Advances in Intelligent Systems and Computing 193

Editor-in-Chief

Prof. Janusz Kacprzyk
Systems Research Institute
Polish Academy of Sciences
ul. Newelska 6
01-447 Warsaw
Poland
E-mail: kacprzyk@ibspan.waw.pl

For further volumes:
http://www.springer.com/series/11156

Sukhan Lee, Hyungsuck Cho, Kwang-Joon Yoon,
and Jangmyung Lee (Eds.)

Intelligent Autonomous Systems 12

Volume 1: Proceedings of the 12th International Conference IAS-12, Held June 26–29, 2012 Jeju Island, Korea

Springer

Editors
Sukhan Lee
College of Information and
Communication Engineering
Sungkyunkwan University
Gyeonggi-Do
Korea

Hyungsuck Cho
Daegu Gyeongbuk Institute of Science
and Technology
Daegu
Korea

Kwang-Joon Yoon
Konkuk University
Seoul
Korea

Jangmyung Lee
Department of Electronics Engineering
Pusan National University
Pusan
Korea

ISSN 2194-5357
e-ISSN 2194-5365
ISBN 978-3-642-33925-7
e-ISBN 978-3-642-33926-4
DOI 10.1007/978-3-642-33926-4
Springer Heidelberg New York Dordrecht London

Library of Congress Control Number: 2012948092

Printed on acid-free paper

Springer is part of Springer Science+Business Media (www.springer.com)

Preface

Intelligent autonomous systems are emerged as a key enabler for the creation of a new paradigm of services to humankind, as seen by the recent advancement of autonomous cars licensed for driving in our streets, of unmanned aerial and underwater vehicles carrying out hazardous tasks on-site, and of space robots engaged in scientific as well as operational missions, to list only a few. In the foreseeable future, the services intelligent autonomous systems provide us with are expected to spread into every fabric of our society, beyond the aforementioned transportation, security and space exploration, to include manufacturing, healthcare, environment and energy, and, ultimately, personal/domestic assistance for our daily lives. This edition aims at serving the researchers and practitioners in related fields with a timely dissemination of the recent progress on intelligent autonomous systems, based on a collection of papers presented at the 12th International Conference on Intelligent Autonomous Systems, held in Jeju, Korea, June 26–29, 2012. With the theme of "Intelligence and Autonomy for the Service to Humankind, the conference has covered such diverse areas as autonomous ground, aerial, and underwater vehicles, intelligent transportation systems, personal/domestic service robots, professional service robots for surgery/rehabilitation, rescue/security and space applications, and intelligent autonomous systems for manufacturing and healthcare. The papers presented at the conference are not only rich in their technical contents, including autonomous navigation, SLAM, machine vision/perception, cognitive systems, human-robot interaction, biomimetic systems, decision-support systems, web-based and networked robotics, mobiligence, life engineering, and micro/nano robots for intelligent autonomous systems. But, they are in high quality as more than 50% of the papers presented were from the 24 special invited sessions. For better readability, this edition has the total 168 papers grouped into 4 chapters: Chapter I: Autonomous Ground Vehicles and Mobile Manipulators, Chapter II: Unmanned Aerial and Underwater Vehicles and Bio-inspired Robotics, Chapter III: Service Robotics and Human-Robot Interaction,

Chapter IV: Autonomous Multi-Agent Systems and Life Engineering, where individual chapters, edited respectively by Sukhan Lee, Kwang Joon Yoon, Jangmyoung Lee and Hyungsuk Cho, begin with a brief introduction written by the respective chapter editors. I do hope that readers find Intelligent Autonomous System 12 stimulating and enjoyable.

Sukhan Lee
General Chair, IAS-12

Contents

I: Autonomous Ground Vehicles and Mobile Manipulators

I: Autonomous Ground Vehicles and Mobile Manipulators

Introduction by Sukhan Lee

Chapter I, with 44 papers in total, covers the recent advancement in autonomous vehicles and mobile robots/manipulators, including the technologies for autonomous navigation, motion control, SLAM, vision and cognitive systems and web-based/ networked robotics.

For autonomous navigation, the real-time path/trajectory generation for safe navigation in a cluttered environment, infested with static as well as dynamic obstacles such as moving vehicles and pedestrians, poses as an important yet difficult problem for a viable solution. There are 11 papers presented for dealing with the above problem of autonomous navigation, as introduced briefly in the following: 1) reactive yet smooth trajectory generation for safe navigation in a cluttered environment, 2) a 3D environmental map generated with an omnidirectional camera for obstacle avoidance, 3) a general method to interpret navigational data for neural network based navigation, 4) drivable road identification by integrating LiDAR and vision, 5) autonomous crossing of a street by identifying pedestrian sidewalks, push-buttons, zebra stripes and pedestrian crossing lights, 6) trajectory generation under time constraints with the vision-based estimation of relative target location, 7) reactive control of a personal transporter for safe navigation under a group of moving pedestrians, 8) real-time collision avoidance of an omnidirectional mobile robot in consideration of robot shape and movable area, 9) an autonomous vehicle for high speed on/off road navigation avoiding diverse patterns of obstacles, 10) mobile robot navigation in a closed world environment based on reliable, fast, and efficient omnidirectional vision, and 11) an android remote call service for an OSGi-based autonomous vehicle with a mobile device.

SLAM, with 6 papers presented, covers the issues involved in map building and localization in navigation, dealing, especially, with the problems of updating a visual memory and of establishing the efficiency in handling a landmark database. They include 1) updating a visual memory based on the concept of short- and long-term memories for visual memory based navigation, 2) building a sparse topological map based on image sequence segmentation, 3) a Kinect enabled 3D region descriptor for Monte **Carlo localization, 4) a succinct random accessible database for a large collection of point landmarks based on data compression techniques, 5) 3D thermal mapping of buildings based on a 3D laser scanner, a color camera and a thermal camera, and 6) moving object detection using monocular vision for SLAM.** On the other hand, we have 5 papers presented in motion control, addressing such issues as state estimation, stability and motion generation, as follows: 1) state estimation under the measurement signals with dead time based on a linear Kalman filter, 2) balancing a two wheeled mobile manipulator of inverted pendulum with the variable centroid based on compensatory angles, 3) stability analysis of a discrete non-linear truck-trailer mobile robot system via discrete linear parameter varying and fuzzy modeling, 4) discovering novel motor/movement primitives by the transition graph constructed from a database of example trajectories, and 5) design of a space robot system to simulate climbing of astronauts based on a binocular vision system.

For machine vision systems, 12 papers are included to show the progress in capturing and registering 3D point clouds, 6D visual odometry, motion detection and people tracking: among them, 5 papers are dedicated to omnidirectional visions for their applications to depth estimation, visual odometry and 3D environmental modeling. They are: 1) fast 6D odometry based on the visual features and depth from an affordable RGB-D camera, 2) a method of pixel-wise segmentation of moving regions based on sparse motion cues, 3) fast and robust multi-people tracking using RGB-D data for point cloud depth clustering and HOG-like classification, 4) an algorithm for computing optical flows with the multi-agent based feature point tracking implemented in parallel, 5) recognition of faces with varying expression based on 3D curvature features, 6) a comparative study of 3D cameras: time-of-flight vs. active triangulation, 7) a novel 3D free-form registration algorithm using parallel-light beam data association and 8) real-time model based visual servo for the self-localization and manipulation of a humanoid robot. In addition, we have 6 papers dedicated to omnidirectional vision: 9) a visual gyroscope for general attitude estimation using omnidirectional cameras, 10) panoramic 3D reconstruction with three catadioptric cameras for an autonomous vehicle, 11) recovery of an indoor spatial layout from a collection of lines using omnidirectional vision, and 12) Depth estimation from a moving fish-eye camera based on epipolar-plane image analysis.

Cognitive systems are to care for intelligent decision-making toward the situation analysis and behavior control for mobile robots, as a means of dealing with complexity, as well as for generating emergent social behaviors in human-robot and robot-robot interactions. There are 7 papers presented here for such cognitive systems, including 1) a Fuzzy based system for adaptive real-time behavior control and improvement, 2) building an ontology based semantic map for robust instantiation to detect false recognition, 3) recognizing hardware faults on mobile robots using situation analysis, 4) an intelligent decision system for fault diagnosis based on learning decision trees, 5) a theoretical foundation of cognition regarding a sociology of human and robot, 6) a model of emotional conversion process based on the Gross cognitive reappraisal strategy, and 7) an actor studio for the development of user friendly action editing system for performance robots.

Last but not least, with 3 papers presented, web-based and/or networked robotics, being emerged as significant, deals with such issues as the web-based exchange of information, or the reuse of task execution plans, among robots, as follows: 1) RoboEarth action recipe execution: storing and reusing task plans on a global accessible database, 2) exchanging action-related information such as environmental maps and object properties among autonomous robots, and 3) a ubiquitous network robot platform for realizing integrated robotic applications.

On-Line Obstacle Detection Using Data Range for Reactive Obstacle Avoidance

José Miguel Vilca, Lounis Adouane, and Youcef Mezouar

Pascal
UBP – UMR CNRS 6602
Clermont-Ferrand, France
Jose_Miguel.VILCA_VENTURA@univ-bpclermont.fr

Abstract. This paper deals with the reactive navigation in clustered environment. It proposes an online and adaptive elliptic trajectory to perform smooth and safe mobile robot navigation. These trajectories use limit-cycle principle already applied in the literature [3]. The main contribution proposed here is to perform this navigation in a completely reactive way while using only range sensor data. At this aim, each obstacle to avoid is surrounded by an ellipse and its parameters are obtained online while using the sequential range data and appropriate method to identify the enclosed ellipse. Different methods to obtain these ellipse parameters are presented and implemented. A specific criterion is taken into account to obtain always smooth change in these parameters. A large number of simulations permit to show the efficiency of our proposal for the navigation in cluttered environment.

Keywords: Mobile robot navigation, Obstacle detection and avoidance, Telemetry, Parameter identification, Least square.

1 Introduction

An important issue for successful mobile robot navigation is obstacle avoidance. In fact, this function permits to prevent robot collision and to insure thus robot safety. One area of research in obstacle avoidance is focused on reactive methods, where only local sensors information is used rather than a prior knowledge of the environment [1], [9], [13]. In [11], the author proposes a real-time obstacle avoidance approach based on the principle of artificial potential fields. In this work, it is assumed that the robot actions are guided by the sum of attractive and repulsive fields.

Several other approaches can be found in the literature, such as the obstacle avoidance based on orbital trajectories. This approach is described by circular limit-cycle differential equations [2], [10] and [12]. Our work uses elliptical trajectories that was presented in [3]. Therefore, more generic and efficient obstacle avoidance is performed and this even with different obstacle shapes, for instance long walls. In fact, an ellipse fit better this kind of obstacles than a circle.

For this purpose, different techniques have been proposed in the literature to enclose the data with an ellipse [7], [14] and [15]. In [15], the author presents a summary of the methods to fit a set of data with an ellipse. The presented methods are the least square

S. Lee et al. (Eds.): Intelligent Autonomous Systems 12, AISC 193, pp. 3–13.
springerlink.com

fitting based on algebraic and Euclidean distance, Kalman filtering method and robust estimation. In [14], the author proposed a technique to obtain the smallest enclosing ellipse by a set of data improving thus the Welzl's algorithm [14] with linear increasing time with regards to data dimension. In [7], the authors constructed an ellipse using the mean and covariance of the data and the Mahalanobis distance to analyze the relationship among them. The maximum Mahalanobis distance is used for different purposes, namely for detection of outliers and for investigating the representativity between two data sets. This work proposes an on-line heuristic method based on the distance between the data to compute the parameters of ellipse.

The rest of the paper is organized as follows: in the next section, the task of navigation using elliptic trajectories is presented. In section 3, the details of the control architecture are introduced and obstacle avoidance algorithm is given. Section 4 presents the methods for enclosing the range data with an ellipse. Simulation results are given in section 5. Finally, conclusion and future works are given in section 6.

2 Navigation in Cluttered Environment

First, let us assume that the obstacle O can be surrounded by elliptical box (cf. Fig. 1). The elliptical shape is represented by its Cartesian form:

$$(x-h)^2 / a^2 + (y-k)^2 / b^2 = 1 \tag{1}$$

where $(h,k) \in \Re^2$ are the ellipse center coordinates and $a,b \in \Re^+$ are the semi-axes with $a \geq b$.

In section 4, it will be shown how the ellipse parameters can be efficiently computed from range data. Let us also represents the robot and the target by circles C_R and C_T of radius R_R and R_T respectively (cf. Fig. 1). One can define:

1. D_{RO} as the minimal distance between the robot and the obstacle "O" [8].
2. *Ellipse of influence* (E_f) as an ellipse that has the same center (h,k) and tilt angle Ω as the ellipse which surround the obstacle (1) while its major and minor semi-axis, a_{lc} and (b_{lc}), are defined as follows

$$\begin{cases} a_{lc} = a + R_R + Margin \\ b_{lc} = b + R_R + Margin \end{cases}$$

   where *Margin* represents safety tolerances encapsuling: perception uncertainties, control reliability and accuracy.
3. "l" as the line passing through the center of C_R and C_T. As we will see in the sequel, our method only needs to know if it exists intersection points between l and E_f (cf. Fig. 1).

The objective of this navigation task in cluttered environment is to lead a mobile robot towards a specific target in an unstructured environment. This task must be achieved while avoiding static and dynamic obstacles O which can have different shapes. In this work, we consider that the cluttered environment has only static obstacles. Furthermore, the presence of dynamic obstacles will be developed in future work.

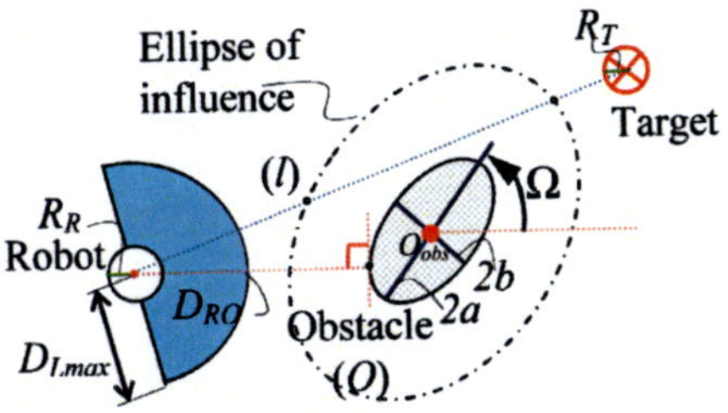

Fig. 1. The used perceptions for mobile robot navigation

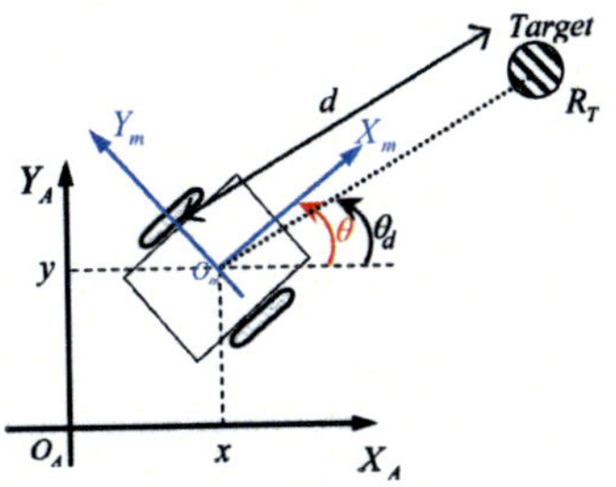

Fig. 2. Robot configuration in a Cartesian reference frame

3 Control Architecture

The used control structure is based on [3] (cf. Fig. 3). It aims to manage the interactions between elementary controllers while guaranteeing the stability of the overall control as proposed in [1]. Its objective is also to insure safe, smooth and fast robot navigation. The specific blocks composing the global controller are detailed below.

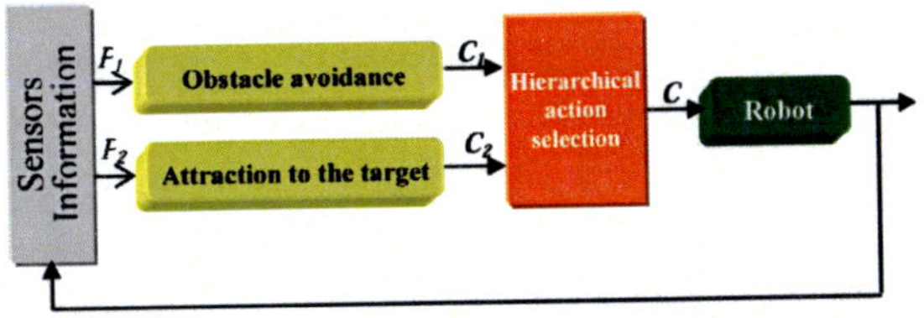

Fig. 3. Control architecture for mobile robot navigation

This control architecture uses a hierarchical action selection mechanism to manage the switches between the controllers, according to environment perception. It has a mechanism that activates the obstacle avoidance controller as soon as it exists at least one obstacle which can obstruct the future robot movement toward its target [2]. This permits to anticipate the activation of obstacle avoidance controller unlike what is proposed in [6] or [4], which wait until the robot is in the immediate vicinity of the obstacle (i.e. $D_{RO} \leq R$ "a certain radius value").

Before describing each elementary controller, let us briefly recall the kinematic model of an unicycle robot (cf. Fig. 2)

$$\begin{bmatrix} \dot{x} \\ \dot{y} \\ \dot{\theta} \end{bmatrix} = \begin{bmatrix} cos(\theta) & 0 \\ sin(\theta) & 0 \\ 0 & 1 \end{bmatrix} \begin{bmatrix} v \\ \omega \end{bmatrix} \qquad (2)$$

where x, y, θ are configuration state of the unicycle at the point O_m, v and ω are respectively, the linear and angular velocity of the robot at the point O_m.

The attraction to target controller guides the robot toward the target. This controller is based on the control position of the robot to the target. In this paper, we focus on the obstacle avoidance controller, more details about attraction to target controller are given in [5].

3.1 Obstacle Avoidance Controller

To perform the obstacle avoidance behavior, the robot needs to follow accurately limit-cycle trajectories as what is given in [2] and [12]. In these works authors use a circular limit-cycle characterized by a circle of influence of R_I radius. In [3], it is proposed to extend this methodology for more flexible limit-cycle shape (an ellipse). The main ideas of this controller are detailed below.

The differential equations giving elliptic limit-cycles are:

$$\dot{x}_s = my_s + x_s(1 - x_s^2/{a_{lc}}^2 - y_s^2/{b_{lc}}^2 - cx_sy_s) \quad (3)$$

$$\dot{y}_s = -mx_s + y_s(1 - x_s^2/{a_{lc}}^2 - y_s^2/{b_{lc}}^2 - cx_sy_s) \quad (4)$$

with $m = \pm 1$ according to the direction of avoidance (clockwise or counter-clockwise, cf. Fig. 4). (x_s, y_s) corresponds to the position of the robot according to the center of the ellipse; a_{lc} and b_{lc} characterize respectively major and minor elliptic semi-axis (cf. Fig. 1); c if $\neq 0$ gives the Ω ellipse angle.

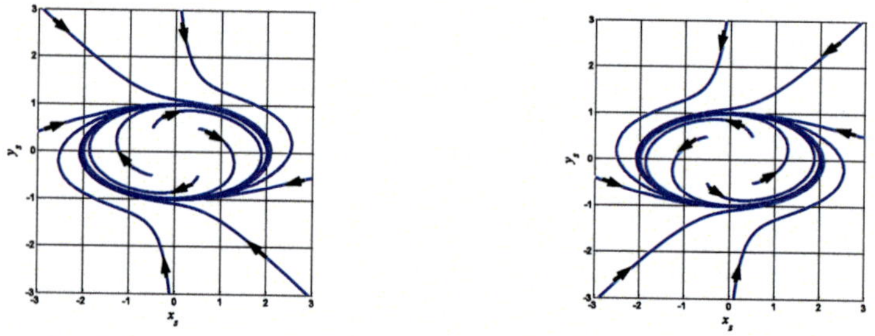

Fig. 4. Clockwise and counter-clockwise shape for the elliptic limit-cycles

The algorithm for obstacle avoidance is summarized in the following main steps:

- The nearest hindering obstacle is detected,
- The direction of avoidance is chosen according to the position of the robot with regards to the position of obstacle and the target,
- The robot avoids the obstacle while following an elliptic limit cycle which has the semi-axes a_{lc} and b_{lc}, with $a_{lc} \geq b_{lc}$.

More details of the used obstacle avoidance algorithm is given in [3].

The desired robot orientation is thus given by the differential equation of the limit-cycle (3) and (4) as:

$$\theta_d = arctg\,(\dot{y}_s/\dot{x}_s) \quad (5)$$

and the error by:

$$\tilde{\theta} = \theta_d - \theta \tag{6}$$

3.2 Control Law

The used control law is expressed as follows [5]:

$$v = v_{max} e^{-1/d} cos(\tilde{\theta}) \tag{7}$$
$$\omega = \omega_r + K_p \tilde{\theta} \tag{8}$$

where v_{max} is the maximum linear velocity, K_p is a constant such that $K_p > 0$ and if the *obstacle avoidance* controller is activated, d is equal to D_{RO} (cf. Fig. 1) else d is the distance between the robot and the target when the *attraction to the target* controller is activated. The robot reaches the target when $0 < d \leq R_T$, where R_T is the radius of the target (cf. Fig. 1).

It is interesting to notice that only one control law is applied to the robot even if its architecture of control contains two different controllers. Only the set points change according to the applied controller [5].

4 Enclosing Data Range with an Ellipse

During the robot movement, it is important for the robot to detect on-line and to avoid the hinder obstacle. The robot will try to surround the observed range data with the closest ellipse to apply elliptic limit-cycle approach.

For this purpose, let us consider a set of n points in $\Re^2$ with coordinates $P_i(x_i, y_i)$ (cf. Fig. 5). These points are computed from the data range of the robot, and it is considered that it was previously filtered, i.e, without noisy and outliers. In this section, it will be shown how to compute the ellipse that encloses all points. An important condition in this work is that the methods need to start at least with three different points.

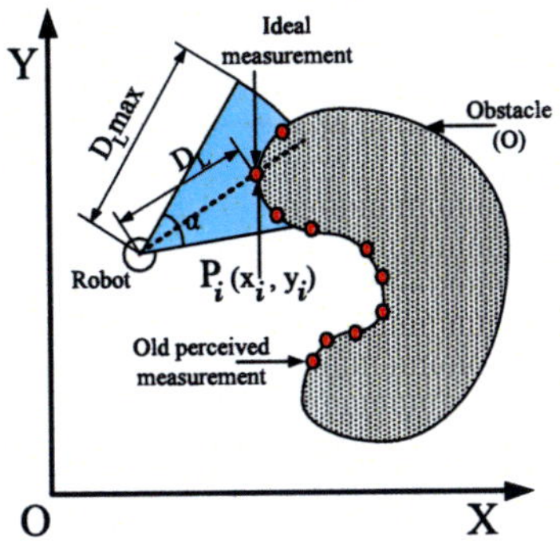

Fig. 5. Data set of *n* points

This paper considers that the robot detects one obstacle at a time. The segmentation method of the set of points will be used in future works to the detection of more than one obstacle at a time and for big obstacles.

The following sub-sections will present different methods to enclose the data points with an ellipse. Sub-sections 4.1 and 4.2 are applied and (more or less) modified approach with regards to what is proposed in the literature and the method given in sub-section 4.3 is a completely new approach to deal with this problem.

4.1 Fitting Using Least Square Approach

This approach uses the general conic equation which is given by:

$$f(x,y) = Ax^2 + 2Bxy + Cy^2 + 2Dx + 2Ey + F = 0 \tag{9}$$

According to the real constants A, B, C, D, E anf F, we obtain the analytic equation of the different kind of conics (parabola, ellipse and hyperbole). An ellipse is defined if the conic parameters (9) satisfy the following condition $B^2 - AC < 0$.

The problem is to fit a conic section (9) with a set of n points $\{p_i\} = \{(x_i, y_i)\} \mid i = 1, \ldots, n$. As the data are noisy, it is unlikely to find a set of parameters (A,B,C,D,E,F) (except for the trivial solution $A = B = C = D = E = F = 0$) such that $f(x_i, y_i) = 0$. Instead, we will try to estimate them by minimizing some objective function J.

A common practice is to minimize the algebraic distance $f(x_i, y_i)$, we use the least-square fitting based on algebraic distance [15] to minimize the following function:

$$J = \sum_{i=1}^{n} f^2(x_i, y_i) \tag{10}$$

Note that there is no justification for using algebraic distance apart from easy implementation. To avoid the trivial solution, $f(x_i, y_i)$ should be normalized. We use the normalization $A + C = 1$ because it is one of the most popular methods proposed in the literature [15], and its implementation is easy.

Using the normalization in (9), the set of equation $f(x_i, y_i)$ can be written as:

$$f_i \equiv \mathbf{a}_i\mathbf{s} - \mathbf{b}_i = 0 \tag{11}$$

where $\mathbf{s} = [A, B, D, E, F]^T$, $\mathbf{a}_i = [x_i^2 - y_i^2, 2x_iy_i, 2x_i, 2y_i, 1]$ and $\mathbf{b}_i = -y_i^2$. The solution that minimize the functional J is given by

$$\mathbf{s} = (\mathbf{A}^T\mathbf{A})^{-1}\mathbf{A}^T\mathbf{b} \tag{12}$$

where $\mathbf{A} = [\mathbf{a}_1, \mathbf{a}_2, \ldots, \mathbf{a}_n]^T$ and $\mathbf{b} = [\mathbf{b}_1, \mathbf{b}_2, \ldots, \mathbf{b}_n]^T$. This method is known as the pseudo inverse technique.

4.2 Covariance Approach

This method is based on the analysis of the relationship between the n points. We use the covariance of the data and the Mahalanobis Distance (*MD*) [7]. In the field of multivariate calibration, the *MD* is used for different purposes, namely: for the detection of outliers, the selection of calibration samples from a large set of measurements and for investigating the representativity between two data sets. In the original variable space, the *MD* takes into account the correlation in the data, since it is calculated using the inverse of the covariance matrix of the data set of interest.

The principle of covariance approach is given below: First the covariance matrix $\mathbf{C}_x$ is constructed:

$$\mathbf{C}_x = \frac{1}{(n-1)}(\mathbf{X}_c)^T(\mathbf{X}_c) \tag{13}$$

where $\mathbf{X}_c$ is the column-centered data matrix $(\mathbf{X}-\bar{\mathbf{X}})$. $\mathbf{X}$ is the data matrix containing n objects in the rows measured for p variables and $\bar{\mathbf{X}}$ is the data mean.

Now, we analyze the covariance matrix $\mathbf{C}_x$ using the eigenvalues and eigenvectors to obtain the parameters of the ellipse $f(x', y')$ (1). Furthermore, we use the Mahalanobis distance to select the dimension of the semi-axes (percentage the data inside the ellipse) [7]. The eigenvalues λ_1, λ_2 $(\lambda_1 > \lambda_2)$ are related to the semi-axes a and b as follows:

$$\begin{aligned} a &= MD_{max}\sqrt{\lambda_1} \\ b &= MD_{max}\sqrt{\lambda_2} \end{aligned} \tag{14}$$

where MD_{max} is the maximum Mahalanobis distance to ensure that all the set of data is enclosed by the ellipse. The eigenvectors $\mathbf{v}_1$, $\mathbf{v}_2$ are related with the orientation Ω.

$$\Omega = \arctan(\mathbf{v}_2/\mathbf{v}_1) \tag{15}$$

The center of the ellipse is the mean of the data $\bar{\mathbf{X}}$ and the maximum semi-axes a is the direction where there are the most important amounts of data.

4.3 Heuristic Approach

The proposed heuristic approach uses the distance between the points to obtain one of the axes.

This method computes the distance between all the points $d_{ij} = \|p_i - p_j\|$ with $i, j = 1, \dots, n$; and select the maximum distance d_{max}. We have thus, $d_{ij} \le d_{max}$. With this manner, this maximum distance is not decreasing. The line connecting the corresponding two points (the points with the maximum distance) is one of the ellipses axes, where the ellipse center C_O is the middle point between the points with maximum distances and the first semi-axes is $a_1 = d_{max}/2$.

Now, we work in the new coordinates system $X'Y'$ to obtain the second ellipse semi-axes a_2. We transform the n points to new coordinates system using (16).

$$\mathbf{P}'_i = \begin{bmatrix} cos(\Omega) & sin(\Omega) \\ -sin(\Omega) & cos(\Omega) \end{bmatrix} (\mathbf{P}_i - \mathbf{C}_O) \tag{16}$$

Where Ω is the orientation of the line between the two points that have the maximum distance. $\mathbf{P}'_i(x'_i, y'_i)$ is the coordinate in the new system, $\mathbf{P}_i(x_i, y_i)$ is the coordinate in the initial system and $\mathbf{C}_O$ is the coordinate of the ellipse center in the initial system.

If the coordinate y'_i of the points is greater than the threshold $\varepsilon > 0$, we compute the distance of $\mathbf{P}'_i$ to the origin O', i.e., $|y'_i| > \varepsilon \Rightarrow d'_i = \|\mathbf{P}'_i\|$. The threshold is used to eliminate the points that are collinear with the two points that have the maximum distance (axes).

We know that all points inside the ellipse have the distance $d'_i < max\{a_1, a_2\}$. As we do not know the spatial distribution of the set of points and to ensure that all the points are inside the ellipse, we choose that $a_2 = max\{b_i\}$, where b_i is the computed semi-axes using $\mathbf{P}'_i$ in (1). In other words, $b_i \leq a_2 \leq max\{a_1, a_2\} \Rightarrow \mathbf{P}_i \in Ellipse$. Therefore, the ellipse will enclose all points without regard of the obstacle shape.

Finally, we obtain the semi-axes of the ellipse (1) such as:

$$\begin{aligned} a &= max\{a_1, a_2\} \\ b &= min\{a_1, a_2\} \end{aligned} \tag{17}$$

5 Simulations Results

To demonstrate the efficiency of the proposed approach to enclosing the obstacle with the ellipse, a two statistical survey was made. We consider a mobile robot with a radius of $R_R = 0.5\ m$ and a $\alpha = 180°$ laser range sensor with the maximum detected range is equal to $D_{Lmax} = 3.0\ m$ (cf. Fig. 5). The maximum linear velocity of the robot is $1.5\ m/s$ and the sample time is $0.01\ s$. For each simulation the robot starts at the same configuration and reaches the same final configuration. We do not start to use any method until we have the enough data range $n_{data} \geq 3$. The first survey is used to compare the methods presented above to identify the closer ellipse shape which surround the detected obstacle (cf. section 4).

Fig. 6 shows the identified ellipse parameters using the different methods. We observe that the heuristic parameters change more smoothly if they are compared with the others.

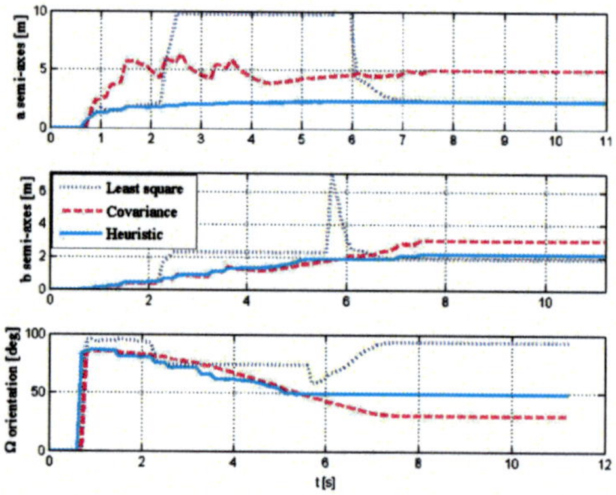

Fig. 6. Parameter of the estimated ellipses

Fig. 7, 8, 9 show the evolution of the identified ellipse for the least square, covariance and heuristic methods, respectively. We observe that the least square method and covariance change abruptly.

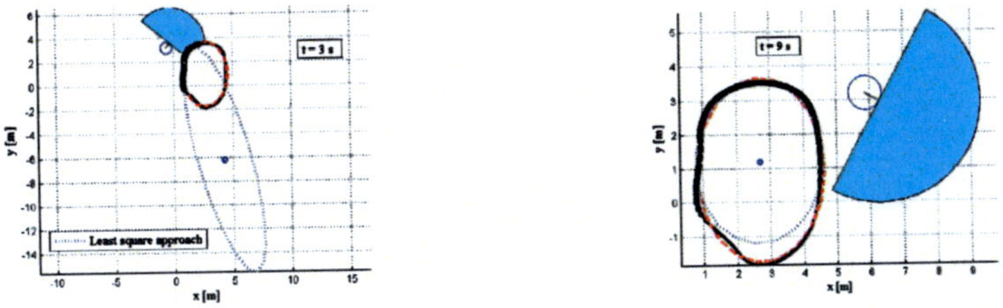

Fig. 7. Evolution of the obtained ellipse using least square approach

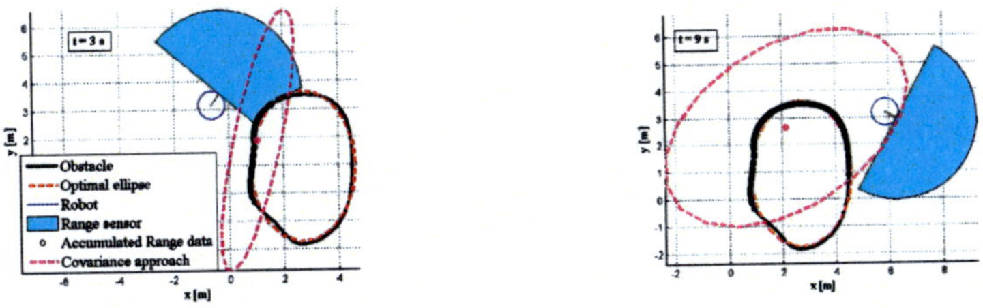

Fig. 8. Evolution of the ellipse using covariance approach

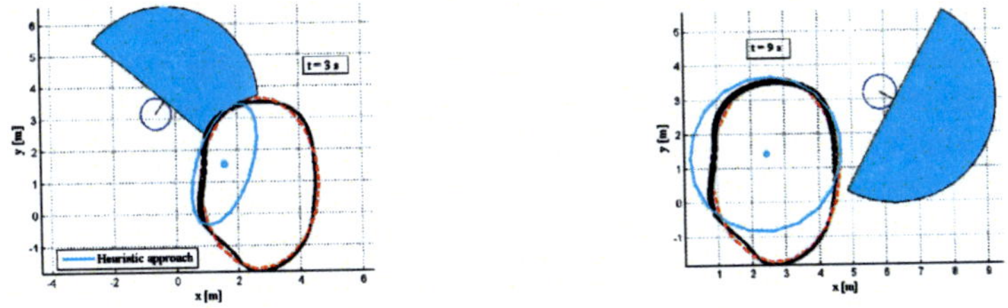

Fig. 9. Evolution of the ellipse using heuristic approach

The second survey is used to make a focus only around the proposed heuristic method which gives satisfactory results and this while making an on-line navigation in cluttered environment (cf. Fig 10).

Fig. 10 shows the trajectory of the robot in the environment with three obstacles. We observe that the robot avoid the obstacles with a smooth trajectory. This trajectory was obtained while using the on-line obstacle avoidance algorithm [3] which takes its parameters (elliptical limit-cycle to follows) from the proposed heuristic approach.

Fig. 11 shows the distance between the optimal ellipse (obtained while knowing all the range data which surround the obstacle) and the position of the robot (red dotted

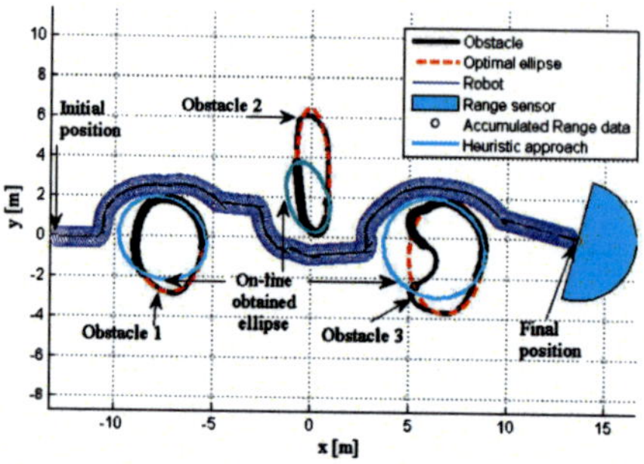

Fig. 10. Robot trajectory using the heuristic approach to enclosing the obstacle

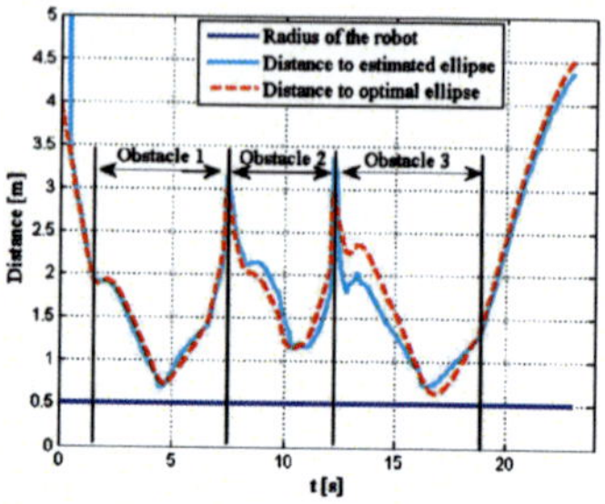

Fig. 11. Distance from the robot to the optimal ellipse and to on-line estimated ellipse using the heuristic approach

line), and the distance between the ellipse using the heuristic method and the robot. This figure shows that the robot does never collide with any obstacles, thus the proposed on-line approaches is efficient to deal with cluttered unstructured environment. Future work will consist to implement this approach in real robots.

6 Conclusion

This paper proposes an on-line elliptic limit-cycle trajectory to perform smooth and safe mobile robot navigation. This elliptic trajectory is obtained while using only local range data. The later, permits to obtain the parameters of the ellipse which surround the obstacle to avoid. After giving a review of different methods to enclosing an ellipse, an appropriate heuristic method was proposed and compared to the existing approaches (based on least square or covariance). The proposed heuristic permits to obtain a smooth changes of the ellipse parameters. The overall obstacle avoidance strategy permits us to obtain generic and flexible navigation in very cluttered environments. The proposed reactive navigation was embedded in multi-controller architecture and permits for a mobile robot to efficiently navigate in environments with different obstacles shapes. In future works, the problem of dynamic obstacles, outliers detection and uncertainty of the range data will be considered. Furthermore, we will use the Kalman filter to deal with the uncertainty of the range data.

References

1. Adouane, L.: Hybrid and safe control architecture for mobile robot navigation. In: 9th Conference on Autonomous Robot Systems and Competitions, Portugal (2009)
2. Adouane, L.: Orbital obstacle avoidance algorithm for reliable and on-line mobile robot navigation. In: 9th Conference on Autonomous Robot Systems and Competitions, Portugal (2009)
3. Adouane, L., Benzerrouk, A., Martinet, P.: Mobile robot navigation in cluttered environment using reactive elliptic trajectories. In: 18th IFAC World Congress (2011)

4. Adouane, L., Le Fort-Piat, N.: Behavioral and distributed control architecture of control for minimalist mobile robots. Journal Europen des Systémes Automatisés 40(2), 177–196 (2006)
5. Benzerrouk, A., Adouane, L., Martinet, P.: Lyapunov global stability for a reactive mobile robot navigation in presence of obstacles. In: ICRA 2010 International Workshop on Robotics and Intelligent Transportation System (2010)
6. Brooks, R.A.: A robust layered control system for a mobile robot. IEEE Journal of Robotics and Automation RA-2, 14–23 (1986)
7. De Maesschalck, R., Jouan-Rimbaud, D., Massart, D.: The mahalanobis distance. Chemometrics and Intelligent Laboratory Systems 50(1), 1–18 (2000)
8. Eberly, D.: Distance from a point to an ellipse in 2d. In: Geometric Tools, LLC (2008), http://www.geometrictools.com/
9. Egerstedt, M., Hu, X.: A hybrid control approach to action coordination for mobile robots. Automatica 38(1), 125–130 (2002)
10. Jie, M.S., Baek, J.H., Hong, Y.S., Lee, K.W.: Real time obstacle avoidance for mobile robot using limit-cycle and vector field method. Knowledge-Based Intelligent Information and Engineering Systems, 866–873 (2006)
11. Khatib, O.: Real-time obstacle avoidance for manipulators and mobile robots. The International Journal of Robotics Research 5, 90–99 (1986)
12. Kim, D.H., Kim, J.H.: A real-time limit-cycle navigation method for fast mobile robots and its application to robot soccer. Robotics and Autonomous Systems 42(1), 17–30 (2003)
13. Toibero, J., Carelli, R., Kuchen, B.: Switching control of mobile robots for autonomous navigation in unknown environments. In: IEEE International Conference on Robotics and Automation, pp. 1974–1979 (2007)
14. Welzl, E.: Smallest Enclosing Disks (Balls and Ellipsoids). In: Maurer, H.A. (ed.) New Results and New Trends in Computer Science. LNCS, vol. 555, pp. 359–370. Springer, Heidelberg (1991)
15. Zhang, Z.: Parameter estimation techniques: A tutorial with application to conic fitting. Image and Vision Computing 15, 59–76 (1997)

Obstacle Avoidance Based on Plane Estimation from 3D Edge Points by Mobile Robot Equipped with Omni-directional Camera

Kazushi Watanabe[1], Ryosuke Kawanishi[1], Toru Kaneko[1], Atsushi Yamashita[2], and Hajime Asama[2]

[1] Dept. of Mechanical Engineering, Shizuoka University
3-5-1 Johoku, Naka-ku, Hamamatsu-shi, Shizuoka 432-8561, Japan
{f0130076,f5945015,tmtkane}@ipc.shizuoka.ac.jp
[2] Dept. of Precision Engineering, The University of Tokyo
7-3-1 Hongo, Bunkyo-ku, Tokyo 113-8656, Japan
{yamashita,asama}@robot.t.u-tokyo.ac.jp

Abstract. In this paper, we propose a method for a mobile robot to avoid obstacles in its environment using an omni-directional camera. The method makes an environment map consisting of 3D edge points obtained from omni-directional camera images and estimates the locations of planes by analysing these 3D edge points so that the robot can avoid walls as obstacles. The method has the advantage that it can generate a 3D map in environments constructed by textureless planes. Experimental results show the effectiveness of the proposed method.

Keywords: navigation, environment recognition, omni-directional camera.

1 Introduction

Localization and path planning are indispensable functions for an autonomous mobile robot. Since the robot does not have an environment map in unknown environment, the robot needs to generate a map by itself by measuring surrounding environment. In addition to map generation, path planning, movement, and localization are needed for the robot to generate the environment map over a wide area. Therefore, in order for the robot to work in unknown environment, the function of generating a map while autonomously navigating is required.

As a sensor for environment measurement in map generation, an omni-directional camera is sometimes used (Fig. 1(a)). An omni-directional camera is suitable for environment measurement [1], because the acquired image has a panoramic field of view (Fig. 1(b)).

There are methods for map generation and path planning in unknown environment of a mobile robot equipped with an omni-directional camera [2-4]. If we use multiple sensors for measurement, measurement result is more accurate and denser than using only an omni-directional camera. Then the robot can avoid the surface of a wall and obstacles, and it can run autonomously in unknown environment. However, this

S. Lee et al. (Eds.): Intelligent Autonomous Systems 12, AISC 193, pp. 15–24.
springerlink.com

method needs other sensors in addition to a camera. Therefore this method has disadvantages in terms of cost and weight. If the robot can work in unknown environment using only an omni-directional camera, there are significant benefits in terms of cost and weight compared with using other sensors such as laser scanner, ultrasound sensor, etc.

Methods for generating map and path planning using a single omni-directional camera have been already proposed. However, generating a map in an environment with few textures is difficult by use of only a single omni-directional camera. Thus running in such environment is difficult for the robot. Therefore in [5][6], operation of robot is manual.

As a method for reconstructing an environment map denoted by a set of dense points, reference [7] is proposed. This method generates planes consistently fitting to 3-D edge points reconstructed from the image edge points. However, in [7], obstacle avoidance of robot using the result of plane reconstruction is not mentioned.

Therefore, in this paper, we propose a method for avoiding obstacles in an environment which has few feature points. The proposed method reconstructs planes from a 3-D map generated by tracking the edge points in omni-directional image sequences. By interpolating the area which has few feature points, a dense 3-D map can be generated in environment with few textures. Since our method generates a dense 3-D map, the robot can run without a map generated beforehand.

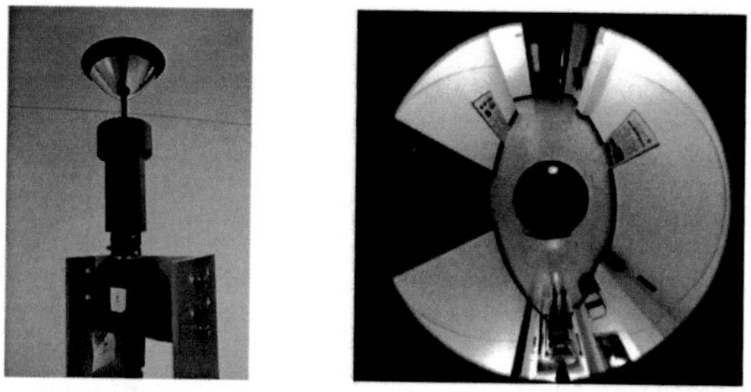

(a) Omnidirectional camera. (b) Acquired image.

Fig. 1. Omnidirectional camera and acquired image

2 Process Outline

In the proposed method, we use a mobile robot equipped with an omni-directional camera as shown in Fig. 1(a). The camera attaches a hyperboloid mirror in front of its lens. An acquired image is shown in Fig. 1(b). Internal parameters of camera and distortion of lens are determined in advance.

Figure 2 describes an overview of the method for 3-D measurement. An omni-directional camera is mounted on the mobile robot. The robot moves constantly and captures omni-directional images simultaneously. When the robot moves along a fixed distance, it stops capturing images. Then the robot carries out the camera motion estimation and 3-D measurement.

Next, the robot reconstructs planes around itself from the result of 3-D measurement. From the result of plane estimation, the robot determines the direction which has no plane from its view point. The robot turns to the determined direction, and resumes moving and capturing images.

By repeating the above behavior, the robot can move autonomously and generates an environment map.

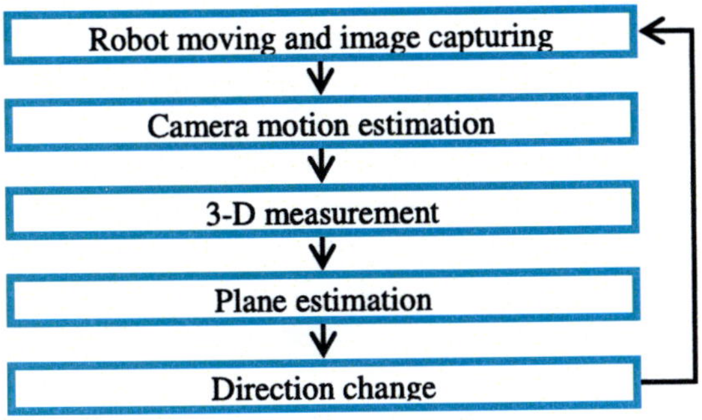

Fig. 2. Procedure of our proposed method

3 3-D Measurement

Our method of 3-D measurement is based on reference[8]. In this paper, the camera motion is estimated based on the 8 point algorithm which uses correspondence relationship between the tracked feature points. We use the Lucas Kanade tracker algorithm for tracking feature points [9]. Simultaneously, the method of RANSAC (RANdom SAmple Consensus) [10] is applied to remove outliers. Finally, the reprojection errors are minimized by using bundle adjustment[11].

We employ a 3-D measurement method using estimated camera motion[12]. This method measures edge points which are tracked through the image sequences. Edge points are obtained from the captured image by the Canny edge detector [13].

3.1 Camera Motion Estimation

An omni-directional camera we use has a hyperboloid mirror in front of the camera lens. A ray vector $\mathbf{r} = [x, y, z]^T$ is defined as a unit vector heading to the reflection point on the mirror surface from the focus of the hyperboloid mirror as shown in Fig. 3. A ray vector $\mathbf{r}$ is calculated by Eq. (1) from image coordinates $[u, v]^T$.

$$\mathbf{r} = \begin{bmatrix} su \\ sv \\ sf - 2c \end{bmatrix} \tag{1}$$

Here, s is defined as Eq. (2). In Eq. (2), a, b and c are the hyperboloid parameters and f is the focal length of the lens.

$$s = \frac{a^s\left(f\sqrt{a^2+b^2}+b\sqrt{u^2+v^2+f^2}\right)}{a^2f^2-b^2\left(u^2+v^2\right)} \tag{2}$$

The proposed method calculates a ray vector of image coordinates $[u, v]^T$ at the arbitrary viewpoint.

Matrix $\mathbf{E}$ which satisfies Eq. (3) is called an essential matrix.

$$\mathbf{r}_i'^{\mathrm{T}}\mathbf{E}\mathbf{r}_i = 0 \tag{3}$$

where ray vectors $\mathbf{r}_i^{\mathrm{T}} = [x_i, y_i, z_i]^{\mathrm{T}}, \mathbf{r}_i'^{\mathrm{T}} = [x_i', y_i', z_i']^{\mathrm{T}}$ are those of the corresponding point in two images, respectively. By decomposing $\mathbf{E}$ into translational element and rotational element, camera motion is estimated.

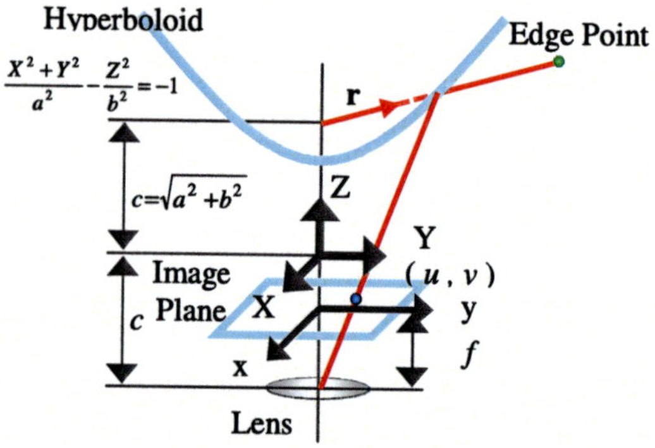

Fig. 3. Calculation of ray vector

3.2 Outlier Rejection

Not all feature points tracked in the image sequence correspond satisfactorily due to image noise, etc. Mistracked feature points should be rejected, by using the RANSAC (RANdom SAmple Consensus) algorithm [10].

In reference [8], determining essential matrix $\mathbf{E}_{rand}$ is repeated for a predetermined number of times, then we choose the case with the maximum number of feature points satisfying Eq. (4), and calculate essential matrix $\mathbf{E}$ again using remaining feature points.

$$\mathbf{r}_i'^{\mathrm{T}}\mathbf{E}_{rand}\mathbf{r}_i < q \tag{4}$$

where q is a given threshold.

3.3 Decision of Feature Point Number

In [8], the optimum number of feature points to use in measurement is automatically decided by defining it as the number of feature points including the maximum number of outliers in a range in which camera movement is still estimated precisely enough. From the above, Eq. (5) and (6) are given as follows:

$$\zeta(k) = \left| \frac{\omega \sum_{i=k}^{k+\omega} i \cdot O_i - \sum_{i=k}^{k+\omega} i \sum_{i=k}^{k+\omega} O_i}{\omega \sum_{i=k}^{k+\omega} i^2 - \left(\sum_{i=k}^{k+\omega} i \right)^2} \right| \tag{5}$$

$$\zeta(k) < t \tag{6}$$

where O_i is the number of outliers and i is the number of extracted feature points. Equation (5) represents the slope of the straight line calculated by the least squares method, so $\zeta(k)$ shows the increase in the number of outliers when the number of the extracted feature points changes from k to $(\omega + k)$. $\zeta(k)$ is calculated stably by setting ω appropriately. In Eq. (6), t is a given threshold. The largest value of k which satisfies Eq. (6) is determined as the optimal number of extraction to camera motion estimation.

4 Plane Reconstruction

We employ the method in [7] for reconstructing planes. In [7], planes fitting to edge points are reconstructed if they satisfy visibility constraints.

A triangle is generated by choosing arbitrary combination of three edge points. The planes which satisfy visibility constraint and fitting to edge points are chosen as correctly estimated planes. Visibility constraint is as follows.

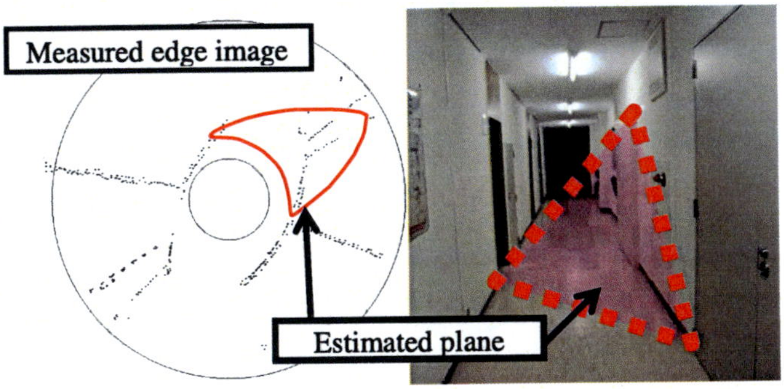

Fig. 4. Visibility constraint

If an estimated plane exists in real environment, objects behind the plane are invisible. Contrarily, if there exists an object visible through an estimated plane, the plane is determined as a plane which is inconsistent with the real environment (Fig. 4). This plane is said to offence visibility constraint. Estimated planes which are inconsistent with the real environment are excluded by applying visibility constraint.

5 Obstacle Avoidance

When the robot finishes plane reconstruction, it determines the direction which has no plane from its view point. In this paper, the area which has no plane is called an empty area. The robot runs toward the center of an empty area.

Since the purpose of our method is determining an advancing direction of the robot, it is enough to search for empty areas on the horizontal 2-D plane where the height of point is not considered. A plane outputted in the 3-D map is a triangle whose sides consist of edge points. Many edge points are on each side in the 3-D map, and a straight line segment consisting of these edge points appear in the 2-D map (Fig. 5).

To detect empty areas, our method draw lines each of which connects the present point of the robot to each edge point in the 2-D map. Then it determines an area containing two adjacent lines as an empty area if the angle between the lines exceeds a threshold (Fig. 6). Because the method uses the result of plane reconstruction instead of 3-D measurement point data which sometimes include outliers, robust empty area detection is realized.

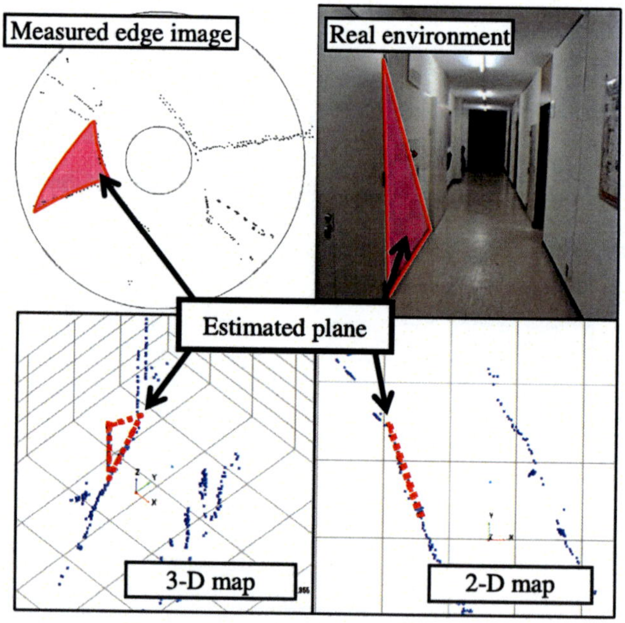

Fig. 5. Estimated plane

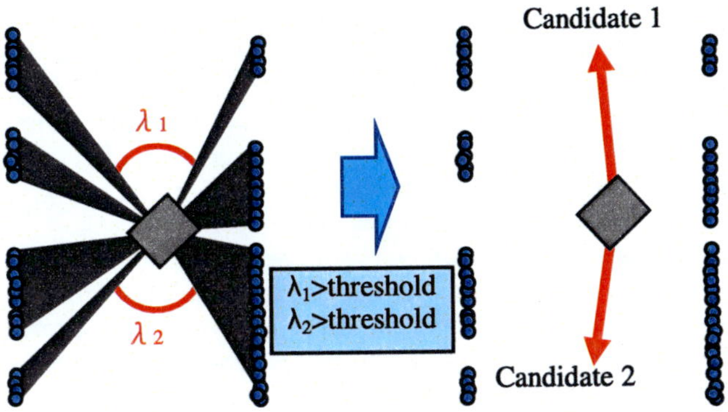

Fig. 6. Detection of empty areas

If there are several directions of empty area, the direction closest to the current direction is chosen as a new direction (Fig. 7). By considering a new direction every time the robot stops at a fixed distance, the robot can generate a map of unknown environment while avoiding obstacles.

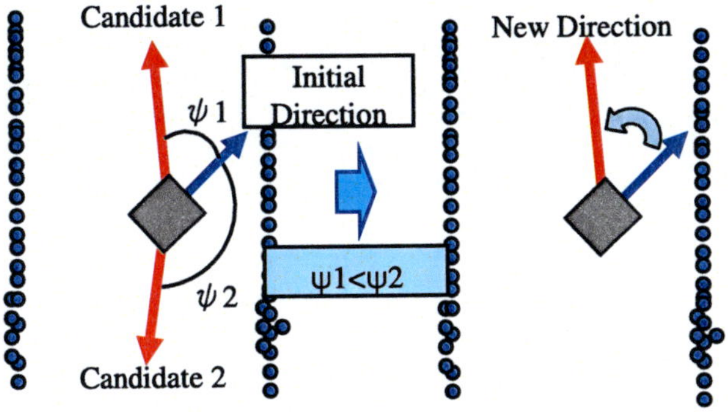

Fig. 7. Selection of a new direction from candidates

6 Experiment

An experiment was carried out at a corridor (Fig. 8(a)). A mobile robot equipped with omni-directional camera for the experiment is shown in Fig. 8(b). The robot had a single omni-directional camera. The robot had two types of behavioral patterns, advance and rotation. Rotation was only used for change of direction when the robot detects obstacles existing in the forward direction.

A result of 3-D measurement in experimental environment is shown in Fig. 9(a), 10(a) and11(a). In these figures, the initial position of the robot is drawn as a triangle. A rectangle connected to the triangle by a line is the position of the robot after moving. The points existing around the triangle and the rectangle are measured edge points.

When we compare the experimental environment (Fig. 8(a)) with the experimental result (Fig. 9(a), 10(a) and 11(a)), the latter reconstructs only a part of the real environment.

The result of plane reconstruction using the result of 3-D measurement is shown in Fig. 9(b), 10(b) and 11(b). In these figures, the new direction of the robot determined from the result of plane reconstruction is represented by an arrow. Each plane is represented by 3 sides which are drawn by points in yellow.

From the result of plane reconstruction, a dense 3-D map is obtained. The viewing angle in Fig. 11(b) is almost same with that of Fig. 8(a). Reconstructed planes are in position of the plane in real environment. Since a dense 3-D map was obtained, the robot turned to the direction which had no plane. Herewith, the robot autonomously generated the map and avoided the wall while moving.

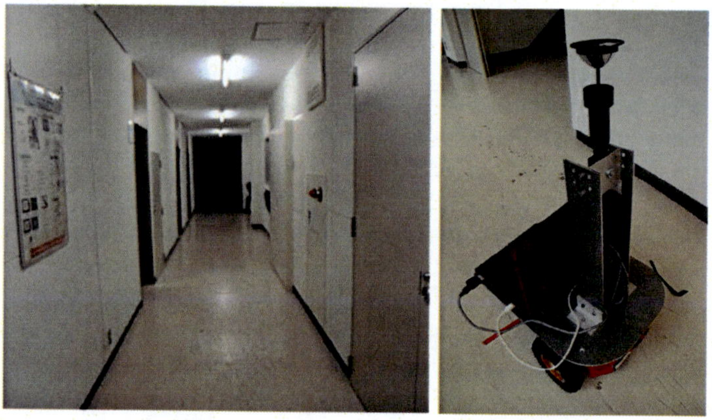

(a) Experimental environment. (b) Mobile robot.

Fig. 8. Experimental environment and mobile robot

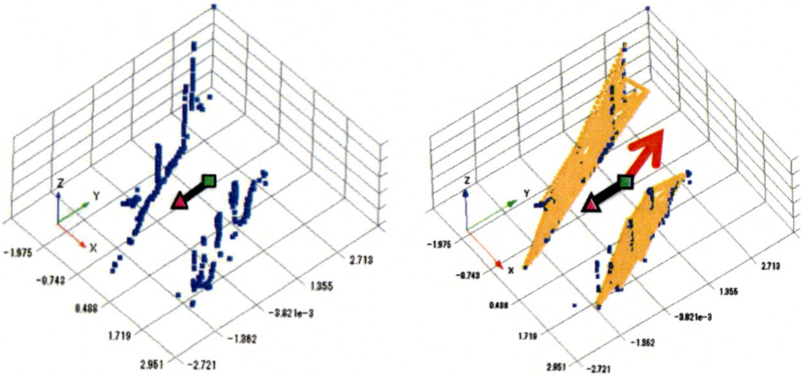

(a) Result of 3-D measurement. (b) Result of plane estimation.

Fig. 9. Bird's eye view of a result of 3-D measurement and plane estimation

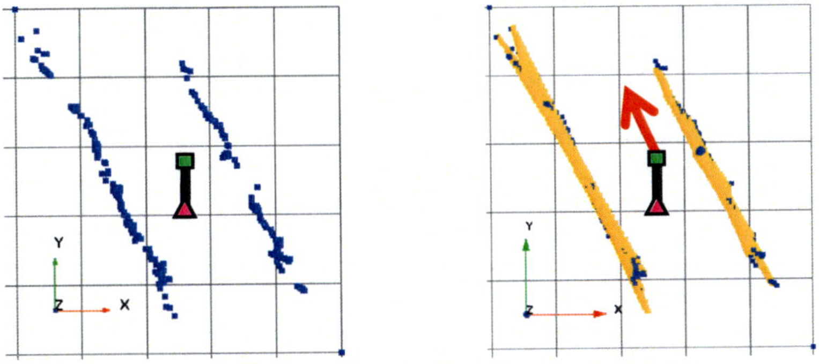

(a) Result of 3-D measurement. (b) Result of plane estimation.

Fig. 10. Top view of a result of 3-D measurement and plane estimation

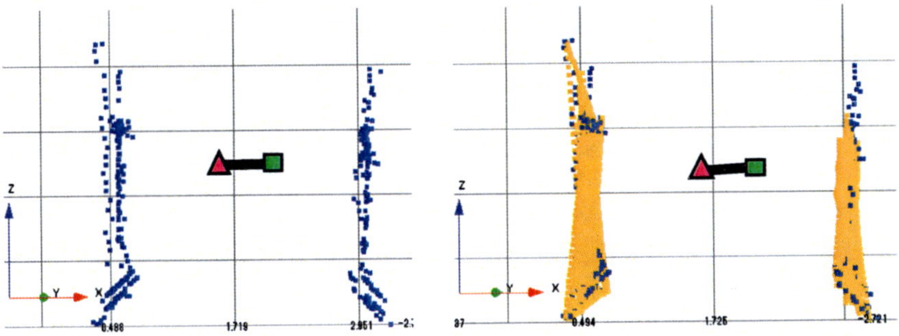

(a) Result of 3-D measurement. (b) Result of plane estimation.

Fig. 11. Side view of a result of 3-D measurement and plane estimation

7 Conclusion

In this paper, we proposed a method of avoiding obstacles for a mobile robot equipped with omni-directional camera. The proposed method makes the robot possible to autonomously move in unknown environment with few textures.

As future works, navigation in dynamic environment, and avoiding obstacles consisting of not only planes should be studied.

Acknowledgements. This work was in part supported by MEXT KAKENHI, Grant-in-Aid for Young Scientist (A), 22680017.

References

1. Geyer, C., Daniilidis, K.: Omnidirectional Video. The Visual Computer 19(6), 405–416 (2003)
2. Wang, C., Thorpe, C.: Simultaneous Localization and Mapping with Detection and Tracking of Moving Objects. In: Proceedings of the 2002 IEEE International Conference on Robotics and Automation, pp. 2918–3004 (2002)
3. Kim, D., Sun, J., Rehg, S.M., Bobick, A.F.: Traversability Classification Using Unsupervised On-line Visual Learning for Outdoor Robot Navigation. In: Proceedings of the 2006 IEEE International Conference on Robotics and Automation, pp. 518–525 (2006)
4. Konolige, K., Agrawal, M., Bolles, R.C., Cowan, C., Fischler, M., Gerkey, B.P.: Outdoor Mapping and Navigation Using Stereo Vision. In: Proceedings of the International Symposium on Experimental Robotics, pp. 1–12 (2006)
5. Morioka, H., Sangkyu, Y., Hasegawa, O.: Vision-based Mobile Robot's SLAM and Navigation in Crowded Environments. In: Proceeding of the 2010 IEEE/RSJ International Conference on Intelligent Robots and Systems, pp. 1–4 (2010)
6. Iwasa, H., Aihara, N., Yokoya, N., Takemura, H.: Memory-based Self-Localization Using Omnidirectional Images. Systems and Computers in Japan 34(5), 56–68 (2003)
7. Tomono, M.: Plane Reconstruction with a Stereo Camera in Non-textured Envirionments. In: Proceeding of the 17th Robotics Symposia, pp. 463–468 (2011) (in Japanese)
8. Kawanishi, R., Yamashita, A., Kaneko, T.: Three-Dimensional Environment Model Construction from an Omnidirectional Image Sequence. Journal of Robotics and Mechatronics 21(5), 574–582 (2009)
9. Shi, J., Tomasi, C.: Good Features to Track. In: Proceedings of the 1994 IEEE Computer Society Conference on Computer Vision and Pattern Recognition, pp. 593–600 (1994)
10. Fischler, M.A., Bolles, R.C.: Random Sample Consensus: A Paradigm for Model Fitting with Applications to Image Analysis and Automated Cartography. Communications of the ACM 24(6), 381–395 (1981)
11. Triggs, B., McLauchlan, P.F., Hartley, R.I., Fitzgibbon, A.W.: Bundle Adjustment – A Modern Synthesis. In: Triggs, B., Zisserman, A., Szeliski, R. (eds.) ICCV-WS 1999. LNCS, vol. 1883, pp. 298–372. Springer, Heidelberg (2000)
12. Tomono, M.: Dense Object Modeling for 3-D Map Building Using Segment-based Surface Interpolation. In: Proceedings of the 2006 IEEE International Conference on Robotics and Automation, pp. 2609–2614 (2006)
13. Canny, J.F.: A Computational Approach to Edge Detection. IEEE Transactions on Pattern Analysis and Machine Intelligence PAMI-8(6), 679–698 (1986)

A Generalized Neural Network Approach to Mobile Robot Navigation and Obstacle Avoidance

S. Hamid Dezfoulian, Dan Wu, and Imran Shafiq Ahmad

School of Computer Science, University of Windsor
401 Sunset Avenue, Windsor, Ontario N9B 3P4
{dezfoul,danwu,imran}@uwindsor.ca

Abstract. Navigation is one of the most important problems in developing and designing intelligent mobile robots. To locally navigate and autonomously plan a path to arrive to a desired destination, Artificial Neural Networks (ANNs) are employed to model complex relationships between inputs and outputs or to find patterns in data as they provide more suitable solutions than the traditional methods. However, current neural network navigation approaches are limited to one kind of robot platform and range sensor, and usually are not extendable to other types of robots with different range sensors without the need to change the network structures. In this paper, we propose a general method to interpret the data from various types of 2-dimensional range sensors and a neural network algorithm to perform the navigation task. Our approach can yield a global navigation algorithm which can be applied to various types of range sensors and robot platforms. Moreover, this method contributes positively to reducing the time required for training the networks.

Keywords: Robot Navigation, Obstacle Avoidance, Artificial Neural Networks, Sensor Visualization, Generalization.

1 Introduction

Mobile robot navigation field involves the assets such as “nonlinear mapping, ability to learn from examples, good generalization performance, massively parallel processing, and the ability to approximate any function given adequate number of neurons” [1]. These involvements have caused extensive use of neural networks in this field. The different kinds of neural nets used, include: feedforward neural network, self-organizing neural network, principal component analysis (PCA), dynamic neural network, support vector machines (SVM), etc. [2], [3]. Training a feedforward multi-layer neural network using the back-propagation algorithm can be used to approximate arbitrary functions and to classify patterns. In [4], Thrun has "translated" the readings of sonar sensors into occupancy values of each grid cell for building metric maps by employing a feedforward neural network.

Working with the High-dimension data, requires effective techniques for reducing the dimension of the data and extracting the principal features. In the different fields of machine learning such as data compression and pattern recognition, PCA is one of

S. Lee et al. (Eds.): Intelligent Autonomous Systems 12, AISC 193, pp. 25–42.
springerlink.com

the most effective and known methods [5]. To localize a mobile robot, Vlassis et al. [6] proposed a method using PCA to reduce the dimensions of sonar sensor data. In [7] Crowley et al. proposed a method employing the reduced data from laser range sensor using PCA to estimate the position of a mobile robot.

Avoiding obstacles by the robots in their navigation environment is a necessity. The data gathered from the environment by the robot's sensors, are fed in neural networks designed for this matter to predict the driving direction for the robot to proceed to its target. In [8], Fujii et al. used reinforcement learning algorithm to train a multilayered neural network navigate a mobile robot by avoiding any collisions. In [9], to avoid and detect obstacles in unknown environments, Silva et al. proposed a modular network called MONODA (MOdular Network for Obstacle Detection and Avoidance).

However, most of the current researches address only one particular type of sensor or robot platform. The main issue with neural network approaches is the training of the network. Collecting sufficient, yet valuable, samples from the environment to train the network can sometimes be frustrating and very time consuming [10]. In addition, apart from the effort that has to be put to collect valuable samples, the training time of a network can be significantly high [10]. In any neural network navigation algorithm, when a network structure is designed for a specific type of sensor, it cannot be used for other types or different numbers of sensors. By changing the structure of the network, thus, new training samples are required and the network needs to be trained from the beginning. This presents an opportunity to develop a general method to interpret sensor data from different types of sensors that can yield a global navigation algorithm, applicable to various types of range sensors and robot platforms.

In this paper, we make several contributions that address the challenge involved in applying neural network navigation algorithms to various types of range sensors and robot platforms. First we propose a general method for interpreting different types of range sensors, such as laser and ultrasonic sensors. This approach extends the work done by Janglová [11] for determining the free-spaces by applying PCA Neural Network (PCNN) [12]. Then we introduce an algorithm to perform the navigation task using our interpretation of sensory data which has been inspired by the works done by Parhi and Singh [13] for neural network robot navigation.

Then, we carry out experiments with simulated robots, to demonstrate the usefulness of the sensor data interpretation for various range sensors. We empirically demonstrate that even though we have trained our neural networks by only employing readings from laser range sensor, the robots are able to perform navigation tasks using other range sensors as well.

2 The Proposed Method

This work is concerned with the problem of generalizing the interpretation of sensory data and mobile robot navigation. The contributions of our work are as follows:

The primary contribution of this paper is to develop a general method for interpretation of different types of sensors, such as laser and ultrasonic sensors. Our approach extends the work done by Janglová [11] for determining the free-spaces by applying PCA Neural Network (PCNN). We study the problem of how current neural network navigation approaches are limited to one type of sensor and the kinematic constraints of a mobile robot.

The second important contribution of this paper is to implement an algorithm to perform the navigation task using our interpretation of sensory data. Our approach is inspired by the works done by Parhi and Singh [13–15] for neural network robot navigation.

The third contribution of this work is that we develop a software application to carry out the proposed approach. The experimental results obtained through this application indicate feasibility of our approach in simulation robots.

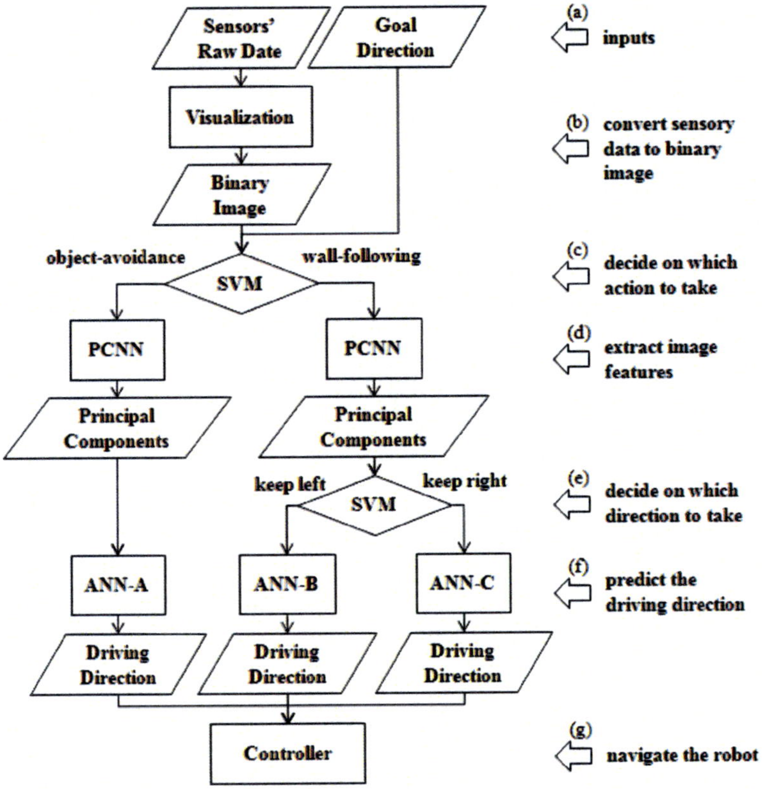

Fig. 1. Flowchart of proposed method

In this section, we propose an approach to address the problem of extending neural network navigation algorithms for various robots and range sensors. In order to overcome this problem, we introduce a new structure as illustrated in Fig. 1, to interpret different types of 2-dimensional sensor data and provide a global navigation algorithm that can be applied to various types of sensors and robot platforms.

We can divide our proposed method (Fig. 1) into four main sections; Visualization, Dimensionality Reduction - Feature Extraction, Classification and Prediction.

2.1 Visualization

The *visualization* section (Fig. 1(b)) is the first and most essential section in our proposed method. It is very important to have a general representation of the raw sensory data for all types of 2-dimensional range sensors so that different types, or numbers, of range sensors can be processed in the same manner. For this purpose the data from different sensors are visualized onto 50×50 pixels bitmaps covering a 25 meter area (5×5 meters) in front of the mobile robot. Fig. 2 illustrates examples of how readings from two different range sensors are visualized onto images.

For the laser range sensor (SICK) [16], the server returns 181 values at every sensor reading, representing the distance of obstacles in 180 degrees in front of the robot (0°-180°) with a maximum range of 6 meters. However, visualization of the sonar sensor is somewhat different than the laser range sensor. Each value from the sonar sensors represents the distance from the closest object in a 20° cone in the orientation of that sensor with a maximum range of 6 meters. Fig. 2(b) depicts visualizations of 8 sonar sensors mounted in front of the robot.

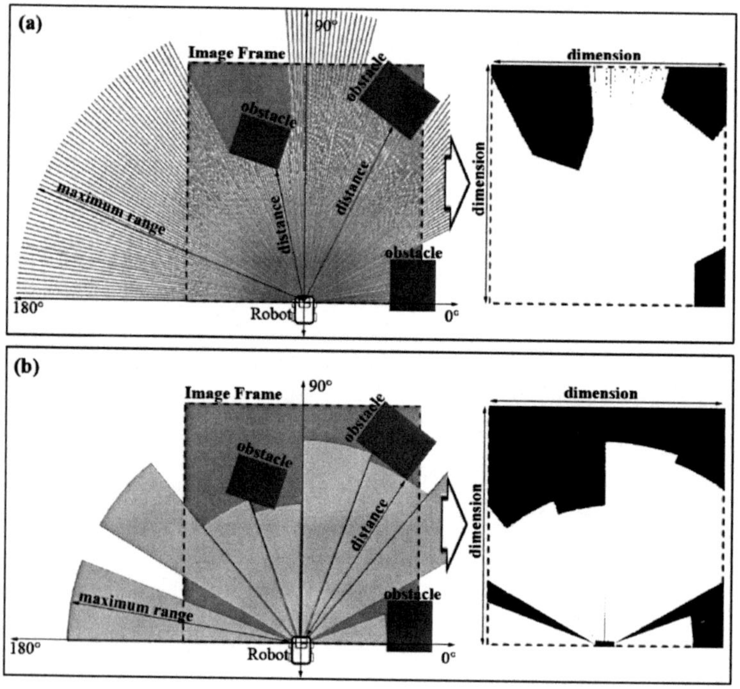

Fig. 2. Example of visualizing the data from a) a laser range finder b) 8 sonar sensors, into a binary image

2.2 Feature Extraction

This section (Fig. 1(d)) consists of a *dimensionality reduction* algorithm to reduce the dimensions of the images. If we apply the plain images as the inputs by means of using each pixel as an input unit to the networks, the training process could be very slow due to the large number of pixels. Since, multilayer neural networks can be very slow when introduced with too many input data [17], it is necessary to reduce the dimensions of the inputs. This can be done by extracting the most relevant and important features from the images using feature extraction methods [18–24].

One of the best known feature extraction learning algorithms is the Generalized Hebbian Algorithm (GHA) [25], which can extract a selected number of principal components from a stationary or quasi-stationary multivariate random process. It applies to a single-layered feed-forward neural network described by $z(t) = Q^T(t)x(t)$, where $x(t) \in R^p$, $z(t) \in R^m$, thus $Q(t) \in R^{p \times m}$. The GHA rule writes:

$$Q(t+1) = \mu x(t) z^T(t) + Q(t)(I_m - \mu LT[z(t) z^T(t)]) \quad (1)$$

where μ is a small learning rate (empirically chosen as 1.0E-05), m is a small positive learning step size, the operator $LT[\cdot]$ returns the lower-triangular part of the matrix contained within, and I_m denotes the identity matrix of size m. In our experiments we have used this method to reduce the dimension of the images from 2500 features to 100 features.

However, there is always a trade-off between performance and speed when selecting the number of extracted features. Through experiments, we came to believe that 100 features are sufficient for our proposed method, which maintains a reasonable training speed and at the same time a high accuracy.

2.3 Classification

In the *classification* sections, the first support vector machine (SVM) (Fig. 1(c)) is trained by introducing it with 150 training samples gathered from the simulation environment using the laser range sensor data. A Radial Basis Function (RBF) with sigma value equal to 10 is used for the kernel of this SVM. This SVM classifies any new sample to either object-avoidance or wall-following class, given an image and the direction of the target.

When a mobile robot encounters a wide object, or a wall, while moving towards the target, it needs to decide which direction to take (left or right), to get past the object. In addition, to safely navigate around U-shaped objects, the robot requires to keep to only one direction. Hence, the second SVM (Fig. 1(e)), which is used only in case of wall-following, with similar parameters as the previous SVM, is trained with 200 training patterns to classify the new samples into *keep-left* or *keep-right* classes.

2.4 Prediction

In this section, three multilayer Artificial Neural Networks (ANN - A, B and C) are trained by using supervised learning with backpropagation algorithm to provide driving directions. The chosen number of layers and units in each layer were found empirically to facilitate training. Moreover, these networks are trained only by samples generated using the laser range scanner. The prediction step of our algorithm consists of two main modules; object-avoidance and wall-following.

2.4.1 Object-Avoidance

In the object-avoidance module, an artificial neural network (ANN-A in Fig. 3) is trained by introducing it with 3000 samples gathered from the training environment shown in Fig. 6. This network is used to produce driving directions to navigate the mobile robot towards the target.

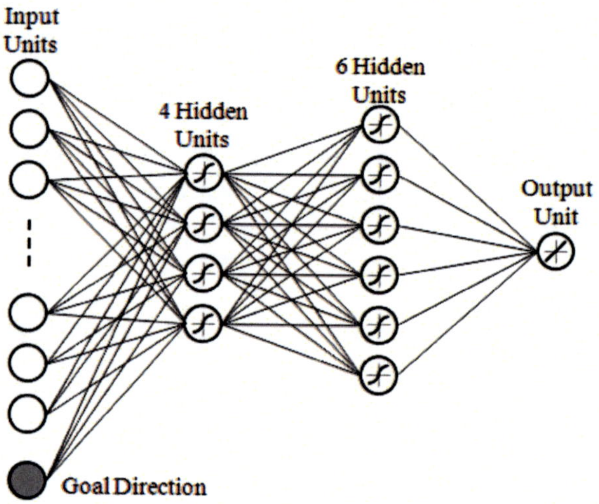

Fig. 3. Proposed neural network architecture for ANN-A

During training and normal operation, the input patterns provided to neural network A comprise the following components:

$y_n^{\{1\}}$ = Extracted features from the binary image.

$y_{n+1}^{\{1\}}$ = Target bearing from the robot's location.

where n is the number of extracted features. These input values are distributed to the hidden units which generate outputs by

$$y^{\{l\}} = f\left(\sum_i W_{ji}^{\{l\}} . y_i^{\{l-1\}}\right) \tag{2}$$

where l is layer number (2 or 3); j is label for j th unit in hidden layer $\{l\}$; i is label for i th unit in hidden layer $\{l-1\}$;

$W_{ji}^{\{l\}}$ is the weight of the connection from unit i in layer $\{l-1\}$ to unit j in layer $\{l\}$; $f(\cdot)$ is the activation function, chosen in this work as hyperbolic tangent sigmoid:

$$n = \frac{2}{1+e^{-2\times n}} - 1 \tag{3}$$

The mean squared difference between $\theta_{desired}$ and θ_{actual} for the set of presented training patterns is used as a measure of performance for the network.

$$Err = \frac{1}{n}\sum_{i=1}^{n}\left(\theta_{desired} - \theta_{actual}\right)^2 \tag{4}$$

where n is the number of training patterns. The error is then back propagated to the previous layers using the backpropagation method to train the network. In order to determine the appropriate weight adjustments to reduce error, this method requires the computation of local error gradients. For the output layer, the error gradient, $\delta^{\{4\}}$, is

$$\delta^{\{4\}} = f'\left(\sum_i W_{1i}^{\{l\}}.y_i^{\{l-1\}}\right)\left(\theta_{desired} - \theta_{actual}\right) \tag{5}$$

The local gradient for units in the hidden layer $\{l\}$ is provided by

$$\delta_j^{\{l\}} = f'\left(\sum_i W_{ji}^{\{l\}}.y_i^{\{l-1\}}\right)\sum_k \delta_k^{\{l+1\}}.W_{kj}^{\{l+1\}} \tag{6}$$

The synaptic weights are updated according to the following expressions

$$W_{ij}(t+1) = W_{ij}(t) + \Delta W_{ij}(t+1) \tag{7}$$

where

$$\Delta W_{ij}(t+1) = \alpha\Delta W_{ij}(t) + \Delta\eta\delta_j^{\{l\}}y_i^{\{l-1\}} \tag{8}$$

where η = learning rate and α = momentum were found empirically to be 0.3 and 0.06 respectively and t is the iteration number. Every iteration is composed of presentation of a training pattern and correction of the weights. The final output from the neural network is

$$\theta_{actual} = \sum_i W_{1i}^{\{4\}}.y_i^{\{3\}} \tag{9}$$

However, this will only ensure a collision free path to the target, which does not provide a solution for navigating around deadlocks and U-shaped objects or for instance going from one room to another. Therefore, a wall-following algorithm is a necessity.

2.4.2 Wall-Following

In the wall-following section of the algorithm, an SVM classifier (Fig. 1(e)) is trained using 200 patterns to select a wall-following direction (keep-left or keep-right). In this section, two identical networks (Fig. 4) are separately trained by introducing 1500 samples gathered from the environment shown in Fig. 7. These networks provide driving directions for keeping left or keeping right at each time step of wall following. During training and normal operation, the input patterns provided to the *wall-following* neural networks consist of only the extracted features from the binary image ($y_n^{\{1\}}$, n being the number of extracted features). In other words, these networks provide an output regardless of the direction of the goal. Therefore, allowing the robot to safely move out of U-shaped objects and other situations such as deadlocks without getting stuck. The performance measurement of the whole network and the transfer functions in different layers are the same as ANN-A.

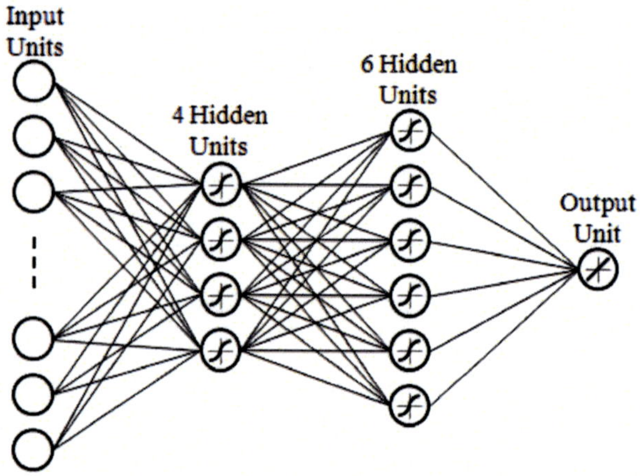

Fig. 4. Proposed neural network architecture for ANN-B and ANN-C

3 Experimental Results

In the following experiments we use USARSim server [26], [27], which is installed on top of the Unreal Tournament [28] game engine. Under the help of this tool, we can test our proposed method in a variety of scenarios.

3.1 Training

Our experiments are conducted with three different simulated robots; P2AT [29], Zerg [26], and Talon [26] with three different configurations of two types of sensors (laser range sensor and sonar sensor) shown in Fig. 5.

For training purposes, only the laser range sensor readings of the P2AT mobile robot have been used. To train the object-avoidance network (ANN-A), 3000 training patterns are gathered from the environment shown in Fig. 6. This is done by manually navigating the robot from the starting positions to the target. To train the wall-following network, ANN-B, the robot is driven manually in the environment shown in Fig. 7 only following the wall and keeping-left. The same environment, but mirrored, is used for collecting training data for ANN-C (keep-right).

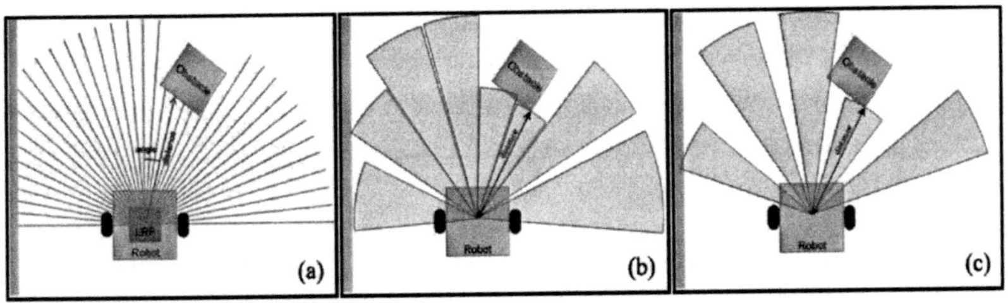

Fig. 5. Sensor distributions. a) laser range scanner b) 8 sonar sensors c) 5 sonar sensors.

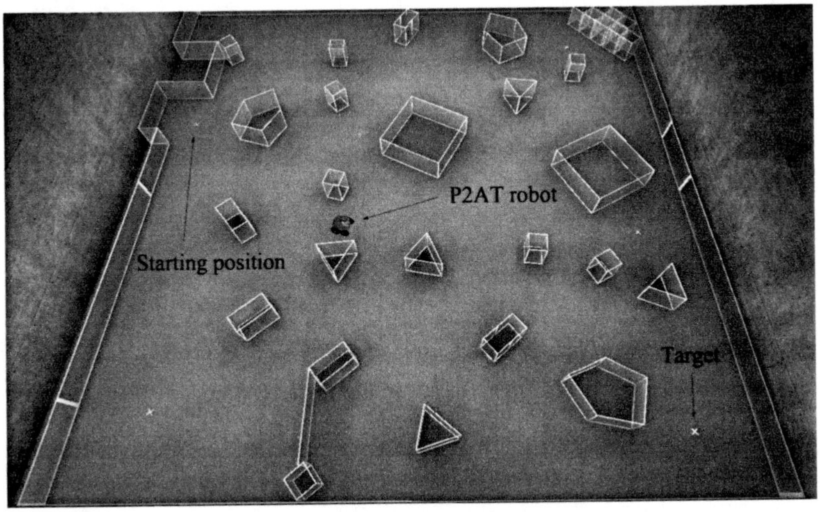

Fig. 6. Training environment for object avoidance

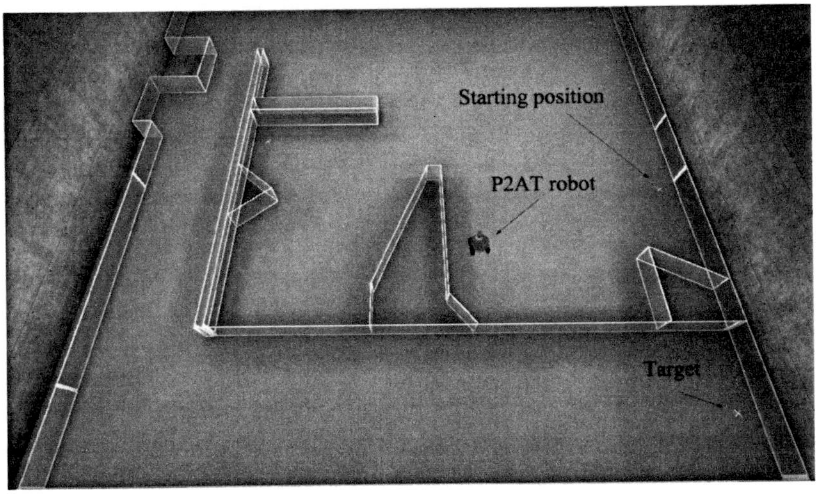

Fig. 7. Training environment for wall following

Table 1 compares performances and training durations of three feature extraction methods, GHA, APEX [13] and NMF [30], for training, object avoidance and wall following neural networks, ANN-A, ANN-B and ANN-C. The values in the table illustrate the mean of 10 performance tests carried out for each network and each feature extraction method. The performance columns show the mean squared error for training the neural networks. Also the regression columns describe the relationship between predictor and response variables. The bold values in the table highlight the best performances and training times for the networks using different feature extraction methods.

Table 1. Comparison of three feature extraction methods for training ANN-A, ANN-B and ANN-C using the average value of 10 performance trainings

		Training Time		Performance	Regression
ANN	F. E. Method	F. E.	NN Train	Overall	Overall
A	GHA	114.190s	14.820s	**410.8**	**0.7669**
	APEX	164.240s	12.310s	424.12	0.7597
	NMF	**20.254s**	**10.180s**	821.79	0.4257
B	GHA	59.511s	8.076s	**210.57**	**0.8591**
	APEX	80.160s	**6.956s**	215.37	0.8538
	NMF	**20.017s**	7.381s	326.12	0.7498
C	GHA	61.717s	8.186s	**220.82**	**0.8523**
	APEX	89.854s	7.762s	236.44	0.8379
	NMF	**13.799s**	**6.856s**	315.18	0.7787

As shown in the table, the training speed of NMF is noticeably faster than the other two methods. The extracted features, using this method, even have some affects on reducing the training time of the neural networks. However, compared to the other methods, the NMF method has the lowest training performance. In our case the training time does not affect the overall performance of our proposed method. Therefore, based on our experiments shown in Table 4, we have chosen GHA for testing our algorithm in different environments. As it can be seen from the table, the GHA method has the highest performance in training the networks and generalizes very well to the validation set. The APEX method is not considered due to the fact that its performance is not as good as GHA and it also has a higher training time.

Fig. 8 and Fig. 9 depict the regression and error plots for training ANN-A from our algorithm. As it can be seen from Fig. 9, the mean squared error of the validation and test samples start to increase after epoch 13. Therefore to prevent the network from over fitting the training samples and to be able to generalize to new samples, training is halted at epoch 13. The regression plots in Fig. 8 show the results of the networks outputs for the training patterns compared to the actual targets at step 13.

Regression and performance plots of training results for keep-left and keep-right wall following networks are shown in Figure 45 through Figure 48. Based on the performance plots we can see that the networks have obtained the best validation performance for training ANN-B and ANN-C at epochs 16 and 12 respectively. The plots show very good results for the laser scanner patterns (training samples).

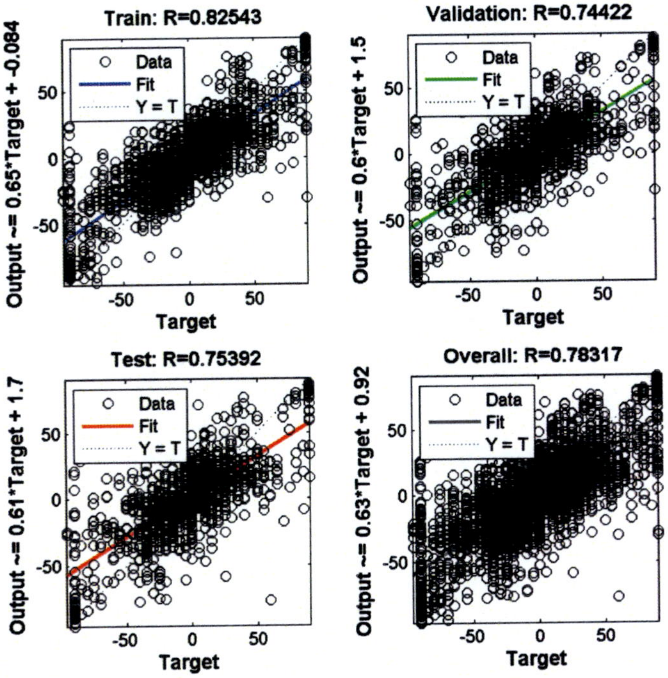

Fig. 8. Regression plots for training ANN-A with 3000 training samples and 3000 samples for testing and validation

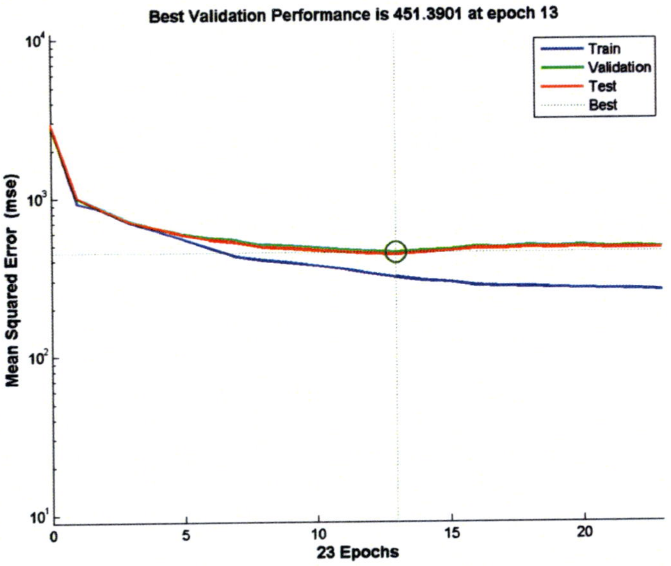

Fig. 9. Performance plot for training ANN-A

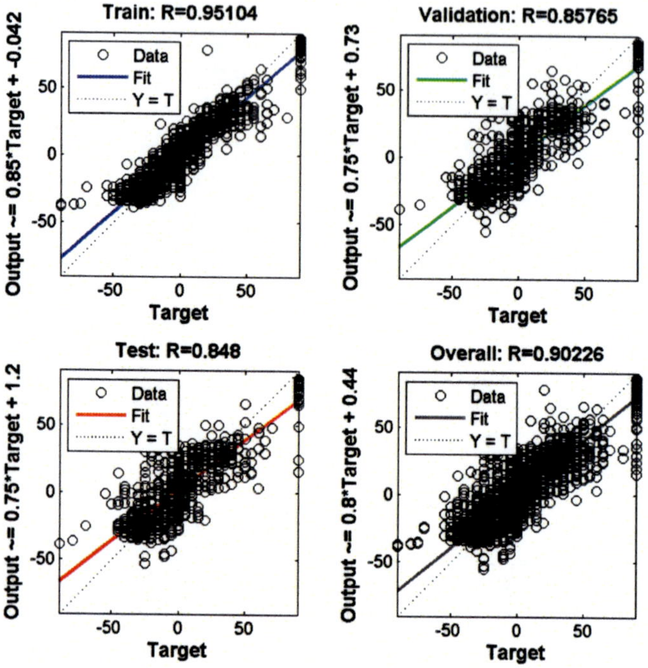

Fig. 10. Regression plots for training ANN-B with 1500 training samples and 1500 samples for testing and validation

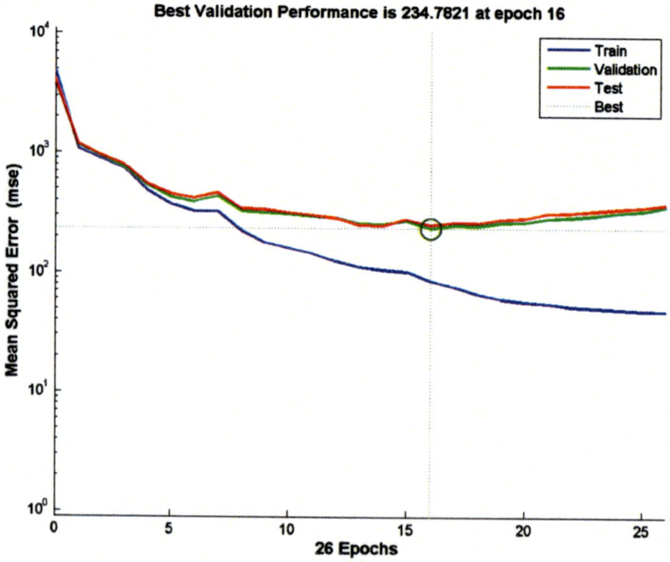

Fig. 11. Performance plot for training ANN-B

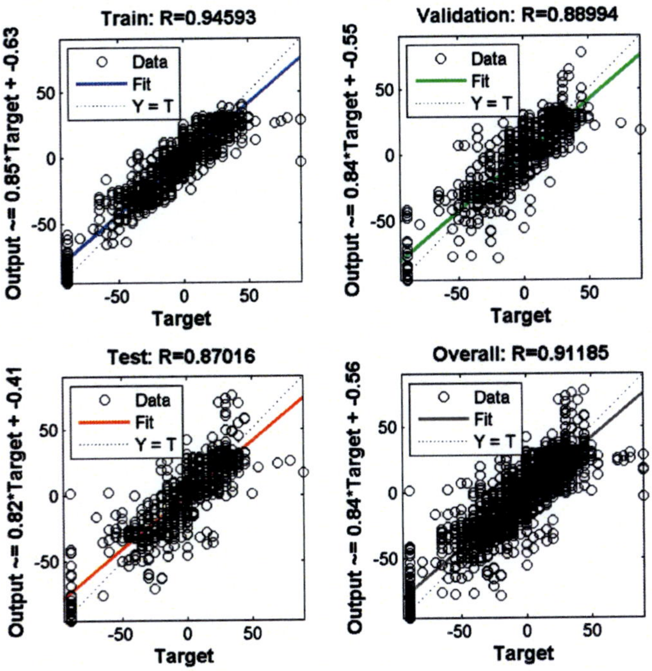

Fig. 12. Regression plots for training ANN-C with 1500 training samples and 1500 samples for testing and validation

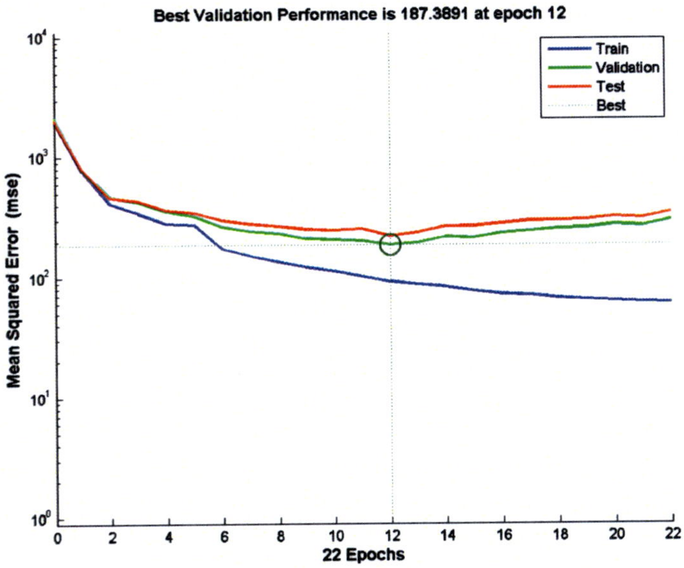

Fig. 13. Performance plot for training ANN-C

Although the accuracy in comparison to the validation and testing patterns are a bit lower than the training samples, the final result is satisfying.

3.2 Experiments

To test our algorithm, we conducted experiments in several environments replicated from [13]. Fig. 15(a), Fig. 16(a) and Fig. 17(a) show the experimental result and simulation result conducted by Parhi and Singh in [13] in these three different environments. We simulated these environments in Unreal Tournament's engine. Fig. 15(b-d), Fig. 16(b-d) and Fig. 17(b-d) show our simulation results for the three robots in these three environments respectively using three different range sensors and settings.

We can see the some differences between the paths traversed using different range sensors using the color scheme shown in Fig. 14. However, all three mobile robots can successfully navigate from the starting position to the target's location using the three different range sensor configurations.

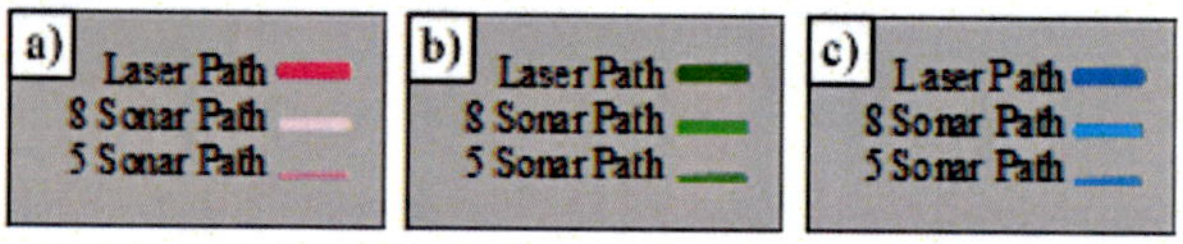

Fig. 14. Legend for simulation results in Fig. 10, Fig. 11 and Fig. 12 for a) P2AT, b) Zerg and c) Talon using three different sensors

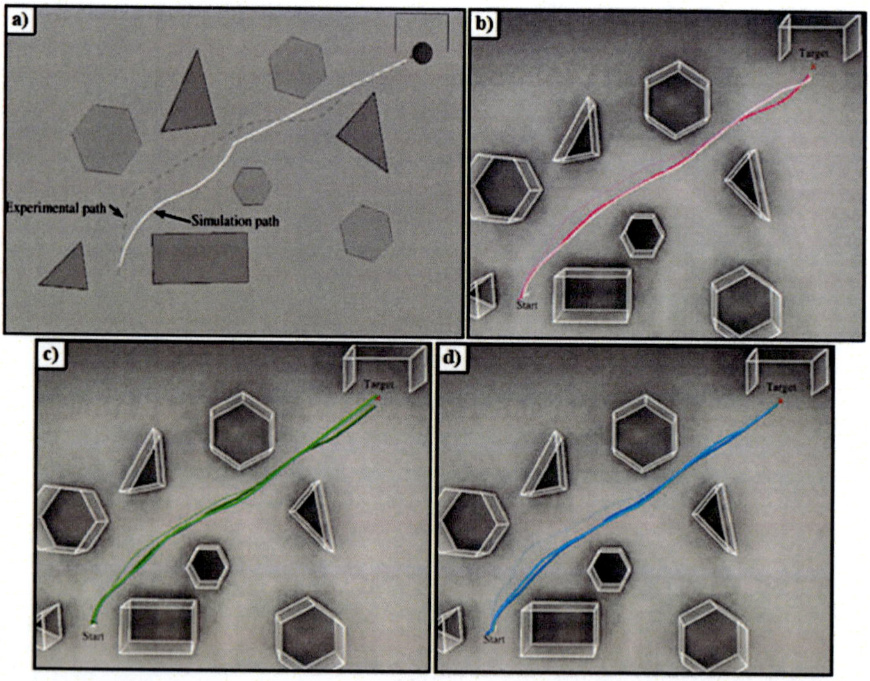

Fig. 15. Simulation results for b) P2AT, c) Zerg and d) Talon using three different sensors in environment 1 (a)

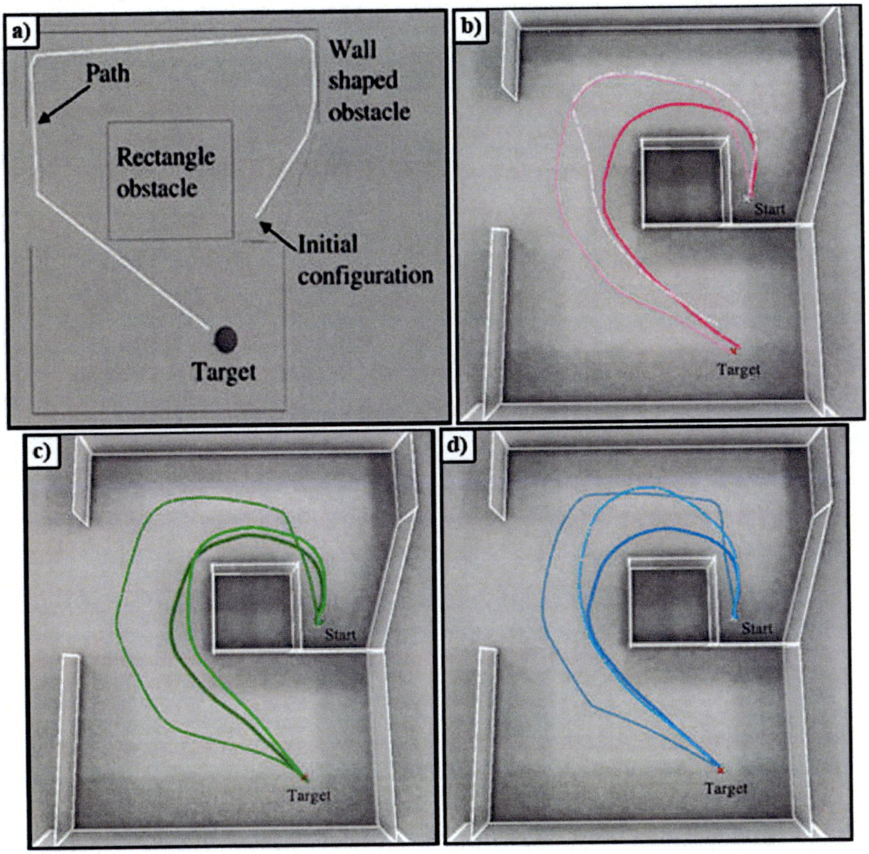

Fig. 16. Simulation results for b) P2AT, c) Zerg and d) Talon using three different sensors in environment 2 (a)

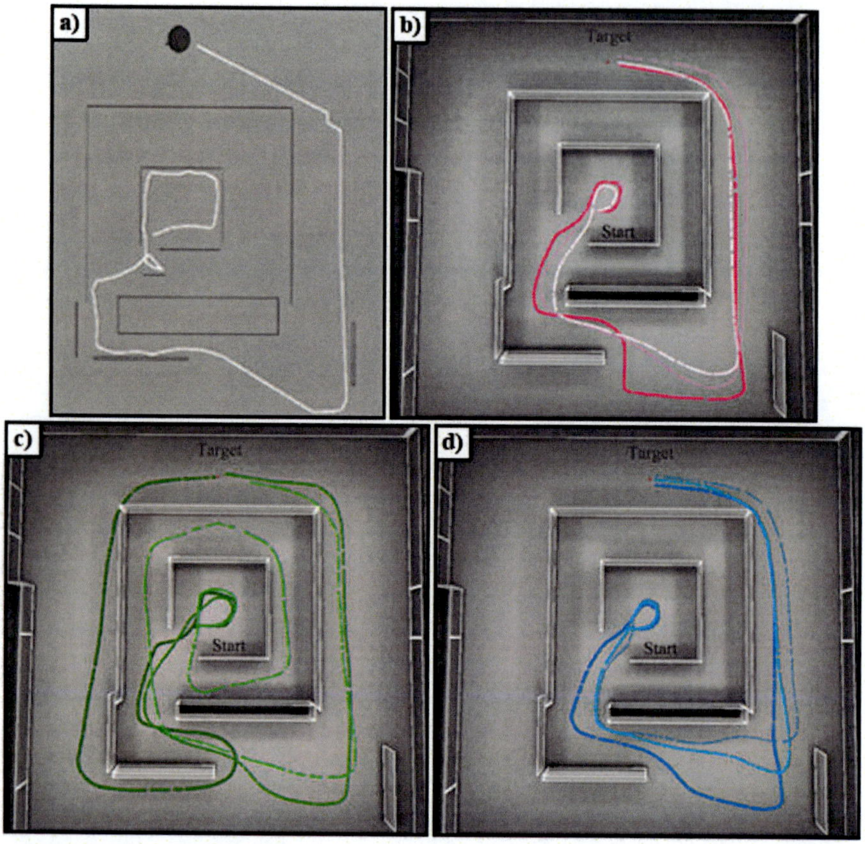

Fig. 17. Simulation results for b) P2AT, c) Zerg and d) Talon using three different sensors in environment 3 (a)

4 Conclusion and Future Work

In this paper, a general approach for interpretation of different types of sensors, such as laser range scanner and ultrasonic sensors has been developed. A system for navigation of mobile robots using these interpretations has also been developed. In our approach we propose a generalized method to interpret different kinds of 2-dimensional- sensors. Instead of training the networks for all types of sensors, our approach only trains the networks using one of the most accurate 2-dimensional sensors. Therefore, it avoids unnecessary training periods and the necessity to gather training samples for different kinds of sensors. Experiments carried in simulated environments demonstrate that our approach is affective in mobile robot navigation for different kinds of robots and sensors. Therefore, our proposed globalized navigation algorithm can yield significant improvements in training time and to lower sensor costs.

In future, we would like to employ online or reinforcement learning algorithms [31–33] to reduce the effort of gathering training samples and to improve our algorithms adaptability to different environments.

In this paper we have only used three kinds of simulated robots and three types of range sensors and configurations. Therefore, another objective for future research is to extend the algorithm to real world environments using real robots and other types of range sensors.

References

[1] Zou, A.-M., Hou, Z.-G., Fu, S.-Y., Tan, M.: Neural Networks for Mobile Robot Navigation: A Survey. Science and Technology, 1218–1226 (2006)

[2] Bishop, C.M.: Pattern recognition and machine learning, vol. 4. Springer, New York (2006)

[3] Mitchell, T.M.: Machine learning. McGraw Hill, Burr Ridge (1997)

[4] Thrun, S.B.: Exploration and model building in mobile robot domains. In: IEEE International Conference on Neural Networks, pp. 175–180 (1993)

[5] Jolliffe, I.: Principal component analysis, pp. 29–62. Springer, New York (2002)

[6] Vlassis, N., Motomura, Y., Kröse, B.: Supervised dimension reduction of intrinsically low-dimensional data. Neural Computation 14(1), 191–215 (2002)

[7] Crowley, J.L., Wallner, F., Schiele, B.: Position estimation using principal components of range data. Robotics and Autonomous Systems 23(4), 267–276 (1998)

[8] Fujii, T., Arai, Y., Asama, H., Endo, I.: Multilayered reinforcement learning for complicated collision avoidance problems. In: Proceedings of the 1998 IEEE International Conference on Robotics and Automation, vol. 3, pp. 2186–2191 (1998)

[9] Silva, C., Crisostomo, M., Ribeiro, B.: MONODA: a neural modular architecture for obstacle avoidance without knowledge of the environment. In: Proceedings of the IEEE-INNS-ENNS International Joint Conference on Neural Networks, IJCNN 2000. Neural Computing: New Challenges and Perspectives for the New Millennium, vol. 6, pp. 334–339 (2000)

[10] Fausett, L.V.: Fundamentals of neural networks: architectures, algorithms, and applications. Prentice-Hall, Englewood Cliffs (1994)

[11] Janglová, D.: Neural Networks in Mobile Robot Motion. Advanced Robotic 1(1), 15–22 (2004)

[12] Diamantaras, K.I., Kung, S.Y.: Principal component neural networks: theory and applications. John Wiley & Sons, Inc. (1996)

[13] Singh, M.K., Parhi, D.R.: Path optimisation of a mobile robot using an artificial neural network controller. International Journal of Systems Science 42(1), 107–120 (2011)

[14] Singh, M.K., Parhi, D.R.: Intelligent neuro-controller for navigation of mobile robot. In: Proceedings of the International Conference on Advances in Computing, Communication and Control, ICAC3 2009, pp. 123–128 (2009)

[15] Parhi, D.R., Singh, M.K.: Real-time navigational control of mobile robots using an artificial neural network. Proceedings of the Institution of Mechanical Engineers, Part C: Journal of Mechanical Engineering Science 223(7), 1713–1725 (2009)

[16] SICK Group | Sensor Intelligence, `http://www.sick.com` (accessed: October 11, 2011)

[17] Haykin, S.: Neural networks: a comprehensive foundation. Prentice-Hall (1999)

[18] Wold, H.: Partial least squares (1985)

[19] Weinberger, K.Q., Saul, L.K.: Unsupervised learning of image manifolds by semidefinite programming. International Journal of Computer Vision 70(1), 77–90 (2006)
[20] Lu, H., Plataniotis, K.N., Venetsanopoulos, A.N.: A survey of multilinear subspace learning for tensor data. Pattern Recognition 44, 1540–1551 (2011)
[21] Lee, J.A., Verleysen, M.: Nonlinear dimensionality reduction. Springer (2007)
[22] HyvArinen, A., Karhunen, J., Oja, E.: Independent component analysis, vol. 26. Wiley-interscience (2001)
[23] Dumais, S.T.: Latent semantic analysis. Annual Review of Information Science and Technology 38(1), 188–230 (2004)
[24] Joliffe, I.T.: Principal component analysis. Springer, New York (1986)
[25] Sanger, T.D.: Optimal unsupervised learning in a single-layer linear feedforward neural network. Neural networks 2(6), 459–473 (1989)
[26] Balakirsky, S., Proctor, F.: USARSim
[27] Carpin, S., Lewis, M., Wang, J.: USARSim: a robot simulator for research and education. Robotics and, 1400–1405 (2007)
[28] Epic-Games, Unreal Tournament 2004, `http://www.gog.com/en/gamecard/unreal_tournament_2004_ece`
[29] ActivMedia, ADEPT mobile robots, `http://www.activrobots.com`
[30] Lee, D.D., Seung, H.S.: Algorithms for non-negative matrix factorization. In: Advances in Neural Information Processing Systems, vol. 13 (2001)
[31] Saad, D.: On-line learning in neural networks, vol. 17. Cambridge Univ. Pr. (1998)
[32] Borodin, A., El-Yaniv, R.: Online computation and competitive analysis, vol. 53. Cambridge University Press, New York (1998)
[33] Sutton, R.S., Barto, A.G.: Reinforcement learning: An introduction, vol. 116. Cambridge Univ. Press (1998)

Drivable Road Recognition by Multilayered LiDAR and Vision

Suhyeon Gim[1], Ilyas Meo[1], Yongjin Park[1], and Sukhan Lee[1,2]

Intelligent Systems Research Institute,
SungKyunKwan University[1], Dept. of Interaction Science,
School of Information and Communication Eng.[2]
Suwon 440-746, Rep. of Korea
{soohyunkim,isri}@skku.edu, snulove@yahoo.com,
lsh@ece.skku.ac.kr

Abstract. This paper presents the processing of the drivable road recognition by multilayered LiDAR and vision. Multilayered LiDAR information used for detecting a planetary region with boundaries and vision processing gives colored lane information for safe driving control. During navigating the road, EKF result on two different information are fused for robust and reliable navigation. This sensor fusing technique makes the autonomous navigation system to be robust and useful in real environment not only on regular road intersection but also unpaved ground way.

Keywords: Drivable road recognition, multilayer LiDAR, computer vision, Extended Kalman Filter Fusion.

1 Introduction

The recognition of human activity is a concerning problem to provide interactive service with the user in various environments. Research about activity recognition is emerging recently, using various sensory resources. There are studies using cameras or GPS based on the user's movement by using pattern recognition techniques. Many of the other research use accelerometers as the main sensor placing the accelerometer on the user's arm, leg or waist. They recognize activities like walking or running, which has movements in the activity and can be optimized for the recognition based on the accelerometer.

Accelerometers show a high accuracy in recognizing activities with lots of movements. But they are weak to recognize the activities with little movements. Accelerometers also provide continuous sensory data which is hard to separate to the exact boundary of several states. Other research use physiological sensors to recognize the user's context, as the user's body status can represent the user's activity and also the user's emotion which depends on the activity. Therefore using the physiological sensor with accelerometers will help to recognize the user's activity.

In this paper, we use not only an accelerometer but also physiological sensors to help the recognition of activities. Accelerometers have advantages to measure the

S. Lee et al. (Eds.): Intelligent Autonomous Systems 12, AISC 193, pp. 43–56.
springerlink.com

user's movement and the physiological signals have advantages to measure the user's status of the body. Most of these sensors calibrate continuous sensor data, and if the data is near the boundary of a specific state, the evidence variable will show a radical difference in small changes. To lessen these changes in this situation we propose to preprocess the data with fuzzy logic. Fuzzy logic can represent ambiguous states in linguistic symbols, which is good in continuous sensory data. As sensory data include uncertainty of calibrating data and also human activity itself has it too, we use Bayesian network to do the inference, and modify the learning and inference methods that fit to the fuzzy preprocessing.

2 Sensor Configuration

2.1 Multi Layered LiDAR Sensor

From the control point of view, the drivable road region is a certain frontal area from the vehicle head since each sensor has limited effective range. The LiDAR sensor has range information intuitively at the scanned line which makes cumulative data set definitely flat or not as the vehicle proceeds. Generally some curbs or obstacles are detected and positional information has been used for the vehicle navigation. However, these schematics may have some risky conditions in case of intersection or roads where there is no curb. Since it is considered that the vehicle navigates both the paved lane road and open terrain, the accumulated summing on each scanned line and registering by some polygonal area in memory are useful for identification.

The LiDAR sensor was developed for the distance warning system in 1996 and improved for Adaptive Cruise Control functionality in 1997[7]. The curb was used as one of the most useful targets for road detection and track following control since the curb is generally located at the road boundary and its distinction between road and footpath makes clear for detecting the road [9]. Some research calculated the lane parameters such as the curvature, yaw angle, i.e. angle between the lane direction and ego-vehicle motion, relative offset between the vehicle and the road boundary and the lane width by separating lane from the road of different reflectivity [5-6].

In this work, the ibeo LUX LiDAR system was used for range measurement and its specification is as follows.

Table 1. The average results of the experiment

Horizontal FOV	85 °
Vertical FOV	3.2 °
Update rate	12.5 Hz
Max. Position errors	10 cm
Range	200 m
Angular resolution	H: 0.125 ° , V: 0.8 °

a. Field of View

The system has four layers which allows for angle compensation by means of multi scan levels with different vertical angles and enables the vehicle to detect the object robustly even under a bumped ground. Two LiDAR sensors are installed in the head of the vehicle and the position and range for left one is presented as below.

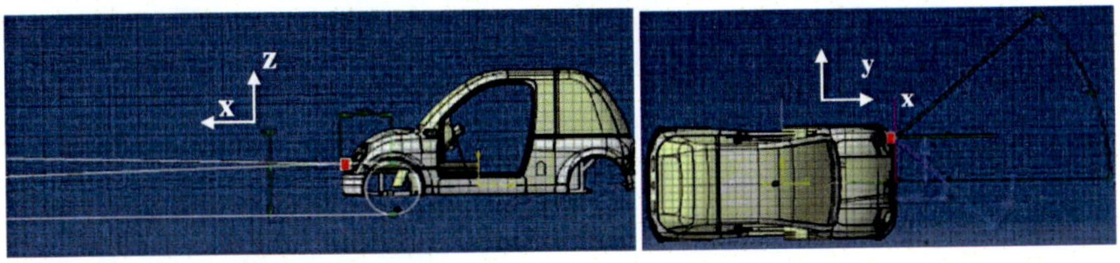

Fig. 1. Position and range of the LiDAR sensor

The sensor is mounted in 12 ° downwardly to detect road plane where the lower three layers take part in the road plane detection and upper layer data is analyzed for obstacle detection as this layer is emitted and reflected almost horizontally.

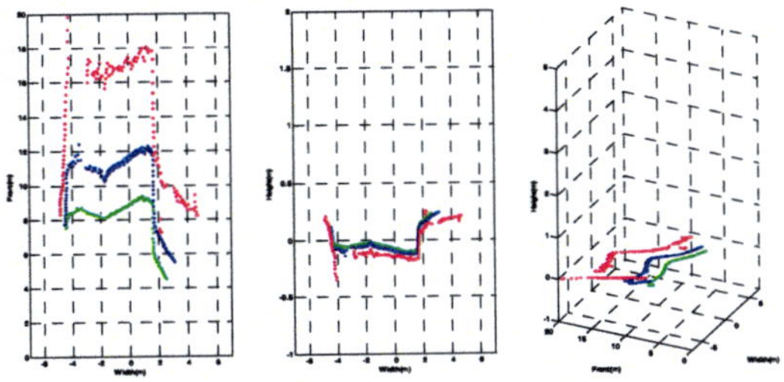

Fig. 2. LiDAR scanned results of the road

Fig. 2 shows a result of LiDAR scanning by the vehicle driving. The left one is plotting the X-Y plane data where X axis is corresponding to the road/driving direction and the Y axis is road width. Middle plot of the figure gives depth information of the road and the right shows the three dimensional construction by sensor coordinate transformation.

Extracting the drivable road boundary using LiDAR sensor is most important task in this research. In our work, histogram analysis is applied for robust information as following figure.

In Fig. 3 the histogram is generated for first laser scan layer data on driving direction, depth and road width axis. Each histogram gives the highest bin which has the greatest number of points. By integrating three results, road boundary can be determined for each layer. The road boundary has two end points which are calculated from analysis of the number of points at each bin. For driving direction axis, there are most reflective points around the arrival distance on the road plane. As shown in the first image in Fig. 2, there could be some fluctuations due to the road shape. In such

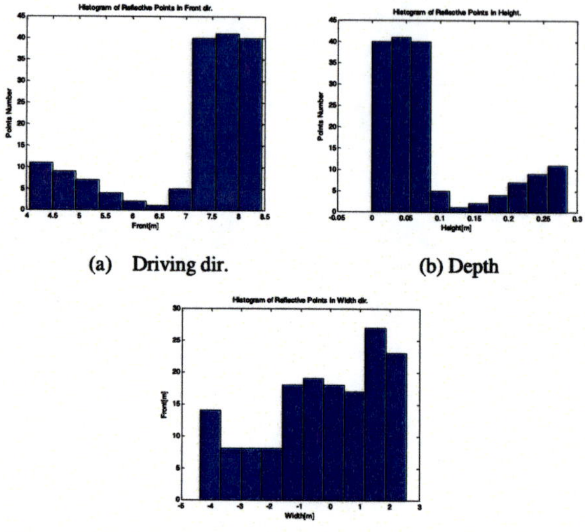

(a) Driving dir. (b) Depth

(c) Width

Fig. 3. Histogram analysis on each axis

cases, it makes to have powerful performance by using three axis analyses and analyzing for each histogram bin. Drivable road recognition by multilayer data is simply using individual results of each layer and they are connected to make some tetragonal polygons. Following figure presents the finally acquired drivable road regions.

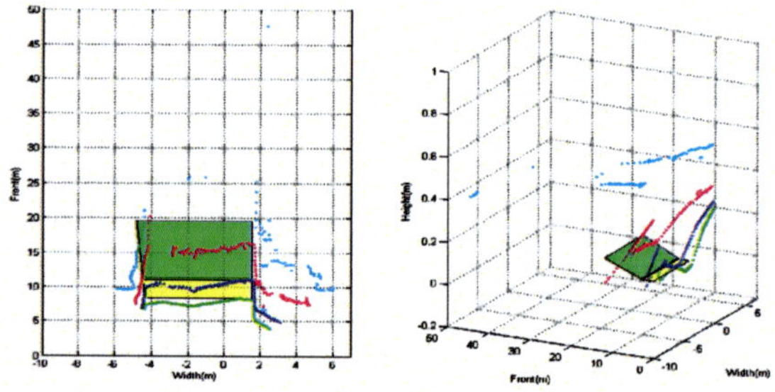

Fig. 4. Drivable road recognition by LiDAR

The first in Fig. 4 represents the top-viewed result which has two polygons with different colors from analyzing three layers where each polygonal area is the result from two consecutive layers. Nearest polygon region from the bottom signifies the more important region to be considered than the upper area. Three layers makes two drivable areas and all detected points are applied to fusion process with vision result as can be described in the latter section. For all axes in each layer, two bins in highest order are selected for candidate positions and if the positions for selected two bins are close each other, resultant position is the average sum of two bins.

2.2 Vision System

One of the most important image features with respect to the recognition of the drivable road is a stepped difference at the gradient intensities near road boundaries. Even in the unpaved road or cluttered open area, each point in the image for the drivable region has similar intensity values around all directions [8]. Detected boundary can be approximated by clothoidal segment which presents continuous change of the curvature along the road. The information from vision gives only colorful difference on the road and environment on the image plane. As the paved road is universally dark grayed and lane or road markers are bright, the drivable road can be recognized from discriminating the color intensities. Even in the dirt road where detected edges are seriously cluttered, using the property is useful as well.

Fig. 5. Drivable road recognition by LiDAR

For processing discrimination of the drivable road, 5 pixels shifted image in left is subtracted from the original image using the characteristic that drivable area has uniform color variation even though there is a lane or not. As each pixel in the drivable area has uniform intensities, the area in the subtracted result has almost zero intensity, i.e. black color. This procedure is also useful for identifying shadow region or small pebbles which have darker intensity as well as in recognizing a dirt road. And then, with a threshold value, the resulting image presents the black and white image for efficient edge detection and further processing.

The edge detection and corner detection are then performed. The result has unnecessary points especially in the upper region of the horizon, hence trapezoidal area is masked and filtered out, i.e. forced to be transformed zero intensity for the exterior area. The mask area is assigned sufficiently large to cover the turning section or intersection as well.

By the Hough Transform (HT) in the masked area, the lanes are recognized as follows. HT is well-known method for detecting line or circle and it is not described more in detail.

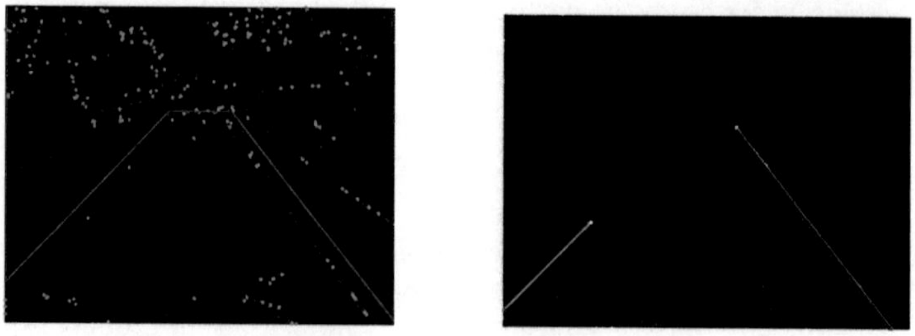

Fig. 6. Edge and corner detection (left) and road lane detection by HT (right)

The first result in Fig. 6 has a lot of detected corner points most of which are trees or some building structures and which should better be filtered out as analysis on such area is time-consuming. The other area to be ignored is the bottom part in the image which contains vehicle head. Since it spends some milliseconds to acquire the result while vehicle moves forward some meters until taking next snapshot at the sampling time, namely it is advantageous to consider forward distant region some meters ahead the vehicle. The second image shows the detected lane by HT of which highest two detections are chosen. It is much time-consuming and computationally complex that the Inverse Perspective Mapping (IPM) is performed for every acquired image frame. In the process, the lane or curve detection by HT is done in advance and then the IPM for a small number of points on the detected geometrically meaningful ends which can reduce about 80% of the whole computing time.

In order to obtain the IPM images, the intrinsic (focal length and optical center) and the extrinsic (pitch angle, yaw angle, and height above ground) camera parameters are required. The relation between the world frame P_w, image frame I_{img} is defined as

$$P_w = P_{IPM} I_{img} \tag{1}$$

Eqn. (1) is for converting the original image to the world point by IPM matrix [10].

$$I_{img} = P_{IPM}^{-1} P_W \tag{2}$$

Eqn. (2) is a relatoin that converts the World Point to Image Point. Parallel straight lines on the road can be created because the perspective is removed from the converted image. First of all, it is necessary to know the geometry of the camera to complete above equations. The basic geometry of the camera is assumed to be known. The position of the standard of the XY axis of the camera becomes a horizontal plane against the position of ZY axis, center axis of the camera in the space is also supposed to be known.

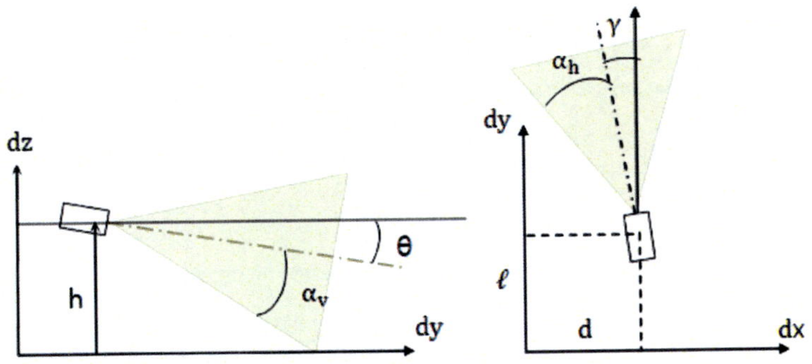

Fig. 7. Camera geometry configuration

In the above, the yaw angle is dx over dy, and dz and dy means the pitch angle, and θ_0 is the angle from the horizontal plane, and w and h are the width and height of the image respectively. The equation for converting image frame to the world frame is as follows.

$$x(u,v) = h \cdot \cot\left[(\theta - \alpha_v) + u \cdot \frac{2\alpha_v}{h-1}\right] \cdot \cos\left((\gamma - \alpha_h) + v \cdot \frac{2\alpha_h}{w-1}\right) + d \tag{3}$$

$$y(u,v) = h \cdot \cot\left[(\theta - \alpha_v) + u \cdot \frac{2\alpha_v}{h-1}\right] \cdot \sin\left((\gamma - \alpha_h) + v \cdot \frac{2\alpha_h}{w-1}\right) + \ell \tag{3}$$

For the control of vehicle steering, the orientation for the detected lanes is necessary. The Inverse Perspective Mapping is conducted as follows and the transformed image could give the lane orientation which informs the vehicle heads. The following two figures present the IPM results and lane detection results for straight and crossing roads respectively.

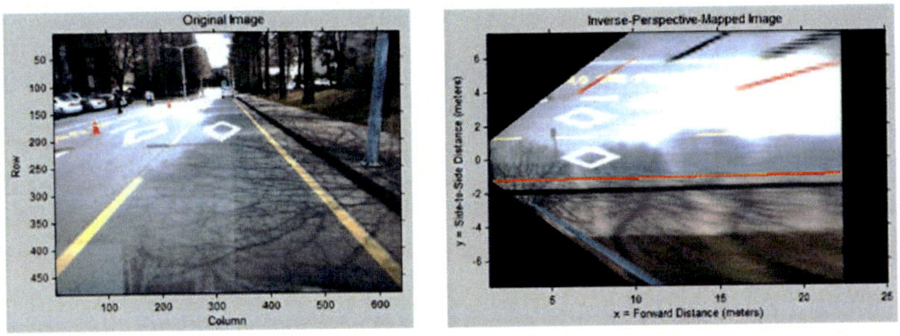

Fig. 8. IPM result and lane detection on straight road

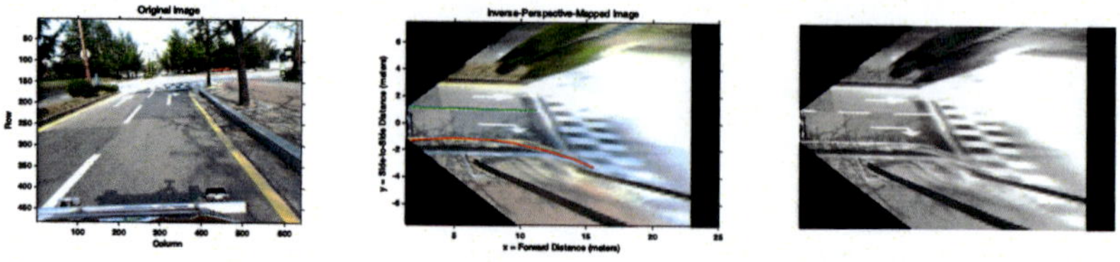

Fig. 9. IPM result and lane detection on crossing road

The detected lanes are approximated by clothoid segments as in the next section with fusion process. Specified pseudo algorithm is described as following diagram.

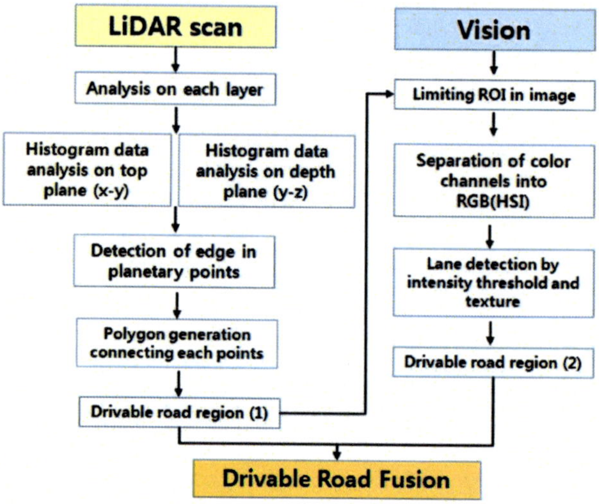

Fig. 10. Algorithm flow for the drivable road fusion

3 Sensor Fusion

Main purpose of the sensor fusion is to increase robustness and safety regardless environmental dynamic changes or uncertainties for real application. For the robust recognition on the drivable road which a vehicle can navigate or passing by, LiDAR for discriminating planetary area and vision for road lane detection are fused. In the most cases, paved road has some guided lanes by vivid color difference and it also has property of flatness between stepped curb in both sides.

However, there are many irregular cases such as where there is no lane, no curb, some dump and damages surface. Assuming that obstacle detection or identification is performed in another part, it is focused on drivable road by sensor fusion in the paper. Sensor fusion could give robust and safe information to the control system in nearly real time during vehicle navigation. To reduce the computational procedure and synchronize two system processes, the fusion algorithm chooses each result in alternative manner by comparing the road geometry according to whether it is straight or not. From our experimental experiences it is most useful to use left side data of laser in the road section during right-turning and vice versa. And our filtering process could give better results when using vision data analysis for straight road and using laser data analysis on curved way. This process is implemented by a selective manner in using only HT for vision information and clothoidal segment modeling for laser data analysis in the filtering process, especially for the computational speed performance.

We integrated the sensor information of two sensor systems and apply a filtering process for automatic steering control while following the designated track. We consider that the vehicle navigates in less than 30 Km/h (8.3 m/s) and standard Korean city street for experimental environments. The amount of data to be processed is regulated, the region of interest for each data and sampling rate while assuring its safety and robustness.

Some curved road can be represented as a sequence of successive clothoidal segments. In [1], road boarders are described by a clothoid which is approximated by 3rd polynomial equations and road geometry are extracted and estimated from vision and laser scanner and then fused by fuzzy logic inference algorithm.

In [8], the work try to generate a smooth, natural and easily drivable path between two points and construct continuous trajectories where the curvature k, and sharpness α are the smallest values needed to achieve a given change of heading over a given distance traveled. The design parameters were changed in terms of the total deflection angle and maximum curvature at the end of the curve since originally the parameters for clothoid curve generation are curve position s and sharpness α.

$$k(s) = k_0 + \alpha s \quad (7)$$

$$\theta(s) = \int_0^s k(u)\,du = \frac{\alpha}{2}s^2 + k_0 s + \theta_0 \quad (7)$$

For control with consideration of various cases which could occur in real environments, differential values acquired through two sampling times are used for offset and orientation.

3.1 Road Modeling

Normally roads are not straight lines only but have significant curves depending upon the terrain underneath. For road modeling, civil engineers use clothoids i.e curves whose curvatures linearly dependent on their arc length(l). Mathematically, a clothoid may be expressed as:

$$c(l) = c_0 + c_1 l \quad (7)$$

where c_0 is regarded as initial curvature and c_1 as rate of change of curvature w.r.t length. Road borders are represented by clothoids which can mathematically be represented by 3rd order polynomial. This polynomial has, in general, representation as:

$$y(x) = y_0 + c_0 \frac{x^2}{2} + c_1 \frac{x^3}{6} \tag{8}$$

y_0 is offset from centerline from vehicle's position on road.

This road curvature may also be estimated approximately using vehicles' motion parameters such as velocity(v) and turn rate(ω), but it gives faithful results only when the vehicles is under smooth curvatures. In case of, for example, vehicle's overtaking scenario, although the road is straight, but vehicles data based curvature calculation will exhibit some curvatures because of change in velocity and/or turn rate. In such and more complex turn case and for greater distances, vehicle data based curvature calculation becomes vague. In this kind of situation we need additionally some other maneuvering parameters of the vehicle e.g lateral velocity, lateral acceleration, offset distance etc. Another approach of road curvature estimation is to use road model as given in Eqn. (8). The Clothoid given above can describe straight edge of a road if c_1=0 and a curved road if $c_1 \neq 0$. Hence if we want to represent two edges/borders of a road, we can use two clothoids with same curvature and curvature rate but different offsets to the left and right of the vehicle on the road. Hence we can choose an appropriative state vector which can describe a road shape sufficiently as,

$$X = [c_0 \; c_1 \; y_{0l} \; y_{0r}] \; . \tag{9}$$

A discrete time state space model for road border's dynamics is given as,

$$\begin{aligned}
c_0(k+1) &= c_0(k) + c_1(k)\Delta x \\
c_1(k+1) &= c_1(k) \\
y_l(x) &= y_l(k) + c_0(k)\frac{\Delta x^2}{2} + c_1(k)\frac{\Delta x^3}{6} + (\Delta y)w_{\Delta y} \\
y_r(x) &= y_r(k) + c_0(k)\frac{\Delta x^2}{2} + c_1(k)\frac{\Delta x^3}{6} + (-\Delta y)w_{-\Delta y}
\end{aligned} \tag{10}$$

In vector-matrix form, discrete time state space road dynamic model is given by following equation set,

$$X_{k+1} = \mathrm{A}_k X_k + \mathrm{B}_k W_k \, , \tag{11}$$

where A_k is state transition matrix (STM) and B_k is process noise gain matrix respectively. Note that STM depends mainly on the distance covered (Δx) in x-direction (direction of travel) and offset displacements here treated as process noise. For Extended Kalman filter design to estimate road parameters state vector given above, we need additional parameters to be calculated e.g. process noise covariance matrix which exhibits uncertainties in modeling as well as unseen errors, measurement covariance matrix etc.

We can extract road end points in successive layers and use geometrical relationship to calculate curve parameters such as c_0 and c_1. Considering vehicle in the center of the road or lane, we can extract 4 pairs of data points each for left and right side road boundary using this 4 layered data set. However the furthermost layer has more errors and scattered data points, so we use data only from first two layers for calculation of clothoid parameters. Setting up measurement model this way, we can get 2 pairs of data points for left and right side road boundary i.e our measurement vector will consist of:

$$y = [(x_{1l}, y_{1l})\ (x_{2l}, y_{2l})\ (x_{1r}, y_{1r})\ (x_{2r}, y_{2r})]^T \tag{12}$$

However if we take y_{ij} as measurements of road offsets, we can express our measurement model in the following form.

$$\begin{aligned} y_{1l}(k) &= y_l(k) + c_0(k)\frac{x_{1l}^2}{2} + c_1(k)\frac{x_{1l}^3}{6} + v_{1l} \\ y_{2l}(k) &= y_l(k) + c_0(k)\frac{x_{2l}^2}{2} + c_1(k)\frac{x_{2l}^3}{6} + v_{2l} \\ y_{1r}(k) &= y_r(k) + c_0(k)\frac{x_{1r}^2}{2} + c_1(k)\frac{x_{1r}^3}{6} + v_{1r} \\ y_{2r}(k) &= y_r(k) + c_0(k)\frac{x_{2r}^2}{2} + c_1(k)\frac{x_{2r}^3}{6} + v_{2r} \end{aligned} \tag{13}$$

where v_{ij} is measurement noise with covariance R which incorporates the uncertainty in measurement sensor reading.

Measurement Jacobian matrix corresponding to the above measurement model is:

$$H = \begin{bmatrix} \frac{x_{1l}^2}{2} & \frac{x_{1l}^3}{6} & 1 & 0 \\ \frac{x_{2l}^2}{2} & \frac{x_{2l}^3}{6} & 1 & 0 \\ \frac{x_{1r}^2}{2} & \frac{x_{1r}^3}{6} & 0 & 1 \\ \frac{x_{2r}^2}{2} & \frac{x_{2r}^3}{6} & 0 & 1 \end{bmatrix}. \tag{14}$$

Equations (9~15) constitute framework for use of Extended Kalman filter as road parameter estimator in this paper. We validated our work through simulations and experimental data acquired by manually driving the experimental vehicle, SSKU ARORA.

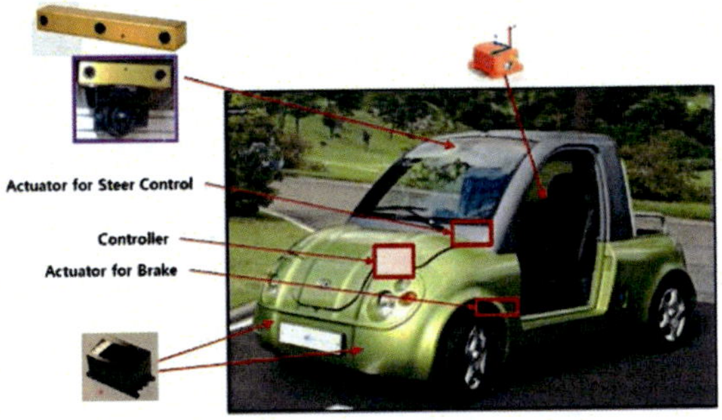

Fig. 11. Electric vehicle system for road recognition

The vehicle is equipped with 4-layer LiDAR Ibeo's LUX 2010™, Camera POINT GREY™ Bumblebee2, IMU/GPS module (xsens™ MTi-G) and onboard computing recourses etc. Simulations and experimental results using above designed EKF for estimation of road parameters are given below.

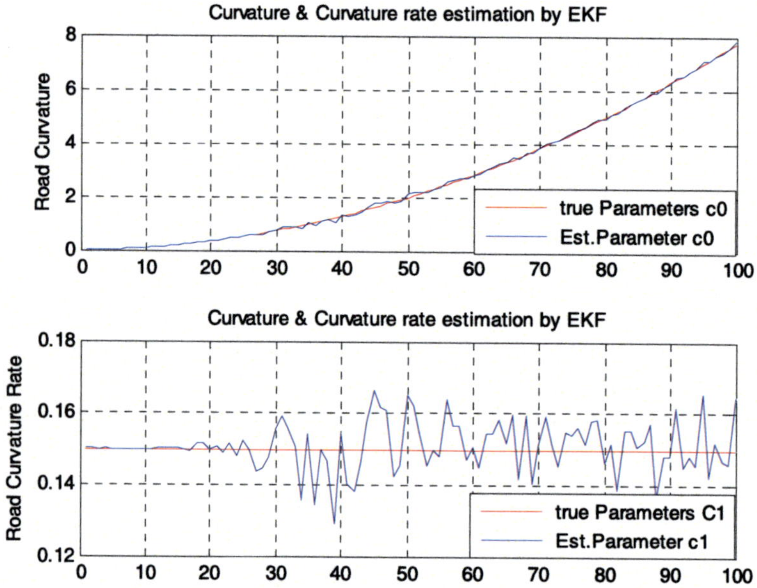

Fig. 12. Road Parameters Estimation result by EKF

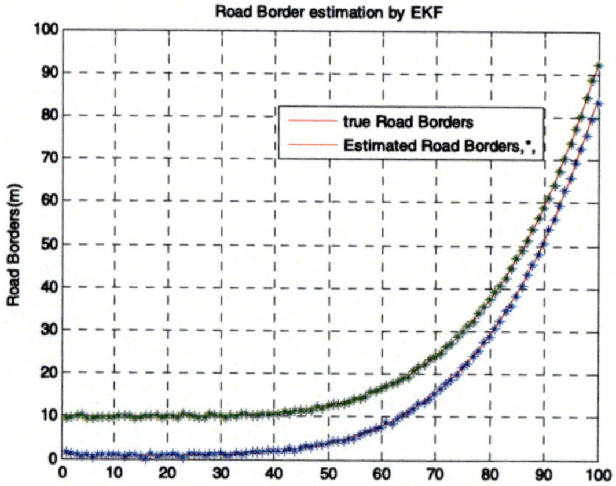

Fig. 13. Road detection result by EKF

Above Fig. 11 shows results of EKF road parameter estimation,namely curvature and curvature-rate respectively. Fig. 12 exibits estimation of road borders. As apparent from two figures above that initially the road shape is almost straight so constant curvature and its rate are estimated faithfully during t=0~30. However during constant turns between t=40~90 there is degraded estimation in curvature rate shown in Figure 11. This is due to nonlinear nature of model and its approximation used in EKF. However the road shape estimation,overall, is quite promising as shown in figure 12. This road shape estimation may be enhanced by using modern nonlinear estimators e.g UKF and PF, which is our future work plan.

4 Experimental Results and Conclusions

Experiments are conducted at road in SungKyunKwan University in Suwon, Korea. The road has almost single lane with pavement and some sections have intersections and curved sections. The vehicle is driven manually at the speed of around 10 km/h for stable acquisition of sensor data and fusion process. The POINT GREY™ Bumblebee 2 was used for vision of which focal length is 3.8 *mm* and HFOV is set up as 66 °. The Ibeo's LUX 2010™ model having four layers was used for LiDAR sensor. The minimum measuring distance of LiDAR is 0.5 *m*, the maximum measuring distance arrives at 200 *m* and accuracy is less than 10 *cm*. The horizontal and vertical angle resolution are 0.125 ° and 0.8 ° respectively and the distance resolution is 4 *cm* in the experiment. The other important sensor is IMU/GPS where the Xsens™ MTi-G model is used which gives EKF filtered Integrated INS/GPS navigation solution alongwith raw and calibrated data about magnetic orientation, acceleration, tilt angle and GPS longitude/latitude position. LiDAR data is acquired at 25 Hz and vision image is updated at 5 Hz. However some synchronization was carried out as the vehicle odometry, rear-wheel encoders and steering angles are acquired every 3 *Hz* through CAN communication.

The specific results of the process give some information of how much the vehicle is distant from the right side margin in the road and what direction the vehicle heads for. This process will be implemented in the autonomous navigation framework, especially for locally safe navigation abilities in the future.

Acknowledgements. This research was supported by a grant (I 091003) from Gyeonggi Technology Development Program funded by Gyeonggi Province.

References

1. Tsogas, M., Floudas, N., Lytrivis, P., Polychronopoulos, A.: Combined lane and road attributes extraction by fusing data from digital map, laser scanner and camera. Information Fusion 12, 28–36 (2011)
2. Gupta, R.A., Snyder, W.S., Pitts, W.S.: Concurrent visual multiple lane detection for autonomous vehicles. In: IEEE Int. Conf. on Robotics and Automation, Anchorage, USA, May 3-8, pp. 2416–2422 (2010)

3. Kim, Z.: Realtime lane tracking of curved local road. In: Proc. of IEEE Int. Transportation Sys. Conf., Toronto, Canada, pp. 17–20 (September 2006)
4. Lipski, C., Scholz, B., Berger, K., Linz, C., Stich, T.: A fast and robust approach to lane marking detection and lane tracking. In: Southwestern Symp. on Image Analysis and Interpretation, pp. 57–60 (2008)
5. Peterson, K., Ziglar, J., Rybski, P.E.: Fast feature detection and stochastic parameter estimation of road shape using muliple LIDAR. In: IEEE Int. Conf. on Int. Robots and Systems, Nice, France, September 22-26, pp. 612–619 (2008)
6. Ogawa, T., Takagi, K.: Lane recognition using on-vehicle LIDAR. In: Int. Vehicles Symposium, Tokyo, Japan, June 13-15, pp. 740–741 (2006)
7. Lindner, P., Rochter, E., Wanielik, G., Takagi, K., Isogai, A.: Multi channel LiDAR processing for lane detection and estimation. In: Proc. of Int. IEEE Conf. on Intelligent Transportation Systems, St.Louis, USA, October 3-7, pp. 202–207 (2009)
8. Manz, M., von Hundelshausen, F., Wuensche, H.-J.: A hybrid estimation approach for autonomous dirt road following using multiple clothoid segments. In: IEEE Int. Conf. on Robotics and Automation, Anchorage, USA, May 3-8, pp. 2410–2415 (2010)
9. Chun, C., Suh, S., Lee, S., Roh, C.-W., Kang, S., Kang, Y.: Autonomous navigation of KUVE. Journal of Institute of Control, Robotics and Systems 16(7) (July 2010)
10. Hartley, R., Zisserman, A.: Multiple View Geometry in Computer Vision. Cambridge Univ. Press (2000)

Field Trial Results of Autonomous Road Crossing Mobile Robot

Aneesh Neeschal Chand[1] and Shin'ichi Yuta[2]

[1] School of Electrical Engineering, Fiji National University
P.O. Box 3722, Samabula, Fiji
aneesh@roboken.esys.tsukuba.ac.jp
[2] Department of Electrical Engineering, Shibaura Institute of Technology
3-7-5, Toyosu, Koto-ku, Tokyo 135-8548, Japan
yuta@shibaura-it.ac.jp

Abstract. We have developed a fully-integrated outdoor mobile robot that is capable of crossing roads autonomously in real world urban environments. To do this, the robot travels along pedestrian sidewalks autonomously, continually detects pedestrian push button boxes and navigates to it when one is detected. It then activates the push-button using an onboard-finger, moves to the crossing zone and crosses the road after detecting the zebra stripes and pedestrian lights. In this paper, we report the results of preliminary field trial experiments where the robot was deployed in a real world environment and its performance was evaluated.

Keywords: road-crossing mobile robot, real world outdoor environment.

1 Introduction

There are mobile robots already existing that can travel in outdoor urban environments. They are capable of navigating along sidewalks and footpaths thereby emulating pedestrians e.g [1-2]. Typically, these robots travel along a straight and continuous path segment and cannot cross roads independently unless explicitly assisted by humans.

Consider one such pedestrian-type robot. Suppose that it is pre-supplied with a navigational map. This map consists of a set of sparsely defined navigational waypoints $W = \{w_1, w_{2,} w_3 \ldots\}$ that loosely defines the path of the robot. The waypoints may be in the form of Global Positioning Systems (GPS) coordinates, or coordinates of a number of navigational landmarks or any other navigational forms of waypoints. The robot then starts navigating from a starting point on the sidewalk and travels according to the supplied waypoints. Most of the time, the waypoints are deliberately chosen so that the robot path spans a narrow area and the robot just needs to navigate along a continuous sidewalk with ever having to cross the road. Or if it needs to cross, then the robot is assisted by a human. Now what if it were to be that the supplied waypoints were distributed over a much wider area and spanned a whole network of roads and sidewalks and the robot is required to navigate the full journey

S. Lee et al. (Eds.): Intelligent Autonomous Systems 12, AISC 193, pp. 57–66.
springerlink.com

without any human intervention? In this scenario, one waypoint w_x is located on a sidewalk which is on one side of the road but the next waypoint w_{x+1} is located on the other side of the road. In such a case, the waypoints lead the robot to the vicinity of a pedestrian crossing while it is traveling along a sidewalk but the next waypoint (and the final destination) happens to be located on the opposite side of the road. Inevitably, the robot needs to cross the road in order continue the journey and reach the destination. How the mobile robot can autonomously cross the road was the specific problem of interest in this research. Specifically, what the robot needs is an autonomous crossing function so that whenever it is traveling on a sidewalk and needs to cross the road, it can do so at the next detected crossing zone.

If an outdoor mobile robot could be endowed with such an ability to autonomously cross a street or road, the scope of the robot would be greatly enhanced. Accordingly, we have developed such an autonomous road-crossing robot.

2 Overview of Road Crossing Mobile Robot

2.1 Related Works

First the target environment, a real world intersection, where the robot would be deployed was selected. The motion of the robot in attaining the intended street-crossing behavior in the environment was systematically analyzed and the requirements to realize the street-crossing robotic system were elicited. Based on the identified requirements, the road-crossing motion and design of the robot was formalized.

The developed robot, shown in figure 1, first autonomously travels along pedestrian sidewalks. For this, the robot uses a single forward-looking, wide-angle lens monocular camera to spontaneously detect sidewalks. From the detected sidewalk, the robot acquires the trajectory along which to travel. After validating the acquired trajectory by approximating it with a straight line, the robot starts traveling along the validated trajectory. While traveling, continuous detection of pedestrian push-button boxes is performed using real time object detection techniques. i.e a Haar classifier [3]. Filters are used to filter out false positives. When a button box is detected, the robot acquires its position by fusing data from the monocular camera and a laser range scanner (LRS) so that the robot can calculate the trajectory required to approach the button box [4]. After approaching, the robot detects the circular push button and precisely activates it with an on-board finger. To activate the button, the finger is pre-calibrated so that it is installed at the same height at the button and the distance that the LRS returns when the finger has pressed the button is noted and it becomes the reference distance during an actual button activation operation. Next, the robot approaches the pedestrian crossing using the detected button box as a landmark. At this point, the robot robustly detects pedestrian crossings and is capable of detecting a crosswalk in varying environmental conditions. In doing this, we also consider potential real-world challenges like shadow-induced occlusions using a combined Hue-color extraction and Gaussian pyramid-based segmentation method. Moreover, reduced detection due to diminution of the pedestrian zebra lines is also considered and the orientation of the detected crossing is determined next. The trajectory to cross the road is then perceived from the detected crosswalk. Finally, color thresholding using spatial and temporal filters determines the existence of pedestrian lights and identifies the lit signal light and the robot crosses the road on a green signal. Further details of the developed robot and the underlying technologies can be found in [5].

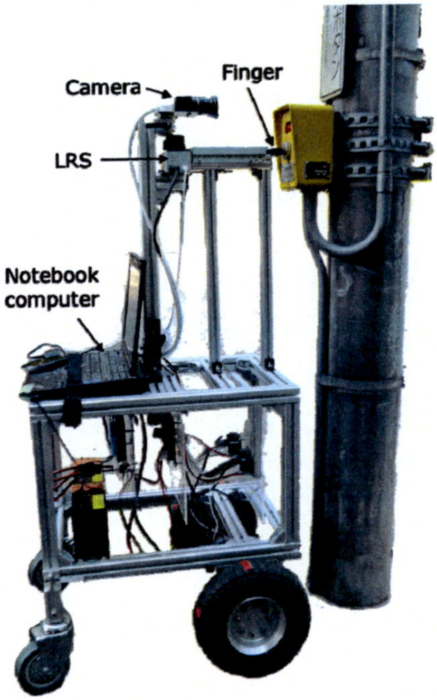

Fig. 1. The developed autonomous road-crossing robot

3 Validation by Real World Experiments

A major objective of this work was not just to implement a road-crossing robot but to implement one in a real world environment. So notably a salient feature of our work is that the design was implemented on a real robot and all experiments were conducted in a completely real world outdoor environment in line with our goal of developing real robots for the real world.

That the experiments were conducted in a real world environment has further validated and lent credibility to the design of the robot and the road crossing algorithm developed. The following sections report the preliminary field trial results and the experiences we gained, which is the overall focus of this paper.

4 Experimental Results and Discussion

4.1 Experimental Environment

Figure 2 shows the satellite image of the experimental environment together with its simplified schematic. The geographical coordinates of the experimental environment in decimal degrees are 36.115346°N, 140.096727°W. This is a site located in the campus of the University of Tsukuba, Japan. The yellow boxes in figure 2(b) represent the button boxes and the approximate route of the robot is indicated in red dashed lines. Figure 3 shows in greater detail the section of the selected intersection where the experiment was done. The relevant objects of interest that the robot would sense or interact with are highlighted: sidewalks, push button box, zebra stripes representing crossing zone and pedestrian lights.

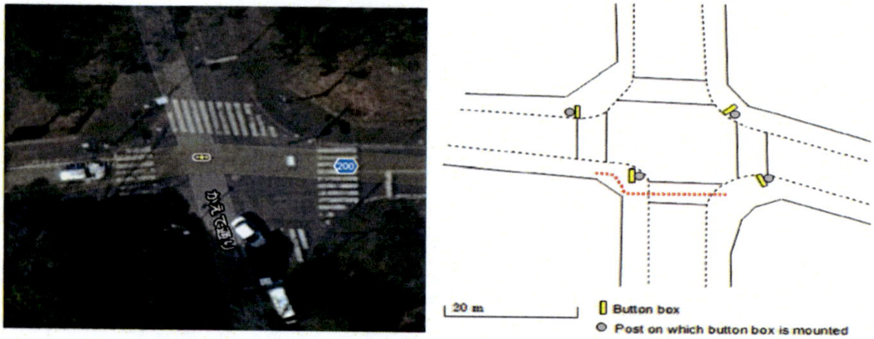

Fig. 2. Selected experimental environment. (a) GoogleMaps image. (b) Simplified schematic with robot route indicated.

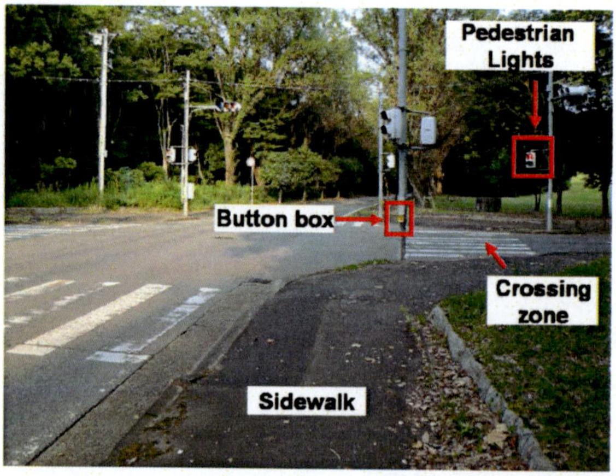

Fig. 3. Experimental environment with objects of interest indicated

4.2 Results

The experimental results are presented in the same sequence as the robot operation. Videos of experiments are available for viewing online[1].

4.2.1 Navigation along Sidewalks

For navigation on the sidewalk, the robot could detect the sidewalk, generate the trajectory and navigate forward reasonably accurately. Figure 4 shows a selected frame of the sidewalk detection and trajectory generation result. However, there are some instances when the sidewalk is not detected accurately and the resulting trajectory generation was then incorrect. This occurred in approximately 30% of the total frames processed. This occurred for sidewalk sections that were less defined and in a state of disrepair; the detection results were then poor. However, the trajectory validation method used proved extremely beneficial as all faulty trajectories were detected and discarded.

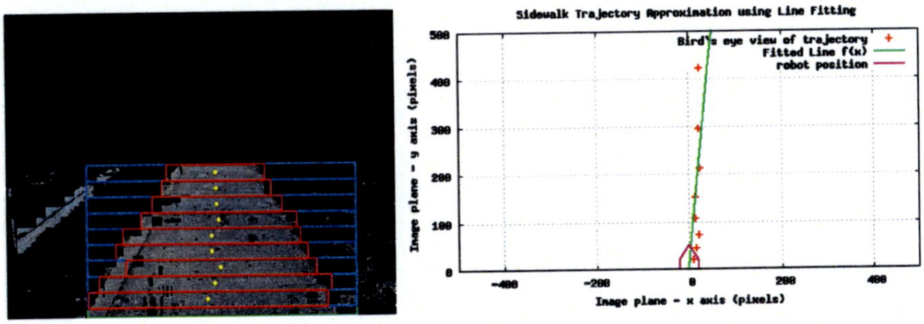

(a) (b)

Fig. 4. Selected sidewalk detection (a) and trajectory generation result (b)

4.2.2 Navigation towards Button Box

The Haar classifier constructed for button box detection provided a varying detection rate for the reason that not as many training images were used for its construction. Consequently, the best detection rate attained was 85% with a corresponding accuracy rate of 99%. This means that whenever a button box was detected (which was 85% of the time), the detected box turned out to be a correct detection 99% of the time. Attributed to the filters used, hardly any false positives were encountered. The highest detection rate was obtained in conditions of sufficient lighting. In other conditions, these rates may vary.

After vision-based detection, the estimation of the push button box angular position relative to robot is attained with less than two degrees of error. This magnitude of error is not fatal since the angular resolution of the URG-04LX sensor is 0.36 degrees and the diameter of the post on which the button-box is housed is 19.2cm, there are

[1] `http://www.roboken.esys.tsukuba.ac.jp/~aneesh/phd_videoresults.html`

than one laser beam striking the button box post. Since the button box position estimation method yields at least one individual laser beam and as long as one laser beam can be identified, the entire cluster can also be determined. Figure 5 shows the button box being detected and the trajectory generated to approach it from a distance of about 1.9m.

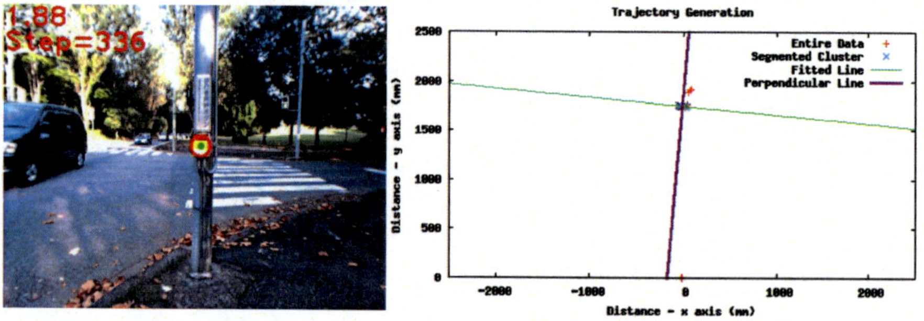

(a) (b)

Fig. 5. Detecting button box (a) and generating trajectory to approach it (b)

(a) Button detection using HoughCircle (b) Robot p.ov. image showing button activation

Fig. 6. Detection and activation of pedestrian push-button

The detection and perpendicular trajectory generation time for a single frame of size 640x480 pixels ranges from 200ms to 900ms. This means that the overall process is fast enough for real time operation for a moderate robot speed of 150mm/s.

4.2.3 Button Activation

The results of an autonomous button alignment and activation operation are presented here. In figure 6(a), the robot point-of-view image shows the button detection result using the HoughCircle detection algorithm [6]. Figure 6(b) shows the robot point-of-view after a button activation operation. During experiments, the button activation function was successfully tested on multiple button boxes located on other sections of the 4-way intersection. Provided that the finger is calibrated accurately, autonomous activation was always achieved. Some limitations of this function are discussed in subsequent sections. In the current implementation, there is no feedback confirming the button activation. The robot simply moves forward by a fixed distance (obtained in calibration stage) in order to activate the button.

4.2.4 Navigation towards Crossing Zone

The robot uses the button box position as a landmark to navigate to the front of the crossing zone, in figure 7. It is capable of correctly navigating from the button box to crossing zone accurately all the time. In the current implementation, there is no obstacle avoidance incorporated. When an obstacle is encountered, the robot halts and waits for the obstacle to pass.

Fig. 7. Superimposed images showing robot moving to crossing zone after button activation

4.2.5 Crossing the Road

The final stage is the identification of the pedestrian lights and detection of the crossing zone. Application of the color thresholding and filtering algorithms detects and identifies the pedestrian lights correctly. It is seen in figure 8 that the green pedestrian signal is correctly detected even with the green vegetation in the background. Only in cases when the pedestrian light is under direct sun-glare is the green signal undetected. This is because in sun-glare conditions, the green light becomes indistinguishable from the surrounding green vegetation. Overall, the green signal is detected about 90% of the time. The red signal however is detectable more robustly than the green signal, even under sun-glare, since it is the only red colored object in the environment. Consequently, to alleviate the 10% non-detection rate of the green signal, additional logic was added to the color thresholding conditions.

(a) Green signal identified (b) Red signal identified

Fig. 8. Identification of pedestrian lights

Fig. 9. Crossing trajectory generation

Since the red signal is detected more robustly and since it is known that the red and green signal cannot mutually coexist together, we deem the green signal to be lit when a red signal is *not* detected. This greatly increases the detection of the green signal.

The detection of the crossing zone is also slightly modified here since the zebra stripes are badly faded. Instead of detecting the zebra stripes from the perspective view, the perspective view is first converted to a bird's eye view and *then* the stripes are detected. The result is shown in figure 9. The first zebra stripe is ignored in the detection process as it length is shorter compared to the remaining stripes. Similarly, some of the more seriously faded stripes are also discarded in the blob extraction phase. However, the stripes that are detected generate the correct trajectory that is representative of the entire crossing zone.

4.3 Discussion on Experimental Results

The robot is capable of crossing the road in general, however, there are some instances in which it fails. The single most fatal source of failure is non detection of button boxes. When using vision in too bright or cloudy conditions, the Haar classifier failed to detect the button box. It is suggested in [7] that 3000 training images be used for the classifier construction. Likewise, [8] cites a similar quantity of 10,000 images. To generalize, training images in the order of magnitude of thousands are needed to construct a sufficiently robust classifier. However, only 100 training images were used in this work since a (weak) classifier was required for proof of concept purposes only. As long as there is at least *one* positive detection, the robot generates the path to the button box correctly. But a complete non-detection leads to an outright failure.

Similarly, the Hough circle detection method does not detect a circle in some video frames at times. Although the shutter speed, gain and exposure parameters of the Flea2 camera are automatically adjusted, the iris of the lens requires manual adjustment to control the amount of light entering. Without this manual adjustment, the button goes undetected in too bright or dark conditions. However such a case does not lead to failure but just results in a prolonged operation time as the robot waits until the button has been detected.

The overall lesson gained from this work is that while the implementation of a robotic system for real world environments can be realized, the task of validating through real world experiments requires numerous trials and consists of improvement by discovery. That is, one trial reveals some shortfall which is fixed and the next trial exposes some other shortfall and so on. Also, while real world robotic systems can be realized using a simple sensor configuration, such as in this case, a multiple redundant sensor system is probably imperative to counter failure.

5 Conclusion

The results of preliminary field trial experiments of an autonomous road-crossing mobile robot were reported in this paper. The robot was deployed in a real world environment and its performance was evaluated. Despite some instances of failure, there is overall a great potential for realizing a real road-crossing mobile robot. It is expected that future works will include improving the system for increased reliability and success. Such future works include tackling the sources of failure and expanding the scope of the robot to deal with even more likely real world events.

Acknowledgements. This research was done at and funded by the Intelligent Robot Laboratory of the University of Tsukuba, Japan.

References

1. Morales, Y., Takeuchi, E., Carballo, A., Tokunaga, W., Kuniyoshi, H., Aburadani, A., Hirosawa, A., Nagasaka, Y., Suzuki, Y., Tsubouchi, T.: 1Km Autonomous Robot Navigation on Outdoor Pedestrian Paths Running the Tsukuba Challenge 2007. In: 2008 IEEE/RSJ International Conference on Intelligent Robots and Systems, France, September 22-26, pp. 219–225 (2008)
2. Trulls, E., Corominas Murtra, A., Pérez-Ibarz, J., Ferrer, G., Vasquez, D., Mirats-Tur, J.M., Sanfeliu, A.: Autonomous navigation for mobile service robots in urban pedestrian environments. Journal of Field Robotics 28, 329–354, doi:10.1002/rob.20386

3. Viola, P., Jones, M.J.: Rapid Object Detection Using a Boosted Cascade of Simple Features. In: IEEE Conf. Computer Vision and Pattern Recognition, CVPR (2001)
4. Chand, A.N., Yuta, S.: Vision and Laser Sensor Data Fusion Technique for Target Approaching by Outdoor Mobile Robot. In: IEEE International Conference on Robotics and Biomemitics, Tianjin, China, December 14-18, pp. 1624–1629 (2010)
5. Chand, A.N., Yuta, S.: Design of an Intelligent Outdoor Mobile Robot with Autonomous Road-Crossing Function for Urban Environments. In: 2012 IEEE/ASME International Conference on Advanced Intelligent Mechatronics (2012)
6. Duda, R.O., Hart, P.E.: Use of the hough transformation to detect lines and curves in pictures. Communications of the ACM 15(1), 11–15 (1972)
7. Kuranov, A., Lienhart, R., Pisarevsky, V.: An Empirical Analysis of Boosting Algorithms for Rapid Objects Withan Extended Set of Haar-like Features. Tech. Rep. MR LTR, Intel Technical Report (July 2002)
8. Lidoris, G., Rohrmuller, F., Wollherr, D., Buss, M.: The Autonomous City Explorer Project – Mobile robot navigation in highly populated urban environments. In: IEEE International Conference on Robotics and Automation, Kobe, May 12-17, pp. 1416–1422 (2009)

Timed Trajectory Generation Combined with an Extended Kalman Filter for a Vision-Based Autonomous Mobile Robot

Jorge Bruno Silva[1], Cristina P. Santos[1], and João Sequeira[2]

[1] University of Minho, Guimarães, Portugal
{jbruno,cristina}@dei.uminho.pt
[2] Instituto Superior Técnico, Lisbon, Portugal
jseq@irs.ist.utl.pt

Abstract. Planning collision-free trajectories requires the combination of generation and modulation techniques. This is especially important if temporal stabilization of the generated trajectories is considered. Temporal stabilization means to conform to the planned movement time, in spite of environmental conditions or perturbations. This timing problem has not been addressed in most current robotic systems, and it is critical in several robotic tasks such as sequentially structured actions or human-robot interaction. This work focuses on generating trajectories for a mobile robot, whose goal is to reach a target within a constant time, independently of the world complexity. Trajectories are generated by nonlinear dynamical systems. Herein, we extend our previous work by including an Extended Kalman Filter (EKF) to estimate the target location relative to the robot. A simulated hospital environment and a Pioneer 3-AT robot are used to demonstrate the robustness and reliability of the proposed approach in cluttered, dynamic and uncontrolled scenarios. Multiple experiments confirm that the inclusion of the EKF preserves the timing properties of the overall architecture.

1 Introduction

The ability to plan collision-free trajectories for autonomous mobile robots includes fundamental issues, such as trajectory modulation and generation. This issue becomes more relevant in the path planning problem if temporal stabilization of the generated movements is considered, particularly in discrete movements.

Temporal stabilization means to conform to the planned movement time, despite varying environmental conditions or perturbations. Thus, robot movements must compensate for disturbances, accelerating or decelerating. This timing problem has not been addressed in most current robotic systems. For instance, the classical problem of autonomous navigation, tasks performed by the robot only require a convenient overall speed and when a robot takes longer to arrive at a goal because of some unforeseen disturbance such as obstacle circumnavigation, this change of timing is not compensated for by accelerating the vehicle along its trajectory.

Timed actions, involve stable temporal relationships and this temporal stabilization is critical in multiple robotic tasks including avoidance of moving objects, catching,

S. Lee et al. (Eds.): Intelligent Autonomous Systems 12, AISC 193, pp. 67–79.
springerlink.com

hitting, drumming, juggling, coordination between degrees of freedom, human-robot scenarios and generating sequentially structured actions.

In this contribution we continue previous work attempts to tackle these spatial-temporal problems. Herein, we extend our previous work aiming that the robot reaches a goal location while avoiding obstacles within a specified movement time [1]. However, the target which identifies the goal location was defined geographically. To make the application more realistic, in [2], the sensory loop was closed by online acquiring the location of the target through the built-in robot's visual system, such that action was steered by sensory information. However, when the localization of a target is obtained from a vision system there may be situations in which the robot's visual system is unable to detect the respective target. Obstacles in the environment occluding the target are an example of such situations.

An Extended Kalman Filter (EKF) is used to estimate the location of a ball relative to the current position of the mobile robot. We have chosen to apply an EKF because this filter has a low computational cost, and it is an optimal estimator that linearizes the nonlinear system around the estimation. Note that our focus is to develop an architecture that integrates the dynamical systems theory and an EKF while providing for the timed control of an autonomous mobile robot. The resulting architecture is useful in robotics applications that have to generate rhythmic movements and simultaneously to deal with targets, which can be visible or undetectable during periods of time.

However, an important question which has to be considered is whether the inclusion of an EKF affects the temporal nature of the previously proposed dynamical system architecture, *i.e.*, the robot has to keep reaching the target location within the specified time.

Comparative examples of temporal solutions with and without the EKF are presented which show the intrinsic robustness of the system even in the presence of non-modeled dynamics (Brownian motion of the target).

The proposed architecture is verified in a hospital environment. In such application, the robot is expected to carry out a task or a sequence of tasks within a specified timing. Specifically, if we consider tasks that require a human-robot interaction, or if the mobile robot has to delivery or carry goods between different locations of the hospital.

In such case, it is required to guarantee that the response time will be inferior to a certain value, which depends on the type of interaction, and which can be specified by higher levels of the modular architecture herein presented.

The experiments were performed in a Webots simulation environment using a Pioneer 3-AT robot. The EKF was able to robustly estimate the location of the target even when the path covered by the robot is subjected to variations not known *a priori*, and to converge to the reference path after long periods of time without observations. This is especially important if the robot has to navigate in cluttered environments.

2 Related Work

The generation of timed trajectories embedded in feedback structures remains, to the best of our knowledge, an issue currently receiving attention both in robotics and in neurosciences. Usually the proposed control approaches are based on nonlinear dynamical systems.

Research has been revolving around related themes considering learning by demonstration, including the dynamical motor primitives [3], combining dynamics for rhythmic and discrete motion, generation of synchronized movement for robot with multiple degrees of freedom to achieve tasks which involve rhythmic motions such as biped and quadruped autonomous adaptive locomotion over irregular terrain [4], juggling [5], drumming [3], playing with a slinky toy, basis field approaches for limb movements [6] and locomotion [7] for instance, as well as superposing movement primitives.

The proposed systems are robust against disturbances but temporal stabilization is not effectively attempted. These ideas were previously extended to signal processing in biped and quadruped locomotion [8], using a frequency-learning mechanism for nonlinear oscillators in the presence of external signals.

In [9] the drumming problem is also addressed but it does not take into account any kind of disturbance which may affect the rhythmic movement. Also [10], combined rhythmic and discrete movements were able to generate movements in a Ball-Paddling robotic application.

Human-like reaching has been implemented in a robot manipulator [11] to reach a tracked ball by applying VITE-like dynamical systems. Recent studies include robust manipulation through the learning of motion primitive goal [12].

Nevertheless, works addressing the generation of rhythmic movements in wheeled mobile robots has not been addressed in most current works. In [13], rhythmic movement is generated using limit cycle attractors for a wheeled mobile robot that is able to perform a sequence of basic behaviors. However, an inherent limitation of this approach is that it only generates a single continuing pattern of rhythmic movement, remaining still limited with respect to the integration of multiple constraints and discrete movements.

These works have used nonlinear dynamical systems, including oscillators, and the oscillatory behavior has been explored for responsiveness and flexibility as well as for timing. The dynamical systems theory provides several advantages to address the temporal stabilization problem such as low computational cost, smooth online modulation while keeping the general features of the original movements, robustness against small perturbations, and possibility to integrate sensory feedback.

The work in [14] has initiated the dynamical system approach to understand movement coordination to generate timed actions, in which time is a controlled variable against perturbations when a particular action is initiated and terminated. In [15], this architecture was implemented in a real vehicle, where a temporal stabilization mechanism was proposed. The same framework was modified in [16], in which some changes to the approach were proposed, since they argued that the included temporal stabilization mechanism was not able to preserve the temporal constraint.

We continued to tackle this challenge and proposed a few novel adaptations to the system in [1], by exploring the intrinsic properties of the used oscillator. The main novelty was in exploring bifurcation theory to switch the qualitative dynamics of the oscillators, instead of switching among different dynamical systems. Thus, we generated sequences of both discrete and rhythmic movements through the same system similarly to work in locomotion [17] and [4], but applied to a wheeled robot. Further, an adaptive mechanism for frequency modulation of the velocity profile based on the oscillators current state was proposed, allowing to set different times for acceleration and deceleration.

The applied type of movements, discrete and rhythmic, is strongly supported by research on human motor control [18].

In mobile robotics the EKF has been extensively used *e.g.*, to smooth data from ultrasonic sensors in order to estimate the location of objects [19] or in robotic arms catching a flying ball [20],[21] where a ball is launched towards to the robotic arm, and an EKF is used to track and to predict the position of the ball at each instant of time. EKFs have been also widely spread in the literature addressing target tracking. However, its effects in the overall dynamics have not received much attention, *e.g.*, related to target occlusion during large periods of time. This is an aspect enforced by the experiments presented here.

3 System Overview

The overall architecture of our system is depicted in fig 1. The system is hierarchically divided in two functional layers according to their level of abstraction. These two layers implemented through a mesh of interacting dynamical systems, are mutually coupled, however they work independently of one another, and each layer sends sets of parameters to the other layer. The second layer mimics the role of supra-spinal structures in

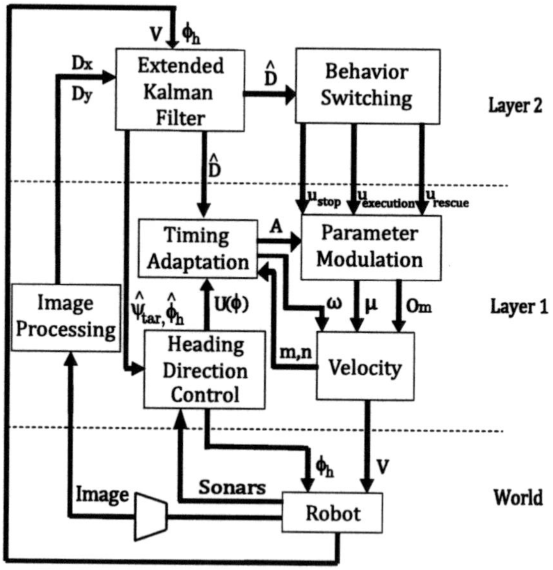

Fig. 1. System overall schematic

biological systems. Basically, it is responsible for selecting the most adequate robot motor behavior accordingly to external conditions and current states of the task at hand, map it onto the corresponding set of dynamical parameters and send them to the first layer at the right timing.

The first layer focuses its operation on controlling the 2D motion of the robot according to the dynamics of two behavioral variables: the heading direction ϕ_h and the forward velocity v.

In the following, we detail each layer of the overall architecture. For a complete explanation please see [2,1].

3.1 First Layer

For heading control, we motivate the choice of the dynamical systems approach [22], because its properties allow for the integration of sensory-motor feedback and thus for closed-loop control. The directional dynamics rules the robot's heading direction ϕ_h by avoiding obstacles and following towards to the target.

The velocity v of the robot is the major determinant to the success of the movement task within the specified timing constraints. Velocity is set as the solution of a stable limit-cycle generated by a Landau-Stuart oscillator [2,15,1], which is modulated by the *parameter modulation*. The performed velocity consists of a single oscillation cycle, adapted in order to accelerate or decelerate in the presence of disturbances.

The proposed oscillator is able to generate discrete, rhythmic or a superimposition of both movements, resulting in a more complex movement. The choice and switch among these different movements depends explicitly on the parameters values.

Timing adaptation is achieved by computing online an updating rule for the oscillator parameters. It is explained detail in [2] as well as obstacle avoidance.

This layer receives from the second layer at the right timing, sets of parameters that specify and modulate in a simple and straightforward manner the generated trajectories.

3.2 Second Layer

Commands are sent from the second layer, changing qualitatively the dynamic solution by taking advantage of the Hopf bifurcation of the Landau-Stuart oscillator, resulting in three motor behaviors, available in the level's repertoire: stop, the robot does not move; execution, the robot performs the rhythmic movement; rescue, discrete movement.

The *Behavior Switching* module autonomously selects and sequences these motor behaviors accordingly to external conditions and current states of the task at hand, such that action itself is elicited by perception.

The *Extended Kalman filter* module, employs an EKF in order to estimate the localization of the target relatively to the robot's position, *i.e.* the $\hat{D}_x$, $\hat{D}_y$ projections onto the robot X and Y-axes of the remaining distance to reach the target, and consequently the distance $\hat{D}$ to the target and the direction $\hat{\psi}_{\text{tar}}$ where the target lies[1].

It receives from the *Image processing* module the measured D_x and D_y.

[1] Along the paper this robot frame refers to a reference frame that has origin located at the position of the robot and is always aligned with an inertial world frame.

Note that the timing adaptation is directly proportional to the estimation of $\hat{D}$ and thus the EKF has a very important role in the overall architecture.

3.3 Timing Constraints

This timing architecture is very useful for robotic applications that need to generate robust timed movements. Moreover, the architecture is able to generate rhythmic movements that meet the requirements imposed by the dynamical parameters. This means that the proposed dynamics are to a large extent independent from the physical implementation. The necessary modifications are only required at the level of the heading direction dynamics.

However, these timing movements are bounded due to the robot physical constraints. Herein, we can identify physical limitations such as accelerations or decelerations of the robot or even the maximum velocity reached by the robot. Let us consider a complex and cluttered environment, for instance a hospital environment, in which a mobile robot has to perform timed sequentially delivery tasks. The timing movements are constrained by the maximum velocity that the robot can reach in such environments. This is dependent on the velocity that the users move in the environment, in the degree of clustering and also on its own achievable velocity. This is a typical physical limitation of the robotic application and not from the proposed timing architecture.

4 Extended Kalman Filter Formulation

The EKF needs a state space model of the dynamic system describing the time evolution of the state to be estimated, herein the target position in the robot coordinate system, expressed by the system state discrete variable $x_k = [Dx_k, Dy_k, \phi_{\mathrm{h}_k}]^T$.

In order to calculate the localization of a target, we need to calculate the distance D between the robot and the target, and the angle ψ_{tar} from the robot X-axes to the target around the robot Z-axes. Figure 2 details the variables involved when representing the motion of the robot relative to the target. All variables are defined in the robot coordinate system.

The following relationships may be established among these variables:

$$Dx_k = Dx_{k-1} - v_k \cos(\phi_{\mathrm{h}_k})\mathrm{dt}, \tag{1}$$

$$Dy_k = Dy_{k-1} - v_k \sin(\phi_{\mathrm{h}_k})\mathrm{dt}, \tag{2}$$

$$\phi_{\mathrm{h}_k} = \phi_{\mathrm{h}_{k-1}} + \omega_k \mathrm{dt}, \tag{3}$$

where Dx_k, Dy_k are the projections onto the robot X and Y-axes of the distance between the robot and the target, ϕ_{h_k} is the current heading direction of the robot, v_k is the robot forward velocity and ω_k is the robot angular speed.

The angle that the robot has to turn in order to face the target ψ_{tar_k} and the distance D_k that the robot has to travel to reach the target at instant k are given by:

$$\psi_{\mathrm{tar}_k} = \arctan\left(\frac{Dx_k}{Dy_k}\right), \tag{4}$$

$$D_k = \sqrt{Dx_k^2 + Dy_k^2}. \tag{5}$$

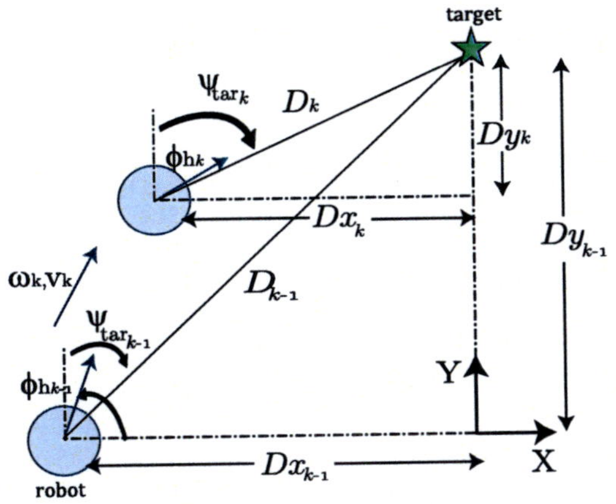

Fig. 2. Representation of the variables involved in the motion of the robot relative to the target. Variables are described in the robot coordinate system. The robot has linear and angular velocity in both instants of time.

A nonlinear system describing equations 1,2,3 can be expressed as follows:

$$x_k = f(x_{k-1}, u_{k-1}) + w_k, \tag{6}$$

$$y_k = h(x_k) + z_k, \tag{7}$$

where $u_{k-1} = [v_k, \omega_k]^T$ are the control variables.

The random variables w_k and z_k represent the process and measurement noise respectively. They are assumed to be white and independent of each other, and with normal probability distributions.

Let us write the new governing equations that linearize an estimate about eqs. 6 and 7:

$$x_k \approx \hat{x}_k^- + F_k(x_{k-1} - \hat{x}_{k-1}) + W_k w_{k-1}, \tag{8}$$

$$y_k \approx \hat{y}_k^- + H_k(x_{k-1} - \hat{x}_{k-1}) + Z_k z_{k-1}, \tag{9}$$

where F_k and H_k are the Jacobian matrices of partial derivatives of f and h with respect to x. W_k and Z_k are the Jacobian matrices of partial derivatives of f and h with respect to w and z respectively.

The EKF time update equations are:

$$\hat{x}_k^- = f\left(Dx_{k-1}, Dy_{k-1}, \phi_{h_{k-1}, u_{k-1}}\right) \tag{10}$$

$$P_k^- = F_k P_{k-1} F_k^T + Q, \tag{11}$$

where P_k^- is the *a priori* error covariance, P_k is the *a posteriori* error covariance, and $\hat{x}_k^-$ is the state estimated *a priori*. The EKF measurement update equations:

$$K_k = P_k^- H_k^T (H_k P_k^- H_k^T + R)^{-1} \tag{12}$$

$$N_k = y_k - h\left(\hat{x}_k^-\right) - y_k - h\left(\hat{x}_k^-\right) \tag{13}$$

$$\hat{x}_k = \hat{x}_k^- + K_k N_k \tag{14}$$

$$P_k = (I - K_k H_k) P_k^- \tag{15}$$

where K_k is the Kalman gain, N_k is the innovation, and $\hat{x}_k$ represents the state estimated *a posteriori*. The initial error covariance P_0 is set equal to R.

R was obtained through some off-line samples of the measurements (Dx_k, Dy_k, ϕ_{h_k}), from which the respective covariance was computed.

5 Performance of the Extended Kalman Filter

In order to analyze the impact of the EKF in the timing properties of the architecture and on the planned movement time, comparative examples of the robot moving towards a target with and without the EKF are evaluated.

In these experiments the robot has to reach a target at a distance of 4 m within 30 s, no obstacles appear during its direct path and the target is always visible to the robot's vision system. Table 1 shows the obtained results for several performance indexes defined ahead. The target position is perceived by the robot's visual system as if the target is moving inside an hypothetical circumference, whose radius is varied in the different experiments.

Due to the robot's physical constraints, we consider that the noise in the target position is perceived only during the first 27 s of each experiment. After this instant of time the target will remain in the same position in order for the robot to have enough time to reach it.

The value ΔP_f indicates the Euclidean distance between the target position and the final position of the robot. ΔP is the Euclidean distance between the robot and the target at $t = 27$s. In the same experiment without noise, the value of ΔP is 0.197 m. ΔT indicates if the time constraints were verified. It is calculated as the ratio between the time needed by the robot to reach the target and the specified movement time. If this value is smaller than 1 the timing properties are verified.

Table 1. Performance of the system resulting from different perceptions of the target noise

Radius		ΔP (m)	$\Delta P_f(m)$	ΔT (s)	E
5 cm	without Kalman	0.222	0.034	0.970	0.1155
5 cm	with Kalman	0.181	0.023	0.957	0.1148
10 cm	without Kalman	0.231	0.036	0.966	0.1163
10 cm	with Kalman	0.185	0.011	0.93	0.1140
20 cm	without Kalman	0.195	0.039	0.960	0.1166
20 cm	with Kalman	0.196	0.007	0.921	0.1104
30 cm	without Kalman	0.184	0.040	0.957	0.1166
30 cm	with Kalman	0.189	0.031	0.939	0.1105

From table 1, when the EKF is used the error in the final position of the robot, ΔP_f, is smaller. Regarding ΔP also we can verify that when the noise perception is disabled, at $t = 27\ s$, the difference between this value and the ideal $\Delta\mathrm{P} = 0.197$ is smaller when the EKF is used.

Also, it can be verified that $\Delta\mathrm{T}$ is smaller than 1. Therefore, the inclusion of the EKF besides reducing the error in the target perception did not compromise the timing properties.

The amount of energy required by the robot to move is estimated as $E = \sqrt{v^2 + \omega^2}$. These results clearly show that the average energy required for the robot to successfully complete the missions is lower when the EKF is used.

6 Experimental Results

Here we describe some experiments realized in a simulated hospital environment using a Pioneer 3-AT robot. The robot has no previous knowledge about the position of any obstacle in the world, or other disturbance that possibly may occur. If the robot stops detecting the ball by any reason, the EKF keeps providing at each instant of time the direction to the ball, as well as the current distance between the ball and the robot.

At each sensorial cycle, information is acquired, dynamic equations are calculated and integrated using an Euler method with a time step of 200 ms. The robot is able to reach a maximum velocity of approximately 0.75 m/s.

In all the presented experiments the robot always looses visual contact with the ball when the distance between both is small (≈ 0.4 m), because the ball's size is larger than the camera's field of view and the visual system of the robot is automatically disabled.

6.1 Experiment 1

This first experiment aims to demonstrate the EKF performance when the robot is compensating the disturbance caused by two obstacles that force an obstacle circumnavigation and to cover more distance (fig. 3). Thus, the robot has to increase its velocity

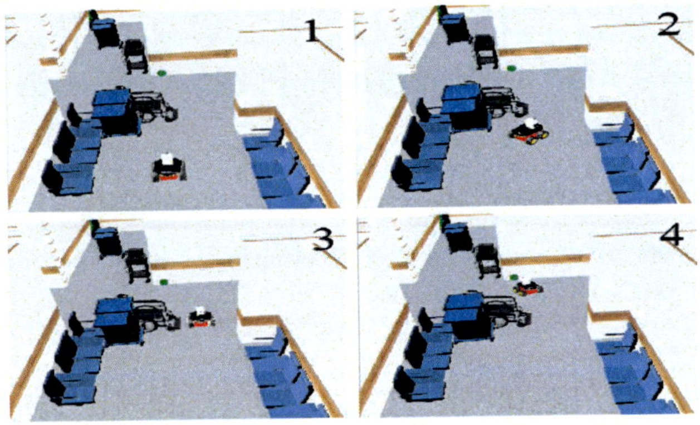

Fig. 3. Snapshots of robot motion when it is moving towards to the ball and has to compensate for the disturbance created by obstacles

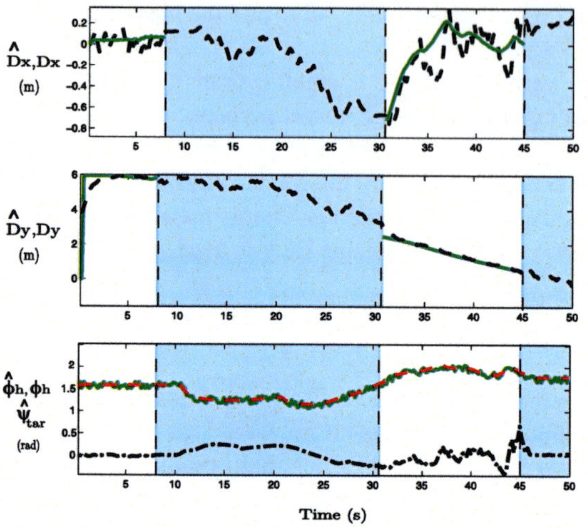

Fig. 4. Top panel (Middle): Measured D_x (D_y) by the robot's visual system (green continuous line), and estimated $\hat{D}_x$ ($\hat{D}_y$) by the EKF (black dashed line). Bottom: Estimated robot's heading direction $\hat{\phi}_h$ (dashed red line), the one calculated by the dead-reckoning ϕ_h (continuous green line), and the angle $\hat{\psi_{tar}}$ (continuous-dashed black line) that the robot has to update its heading direction in order to reach the target.

in order to compensate for this delay. Additionally, when the robot is overcoming the obstacles it looses visual contact with the ball.

Fig. 4 shows the D_x (top) and D_y (middle) measured (green continuous line) and estimated (black dashed line). The bottom panel depicts the estimated $\hat{\phi}_h$, the measured robot's heading direction ϕ_h, and the angle $\hat{\psi}_{tar}$, provided by eq 4. Shadow areas represent intervals of time when the robot does not detect the target, $8\ s \leq t < 31\ s$ due to the obstacle circumnavigation and after $t > 45\ s$ due to the target proximity. The EKF keeps providing good estimates even after the long period without observations, showing a fast convergence to the real value, despite the time spent without new sensorial information. This is a specially important behavior if the robot has to navigate in cluttered environments.

6.2 Experiment 2

In this experiment we want to verify the EKF behavior when the target introduces non-modeled dynamics. Initially the distance between the robot and the ball is approximately 4 m, and the robot starts moving towards to it. However, as the robot is moving the position of the ball is disturbed.

The velocity profile is presented in the first panel of fig. 5. Blue continuous line identifies the measured robot's velocity through odometry, red dashed line depicts the generated velocity by the oscillator and the continuous-dashed black lines describes the amplitude of the respective oscillator. The second, third and bottom panels depict the same variables identified in fig. 4. At $t = 3$ s, when the target moves the amplitude of

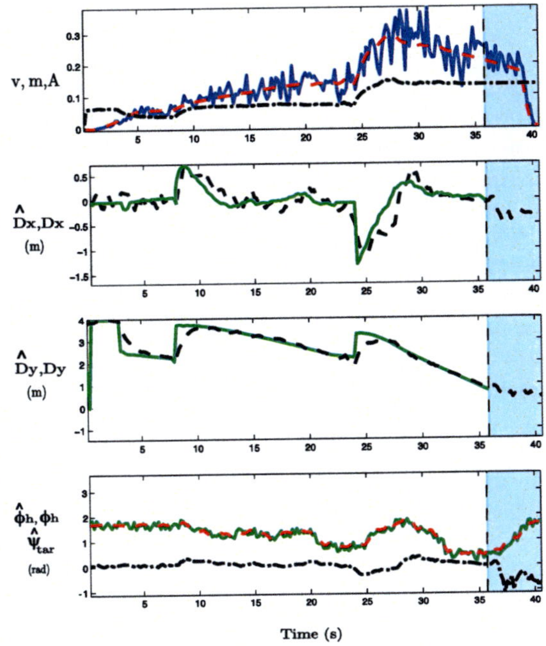

Fig. 5. Similar to fig. 4 but for the experiment 2

the oscillator is decreased according and consequently decreased the robot's velocity. At $t = 8$ s and at $t = 25$ s, a new displacement of the target happens and a new adjustment of the robot's velocity is required. With these adjustments the robot can reach the target within the specified movement time.

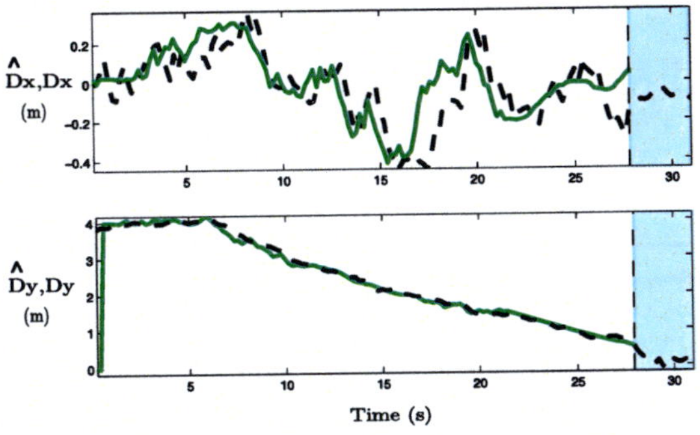

Fig. 6. Similar to fig. 4 but for the experiment 3

6.3 Experiment 3

In the final experiment we aim to verify if the noise resultant from non-modeled dynamics can be removed by the EKF. The ball presents a Brownian movement.

In fig. 6 we can verify that the EKF is able to cope with the unmodeled Brownian motion dynamics of the target within the required 30 s. These experiments were performed for a noise standard deviation of 0.01. Above this value, the target moves too fast for the robot to follow.

7 Conclusions

The paper addresses the problem of generating timed trajectories for an autonomous robot that has to reach a location, within a specified time. While the robot is performing the task of reaching the target, it has to simultaneously compensate any delay or rush as a result of any obstacle circumnavigation or target displacements. The target (ball) is robustly detect by the robot's visual system through its color. Moreover, an EKF was developed and applied to estimate the location of the ball in cases where the visual system does not provide visual information.

We have successfully demonstrated the integration of a standard EKF and a nonlinear dynamical system to robotics in the same approach, without degrading the temporal properties of the proposed architecture. The inclusion of the EKF allowed the reduction of the error between the final position of the robot and that of the target.

Future work includes addressing more complex and cluttered environments, as well as to address sequentially tasks.

Acknowledgement. Work supported by the Portuguese Science Foundation (grant PTDC/EEA-CRO/100655/2008), and by project FCT PEst-OE/EEI/LA0009/2011. Jorge B. Silva is supported by PhD Grant SFRH/BD/68805/2010, granted by the Portuguese Science Foundation.

References

1. Silva, J., Santos, C., Matos, V.: Generating trajectories with temporal constraints for an autonomous robot. In: 8th IEEE International Workshop on Safety, Security & Rescue, Bremen, Germany, July 26-30 (2010)
2. Silva, J., Santos, C., Matos, V.: Timed trajectory generation for a toy-like wheeled robot. In: 36th Annual Conference of the IEEE Industrial Electronics Society, Glendale, USA, November 07-10, pp. 1645–1650 (2010)
3. Ijspeert, A., Nakanishi, J., Schaal, S.: Learning attractor landscapes for learning motor primitives. In: Advances in Neural Information Processing Systems, vol. 15, pp. 1547–1554. MIT Press (2002)
4. Santos, C., Matos, V.: Gait transition and modulation in a quadruped robot: A brainstem-like modulation approach. Robot. Auton. Syst. 59, 620–634 (2011)
5. Buhler, M., Koditscheck, S.: Planning and control of a juggling robot. International Journal of Robotics Research 13(2), 101–118 (1994)

6. Bizzi, E., Saltiel, P., D'Avella, A., Tresch, M.: Modular organization of spinal motor systems. Neuroscientist 8(5), 437–442 (2002)
7. Maufroy, C., Kimura, H., Takase, K.: Towards a general neural controller for quadrupedal locomotion. Neural Networks 21(4), 667–681 (2008)
8. Righetti, L., Buchli, J., Ijspeert, A.J.: Dynamic hebbian learning in adaptive frequency oscillators. Physica D: Nonlinear Phenomena 216(2), 269–281 (2006)
9. Crick, C., Munz, M., Scassellati, B.: Synchronization in social tasks: Robotic drumming. In: Proceedings of the 15th IEEE International Symposium on Robot and Human Interactive Communication, pp. 97–102 (2006)
10. Kober, J., Peters, J.: Learning motor primitives for robotics. In: Proceedings of the IEEE International Conference on Robotics and Automation, Kobe International Conference Center, Kobe, Japan, pp. 2112–2118 (2009)
11. Hersch, M., Billard, A.G.: Reaching with multi-referential dynamical systems. Autonomous Robots 25, 71–83 (2008)
12. Stulp, F., Theodorou, E., Kalakrishnan, M., Pastor, P., Righetti, L., Schaal, S.: Learning motion primitive goals for robust manipulation. In: Proceedings of the IEEE International Conference on Intelligent Robots and Systems (IROS), pp. 325–331 (2011)
13. Clark, M.R., Anderson, G.T., Skinner, R.D.: Coupled oscillator control of autonomous mobile robots. Autonomous Robots 9, 189–198 (2000)
14. Schöner, G.: A dynamic theory of coordination of discrete movement. Biological Cybernetics 63, 257–270 (1990)
15. Santos, C.: Generating timed trajectories for an autonomous vehicle: a non-linear dynamical systems approach. In: Proc. of the IEEE Int. Conf. on Robotics and Automation, ICRA (2004)
16. Tuma, M., Iossifidis, I., Schöner, G.: Temporal stabilization of discrete movement in variable environments: an attractor dynamics approach. In: IEEE International Conference on Robotics and Automation, Kobe International Conference Center, Kobe, Japan, pp. 863–868 (2009)
17. Degallier, S., Ijspeert, A.: Modeling discrete and rhythmic movements through motor primitives: A review. Biological Cybernetics 103(4), 319–338 (2010)
18. Schaal, S., Kotosaka, S., Sternad, D.: Nonlinear dynamical systems as movement primitives. In: International Conference on Humanoid Robotics, Cambridge, MA, pp. 117–124 (2001)
19. Sekmen, A., Barshan, B.: Estimaton of object location and radius of curvature using ultrasonic sonar. IEEE Transactions on Pattern Analysis and Machine Intelligence 62, 841–865 (2001)
20. Smith, C., Christensen, H.I.: Using cots to construct a high performance robot arm. In: Proceedings of the IEEE International Conference on Robotics and Automation, Roma, Italy, pp. 4056–4063 (2007)
21. Wimböck, T., Bäuml, B., Hirzinger, G.: Kinematically optimal catching a flying ball with a hand-arm-system. In: Proceedings of the IEEE/RSJ International Conference on Intelligent Robots and Systems (IROS), pp. 2592–2599 (2010)
22. Bicho, E., Mallet, P., Schöner, G.: Target representation on an autonomous vehicle with low-level sensors. The International Journal of Robotics Research (210), 424–447 (2000)

Autonomous Navigation of a Personal Transporter within Moving Human Groups Using Reactive Control

Florian Steinhardt, Marcus Strand, and J. Marius Zöllner

FZI Research Center for Information Technology, Karlsruhe, Germany
{Steinhardt,Strand,Zoellner}@fzi.de

Abstract. If a vehicle has to drive autonomously in an area with mostly pedestrian traffic and thus needs to share the free space with pedestrians it has to safely navigate in this environment. A pedestrian area lacks fixed lanes and definite traffic rules which leads to more complex paths. Classical path planning in a partially known environment with multiple moving pedestrians is a hard planning problem. Just the task of accurately predicting the individual path of multiple pedestrians in a group accounting for all interactions between the different persons over more than a few seconds is as good as impossible. Using this faulty prediction as a foundation for a path planning algorithm will produce a flawed plan. However such an environment is easy for humans or animals to navigate, avoidance of other persons and moving in a group of pedestrians happens almost unconsciously. In this paper a simple way of autonomously navigating such an environment by using a pure reactive control based on a few simple rules is explored. This method doesn't need a completely known environment, doesn't plan ahead, is simple to implement and can react quick to changes in the environment.

1 Introduction

Individual public transportation in urban environments is a future key element of mobility solutions. The research directions follow assisted driving solutions for dedicated lanes up to autonomoulsy navigating little mobile elements for person or freight transportation. Due to the development of new robotic vehicle concepts in the last years, the trend for autonomy is gaining more and more impact. The possibility to call a personal transporter, ride it in an autonomous or semi-autonomous manner and leave it at any point is not given yet, but reserach is working on possbile solutions. A vehicle like this can be easily integrated into a multimodal mobility chain, picking up people at specific points and navigate them just in time to the next transportation system. This holistic approach for urban navigation using different kinds of transport modi is depicted in fig. 1. One key element in this chain is the navigation in inner-city areas especially within moving crowds of humans. Therefore this paper deals with the problem of navigation of autonomous systems within human groups.

1.1 Related Work

C.W. Reynolds generated flocking of moving 'boids' (simulated bird-like objects) [6] by controlling each individual member of the swarm using three simple behaviours:

S. Lee et al. (Eds.): Intelligent Autonomous Systems 12, AISC 193, pp. 81–90.
springerlink.com

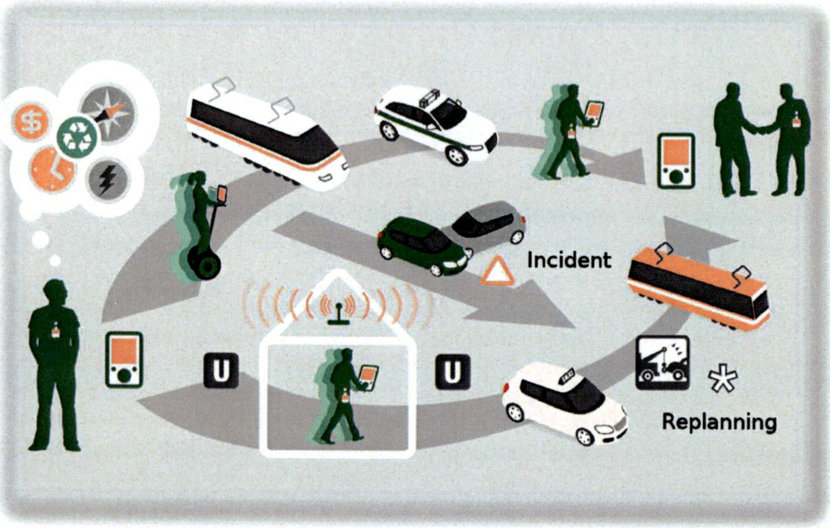

Fig. 1. Urban navigation using different kinds of transport modi

separation, alignment and cohesion. The complexity of a boid swarm is the result of interaction between single boids, splitting and merging of flocks after avoiding obstacles is an emergent behaviour.

The social force model proposed by Helbing and Molnar [4] to simulate human crowds is based on three effects: acceleration towards a desired velocity, attractive effects (e.g. a group of friends or a window display) and repulsive effects (keeping clear of the personal space close to other pedestrians). It is used to create a microsimulation of the movement of large crowds by simulating all individual pedestrians.

Hoogendoorn bases his model of pedestrian movement [5] on several behavioural hypotheses, including: Pedestrians reconsider their walking choices by using current observations and the resulting predictions into the subjective utility optimization. The walking behaviour of other pedestrians is predicted according to non-cooperative or cooperative strategies. Pedestrians only have limited prediction possibilities and mainly consider pedestrians in their direct environment and they minimize predicted discounted costs resulting from straying from the planned trajectory, vicinity of other pedestrians and obstacles, and applying control.

A theory of self-organisation in pedestrian flow is proposed by Daamen and Hoogendoorn [1] to explain emergent patterns of crowd movement that can be observed in real human crowds.

Not all members of a swarm have to belong to the same species, even a swarm consisting of animals and robots is possible. Faria et al. use a robotic fish [2] to examine the behaviour of fish swarms. This fish is a simple dummy on top of a magnet which can be moved by a robotic arm with another magnet under the tank. The robotic fish is not moving in reaction to the movement or position of the other fish, so it doesn't try to move like it is part of the swarm.

Although this robofish only has minimal abilities compared to a real fish its simple movement and appearance seem to be enough to become an accepted member of the swarm.

2 The Personal Transporter *CityPod*

CityPod is a small and flexible personal transporter for the inner city transportation of humans (fig. 2). To allow for a small footprint and the possibility to turn on the spot this vehicle is using a balancing two wheel differential drive while moving around. The vehicle is built on top of a robotic mobile platform, the Segway RMP 200 ATV[1].

Fig. 2. The new version of CityPod with additional sensors in an elevated position on top of the new cabin

[1] http://rmp.segway.com/rmp-200/

It has a range of up to 20 km. To allow a passenger to board or dismount the vehicle can be lowered down to rest on two additional support wheels in the front. These wheels serve a second purpose: if the emergency stop is activated and the segway platform is turned off the whole vehicle is kept standing instead of falling over.

The installed sensor equipment consists of two planar laser scanners (both with a scanning angle of 270°, mounted on two opposing corners of the vehicle allowing for a complete circumferential view), floor contact sensors of the support wheels, the internal sensors of the segway platform (e.g. acceleration sensors for tilt measurement, wheel encoders, battery voltage sensor), a seat-occupied switch and a gyroscope. A Kinect 3D sensor is installed on top of the cabin (eye level of a pedestrian) to simplify the detection and tracking of pedestrians and other obstacles and features in the environment. An embedded PC running Linux and the MCA2 robot control framework (Modular Controller Architecture, see [7], [8]), a WLAN router, payload batteries and hardware interface electronics complete the control hardware.

CityPod supports different modes of operation: directly steered by a driver, autonomous driving (with or without a passenger), driving remote controlled or autonomously. A tablet computer or smartphone is used as the main user interface and allows controlling the vehicle using the integrated acceleration sensors or touchscreen. For testing purposes a gamepad can be used to steer the vehicle. The user interface can be connected via WLAN or an internet connection to the vehicle control computer, it shows a HTML5 interface using websockets to send and receive data to the vehicle. This allows a passenger to use his own smartphone to control the vehicle without having to install software or apps first.

3 Moving with Human Groups

The control algorithm needs information about close moving obstacles in the vehicles environment. These obstacles need to be detected and tracked to provide the current position, direction and velocities of the pedestrians. Static obstacles (e.g. walls) are detected to avoid them as well. For experiments with simulated moving obstacles this detection and tracking step is easy as the position, moving direction and velocities of the obstacles are already known.

The reactive control is based on three different forces or influences: the calculated center of the group of tracked obstacles close to the vehicle, average motion direction and velocities of the moving obstacles in this group, and a drive to avoid collisions with static and moving obstacles.

Attraction towards the center of the close group is realized by first calculating the centroid point of all detected obstacles closer than a specific threshold to the vehicle. The influence of obstacles that move into this region of interest is weighted to avoid a sudden change of this mean position, the obstacles weight is gently faded in from 0 to 1 at the border of the region. The resulting calculated force is the vector to this centroid point (**c**).

The average motion vector (**am**) of the close group of neighbouring objects is used as the second influence to control the vehicles movement.

(a) Escape (b) Evade

Fig. 3. Virtual force field of the evade and escape reflexes

The collision avoidance uses two reflexes to avoid collisions with obstacles and other members of the swarm. These reflexes generate a virtual force field around obstacles to guide the vehicle away from dangerous areas (see fig. 3).

The first of these reflexes is the *escape reflex*, it generates a velocity vector away from an object if it is closer than one meter. The closer the object the stronger the repulsive influence. A second reflex, the *evade reflex*, is used to move the vehicle out of the predicted way (in the vehicles coordinate system thus using the relative movement between vehicle and object) of an approaching obstacle. It's force vector is always orthogonal to and points away from the predicted path of an obstacle. The obstacles movement prediction is a simple one: it is assumed that the obstacles direction and velocity remains unchanged. These two reflexes comprise the reactive part of the obstacle avoidance system already used in [3] (this paper includes an explanation of these reflexes in more detail).

Both two reflexes generate a velocity vector (**escape** and **evade**) following the virtual force field of the obstacles which are combined to the *obstacle avoidance* vector **oa** (equation 1).

$$\mathbf{oa} = \frac{\mathbf{escape} + \mathbf{evade}}{2} \tag{1}$$

$$\mathbf{r} = \frac{\mathbf{am} + \mathbf{c} + w\mathbf{oa}}{3} \tag{2}$$

All tree vectors of the different influences are fused using a simple weighted vector addition (equation 2) where the collision avoidance vector is weighted stronger than the other vectors. A value of $w = 2$ has proven to be high enough to strenghten **oa** to avoid obstacles even in centroid direction. The resulting vector **r** is used to control the vehicle.

As our vehicle uses a differential drive it is not possible to directly use this velocity vector to control the movement because a differential drive can't move sideways. This vector only could be used if the robot turned on the spot until it faced into the right direction and then moving with the calculated speed, but this would lead to repeated stops to change the vehicles direction.

Instead the velocity vector has to be separated into a translative velocity (the portion of the vector along the x-axis) and an angular velocity (proportianal to the vector part in y-direction, scaled to generate favorable turning velocities). These new velocities are used to control the differential drive.

4 Experimental Results

As the obstacle tracking software and the enhanced sensor infrastructure on the vehicle is not finished yet the experiments of the motion control system were carried out in a simulated environment.

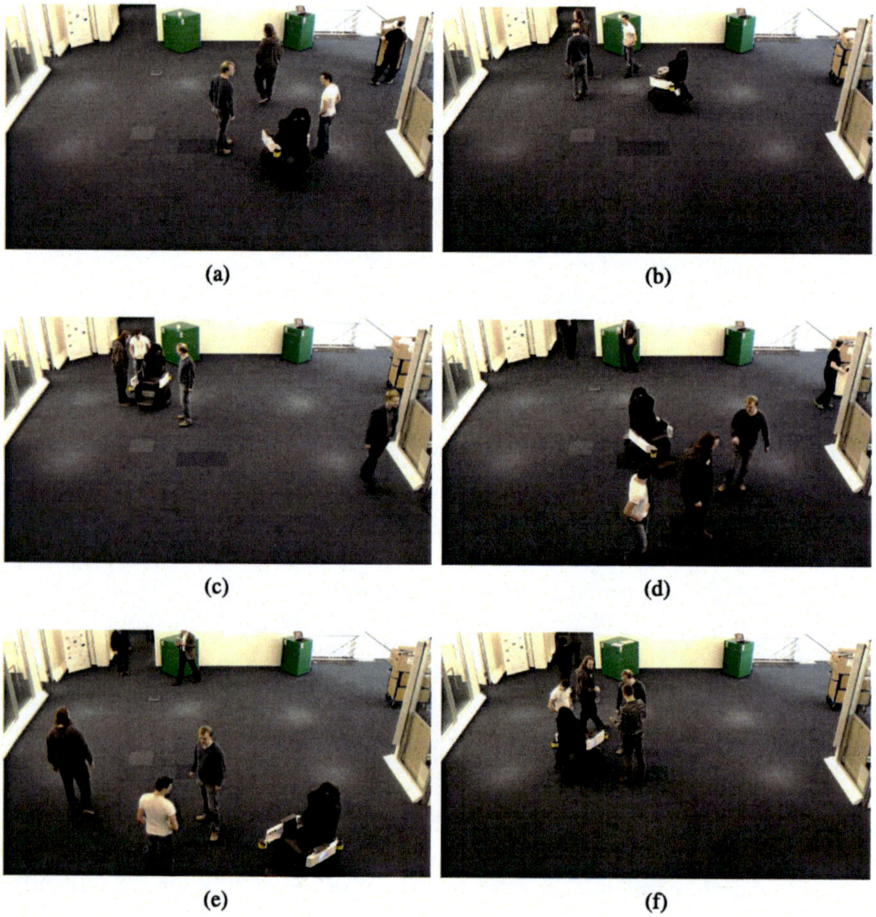

Fig. 4. CityPod as part of a pedestrian group. An older version of the vehicle was used for this experiment.

(a) t = 31 s

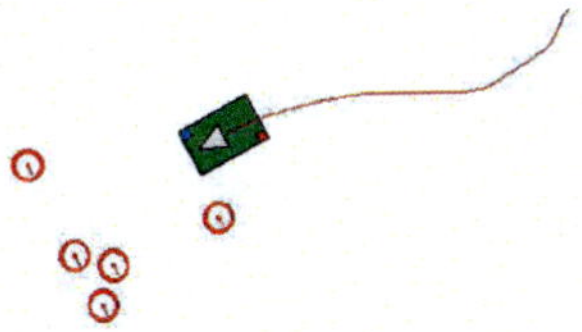

(b) t = 36 s

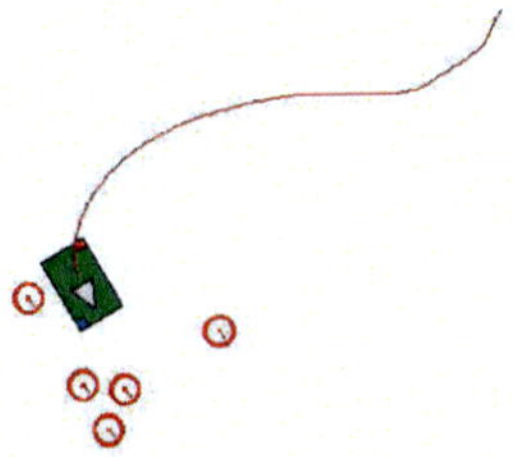

(c) t = 41 s

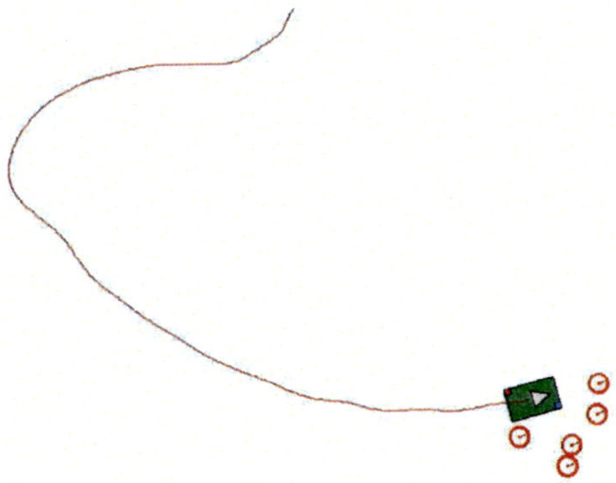

(d) t = 71 s

Fig. 5. Screenshots of the simulation at different points in time. The vehicle (green rectangle, triangle points to the front) closes in on the group and follows close to the calculated centroid point of the group. The red line shows the path of the vehicle. These screenshots show the same experiment as fig. 6.

4.1 Simulation Experiments

The vehicle, its acceleration behaviour and the surrounding moving obstacles were simulated in a 2D environment. The position and current motion of the obstacles (pedestrians) used by the motion system is read directly from this simulation. Each single obstacle is moving on a fixed path. This path and the velocities of the single obstacles are offset by a small value to create a changing group arrangement over time. All simulated obstacles don't react to their environment or deviate from their fixed path.

The vehicles region of interest is a circle with a radius of 6 meters. From the center up to a distance of 5 meters from the vehicle an obstacle is assigned the full weight of 1, from 5 to 6 meters this weight is decreased to 0.

The mean distance to the closest obstacle, the smoothness of the path and the distance to the calculated center of the surrounding obstacles can be used as indicators of the performance of the steering system. The distance to the closest object is calculated from the closest point of the vehicle hull to the circular obstacles center.

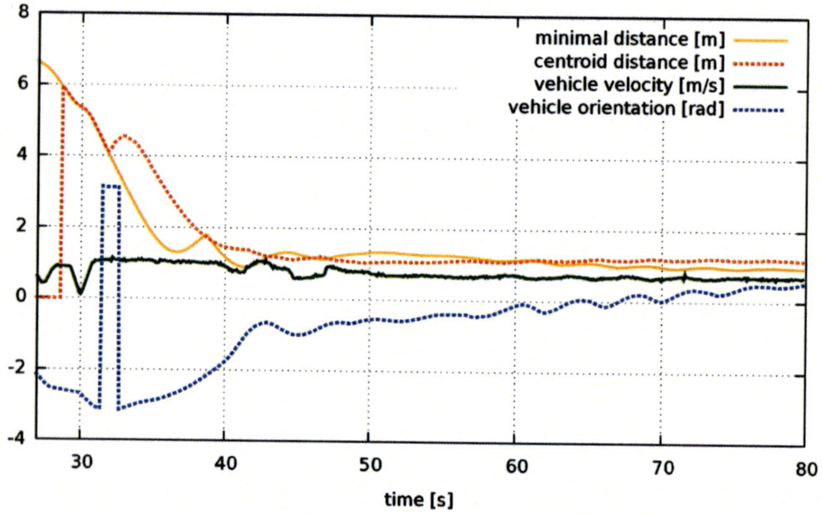

Fig. 6. Plotted data from the same simulation experiment as shown in fig. 5. CityPod detects moving obstacles close to it's position, drives towards it and swims along this group. Shown are the distance to the closest obstacle, the distance to the groups centroid point, the vehicles velocity and orientation. The distance to the calculated centroid point decreases, the minimal distance (orange) to obstacles is kept. The vehicles movement (the orientation changed and the resulting path) is smooth, the velocity is kept almost constant.

Figures 5 and 6 show the results of the same simulation run. The vehicle (green rectangle, a small triangle points to the front) and five simulated pedestrians (red circles with a line from the center showing the current movement direction) are shown. After driving closer to the group the vehicle follows the calculated centroid point of the group. It's path and velocity are smooth and it is keeping it's distance from the obstales.

4.2 Experiments with Real Pedestrian Groups

The main challenge on using this movement control system on the real vehicle is to obtain good pedestrian position and movement data, the detection and tracking of surrounding obstacles can be tricky on a moving platform. A slightly easier tracking could be done by using fixed external sensors (e.g. a topview camera) in an 'intelligent envirionment'.

However using only a simple person detection approach the real vehicle is already able to accomplish the results of the simulation experiments: figure 4 shows CityPod following a group of pedestrians and moving as part of their group.

5 Conclusion and Outlook

Using the proposed system the vehicle has the new ability to move safely in the vicinity of pedestrians. The system allows following groups of pedestrians, avoiding obstacles and adapting the vehicles movement to the surrounding pedestrian group. It enables a vehicle to move like it is a part of a human crowd, thus possibly benefitting from the emergent movement patterns in human crowds (e.g. self separating pedestrian groups with more or less similar moving directions which decreases the number of head-on encounters with obstacles). But problematic behaviour of the system arises in such an inhomogen group of surrounding pedestrians (e.g. two different streams of pedestrians in opposite directions). A differentation between different pedestrian groups in a crowd and choosing and following a single group which moves closest to the own desired direction and speed would increase the effectiveness of this system.

This movement control system can be used as a low level movement control of an autonomous or comandeered vehicle movement in an environment filled with pedestrians.

References

1. Daamen, W., Hoogendoorn, S.P.: Experimental research on pedestrian walking behavior. In: Transportation Research Board Annual Meeting, pp. 1–16. National Academy Press, Washington DC (2003)
2. Faria, J.J., et al.: A novel method for investigating the collective behaviour of fish: introducing "Robofish". Behavioral Ecology and Sociobiology 64, 1211–1218 (2004), doi:10.1007/s00265-010-0988-y
3. Göller, M., Steinhardt, F., Kerscher, T., Zöllner, J.M., Dillmann, R.: Proactive avoidance of moving obstacles for a service robot utilizing a behavior-based control. In: 2010 IEEE/RSJ International Conference on Intelligent Robots and Systems (IROS), October 18-22, pp. 5984–5989 (2010), doi:10.1109/IROS.2010.5651506
4. Helbing, D., Molnàr, P.: Social force model for pedestrian dynamics. Physical Review E, 4282–4286 (1995)
5. Hoogendoorn, S.P.: Pedestrian flow modeling by adaptive control. Transportation Research Record 1878(1), 95–103 (2004)

6. Reynolds, C.W.: Flocks, Herds, and Schools: A Distributed Behavioral Model. Computer Graphics 21(4), 25–34 (1987); SIGGRAPH Conf. Proc.
7. Scholl, K.U., Albiez, J., Gassmann, B.: MCA2 - an expandable modular controller architecture. In: 3rd Real-Time Linux Workshop (2001)
8. Uhl, K., Ziegenmeyer, M.: MCA2 - an extensible modular framework for robot control applications. In: Advances in Climbing and Walking Robots, vol. 2, pp. 680–689 (2007)

Simultaneous Control of Translational and Rotational Motion for Autonomous Omnidirectional Mobile Robot Considering Shape of the Robot and Movable Area by Heights

Ayanori Yorozu, Takafumi Suzuki, Tetsuya Matsumura, and Masaki Takahashi

Dept. of System Design Engineering, Keio University
3-14-1 Hiyoshi, Kohoku-ku, Yokohama 223-8522, Japan
takahashi@sd.keio.ac.jp

Abstract. This paper presents a real time collision avoidance method for an autonomous omnidirectional mobile robot considering shape of the robot and movable area by heights based on simultaneous control of translational and rotational motion. Service robots which have been developed in recent years have arms to work and execute tasks. In these robots, the size of width is sometimes not equal to that of depth by heights. In order to avoid obstacles considering safety and mobility for the robots, it is necessary to evaluate shape of the robot and movable area by heights. To evaluate them, the robot model is defined by heights. Evaluating of the robot model and the movable area for each height, if the robot is unable to move keeping a safe distance from the obstacles, the robot determines the suitable orientation angle considering the minimum length from the center of the robot model to that outer shape. In this paper, the novel control method based on the fuzzy potential method is presented. To verify the effectiveness of the proposed method, several numerical simulations are carried out.

Keywords: Autonomous mobile robot, Obstacle avoidance, Omnidirectional platform, Fuzzy potential method.

1 Introduction

Recently, autonomous mobile robots work in human living space have been studied and developed. Some cases of these robots installation to public facilities have been reported [1-3]. These robots sometimes have two arms mounted on a torso so these robots can be used for manipulation, whole-body activities, and human-robot interaction [4, 5]. In these robots, the size of width is not equal to that of depth by heights. In order to avoid obstacles considering safety and mobility, it is necessary to evaluate shape of the robot and movable area by heights.

Various obstacle avoidance methods and their availabilities for mobile robots have described [1-3, 6-8]. Most of these studies regard the robots as points or circles and control methods of the translational movements in two-dimensional plane are

S. Lee et al. (Eds.): Intelligent Autonomous Systems 12, AISC 193, pp. 91–100.
springerlink.com

discussed. In these studies, a non-circle robot is regarded as a circle robot with consideration of maximum size of the robot. However, depending on the shape of the robot such as horizontally long shape, this approach reduces and wastes available free space and can decrease the possibility that the robot reaches the goal. To enable wide robots to avoid obstacles safely and efficiently, it is necessary to control not only a translational motion but also a rotational motion. Several studies have focused on the orientation angle of the robot [9, 10]. However, these methods require an environmental map and the studies have not shown the effectiveness for avoidance of unknown obstacles by autonomous mobile robots. Therefore, in our current research, to avoid unknown obstacles reactively for wide robots, simultaneous translational and rotational motion control method is presented [11]. In addition, there are obstacles of various heights in the human living space and the relation between the robot and the surrounding environment varies depending on shape of the robot. Therefore, in order to avoid obstacles more safely and efficiently, it is necessary to evaluate cross-sectional shape of the robot and movable area by heights.

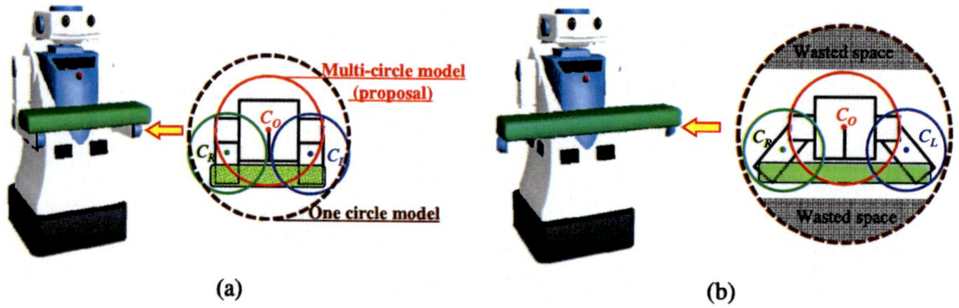

Fig. 1. Two robot models considering cross-sectional shape of the robot at the height of arms: (a) robot holds a short baggage, (b) robot holds a long baggage

Consequently, this study proposes the following two points.

I. Robot model considering cross-sectional shape of the robot.
II. Simultaneous control of translational and rotational motion considering shape of the robot and movable area by heights.

With the proposed method, if the robot is unable to move keeping a safe distance from the obstacles, the robot determines the suitable orientation angle considering the minimum length from the center of the robot to that outer shape in real time. To verify the effectiveness of the proposed method, several simulations are carried out.

2 Robot Model Considering Cross-Sectional Shape of the Robot

Service robots sometimes use the arms to transport baggage. In that case, the robot size of width is not equal to that of depth by heights. If these wide robots are modeled as a circle, the robots decrease the possibility to reach the goal because of wasting available free space. In order to achieve obstacle avoidance considering safety and mobility, it is necessary to introduce a robot model considering cross-sectional shape. In this study, to consider the changes of the shape of the robot as shown in Fig. 1, a new robot model with multi-circle is proposed. The modeling method is as follows.

- Robot body is enclosed in a circle.
- If the parts such as arms that cannot be enclosed in the body circle exist, these are enclosed in each circle.

As shown in Fig. 1, the proposed robot model is available to change the angle of the arms according to the size of baggage that the robot holds.

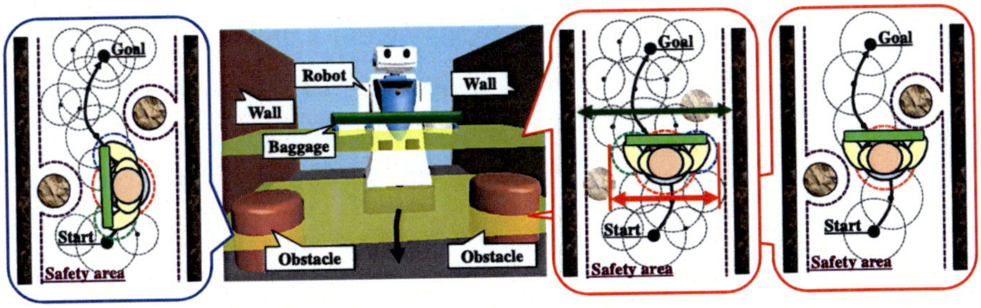

Fig. 2. Trajectories of the robot with two motion control method: (*left*) considering footprint at the height of wheeled platform, (*right*) evaluating cross-sectional shape of the robot and movable area by heights

3 Simultaneous Control of Translational and Rotational Motion Considering Shape of the Robot and Movable Area by Heights

For the robot that the size of width is not equal to that of depth by heights, simultaneous translational and rotational motion control method is presented [11]. In addition to the shape of the robot is different by heights, there are obstacles of various heights the human living space. For example, in situation as shown in Fig. 2, considering the shape of the robot by heights, it is able to move keeping a safe distance from obstacles without rotational motion. Therefore, in order to avoid obstacles safely and efficiently, it is necessary to evaluate cross-sectional shape of the robot and the movable area by heights. In this study, evaluating shape of the robot and movable area by heights, if the robot is unable to move keeping a safe distance from the obstacles, the robot determines the suitable orientation angle considering the minimum length from the center of the robot to that outer shape. Then, evaluation of the shape of the robot and the movable area is used the width of the robot model and the movable area measured with range sensor like Laser Range Finder (LRF) by heights.

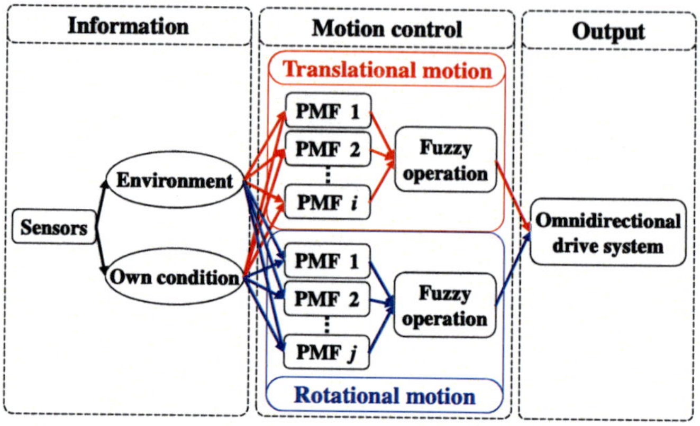

Fig. 3. Concept of fuzzy potential method for simultaneous translational and rotational motion with an omnidirectional platform

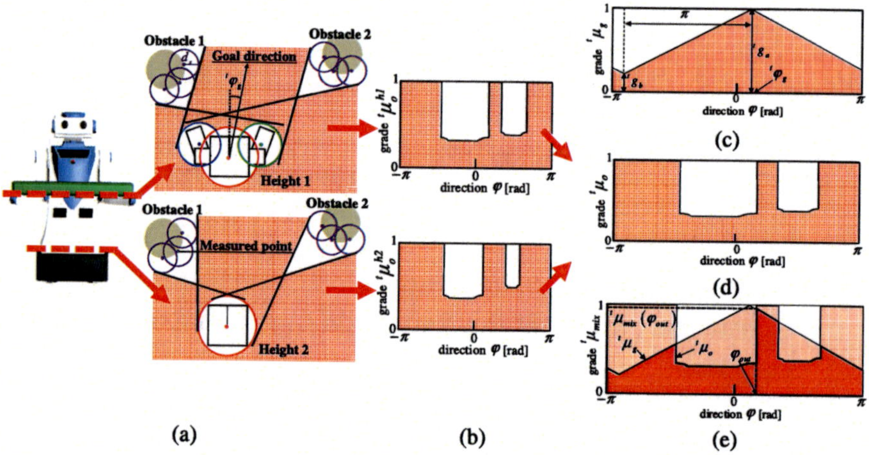

Fig. 4. An example of determination of translational motion: (a) situation of the robot and the environment, (b) PMFs for obstacles in each height, (c) PMF for a goal, (d) integrated PMF for obstacles, (e) mixed PMF for translational motion

In this study, the novel control method based on the fuzzy potential method (FPM) [12] with proposed model is presented. Fig. 3 shows a concept of the FPM that takes into consideration both translational and rotational motion by heights. In the FPM, a command velocity vector that takes into consideration element actions is decided. Element actions are represented as potential membership functions (PMFs), and then they are integrated by means of fuzzy inference. The horizontal axis of PMF is directions which are from $-\pi$ to π radians measured clockwise from the front direction of the robot. The vertical axis of PMF is the grade for the direction. The grade, direction, and configured maximum and minimum speeds, are used to calculate the command velocity vector. Finally, translational and rotational velocity commands, which are calculated by defuzzification of mixed PMFs, are realized by an omnidirectional drive system.

3.1 Translational Motion Considering Shape of the Robot and Environment by Heights

For the robot shown in Fig. 4(a), to generate translational PMF for moving toward the goal keeping the safe distance d_s from obstacles that detected in each height.

3.1.1 PMF for a Goal

To head to the goal, a triangular PMF ${}^t\mu_g$ is generated, as shown in Fig. 4(c). As a measure to decide how close to the goal the robot should go, ${}^tg_a$ is defined as the height of the triangular PMF. As a measure to decide how much the robot can back away from obstacles, ${}^tg_b$ is defined. ${}^t\mu_g$ reaches the maximum value as ${}^tg_a$ at an angle of the goal direction relative to the front direction of the robot ${}^t\varphi_g$.

$$ {}^tg_a = \begin{cases} \dfrac{\left\|\mathbf{R}_{r,g}\right\|}{\varepsilon} & if \ \left\|\mathbf{R}_{r,g}\right\| \le \varepsilon \\ 1.0 & otherwise \end{cases}. \tag{1} $$

$$ {}^tg_b = \eta\,{}^tg_a \qquad (0.0 \le \eta < 1.0). \tag{2} $$

where $\left\|\mathbf{R}_{r,g}\right\|$ is an absolute value of the position vector of the goal relative to the robot. ε and η are constants. If $\left\|\mathbf{R}_{r,g}\right\|$ is below ε, ${}^tg_a$ is defined.

3.1.2 PMF for Obstacles

To avoid obstacles safely and efficiently in real time, a concave shaped PMF ${}^t\mu_o^{hi}\ (hi : height\ i = 1, 2, \cdots, n)$ are generated. These PMFs are specified by depth and width, which are calculated based on the geometrical relation between obstacles and the robot in each heights as shown in Fig. 4(b). Then, they are all integrated to ${}^t\mu_o$ shown in Fig. 4(d) by calculating the logical product.

$$ {}^t\mu_o = {}^t\mu_o^{h1} \wedge {}^t\mu_o^{h2} \wedge \cdots \wedge {}^t\mu_o^{hn}. \tag{3} $$

3.1.3 Calculation of a Translational Command Velocity Vector

The PMF ${}^t\mu_{mix}$ for translational motion is generated and calculate the command velocity vector. ${}^t\mu_{mix}$ in Fig. 4(e) is an logical product of ${}^t\mu_g$ and ${}^t\mu_o$. By defuzzifier, a velocity command vector is calculated as a traveling direction φ_{out} and an absolute value of the reference speed of the robot based on the mixed PMF ${}^{t}\mu_{mix}$. φ_{out} is decided as the direction that makes the PMF ${}^{t}\mu_{mix}(\varphi)$ maximum. Based on φ_{out}, v_{out} is calculated as follows.

$$v_{out} = {}^{t}\mu_{mix}(\varphi_{out})(v_{max} - v_{min}) + v_{min}. \tag{4}$$

where ${}^{t}\mu_{mix}(\varphi_{out})$ is the mixed PMF for translational motion correspond to the φ_{out}. v_{max} and v_{min} are respectively the upper and lower limits of the robot speed.

3.2 Rotational Motion Based on the Evaluation of the Shape of the Robot and the Movable Area by Heights

PMF for rotational motion based on the evaluation of the shape of the robot and the movable area by heights is generated.

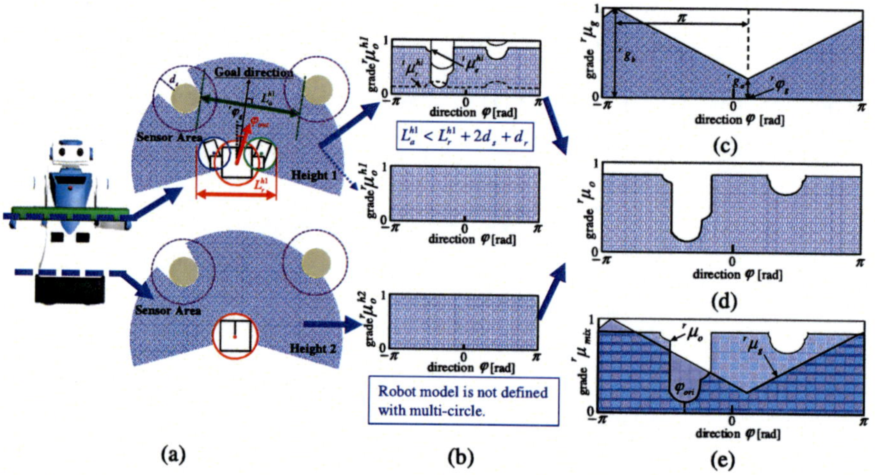

Fig. 5. An example of determination of rotational motion: (a) situation of the robot and the environment, (b) PMFs for obstacles in each height, (c) PMF for a goal, (d) integrated PMF for obstacles, (e) mixed PMF for rotational motion

3.2.1 PMF for a Goal

In order to turn the front of the robot toward the goal direction or the travelling direction if there is no obstacle to avoid, PMF for a goal is generated as ${}^{r}\mu_{g}$ shown in Fig. 5(c). This shape is decided in same way as ${}^{t}\mu_{g}$.

3.2.2 PMF for Obstacles

If the robot is unable to move keeping the safe distance d_s from the obstacles with facing the goal direction or the travelling direction, the PMF ${}^{r}\mu_{o}$ to determine the suitable orientation angle considering the minimum length from the center of the robot model to that outer shape is generated. To evaluate the shape of the robot and the movable area, the robot width L_r^{hi} and the movable area width L_a^{hi} are calculated at the height that the robot model defined multi-circle. As shown in Fig. 5(a), L_a^{hi} are the vertical distances between the obstacles detected in the travelling direction. To determine whether the robot is able to move keeping the distance d_s from the obstacles with facing the goal direction is used follow equation.

$$L_a^{hi} < L_r^{hi} + 2d_s + d_r. \tag{5}$$

where d_r is the parameter to avoid determining to be unnecessary rotational motion for obstacles even if that is required due to measurement error. Accordingly, the PMF for rotational motion considering obstacles is as follows.

$${}^r\mu_o^{hi}(\varphi) = \begin{cases} {}^r\mu_e^{hi}(\varphi) - {}^r\mu_r^{hi}(\varphi) & \textit{if } \text{Multi-circle model } \textit{and } \text{Formula (5)} \\ 1.0 & \textit{otherwise} \end{cases}. \tag{6}$$

${}^r\mu_e^{hi}(\varphi)$ are generated based on the distance from the center of the robot to obstacles, as shown in Fig. 5(b). The relative distances are obtained with range sensor such as LRF. Then, they are all integrated to ${}^r\mu_o$ shown in Fig. 5(d) by calculating the logical product in same way as ${}^t\mu_o$.

3.2.3 Calculation of a Rotational Command Velocity Vector

For the rotational motion, like the translational motion, the rotational command velocity vector is derived. The PMFs ${}^r\mu_g$ and ${}^r\mu_o$ are integrated by fuzzy operation into a mixed PMF ${}^r\mu_{mix}$, as shown in Fig. 5(e). By defuzzifier, the command velocity vector is calculated as a rotational direction φ_{ori} and an absolute value of the reference speed of the robot. φ_{ori} is decided as the direction that makes the following function $h(\varphi)$ minimum.

$$h(\varphi) = \int_{\varphi-\zeta}^{\varphi+\zeta} {}^r\mu_{mix}(\psi)\, d\psi. \tag{7}$$

where ζ is the parameter to avoid choosing an uncertainty φ caused by, for example, noise on the sensor data. Based on φ_{ori}, ω is calculated as follows.

$$\omega = \frac{\varphi_{ori}}{\pi}(\omega_{max} - \omega_{min}) + \omega_{min}. \tag{8}$$

ω_{max} and ω_{min} are the upper and lower limits of the robot rotational speed.

Finally, translational and rotational velocity commands are realized simultaneously by an omnidirectional drive system [11].

4 Simulation

In simulation, to verify the effectiveness of the proposed robot model and motion control method, the situation that the robot transports baggage with arms was assumed.

4.1 Situation of Simulation

The robot has an omnidirectional drive system, and can measure 4.0 m in ±120 deg range at 0.23 m height of the wheel platform and 0.60 m height of arms with two LRFs. A safe distance d_s is 0.20 m and complement measurement error parameter d_r is 0.20 m. As shown in Fig. 6, there are two static obstacles (obstacle 1: 0.50 × 0.60 × 0.80 m, obstacle 2: 0.50 × 0.60 × 0.40m) in the passage of 2.3 m width. However, the robot is assumed to have no priori information about these obstacles.

In situation A, the robot determines the translational motion with one circle robot model. In situation B, the robot determines translational and rotational motion considering the footprint at the height 2. In situation C, the robot determines the suitable orientation angle based on the evaluation of the shape of the robot and

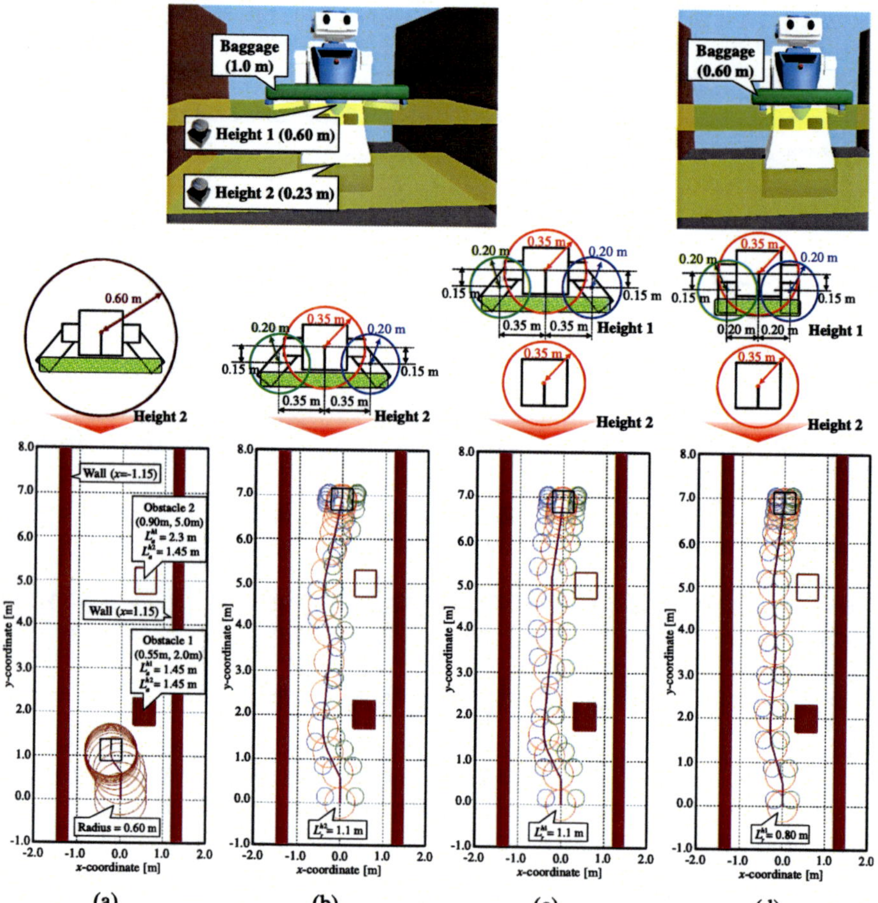

Fig. 6. Simulation results; robot models and trajectories of the robot: (a) situation A, (b) situation B, (c) situation C, (d) situation D

movable area by heights (proposal). In situation D, the position of the arms is different from other situations A-C according to the size of baggage.

4.2 Performance of Multi-circle Robot Model

In situation A shown in Fig. 6(a), the robot regarded as one circle did not get to the goal because the robot model is larger than the movable area next to the obstacle 1 $L_a^{h2}(=1.45\,\text{m}) < L_r^{h2} + 2d_s(=1.6\,\text{m})$. On the other hand, in situation B shown in Figs. 6(b) and 7, the robot modeled with multi-circle was able to pass the obstacle 1 with rotational motion considering the minimum length from the center of the robot to that outer shape. Moreover, in situation D shown in Figs. 6(d) and 7, the robot was able to pass the obstacle 1 without rotational motion because the relation between the robot and the movable area is $L_a^{h2}(=1.45\,\text{m}) > L_r^{h2} + 2d_s + d_r(=1.4\,\text{m})$.

These results show that defining the robot model with multi-circle, it is possible to respond flexibly to the changes of the cross-sectional shape of the robot and design simultaneous control of translational and rotational motion based on the evaluation of the cross-sectional shape of the robot and movable area.

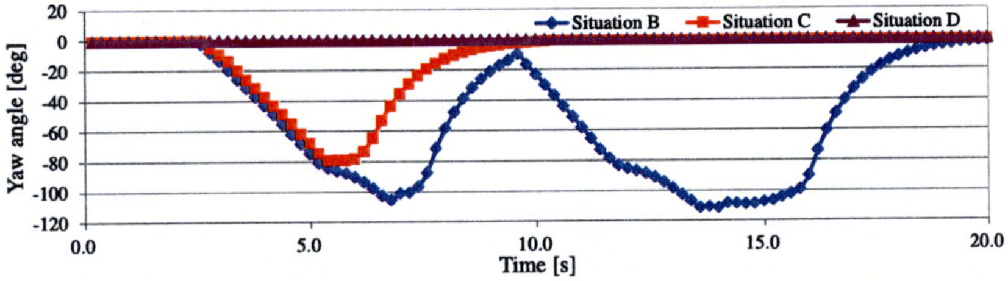

Fig. 7. Time history of yaw angle of the robot in situations B-D

4.3 Performance of Motion Control Based on the Evaluation of the Shape of the Robot and the Movable Area by Heights.

In situation B shown in Figs. 6(b) and 7, the robot changed yaw angle for the obstacle 2 which can be avoided without rotational motion because the robot does not evaluate the shape of the robot and the movable area by heights. On the other hands, in situation C shown in Figs. 6(c) and 7, the robot has achieved to avoid the obstacle 2 without rotational motion depend on the evaluating of the shape of the robot and the movable area by heights $L_a^{h1}(=2.3\,\text{m}) > L_r^{h1} + 2d_s + d_r(=1.7\,\text{m})$. In addition, as shown in Fig. 7, after the rotational motion for the obstacle 1, the robot was able to quickly return to the goal direction, so it is more expected to respond quickly to obstacles in proposed method.

These results show that the evaluation of the cross-sectional shape of the robot and the movable area by heights to the robot which the size of width is not equal to that of depth by heights, the robot can decide the suitable orientation angle according to the situation. As a result, the robot can get to the goal point safely and smoothly.

5 Conclusion

In this paper, toward the realization of motion control for autonomous mobile robots that can be flexible in various situations, the real time collision avoidance method with simultaneous control of translational and rotational motion considering the shape of the robot and the movable area by heights has been proposed. This method used an omnidirectional platform for the drive system and was based on the fuzzy potential method. Evaluating of the cross-sectional shape of the robot and the movable area by heights, if the robot is unable to move keeping a safe distance from the obstacles, the robot determines the suitable orientation angle considering the minimum length from the center of the robot model to that outer shape. The effectiveness has been verified by numerical simulations. It has been shown that the proposed method performs translational and rotational movement simultaneously according to the situation.

Acknowledgements. This work was supported in part by Grant in Aid for the Global Center of Excellence Program for "Center for Education and Research of Symbiotic, Safe and Secure System Design" from the Ministry of Education, Culture, Sport, and Technology in Japan.

References

1. Tiejun, Z., Dalong, T., Mingyang, Z.: The Development of a Mobile Humanoid Robot with Varying Joint Stiffness Waist. In: Proc. IEEE Int. Conf. Mechatronics and Automation, vol. 3, pp. 1402–1407 (2005)
2. Evans, J.M.: HelpMate: An Autonomous Mobile Robot Courier for Hospitals. In: IEEE/RSJ/GI Int. Conf. on Intelligent Robots and Systems, vol. 3, pp. 1695–1700 (1994)
3. Krishnamurthy, B., Evans, J.: HelpMate: A Robotic Courier for Hospital Use. In: IEEE Int. Conf. on Systems, Man and Cybernetics, vol. 2, pp. 1630–1634 (1992)
4. Kuindersma, S., Hannigan, E., Ruiken, D., Grupen, R.: MINERVA: Dexterous Mobility with the uBot-5 Mobile Manipulator. In: Proc. IEEE Int. Conf. on Robotics and Automation, vol. 3, pp. 1999–2005 (1999)
5. Mehling, J.S., Strawser, P., Bridgwater, L., Verdeyen, W.K., Rovekamp, R.: Centaur: NASA's Mobile Humanoid Designed for Field Work. In: IEEE Int. Conf. on Robotics and Automation, pp. 2928–2933 (2007)
6. Borenstein, J., Koren, Y.: The Vector Field Histogram Fast Obstacle Avoidance for Mobile Robots. IEEE Trans. on Robotics and Automation 7(3), 278–288 (1991)
7. Fox, D., Burgard, W., Thrun, S.: The Dynamic Window Approach to Collision Avoidance. IEEE Robotics and Automation 4(1), 23–33 (1997)
8. Minguez, J., Montano, L.: Nearness Diagram (ND) Navigation: Collision Avoidance in Troublesome Scenarios. IEEE Trans. on Robotics and Automation 20(1), 45–59 (2004)
9. Kavraki, L.: Computation of Configuration Space Obstacles Using the Fast Fourier Transform. IEEE Trans. on Robotics and Automation 11(3), 408–413 (1995)
10. Wang, Y., Chirikjian, G.S.: A New Potential Field Method for Robot Path Planning. In: Proc. IEEE Int. Conf. on Robotics and Automation, pp. 977–982 (2000)
11. Suzuki, T., Takahashi, M.: Translational and Rotational Motion Control Considering Width for Autonomous Mobile Robots Using Fuzzy Interence. Numerical Analysis – Theory and Application. InTech Book (2011) ISBN 978-953-307-389-7
12. Tsuzaki, R., Yoshida, K.: Motion Control Based on Fuzzy Potential Method for Autonomous Mobile Robot with Omnidirectional Vision. J. of the Robotics Society of Japan 21(6), 656–662 (2003)

Development of an Autonomous Vehicle for High-Speed Navigation and Obstacle Avoidance

Jee-Hwan Ryu, Dmitriy Ogay, Sergey Bulavintsev, Hyuk Kim, and Jang-Sik Park

Korea University of Technology and Education, Cheonan, South Korea
{jhryu,cactus,perseus,ganggai}@koreatech.ac.kr,
bulavinstevs@gmail.com

Abstract. This paper introduces the autonomous vehicle named Pharos that participated in 2010 Autonomous Vehicle Competition organized by Hyundai-Kia Automotive Group. Pharos was developed for high-speed on/off-road unmanned driving avoiding diverse patterns of obstacles. For the high speed travelling up to 60 Km/h, long range terrain perception, real-time path planning and high speed vehicle motion control algorithms are developed. This paper describes the major hardware and software components of our vehicle.

1 Introduction

The first autonomous vehicle competition in South Korea organized by Hyundai-Kia Automotive Group took place on November, 2010 [11]. The mission of the competition was unmanned travelling of about 4 Km on/off road, passing 7 different patterns of obstacles. Missions also included lane keeping and stopping within 1 m of cross-walk. Out of eleven universities from South Korea, participating in the competition, "Pharos" has taken 4th place. It finished the course in 8 min 52 sec clearing all the missions except cross-walk stopping, which gave 5 min penalty.

Recently there have been many research activities in autonomous vehicle area. Especially, DARPA Grand Challenge [1], [8] and Urban Challenge [10] have made a big progress in the area of autonomous vehicles. However, there are still many open issues for high speed unmanned driving such as long range terrain perception [9], real-time obstacle avoidance and trajectory planning [3], [6], and high speed vehicle motion control [5] et al.

In this paper autonomous vehicle based on real Sport Utility Vehicle (SUV) is introduced, motivated by the competition. The main challenging issue in the development of the vehicle was to build a reliable system, able to travel at high speed up to 60 Km/h through on-road and unstructured off-road while avoiding different types of obstacles. To satisfy these requirements, new methods were developed and extended based on existing methods in the field of autonomous navigation, such as long-range obstacle detection and mapping, real-time collision avoidance and trajectory planning, and stable vehicle control on slippery and rugged terrain.

S. Lee et al. (Eds.): Intelligent Autonomous Systems 12, AISC 193, pp. 101–109.
springerlink.com

2 Hardware Design

Pharos' hardware design consists of several blocks, which are roof platform, which holds sensors, steering and braking actuators, power and data processing systems, and wiring.

Actuation is done by rotating the steering wheel with a controlled motor. A geared DC motor is attached to the steering column. After analysing required torque and rotational speed the choice was given to Smart motor from Animatic Co. with 1:4 gear ratio to control the steering wheel.

Fig. 2 shows the assembled 3D CAD model of the steering actuation mechanism. one-to-one pulley powered transmission mechanism was used, and steering angle sensor was moved to the motor output axis to measure the absolute steering angle.

Braking is actuated by the same type motor with gear, which pushes the brake pedal. Throttle is controlled through internal vehicle electronic system. Vehicle data, such as steering angle, vehicle speed, APS signal, etc are fed to the computing system through CAN interface.

Shown in the Fig. 1 is the custom-made roof platform, holding four SICK laser scanners are installed there. All scanners are facing forward along the driving direction of the vehicle, but with slightly different tilting. A single CMOS camera is installed on the roof for cross-walk detection. GPS receiver, DGPS RF modem and a radio antenna for E-stop (emergency stop system) are also installed on the roof platform. The E-stop system was provided by the organizers of the competition to allow the following patrol car safely stop an autonomous vehicle. Three physical E-stop buttons are also available on both sides of the vehicle and in the rear side.

Pharos' controllers and communication system are located inside of the trunk. Six Compact PCI computers, NI-CompactRIO, two Gigabit Ethernet switches and various signal converters are installed in a rack. Custom-made power system with backup batteries is installed underneath of the rack to supply power. A six degree-of-freedom inertial measurement unit is attached close to the gravity center of vehicle.

Fig. 1. Pharos standing at the finish line of competition track

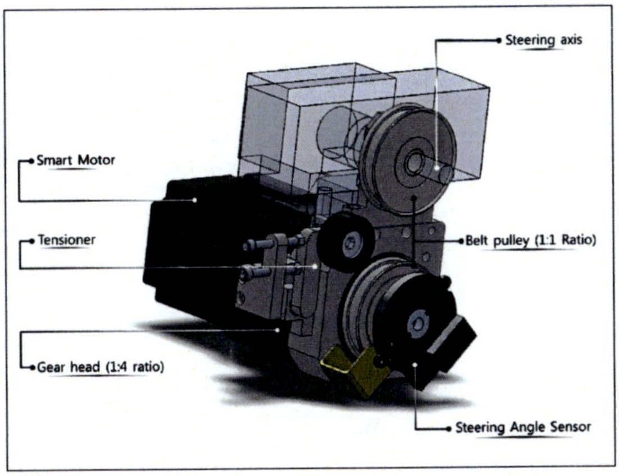

Fig. 2. Assembled 3D CAD model of the steering actuation mechanism

3 Software Architecture

3.1 Overview

The main principles that we wanted to follow when developing the system architecture were: reliability of a system, ability to distribute software over several independent computers, easy configuration, simulation and development. To reach these requirements we have chosen an architecture, where each module was an independent program that could be run stand alone in a separate process of an operating system. Communication between modules is done through publish-subscribe mechanism.

There is no central process, which encapsulates modules, as it could reduce reliability of the system, if such a process dies. All communication is done through a lightweight messaging server, which supports connectivity through network protocols. It allowed us to easily distribute modules over several computers, and even gave opportunity to run different modules on different operating systems.

In a running integrated system all processes are put on watchdogs, so they are restarted if fail, or they can be controlled remotely from the user interface or health monitoring system.

3.2 System Components

Our system has multiple processing layers, where data flow from sensors through consecutive layers, until it reaches steering and braking-acceleration actuators. These layers are: sensor interface, perception layer, planning layer, and control layer. Each layer consists of several modules, as shown in Fig. 3.

Sensor interface layer consists of modules that receive data directly from sensors, using sensor protocols, and then publishes the data in a format used inside our system. They run on frequency determined by corresponding sensors, which are 75Hz for laser scanners and 100 Hz for GPS/IMU.

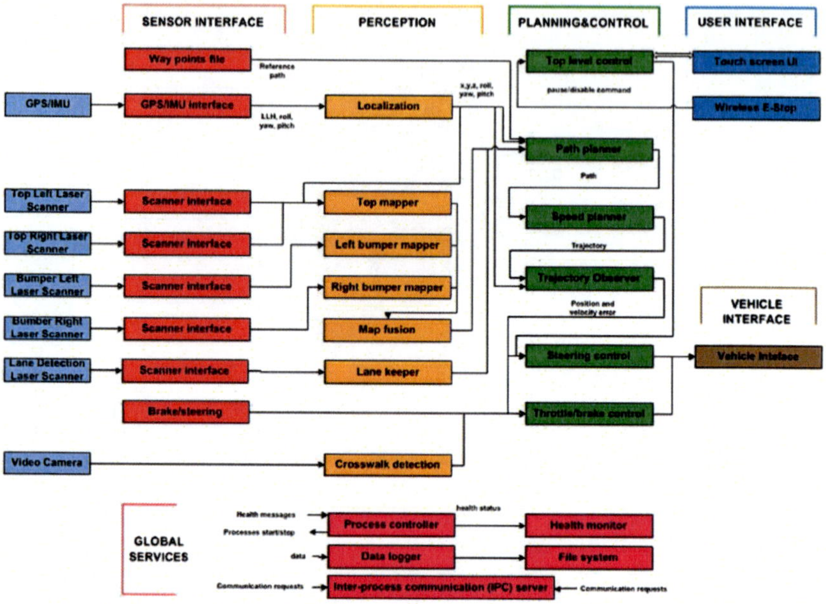

Fig. 3. Overview of the software architecture

Perception layer has several independent parts that are: localization for estimating vehicle state, laser mapping for generating obstacle occupancy grid map from laser scanner data, and vision mapping for detecting road markings. Data to the path planner are sent at 10 Hz frequency.

Planning layer consists of path planning module that uses 2D obstacle map to generate traversable path, speed planning layer that plans speed taking into account current vehicle speed, speed limits imposed by competition requirements, and shape of the generated path.

Control layer consists of a Trajectory Observer, which calculates difference between planned path and current vehicle position and sends it to the motion control module, which is run in a separate real-time system. Unlike path planner, which is nondeterministic, Trajectory Observer is run at 100 Hz frequency, monitors vehicle position and stops vehicle if no data from path planner come, which can occur in case of a process failure.

All modules also generate health monitoring messages, which are listened by a health monitoring block.

4 Perception

4.1 Obstacle Detection and Mapping

To make safely avoid obstacles, Pharos was to be able to create a precise terrain map with sufficient range of view for undertaking appropriate actions. The map should

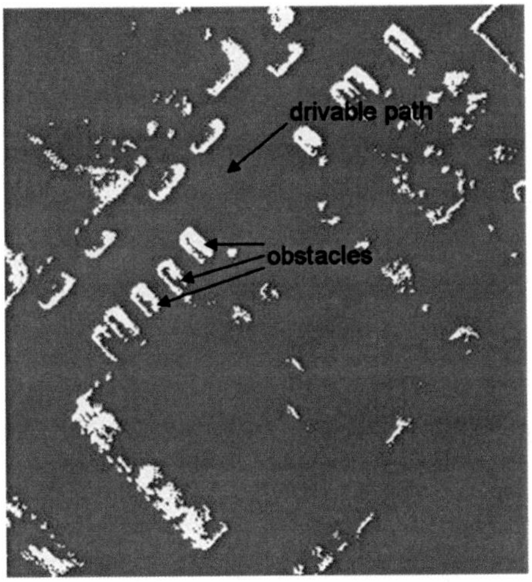

Fig. 4. An example of occupancy grid map

contain accurate position of obstacles to maintain the movement of the vehicle even in narrow paths with widths, comparable to the width of a vehicle. Detection range of the obstacles is also important for driving at higher speeds. To satisfy these main criteria, medium range laser scanners were chosen. A laser scanner has a 75 Hz update rate with accuracy $1cm$ at the 30 m distance. Laser scanner system consists of three laser scanners, mounted on the roof, tilted downward and two scanners, installed on the front bumper with 45 degree yaw angle with respect to the vehicle to provide wider area of view. Different obstacle classification algorithms are used for scanners installed on the roof and on the bumper. Each laser scanner acquires distance information in its own local coordinate frame. By using estimated position and orientation of the vehicle, data from scanners are transformed into points in global coordinate frame and represent 3D point cloud. Each point in 3D point cloud can be described as (X_m^i, Y_m^i, Z_m^i) where m is the index of the individual scanner and i is the index of each measurement in one scan.

The classification method is based on the occupancy grid map [2]. The map is represented by a 2D array, where each cell contains information about its occupancy. To classify a cell as obstacle, two nearby points in 3D point cloud must be found and their vertical distance must satisfy condition $|Z_k^i - Z_l^j| > \varepsilon$, where i and j are indices of two closest points in 3D cloud point from two different laser scanners and ε is the threshold, assumed to being obstacle. If no such points can be found or no value is acquired, this cell is assigned as unknown. Unknown zone considered as drivable. Fig. 4 shows the occupancy grid map, drawn in our campus.

The advantage of this algorithm is that the measurements from two different scanners are used to detect obstacles. This excludes possibility to detect fake obstacles due to the error in the orientation estimation of the vehicle. In addition, it is possible to make six combinations when three scanners are used, therefore reliability of the algorithm can be increased. Bumper laser scanners use a different way to classify obstacles. They are mounted at some distance above the ground and scan in parallel with the driving surface. Cell is assigned to be an obstacle if points with same X and Y location satisfies condition $\sum_{j=1}^{n} p_j(Z_j > |h - \delta|) \cdot w > 0.5$, where j is the time index for the series of height measurements acquired for same cell, w is weight of the possibility that point is obstacle and δ is vertical distance that determines size of the obstacle. If same point has been classified as obstacle several times, likelihood of the obstacle presence in this location increases and obstacle is detected.

In practice both algorithms have shown reliable performance and allow our vehicle to avoid obstacles at the distance of around 30 m ahead, with speed up to 60 km/h.

4.2 Road Boundary Estimation

One of the problems that can occur during autonomous driving is reference path drifting, due to the inaccuracy of GPS system. To avoid it, we developed an algorithm, which helps to determine the position of the vehicle with respect to the road boundary. On the asphalt road, where static obstacles usually aren't present, vehicle had to keep travelling by he right side of the road. Therefore our task simply was to define lateral distance between car and one of the road sides. To distinguish actual road from off-road, we use reflection properties of the road. Laser scanners can provide reflectivity index of the surface. Based on this reflection, occupancy grid can be built. The reference path, is assumed to be parallel to the road side. After the road boundary is detected, drifting error can be calculated. In real driving we had to smooth the drifting correction with one-dimensional Kalman filter.

5 Path and Speed Planning

The main mission that car should have accomplished is the following of a pre-recorded reference path. Artificial and natural obstacles can be present on a reference path, so vehicle should avoid them, and also edges of a road can be natural obstacle, because of GPS localization drifting.

The reference path is stored in a text file. Before being given as an input to the path planner, the reference path is smoothed, and such parameters as tangential vector and curvature of the path are calculated for each reference point.

5.1 Path Planner

For our implementation, we have chosen a sampling based approach, as it can drive vehicle through a narrow path and can take kinematic constraints of a vehicle into account. Algorithm represents a wave front propagation. But unlike in discrete Dijkstra based implementation, it uses sampling in continuous space with a nearest neighbour node search for expansion, like in RRT [7]. But computation speed is a challenge for the development of real-time algorithms [4]. We wanted our vehicle to response faster to the environment changes, because the range of car's perception was limited.

We have developed a new method to increase computing time efficiency of the path planner. It should be mentioned that our path planner is run in a single thread and has re-planning time about 100 ms, which is enough to drive vehicle at speeds up to 60 km/h, given car's perception range.

For the simplicity of implementation and computation efficiency we assumed that car is travelling along short arcs. Then we used a curvature and curvature-difference based distance measure in the configuration space for the nearest neighbor search to provide smooth and fast motion.

After planned path is calculated it is processed by a Savitzky-Golay low-pass filter to calculate smooth tangential vectors and curvatures of the path at every point.

5.2 Speed Planner

The main task for a speed planner is to follow maximum safe speed, while taking into account speed limit zones, that are recorded together with the reference path. Safe speed is calculated based on the curvature of a path. If there is a sudden change in a path curvature, speed planner reacts in advance.

5.3 Trajectory Observer

Trajectory Observer is a special block, running at a high frequency, it measures difference between current state of a vehicle and a planned path. Then it sends the data to the control block. There were two reasons to introduce this block. At first it is an interface between the whole system run on a general purpose operating system, which is linux in our case and a real time system. At second it improves safety of a vehicle, as it is deterministic and doesn't have delays or failures like path planner.

6 Real-Time Vehicle Motion Control

Once the errors to the reference trajectory of the vehicle is received from the trajectory observer, the motion controller produces appropriate steering, throttle and brake commands to follow the given trajectory.

6.1 Steering Control

Lateral offset, angle difference and curvature of the reference path are used as an input to the controller. Control law is similar to the one in Stanley's [8] implementation with some modifications. It is basically a PD controller with curvature feed-forward and speed based adaptive look-ahead point selection. The steering controller gives steering commands to the steering motor at a rate of 100 Hz.

Fig. 5 shows steering performance while Pharos was making a corner of 40 m radius curved road. Steering controller maintained within 30 cm lateral offset around 60 km/h.

6.2 Speed Control

Once the desired speed is determined, low-level speed controller control the brake level and throttle level exclusively. It treats the break and throttle as two opposing single-acting actuators that produce longitudinal force on the vehicle. The controller computes

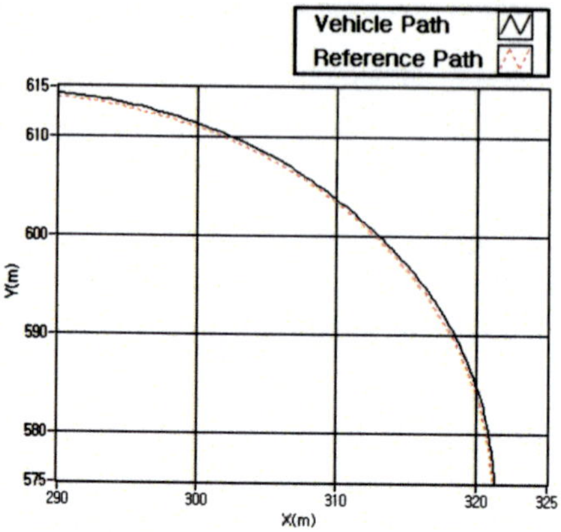

(a) Tracking performance of the steering controller

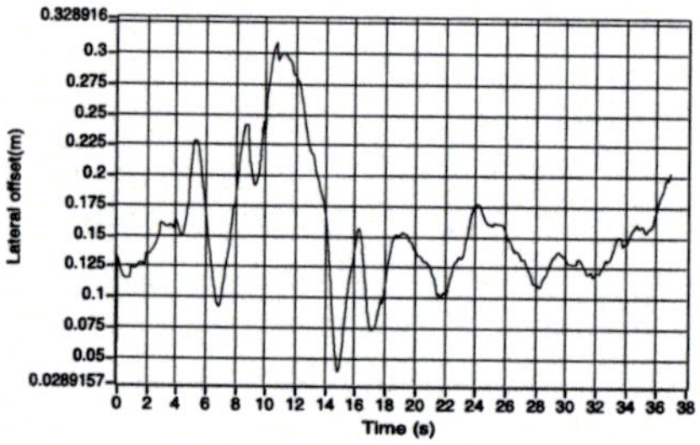

(b) Lateral offset error of the vehicle

Fig. 5. Steering performance of the vehicle on 40m radius curved road

weighted sum plus weighted integration of the speed error. Based on the sign of the computed value, either brake or throttle level is controlled. By considering dead-zone, the controller is able to avoid the chattering behaviour.

7 Discussion

This paper provides an overview of the autonomous vehicle Pharos, developed to participate in the 2010 Autonomous Vehicle Competition in South Korea.

From a broad perspective point of view, Pharos' control software mirrors common methodology in the area of autonomous vehicle. However many of the individual modules have been developed to achieve the high speed unmanned driving. Long range terrain perception and obstacle detection algorithm, real-time trajectory planning algorithm, and high speed vehicle motion control method are developed and integrated all together. All the developed algorithms are extensively tested in the field to prove the reliability.

Even though our vehicle could successfully finish the race, there are many issues to be improved, such as extending perception range, perceiving dynamic environment and advanced localization.

References

1. Buehler, M., Iagnemma, K., Singh, S.: The 2005 DARPA grand challenge: the great robot race. Springer Trats in Advnaced Robotics, vol. 36. Springer, Berlin (2007)
2. Elfes, A.: Using occupancy grids for mobile robot perception and navigation. Computer 22(6), 46–57 (1989)
3. Dolgov, D., Thrun, S., Montemerlo, M., Diebel, J.: Practical Search Techniques in Path Planning for Autonomous Driving. In: Proceedings of the First International Symposium on Search Techniques in Artificial Intelligence and Robotics (2008)
4. Frazzoli, E., Dahleh, M.A., Feron, E.: Real-time motion planning for agile autonomous vehicles. AIAA Journal of Guidance and Control 25(1), 116–129 (2002)
5. Kelly, A.: A partial analysis of the high speed autonomous navigation problem, Project report of "Perception for Outdoor Navigation" and "Unmanned Ground Vehicle System," Carnegie Mellon University (1994)
6. Kuwata, Y., Fiore, G.A., Teo, J., Frazzoli, E., How, J.P.: Motion planning for urban driving using RRT. In: Proc. IROS, pp. 1681–1686 (2008)
7. LaValle, S.M., Kuffner, J.J.: Randomized kinodynamic planning. International Journal of Robotics Research 20(5), 378–400 (2001)
8. Thrun, S., et al.: Stanley: The Robot that Won the DARPA Grand Challenge. Journal of Field Robotics 23(9), 661–692 (2006)
9. Urmson, C., et al.: A robust approach to high-speed navigation for unrehearsed desert terrain. Journal of Field Robotics 23(8), 427–508 (2006)
10. Urmson, C., et al.: Autonomous driving in urban environments: Boss and the Urban Challenge. Journal of Field Robotics 25(8), 425–466 (2008)
11. (in Korean), http://www.hyundai-ngv.com/techcontest/

Advanced Perception for Robots in a Closed World Environment

Mahmoud El Shaikh[1], Andreas Koch[2], Bernd Eckstein[2], Kai Häussermann[2], Oliver Zweigle[2], and Paul Levi[2]

[1] German University of Cairo
mahmoud.shaikh@gmail.com
[2] University of Stuttgart, Universitaetsstr. 38, 70569 Stuttgart, Germany
{kochas,eckstein,haeussermann,zweigle,levi}@ipvs.uni-stuttgart.de

Abstract. A basic requirement for all mobile robots is a precise, accurate and fast perception of the environment to allow intelligent behavior through interaction with its surrounding. This paper introduces a new method for a reliable, fast and efficient purely omni-directional vision based mobile robot navigation for a closed world environment. The proposed method enables a robot that is driving at a high-speed to detect and classify different objects of different colors by only calibrating one dominant environment color. The correctness and the effectiveness of the introduced approach was evaluated successfully in the highly dynamical RoboCup Middle Size Soccer scenario.

1 Introduction

To enable the ability of navigation, object recognition, reasoning and interaction for autonomous mobile robots a precise, accurate and fast perception of the environment [1] is required. In highly dynamical environments the perception is especially important due to the direct feedback on the decision making process and the overall intelligence of the robot. Slow processing or even a small percentage of failure will have major negative effects on the reliability of the robot to accomplish its task. Thus the perception has to be fast in order to acquire quick reaction times. Moreover, avoiding bottlenecks in the perception level is desirable to allow more computation time for strategies, behavior selection and role assignment, e.g. in a multi-agent system. Even though there exists several sensors for perception, computer vision is one of the most promising sensor for perception; after all humans heavily depend on their eyes to perceive their environment. In order to the underlying highly dynamic real-time scenario the methods have to meet the real-time constraint. Furthermore the robots are moving at high speeds of up to 5,5m/s. Consequently in this work we will present a novel purely vision based classification and detection algorithm for a set of closed world computer vision problems. This paper is organized as follows. Section 2 outlines some related work. Section 3 describes the proposed framework with the corresponding idea, the data structures, the classification approach and the adaption method. Section 4 shows briefly the evaluation and the results of the approach. In Section 5 we sum up the presented method and discuss further work.

S. Lee et al. (Eds.): Intelligent Autonomous Systems 12, AISC 193, pp. 111–122.
springerlink.com

2 Related Work

The most widely used color segmentation and -classification methods in practice can be divided into the following classes: *Edge-based methods* exploit the change of pixel values at the edge between two different regions [2]. To obtain the closed region boundaries usually Canny or Sobel filters were used. Although the approach achieves good results the computation of the filters are expensive. In contrary to the above class *region-based methods* base on criteria that take into account color similarity and spatial proximity. Thus it is assumed that adjacent pixels in the same region have also similar color value [3], [4]. Therefore the *seeded region growing (SRG) method* based on watershed segmentation is usually applied. Unfortunately the success of the method highly depends on the inital seed [4] and the unsupervised selection of the seeds is a difficult task [5]. The *threshold technique methods* assume that pixels with similar color value belong to the same class [6], [7], [8], [9], [10]. In comparison with the other approaches this method is very efficient and allows a fast classification. After the training or calibration-phase the classification is usually done using a look-up table that divides various color volumes from the entire color space into ranges such that a group of different color intensities within different channels represents a specific object of interest. Thus, the color calibration is of key importance and must be done with extreme care. Because the methods are extremely sensitive to changes in lighting conditions a carefully tuning of the thresholds in the calibration-phase is required. The wrong inclusion of wrong color intensities in the calibration may result in wrong classifications. Therefore, a new method for a reliable and efficient perception adopted solely through computer vision for a closed world environment is proposed that allows both a simpler calibration approach and even a more precise, accurate and fast classification and detection system.

3 Adaptive Color Segmentation

Recognition of objects can be done with computer vision either through shape or color. In order to depend purely on color information it is necessary to have a very reliable classification algorithm because the detection raises many challenges. The light is usually inhomogeneously spread across the environment and different light intensities yield variations in color that might conflict with the initial calibration [11]. Moreover, with the existence of several classes of objects it is possible to interchange dynamically between different objects with slight color variations within the same class. Thus the independence from any color intensity threshold in a specific range is an advantage. Another advantage would be the reduction of the number of objects of interest that need to be calibrated. The proposed color classification algorithm meets the above challenges with precision, accuracy, proved robustness and stability without the use of any look-up tables by only calibrating one of the objects of interest, which is the most dominant one. The algorithm assumes the existence of a specific dominant color that is part of the color-coding of the closed world environment. The algorithm works in three stages that will be further discussed.

3.1 Data Structures

In computer vision the order in which pixels are being processed is of extreme importance and has a large impact on how algorithms are being shaped. Due to this significance it is necessary to have smart data structures that facilitate the retrieval of information. Since in omni-directional vision all the information can be found in a circular 360° reflection of the environment, the data structure is then also very crucial for excluding the useless parts of the rectangular image from the processing. Similarly the data structure can manipulate how much pixels within the circle are allowed for processing. This is usually a trade-off between precision and even more speed.

The proposed data structure is based on the principle of radial search lines that are constructed from the center of the circular reflection with length equal to the radius of the circle using Bresenham's line algorithm [12]. The whole search space will contain 360 lines (see figure 1) if 1° difference is left between each two consecutive lines. This approach significantly reduces the amount of information needed to process while maintaining a good balance for precision. Therefore, it is not needed to reduce the number of lines in the search space in order to be able to detect far objects.

Processing the image will then be reduced to accessing the search space line by line. Then in each line the color information of the pixels is accessed. This can be done either inside-out or outside-in depending on the application. Of course, even parts of the data

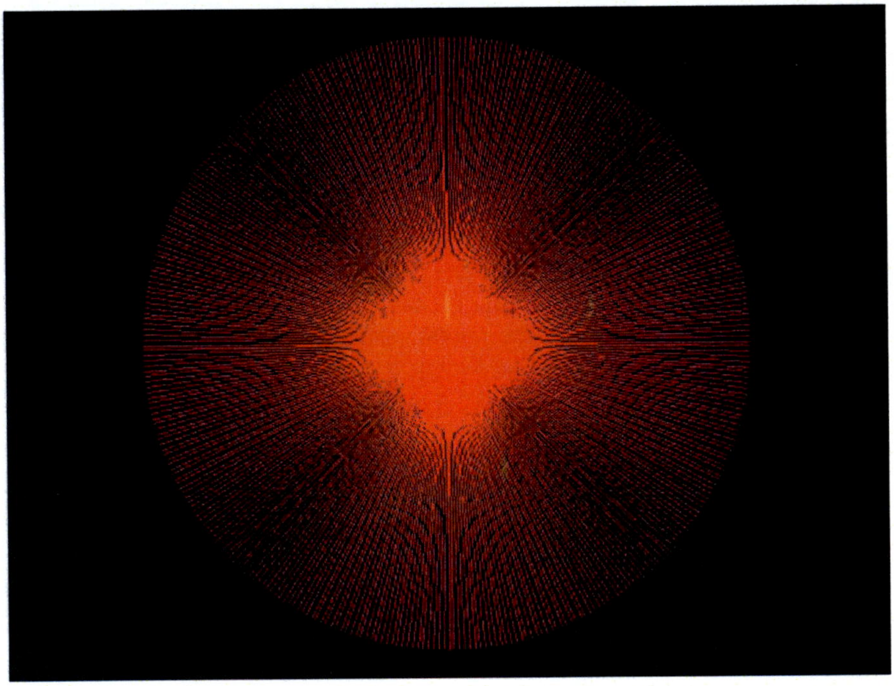

Fig. 1. 360 radial search lines constructed using Bresenham's line algorithm

structure should be excluded from the search space such as the reflection of the camera itself in the center of the circle. Another important data structure that is basically the result of the transition from classifying different objects of interest to actually detecting them is the "segment" data structure. A new instance of this data structure is created whenever a new class of objects is classified. Important information is stored in this data structure like the start of the segment on the radial line and its length. The X and Y coordinates of the pixels within each radial line are precomputed and stored in the search space data structure prior to running the algorithm.

3.2 Basic Idea and Approach

Even in closed world environments the dependency on specific conditions such as information about the field where the robots are located is considered as a weak point. Applications should be robust enough to adapt to different environments within the same domain and different environments have different color intensities due to the lighting conditions even if the human eye cannot see it [11].

This is exactly why a calibration phase is essential. Without calibration, robots will be environment specific and will have to be programmed for each and every environment. Several questions rise under this matter. Which objects of interest should be calibrated and which color space is the most practical?

The combination of luminance with RGB intensities is really powerful. Luminance -or brightness- is very sensitive for human perception [13]. On the other hand, slight RGB variations are nearly insensitive for the human eye but very sensitive in image processing. The transformation from the YUV color space to the RGB color space is linear. Therefore, it is possible to acquire images in YUV in order to have control over the luminance and in the meantime transform all RGB variables of the logical equations to YUV variables with the help of mathematical linear transformations. Consequently the new equations are used to maintain the runtime of the application. Such equations will be further discussed.

As previously mentioned, instead of relying on any color intensity for either the calibration or detection, the robot in the first phase is manually calibrated for the ratios between the different channels of the most dominant color. These ratios are then used in the second and third phases where the processing of the search space is done to classify and detect different objects of interest. An important keynote is that in the classification and later in the detection phases the traversal of the whole search space is done only once. Both phases are using the same traversal per each frame due to the very strong key advantage gained from the ratio approach, which is its precision. Once a pixel is classified, it directly proceeds to the detection phase where it will be clustered along with other similar neighboring pixels into a segment in the radial line and later grouped into specific objects of interest within the whole search space. In the meantime, the automatic adaption algorithm is running cooperatively in the background. This approach eliminates the use of any look-up tables that include the severe choice of the bounds of the color volumes discussed previously.

3.3 Manual Color Calibration

The calibration phase is only done once for each environment. Of course, after some significant amount of time the color intensity of the most dominant color is prone to change, e.g. due to the continuous flickering of lamps. Also, the severe treatment of the field by the robots' wheels might slightly erode the ground thus affecting the color. Other unknown factors also exist that contribute to the fact that color intensities are varying even slightly.

Due to these factors it is sometimes essential to re-calibrate the vision to cope with those variations. Instead of calibrating the exact color intensity, the vision is calibrated for the ratios between the different channels of the RGB color space. This way, the need for re-calibration is alleviated.

There are two approaches when it comes to calibration. Either choose in an interactive window (e.g. by dragging with the mouse) a relatively large area of the most dominant color that should be calibrated or by placing the camera (e.g. mounted on a robot) in a place where that color is surrounding the camera with a significant amount of distance. Afterwards probably choose the partition where the calibration should take place.

After selecting a big portion of the color, the average intensity of each channel is computed. Then, the ratios between the channels are calculated. This way, it is possible to maintain the calibration information without the need for re-calibrating the colors, because slight variations of the color intensity do not affect the ratios significantly. The results from the calibration phase are six values: red average, green average, blue average, green/red ratio, green/blue ratio and red/blue ratio (or other possible permutations of the ratios). After the calibration phase, these numbers should be saved.

However, even though re-calibration is not needed, it will not do any harm. Therefore, the calibration phase should be very fast and efficient and should not be thought of as a hard task. Especially when moving to new environments – of course in the same domain – a new calibration process should not be compulsory.

3.4 Color Classification

Classifying individual pixels is an important aspect of the perceiving capabilities. These individual pixels will later be clustered into segments of specific objects of interest during the detection phase. Therefore, precision is of high importance, as it will later in the detection phase contribute to the simplicity of the approach of clustering the individual pixels to form segments of different objects of interest.

Depending on the closed world environment domain, one or more inequalities will be responsible for the classification of pixels to their respective objects of interest. This mapping between inequalities and objects of interest should be determined by the color intensities of the different objects of interest in the domain.

The plot in Fig. 2 shows a *ratio plot* for red-green. Corresponding plots exist for the green-blue and red-blue ratios calculated in the calibration phase. Both the average green and red intensities (from the calibration phase) are marked in Fig. 2 with an *X*. It is important to note that the above graph is not a representation of channel intensities. Instead, it represents all possible ratios between the two channels. The above example

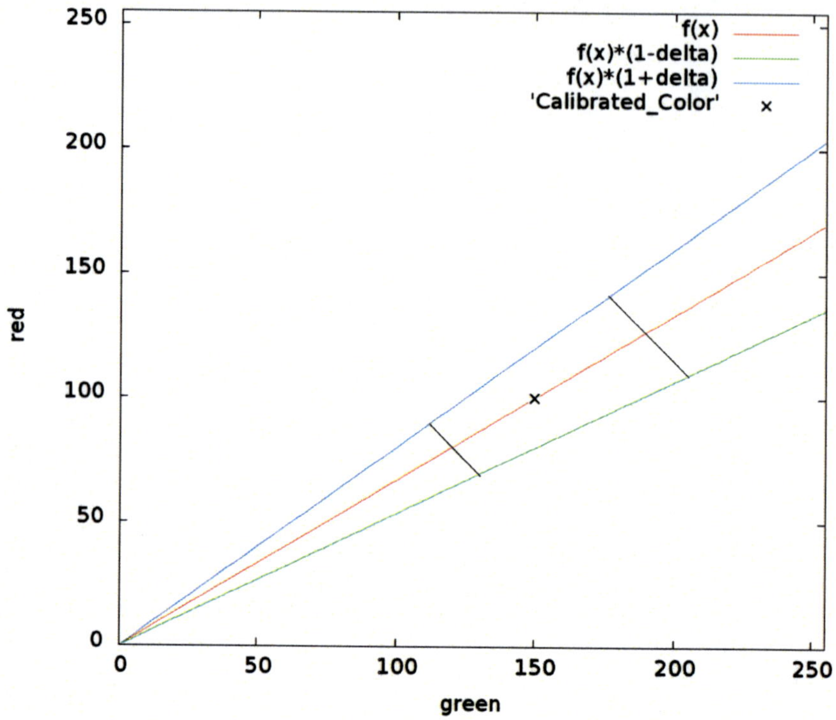

Fig. 2. The minimum and maximum bounds of the initially calibrated green - red ratio marked by a X

shows that the average intensities are 150 and 100 for both the green and red channels respectively. Therefore, the marked X has the value $150/100 = 1.5$.

In order to successfully classify a pixel as a member of the most dominant color that was initially calibrated, the pixel must be within the maximum and minimum bounds of the corresponding ratio plot. On the other hand, depending on which bound the pixel broke (from all the plots) it is classified to its respective object of interest. These boundaries are represented mathematically in the form of inequalities. Generally speaking, with the existence of n inequalities there will be $n+1$ possible classifications of objects of interest. This additional classification occurs when the pixel lies within all boundaries of all plots. Thus, should be classified as the most dominant color that was initially calibrated.

The middle red line $f(x)$ passing through the X mark (see Fig. 2) represents all the points in the green-red space with green-red ratios equal to the calibrated green-red ratio. The bottom green line $f(x) * (1 - delta)$ represents the minimum bound and finally, the top blue line $f(x) * (1 + delta)$ represents the maximum bound. Another two lines could be added to filter out some of the values that have the same or very close ratios but very high or very low intensities. However, their existence is very unlikely to occur in closed world environments where everything is color-coded. Nevertheless, they are shown in Fig. 2 as two black lines perpendicular to the maximum bound.

In Fig. 2, two objects of interest can be classified. All pixels that will lie below the minimum boundary could be assigned to a specific object of interest. Similarly, all pixels lying above the maximum boundary could be assigned to another object of interest. In addition to the two black lines, additional classifications could exist within the same figure by having additional boundaries.

For each pixel this inequality is calculated for different values of X:

$$(1-\delta_X)\frac{C_{X_1}}{C_{X_2}} < \frac{V_{X_1}}{V_{X_2}} < (1+\delta_X)\frac{C_{X_1}}{C_{X_2}} \quad \delta \in [0,1] \tag{1}$$

Where $X \in \{\{green, red\}, \{green, blue\}, \{red, blue\}\}$, C is the initially calibrated value of the respective channel, V is the value (intensity) of the respective channel of the current pixel being processed and δ is a threshold of the ratio between the members of X and their corresponding calibrated values.

Often camera images are retrieved in other color spaces than RGB. In this case it would be inefficient to transform each pixel into the respective RGB values and use those new values in the above inequalities. If the image is being retrieved in YUV from the camera, we can take advantage of the fact that the transformation from YUV to RGB is linear and transform the inequalities to YUV variables instead of transforming each and every pixel before beginning the processing. Furthermore we can also use the luminance (previously discussed) in the classification of objects. It is then possible to gain very satisfying speedups.

The transformation from YUV to RGB is as follows [14]:

$$r = \frac{y + (v * 1436)}{1024} \tag{2}$$

$$g = \frac{y - (u * 352 + v * 731)}{1024} \tag{3}$$

$$b = \frac{y + (u * 1814)}{1024} \tag{4}$$

As it is apparent from the equation, every floating-point operation in the transformation from YUV to RGB is replaced by integer operations. With a large number of pixels, speedups can be gained because the division by 1024 can be done using the the low cost shifting operation. Later on, in the linear transformation of the equations all divisions will cancel each other out.

By proceeding with the left hand side of the green-red ratio inequality example in eq. 1, only with direct substitution from the green and red variables to the y, u and v variables in eq. 3 and eq. 4 the inequality will look like this:

$$(1-\delta_X)\frac{C_{X_1}}{C_{X_2}} < \frac{V_{X_1}}{V_{X_2}} < (1+\delta_X)\frac{C_{X_1}}{C_{X_2}} \quad \delta \in [0,1] \tag{5}$$

After some linear mathematical transformation the above equation will look like this:

$$y * (CGR - \delta) + v * (1436 * CGR + 731 * \delta) \\ + u * (\delta * 352) > 0 \quad \delta \in [0, 1] \tag{6}$$

Where Y, U and V are the luminance and chrominance channels, δ is the percentage threshold and CGR is the initially calibrated *Green / Red* ratio which is a constant number. This transformation should be done to all the inequalities.

Therefore calculating the inequality in eq. 6 is in other words checking whether the green-red ratio of a pixel lies above the lower bound of the calibrated green-red ratio calculated in the calibration phase. It could be enough to directly classify the pixel if it dissatisfies the inequality or it could be a part of a logical expression and needs further computations of other inequalities to successfully classify a pixel as a member of an object of interest.

3.5 Object Detection

The data structure contents of the search space should not only be limited to the lines and the X and Y coordinates of their pixels. Sets of different objects of interest can keep track of the objects so far detected.

Phase two of the algorithm (the object detection mechanism) is rather straightforward due to the precise classification from the previous phase. A new segment is created for the first classified pixel for as long as the next pixels are equally classified throughout the iteration of the pixels on the line. A new segment is only created when a different classification takes place. At this point, the last pixel of the previous segment is stored and the first pixel of the new segment is also stored but in the new segment. This is done for each line.

As the traversal of the search space proceeds and the next lines are being processed the algorithm starts looking for overlapping segments starting from the traversal of the line on the second degree of the search space. Basically, all the segments of the first line are considered as parent segments and if the next line contains overlapping equally classified segments with one of the segments of the previous line, this segment is then considered as the parent segment and the old parent segment will be considered as a child segment pointing to the new parent segment.

The sets of the different objects of interest within the search space contain the parents of all the different objects of interest detected. This set is cleared at the start of the traversal of the search space for each frame.

This way, singly classified pixels are clustered into groups of segments within each line and further grouped into different objects of interest contained in the different sets within the search space.

3.6 Automatic Adaption

In order to adapt to the different lighting conditions (e.g. across a typical RoboCup playing field), the image is divided into 6 or more partitions with one or more radars for each partition. The radar is a randomly chosen radial search line within the partition and must contain a minimum number of pixels from the dominant color in order to be taken into account. Otherwise, another random line within the partition is picked and if that line also does not contain the minimum number of pixels possibly due to an obstacle the automatic adaption for this partition is deactivated for this frame and will be re-activated in the next frame. The radar is used to calculate independently from the rest of the partitions the average intensity of the most dominant color.

Although the minimum and maximum bounds in the classification phase have been parameterized by δ and can be manipulated in real time to check which values suit the current environment. The radar is very useful to detect the edge of the world very precisely especially when the system is located far away from that boundary as they are "learning" the nature of the color in real time. The process should be separated for each partition due to the inhomogeneous lighting conditions across the environment.

The automatic adaption algorithm works as follows: the traversal is moving from on radial lines from the outside towards the center of the data structure. For each pixel a on the radial line, proceed with the rest of the pixels after pixel a searching for the pixel with the maximum difference of the most dominant color intensity between itself and pixel a (call it pixel b). Then check, if pixel b is certainly the color being searched for (because white pixels also have maximum values in all channels, not mentioning other noise possibilities). Calculating the ratio of all channels to their respective calibrated ones does this. Pixel b is only considered if all ratios are within a certain level of tolerance, e.g. within a margin of 20%. The traversal is done for each pixel on the line but the search will only be done for the pixels after the pixel being considered.

The process is running automatically and is computed online during runtime of the vision component without any manual interference. This way it is possible to compensate for the different lighting conditions due to the independence of each partition.

Several optimizations have been implemented to speed up this compute intensive approach. It is not necessary to start from the beginning of the search line. Of course it would be much more accurate but starting from the middle of the line did not show significant variations in the results. Another optimization would be increasing the incrementation step. Instead of applying the algorithm on every pixel of the line, it is also possible to do it for every second pixel. Furthermore, the search of the pixel with maximum difference (pixel b) can be done for every second or even every third pixel.

3.7 Drawbacks

With omni-directional vision it is very hard to detect or even classify far objects since the images captured by the camera are just the reflection of the surrounding environment. It has been previously mentioned that leaving 1° difference between the lines maintains a reasonable balance between precision and speed. But nevertheless, with a 640×480 resolution eight to nine meters from the center of the camera is considered as a significantly far distance even with 1° difference between the radial search lines.

4 Evaluation and Results

All approaches presented in this paper were successfully evaluated and applied in the Middle Size League of RoboCup [15]. In this scenario the robots need to classify and detect various objects of interest: the green field which is the border of the world, the white-lines that are important for self localization, the ball and other objects for obstacle avoidance. These objects of interest are crucial to successfully coordinate the robots for a specific target, which is to score a goal against the opponent's goalkeeper. In this environment there are many challenges that the robots must overcome, i.e. different lighting conditions, robust detection and a fast processing. The proposed architecture has proven robustness in the Middle Size Team of the University of Stuttgart. The processing of the search space involves the classification and detection of the edge of the green field, the white lines on the field as well as different robots and the ball. Balls were being successfully detected from distances as far as 7 meters. With Intel dual core processors running on 2.16 GHz frequency, the processing of the whole search space (classification and detection) with 1° difference between the radial search lines took on average 2.67 ms. In addition to the speed and precision of the perception capabilities of this vision approach, it also showed stability of the results and overall robustness during the RoboCup matches. Fig. 3 shows the precise classification of four objects of interest namely the ball, white lines, obstacles (other robots) and the green field while only calibrating the most dominant color in the environment, which is of course in this case the green field. Furthermore Fig. 4, shows the detection of the ball along with the precise edge detection of the green field by the automatic adaption algorithm.

Fig. 3. Classification of individual pixels into four different objects of interest

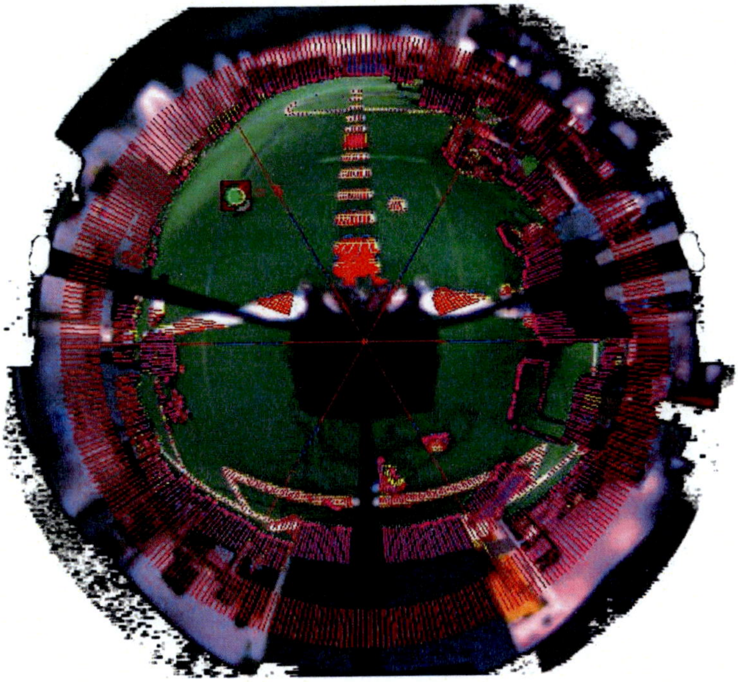

Fig. 4. Detection and clustering of segments

5 Conclusion

In order to have intelligent robots with smart behavioral patterns, there should be no doubts about the robot's perception abilities in any possible situation that might occur at any place of the environment in any instant of time. Perception in general could be broken down into several types of sensors each with intelligence behind it. But computer vision in general and omni-directional vision in specific proves its trustworthiness in environments where robots are free to move in a given area and act according to different situations.

With the use of computer vision, several challenges have been encountered such as the inevitable trade-off between precision and speed, the ability to adapt to different environments within the same domain and of course the inherent question about stability that rises when the color information is the primary candidate for detection. But, with the use of simple calibration, smart data structures, automatic adaption and with the help of some linear mathematical transformations, it was possible to mitigate the undesirable trade-off and reach very stable results even under different lighting conditions across the environment algorithm. The future works of this approach is to gain more speed-ups with the help of parallel programming. Since the whole search space is divided into different independent partitions the computation can be parallelized and might benefit from GPU acceleration.

References

1. Zhang, J., Weser, M., Jockel, S.: Fuzzy multisensor fusion for autonomous proactive robot perception
2. Ruzon, M.A., Tomasi, C.: Edge, junction, and corner detection using color distributions. IEEE Transactions on Pattern Analysis and Machine Intelligence 23(11), 1281–1295 (2001)
3. Altunbasak, Y., Eren, P.E., Tekalp, A.M.: Region-based parametric motion segmentation using color information. Graphical Models and Image Processing 60(1), 13–23 (1998)
4. Ikonomakis, N., Plataniotis, K.N., Venetsanopoulos, A.N.: A region-based color image segmentation scheme. In: Proc. of the SPIE: Visual Communications and Image Processing, Citeseer, vol. 3653, pp. 1202–1209 (1999)
5. Ikonomatakis, N., Plataniotis, K.N., Zervakis, M., Venetsanopoulos, A.N.: Region growing and region merging image segmentation. In: 1997 13th International Conference on Digital Signal Processing Proceedings, DSP 1997, vol. 1, pp. 299–302. IEEE (1997)
6. Quan, Z.G., Ming, L.Z.: Research on preprocessing of color image for vision based mobile robot navigation. Inform. Technol. J. 10, 597–601 (2011)
7. Posada, L.F., Narayanan, K.K., Hoffmann, F., Bertram, T.: Floor segmentation of omnidirectional images for mobile robot visual navigation. In: 2010 IEEE/RSJ International Conference on Intelligent Robots and Systems (IROS), pp. 804–809. IEEE (2010)
8. Li, Y., Feng, Z., Sheng, H.: A method of recognizing objects specified by color and shape in autonomous robot. In: 2010 IEEE International Conference on Information and Automation (ICIA), pp. 2466–2470. IEEE (2010)
9. Hengst, B., Ibbotson, D., Pham, S.B., Sammut, C.: The unsw united 2000 sony legged robot software system. In: RoboCup 2000: Robot Soccer World Cup IV. Citeseer (2001)
10. Bruce, J., Balch, T., Veloso, M.: Fast and inexpensive color image segmentation for interactive robots. In: Proceedings of the 2000 IEEE/RSJ International Conference on Intelligent Robots and Systems (IROS 2000), vol. 3, pp. 2061–2066. IEEE (2000)
11. Oberasche, L.: Interaction of colour and light: Simulation and assessment under artificial and natural light (2011)
12. Bresenham, J.E.: Algorithm for computer control of a digital plotter. IBM Syst. J. 4, 25–30 (1965)
13. Halsted, C.P.: Halsted. Brightness, luminance, and Confusion (1993)
14. Damien Douxchamps. libdc1394: Api version 2: Conversions, `http://damien.douxchamps.net/ieee1394/libdc1394/v2.x/api/conversions` (accessed September 13, 2011)
15. RoboCup. Soccer middle size league, `http://www.robocup.org/robocup-soccer/middle-size` (accessed September 13, 2011)

An Android Remote Call Vehicle Service for OSGi-Based Unmanned Vehicle Using by a Mobile Device

Choon-Sung Nam, Sukhan Lee, and Dong-Ryeol Shin

School of Information and Communication Engineering, Sunkyunkwan University
300 CheonCheon-Dong, JangAn-Gu, Suwon, Gyenggi-Do, 440-746, Korea
{namgun99,drshin}@skku.edu, lsh@ece.skku.ac.kr

Abstract. Almost every country is interested in electronic vehicle uses of green energy. Smart device appeared by developing mobile devices. Thus, the desire to overcoming our rapid changeable world, we believe that the study of balancing way between Smart Technology and Vehicle must be necessary. To achieve this, we should support open interface by adopting OSGi framework in a vehicle. And for mobile device user or caller, we should design and implement android application of mobile devices. This helps cooperation to the OSGi framework and Android. Especially, we propose the Android-based Remote Call Vehicle Service (ARC-VS) for mobile user or caller.

Keywords: OSGi, unmanned vehicle, remoter call service, android, mobile device procedure call.

1 Introduction

Recently, at last 100 years, electronic vehicle technology, especially environment protection has been issued among people. Green energy technology, one of the most common issues in IT world, leads people to realize our serious energy problem, diminishing fuel resources and also leads people to be interested in electronic vehicle developments. To be specific, because of its amazing advantage that electronic car gets its resources from electricity, not fuel sources, this technology have been one of the most debatable issue among IT and vehicle industrial fields [1][2].

Additional function, monitoring vehicle conditions timely, has been required by the customers because of requiring more comfortable service. Fortunately, these requests would be satisfied as installing initial sensor and motor devices into the car. Many people are able to use smart devices because of rapid smart technologies development and especially it leads people to live more easily and continent than before. For example, people are able to check their e-mail and are surfing the Internet as using various mobile applications. It means people can work their computer desk jobs without their laptop. It must be breakthrough news. Under this environment, vehicle industries also have made an effort to develop their smart applications and put that

S. Lee et al. (Eds.): Intelligent Autonomous Systems 12, AISC 193, pp. 123–132.
springerlink.com

into their vehicles (products). However, unfortunately, so far, these applications are designed and operated for their marketing or advertisement. As one of its evidences, nowadays, Maserati, one of the major vehicle vendors provides their customers several advertising services- vehicle engine sounds, Ads videos, telephone services, GCP- though their smart applications. In Korea, Ssang-Young also provides the basic guideline manual for Korando (their product) and GPS services though their applications.

For following facts, the desire to overcoming our rapid changeable world, we believe that the study of balancing way between smart technology and vehicle must be necessary. In this paper, we will develop and propose smart application, which is able to remote and control vehicle as possible. To do that, basically we will use OSGi[3] platform to implement the user-interfaces between users (drivers) and smart devices (installed in vehicle). To be specific, firstly, we will design and implement electronic vehicle architecture, which is based in OSGi. Secondly, to communicate with Android devices, we will also design and implement the remote call vehicle applications as using Android-developing platforms.

The structure of this paper is like that. At section 2, we will mention the description and advantages of OSGi briefly and then at section 3 will describe our proposal-Android remote call vehicle services. Section 4 describes our test mythology and its result data. Finally we will mention the conclusion and discussion which includes future works.

2 Related Works

2.1 OSGi(Open Service Gateway Initiative)

OSGi (Open Service Gateway Initiative) in March 1999 that Sun, IBM, including the 15 companies founded together with a group of about 80 over the world was composed of a consortium of companies, middleware and application services by providing the API with service provider for the consumer's home environment, easy to install the service and maintenance to allow Java-based open service platform specification is to standardize the work [3]. OSGi between the various local network devices by providing a standard specification to ensure interoperability and connectivity with the outside world through them, or a variety of clients, administrators can support. Scalability of OSGi has the properties that it supports a variety of device's communications.

The OSGi framework is a module system and service platform for java programming language that environment managed to run the one bundle at least. OSGi service based on framework is defined and users are possible to develop a service. The core of the framework is life-cycle. All Bundles has life-cycle and framework is control the bundle. Due to these technical, OSGi has very useful properties. Application or a bundle of components are performed without a reboot to install, start, stop, update, remove, and allow it to do. OSGi can be included in the application of SOA has been implemented. This means that the combined application components without restarting the system to dynamically change the structure of the connections can be said that, and the dynamic connection structure is called the "SOA" (Service Oriented Architecture). Each of the components / applications in the Service Registry to register their services through the OSGi service export can do. The components you want to use this service from a service registry easily import rather than the strong coupling between the components connected to "Loosely Coupling" and can be used.

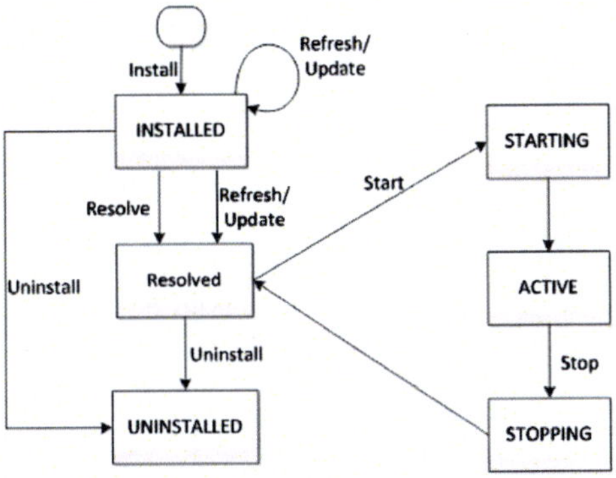

Fig. 1. Bundle's life cycle of OSGi

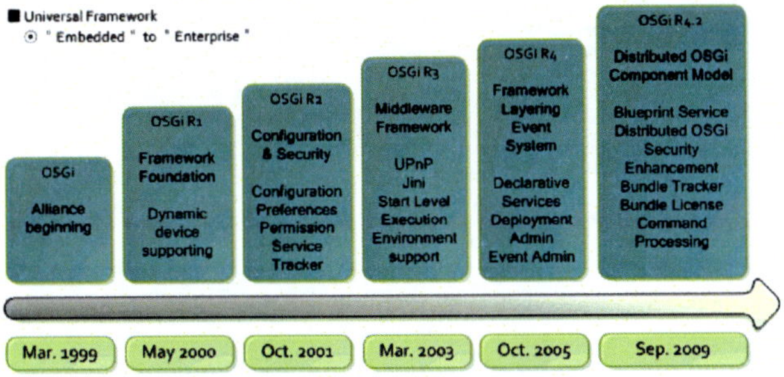

Fig. 2. Expansion of OSGi service

This template, modified in MS word 2003 and saved as "word 97-2003 & 6.0/95 – RTF" for the PC, provides authors with most of the formatting specifications needed for preparing electronic versions of their papers. All standard paper components have been specified for three reasons: (1) ease of use when formatting individual papers, (2) automatic compliance to electronic requirements that facilitate the concurrent or later production of electronic products, and (3) conformity of style throughout conference proceedings. margins, column widths, line spacing, and type styles are built-in; examples of the type styles are provided throughout this document and are identified in italic type, within parentheses, following the example. Some components, such as multi-leveled equations, graphics, and tables are not prescribed, although the various table text styles are provided. The formatter will need to create these components, incorporating the applicable criteria that follow.

The May 2000 OSGi Release 1.0 (R1) specification was created and R2 Release (2001. Oct), R3 Release (2003. Mar), the current version of the specification R4.3 was released. R1 defines the basic framework and support method for the dynamic devices. R2 were added the services that configuration service for storing application setting and preference service and permission admin service to ensure security. Then later, R3 was led to major changes and expansion. The UPnP and Jini standard compatible with the household middleware has been added to the service, start level and execution environment support, XML parser and the addition of services such as URL handler as the shape became middleware framework to support a variety of environments. Figure 2 shows expansion of OSGi service.

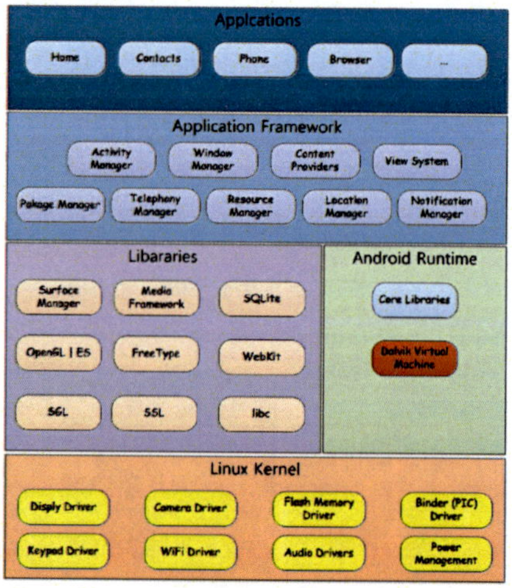

Fig. 3. Android system architecture

2.2 Android System

The Android operating system software stack consists of several layers which are application layer, application framework layer, Android runtime and libraries, and Linux kernel [4]. At the core of the Android platform is Linux kernel version 2.6.29, responsible for device drivers, resource access, power management, and other OS duties. Sitting at the next level, on top of the kernel, are a number of C/C++ libraries such as Secure Socket Layer (SSL), WebKit, etc. Most of the application framework accesses these core libraries through the DVM(Dalvik Virtual Machine), the gateway to the Android platform. The Android Java applications runs on a Java based object oriented application framework which includes activity manager, package manager, window manager, etc.

3 The Android Remote Call Vehicle Service System

ARC-VS(Android Remote Call Vehicle Service) is to provide a vehicle service that an user(caller), who has a smart device based on Android platform, sends his location(GPS) to an electric vehicle based on OSGi framework and then, the vehicle responses its location and information back to user and moves to the caller. To be possible for this service, we need the architecture like a following figure 4.

3.1 Architecture of ARC-AV

In figure 4, caller receives his location from GPS service and calls a vehicle using caller management server through WiFi or 3G/LTE. caller management server receives information and location of caller and sends that to first place vehicle in the vehicle rest area. The vehicle moves to caller. We need two mechanisms to implement figure 4. First mechanism is to control data of caller or vehicle using an interface based on OSGi inside the vehicle. Second, caller must be able to access the interface using Android smart phone application. For these works, we need to use or modify OSGi-based unmanned vehicle like a former work [5]. In addition, it's necessary to design new mechanism using not only WiFi but also 3G/LTE for direct communication with vehicle.

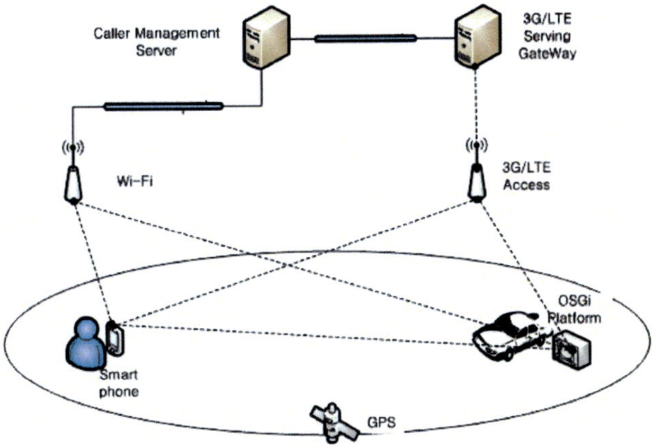

Fig. 4. ARC-VS architecture

3.2 OSGi-Based Unmanned Vehicle System

OSGi-based platform is required for communication between inside and outside vehicle. Therefore, in this paper we define, modify and install following OSGi bundles into a vehicle. In figure 5, there are 3 types of communication services, (1)Passenger Interaction Service, (2)Vehicle Control Module, (3)Android-based Remote Call Service. In-vehicle OSGi bundle is as follows

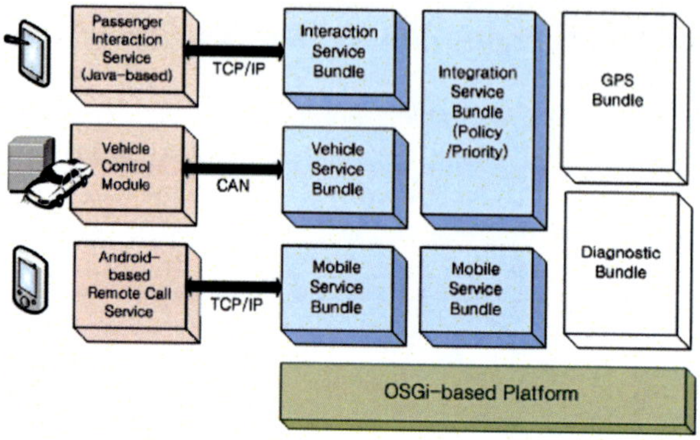

Fig. 5. OSGi-based unmanned vehicle system architecture

- Interaction Service: Interaction Service: This part is for interaction with driver or user. In-vehicle OSGi permits communication data through TCP/IP after passenger inside a vehicle inputs his requirement. The vehicle, then, can move to destination based on information that passenger wants goal and velocity using Interaction Service in OSGi.
- Vehicle Service: This part is for instruction of vehicle control with CAN communication. The instruction is made by in-vehicle integration service bundle. This instruction is sent to vehicle controller system through vehicle service bundle and controls the vehicle.
- Android-based Remote Call Service: This part is for communication with Android-based mobile device. This sends a remote call of the Android application to caller management server through WiFi or 3G/LTE. For connection with in-vehicle mobile service bundle, then caller management server checks state of vehicle and connects caller with the vehicle. Afterwards, requirements of Android-based remote call service are sent to the vehicle, then location and information of caller is sent to interaction service bundle and vehicle Service bundle through integration service bundle. After that, the vehicle moves to caller according to road environment..
- Integration Service Bundle: This part is to process information inside and outside car. This supports information collected about GPS, requirement, etc.
- GPS Bundle: This part is to show location of vehicle. OSGi service that uses default location information is transmitted to integration service bundle that shows current location of vehicle. Not only this, but caller is able to check location of vehicle on the mobile screen..
- Basic Services Bundle: Three services of default services are used. (1)Configuration service is used for initial setup at the beginning of system. (2)Log service records how OSGi performance is improved like black box. (3)Reference service is connected with independent service bundles. Each bundle is used for loading or saving of preference service function.
- Diagnostic Bundle: This is equipped inside and outside vehicle sensor. These sensors are related with inside information of vehicle. We define some data related with driving environment for that, such as brightness, temperature and humidity. These data are also needed for driving and these data can be extended.

3.3 Android-Based Mobile Device System

After OSGi-based platform is built inside vehicle, it should be connected with Android-based mobile device using mobile service bundle. Thus, Android-based mobile device is needed for caller to call and control vehicle. Three applications are needed to implement mobile device. (1)Caller Android-based application for remote call, (2)Vehicle Android-based application or mobile service bundle to transmit/receive call, (3)Call management server to control require information between caller and vehicle.

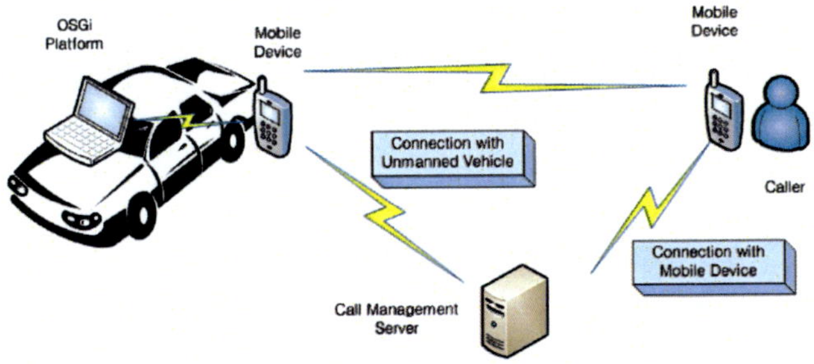

Fig. 6. Android-based mobile device service architecture

- Call Management Server: This is a server that manages requirement to connect available vehicle after checking state of managing vehicles if caller or user calls using Android-based mobile device.
- Caller Android-based Application: This is an application for caller to call and control vehicle. Caller sends his location and state to call management server using application, and call management server helps connection between caller and vehicle sending information of caller to available vehicle. Besides, it updates GPS data through constant exchange of information.
- Vehicle Android-based Application or Mobile Service Bundle: This checks state of its vehicle receiving requirement of caller from call management server. If found state of vehicle suited for requirement, call management server has vehicle move to caller using integration service bundle inside vehicle. And this is linked to caller Android-based application based on available communication of WiFi, 3G and LTE depend on network state.

4 Implementation of AVR-VS

The experimented vehicle is made by the help of Ad-TECH Company. The vehicle is consist of electric systems and possible to use various controller. Whole system is comprised of three parts, vehicle control system, OSGi System, Android application [6,7].

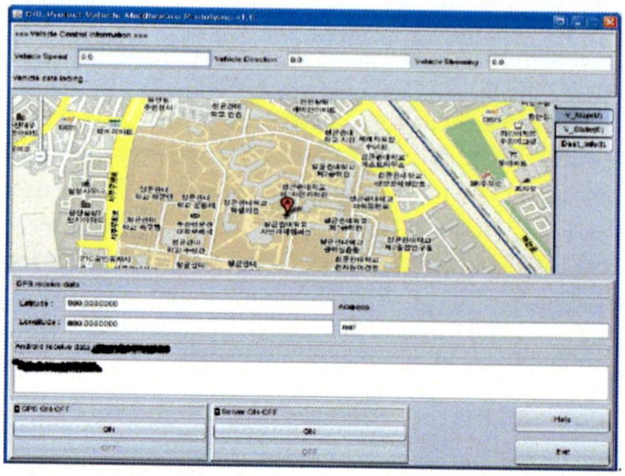

Fig. 7. User interface of Interaction service

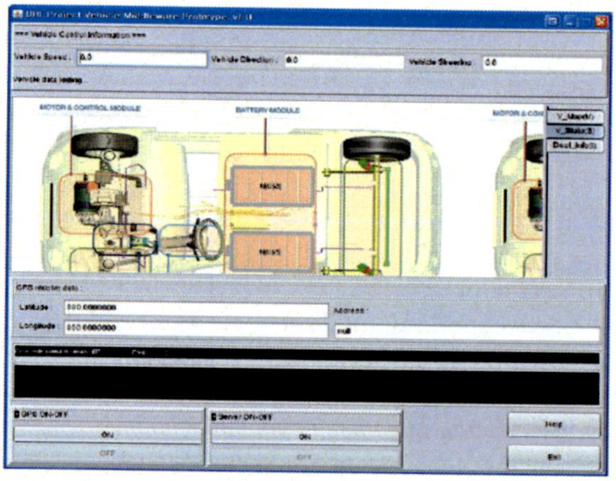

Fig. 8. User interface of sensors in a vehicle

4.1 Vehicle Control System

This system is ordered through communication between vehicle control and other system. Controller carries value to accuator motor when transmit communication related with protocol. The hardware is as a follow.

- Steering Actuator Module: EPS type steering module, smart motor controlled by RS232c.
- Brake Actuator Module: mechanical link type brake module, smart motor controlled by RS232c.
- Accelerator Actuator Module: wire type accelerator module, DC servo motor with encoder & controller
- Controller: Real Time Controller, Message Protocol: CAN (high speed).

4.2 OSGi Framework System

This system receives outside information or transmits it to vehicle control. The main role of this part is to process command control of communication module. It's based on JVM and OSGi(R.4.2). In-vehicle user interface is as figure 7. This program is to receive user event and show state of vehicle. This shows driving location of vehicle after receiving current location from OSGi integration service. In addition, this receives environment of vehicle from sensor and shows. The user interface is as figure 8.

Figure 9 shows Android application for remote call vehicle service. The left picture in figure 9 shows that location of the caller(Galaxy S-II) is transmitted to call management server. The right picture in figure 9 shows that call management server transmits information to vehicle around that.

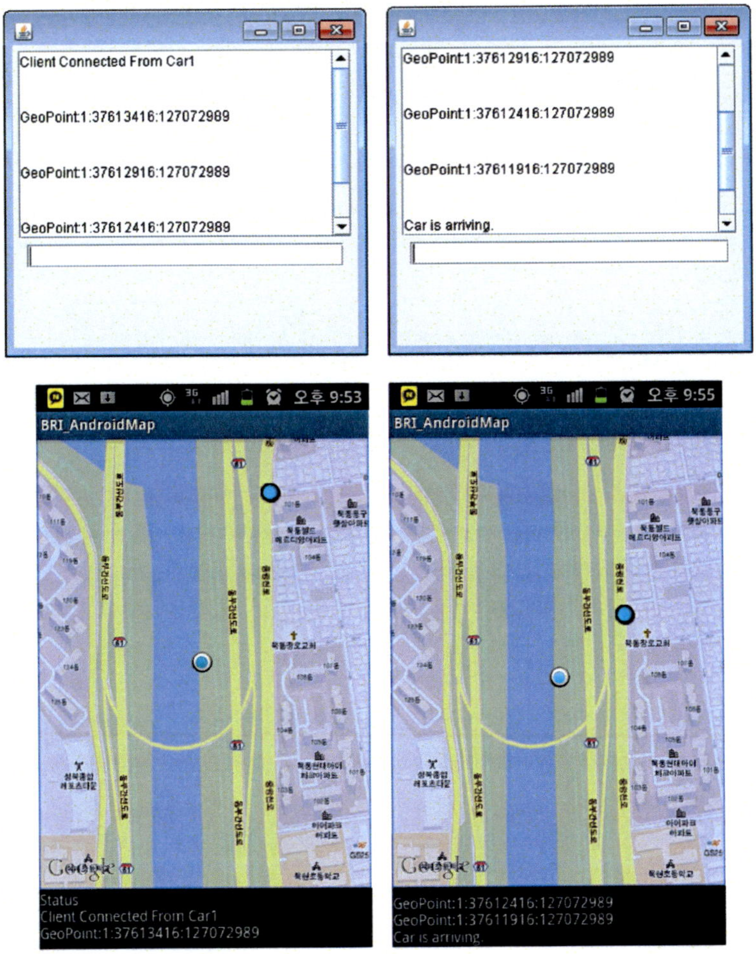

Fig. 9. Android application for remote call vehicle service by callers and a server

4.3 Android Application for Caller

We implemented android application for communication between user and vehicle. We used Samsung Galaxy S and Galaxy S-II as mobile device. Galaxy S is based on Android 2.2 and Galaxy S-II is based on Android 2.3.

5 Conclusion

This paper designed and implemented the Android-based remote call vehicle service using by OSGi-based platform. We made additional bundles of OSGi in the vehicle and revised basic bundles to support unmanned vehicle service. These results help the use of electronic vehicle and mobile device services by changing future energy environment. To implement this service, we set the OSGi-based platform up the electronic vehicle and set up mobile application of smart phone using by Android platform. In future works, we plan to improve services on requirement of user(caller) and vehicle operation and to make new service and operation control method.

Acknowledgements. This research was supported by a grant (I 091003) from Gyeonggi Technology Development Program funded by Gyeonggi Province.

References

1. Dunning, J.S.: Electric vehicles and battery technology. In: Proc. Int. Symp. Low-Pollution Vehicles, p. 13 (1990)
2. Irie, N., Fukino, M., Horie, H.: Nissan FEV—A new concept electric vehicle. In: Proc. Int. Electric Vehicle Symp., vol. 13(02) (1992)
3. OSGi Service Platform, Release 4, OSGi Alliance, http://www.osgi.org
4. Android SDK, Android Developers, http://developer.android.com/sdk/index.html
5. Lee, B.H., Ahn, S.H., Shin, D.R.: Remote Control Service for OSGi-based Unmanned Vehicle using SmartPhone in Ubiquitous Environment. In: Proc. CICSN 2011, pp. 158–163 (2011)
6. Eclipse IDE for Java EE Developers, The Eclipse Foundation, http://www.eclipse.org/downloads/
7. ADT Plugin 16.0.1 for Eclipse, Android Developers, https://dl-ssl.google.com/android/eclipse/

Visual Memory Update for Life-Long Mobile Robot Navigation

Jonathan Courbon[1], Hemanth Korrapati[1], and Youcef Mezouar[1,2]

[1] Clermont Université, Université Blaise Pascal, Institut Pascal, BP 10448, CLERMONT-FERRAND, F-63000
[2] CNRS, UMR 6602, IP, Aubière, F-63171

Abstract. A central clue for implementation of visual memory based navigation strategies relies on efficient point matching between the current image and the key images of the memory. However, the visual memory may become out of date after some times because the appearance of real-world environments keeps changing. It is thus necessary to remove obsolete information and to add new data to the visual memory over time. In this paper, we propose a method based on short-term and long term memory concepts to update the visual memory of mobile robots during navigation. The results of our experiments show that using this method improves the robustness of the localization and path-following steps.

Keywords: Visual memory, life-long mapping, mobile robots.

1 Introduction

Visual memory-based navigation approaches have gained increasing interest in the last few years. They consist of representing the mobile robot environment with visual features gathered in a database. Basically, the navigation process from this visual memory can be split in three stages: 1) visual-memory acquisition, 2) initial localization and 3) path following (refer to Figure 1). In the sequel, we will focus on strategies where key images are stored and used as references during the on-line steps whether with metrical or topological map. The initial localization is usually based on features matching to find the visual image of the memory the most similar to the current image. During the path following step, features matched between the current image and images of the visual memory can be exploited to provide the necessary input for the state estimation supplying the control law (refer to [1] for instance).

Most navigation strategies proposed in the literature assume that the environment where the robot works is static. However, this assumption does not hold for many real environments. Following the taxonomy proposed in [2], changes in the environment may be *transient* or *lasting*. Lasting *perceptual* changes will deteriorate the feature matching process and then the performance of vision-based navigation strategies. To improve the performances of the localization and of the

S. Lee et al. (Eds.): Intelligent Autonomous Systems 12, AISC 193, pp. 133–142.
springerlink.com

path following steps, it seems promising to modify the reference views as proposed in [3,4] using an information model based on the human memory model [5]. Basically, reference views are stored in a Long Term Memory. When features have been seen in many views during the localization step, they are transfered from the Short Term Memory to the long-term memory and missing features are forgotten. It is reported in [3,4] that localization performances are improved with respect to a static map.

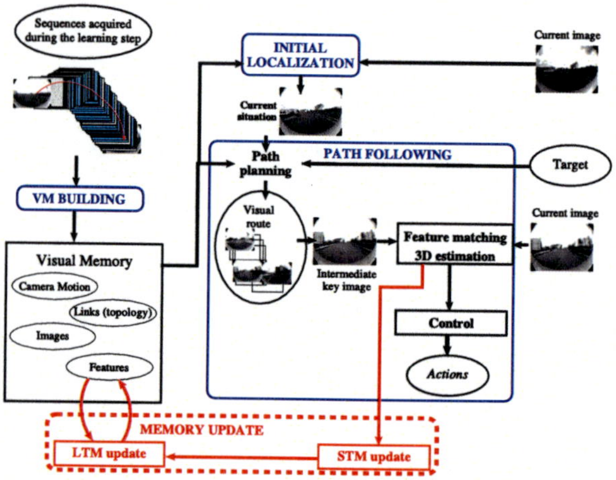

Fig. 1. Navigation process from a visual memory. Visual-memory acquisition, initial localization and path following are the classical stages of such a process. Our contribution is the update of the memory to take into account lasting changes. The concepts of Short-Term Memory and Long-Term Memory are used in that aim.

This paper proposes a strategy to deal with the lasting changes within visual memory-based navigation frameworks using reference images. Each time the robot is realizing a navigation task, the content of its visual memory is updated to take into account relevant changes for future runs. The proposed approach well suits a large class of visual-memory based navigation strategies based on topological or metrical maps. Furthermore, it is not restricted to point features. The remaining of this paper is organized as follows. The problem is stated in Section 2. In Section 3, the principles of our approach are detailed. The Section 4 reports experimental results with a robot navigating in an indoor environment.

2 Problem Statement

We suppose that the environment contains a set of 3D features $\{\mathcal{Q}_l \mid l = 1, 2, \ldots n\}$. The observation (or projection) of a 3D feature $\mathcal{Q}_l$ in an image $\mathcal{I}^{i_a}$ is a visual feature noted $\mathcal{P}_l^*$. We assume that visual features can be located/detected from images and that they are described by feature vectors. Two features $\mathcal{P}^{i1}_{l_1}$ and $\mathcal{P}^{i2}_{l_2}$ from two images $\mathcal{I}^{i1}$ and $\mathcal{I}^{i2}$ are said to be *matched* or *in correspondence* if they are supposed to be the projections of a same 3D feature (i.e. $l_1 = l_2$).

Visual memory. The Visual Memory (VM) of the robot can store different features:

a) n_{VM} key images $\{\mathcal{I}^i \,|\, i = \{1, 2, \ldots, n_{VM}\}\}$ extracted from the video sequence initially acquired,
b) for each key image $\mathcal{I}^i$, a set P^i of n^i descriptive image features $P^i = \{\mathcal{P}^i_{l_j} \,|\, j = \{1, 2, \ldots, n^i\}, l_j \in \{1, 2, \ldots n\}\}$,
c) a set of links between adjacent places $\{(\mathcal{I}^{i_a}, \mathcal{I}^{i_b}), (i_a, i_b) \in \{1, 2, \ldots, n_{VM}\}^2, i_a \neq i_b\}$,
d) a set of 3D features $\{\mathcal{Q}_l \,|\, l = 1, 2, \ldots n\}$,
e) the motion of the camera between acquisitions of the key images.

Initial localization. Let $\mathcal{I}^c$ be the image acquired at the current time during the autonomous navigation stage. We suppose that the initial localization of the robot in its internal representation of the environment starts with finding the index i such that $(\mathcal{I}^i, P^i)$ is the most similar to $(\mathcal{I}^c, P^c)$. A measure of similarity between images and/or features matching can be used in that aim.

Path following. Once the robot is localized and a target is defined, we assume that it is then possible to extract a path from the current situation to the target. This path can be autonomously followed using a set of key images (called *visual route* in the sequel) defining successive references: $\Psi = \{\mathcal{I}^{i_a} \mid a \in \{1, 2, \ldots, n_\Psi\}, i_a \in \{1, 2, \ldots n_{VM}\}\}$. Let $\mathcal{I}^{i_a}$ be the current reference image of the visual route Ψ. A usual way to compute the state of the robot which supplies the control law can be summarized as follows. First, a set of image features P^c are extracted on the current image and matched with those of $\mathcal{I}^{i_a}$ (P^{i_a}). Then, 3D information are estimated and the state of the robot is xtracted from the 3D information and the control law computed.

Problem statement. To deal with perceptual lasting changes, it is necessary to update the content of the visual memory. According to Sections 2 and 2, visual features have to be fulfilled for the initial localization and during the path following stage. 3D features positions can also be required during path following. Our objective is thus to modify the set of features (visual features $\{P^i\}$ and 3D features $\{\mathcal{Q}_l\}$ if required) to improve the performance of the visual navigation process.

3 Map Update Process

To update the visual memory, we propose to use two memories: a Short-Term Memory (STM) and a Long-Term Memory (LTM) [5].

3.1 Long-Term and Short-Term Memories

The Long-Term Memory stores the set of 3D features as well as 2D features. Each visual feature is associated with a state, representing its status at the current time. During the LTM update, states will be modified, resulting to the elimination of obsolete visual features.

The LTM is defined as:

$$LTM = \{\{\mathcal{Q}_l \,|\, l = 1,2,\dots n\}\ ;\ \{(\mathcal{P}^i_{l_j}, s^i_{l_j}) \,|\, j = \{1,2,\dots,n^i\},\ l_j \in \{1,2,\dots n\}, i = \{1,2,\dots n_{VM}\}\}\}$$

where $s^i_{l_j}$ is the state of the visual feature $\mathcal{P}^i_{l_j}$ (the projection of $\mathcal{Q}_{l_j}$ in the image $\mathcal{I}^i$ of the Visual Memory).

Let $\mathcal{I}^{i_a}$ be the temporary reference image of the visual route. A Short-Term Memory (STM) is associated with this image. It stores 2D and 3D features extracted from the N images $\{\mathcal{I}^{c_i} \,|\, 1 \leq i \leq N\}$ acquired during the path following step and such that their visual features have been matched with P^{i_a} for state estimation. More precisely, the STM contains three lists:

- Q^{mem} contains the n_{mem} 3D features whose projections in $\mathcal{I}^{i_a}$ have been matched with visual features of $\{\mathcal{I}^{c_i}\}$:

$$\mathsf{Q}^{mem} = \{\mathcal{Q}_{l_j} \,|\, j = \{1,2,\dots n_{mem}\}, 1 \leq l_j \leq n\}$$

- T^{new} lists the N 3D transformation matrices $\mathsf{T}^{new} = \{{}^{i_a}\mathbf{T}_{c_1}, {}^{i_a}\mathbf{T}_{c_2}, \dots, {}^{i_a}\mathbf{T}_{c_N}\}$ which give the pose of the N cameras' frames in the camera frame corresponding to the acquisition of $\mathcal{I}^{i_a}$. These transformations can be extracted from the 3D reconstruction estimated during path following.
- Q^{new} corresponds to *potentially new* features. A feature of Q^{new} is projected on one image of $\{\mathcal{I}^{c_i}\}$ at least but not in the reference image $\mathcal{I}^{i_a}$. The set Q^{new} is noted:

$$\mathsf{Q}^{new} = \begin{cases} (\mathcal{Q}^{new}_1, \{\mathcal{P}_1^{c_{l_{1,1}}}, \mathcal{P}_1^{c_{l_{1,2}}}, \dots \mathcal{P}_1^{c_{l_{1,n_1}}}\}) \\ (\mathcal{Q}^{new}_2, \{\mathcal{P}_2^{c_{l_{2,1}}}, \mathcal{P}_2^{c_{l_{2,2}}}, \dots \mathcal{P}_2^{c_{l_{2,n_2}}}\}) \\ \dots \\ (\mathcal{Q}^{new}_n, \{\mathcal{P}_n^{c_{l_{n,1}}}, \mathcal{P}_n^{c_{l_{n,2}}}, \dots \mathcal{P}_n^{c_{l_{n,n_n}}}\}) \end{cases}$$

where $\mathcal{P}_j^{l_{i,j}}$ is the projection of the new feature Q^{new}_j in the image $\mathcal{I}^{c_{l_{i,j}}}$ and n_j is the number of images where this 3D feature is projected.

3.2 Overview of the Update Process

The features of the LTM associated to the key images of the visual route Ψ are updated during the path following stage. The first key image of Ψ ($\mathcal{I}^{i_1}$) is initially considered as the reference image $\mathcal{I}^{i_a}$. An empty STM is associated to this key image. Features are detected in the current image $\mathcal{I}^{c_i}$ acquired by the embedded sensor and the STM is updated (refer to Section 3.3). Once the reference image $\mathcal{I}^{i_a}$ is reached, the features of the Long Term Memory associated to this key image is updated using the content of the STM. Obsolete 3D features are deleted and new features are added and can be used in future runs for initial localization and state estimation. The LTM update process is detailed in Section 3.4. After this update, the STM associated to the key image is deleted and the update process is repeated for the next reference image $\mathcal{I}^{i_a}$ until the end of the visual route is reached.

3.3 STM Update

Q^{mem} is updated using 3D features whose projections are in the current image $\mathcal{I}^{c_i}$ and in the reference image $\mathcal{I}^{i_a}$. 3D features not still present in Q^{mem} are added to this set. Each visual feature $\mathcal{P}_r^{c_i}$ not matched with any visual feature of P^{i_a} is supposed to be the projection of a potentially new 3D feature. If $\mathcal{P}_r^{c_i}$ is matched with the projection of a feature $\mathcal{Q}_j^{new}$ of Q^{new} (i.e. matched with a visual feature of the set $\{\mathcal{P}_j^{c_{l_{i,j}}}\}$), then $\mathcal{P}_r^{c_i}$ is added to the list of features associated to $\mathcal{Q}_j^{new}$. Otherwise, a potentially new feature is added to Q^{new} and $\mathcal{P}_r^{c_i}$ is associated to it.

3.4 LTM Update

The LTM is updated when the reference image $\mathcal{I}^{i_a}$ is reached. This process can be split into two steps: feature update and feature addition.

Feature update. The elements of Q^{mem} are used to update the state of features of the LTM. When a 3D feature $\mathcal{Q}_{l_j}$ belongs to Q^{mem} then the state of the corresponding projection $\mathcal{P}_{l_j}$ is reset to the first state otherwise the state progresses (refer to Figure 2). A visual feature that passes through all the states without a "hit" is forgotten *i.e.* eliminated from the LTM. If any projection of the 3D feature $\mathcal{Q}_{l_j}$ exists, then $\mathcal{Q}_{l_j}$ is eliminated.

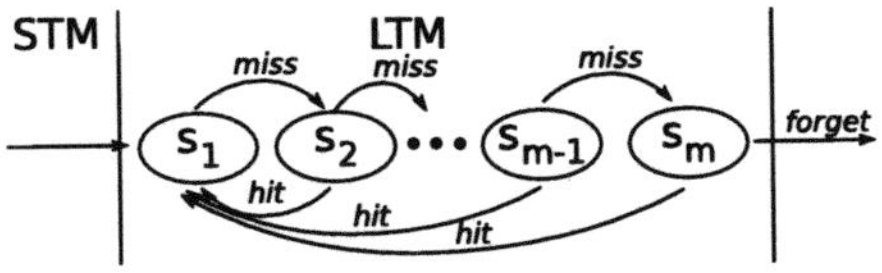

Fig. 2. LTM update, represented as a finite state machine

Feature addition. A feature $\mathcal{Q}_i^{new}$ of the set Q^{new} is added to the LTM if n_i is upper than a threshold. As the number of images N acquired by the embedded camera and matched with a given key image may vary (depending on the acquisition framerate and on the longitudinal velocity of the robot among others), the threshold we propose is function of N: $N\tau_{tf}$ with $0 < \tau_{tf} \leq 1$. It is supposed that this feature is projected on a sufficient number of images such that its 3D position can be computed in the frame attached to $\mathcal{I}^{i_a}$ using the set T^{new} and the positions of its projections. The geometry of $\mathcal{P}_i^{new}$ in the reference image $\mathcal{I}^{i_a}$ can then be obtained by projecting this 3D feature. If the feature is not visible (i.e. it is outside the image), this feature is not added to the LTM.

4 Implementation and Experiments

We use the complete framework for autonomous vehicle navigation proposed in [6]. The robot's internal representation of the environment is a visual memory topologically organized. When running autonomously, the vehicle is guided along the reference visual route with a vision-based control law adapted to the nonholonomic constraint. The estimation of the input of the control law are computed from the camera motion parameters estimated between the current image and the reference image using point as visual features and a local 3D reconstruction.

4.1 Set-Up and Implementation

Our experimental vehicle is a Pioneer AT3 robot equipped with a calibrated camera looking forward. Grey level images of resolution 800×600 pixels are acquired at a rate of 7.5 fps. Vision and guidance algorithms are implemented in C^{++} language on a laptop using RTAI-Linux OS with a 2GHz Centrino processor. The considered visual features are Harris corners described by their neighborhood. The similarity score between two patches is based on the Zero Normalized Cross Correlation (ZNCC). The initial localization of the robot is the image of the visual memory with the highest number of features matched with the features P^c of the current image $\mathcal{I}^c$. During path following, features of the current image $\mathcal{I}^c$ are matched with features of the reference image $\mathcal{I}^{i_a}$ (*matched features*). The camera motion parameters are then determined from the essential matrix between $\mathcal{I}^c$ and $\mathcal{I}^{i_a}$ estimated using five couples of matched points and a RANSAC process. Once the robot reaches the reference image, the next image of the visual route is set as the reference image.

During the learning step, the robot is manually teleoperated in an warehouse-like indoor environment. The LTM built with the 22 selected key images contains 3039 3D points and 73132 visual features. During the navigation task, the robot is approximately situated where the initial image was acquired and the target is the last key image. The translational velocity is $V = 80mm/s$. A navigation task is first realized with a static visual memory. Images are stored and used to compare initial localization results with static and adaptive visual memories. A navigation task is then realized with the adaptive LTM. The environment is quiet similar during both navigation tasks. Later, modifications have been done manually and the same experimentations are repeated 8 times. Some changes in appearance (objects or illumination) for two places at Run1 and Run8 are drawn in Figure 3.

For the memory update, the number of states of the LTM is set to $m = 3$. The threshold for feature addition is $\tau_{tf} = 0.33$. The 3D position of a new feature is computed using a triangulation process with 2 views and a RANSAC technique. The descriptor of its reprojection is set to be similar to the descriptor of the feature with the smallest reprojection error.

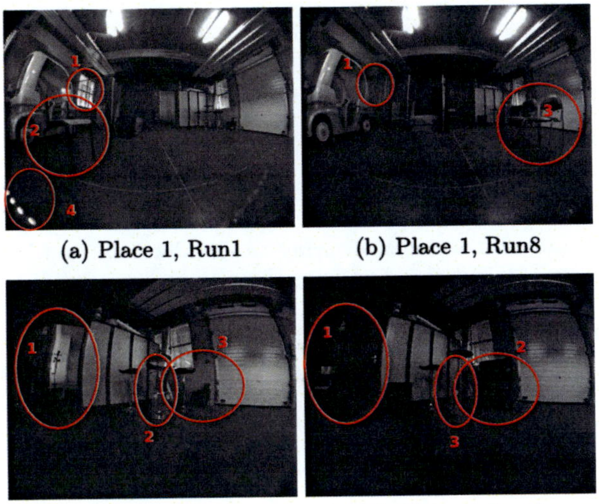

(a) Place 1, Run1 (b) Place 1, Run8

(c) Place 2, Run1 (d) Place 2, Run8

Fig. 3. Changes in appearance between Run1 and Run8 for two different places of the environment

4.2 Memory Content

The number of 3D features and the mean number of point features by key image are represented in Figure 4. The number of 3D features is clearly increasing over time using an adaptive memory. Between Run3 and Run4, the number of image features highly decreases. This is mainly due to the elimination of features detected in the images acquired during the learning stage but never used during the navigation tasks. The number of features then slowly varies, depending on the content of the environment. We observed that using an adaptive visual memory reduces the required resources in terms of memory.

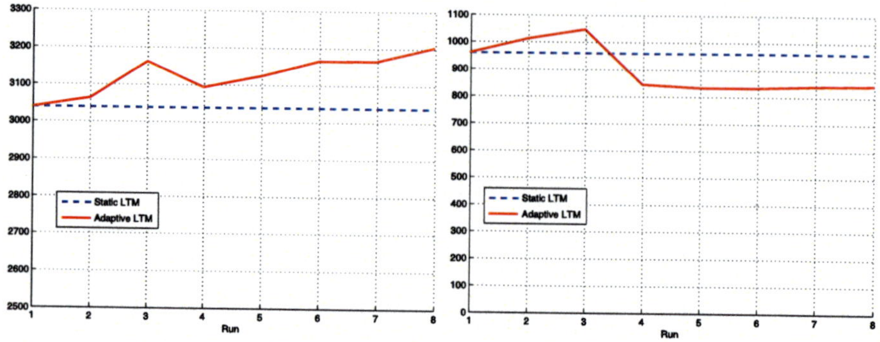

(a) 3D features

(b) Visual features

Fig. 4. Content of the LTM: (a) number of 3D features and (b) mean number of image features by key image versus run number

4.3 Initial Localization

The mean number of features matched between the current image and the most similar key image of the memory during the localization step is represented in Figure 5 (a) while the mean computational time to process it is given in Figure 5 (b). The number of matched features is decreasing over time for both approache but it decreases slower when using an adaptive memory. For Run8 for instance, 33% more features are matched when using adaptive memory. Moreover, after Run3, the computational cost is smaller when using an adaptive memory (7% of amount of time saved for Run8).

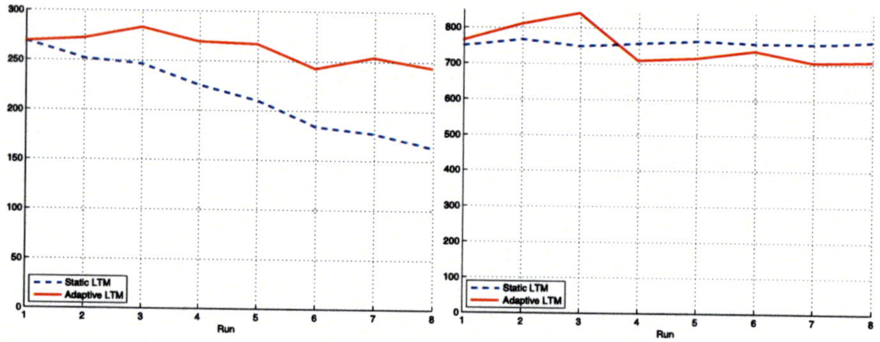

(a) Similarity score

(b) Computational time

Fig. 5. Initial localization: (a) mean number of matched features and (b) mean computational time (in ms) versus run number

4.4 Autonomous Navigation

For each run, the navigation tasks have been successfully executed using static and adaptive memories. Around 900 points are detected in each image. The mean number of features matched between the current image and the reference image as well as the number of features robustly matched are represented in Figure 6. It can be noticed that the percentage of robust matches over visual matches is decreasing (52% for Run1, under 40% for Run8). However, this percentage is higher when using an adaptive memory rather than a static memory (38% vs 26% for Run8). We are currently testing a L2-optimal triangulation technique for three views for the position of the feature in the reference image estimation in order to improve the gain in term of number of robust matches.

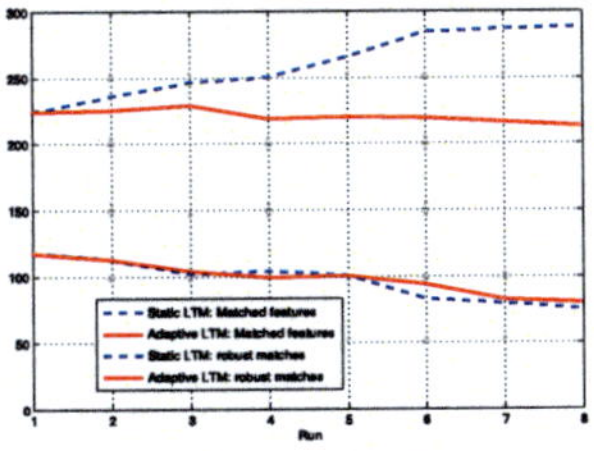

Fig. 6. Path following stage: mean number of matched features versus run number

5 Conclusion

We have presented a strategy to deal with perceptual lasting changes within visual memory-based navigation frameworks. This strategy is based on the concepts of short-term and long-term memories. During the path-following stage, visual features extracted in the current image is used to update a short-term memory. Once a key image is reached, features associated to it in the long-term memory are updated from the STM content. The results show that the performances of the initial localization and of the path following stages are improved. Current works deals with the inclusion of uncertainty propagation within the update process.

Acknowledgment. This work has been supported by the National Research Agency (ANR) of France (under the projects CityVip and R-Discover) and a grant (code I09200) from Gyeonggi Technology Development Program funded by Gyeonggi Province of Korea.

References

1. Royer, E., Lhuillier, M., Dhome, M., Lavest, J.-M.: Monocular vision for mobile robot localization and autonomous navigation. International Journal of Computer Vision, Special Joint Issue on Vision and Robotics 74, 237–260 (2007)
2. Yamauchi, B., Langley, P.: Spatial learning for navigation in dynamic environments. IEEE Transactions on Systems, Man and Cybernetics 26(3), 496–505 (1997)
3. Bacca, B., Salvi, J., Batlle, J., Cufi, X.: Appearance-based mapping and localization using feature stability histograms. Electronic Letters 46(16), 1120–1121 (2010)
4. Dayoub, F., Cielniak, G., Duckett, T.: Long-term experiments with an adaptive spherical view representation for navigation in changing environments. Robotics and Autonomous Systems 59(5), 285–295 (2011)
5. Atkinson, R., Shiffrin, R.: Human memory: a Proposed System and its Control Processes. In: Spence, K.W., Spence, J.T. (eds.) The Psychology of Learning and Motivation. Academic Press, New York (1968)
6. Courbon, J., Mezouar, Y., Martinet, P.: Autonomous navigation of vehicles from a visual memory using a generic camera model. Intelligent Transport System (ITS) 10, 392–402 (2009)

Topological Mapping with Image Sequence Partitioning

Hemanth Korrapati[1], Jonathan Courbon[1], and Youcef Mezouar[1,2]

[1] Clermont Université, Université Blaise Pascal,
Institut Pascal, BP 10448, CLERMONT-FERRAND, F-63000
[2] CNRS, UMR 6602, IP, Aubiére, F-63171
firstname.lastname@univ-bpclermont.fr

Abstract. Topological maps are vital for fast and accurate localization in large environments. Sparse topological maps can be constructed by partitioning a sequence of images acquired by a robot, according to their appearance. All images in a partition have similar appearance and are represented by a node in a topological map. In this paper, we present a topological mapping framework which makes use of image sequence partitioning (ISP) to produce sparse maps. The framework facilitates coarse loop closure at node level and a finer loop closure at image level. Hierarchical inverted files (HIF) are proposed which are naturally adaptable to our sparse topological mapping framework and enable efficient loop closure. Computational gain attained in loop closure with HIF over sparse topological maps is demonstrated. Experiments are performed on outdoor environments using an omni-directional camera.

Keywords: Topological Mapping, Omni-directional Vision, Loop Closure.

1 Introduction

Powerful loop closing techniques for topological maps have been introduced recently [2,3]. Most of them produce dense topological maps, in which every acquired image stands as a node in the topological graph. Sparser topological maps can be built by representing sets of contiguous images with similar appearance features as places. Each place is represented by a node in the topological graph. We refer to the partitioning of an image sequence into places as Image Sequence Partitioning (ISP). Maps with fewer nodes reduce computational complexity involved in loop closure and map merging.

We use a hierarchical topological mapping framework which represents places as nodes of a graph and the adjacency of places by edges. Each node/place consists of all the images acquired in that location. Each node contains a set of representative features extracted from all the member images of the node (place). The representative features are used in evaluating node-image similarity during ISP.

Another contribution of this paper is the proposal of Hierarchical inverted files (HIF) which combine the power of regular inverted files with our map structure and enable efficient loop closure at both node and image levels. Omni-directional image data is used in experimentation as it enables heading invariant loop closure. A spatial constraint demanding the features to stay in exactly the same vertical space of the image is used to improve matching precision and false positive reduction.

S. Lee et al. (Eds.): Intelligent Autonomous Systems 12, AISC 193, pp. 143–151.
springerlink.com

2 Related Work

Many loop closure algorithms for topological mapping have been proposed and tested in both indoor [10], [2], [5] and outdoor environments [3], [4], [1], [7].

In [10], topological maps for indoor environments are built by segmenting the topological graph of the environment using normalized graph-cuts algorithm resulting in subgraphs corresponding to convex areas in the environment. In [5] SIFT features were used to perform matching over a sequence of images by detecting transitions between individual indoor locations depending on the number of SIFT features which can be successfully matched between the successive frames. In [8] fingerprint of an acquired image is generated using omni-directional image and laser readings, and these fingerprints are compared to those of the previous images. All of these works were focused on indoor environments which are simpler to partition due to convex spaces.

A topological mapping framework using incremental spectral clustering has been presented in [9]. Nodes containing similar images are constructed using incremental spectral clustering over the affinity matrix of the images, thereby producing a topological graph. An optical flow based ISP technique was presented in [7] for topological mapping in outdoor environments using a quad rotor robot. Optical flow is used to discover change in environmental appearance. In [1], gist features were used to cluster images with similar appearance for topological map construction.

3 Image Sequence Partitioning

Image Sequence Partitioning (ISP) consists of three main modules: node level loop closure, evaluation of similarity to current place and new node formation. Given a query image, initially SURF features are extracted from the image. Using the SURF features, we evaluate the node-image similarity of the query image with all the nodes in the graph except the current place node and pick out the top k similar nodes. The top k similar nodes are assigned to the set of winning nodes $\mathbf{N_w}$. This process is called node level loop closure as it finds the previously visited places (nodes) most similar to the query image. Only the representative feature sets of the nodes are used to compute the node-image similarities during node level loop closure. In our framework, the representative features of a node are the SURF features of the first image augmented to the node.

An empty $\mathbf{N_w}$ indicates loop closure failure ; that is, query image is not similar to any of the previously visited places. In that case, query image similarity to the current node is evaluated. If the similarity of query image with current place node is above a certain threshold, current node is augmented with the query image. If the query image is not similar to current node also, a new node is created with the query image. But if $\mathbf{N_w}$ is not empty indicating a loop closure, then the set of winning nodes can be considered for a thorough image level loop closure.

Algorithm 1 shows the steps involved in ISP. The 'node_level_loop_closure' function in lines 4 is discussed in detail in sections 4. The function 'current_node_image_similarity' evaluates the similarity of the current node with that of the query image. This is done by matching the SURF features of the query image to that of the features of the current place node. Feature matching is performed as proposed in [6]. Two features are considered to be matched if the ratio of the best match and the second best match is greater than 0.6.

Algorithm 1 . Image Sequence Partitioning Algorithm

procedure PROCESS_QUERY_IMAGE($\mathbf{T}$, I_q, n_c) ▷ $\mathbf{T}$, I_q, n_c, $\mathbf{T.N}$ are the Topological graph, query image, current node in topological graph & node-set of topological graph respectively.
  $\mathbf{N}' = \mathbf{T.N} - \{n_c\}$ ▷ Reference node set excluding current node.
  $\mathbf{N_w} = \{\}$ ▷ Set of winning nodes.
  $\mathbf{N_w}$=Node_level_Loop_Closure($\mathbf{N}'$, I_q, Th_s)
  if is_empty($\mathbf{N_w}$) **then**
    if current_node_image_similarity(n_c,I_q) $> Th_t$ **then**
      n_c.add_image(I_q)
    else
      n=new_node()
      n.add_image(I_q)
      update_map($\mathbf{T}$, n)
    end if
  else
    $\mathbf{N_w}$=get_top_k_similar_nodes($\mathbf{N_w}$)
    $\mathbf{I_w}$ =get_images_of_nodes($\mathbf{N_w}$)
    n^*=Image_Level_Loop_Closure($\mathbf{I_w}$,$\mathbf{N_w}$,I_q)
    add_image_to_node(I_q, n^*)
    update_map($\mathbf{T}$, n^*)
  end if
end procedure

4 Loop Closure and Hierarchical Inverted Files

Node and Image level loop closures are performed at image level using visual words corresponding to the SURF features of the image. We use a variant of inverted files to perform loop closure in a way similar to that of [4,2].

A regular inverted file corresponding to a visual word simply contains a list of all previous images which contained the word. We associate Hierarchical inverted files (HIF) to each visual word. As the name suggests, HIF contain two levels. The first level consists of the ids of the nodes in which the visual word occurred previously. The second level consists of small child inverted files attached to each of the node id. These image ids indicate the images belonging to the parent node in which the visual word has occurred. Each child inverted file corresponding attached to a node of the topological graph, contains the list of all previous images belonging to the parent node in which the word has occurred. A sample HIF is illustrated in figure 1. To perform a node level loop closure using HIF, we do not have to go through the entire HIF, but its sufficient to go through first level (node ids) of the HIFs. For an image level loop closure using HIF, we only have to traverse through those child inverted files corresponding to the winning nodes; which form only a fraction of the total HIF. Thus, HIFs offer computational gain in loop closure when compared to regular inverted files which is demonstrated in section 5. Algorithms 2 and 3 give a clearer picture of the node level and image level loop closures.

It can be observed in the image level loop closure in algorithm 3, that in lines 9 through 12 we impose spatial constraints on the visual word matches. On line 9, we get the best visual word matches - this becomes important in case of unequal number of occurrences of the same visual word in both images. In this case, we select the lower of the two and find their matches in the other images such that they are the closest w.r.t the vertical distance. In lines 10 through 12 we sum up the distances of these multiple occurrences using a gaussian kernel with zero mean and variance σ. This constraint helps in reducing false positives. There can be multiple winning images given by the image level loop closure and hence multiple corresponding nodes, but for the sake of simplicity we do not represent that in the algorithm. In such cases we use RANSAC based geometric verification to find the right match.

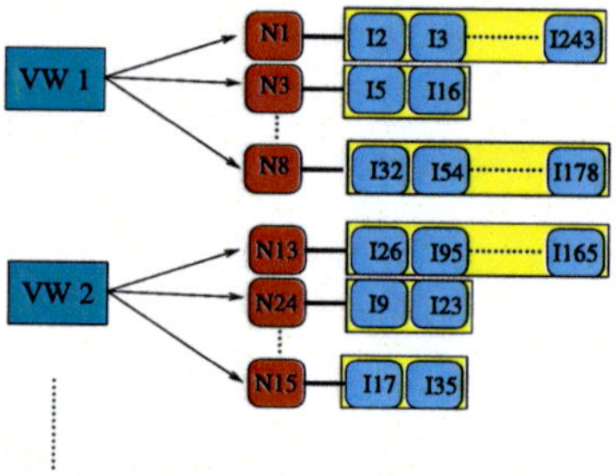

Fig. 1. A hierarchical inverted file. VW1, VW2,... represent visual words. N1, N2,... represent node ids in the first level of HIF and I1, I2, I3, ... represent the image ids in the inverted file.

Algorithm 2 . Node Level Loop Closure Algorithm

procedure NODE_LEVEL_LOOP_CLOSURE($\mathbf{N}'$, I_q, Th_s)
  $\mathbf{N}' = \mathbf{T}.\mathbf{N} - \{n_c\}$
  H=Histogram(size_of($\mathbf{N}'$))
  W=extract_and_quantize_Features(I_q)
  for each word w_i in **W do**
    HIF_i=get_hierarchical_inverted_file(w_i)
    for each node n_j in HIF_i **do**
      $H[n_j] = H[n_j] + 1$
    end for
  end for
  $\mathbf{N}^*$=get_winners_from_histogram(H, Th_s)
  return $\mathbf{N}^*$
end procedure

Algorithm 3 . Image Level Loop Closure Algorithm

```
procedure IMAGE_LEVEL_LOOP_CLOSURE(I_w, N_w, I_q)
    S=Array(M)
    W=extract_and_quantize_Features(I_q)
    for each word w_i in W do
        HIF_i=get_hierarchical_inverted_file(w_i)
        for each node n_j in N_w do
            IF_i^{n_j}=get_inverted_file_of_node(HIF_i, n_j)
            for each entry f_i in IF_i^{n_j} do
                D=get_matching_word_distances()
                sim = 0
                for each distance d_j in D do
                    sim = sim + g(d_j, σ)
                end for
                H[f_i] = H[f_i] + Tf − Idf(f_i, w_i) * sim
            end for
        end for
    end for
    I*=get_winner_from_histogram(H)
    n*=get_corresponding_node(I*)
    return n*
end procedure
```

5 Experiments

Our experimental setup consists of a Pioneer P3DX robot equipped with an omnidirectional camera. A laptop equipped with an Intel Centrino 2 processor running ubuntu 9.04 is used for data processing. The experiments were carried out in our artificial urban environment - PAVIN. The environment contains roads, artificial buildings, and a variety of real-world road settings like junctions, traffic lights, round abouts, curved roads and dead ends.

Omni-directional images were acquired at a frame rate of 2 fps, as the robot moves along a manually controlled trajectory. Two datasets were constructed - Dataset-6560 and Dataset-11200. Dataset-6560 contains 6560 images with 52 possible loop closures. Another dataset Dataset-11200 contains 11200 images and 71 possible loop closures. The number of loop closures were determined by manually examining the datasets.

5.1 ISP - Sparsity

The number of nodes in a topological map indicate its sparsity. The sparsity obtained in a map differs based on the features used to perform ISP. We have experimented using SURF128, U-SURF128, and SIFT features, out of which we found out that U-SURF128 is the best suitable feature for attaining sparse maps. The problem with SIFT is that it produces huge number of features out of which there are huge number of feature mismatches, which makes it impossible to perform feature matching in real-time and also cannot represent appearance properly. On the other hand SURF128 and U-SURF128 produce lesser number of features with relatively fewer mismatches. However U-SURF128 becomes a better choice as it exhibits more discriminative power on planar environments and also has lower computation time. In our implementation we used a matching threshold of 60% for SURF feature matching and a threshold of 40% for SIFT features.

Another important factor that effects stability of appearance is the image distortion. The features directly extracted from warped omni-directional images are unstable as the appearance of key-point patch changes very much even with a small displacement of the camera. Hence it is likely that the maps produced using warped images contain greater number of place nodes. This happens because due to the feature instability, each place can be understood as multiple adjacent places and hence multiple nodes in the topological graph. The undistorted (unwrapped) images produce relatively sparser maps. Tables 1(a) and 1(b) show the sparsity of maps produced by ISP using different features on warped and unwrapped images. We can see that U-SURF128 is the best performing feature producing the most sparse maps.

Table 1. Sparsity

(a) Sparsity - Warped

	SURF128	U-SURF128	SIFT
DATASET-6560	539	502	1582
DATASET-11200	756	742	2795

(b) Sparsity - Unwarped

	SURF128	U-SURF128	SIFT
DATASET-6560	504	473	1037
DATASET-11200	737	723	1257

5.2 Accuracy

The most sparse map may not guarantee an accurate map. Only those maps with accurate place partitioning are accurate and can lead to accurate loop closures. Thus a good mapping technique is one which provides an optimal combination of sparsity and accuracy.

Given a query image, first we perform loop closure at node level and then at image level if more accuracy is required. Tables 2 shows the number of loop closures detected and the number of false positives obtained on Dataset-6560 and Dataset-11200 respectively by node level loop closure. The results in these tables were obtained using unwrapped omni-directional panoramas. We can see that a few loop closures are missed out, and some false positives are observed. The inaccuracy can be attributed to the imperfections in ISP. There is a direct relation between accuracy of the maps and ISP. The way in which we perform topological mapping does not induce any information loss. In other words we do not perform any kind of sampling or selection of reduced number of features and hence there is no information lost during the mapping process. However inaccurate loop closure detection might occur due to inaccurate partitioning of places. For example a place can be represented by two nodes by partitioning it inaccurately during ISP due to appearance feature instability. As a result, during node level loop closure using a query image, both of the nodes representing that place may not get high similarity scores and hence the loop closure becomes inaccurate. A good ISP algorithm produces maps with minimum number of these situations.

Image level loop closure accuracy depends on the accuracy of node level loop closure. If node level loop closure selects an inaccurate set of winning nodes, then as a result, image level loop closure also becomes inaccurate. However in case of an accurate node level loop closure, we have observed that 99% accuracy was possible in image level loop closure irrespective of the ISP technique.

Table 2. Node Level Loop Closure Accuracy

	#(LC)	#(FP)
Dataset-6560	46	2
Dataset-11200	67	4

5.3 Computational Time

Table 3 shows average computational time (in milliseconds) taken by each stage of our topological mapping framework in processing each query frame on both Dataset-6560 and Dataset-11200. The abbreviations NLLC, LFE+QUANT, ILLC stand for Node Level Loop Closure, Local Feature Extraction & Quantization and Image Level Loop Closure respectively. NLLC is the node level loop closure which involves extracting the most similar nodes using HIFs as mentioned in Algorithm 2. This takes $10ms$ as shown in table 3, and requires additional $50ms$ in order to compare with the current place node whenever needed. Obviously, local feature extraction (LFE) ($200ms$) and quantization (QUANT) ($70ms$) time is constant for every acquired frame. Actually, time required for both of these tasks increases with the number of features in an image.

Computation time of image level loop closure (ILLC) is too low. This low computation time is the result of using hierarchical inverted files(HIF). As we mentioned before, HIFs make the loop closure computation almost independent of the number of reference images and also in our case, nodes of the topological graph.

Table 3. Average Computation Times (in ms)

NLLC	LFE+QUANT	ILLC
$10 + 50$	$200 + 70$	21

Figures 2(a) and 2(b) show graphs comparing the loop closure times of our HIF-based method and without using HIF (non-HIF based). Red curves in the graphs indicate the time taken for feature quantization, node level loop closure and image level loop closure, for each image frame in a sequence. The blue curves represent the time taken by feature quantization and similarity score generation using inverted files by using inverse similarity evaluation methodology. We can see that the loop closure time of non-HIF based method increases relatively more with the increase in the number of images in the map, while our method using our method, loop closure time increases more slowly. Also, the performance gain becomes more prominent in case of huge datasets (huge number of images) as can be seen in the figure 2(b) corresponding to Dataset-11200,

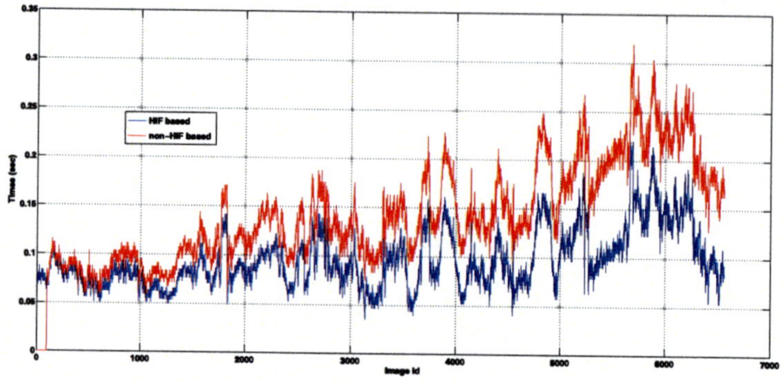

(a) Loop Closure time - Dataset-6560

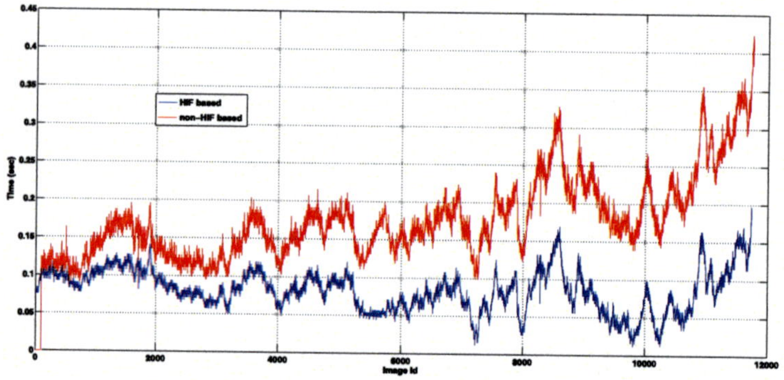

(b) Loop Closure time - Dataset-11200

Fig. 2. Loop closure computation times of non-HIF based loop closure and HIF based loop closure on maps generated on our datasets.

which contains 11200 images. The non-HIF based loop closure time for Dataset-11200 increases less steeply than that of Dataset-6560. This happened because the average number of features of Dataset-11200 is lesser than that of Dataset-6560 and as a result it takes lesser time to process each reference frame. This efficiency of our HIF-based method can be attributed to the combination of sparse topological mapping and HIFs for efficient map storage. The representational power of HIFs saved lot of computation involved in loop closure.

6 Conclusion

We proposed a sparse topological mapping framework involving two levels of representation. Image Sequence Partitioning(ISP) is employed to produce sparse maps. Hierarchical Inverted Files(HIF) were proposed to take advantage of the sparse maps and perform quick loop closure. Sparsity of the maps produced by different features are analysed and the accuracy is evaluated. Finally, our framework is evaluated on computational time required for loop closure.

Acknowledgment. This work has been supported by the National Research Agency (l'ANR) of France (under the projects CityVip and R-Discover) and a grant (I09200) from Gyeonggi Technology Development Program funded by Gyeonggi Province of Korea.

References

1. Kosecka, J., Murillo, A.C., Campos, P., Guerrero, J.J.: Gist vocabularies in omnidirectional images for appearance based mapping and localization. In: 10th IEEE Workshop on Omnidirectional Vision, Camera Networks and Non-Classical Cameras (OMNIVIS), Held with Robotics, Science and Systems (2010)
2. Angeli, A., Filliat, D., Doncieux, S., Meyer, J.-A.: A fast and incremental method for loop-closure detection using bags of visual words. IEEE Transactions on Robotics, Special Issue on Visual SLAM (2008)
3. Cummins, M., Newman, P.: FAB-MAP: Probabilistic Localization and Mapping in the Space of Appearance. The International Journal of Robotics Research 27(6), 647–665 (2008)
4. Cummins, M., Newman, P.: Highly scalable appearance-only slam fab-map 2.0. In: Robotics Science and Systems (RSS), Seattle, USA (June 2009)
5. Koseck, J., Li, F., Yang, X.: Global localization and relative positioning based on scale-invariant keypoints. Robotics and Autonomous Systems 52(1), 27–38 (2005)
6. Lowe, D.G.: Distinctive image features from scale-invariant keypoints. International Journal of Computer Vision 60, 91–110 (2004)
7. Nourani-Vatani, N., Pradalier, C.: Scene change detection for topological localization and mapping. In: IEEE/RSJ Intl. Conf. on Intelligent Robotics and Systems, IROS (2010)
8. Tapus, A., Siegwart, R.: Incremental robot mapping with fingerprints of places. In: IEEE/RSJ Intl. Conf. on Intelligent Robotics and Systems (IROS), pp. 2429–2434 (2005)
9. Valgren, C., Duckett, T., Lilienthal, A.: Incremental spectral clustering and its application to topological mapping. In: Proc. IEEE Int. Conf. on Robotics and Automation, pp. 4283–4288 (2007)
10. Zivkovic, Z., Booij, O., Kröse, B.: From images to rooms. Robot. Auton. Syst. 55, 411–418 (2007)

Kinect Enabled Monte Carlo Localisation for a Robotic Wheelchair

Theodoros Theodoridis, Huosheng Hu, Klaus McDonald-Maier, and Dongbing Gu

School of Computer Science and Electronic Engineering
University of Essex, Wivenhoe Park, Colchester, CO4 3SQ, UK
{ttheod,hhu,kdm,dgu}@essex.ac.uk

Abstract. Proximity sensors and 2D vision methods have shown to work robustly in particle filter-based Monte Carlo Localisation (MCL). It would be interesting however to examine whether modern 3D vision sensors would be equally efficient for localising a robotic wheelchair with MCL. In this work, we introduce a visual Region Locator Descriptor, acquired from a 3D map using the Kinect sensor to conduct localisation. The descriptor segments the Kinect's depth map into a grid of 36 regions, where the depth of each column-cell is being used as a distance range for the measurement model of a particle filter. The experimental work concentrated on a comparison of three different localization cases. (a) an odometry model without MCL, (b) with MCL and sonar sensors only, (c) with MCL and the Kinect sensor only. The comparative study demonstrated the efficiency of a modern 3D depth sensor, such as the Kinect, which can be used reliably for wheelchair localisation.

Keywords: Monte Carlo Localisation, Particle Filter Localisation, Region Locator Descriptors, Kinect sensor.

1 Introduction

The need for reliable wheelchair localisation has become a high priority need for the elderly and disabled. To provide adequate autonomous mobility, robotic wheelchairs can be used for transportation. This would require an efficient localisation method, and a depth sensor for coverage. In our effort to tackle the localisation problem with a robotic wheelchair, the Monte Carlo Localisation (MCL) method was used. The perception was undertaken by the Microsoft's Kinect depth sensor, for the coverage of a 640×480 depth pixels in ~60° wide lens angle. The sensor's resolution along with its low cost, compared to 1D laser range finders, makes it an attractive apparatus for localisation.

MCL is one of the most popular probabilistic methods for online robot localization, which is based on a particle filter algorithm. MCL was primarily introduced by [2], on their work in mobile robot localisation. MCL in its purest form tackles the global position estimation, and the local position tracking problem. To perform localisation, the robot's state space is represented by a sample (particle) density, which approximates the robot's position. The whole process is conducted in two phases; the *prediction* and *update*. In the prediction phase a motion model predicts the robot's

S. Lee et al. (Eds.): Intelligent Autonomous Systems 12, AISC 193, pp. 153–163.
Springerlink.com

current pose with a Probability Density Function (PDF). Similarly, in the update phase a measurement (sensor) model is incorporated to evaluate each particle in the density, and sample with replacement those particles evaluated as good predictors. In [3], an alternative MCL method is proposed based on an adaptive particle density, which showed improved accuracy and less computation effort. A stereo vision based MCL was presented by [4], using a scale invariant feature transform descriptor, given a 3D map. The measurement model incorporated 3D features extracted from landmarks. The proposed method appeared to solve the 6 DOF motion, and perform accurate localisation with a motion model having sensor measurements only. Another MCL method that exploits visual features was introduced by [5], for the localisation of the Aibo dog-robots. Two MCL variants were proposed; a landmark, and a field line-based extraction. The MCL using the visual features performed fast and reactive localisation within a miniaturised football pitch. In [7] a landmark tree model is employed for self-localisation of an autonomous wheelchair. The method utilises an image retrieval technique and the Bayes rule to localise the wheelchair. In addition, a path planning algorithm is introduced, which exploits a treelike structure to locate landmarks and destination locations. Landmarks are recognised through an image by extracting the shape and structure, and localisation is conducted by traversing within the tree nodes to elicit an optimal path. A probabilistic odometry (motion) model was introduced by [8] for an autonomous wheelchair. The method constructs a set of frequency tables of the wheelchair's pose stored in bins. A particle filter advises these tables to make predictions for localisation.

The rest of the paper is organized as follows. Section 2 describes the classical Monte Carlo Localisation method, given a brief algorithm. Section 3 presents the Kinect sensor and a depth-based region locator descriptor used in MCL. The probabilistic models that implement the MCL's structure are given in Section 4. Section 5 demonstrates a tri-case experiment using a mobile robot, and Section 6 points out conclusions and future directions.

2 Monte Carlo Localisation

The general idea of MCL is the deployment of a particle density, designated to predict a robot's position. The particles can be seen as virtual copies of the robot's existence, incorporating 2D kinematics for locomotion (motion model), and a beam array for perception (measurement/sensor model). As stated, MCL is conducted in two main phases to perform localisation. In the *prediction* phase at every time step the particles' kinematics is updated when the robot transits from state x_{k-1} to x_k . Each particle is assigned with a weighting fitness value that assesses how well the actual position of the robot is approximated. This accounts the actual (the robot's) and the expected (the particle's) sensory perception. In the *update* phase a subset of particles is drawn probabilistically with replacement to represent the new particle population. The resampling is based on the Darwinian principle of natural selection. Hence, high fitness particles have higher probability to be selected multiple times. Progressively, the whole particle population converges near the actual position of the robot.

Algorithm 1. Monte Carlo Localisation

1: Generate Particles $X_x \sim N(\mu_x, \sigma_x^2)$

2: Initialise Particle Weights $w_k = 1/Q$

3: Prediction $x_k \sim p(x_k \mid x_{k-1})$

4: Weight Estimation:

$$w_k = p(z_k \mid x_k, u)\, w_{k-1},\quad w_k = w_k / \sum_{q=1}^{Q} w_k^{[q]}$$

5: Update $X_k \sim \left\langle x_k^{[q]}, w_k^{[q]} \right\rangle_{q=1}^{Q}$

Algorithm 1 describes briefly the MCL method in five steps. In the first, a particle distribution X is created with randomly chosen state vectors $\mathbf{x} \equiv \langle\langle x, y, \theta\rangle, w\rangle$ from a Gaussian distribution *N*. The randomly chosen parameters are the robot's pose at coordinates x, y, θ, and the weight factor w. Step 2 initialises the particles weights with the factor $1/Q$, where Q is the total number of particles. In step 3, the prediction takes place of the state x_k given the initial state x_{k-1}. At this step, the robot's kinematics is applied at each particle. The weight estimation in step 4 is done by the product of the measurement probability shown as z_k, given the state $x_k$after applying a control command u, for particle q. Normalisation follows to scale the particle weights in the interval {0, 1}, so all add to 1. The update phase in step 5 samples with replacement a subset of particles based on their weight values w_k. The process is repeated recursively by looping back to step 2.

3 Kinect Depth-Based Perception

3.1 Kinect Sensor

The main sensor apparatus utilised in this work is the Kinect sensor (see Fig. 1), produced by Microsoft. Kinect is a vision sensor providing a RGB and a depth image in a single device. It is being used by Microsoft's game industry for games that do not require physical control devices such as joysticks. Basically, the user becomes the controller by using gestures and body movements. For using the sensor without the X-Box, as it is the primary device to work with, several drivers have been released to access the device from a PC. For the current application we have used the CL-NUI-Platform driver for Windows 7 (version 1.0.0.1121).

Fig. 1. Microsoft's Kinect sensor

The Kinect's video frame rate is up to 30 Hz, with an 8-bit VGA resolution of 640×480 pixels. The device provides 11-bit depth with 2,048 levels of intensity. The sensor's approximate ranging limit is between 0.5 to 6 m distance, while the angular field of view is 57° horizontally, and 43° vertically. There is also an embedded quadruple microphone array with a 16-bit audio and sampling rate of 16 kHz [6].

3.2 Region Locator Descriptor

The sensor coverage of a significant angular range is a crucial factor for autonomous and safe transportation. Kinect's specifications allow us to cover a large field of view from which information can be acquired. Such information regards the extraction of depth in local regions in this field. The local depth regions can play a twofold role; initially to harness them for the MCL's sensor model to conduct localisation. Secondly, for obstacle avoidance and safe navigation.

A local depth region acquisition method is being proposed to work for localisation purposes. For this reason, we generate a 6×6 depth matrix, which discretises the image in 36 depth regions as depicted in Fig. 2. The region matrix is given by matrix **A** in Eq. 1, with entry elements the regions represented by the matrix **B**. The elements of matrix **B** are the individual depth pixels indexing 107×80, as divided by 6 from the overall image resolution. For each region **B** the average depth is estimated, therefore matrix **A** now consists of the mean values for each region. The final region locator descriptor (*rld*) is a vector with 6 elements reflecting to the minimum distance from each column *n* (0, ..., 5) of matrix **A**.

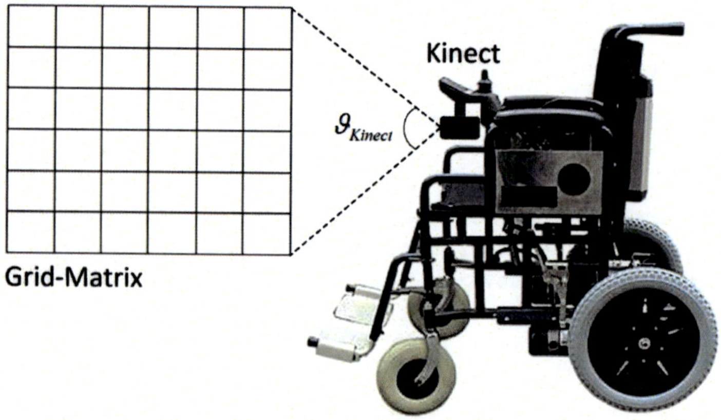

Fig. 2. The Kinect sensor attached on a robotic wheelchair. The grid-matrix is extracted by the visual descriptor from the Kinect's depth image.

$$rld = \begin{bmatrix} \underset{m}{\operatorname{argmin}} A_{m,0} \\ \underset{m}{\operatorname{argmin}} A_{m,1} \\ \underset{m}{\operatorname{argmin}} A_{m,2} \\ \underset{m}{\operatorname{argmin}} A_{m,3} \\ \underset{m}{\operatorname{argmin}} A_{m,4} \\ \underset{m}{\operatorname{argmin}} A_{m,5} \end{bmatrix} \Leftarrow A = \begin{bmatrix} \overline{B}_{0,0} & \overline{B}_{0,1} & \cdots & \overline{B}_{0,5} \\ \overline{B}_{1,0} & \overline{B}_{1,1} & \cdots & \overline{B}_{1,5} \\ \vdots & \vdots & \ddots & \vdots \\ \overline{B}_{5,0} & \overline{B}_{5,1} & \cdots & \overline{B}_{5,5} \end{bmatrix}, \overline{B} = \begin{bmatrix} p_{0,0} & p_{0,1} & \cdots & p_{0,80} \\ p_{1,0} & p_{1,1} & \cdots & p_{1,80} \\ \vdots & \vdots & \ddots & \vdots \\ p_{107,0} & p_{107,1} & \cdots & p_{107,80} \end{bmatrix} \quad (1)$$

4 Probabilistic Models

There are four models required for the implementation of a Markov localisation method such as MCL. These models are being employed to represent mainly the particles' pose and perception. A *motion model* expresses the probability for certain actions to move the robot to certain relative positions. A *sensor model* describes the probability for taking certain measurements at certain locations [5]. A *noise model* adds Gaussian noise to the motion model, so as to spread the particle density. Finally, an *odometry model* can be incorporated to conduct a waypoint navigation. The MCL's particle filter obviously acts on the odometry model, refining its direction towards a predefined location.

4.1 Motion Model

The effects of actions on the robot's pose are represented by a motion model [5]. At every time step the robot's kinematics updates the particle's kinematics. The kinematic update is instrumented by Eqs. 2, 3, and 4, with d denoting the delayed linear displacement, and ϑ the delayed rotational one. The noise terms ε are random Gaussian noise explained in Section 4.3.

$$\begin{bmatrix} x_k \\ y_k \\ \theta_k \end{bmatrix} = \begin{bmatrix} x_{k-1} + (d + \varepsilon_x)\cos\theta \\ y_{k-1} + (d + \varepsilon_y)\sin\theta \\ \theta_{k-1} + (\vartheta + \varepsilon_\theta) \end{bmatrix} \quad (2)$$

$$d = \sqrt{(x_{k-1} - x_k)^2 + (y_{k-1} - y_k)^2} \quad (3)$$

$$\vartheta = |\theta_{k-1} - \theta_k| \quad (4)$$

In the current work, a priory virtual map M is provided for the interaction of the particles with their environment. The ones which violate the map's boundaries are regenerated with a random pose within the map. The map is represented here by a pair of points, each defining the start-end coordinates of a line. Fig. 3 depicts the line representation.

Algorithm 2 illustrates the map validation procedure. Step 1 describes a state estimate for the current particle q. The mapFlag at step 2 is set to true, as it decides later on whether the state of the evaluated particle violates the map's boundaries. The for-loop in step 3 validates the particle's x, y state coordinates with each line in the map. A newly random coordinate state is chosen for the particle which violates the boundaries, or, the particle is left intact otherwise.

Algorithm 2. Particle map (M) validator

1: Estimate $\mathbf{x}_k^{[q]}$

2: mapFlag = **true**

3: **for** $i = 1$ to M **do**

$$j = (i + \mathrm{M} - 1)\,\%\,\mathrm{M}$$

if $$\left(\begin{array}{l} (\mathrm{M}_y^{[i]} > q_y) \neq (\mathrm{M}_y^{[j]} > q_y)\ \& \\ q_x < \dfrac{(\mathrm{M}_x^{[j]} - \mathrm{M}_x^{[i]})(q_y - \mathrm{M}_y^{[i]})}{(\mathrm{M}_y^{[j]} - \mathrm{M}_y^{[i]})} + \mathrm{M}_x^{[i]} \end{array} \right)$$

$$\mathrm{mapFlag} = \overline{\mathrm{mapFlag}}$$

end if

end for

4: **if** mapFlag = **true**

return rand($q^{[x]}$)

end if

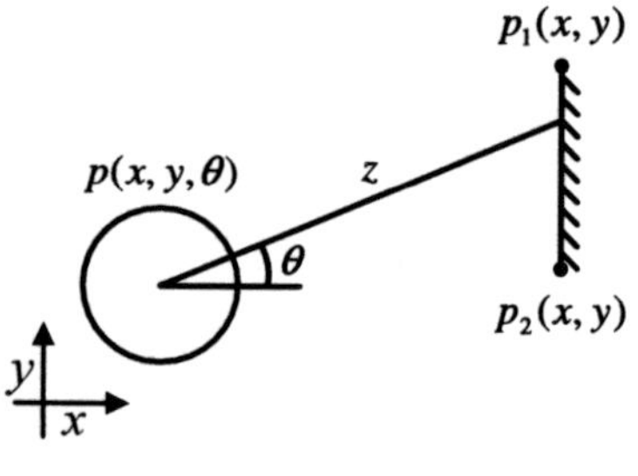

Fig. 3. Beam representation of a particle

4.2 Sensor Model

The sensor model is described by the likelihood probability $p(z|\mathbf{x})$ of Eq. 6. It is a probabilistic error between the particle's virtual beam range z (Eq. 5), as shown in Fig. 1, and the real robot sensor range s, acquired from the visual depth descriptor. The product of the likelihood probabilistic error between virtual and real sensor ranges, for particle q, constitutes the particle's goodness of fit (weight, Eq. 7). The weight is an assessment of how close the particle approximates the robot's state vector $\mathbf{x}$.

$$z = \frac{[p_2(y) - p_1(y)][p_1(x) - x_q] - [p_2(x) - p_1(x)][p_1(y) - y_q]}{[p_2(y) - p_1(y)]\cos\theta - [p_2(x) - p_1(x)]\sin\theta} \tag{5}$$

$$p(z \mid \mathbf{x}) = \begin{cases} 1, & \text{if } |s - \bar{z}| > \sigma_z \\ 1 - \dfrac{|s - \bar{z}|}{\sigma_z}, & \text{otherwise} \end{cases} \tag{6}$$

$$w_q = \prod_{i=1}^{S} p(z^{[i]} \mid \mathbf{x}) \cdot w^{[i]} \tag{7}$$

where $\bar{z}$ is the average of a measurement sample, and σ_z the standard deviation. Ultimately, S is the max number of sensors, which has been defined primarily from the visual descriptor as 6 (region matrix **A**, Eq. 1). Here, for consistency the sensor model assigns six beam sensors to each particle with 10° angular difference.

4.3 Noise Model

The noise model is generated using a zero-mean Gaussian distribution. For the translational noise (see Eq. 8) the delayed distance d is multiplied with the Gaussian N to acquire a random noise scalar, which is added with d in the kinematic update of Eq. 1. The exponent n is chosen for introducing more (=1) or less noise (>1). Similarly for the rotational noise.

$$\begin{aligned} \varepsilon_x &\sim N(0, \sigma_x^2)^n \times d \\ \varepsilon_y &\sim N(0, \sigma_y^2)^n \times d \\ \varepsilon_\theta &\sim N(0, \sigma_\theta^2)^n \times \vartheta \end{aligned} \tag{8}$$

The random Gaussian noise is sampled from a multivariate distribution N, which is characterised by a density function [1] with $\mu = 0$ and $\sigma = 1$ (see Eq. 9).

$$N = \det(2\pi\Sigma)^{-1/2} \exp\left\{-\frac{1}{2}\mathbf{v}^{\mathrm{T}}\Sigma^{-1}\mathbf{v}\right\} \tag{9}$$

where Σ is the covariance matrix, and $\mathbf{v}$ a random vector.

4.4 Odometry Model

An odometry model is particularly useful for waypoint localisation. The model describes the robot's motion in a rotational and a translational displacement. The rotational displacement is estimated by the arctangent between the current (x, y) and the next (x', y') coordinate pairs. The translational displacement is given by the difference of the same coordinate pairs, estimated with the Euclidean distance. The model's displacements are described by Eq. 8

$$\begin{aligned} \delta_{rot1} &= \operatorname{atan2}(y' - y,\ x' - x) \\ \delta_{trans} &= \sqrt{(x - x')^2 - (y - y')^2} \\ \delta_{rot2} &= \theta' - \theta \end{aligned} \tag{8}$$

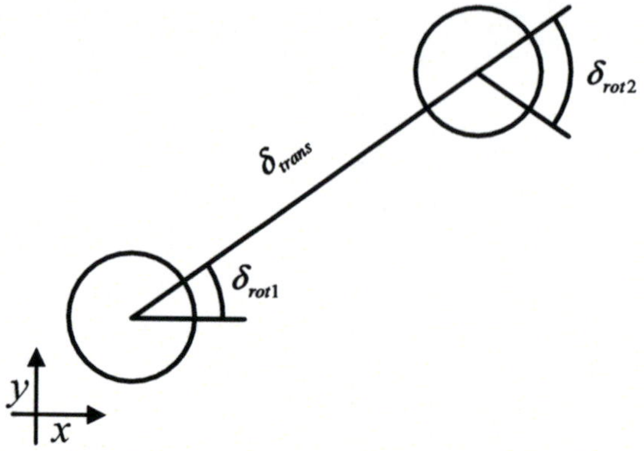

Fig. 4. Representation of the odometry model in two rotational, and one translational transition

The robot's motion at (*t*-1, *t*] is approximated by a sequence of transitions as shown in Fig. 2 [1]. Initially, a rotational transition δ_{rot1} is performed, followed by a translational transition, δ_{trans} and a consequent rotational δ_{rot2} . The robot's transitional displacements over a number of nodes completes the waypoint localisation. The purpose of the MCL's particle filter is to act on the odometry model by refining the robot's heading direction towards a designated node.

5 Experimental Results

5.1 Experimental Setup

The Essex robotic arena was the main experimental hall where the experimental procedure took place. The environmental setup illustrated in Fig. 5(a) was used for the conduction of the experimental cases. The experiments were separated in three

(a) (b) (c)

Fig. 5. (a) The experimental environment, (b) The Kinect-enabled mobile robot, (c) The grid depth map different cases so as to show the efficiency of the visual descriptor when applied for MCL, and the Kinect sensor when compared to proximity sensor localisation. In addition, the initial experiments were taken by an Activmedia Pioneer robot (Fig. 5(b)) for testing purposes before the actual wheelchair is used to carry human beings.

We have used an odometry model to navigate the robot through 19 nodes (waypoints) as Fig. 5(a) also depicts. Lastly, Fig. 5(c) illustrates the region depth map using the robot's onboard Kinect sensor.

5.2 Localisation Cases

a) Odometry Model
Employing an odometry model only, the robot is obviously lost so an incorporated obstacle avoidance undertakes to help the localisation. Fig. 6(a) presents an experimental trial for navigating with an odometry model. The trajectory seems rather unstable due to several maneuvers performed by the robot for the avoidance of the walls. From the poor localisation performance shown in this graph, we see the importance of how useful would be a probabilistic filter for the autonomous transportation of humans.

b) Sonar-Based MCL
The classical MCL method, as initially proposed by [2, 3], employed proximity sensors. In this case, we performed MCL with an array of six sonar sensors. The

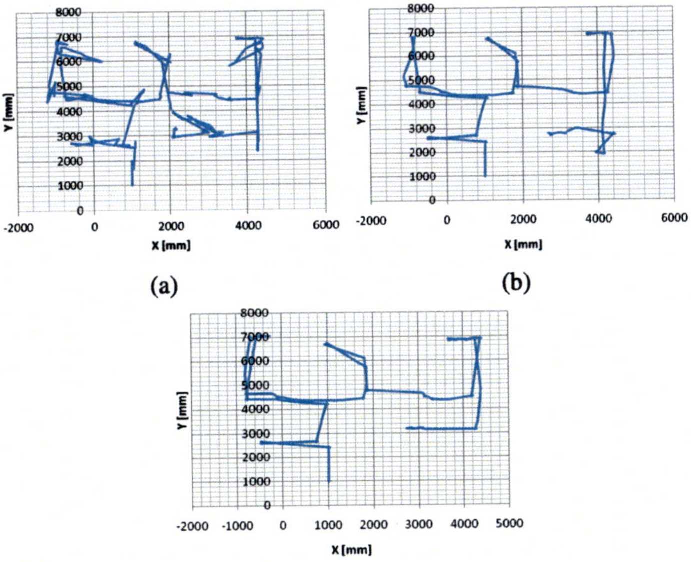

(c)

Fig. 6. Experimental trajectories. (a) Using an odometry model, (b) Using sonar-based MCL, (c) Using Kinect-based MCL.

resultant trajectory of Fig. 6(b) demonstrates a much smoother trajectory when we introduce the MCL method to the odometry model. A diminished use of the obstacle avoidance algorithm is observed, and the path now is drawn clearer.

c) Kinect-Based MCL

The proposed MCL method, enhanced with an incorporated region locator descriptor, shows in Fig. 6(c) to perform better than the sonar-based MCL, and apparently better than the odometry model on its own. The performance enhancement is owed to fact that the Kinect sensor utilises a complete coverage of the frontal view. This turns out to be essentially useful for the sensor model, as now the particles are evaluated more accurately. The Kinect's depth coverage in x and y image coordinates, contrary to lasers acting in x only, is proven also to be safer and more robust for a robotic wheelchair use.

6 Conclusions

In this work an attempt has been made to introduce a region locator descriptor, through a depth image, to Monte Carlo Localisation method. The Microsoft's Kinect sensor was the main apparatus incorporated to enhance the localisation performance. Indeed, the experimental work showed that the suggested descriptor can improve the localisation accuracy when compared with proximity sensors.

Our future directions will focus on testing colour and edge histogram descriptors for MCL. Eventually, to test the performance of the current method on different wheelchair types for indoor and outdoor localisation.

Acknowledgements. This research has been financially supported by the EU Interreg IV A 2 Mers Seas Zeeën Cross-border Cooperation Programme – SYSIASS: Autonomous and Intelligent Healthcare System. More details can be found at the project's website http://www.sysiass.eu/. We would also like to thank Robin Dowling for the technical support.

References

1. Thrun, S., Burgard, W., Fox, D.: Probabilistic Robotics. MIT Press, Cambridge (2005)
2. Dellaert, F., Fox, D., Burgard, W., Thrun, S.: Monte Carlo Localization for Mobile Robots. In: IEEE International Conference on Robotics and Automation, pp. 1322–1328 (1999)
3. Fox, D., Burgard, W., Dellaert, F., Thrun, S.: Monte Carlo Localization: Efficient Position Estimation for Mobile Robots. In: Proceedings of the Sixteenth National Conference on Artificial Intelligence, pp. 343–349 (1999)
4. Elinas, P., Little, J.J.: sMCL: Monte-Carlo Localization for Mobile Robots with Stereo Vision. In: Proceedings of Robotics: Science and Systems, pp. 373–380 (2005)
5. Rofer, T., Jungel, M.: Vision-Based Fast and Reactive Monte-Carlo Localization. In: IEEE International Conference on Robotics and Automation, pp. 856–861 (2003)
6. Wikipedia contributors, Kinect, Wikipedia: The Free Encyclopedia. Wikimedia Foundation (2004), `http://en.wikipedia.org/wiki/Kinect`

7. Zhao, X., Li, X., Tan, T.: A Novel Landmark Tree Based Self-Localization and Path-Planning Method for an Intelligent Wheelchair. In: Proceedings of the 9th IEEE International Workshop on Robot and Human Interactive Communication, pp. 84–89 (2000)
8. Yaqub, T., Tordon, M.J., Katupitiya, J.: A Procedure to Make the Probabilistic Odometry Motion Model of an Autonomous Wheelchair. In: Proceedings of the 2006 IEEE International Conference on Mechatronics and Automation, pp. 526–531 (2006)

Succinct Landmark Database

Kanji Tanaka*

Dept. of Human & Artificial Intelligent Systems, Univ. of Fukui,
3-9-1, Bunkyo, Fukui 910-8507, Japan
tnkknj@u-fukui.ac.jp

Abstract. Recently developed robotic mapping techniques enable the acquisition of large scale landmark databases. This paper explores an approach for succinct landmark database, which memorizes a large collection of point landmarks while allowing to random access the location of i-th landmark. Our approach combines and extends three independent compression techniques: space coding, succinct data structure, and exemplar-based scene compression. Experiments using real datasets evaluate effectiveness of the presented techniques in terms of compactness, access speed, and accuracy of landmark database.

Keywords: landmark database, space coding, succinct data structure, exemplar-based scene compression.

1 Introduction

Landmark database plays an important role in robotic mapping and localization applications [1]-[3]. A landmark database aims to memorize a large collection of locations of point landmarks w.r.t. an environment map, while allowing to random access the location of i-th landmark. Classical data structures such as array of landmark locations allow a modern PC to memorize and random access tens of millions of landmarks. However, it is impossible to scale up to larger collection of landmarks, which is becoming available with recently developed robotic mapping techniques [4]. The goal of our research is to develop *succinct representation* of landmark database, that is, to compress a given landmark database while preserving its random accessibility.

The problem of succinct landmark database might be partially similar to the classical problem of point cloud compression in the field of point-based geometry. Many efforts have focused on compact 3D representation of a given point cloud while preserving details, by employing various types of model-based representations (e.g. meshes [5], depth maps [6], primitives [7], grammars [8], patches [9], hybrids [10]) as well as point-based representations (e.g. spatial subdivision [11], predictive coding [12]), while unfortunately, most of the existing schemes do not preserve the random accessibility [13]. Model-based representations focus on simplification and approximation of surfaces while preserving details, and do not aim to recover the original point cloud. Point-based representations do not support random access to the decompressed points. Predictive coding [12], where points are predicted from previously coded neighbors,

* Corresponding author.

S. Lee et al. (Eds.): Intelligent Autonomous Systems 12, AISC 193, pp. 165–172.
springerlink.com

always relies on previous points in the encoded sequence. Spatial subdivision (e.g. octree)[11], where connectivity information is eliminated for compression, also needs to decode the entire point set.

Our approach combines and extends three independent compression techniques: **space coding, succinct data structure** and **exemplar-based scene compression**. The following subsections briefly review individual techniques, and describe our approach.

1.1 Space Coding

Space coding methods aim at linearization of higher dimensional (e.g. 3D) spaces using **space-filling curves** that have several desirable properties [14], (1) they pass through each point in the space once and only once, (2) two points that are neighbors in space are neighbors along the curve and vice versa, and (3) easy to retrieve neighbor of a point. In particular, we are interested in the **Morton curve**, which uses bit interleaving of binary representation of the individual i-th coordinates of the point. Unlike other curves such as Peano-Hilbert curve, Canto curve as well as Spiral curve, the Morton curve has all of the following desirable properties, (4) it supports **easy converting** between the 3D point locations and the curves, (5) it supports **stable** space ordering i.e. the relative order of the individual 3D locations is maintained when the resolution is doubled or halved, (6) it guarantees **admissible** curves, i.e. at every point on the curve, at least one horizontal and one vertical neighbor of each point have already been encountered.

It is reasonable to assume that difference-codes of Morton code are small integers and compress well by **Gamma codes**. Such an encoding strategy (**Morton+difference+ Gamma**) has been also explored and verified by [15] in the application of nearest-neighbor-graph construction.

1.2 Succinct Data Structure

We now address succinct representation of the Gamma codes. A well-known difficulty in variable length coding (e.g. Gamma code) is that it is not possible to access directly the i-th encoded element, since its position in the encoded sequence depends on the sum of the lengths of the previous codewords. A naive solution to random access a variable-length encoded sequence is to regularly **sample** it and store the positions of the samples in the encoded sequence, which introduces space and time penalty relying on the sequence's length. In contrast, succinct data structure aims to save space cost close to information-theoretic lower bound while allowing random access.

In particular, we adapt **directly addressable variable length codes**, a succinct data structure, recently developed in [16], which consumes only exactly the same space as the original encoded sequence, and also supports fast random access as well as partial sums whose time costs are independent from length of the sequence.

1.3 Exemplar-Based Scene Compression

We further consider exemplar-based compression of the pointset [17]. Its basic idea is to exploit exemplar patterns of local point configuration, such as "walls" and "trees", which should repetitively appear in typical natural scenes as well as in man-made environments. The technique analyzes and detects repetitive patterns in the input pointset, and replaces them with reference to a pattern dictionary. The pattern dictionary is a set of exemplar patterns, and could be learned on-the-fly in the input point cloud, or given in advance in the case of general purpose compressor.

In our approach, we extend the exemplar-based compression scheme so as to allow random access. First, points are reordered in the acsending order of exemplar IDs they belong, in order to eliminate the information of exemplar IDs. Second, points belonging to each exemplar pattern are further reordered and encoded into a space filling code (denoted as **local pointID**), so as to allow random access. In the above approach, overheads introduced by exemplar patterns are composed of the dictionary as well as parameters (e.g. size, location, shape) of each exemplar pattern. Points belonging to exemplar patterns compress well, since total length of their space filling curves should be smaller than that of the input pointset. Other points that do not belong to any examplar pattern are directly Morton-difference-Gamma encoded. Best compression ratio is achieved when all the input points belong to exemplar patterns, and the overheads are practically small.

1.4 Contributions

This paper proposes a succinct representation of landmark database. We describe two different schemes in order to facilitate various applications. The first scheme aims to succinctly represent the landmark database in a lossless manner employing the techniques for space coding and succinct data structure. The second scheme aims at further compaction of the landmark database in a lossy manner combining the above two techniques with the exemplar-based compression technique. In experiments, we investigate compactness of either scheme as well as access speed and accuracy of the presented techniques using real datasets.

2 Approach

2.1 Succinct Data Structure

The compression problem addressed in this paper takes as input a sequence of $N+1$ point locations to be compressed $p_0, \cdots, p_N$, and outputs a succinct representation $S(i)$ of the landmark locations and a sequence of pointIDs (denoted as **global pointIDs**) $i_0, \cdots, i_N$ of the original point locations, so that $S(i)$ returns location of landmark with global pointID i.

As aforementioned, our approach employs the Morton-difference-Gamma encoding scheme. In detail, the input points $p_0, \cdots, p_N$ are encoded into Morton codes, and reordered in Morton order $\alpha_0, \cdots, \alpha_N$. Then, the Morton code sequence is encoded by difference coding $\Delta\alpha_i = \alpha_i - \alpha_{i-1}$ into a length N sequence $\Delta\alpha_1, \cdots, \Delta\alpha_N$. Then, the difference code sequence is encoded into a sequence of Gamma codes $\gamma_1, \cdots, \gamma_N$.

Our scheme for succinct data structure adapts the reordering technique presented in [16] so as to address the Gamma code case. It aims to reorder codes such that it is possible to random access j-th codeword x_j without need of any sampling method. First, it separates the original Gamma code sequence into a pair of sequences of chunks $U_r, \cdots, U_1$ and $B_r, \cdots, B_1$. Each stream U_i (or B_i) contains the i-th significant chunks of the unary part (or the binary part) of a Gamma code. The decoding algorithm recovers at each i-th step, i-th bit of each codeword at once. Since i-th bit of the unary part of a Gamma code indicates whether the codeword terminates at that bit or not, it can identify the number of codewords that have not been terminated yet as well as the concatenation of such non-terminated codewords in an incremental manner. The resulted data structure supports random access to the original Gamma code as well as partial sum $X[j] = x_1 + \cdots + x_j$ for any given i with a constant cost $O(1)$ independent from the length N of original sequence.

2.2 Exemplar-Based Scene Compression

The basic idea of exemplar-based scene compression is to replace a part of input pointset with compact reference to a pattern dictionary. The pattern dictionary is a set of exemplar patterns of local point configuration, each of which is a parametric or non-parametric representation of a pointset, and each point is assigned with a unique local pointID.

The encoding procedure aims to obtain a succinct data structure of a given input pointset (Fig.1a), and proceeds in the following steps:

1. Split points into those which belong to one of exemplar patterns $P_1, \cdots, P_M$ (Fig.1b) or those which do not belong to any exemplar pattern P_0 (Fig.1c).
2. For each exemplar pattern $P_i (i \in [1, M])$, encode each point p with local pointID $j (\in [1, |P_i|])$ by associating it with two-dimensional index value $k = (i, j)$.
3. For the non-exemplar pattern P_0, encode each point p with 3D Morton code j by associating it with two-dimensional index value $k = (0, j)$.

In experiments, we will focus on typical **quasi Manhattan world** environments with **2D extrusion models** [18], mainly composed of planes perpendicular to the floor plane, and employ plane-type exemplar of landmark configuration. A set of plane exemplars are extracted by first projecting the input points onto the floor plane, a set of 2D line segments are extracted from 2D projected points by **Hough transform** algorithm and then those line segments whose score exceeds a preset threshold $C_{score} = 10$ as well as the floor plane are regarded as (2D projections of) exemplar planes, denoted as **DB1**. Given such exemplar planes $P_1, \cdots, P_M$, each point in the input pointset is classified into an exemplar $P_i (i \in [1, M])$ whose distance from the point is smaller than a preset threshold $C_{dist} = 1.0$m if there exists such an exemplar P_i, or the non-exemplar P_0 otherwise. We approximate each point belonging to an exemplar $P_i (i \in [1, M])$ by the perpendicular foot q from the point to the plane, and assign the 2D Morton code j of the 2D location q as the local pointID. Note that 2D Morton code given such a support plane approximates the 3D location with a spatially more efficient manner than the original 3D Morton code.

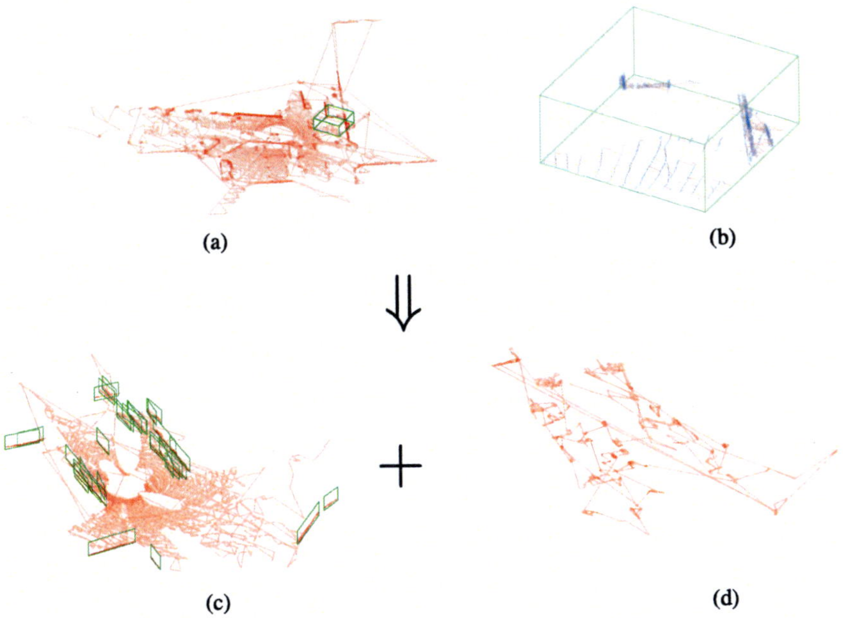

Fig. 1. Overview of our approach. Left: Succinct data structure, where the input pointset is directly encoded and reordered by employing the Morton-difference-Gamma coding scheme, as well as the directly addressable variable length code. Right: Exemplar-based scene compression, where the input pointset is firstly split into those which belong to exemplar patterns and the rest, and then the resulted pointsets are individually encoded and reordered. (a) Input pointset. (b) A close-up. (c) Points belonging to the floor exemplar as well as wall exemplars. (d) Points belonging to non-exemplar patterns.

The exemplars are also encoded with the space coding as well as the succinct data structure. First, points belonging to each cluster are reordered in the ascending order of local pointID, and then encoded by difference-Gamma code as well as the directly addressable variable length code. Second, all the code sequences are concatenated in the ascending order of clusterID, denoted as **DB2**. Note that length of the resulted sequence exactly equals size of the original pointset. We assign global pointID k to k-th point in the concatenated sequence. The information of clusterID is eliminated from each point, and instead, size of each cluster is memorized using the succinct data structure, denoted as **DB3**.

The decoding procedure aims to obtain the location of a point with a given global pointID k, time-space efficiently. It proceeds in the following steps. First, clusterID j of the point is binary searched using the global pointID k as well as the information of cluster size **DB3**. Second, the cluster's 1st point's global pointID k_j is computed using the partial sum operation $X[j]$ over **DB3**. Third, the point's localID k' is computed using the partial sum operations $k' = X[k] - X[k_j]$ over **DB2**, decoded as 2D Morton (if $j > 0$) or 3D Morton (if $j = 0$) code, and then converted to the 3D location using the information of exemplar planes **DB1**, as well as non-exemplar patterns **DB0**.

Obviously, the compression schemes trade time (and accuracy) for space, when compared with the original uncompressed representation. In experiments, we will investigate the tradeoffs using real datasets.

3 Experiments

We evaluate effectiveness of the presented techniques, in terms of compression ratio, access speed, as well as accuracy. Nine pointsets shown in Fig.2 are employed as benchmark dataset. Each pointset is an unorganized pointcloud with up to 70K points acquired by a 3D laser scanner, although we never use the knowledge of sensor's viewpoint for the compression purpose. All experiments are performed on 2.4GHz Intel CPU and 2GB of RAM. The sensor resolution is assumed to be 0.1m. 21bit Morton code per point (+ an offset Morton code) suffice for that resolution. Fig.3 reports size of the original pointset ("SC"), size of the representations presented in 2.1 and in 2.2

Fig. 2. Input pointsets

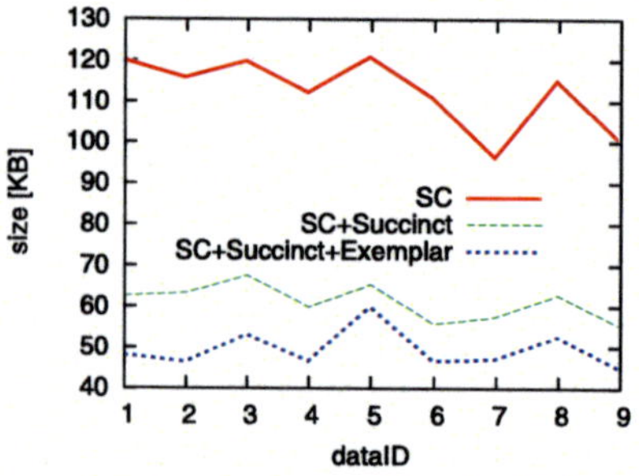

Fig. 3. Datasize

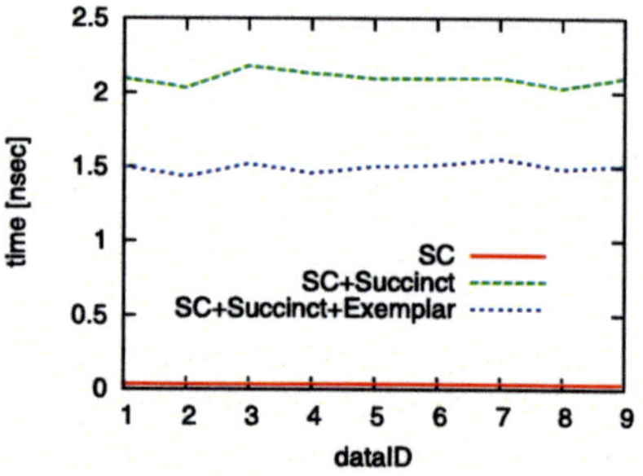

Fig. 4. Access speed

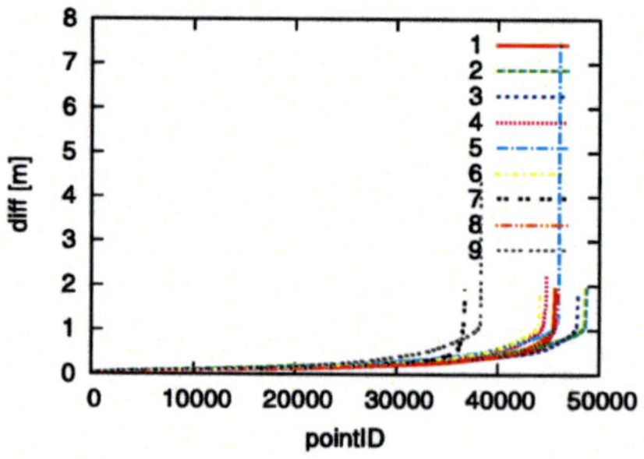

Fig. 5. Accuracy

("SC+succinct" and "SC+succinct+Exemplar"), respectively. For fair comparison, the original pointset is fixed length encoded (i.e. 21bits) and stored in an array. One can see that, the variable length coding techniques ("SC+succinct", "SC+succinct+Exemplar") clearly outperform the naive fixed length coding scheme. "SC+succinct+Exemplar" saves around 10KB-20KB when compared with "SC+succinct". Fig.4 shows average time cost for random access per point. One can see that "SC" beats the other two, and "SC+succinct+Exemplar" is slightly better than that of "SC+succinct". Fig.5 reports approximation error of the "SC+succinct+Exemplar". One can see that around half of points have non-negligible approximation error. This is due to that the exemplar-based compression scheme approximates 3D points with their 2D projection onto exemplar planes.

4 Conclusions and Discussions

The primary contribution of the paper is proposal of succinct representation for landmark database. Our approach combines and extends three independent compression techniques: space coding, succinct data structure, and exemplar-based scene compression. The succinct data structure reorders the sequence in a lossless manner so as to allow random access, which is an essential requirement for landmark database. The exemplar-based compression further compresses the landmark database while preserving the random accessibility. Experiments using real dataset have shown effectiveness of the presented techniques in terms of compactness, access speed, and accuracy.

The tradeoff between accuracy and space of our approach depends on how well the exemplar patterns fit input points. From the viewpoint of accuracy, better fitting between points and exemplars naturally leads to lower approximation error. From the viewpoint of space, better compression ratio is expected when larger portion of the input pointset is compactly explained out by the exemplar patterns. To improve the exemplar-based compression performance is a main direction of our future research.

In this paper, we employed a simple, plane-type exemplar, which is represented by a 2D straight line segment on the floor plane. However, our approach is not restricted to 2D straight line segments, but also applicable to arbitrary shape 2D curved segments, and even general 2D point sequences. Furthermore, a main limitation of segment-based representations is that they essentially require $O(N)$ space for size N input pointset. In our previous studies, we have presented several schemes for compactly representing pointsets (**"landmark map"**) while preserving random accessibility, including **repetitiveness-based compression** [17], **grammar-based compression** [19], **incremental compression** [20], as well as compression using **geometric priors** [21]. In such **"map compression" frameworks**, the coding scheme presented in this paper would serve as an essential building block.

Acknowledgements. This work was partially supported by MECSST Grant in-Aid for Young Scientists (B) (23700229), by KURATA grants and by TATEISI Science And Technology Foundation.

References

[1] Angeli, A., Filliat, D., Doncieux, S., Meyer, J.A.: A fast and incremental method for loop-closure detection using bags of visual words. IEEE Transactions on Robotics, Special Issue on Visual SLAM (2008) (conditionally accpeted for publication)
[2] Cummins, M., Newman, P.: Accelerated appearance-only SLAM. In: Proc. IEEE International Conference on Robotics and Automation (2008)
[3] Williams, B., Klein, G., Reid, I.: Real-time slam relocalisation. In: Proc. IEEE 11th Int. Conf. Computer Vision, pp. 1–8 (2007)
[4] Ila, V., Ni, K., Dellaert, F., Carlson, J., Thorpe, C.E.: Subgraph-preconditioned conjugate gradients for large scale slam. In: Proc. IEEE/RSJ Int. Conf. Intelligent Robots and Systems, IROS (2010)
[5] Zhou, Q.-Y., Neumann, U.: 2.5d building modeling with topology control, pp. 2489–2496 (2011)
[6] Brown, M.Z., Burschka, D., Hager, G.D.: Advances in computational stereo. IEEE Transactions on Pattern Analysis and Machine Intelligence 25, 993–1008 (2003)
[7] Zhou, Q.-Y., Neumann, U.: A streaming framework for seamless building reconstruction from large-scale aerial lidar data, pp. 2759–2766 (2009)
[8] Wonka, P., Wimmer, M., Sillion, F., Ribarsky, W.: Instant architecture. ACM Transaction on Graphics 22(3), 669–677 (2003)
[9] Pfister, H., Zwicker, M., Van Baar, J., Gross, M.: Surfels: Surface elements as rendering primitives. In: Proceedings of the 27th Annual Conference on Computer Graphics and Interactive Techniques, pp. 335–342 (2000)
[10] Lafarge, F., Keriven, R., Brédif, M., Vu, H.: Hybrid multi-view reconstruction by Jump-Diffusion. In: 2010 IEEE Conference on Computer Vision and Pattern Recognition (CVPR), pp. 350–357 (2010)
[11] Schnabel, R., Klein, R.: Octree-based point-cloud compression (2006)
[12] Gumhold, S., Karni, Z., Isenburg, M., Seidel, H.-P.: Predictive point-cloud compression (2005)
[13] Botsch, M., Pajarola, R., (eds), Hubo, E., Mertens,T., Haber, T.: Self-similarity-based compression of point clouds, with application to ray tracing abstract (2007)
[14] Sagan, H.: Space-filling curves. Springer (1994)
[15] Hudson, B.: Succinct representation of well-spaced point clouds. CoRR, abs/0909.3137 (2009)
[16] Brisaboa, N.R., Ladra, S., Navarro, G.: Directly addressable variable-length codes, pp. 122–130 (2009)
[17] Tomoni, N., Kanji, T.: Dictionary-based map compression for sparse feature maps. In: Proc. IEEE Int. Conf. Robotics and Automation, ICRA (2011)
[18] Cham, T.J., Ciptadi, A., Tan, W.C., Pham, M.T., Chia, L.T.: Estimating camera pose from a single urban ground-view omnidirectional image and a 2d building outline map (2010)
[19] Kensuke, K., Kanji, T., Tomomi, N.: Grammar-based map compression using manhattan-world priors. In: Proc. IEEE Int. Conf. Robotics and Biomimetics, ROBIO (2011)
[20] Tomomi, N., KanjiAn, T.: incremental scheme for dictionary-based compressive slam. In: Proc. IEEE Int. Conf. Intelligent Robots and Systems, IROS (2011)
[21] Tomomi, N., Kanji, T.: Dictionary-based map compression using geometric priors. In: Proc. IEEE Int. Conf. Robotics and Biomimetics, ROBIO (2011)

Thermal 3D Mapping of Building Façades*

Dorit Borrmann, Jan Elseberg, and Andreas Nüchter

Jacobs University Bremen gGmbH, Campus Ring 1, 28759 Bremen
{d.borrmann,j.elseberg,a.nuechter}@jacobs-university.de

Abstract. Never before in history were humans as dependant on energy as we are today. But the natural ressources are limited and a waste of energy has drastic influences on the environment. In their Action Plan for Energy Efficiency [6] the European Commission estimates that the largest and cost-effective energy savings potential lies in residential (≈ 27%) and commercial (≈ 30%) buildings. To eliminate heat and air conditioning losses in buildings and factories heat and air leaks need to be localized and identified. Imagine the availability of a complete 3D model of every building that architects can use to analyze the heat insulation of buildings and to identify necessary modifications. In these 3D models temperature peaks are not only detectable but also their extent is visible. A robot equiped with a 3D laser scanner, a thermal camera, and a color camera constitutes the basis for our approach. The data from all three sensors and from different locations are joined into one high-precise 3D model that shows the heat distribution. This paper describes the setup of the hardware and the methods applied to create the 3D model, including the automatic co-calibration of the sensors. Challenges unique to the task of thermal mapping of outdoor environments are discussed.

1 Introduction

Recently a lot of work has been done to capture and reconstruct the world around us. Think of a technology that captures not only the precise 3D representation of buidlings but at the same time thermal information. These models enable architects and construction engineers to inspect existing buildings, to run simulations of heat and air flow, and to use the gained information to increase the energy efficiency of these buildings. Thermal imaging is state of the art in recording energy related issues, while terrestrial laser scanning has been used for years to create 3D models. The combination of these two yield a 3D model that contains precise temperature information including the dimensions of heat and air leaks. In [17] an evaluation on the impact of 3D models onto the user's ab ability to interpret thermal information concludes that a good representation of a 3D model can help the user, e.g., to locate heat sources or to detect thermal bridges. To solve the remaining problem of identifying heat sources an optical camera is

* This work was supported by SEE-ERA.NET project ThermalMapper under the project number ERA 14/01 [14]. We gratefully acknowledge the work of Diana Babiac who collected the photos and performed the image based modeling.

S. Lee et al. (Eds.): Intelligent Autonomous Systems 12, AISC 193, pp. 173–182.
springerlink.com

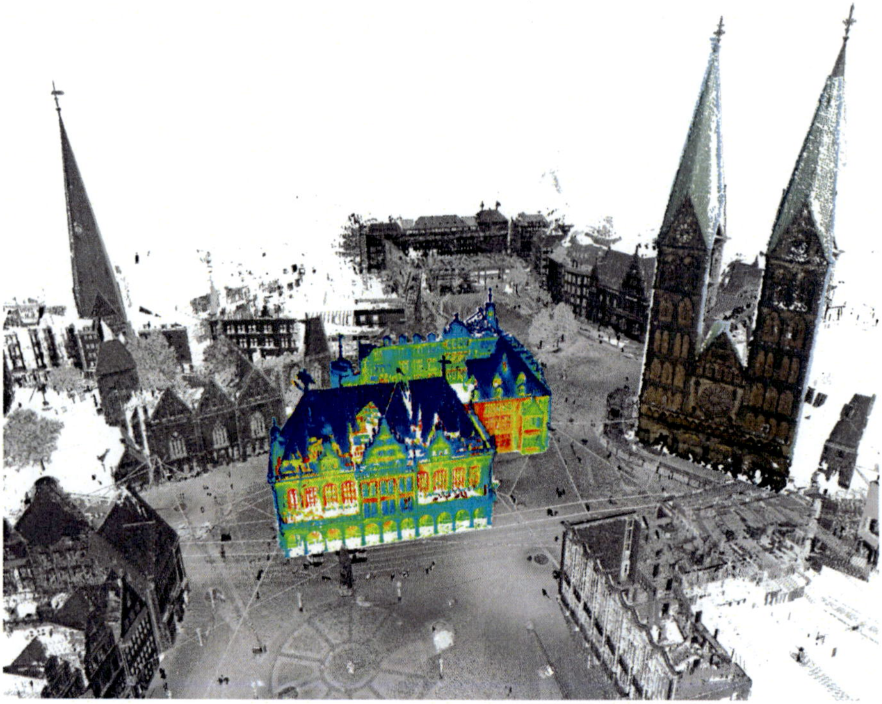

Fig. 1. Laser scan with reflectance, color and thermal information

added to our setup. Laser scan data acquired at different positions is combined into complete models of the environment by use of registration algorithms from the geodesy and robotics community. This paper presents our approach towards the creation of complete 3D thermal models of building façades. Further information including an extensive review of related work can be found on the project webpage [14].

2 Background and State of the Art

To assess the energy efficiency of houses thermal cameras are commonly used. These cameras measure temperatures precisely, but return only 2D images of the environment and therefore the loss of energy can only be roughly quantified. Images are projections to 2D. From a sequence of images it is in principle possible to perform a 3D reconstruction. These approaches are called bundle adjustment or structure from motion (SFM), if the focus lies on solving simultaneous localization and mapping (SLAM), i.e., on recovering the 3D structure of the environment and the sensors poses (position and orientation). Since reliable solutions to image based 3D reconstruction for thermal images have not been presented yet, we use the emerging technology of terrestrial laser scanning. Laser scanning methods are well established in the surveying community and in robotics. Terrestrial 3D laser scanning systems yield precise 3D point clouds. Scanning from different poses enables one to digitize complete indoor environments and to resolve occlusions. Registration algorithms from the geodesy and robotics community are available to automatically align scan views from different poses [4,16].

In related work Högner and Stilla present a modified van as surveying vehicle for acquiring thermal images in urban environments [8]. However, in the focus are image-based techniques like SFM. Prakash et al. present stereo imaging using thermal cameras, but focus on small scale applications [13]. Iwaszczuk et al. suggest an approach to map terrrestrial and airborne infrared images onto existing building models [9]. The model is textured by extracting polygonal parts from the image and mapping those onto the model using standardized masked correlation.

Only a little work has been done for combining 3D scanners and thermal cameras. Carbelles at al. present a methodology to exhaustively record data related to a World Heritage Monument using terrestrial laser scanning, close range photogrammetry and thermal imagery [5]. They use four different sensors for data acquisition: a reflectorless total station, a terrestrial laser rangescanning sensor, a digital photo camera and a thermal camera and use a total of eight natural control points with the help of the total station to relate the geometry between different sensors. In [12] a first automatic approach for multispectral texture mapping is presented. Their method is based on the extraction of a depth map in the form of an image from the model geometry whose pixels maintain an exact correspondence with the vertices of the 3D model. Afterwards the registration with the chosen texture is done which is based on maximization of mutual information. Alba et al. combine the data from a terrestrial laser scan with images acquired by a bi-camera system, i.e., a system consisting of an optical camera and a thermal camera [1]. The fusion of the two data sets is achieved by using some control points that are measured manually with both optical camera and laser scanner. 3D environment mapping using 3D scanners on mobile robots are subject of research [16,11]. Up to our knowledge automatic modeling using 3D scanning and thermal imaging has not been done yet.

3 Advanced Mutual Calibration between the 3D Sensor and the Cameras

3.1 Experimental Setup and Data Acquisition

The setup for simultaneous acquisition of 3D laser scan data, thermal, and optical images is the robot Irma3D (cf. Fig. 2). Irma3D is built of a Volksbot RT-3 chassis. Its main sensor is a Riegl VZ-400 laser scanner from terrestrial laser scanning. Two cameras are mounted on top of the scanner. The Logitech QuickCam Pro 9000 webcam has a video resolution of 1600×1200. The optris PI160 thermal camera has an image resolution of 160×120 pixels and a

Fig. 2. The robot Irma3D, equiped with a 3D laser scanner, a thermal camera and a webcam. In the background the calibration patterns for the optical camera (left) and the thermal camera (right) are displayed.

thermal resolution of 0.1° C. It acquires images at a frame rate of 120 Hz and with an accuracy of 2° C. The laser scanner acquires data with a field of view of $360° \times 100°$. To achieve the full horizontal field of view the scanner head rotates around the vertical scanner axis when acquiring the data. We take advantage of this feature when acquiring image data. Since the cameras are mounted on top of the scanner, they are also rotated. We acquire 9 images with each camera during one scanning process to cover the full 360°.

3.2 Data Processing Procedure

After acquiring the 3D data it has to be merged with the color information. This processing consists of five steps that will be explained in this section.

3.2.1 Intrinsic Calibration of Thermal and Optical Cameras

Each sensor perceives the world in its own local coordinate system. To join the perceived information we need the specific parameters of these coordinate systems. Each camera has unique parameters that define how a point (X, Y, Z) in world coordinates is projected onto the image plane. Given the focal length (f_x, f_y) of the camera and the camera center (c_x, c_y) image coordinates (x, y) are calculated as:

$$\begin{bmatrix} x \\ y \\ 1 \end{bmatrix} = \begin{bmatrix} f_x & 0 & c_x \\ 0 & f_y & c_y \\ 0 & 0 & 1 \end{bmatrix} \begin{bmatrix} X/Z \\ Y/Z \\ 1 \end{bmatrix}. \tag{1}$$

Given the radial distortion coefficients k_1, k_2, k_3 and the tangential distortion coefficients p_1, p_2 and $r = \sqrt{x^2 + y^2}$ the corrected image points (x_c, y_c) are calculated as

$$\begin{pmatrix} x_c \\ y_c \end{pmatrix} = \begin{pmatrix} x(1 + k_1 r^2 + k_2 r^4 + k_3 r^6) + 2p_1 y + p_2(r^2 + 2x^2) \\ y(1 + k_1 r^2 + k_2 r^4 + k_3 r^6) + p_1(r^2 + 2y^2) + 2p_2 x \end{pmatrix} \quad (2)$$

To determine the parameters of optical cameras chessboard patterns are commonly used because the corners are reliably detectable in the images. A number of images showing a chessboard pattern with known number and size of squares are recorded. In each image the internal corners of the pattern are detected and the known distance between those in world coordinates allows to formulate equations (1) and (2) as a non-linear least squares problem and solve for the calibration parameters.

[10] have explored the calibration procedure using different types of thermal cameras. Generally an object with a unique pattern having distinct targets is used which eases labeling and increases accuracy of the calibration process. The points are actively or passively heated. In case of passive heating different material causes the the pattern to show up. For low resolution thermal cameras a chessboard pattern is error-prone even after heating it with an infrared lamp. For pixels that cover the edge of the squares the temperature is averaged over the black and white parts thus blurring the edges. Instead a pattern with clearly defined heat sources such as small lightbulbs is suggested as it shows up nicely in thermal images. Fig. 2 shows our pattern in the background. It is composed of 30 tiny 12 Volt lamps, each with a glass-bulb diameter of 4 mm. The overall size of the board is 500 mm (width) × 570 mm (height). Identifying the heat sources in the image enables us to perform intrinsic calibration in the same way as for optical cameras. To detect the light bulbs in the thermal image a thresholding procedure is applied to create a binary image showing regions of high temperature. A further thresholding step discards effectively all regions that are too large or too small. If the remaining number of regions is equal to the number of lightbulbs in the pattern the regions are sorted according to the pattern. To calculate the exact center of the features with sub-pixel accuracy, the mean is calculated by weighing all the pixels in the region by their temperature.

3.2.2 Extrinsic Calibration – Cameras and Laser Scanner

After calculating the internal parameters of the cameras we need to align the camera images with the scanner coordinate system, i.e., extrinsic calibration. The three rotation and three translation parameters are known as the extrinsic camera parameters and they are unique to every camera view. Once all the points are in the camera coordinate system the projection to the image can be defined up to an factor s using equation [3]:

$$s \begin{bmatrix} x \\ y \\ 1 \end{bmatrix} = \begin{bmatrix} f_x & 0 & c_x \\ 0 & f_y & c_y \\ 0 & 0 & 1 \end{bmatrix} \begin{bmatrix} r_{11} & r_{12} & r_{13} & t_1 \\ r_{21} & r_{22} & r_{23} & t_2 \\ r_{31} & r_{32} & r_{33} & t_3 \end{bmatrix} \begin{bmatrix} X \\ Y \\ Z \\ 1 \end{bmatrix} \quad (3)$$

Suppose there are n images of the calibration pattern and m planar points on the pattern considering the distortions as independent and identically distributed noise than the maximum likelihood estimate of the transformation between the scanner and camera coordinate system is obtained by minimizing

$$\sum_{i=1}^{n}\sum_{j=1}^{m} ||\mathbf{p}_{ij} - \hat{\mathbf{p}}(\mathbf{A}, \mathbf{D}, \mathbf{R}_i, \mathbf{t}_i, \mathbf{P}_j)||^2 \quad (4)$$

where $\mathbf{A}$ is the intrinsic matrix, $\mathbf{R}_i$ the rotation matrix, $\mathbf{t}_i$ the translation vector, and $\mathbf{D}$ the distortion parameters. $\hat{\mathbf{p}}(\mathbf{A}, \mathbf{D}, \mathbf{R}_i, \mathbf{t}_i, \mathbf{P}_j)$ defines the projection of point $\mathbf{P}_j$ in image i, according to equation (3) and (2). This approach assumes that we have a number of points that are identifiable in both the laser scan and the image. For this purpose we attach the calibration pattern onto a board. For the optical camera this is a printed chessboard pattern and for the thermal camera light bulbs arranged in a regular grid pattern. The position of the points of these patterns are known. Algorithm 1 detects the points in a laser scan.

Algorithm 1. Calibration pattern detection in a laser scan

Require: point cloud, specification of calibration pattern
1: discard points outside the area of the expected board
2: find the most prominent plane using RANSAC
3: project a generated plane model into the center of the detected plane
4: use ICP to fit the plane model to the data points
5: **return** position of the lightbulbs according to ICP result

3.2.3 3D to 2D Projection and Color Mapping

During the data acquisition phase laser scans and images are acquired simultaneously. After determining the relations between scanner and cameras in the calibration step this relation is used directly to color the point cloud according to the images.

3.2.4 Projection/Occlusion/Resolution Errors

Due to the different sensor poses and fields of view the sensors perceive different parts of the world. An area that is visible for one sensor might be occluded for the other sensor. When mapping the color information to the point cloud this causes wrong correspondences and therefore faulty colored points. This impact is increased by the low resolution of the thermal camera. With only 120 by 160 pixels per image each pixel corresponds to many 3D points seen by the laser scanner leading to errors at jump edges. Consequently small calibration inaccuracies have a large impact on the results. To solve this problem we take advantage of the fact that if a point belongs to an object there will be more points on that object. We take all points that are are projected onto one pixel and its neighboring pixels. The points are clustered depending on their distance to the scanner. A heuristic based on distance to the 3D scanner and size of the cluster determines effectively which points are considered and enhanced with color information. This removes also some correct color information but the improvement prevails.

3.2.5 Scan Registration

Laser scans acquired at different positions are registered into one common coordinate system using 6D SLAM from *The 3D Toolkit* (3DTK) [2]. The complete model of the environment can be inspected in the viewer from 3DTK enhanced with either reflectance values, thermal data or color from photos (cf. Fig. 1). Switching between the different views enables the user to detect sources of wasted energy and to locate them clearly in the more realistic optical view.

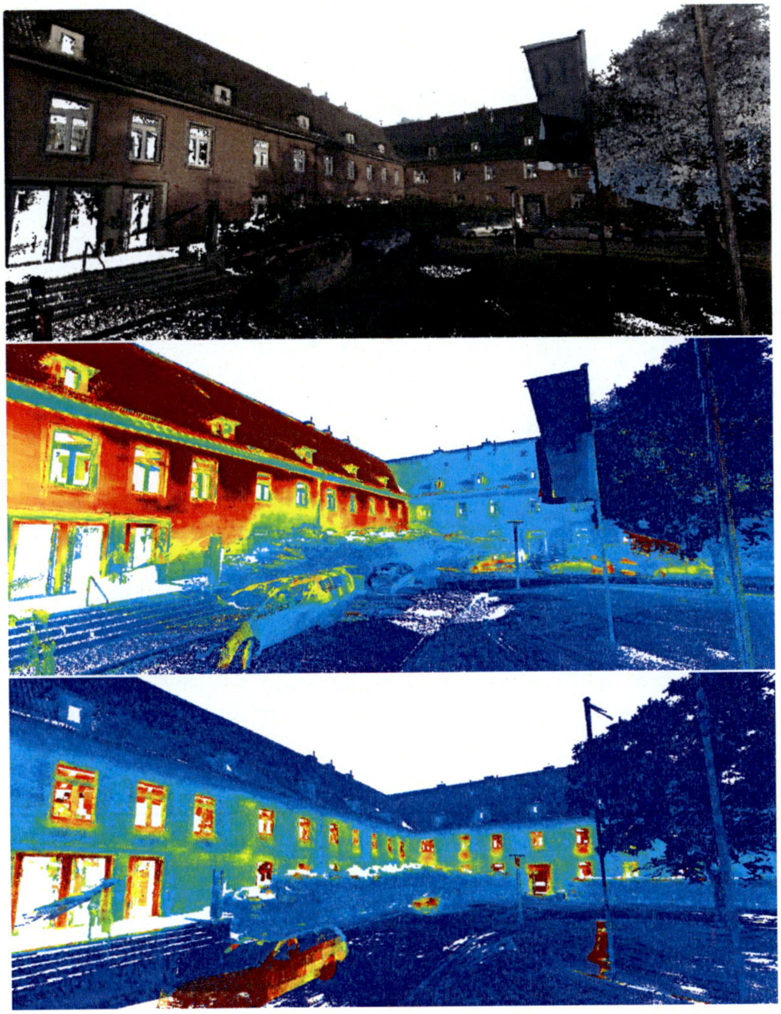

Fig. 3. Colored scans. Top: `Jacobs Day` with color information. Middle: `Jacobs Day` with thermal information. Bottom: `Jacobs Night` with thermal information.

4 Experimental Results

The algorithms were tested on three outdoor data sets each recorded with the robot Irma3D.

Data set `Jacobs Day` consists of laser scans, photos, and thermal images recorded at 13 different positions in the early afternoon. Fig. 3 shows part of the point cloud with color information (top) and thermal information (middle). It strikes immediately that the building to the left appears significantly warmer than the building in the back. Even diffuse sunlight on a cloudy day distorts the measurements in a way that a meaningful analysis becomes impossible.

Fig. 4. Irma3D in front of the Town Hall in Bremen while recording data at night (left). Resulting 3D thermal map of `Bremen` (right).

After the conclusions drawn from data set `Jacobs Day` a second data set `Jacobs Night` was recorded at night consisting of laser scans and thermal images from 15 different positions. The result is shown in Fig. 3 (bottom). It becomes clear that the correct temperature values allow for an analysis of the data.

The experiments on the campus of Jacobs University Bremen showed that reliable temperature measurements are not possible during the day due to the sunlight. However, by collecting data at night the advantage of the photo camera information is mostly lost. To overcome this issue we suggest to collect only data with the laser scanner and the thermal camera at night and combine them with the presented framework. During daylight an independent data set of only photos is recorded. Dataset `Bremen` consists of laser scans and thermal images recorded at night at 11 different positions. In addition to that 143 photos were taken independently during the day. Bundler [15] is used to reconstruct the environment from these photos. The Patch-based Multi-view Stereo Software (PMVS) [7] is used to produce a dense reconstruction of the bundler results. The photos cover only a part of the environment mapped with thermal information, namely the market place. As bundler does not preserve scale of the environment the resulting model was semi-automatically scaled and registered to the point cloud from the laser scanner. The resulting point cloud is shown in Fig. 1. The Town Hall in the center is enhanced with thermal information while the cathedral is part of the result from bundler. A video showing the entire scene with thermal information is available under `http://youtu.be/TPoCebERysc`.

5 Conclusions

We presented a method for multi-modal 3D mapping of building façades by use of a 3D laser scanner, a thermal camera, and a photo camera. We identified problems and suggested solutions distinct to outdoor scenarios. In future work we plan to further develop the suggested methods, i.e., include tree detection and develop automatic registration of data recorded at different times to obtain a multi-modal-spatio-temporal model of the environment.

References

1. Alba, M.I., Barazzetti, L., Scaioni, M., Rosina, E., Previtali, M.: Mapping infrared data on terrestrial laser scanning 3D models of buildings. Remote Sensing 3(9), 1847–1870 (2011)
2. Nüchter, A., et al.: 3DTK – The 3D Toolkit (2011), `http://slam6d.sourceforge.net/`
3. Bradski, G., Kaehler, A.: Learning OpenCV, Computer Vision with OpenCV library, 1st edn. O'Reilly Media (2008)
4. Brenner, C., Dold, C., Ripperda, N.: Coarse orientation of terrestrial laser scans in urban environments. ISPRS Journal of Photogrammetry and Remote Sensing 63(1), 4–18 (2008)
5. Cabrelles, M., Galcera, S., Navarro, S., Lerma, J.L., Akasheh, T., Haddad, N.: Integration of 3D laser scanning, photogrammetry and thermography to record architectural monuments. In: Proceedings of the 22nd CIPA Symposium, Kyoto, Japan (2009)
6. Commission of the European Communities. Addressing the challenge of energy efficiency through Information and Communication Technologies. Communications to the European Parliament, the Council, the European Economic and Social Committee and the Committee of the Regions, COM(2008) 241 final (May 2008)
7. Furukawa, Y., Ponce, J.: Accurate, dense, and robust multi-view stereopsis. IEEE PAMI 32(8), 1362–1376 (2010)
8. Högner, L., Stilla, U.: Texture extraction for building models from ir sequences of urban areas. In: Proc. of 2007 Joint Event: URBAN / URS (2007)
9. Iwaszczuk, D., Hoegner, L., Stilla, U.: Matching of 3D building models with IR images for texture extraction. In: Joint Urban Remote Sensing Event, Munich, Germany (2011)
10. Luhmann, T., Piechel, J., Ohm, J., Roelfs, T.: Geometric calibration of thermographic cameras. Intern. Archives of Photogrammetry, Remote Sensing and Spatial Information 38(pt. 5) (2010)
11. Nüchter, A.: 3D Robotic Mapping: The Simultaneous Localization and Mapping Problem with Six Degrees of Freedom. Tracts in Advanced Robotics, vol. 52. Springer (2009)
12. Pelagottia, A., Del Mastio, A., Uccheddu, F., Remondino, F.: Automated multispectral texture mapping of 3d models. In: EUSIPCO 2009, Glasgow, Scotland (2009)

13. Prakash, S., Pei, L.Y., Caelli, T.: 3d mapping of surface temperature using thermal stereo. In: Proc. ICARCV (2006)
14. Project Webpage. Project ThermalMapper (2011), http://www.faculty.jacobs-university.de/anuechter/thermalmapper.html
15. Snavely, N., Seitz, S.M., Szeliski, R.: Photo tourism: Exploring image collections in 3d. ACM Transactions on Graphics (Proceedings of SIGGRAPH 2006) (2006)
16. Surmann, H., Nüchter, A., Hertzberg, J.: An autonomous mobile robot with a 3D laser range finder for 3D exploration and digitalization of indoor environments. Journal Robotics and Autonomous Systems (JRAS) 45(3-4), 181–198 (2003)
17. Wardlaw, J., Gryka, M., Wanner, F., Brostow, G., Kautz, J.: A new approach to thermal imaging visualisation. EngD Group Project, University College London (2010)

Moving Object Detection Using Monocular Vision

Yin-Tien Wang, Kuo-Wei Chen, and Ming-Jang Chiou

Department of Mechanical and Electro-Mechanical Engineering, Tamkang University
ytwang@mail.tku.edu.tw, 699370218@s99.tku.edu.tw,
698371464@s98.tku.edu.tw

Abstract. This paper presents an algorithm for moving object detection (MOD) in robot visual simultaneous localization and mapping (SLAM). The algorithm is designed based on the defining epipolar constraint for the corresponding feature points on image plane. An essential matrix obtained using the state estimator is utilized to represent the epipolar constraint. Meanwhile, the method of speeded-up robust feature (SURF) is employed in the algorithm to provide a robust detection for image features as well as a better description of landmarks and of moving objects in visual SLAM system. Experiment is carried out on a hand-held monocular camera to validate the performances of the proposed algorithm. The results show that the integration of MOD and SURF is efficient for robot navigating in dynamic environments.

Keywords: Moving object detection, simultaneous localization and mapping, speeded-up robust features, monocular vision.

1 Introduction

In recent years, more and more researchers solve the simultaneous localization and mapping (SLAM) as well as the moving object Tracking (MOT) problems concurrently. Wang et al. [1] developed a consistency-based moving object detector and provided a framework to solve the SLAMMOT problems. Bibby and Reid [2] proposed a method that combines sliding window optimization and least-squares together with expectation maximization to do reversible model selection and data association that allows dynamic objects to be included directly into the SLAM estimation. Zhao et al. [3] used GPS data and control inputs to achieve global consistency in dynamic environments. There are many advantages to cope with SLAM and MOT problems simultaneously: for example, mobile robots might navigate in a dynamic environment crowded with moving objects. In this case the SLAM could be corrupted with the inclusion of moving entities if the information of moving objects is not taken account. Furthermore, the robustness of robot localization and mapping algorithms can be improved if the moving objects are discriminated from the stationary objects in the environment.

Using cameras to implement SLAM is the current trend because of their light weight and low-cost features, as well as containing rich appearance and texture

S. Lee et al. (Eds.): Intelligent Autonomous Systems 12, AISC 193, pp. 183–192.
springerlink.com

information of the surroundings. However, it is still a difficult problem in visual SLAM to discriminate the moving objects from the stationary landmarks in dynamic environments. To deal with this problem, we propose the moving object detection (MOD) algorithm based on the epipolar constraint for the corresponding feature points on image plane. Given an estimated essential matrix it is possible to investigate whether a set of corresponding image points satisfy the defining epipolar constraint in image plane. Therefore, the epipolar constraint can be utilized to distinguish the moving objects from the stationary landmarks in dynamic environments.

For visual SLAM systems, the features in the environment are detected and extracted by analyzing the image taken by the robot vision, and then the data association between the extracted features and the landmarks in the map is investigated. Many researchers [4-5] employed the concept by Harris and Stephens [6] to extract apparent corner features from one image and tracked these point features in the consecutive image. The descriptors of the Harris corner features are rectangle image patches. When the camera translates and rotates, the scale and orientation of the image patches will be changed. The detection and matching of Harris corner might fail in this case, unless the variances in scale and orientation of the image patches are recovered. Instead of detecting corner features, some works [7-8] detect the features by using the scale-invariant feature transform (SIFT) method [9] which provides a robust image feature detector. The unique properties of image features extracted by SIFT method are further described by using a high-dimensional description vector [9]. However, the feature extraction by SIFT requires more computational cost than that by Harris's method [6]. To improve the computational speed, Bay et al. [10] introduced the concept of integral images and box filter to detect and extract the scale-invariant features, which they dubbed Speeded-Up Robust Features (SURF). The extracted SURF must be matched with the landmarks in the map of a SLAM system. The nearest-neighbor (NN) searching method [11] can be utilized to match high-dimensional data sets of description vectors.

In the paper, an online SLAM system with a moving object detector is developed based on the epipolar constraint for the corresponding feature points on image plane. The corresponding image features are obtained using the SURF method [10] and the epipolar constraint is calculated using an estimated essential matrix. Moving object information is detected in image plane and integrated into the MOT process such that the robustness of SLAM algorithm can be considerably improved, particularly in highly dynamic environments where surroundings of robots are dominated by non-stationary objects. The contributions in this paper are two-fold. First, we develop an algorithm to solve the problems for moving object detection in image plane, and then the algorithm is integrated with the robot SLAM to improve the robustness of state estimation and mapping processes. Second, the improved SLAM system is implemented on a hand-held monocular camera which can be utilized as the sensor system for robot navigation in dynamic environments.

2 SLAM with a Free-Moving Monocular Vision

SLAM is a target tracking problem for the robot system during navigating in the environment [12]. The targets to be tracked include the state of the robot itself as well as of the landmarks and moving objects in the environment. The state sequence of the SLAM system at time step k can be expressed as

$$\mathbf{x}_k = f(\mathbf{x}_{k-1}, \mathbf{u}_{k-1}, w_{k-1}) \tag{1}$$

where $\mathbf{x}_k$ is the state vector; $\mathbf{u}_k$ is the input; w_k is the process noise. The objective of the tracking problem is to recursively estimate the state $\mathbf{x}_k$ of the target according to the measurement $\mathbf{z}_k$ at time step k,

$$\mathbf{z}_k = g(\mathbf{x}_k, v_k) \tag{2}$$

where v_k is the measurement noise. A hand-held monocular vision, as shown in Fig. 1, is utilized in this paper as the only sensing device for the measurement in SLAM system. We treat this hand-held vision sensor as a free-moving robot system with unknown inputs. The states of the system are estimated by solving the recursive SLAM problem using the extended Kalman filter (EKF) [12],

$$\mathbf{x}_{k|k-1} = f(\mathbf{x}_{k-1|k-1}, \mathbf{u}_{k-1}, 0) \tag{3a}$$

$$\boldsymbol{P}_{k|k-1} = \boldsymbol{A}_k \boldsymbol{P}_{k-1|k-1} \boldsymbol{A}_k^T + \boldsymbol{W}_k \boldsymbol{Q}_{k-1} \boldsymbol{W}_k^T \tag{3b}$$

$$\boldsymbol{K}_k = \boldsymbol{P}_{k|k-1} \boldsymbol{H}_k^T (\boldsymbol{H}_k \boldsymbol{P}_{k|k-1} \boldsymbol{H}_k^T + \boldsymbol{V}_k \boldsymbol{R}_k \boldsymbol{V}_k^T)^{-1} \tag{3c}$$

$$\mathbf{x}_{k|k} = \mathbf{x}_{k|k-1} + \boldsymbol{K}_k (\mathbf{z}_k - g(\mathbf{x}_{k|k-1}, 0)) \tag{3d}$$

$$\boldsymbol{P}_{k|k} = (\boldsymbol{I} - \boldsymbol{K}_k \boldsymbol{H}_k) \boldsymbol{P}_{k|k-1} \tag{3e}$$

where $\mathbf{x}_{k|k-1}$ and $\mathbf{x}_{k|k}$ represent the predicted and estimated state vectors, respectively; $\boldsymbol{K}_k$ is Kalman gain matrix; $\boldsymbol{P}$ denotes the covariance matrix, respectively; $\boldsymbol{A}_k$ and $\boldsymbol{W}_k$ are the Jacobian matrices of the state equation f with respect to the state vector $\mathbf{x}_k$ and the noise variable w_k, respectively; $\boldsymbol{H}_k$ and $\boldsymbol{V}_k$ are the Jacobian matrices of the measurement g with respect to the state vector $\mathbf{x}_k$ and the noise variable v_k, respectively.

2.1 Motion Model

Two coordinate systems are set at the world frame {W} and the camera frame {C}, as shown in Fig. 1. The state vector of the SLAM system with MOT in Eqn. (1) is arranged as:

$$\mathbf{x} = [\mathbf{x}_C \; \mathbf{m}_1 \; \mathbf{m}_2 \cdots \mathbf{m}_n \; \mathbf{O}_1 \; \mathbf{O}_2 \cdots \mathbf{O}_l]^T \tag{4}$$

$\mathbf{x}_C$ is a 12×1 state vector of the camera including the three-dimensional vectors of position $\boldsymbol{r}$, rotational angle ϕ, linear velocity $\mathbf{v}$, and angular velocity $\boldsymbol{\omega}$, all in world frame; $\mathbf{m}_i$ is the three-dimensional (3D) coordinates of ith stationary landmark in world frame; $\mathbf{O}_j$ is the state vector of jth moving object; n and l are the number of the landmarks and of the moving objects, respectively. The motion of the hand-held camera is presumed to be at constant velocity (CV), and the acceleration is caused by an impulse noise from the external force [4]. The state of ith stationary landmark at time step k is represented by 3D coordinates in space,

$$\mathbf{m}_{ik} = [X_{ik} \quad Y_{ik} \quad Z_{ik}]^T \tag{5}$$

where i is ith landmark, i=1,2, …,n. In the motion model of MOT, the targets to be tracked include the state and motion mode of the moving object in the environment. The state of the MOT system at time step k can be expressed as

$$\mathbf{O}_{jk}=[\boldsymbol{o}_{jk}\ \boldsymbol{s}_{jk}]^T, \qquad \text{for } j=1,2,\ldots,l \tag{6}$$

where $\boldsymbol{o}_{jk}$ and $\boldsymbol{s}_{jk}$ are the state and motion mode of jth moving object, respectively. The MOT problem can be expressed as a probability density function (pdf) in Bayesian probability

$$p(\boldsymbol{o}_k, \boldsymbol{s}_k|\mathbf{z}_{1:k}) = p(\boldsymbol{o}_k|\boldsymbol{s}_k,\ \mathbf{z}_{1:k}) \cdot p(\boldsymbol{s}_k|\mathbf{z}_{1:k}) \tag{7}$$

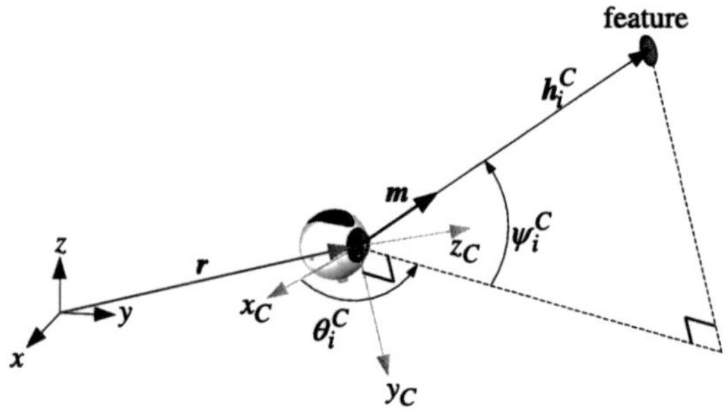

Fig. 1. Coordinate setting for the camera system

where $p(\boldsymbol{o}_k|\boldsymbol{s}_k,\mathbf{z}_{1:k})$ is state inference; $\mathbf{z}_{1:k}$ is the set of measurements for time t=1 to k; $p(\boldsymbol{s}_k|\mathbf{z}_{1:k})$ is the mode learning. The EKF-based interacting multiple model (IMM) estimator [13] can be utilized to estimate the motion mode of a moving object. The state is computed at time step k under each possible current model using r filters, with each filter using a different combination of the previous model-conditioned estimates. Given r motion models, the object state $\boldsymbol{o}_k$ in Eqn. (7) is estimated. Instead of using the IMM estimator, a single CV model is utilized in the paper. That is, the moving objects are also presumed to move at a CV motion model. Their coordinates in 3D space are defined as

$$\boldsymbol{o}_{jk} = \begin{bmatrix} \boldsymbol{p}_{jk-1} + (\mathbf{v}_{jk-1} + \boldsymbol{w}_{jk-1})\Delta t \\ \mathbf{v}_{jk-1} + \boldsymbol{w}_{jk-1} \end{bmatrix} \tag{8}$$

where p_{jk} and $\mathbf{v}_{jk}$ are the vectors of the position and linear velocity of jth moving object at time step k, respectively.

2.2 Vision Sensor Model

The measurement vector of the monocular vision system is expressed as

$$\mathbf{z}_k = [\mathbf{z}_{1k}\ \mathbf{z}_{2k} \cdots \mathbf{z}_{mk}]^T \tag{9}$$

m is the number of the observed image features in current measurement. The perspective projection method [14] is employed to model the transformation from 3D space coordinate system to 2D image plane. The 3D coordinates of ith image feature or landmark in world frame, as shown in Fig. 1, is given as

$$\mathbf{m}_i = [X_i \quad Y_i \quad Z_i]^T = \boldsymbol{r} + \boldsymbol{R}_C^W \boldsymbol{h}_i^C \tag{10}$$

where i indicates ith observed image feature, i=1,2, ...,m; $\boldsymbol{R}_C^W$ is the rotational matrix from world frame to camera frame; $\boldsymbol{h}_i^C = [h_{ix}^C \quad h_{iy}^C \quad h_{iz}^C]^T$ is the ray vector of ith image feature in camera frame.

2.3 Speeded Up Robust Features (SURF)

The basic concept of a scale-invariant method is to detect image features by investigating the determinant of Hessian matrix $\boldsymbol{H}$ in scale space [15]. In order to speed up the detection of image features, Bay et al. [10] utilize integral images and box filters to process on the image instead of calculating the Hessian matrix. The interest points or features are extracted by examining the extreme value of determinant of Hessian matrix. Furthermore, the unique properties of the extracted SURF are described by using a 64-dimensional description vector [9].

Because of the lack of one-dimensional range information in image, how to initialize features becomes an important topic. Some researchers have successfully solved this problem either in time-delayed method [16] or un-delayed method [17]. The undelayed method will be utilized in this research. When an image feature is selected, the spatial coordinates of the image feature are calculated by employing the method of inverse depth parameterization [17].

3 Moving Object Detection and Tracking

The algorithm of moving object detection (MOD) is designed based on the concept of the pixel coordinate constraint of a static object in image plane. An object in space is represented by the corresponding features in two consecutive images. The pixel coordinate constraint equation for these corresponding image features can be expressed as

$$\boldsymbol{h}_d^{C^T} \boldsymbol{E} \boldsymbol{h}_d'^{C} = 0 \tag{11}$$

where $\boldsymbol{h}_d^C$ and $\boldsymbol{h}_d'^C$ are the homogenous normalized image coordinates of the corresponding features abstracted from two consecutive images, image 1 and 2, respectively. They are defined as

$$\boldsymbol{h}_d^C = \boldsymbol{K}_C^{-1} \begin{bmatrix} I_x \\ I_y \\ 1 \end{bmatrix}; \ \boldsymbol{h}_d'^C = \boldsymbol{K}_C^{-1} \begin{bmatrix} I'_x \\ I'_y \\ 1 \end{bmatrix}; \ \boldsymbol{K}_C = \begin{bmatrix} f_u & \alpha_C f_u & u_0 \\ 0 & f_v & v_0 \\ 0 & 0 & 1 \end{bmatrix} \tag{12}$$

where f_u and f_v are the focal lengths of the camera denoting the distance from the camera center to the image plane in u- and v-axis, respectively; (u_0, v_0) is the offset pixel vector of the pixel image plane; α_C is the camera skew coefficient; $\boldsymbol{K}_C$ is the matrix of camera intrinsic parameters which can be obtained from the perspective projection [14]. Note that the image coordinates are defined in camera frame. The essential matrix $\boldsymbol{E}$ in Eqn. (11) is defined as [18]

$$\boldsymbol{E} = [\boldsymbol{t}]_\times \boldsymbol{R} = \begin{bmatrix} 0 & -t_z & t_y \\ t_z & 0 & -t_x \\ -t_y & t_x & 0 \end{bmatrix} \boldsymbol{R} \tag{13}$$

where $\boldsymbol{R}$ is the rotation matrix and $\boldsymbol{t}$ is the translation vector of the camera frame with respect to world frame; $[\boldsymbol{t}]_\times$ is the matrix representation of the cross product with $\boldsymbol{t}$. The rotation matrix and translation vector would be determined using EKF estimator. Therefore, the essential matrix $\boldsymbol{E}$ can be calculated accordingly. Usually, the pixel coordinate constraint in Eqn. (11) is utilized to estimate the state vector and according essential matrix. Given a set of corresponding image points, it is possible to estimate the state vector and the essential matrix which optimally satisfy the pixel coordinate constraint. The most straight-forward approach is to set up a total least squares problem, commonly known as the eight-point algorithm [19]. On the contrary in SLAM problem, the state vector and the essential matrix is obtained using the state estimator. We could further utilize the estimated state vector and the essential matrix to investigate whether a set of corresponding image points satisfy the pixel coordinate constraint in Eqn. (11). First, define the known quantities in Eqn. (11) as a constant vector

$$\begin{bmatrix} a & b & c \end{bmatrix} = \begin{bmatrix} I_x & I_y & 1 \end{bmatrix}^T (\boldsymbol{K}_C^{-1})^T \boldsymbol{E} \boldsymbol{K}_C^{-1} \tag{14}$$

Eqn. (14) is further rearranged as an equation of the epipolar line constraint in image plane for two corresponding points [20],

$$aI'_x + bI'_y + c = 0 \tag{15}$$

Eqn. (15) indicates that the pixel coordinate of the corresponding feature in second image will be constrained on the epipolar line. Motion and measurement noise is involved in the state estimation process. If the measurement of image points is subject to noise, which is the common case in any practical situation, it is not possible to find a corresponding feature point which satisfies the epipolar constraint exactly. That is, the corresponding feature point might not be constrained on the epipolar line because of the presence of noise. Define a distance *D* to represent the pixel deviation of the corresponding point from the epipolar line in image plane [21]. The pixel deviation *D* is induced by the motion and measurement noise in state estimation process. Depending on how the noise related to each constraint is measured, it is possible to design a threshold value which satisfies the epipolar constraint for a given set of corresponding image points.

4 Experimental Results

In this section, the experimental works of the online SLAM with a moving object detector are implemented on a laptop computer running Microsoft Window XP. The laptop computer is Asus U5F with Intel Core 2 Duo T5500 (1.66GHz), Mobile Intel i945GM chipset and 1Gb DDR2. The free-moving monocular camera utilized in this work is Logitech C120 CMOS web-cam with 320×240-pixels resolution and USB 2.0 interface. The camera is calibrated using the Matlab tool provided by Bouquet [22]. The focal lengths are f_u=364.4pixels and f_v=357.4pixels. The offset pixels are u_0=156.0pixels and v_0=112.1pixels, respectively. We carried out three experiments including the SLAM task in a static environment, SLAM with moving object tracking, and people detection and tracking.

In this experiment, the SLAM system with MOD is utilized to detect and track a moving people in the environment. The camera is carried to move around at a corner of our laboratory. The estimated states of the camera pose, the landmarks, and the moving people are illustrated in a 2D top-view plot in the right panel of Figs. 2-5. The captured image is shown in the left panel of each figure. As shown in Fig. 2 for the 350th image frame, the SLAM system builds the environment map and estimates the camera pose concurrently and stably. The ellipses in the right panel indicate the landmarks obtained from the extracted image features. The rectangular box represents the free-moving monocular camera and the thin solid line depicts the trajectory of the camera. One person gets in the scene, as shown in the 365th frame in Fig. 3, and two image features on the human body are detected. These image features are discriminated from the stationary features using the proposed MOD algorithm, and are further initialized in state vector and tracked using Eqn. (8). The SLAM system continuously tracks the moving people until he goes out the scene, as shown in Figs. 4-5. The trajectory of the moving people is depicted as a thick solid line in the right panel of each figure. Hence, the developed system also demonstrates the capability of simultaneous localization, mapping, and moving people tracking.

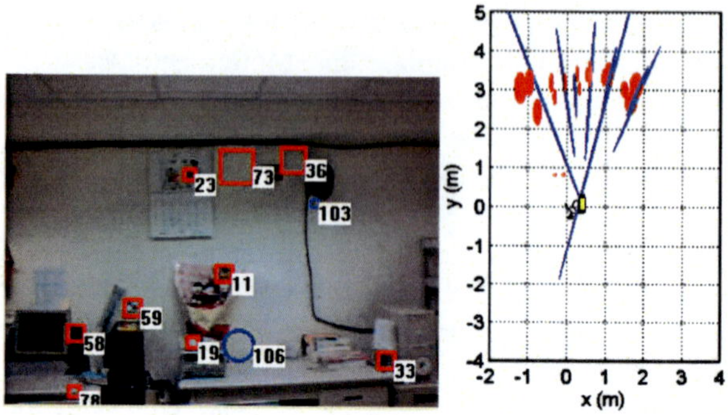

Fig. 2. 350th frame: the system is performing simultaneous localization and mapping

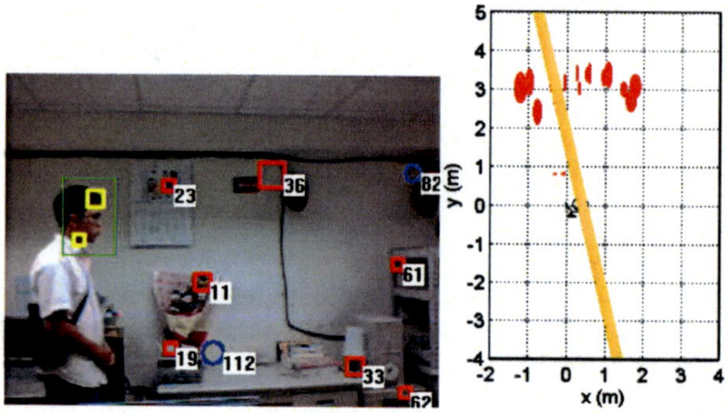

Fig. 3. 365th frame: moving objects are detected

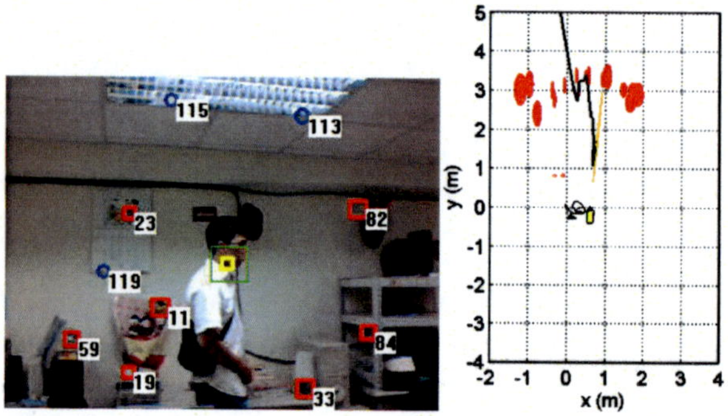

Fig. 4. 386th frame: the detected moving objects are tracked

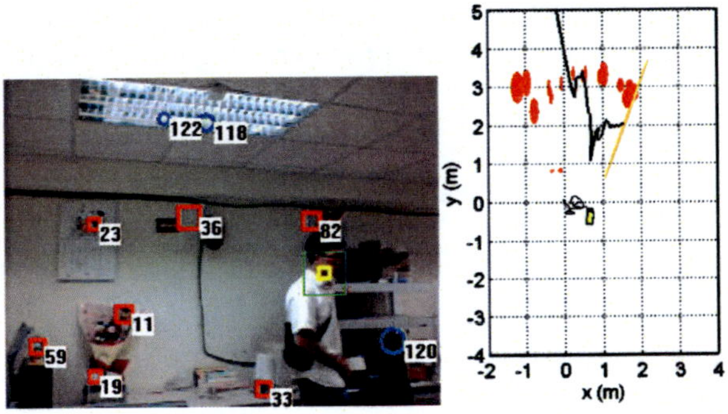

Fig. 5. 410th frame: moving object tracking

5 Conclusions

In this research, we developed an algorithm for detection and tracking of moving objects to improve the robustness of robot visual SLAM system. SURFs are also utilized to provide a robust detection of image features and a stable description of the features. Three experimental works have been carried out on a monocular vision system including SLAM in a static environment, SLAM with moving object tracking, and people detection and tracking. The results showed that the monocular SLAM system with the proposed algorithm has the capability to support robot systems simultaneously navigating and tracking moving objects in dynamic environments.

Acknowledgments. This paper was partially supported by the National Science Council in Taiwan under grant no. NSC100-2221-E-032-008 to Y.T. Wang.

References

1. Wang, C.C., Thorpe, C., Thrun, S., Hebert, M., Durrant-Whyte, H.: Simultaneous localization, mapping and moving object tracking. Int. J. Robot. Res. 26, 889–916 (2007)
2. Bibby, C., Reid, I.: Simultaneous Localisation and Mapping in Dynamic Environments (SLAMIDE) with Reversible Data Association. In: Proceedings of Robotics: Science and Systems III. Georgia Institute of Technology, Atlanta (2007)
3. Zhao, H., Chiba, M., Shibasaki, R., Shao, X., Cui, J., Zha, H.: SLAM in a Dynamic Large Outdoor Environment using a Laser Scanner. In: Proceedings of the IEEE International Conference on Robotics and Automation, Pasadena, California, pp. 1455–1462 (2008)
4. Davison, A.J., Reid, I.D., Molton, N.D., Stasse, O.: MonoSLAM: Real-time single camera SLAM. IEEE T. Pattern Anal. 29, 1052–1067 (2007)
5. Paz, L.M., Pinies, P., Tardos, J.D., Neira, J.: Large-Scale 6-DOF SLAM with Stereo-in-Hand. IEEE T. Robot. 24, 946–957 (2008)
6. Harris, C., Stephens, M.: A combined corner and edge detector. In: Proceedings of the 4th Alvey Vision Conference, pp. 147–151. University of Manchester (1988)

7. Karlsson, N., Bernardo, E.D., Ostrowski, J., Goncalves, L., Pirjanian, P., Munich, M.E.: The vSLAM Algorithm for Robust Localization and Mapping. In: Proceedings of the IEEE International Conference on Robotics and Automation, Barcelona, Spain, pp. 24–29 (2005)
8. Sim, R., Elinas, P., Little, J.J.: A Study of the Rao-Blackwellised Particle Filter for Efficient and Accurate Vision-Based SLAM. Int. J. Comput. Vision 74, 303–318 (2007)
9. Lowe, D.G.: Distinctive image features from scale-invariant keypoints. Int. J. Comput. Vision 60, 91–110 (2004)
10. Bay, H., Tuytelaars, T., Van Gool, L.: SURF: Speeded Up Robust Features. In: Leonardis, A., Bischof, H., Pinz, A. (eds.) ECCV 2006. LNCS, vol. 3951, pp. 404–417. Springer, Heidelberg (2006)
11. Shakhnarovich, G., Darrell, T., Indyk, P.: Nearest-Neighbor Methods in Learning and Vision. MIT Press, Cambridge (2005)
12. Smith, R., Self, M., Cheeseman, P.: Estimating Uncertain Spatial Relationships in Robotics. In: Cox, I.J., Wilfong, G.T. (eds.) Autonomous Robot Vehicles. Springer, New York (1990)
13. Blom, A.P., Bar-Shalom, Y.: The interacting multiple-model algorithm for systems with Markovian switching coefficients. IEEE T. Automat. Control 33, 780–783 (1988)
14. Hutchinson, S., Hager, G.D., Corke, P.I.: A tutorial on visual servo control. IEEE T. Robot. Automat. 12, 651–670 (1996)
15. Lindeberg, T.: Feature detection with automatic scale selection. Int. J. Comput. Vision 30, 79–116 (1998)
16. Wang, Y.T., Lin, M.C., Ju, R.C.: Visual SLAM and Moving Object Detection for a Small-size Humanoid Robot. Int. J. Adv. Robot. Syst. 7, 133–138 (2010)
17. Civera, J., Davison, A.J., Montiel, J.M.M.: Inverse Depth Parametrization for Monocular SLAM. IEEE Trans. Robot. 24, 932–945 (2008)
18. Longuet-Higgins, H.C.: A computer algorithm for reconstructing a scene from two projections. Nature 293, 133–135 (1981)
19. Hartley, R.I.: In Defense of the Eight-Point Algorithm. IEEE T. Pattern Anal. 19, 580–593 (1997)
20. Luong, Q.T., Faugeras, O.D.: The Fundamental Matrix: Theory, Algorithms, and Stability Analysis. Int. J. Comput. Vision 17, 43–75 (1996)
21. Wang, Y.T., Sun, C.H., Chiou, M.J.: Detection of Moving Objects in Image Plane for Robot Navigation using Monocular Vision. EURASIP J. Adv. Sig. Pr. 2012, 29 (2012)
22. Bouguet JY (2012) Camera Calibration Toolbox for Matlab, `http://www.vision.caltech.edu/bouguetj/calib_doc` (accessed March 18, 2012)

Linear Kalman Filter for Dead Time Affected Measurement Signals Implemented in a Small Scale Automated Guided Vehicle

F. Lütteke and J. Franke

Institute for Manufacturing Automation and Production Systems
University of Erlangen-Nuremberg (FAU),
Erlangen, Germany
luetteke@faps-uni-erlangen.de

Abstract. This paper describes an efficient way to calculate state estimations with a linear Kalman filter, if the measurement signal is subject to a fix dead time. In addition the presented solution can handle measurement signals with arbitrarily high sample times. After the derivation of the algorithm an application with high measurement signal dead time is explained. Finally a simulation depicts the result of the algorithm being implemented on the described application.

Keywords: State estimation, dead time, AGV.

1 Introduction

R. E. Kalman described a recursive, discrete, linear filter for state estimation in his famous paper from the year 1960. It is suited to calculate a state estimation for a dynamic system. This technique can for example be used to estimate the position of a vehicle based on onboard accelerometers in combination with noisy (GPS) position data. The so called a-priori estimation is often based on sensor signals, in the described case, the acceleration data. This data is referred to as input data. Afterwards the estimated state is compared with the measured state and corrected in the sense of minimum squared error. This data is called measurement data. The resulting optimized state estimation is called a-posteriori estimation.

This paper describes a technique to effectively update the state estimations, if the measurement data is subject to a fixed dead time. This can be the case, if the position of an object is a-priori-estimated with local acceleration sensors and the a-posteriori-estimation is for example realized by ceiling mounted cameras [1] with time consuming image processing routines.

2 The Kalman Filter

In this paper only linear, time invariant (LTI) systems are considered. They can be described as follows.

S. Lee et al. (Eds.): Intelligent Autonomous Systems 12, AISC 193, pp. 193–200.
springerlink.com

$$\mathbf{x}_{k+1} = \mathbf{A}\mathbf{x}_k + \mathbf{B}\mathbf{u}_k + \mathbf{w}_k \tag{1a}$$

$$\mathbf{y}_k = \mathbf{C}\mathbf{x}_k + \mathbf{v}_k \tag{1b}$$

With state vector **x**, measurement vector **y**, input vector **u**, input noise vector **w**, measurement noise vector **v**, system matrix **A**, input matrix **B** and output matrix **C**. In addition, **w** and **v** describe white noise with known covariance **Q** and **R** and zero cross correlation.

For linear systems, state estimation can be performed with a linear Kalman filter.

$$\mathbf{K}_k = \mathbf{P}_k^- \mathbf{C}' \left(\mathbf{C}\mathbf{P}_k^- \mathbf{C}' + \mathbf{R} \right)^{-1} \tag{2}$$

$$\hat{\mathbf{x}}_k^+ = \hat{\mathbf{x}}_k^- + \mathbf{K}_k \left(\mathbf{y}_k - \mathbf{C}\hat{\mathbf{x}}_k^- \right) \tag{3}$$

$$\mathbf{P}_k^+ = (\mathbf{I} - \mathbf{K}_k \mathbf{C})\mathbf{P}_k^- \tag{4}$$

$$\hat{\mathbf{x}}_{k+1}^- = \mathbf{A}\hat{\mathbf{x}}_k^+ + \mathbf{B}\mathbf{u}_k \tag{5}$$

$$\mathbf{P}_{k+1}^- = \mathbf{A}\mathbf{P}_k^+ \mathbf{A}' + \mathbf{Q} \tag{6}$$

The recursive algorithm given by (2) to (6) contains the correction of an a-priori-estimation resulting from the last timestep $\hat{\mathbf{x}}_k^-$ with the help of a noisy measurement $\mathbf{y}_k$. The resulting a-posteriori-estimation $\hat{\mathbf{x}}_k^+$ is utilized to predict the state of the successive timestep. The Kalmanfilter is explained in detail in [2].

3 Kalman Filter for Linear Models with Dead Time Affected Measurement Signal

In the assumed scenario, the measurement signal $\mathbf{y}_{k\text{-}d}$ can be recorded exactly at the time step *k-d*. However due to data processing and data transmission it takes *d* time steps until the measurement signal can be fed into the state estimation algorithm. In addition, the measurement signal **y** is sampled with a sampletime of s ($s \in \mathbb{N}^+$) time steps while the input signal **u** is sampled every time step. As a consequence the a-priori-estimation described in (5) and (6) is executed more often than the a-posteriori-estimation described in (2), (3) and (4). As long as no measurement is available for time step *k-d*, the corresponding state estimation for this time step is the a-priori-estimation $\hat{\mathbf{x}}_{k-d}^-$ with the covariance $\mathbf{P}_{k-d}^-$. If – at time step k – the measurement $\mathbf{y}_{k\text{-}d}$ is made available to the filter, the state estimation is retroactively corrected to the a-posteriori-estimation $\hat{\mathbf{x}}_{k-d}^+$ with the covariance $\mathbf{P}_{k-d}^+$. At the same time, the current state estimation $\hat{\mathbf{x}}_k^-$ with covariance $\mathbf{P}_k^-$ can be corrected to the updated current state estimation $\hat{\mathbf{x}}_k^*$ with covariance $\mathbf{P}_k^*$. An effective way to calculate $\hat{\mathbf{x}}_k^*$ and $\mathbf{P}_k^*$ is presented in the following.

The a-priori-estimate $\hat{\mathbf{x}}_k^-$ can be expressed in terms of the *d* past state estimations as shown in (7)

$$\hat{\mathbf{x}}_k^- = \mathbf{A}^d \hat{\mathbf{x}}_{k-d}^- + \Delta\hat{\mathbf{\upsilon}}_{k,d} \tag{7}$$

with

$$\Delta\hat{\mathbf{\upsilon}}_{k,d} = \mathbf{A}^{d-1}\mathbf{B}\mathbf{u}_{k-d} + \mathbf{A}^{d-2}\mathbf{B}\mathbf{u}_{k-d+1} + \ldots + \mathbf{A}\mathbf{B}\mathbf{u}_{k-2} + \mathbf{B}\mathbf{u}_{k-1} \tag{8}$$

which results from *d* successive a-priori-estimations, based on the a-priori-estimation at time step *k-d*. It is important to note, that in (8) $\Delta\hat{\mathbf{\upsilon}}_{k,d}$ is not depending on the state vector $\mathbf{x}$.

The change of the state estimation due to $\mathbf{u}$ can be expressed according to (7) as shown in (9)

$$\Delta\hat{\mathbf{\upsilon}}_{k,d} = \hat{\mathbf{x}}_k^- - \mathbf{A}^d \hat{\mathbf{x}}_{k-d}^- \tag{9}$$

Inserting an a-posteriori-estimation in (7) yields

$$\hat{\mathbf{x}}_k^* = \mathbf{A}^d \hat{\mathbf{x}}_{k-d}^+ + \Delta\hat{\mathbf{\upsilon}}_{k,d} \tag{10}$$

and with (9) results

$$\hat{\mathbf{x}}_k^* = \hat{\mathbf{x}}_k^- + \mathbf{A}^d (\hat{\mathbf{x}}_{k-d}^+ - \hat{\mathbf{x}}_{k-d}^-) \tag{11}$$

Equation 11 describes a very efficient way to calculate the a-priori-estimation at time step *k* by a retroactively corrected state estimation $\hat{\mathbf{x}}_{k-d}^+$. To be able to calculate $\hat{\mathbf{x}}_k^*$, the a-priori-estimate $\hat{\mathbf{x}}_{k-d}^-$ as well as the a-priori-covariance $\mathbf{P}_{k-d}^-$ have to be stored when sampling the measurement $\mathbf{y}_{k-d}$.

Analogous to the calculation of $\hat{\mathbf{x}}_k^*$, the corresponding covariance matrix $\mathbf{P}_k^*$ can be obtained. $\mathbf{P}_k^-$ results from *d* times execution of (6)

$$\mathbf{P}_k^- = \mathbf{A}^d \mathbf{P}_{k-d}^- \mathbf{A}'^d + \Delta\mathbf{O}_{k,d} \tag{12}$$

with

$$\Delta\mathbf{O}_{k,d} = \mathbf{A}^{d-1}\mathbf{Q}\mathbf{A}'^{d-1} + \mathbf{A}^{d-2}\mathbf{Q}\mathbf{A}'^{d-2} + \ldots + \mathbf{A}\mathbf{Q}\mathbf{A}' + \mathbf{Q} \tag{13}$$

Solving (12) for $\Delta\mathbf{O}_{k,d}$ yields

$$\Delta\mathbf{O}_{k,d} = \mathbf{P}_k^- - \mathbf{A}^d \mathbf{P}_{k-d}^- \mathbf{A}'^d . \tag{14}$$

The updated covariance $\mathbf{P}_k^*$ can be calculated by inserting the covariance of the a-posteriori-estimation $\mathbf{P}_{k-d}^+$ in (12).

$$\mathbf{P}_k^* = \mathbf{A}^d \mathbf{P}_{k-d}^+ \mathbf{A}'^d + \Delta\mathbf{O}_{k,d} . \tag{15}$$

With (14) results

$$\mathbf{P}_k^* = \mathbf{P}_k^- + \mathbf{A}^d(\mathbf{P}_{k-d}^+ - \mathbf{P}_{k-d}^-)\mathbf{A}'^d. \quad (16)$$

To decrease the computational load for solving (11) and (16), $\mathbf{A}^d$ can be precalculated, if d is known in advance.

In the special case, that the dead time of d time steps exceeds the sample time of s time steps, all stored values of $\mathbf{x}$ and $\mathbf{P}$ have to be corrected, as soon as a new a-posteriori-estimation $\hat{\mathbf{x}}_{k-d}^+$ with covariance $\mathbf{P}_{k-d}^+$ is available. According to (11) and (16) all stored state estimations $\hat{\mathbf{x}}_{k-d+n}^-$ with corresponding covariances $\mathbf{P}_{k-d+n}^-$ have to be modified according to (17) and (18).

$$\hat{\mathbf{x}}_{k-d+n}^* = \hat{\mathbf{x}}_{k-d+n}^- + \mathbf{A}^n(\hat{\mathbf{x}}_{k-d}^+ - \hat{\mathbf{x}}_{k-d}^-) \quad (17)$$

$$\mathbf{P}_{k-d+n}^* = \mathbf{P}_{k-d+n}^- + \mathbf{A}^n(\mathbf{P}_{k-d}^+ - \mathbf{P}_{k-d}^-)\mathbf{A}'^n \quad (18)$$

with

$$n = s, 2s, 3s, \ldots < d.$$

If however d is an integer multiple of s, (11) and (16) can be ignored. Instead (17) and (18) have to be executed for $n = s, 2s, 3s, \ldots <= d$.

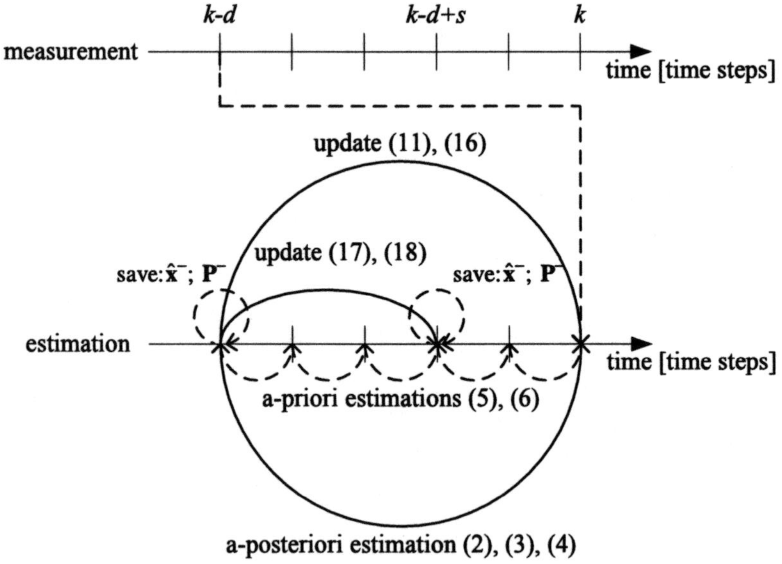

Fig. 1. Kalman filter-based state estimation process with a sample time of the measurement of s=3 time steps. The dead time of the measurement signal amounts to 5 time steps. The a-priori estimation is performed every time step. Solid lines display operations executed at time step k when the measurement from time step k-d has just been received. Dotted lines show operations that took place in the past (like saving state values or taking the measurement).

The Algorithm can be summed up in the following manner

1. Calculate a-priori estimations $\hat{\mathbf{x}}_{k+1}^-$ and $\mathbf{P}_{k+1}^-$ according to (5) and (6) as long as no measurement signal is available. Save state estimations at time steps when measurements are taken.
2. If measurement $\mathbf{y}_{k\text{-}d}$. is available, calculate $\hat{\mathbf{x}}_{k-d}^+$ and $\mathbf{P}_{k-d}^+$ by using the saved values $\hat{\mathbf{x}}_{k-d}^-$ and $\mathbf{P}_{k-d}^-$ with (2), (3), (4)
3. Update all saved a-priori estimations according to (17) and (18) and the current state estimation according to (11) and (16), go on with 1.

Fig. 1 shows the described process of handling measurement signals with a dead time of *d* time steps that are fed into a linear Kalman filter. The dead time amounts to 5 time steps and exceeds the sample time of the measurement which amounts to 3 time steps.

4 Experiment

The algorithm is intended to be implemented in a small scale automated guided vehicle [3] (AGV). It is based on cameras that are mounted on the ceiling, enabling to monitor the whole operational area of the vehicles. The cameras data is processed in a stationary computer to generate a simplified representation of the operational environment of the vehicles.

This so called world model is made available via WLAN to all vehicles of the system. Based on this data that represents the environmental conditions, the vehicles are able to autonomously execute pending transportation tasks. In Fig. 2 the dataflow to generate and continuously update the world model is depicted. It is shown, that the world model distinguishes between three types of world objects.

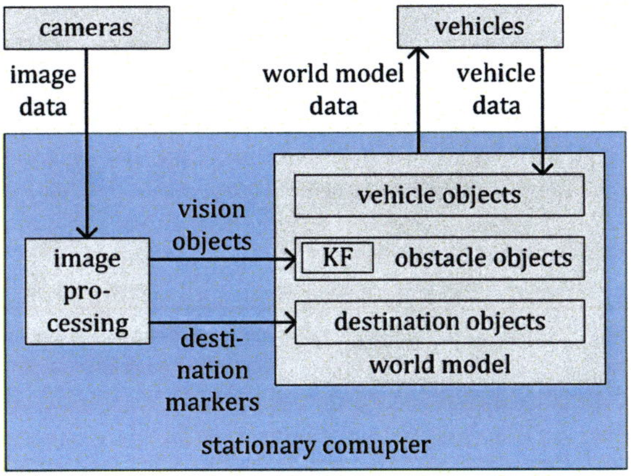

Fig. 2. Dataflow in the stationary computer to generate/update the world model. KF indicates that the state estimation of obstacle objects is performed with a Kalman filter.

Vehicle objects describe operated transportation vehicles. The vehicles are able to perform their position estimations themselves. The state of each vehicle object contains the data each vehicle provides to the stationary computer.

Every docking position of the transportation system is marked with a unique optical marker. These destination markers can be detected by the image processing routine. Destination objects contain all data that is returned by the image processing routine like position, orientation and ID of the corresponding marker.

Obstacle objects represent all objects detected by the image processing routine that are neither assumed to be vehicles nor destinations. For each obstacle object, besides its position, also its velocity is estimated.

The mathematical model of obstacles is given in (19).

$$\begin{bmatrix} x \\ y \\ \dot{x} \\ \dot{y} \end{bmatrix}_{k+1} = \begin{bmatrix} 1 & 0 & dt & 0 \\ 0 & 1 & 0 & dt \\ 0 & 0 & 1 & 0 \\ 0 & 0 & 0 & 1 \end{bmatrix} \begin{bmatrix} x \\ y \\ \dot{x} \\ \dot{y} \end{bmatrix}_{k} \tag{19}$$

where dt describes the cycletime of the a-priori estimation which equates to the duration of a time step k. Comparing (19) with (1α) it can be seen, that no input signal **u** for the state estimation is used. Instead, the current velocity estimation is used to predict the position of the obstacle.

In the following, simulation results for state estimations based on the model described in (19) are discussed. Fig 3 shows a linear movement process of the obstacle. The velocity of the obstacle is constant. In addition the measurement signal **y** (x- and y- coordinate of the obstacle) is delayed by 5 time steps. The sample time s amounts to 3 which means, that only every 3rd time step a measurement is recorded.

Because (19) predicts the next position based on the current longitudinal velocity, the position estimation can only converge against the exact value, if the obstacle moves on a straight line with constant velocity. In Fig. 3 this case of obstacle movement is shown. The first position estimation is wrong by 1 unit of length in x direction. The velocity estimation is initialized with zero. After 5 time steps, position and velocity estimations are corrected due to a measurement signal the first time. The filter calculates an a-posteriori state estimation that is likely to have been valid at time step k=0. The red dots show this a-posteriori position estimation. Because the sample time of the position measurement amounts to 3 time steps, in general no new a-posteriori estimation can be calculated during the following 2 time steps. For every displayed time step, the last a-posteriori estimation value describes the position of the obstacle either 5, 6 or 7 time steps in the past. In addition to the position, the velocity is estimated. It is shown in the second plot of Fig. 3. Also this velocity is assumed to have been valid for the past 5, 6 or 7 time steps. The combination of the most recent position estimation, most recent velocity estimation and the number of time steps, the measurement signal that lead to the last a-posteriori estimation is lying in the past lead to the displayed estimation of the position.

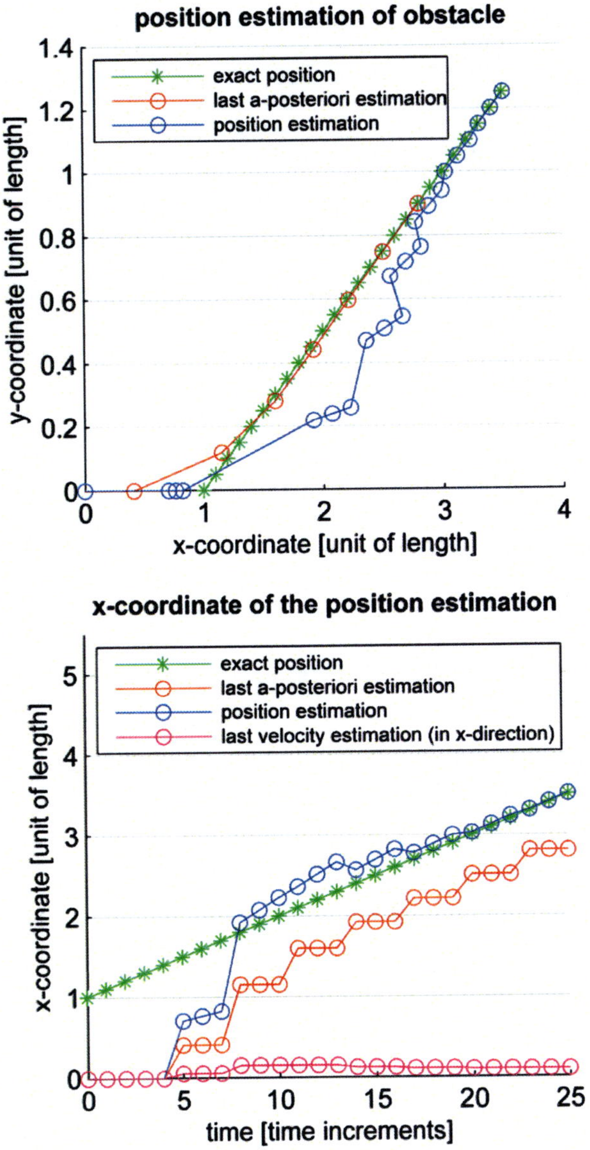

Fig. 3. Simulation of the position estimation of an obstacle object. The exact measurement signal (exact position) is sampled every 3rd time step. In addition, the measurement signal is delayed by 5 timesteps.

5 Conclusion

The proposed algorithm has been motivated because tracking of an object with a camera lead to high estimation errors if a standard linear Kalman filter was used. This error resulted from high dead time and low sample rate of the measurement signal that provided the position of the moving object. If the ordinary Kalman filter that is stated in (2) to (6) was used, (3) could not be executed until measurement data was valid. Therefore the algorithm hang. This resulted in erroneous position estimation of the object.

It has been shown, that in the linear case, fairly simple mathematical operations enable to update the state estimation with a constant sampling rate independently of the dead time of the measurement signal. By implementing the described algorithm according to (11), (16), (17) and (18) in addition to (2) to (6) a mathematically correct solution to the described problem was provided.

References

1. Franke, J., Lütteke, F.: Small Scale Versatile Autonomous Indoor Transportation System. In: Proceedings of International Chemnitz Manufacturing Colloquium (ICMC 2010), Chemnitz, Germany, Wissenschaftliche Scripten, pp. 897–904 (2010)
2. Welsh, G., Bishop, G.: An introduction to the kalman filter. University of North Carolina at Chapel Hill (1995)
3. Ullrich, G.: Fahrerlose Transportsysteme. Vieweg+Teubner, Wiesbaden (2011)

Balance Control of a Variable Centroid Inverted Pendulum Robot

Gang Wang, Sang Yong Lee, Seok Won Kang, and Jang-Myung Lee

Department of Electrical Engineering
Pusan National University
Jangjeon2-dong, Geumjeong-gu, Busan 609-735, Korea
{jlwg142013,Sangyong1696,sucwon1696,jmlee}@pusan.ac.kr

Abstract. A basic problem has been found in two wheels inverted pendulum based mobile manipulator. It is difficult to keep balance as the manipulator operates and it is necessary to find a method to resolve it. This research presents a simple and feasible method for this kind of problem. It involves a common PD control and a method to calculate the compensatory angle for a variable centroid inverted pendulum robot. A 4 DOF mobile manipulator based on inverted pendulum has been developed to analyze the problem, and a simulation is also used to certify the method in this research. The experiment and its results show the performance of the method. It is possible to make the robot keep balance by using the compensatory angle even though the position of centroid is time variant.

Keywords: angle compensation, inverted pendulum, variable centroid, Segway.

1 Introduction

Two wheels inverted pendulum robot has caused more interests in resent ten years. Several kinds of experiments have been finished since Segway was invented as some of them were given to some universities as gifts [1]. Some other kinds of inverted pendulum robot also have been developed in these years. These robots can be divided into two kinds before they get load, one is with a certain centroid and another with a time variant centroid. In the early phase, more experiments were about a constant centroid robot such as Joe[2]. B2 also was a constant centroid robot before it carried a people [3]. Quasimoro was a wheel chair like robot, but its wheel axis was different with drive motors [4]. In recent years, mobile manipulators based on two wheels inverted pendulum have caused more attentions for that this kind of robot can turn around with a smaller zone. These robots are different with those have been introduced above. In fact they are variable centroid robot. uBot was such a robot that was used to pull or push an object[5], it did not concentrate more about balance control. There was also a method provided in other research and this method mainly depends on the horizontal acceleration of the robot by given friction factor. But it is not feasible for different ground, and needs a short time to get the compensatory angle. To make the robot work well in different situation and react fast, it is necessary to find another method.

S. Lee et al. (Eds.): Intelligent Autonomous Systems 12, AISC 193, pp. 201–208.
springerlink.com

This paper includes these parts: In section 2, a method to calculate the position of centroid is introduced. In section 3, dynamic model will be shown. And section 4, centroid simulation and experiment are described.

2 Model of Compensatory Angle

Angle compensation for a variable centroid inverted pendulum robot is important and necessary. It is much more difficult to get the position of the centroid when the robot has more joints. In this section, a method to calculate the position of centroid is introduced. The model of the robot which is used in this paper is shown in Fig1. This robot is mainly consisted of two parts, the chassis is a mini Segway, and the upper part is a mini manipulator. The manipulator has 4 DOF, for each link, it has a DC sever motor, then connect this manipulator to mini Segway by a passive joint. So this robot has one passive joint and four active joints. In Segway, one encoder is installed to get the shift value. The coordinate systems are established as shown in figure 1.

The basic method of this part is coordinate transformation. D-H method is used in this experiment because it is widely used in robot engineering, and also suitable for most of the extant manipulators. D-H parameters of this robot are shown in table1.

Table 1. D-H parameters

L	θ(degree)	d(mm)	a(mm)	α(degree)
1	θ_1	0	136	0
2	θ_2	0	60	0
3	θ_3	0	75	0
4	θ_4	0	60	0
5	θ_5	0	32	0

So transformation matrix can be got from table 1, here c_1 means $\cos\theta_1$and s_1 means $\sin\theta_1$.

$$ {}^{0}H_{1} = \begin{bmatrix} c_1 & -s_1 & 0 & 136*c_1 \\ s_1 & c_1 & 0 & 136*s_1 \\ 0 & 0 & 1 & 0 \\ 0 & 0 & 0 & 1 \end{bmatrix} \tag{1} $$

$$ {}^{1}H_{2} = \begin{bmatrix} c_2 & -s_2 & 0 & 60*c_2 \\ s_2 & c_2 & 0 & 60*s_2 \\ 0 & 0 & 1 & 0 \\ 0 & 0 & 0 & 1 \end{bmatrix} \tag{2} $$

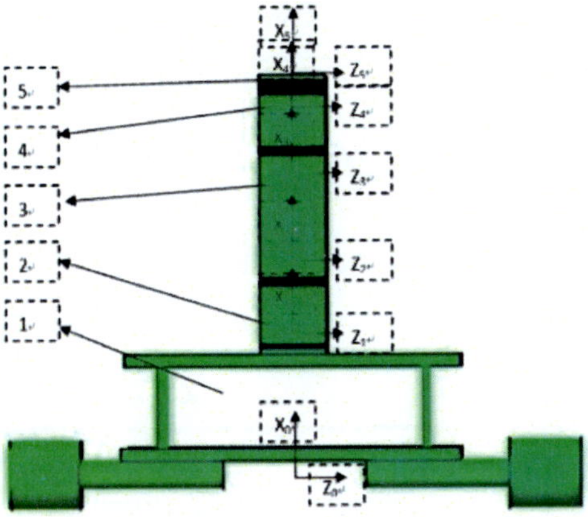

Fig. 1. Model of the robot. As shown in this figure, it includes 5 parts, and also it has 6 coordinate systems.

$$^{2}H_{3} = \begin{bmatrix} c_3 & -s_3 & 0 & 75*c_3 \\ s_3 & c_3 & 0 & 75*s_3 \\ 0 & 0 & 1 & 0 \\ 0 & 0 & 0 & 1 \end{bmatrix} \tag{3}$$

$$^{3}H_{4} = \begin{bmatrix} c_4 & -s_4 & 0 & 60*c_4 \\ s_4 & c_4 & 0 & 60*s_4 \\ 0 & 0 & 1 & 0 \\ 0 & 0 & 0 & 1 \end{bmatrix} \tag{4}$$

$$^{4}H_{5} = \begin{bmatrix} c_5 & -s_5 & 0 & 32*c_5 \\ s_5 & c_5 & 0 & 32*s_5 \\ 0 & 0 & 1 & 0 \\ 0 & 0 & 0 & 1 \end{bmatrix} \tag{5}$$

Transformation matrixes of each coordinate system are shown like this:

$$^{0}H_{2} = {}^{0}H_{1} * {}^{1}H_{2} \tag{6}$$

$$^{0}H_{3} = {}^{0}H_{1} * {}^{1}H_{2} * {}^{2}H_{3} \tag{7}$$

$$^{0}H_{4} = {}^{0}H_{1} * {}^{1}H_{2} * {}^{2}H_{3} * {}^{3}H_{4} \tag{8}$$

$$^{0}H_{5} = {}^{0}H_{1} * {}^{1}H_{2} * {}^{2}H_{3} * {}^{3}H_{4} * {}^{4}H_{5} \tag{9}$$

Table 2. Coordinates

Part number	Mass(g)	Coordinate system	Centroid(mm)
1	m_1(2420)	$X_1Y_1Z_1$	x_1,y_1,z_1(-113,0,0)
2	m_2(110)	$X_2Y_2Z_2$	x_2,y_2,z_2(-44,0,0)
3	m_3(190)	$X_3Y_3Z_3$	x_3,y_3,z_3(-32.5,0,0)
4	m_4(110)	$X_4Y_4Z_4$	x_4,y_4,z_4(-16,0,0)
5	m_5(20)	$X_5Y_5Z_5$	x_5,y_5,z_5(-8,0,0)

Express all the positions of each centroid in $X_0Y_0Z_0$ coordinate system:

$$\begin{bmatrix} {}^0x_i \\ {}^0y_i \\ {}^0z_i \\ 1 \end{bmatrix} = {}^0H_i \begin{bmatrix} x_i \\ y_i \\ z_i \\ 1 \end{bmatrix} \tag{10}$$

Where: $i \in (1,2,3,4,5)$, ${}^0x_i$ means the ith coordinate expressed in $X_0Y_0Z_0$ system. And then the position of centroid of whole body will be like this:

$${}^0x_c = \frac{\Sigma\, {}^0x_i * m_i}{\Sigma m_i},\ {}^0y_c = \frac{\Sigma\, {}^0y_i * m_i}{\Sigma m_i},\ {}^0z_c = \frac{\Sigma\, {}^0z_i * m_i}{\Sigma m_i} \tag{11}$$

Here, $({}^0x_c, {}^0y_c, {}^0z_c)$ means the coordinate of centroid expressed in $X_0Y_0Z_0$ system. So it is very easy to get the compensatory angle:

$$\theta = \arctan\left(\frac{{}^0y_c}{{}^0x_c}\right) \tag{12}$$

3 Dynamic Modeling

In this section, Lagrangian method is used to derive the dynamic model of the robot. In fig.2, M_b means the mass of the whole body without wheels, l means the distance between the centroid and the wheel axis line, θ_w is rotation angle of wheels, J_w is the moment of inertia of wheels, R is the radius of the wheel, $\theta_c=\theta_0+\theta$ is the incline angle of L, $\theta_0$is the incline angle of platform, J_b is moment of inertia of body and M_w means the mass of wheels. XuOYu belongs to the universe coordinate system.

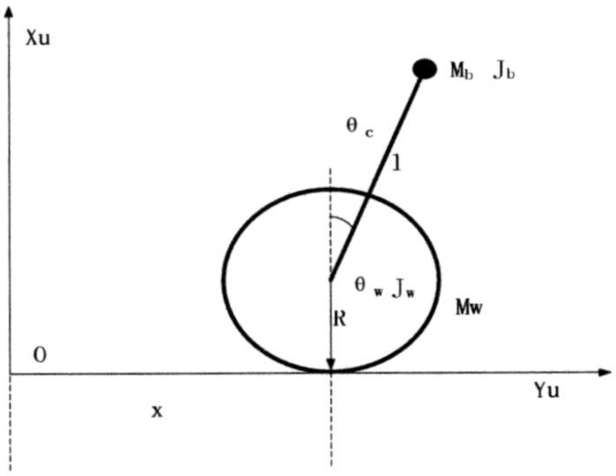

Fig. 2. Dynamic model

where:

$$l = \sqrt{{}^0x_c^2 + {}^0y_c^2} \tag{13}$$

The kinetic energy T of robot is set as the following equation:

$$\begin{aligned} T &= \frac{1}{2}M_b(\dot{\theta}_w R + l\cos\theta_c \dot{\theta}_c)^2 + \frac{1}{2}M_b(-l\sin\theta_c \dot{\theta}_c)^2 \\ &+ \frac{1}{2}M_w(\dot{\theta}_w R)^2 + \frac{1}{2}J_w \dot{\theta}_w^2 + \frac{1}{2}J_b \dot{\theta}_c^2 + \frac{1}{2}n^2 J_m(\dot{\theta}_w - \dot{\theta}_c)^2 \end{aligned} \tag{14}$$

Where: J_m and n represent the moment of inertia of the motor and the gear ratio. The potential energy V of robot is set as the following equation:

$$V = M_w gR + M_b g(R + l\cos\theta_c) \tag{15}$$

Therefore the Lagrange equation L can be got as this:

$$L = T - V \tag{16}$$

Use followed equation to get the dynamic of the robot:

$$\frac{d}{dt}\frac{\partial L}{\partial \dot{\mathbf{q}}} - \frac{\partial L}{\partial \mathbf{q}} = \tau_q \quad where, \mathbf{q} = [\theta_\omega \quad \theta_c]^T \tag{17}$$

Dynamic equations of inverted pendulum model which is obtained by means of Lagrange equation, is described in (18).

$$\tau = M_s(q)\ddot{q} + h_s(q, \dot{q}) + g_s(q) \tag{18}$$

Take (14) and (15) into (16), use (17), write the result in term of (18):

$$M_s = \begin{bmatrix} M_b R^2 + M_w R^2 + J_w + n^2 J_m & M_b Rl\cos\theta_c - n^2 J_m \\ M_b lR\cos\theta_c - n^2 J_m & M_b l^2 + J_b + n^2 J_m \end{bmatrix} \tag{19}$$

$$h_s = \begin{bmatrix} -M_b R l \sin\theta_c \dot{\theta}_c^2 \\ 0 \end{bmatrix} \tag{20}$$

$$g_s = \begin{bmatrix} 0 \\ -M_b g \sin\theta_c \end{bmatrix} \tag{21}$$

4 Simulation and Experiment

4.1 Simulation about Centroid

A simulation was taken out to check part Ⅱ in this paper. The software used in this paper is Wolfram Mathematica 7. And the simulation result is shown as followed:

Fig. 3. A simulation of the robot. The red dot is the centroid of whole body.

4.2 Controller Illustration

PD control is widely used in this kind of problems. In order to certify the idea and make the controller easier, PD control is also used in this paper. The system structure is shown in figure 4.

In this diagram, horizontal acceleration is used as an auxiliary function to decrease the shift. θ_0 and$\theta$are used to control the balance of the robot.

4.3 Experiment and Results

In this experiment, θ_2 to θ_5 were taken out to check the method separately. When θ_2 increased from 0-60 degree, the tilt angle of platform changed from 0-15 degree in order to keep the robot balance and the robot also did not have a significant shift. In order to make comparisons, θ_2 to θ_5 were changed at the shift position around 200mm. From the experiment figure shown in figure 5, it is obvious that θ_2 and θ_3 have more influence to compensatory angle, θ_4 and θ_5 have less influence.

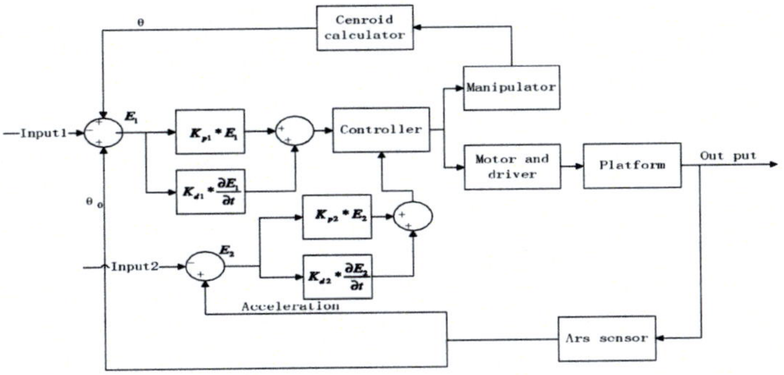

Fig. 4. System structure

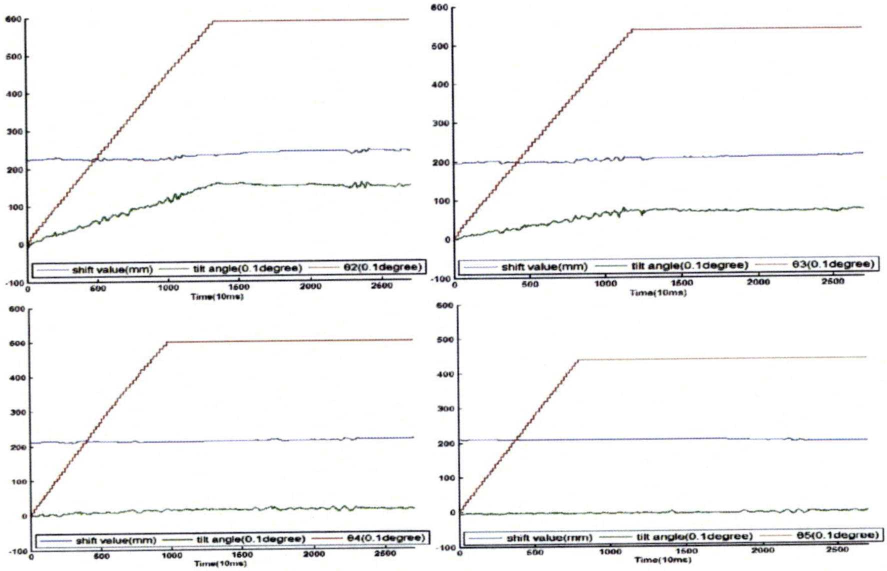

Fig. 5. Experiment data curves

5 Conclusion

In this research, a method of angle compensation is described. The effectiveness of the method has already been proved by experiment. It makes the robot keep the balance by using the compensatory angle even though the centroid of the whole body is changed. According to part 2, it is also useful for dual arm mobile manipulator because it is possible to get the compensatory angle in the same procedure.

Acknowledgements. This research was supported by Basic Science Research Program through the National Research Foundation of Korea (NRF) funded by the Ministry of Education, Science and Technology (2011-0026949).

This research was supported by MKE(The Ministry of Knowledge Economy), Korea, under Specialized Field Navigation/Localization technology research center support program supervised by the NIPA(National IT Industry Promotion Agency).

References

1. Nguyen, H.G., Morrell, J., Mullens, K., Burmeister, A., Miles, S., Farrington, N., Thomas, K., Gage, D.W.: Segway Robotic Mobility Platform. In: SPIE Proc. Mobile Robots XVII, Philadelphia, PA, October 27-28, vol. 5609 (2004)
2. Grasser, F., D'Arrigo, A., Colombi, S.: JOE: A Mobile, Inverted Pendulum. IEEE Transactions on Industrial Electronics 49(1), 107–114 (2002)
3. Baloh, M., Parent, M.: Modeling and Model Verification of an Intelligent Self-Balancing Two-Wheeled Vehicle for an Autonomous Urban Transportation System. In: The Conference on Computational Intelligence, Robotics and Autonomous Systems, Singapore, December 15 (2003)
4. Oryschuk, P., Salerno, A., Al-Husseini, A.M., Angeles, J.: Experimental Validation of an Underactuated Two-Wheeled Mobile Robot. IEEE/ASME Transactions on Mechatronics 14(2), 252–257 (2009)
5. Kuindersma, S.R., Hannigan, E., Ruiken, D., Grupen, R.A.: Dexterous Mobility with the uBot-5 Mobile Manipulator. Advanced Robotics, 1–7 (2009)

Stabilization for Truck-Trailer Mobile Robot System via Discrete LPV T-S Fuzzy Models

Wen-Jer Chang and Po-Hsun Chen

Dept. of Marine Engineering, National Taiwan Ocean University
Keelung 202, Taiwan, R.O.C.
{wjchang,B976A0039}@mail.ntou.edu.tw

Abstract. The stability analysis and synthesis problem of discrete nonlinear Truck-Trailer Mobile Robot (TTMR) system is studied in this paper. The Linear Parameter Varying (LPV) model and Takagi-Sugeno (T-S) fuzzy model are employed to construct the discrete nonlinear TTMR system. There are few researchers discussing the stability analysis and synthesis for the discrete LPV T-S fuzzy models. Therefore, the problem discussed in this paper is worthy to receive the attention of control engineers.

Keywords: fuzzy control, Takagi-Sugeno fuzzy model, discrete linear parameter varying model, linear matrix inequality.

1 Introduction

The TTMR system is often used as a benchmark for new control approaches. Tanaka and Sano [1] presented a model-based fuzzy controller, which realized backward movement control of the TTMR system. According to [1], one can transform the TTMR system into T-S fuzzy model to discuss the stability conditions and the controller design problems [2-3]. In [1-3], the speed of backward movement for the TTMR system is assumed to be constant. This assumption reduces the complex in the computing process of stability analysis and synthesis. However, it is unreasonable to assume constant speed of backward movement for the TTMR system in practical applications.

By assuming the speed of backward movement to be varying and bounded in a range, the nonlinear TTMR system can be modeled by a LPV T-S fuzzy model. It is well known that a nonlinear plant can be reformulated as LPV model via parameterized Jacobian linearization or quasi-LPV method [4]. Based on the LPV model, some control approaches [5-6] have been developed to formulate analysis and synthesis problems as convex optimization problems involving Linear Matrix Inequality (LMI). Besides, the T-S fuzzy modeling technique also provides a proper solution to construct the mathematic model for a nonlinear system. Combining LPV model and T-S fuzzy model, Ge and Huang [7] have studied the analysis and synthesis problems for the continuous LPV T-S fuzzy systems. However, there are few literatures discussing the same problem for the discrete time case. Therefore, the objective of this paper is to derive the stability conditions for the analysis and synthesis of discrete-time LPV T-S fuzzy systems.

S. Lee et al. (Eds.): Intelligent Autonomous Systems 12, AISC 193, pp. 209–217.
springerlink.com

Under the T-S fuzzy modeling technique, several piecewise local LPV models with affine parameter-dependent structure are blended by fuzzy logic to approximate the nonlinear discrete TTMR systems. Based on the Parallel Distributed Compensation (PDC) concept [1-3, 8], a parameter-dependent quadratic Lyapunov function is used to for the synthesis of the discrete LPV T-S fuzzy systems. It can be found that the stability conditions derived in this paper can be transformed to the LMI formulas. Hence, the convex optimization technique [9] can be used to solve these LMI stability conditions.

2 LPV T-S Fuzzy Model of TTMR System

The dynamic equations of nonlinear TTMR system have been introduced in [1-3]. The structure of the TTMR system can be referred to the Figure 1.

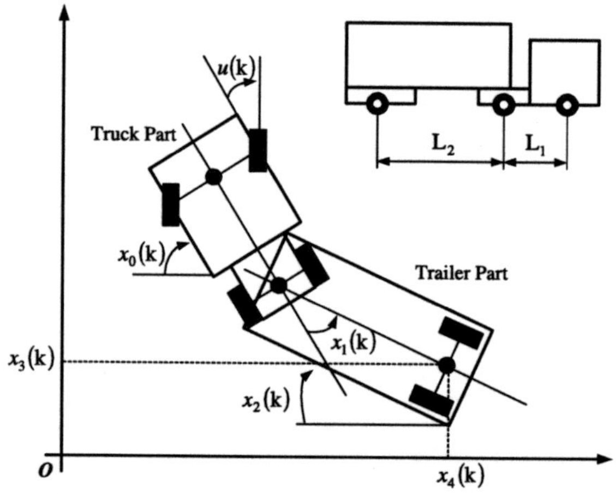

Fig. 1. Structure of truck-trailer mobile robot system

According to [1-3], the simplified TTMR system can be described as follows.

$$x_1(k+1)=\left(1-\frac{\theta(k)\cdot\Delta t}{L_2}\right)x_1(k)+\frac{\theta(k)\cdot\Delta t}{L_1}u(k) \tag{1a}$$

$$x_2(k+1)=\frac{\theta(k)\cdot\Delta t}{L_2}x_1(k)+x_2(k) \tag{1b}$$

$$x_3(k+1)=\theta(k)\cdot\Delta t\cdot sin\left(\frac{\theta(k)\cdot\Delta t}{2L_2}x_1(k)+x_2(k)\right)+x_3(k) \tag{1c}$$

where $x_1(\mathrm{k})$ is the angle difference between truck and trailer, $x_2(\mathrm{k})$ is the angle of trailer, $x_3(\mathrm{k})$ is the vertical position of rear end of trailer, $u(\mathrm{k})$ is the steering angle, L_1 is the length of truck, L_2 is the length of trailer, $\Delta\mathrm{t}$ is the sampling time and $\theta(k)$ is the speed of backward movement.

Assume that $z(k)=\frac{\theta(k)\cdot\Delta t}{2L_2}x_1(k)+x_2(k)$ is the antecedent variable and it is operated between $\pm\pi$. Besides, it is assumed that $\theta(k)\in\left[\underline{\theta},\overline{\theta}\right]$ with constants $\underline{\theta}$ and $\overline{\theta}$. Then the LPV T-S fuzzy model, which approximately represents the dynamic of TTMR system (1), can be described as follows.

$$\begin{aligned}&Rule^1:\ \text{IF } z(k) \text{ is about } 0\\&\qquad \text{THEN } x(k+1)=\mathbf{A}_1(\theta)x(k)+\mathbf{B}_1(\theta)u(k)\end{aligned} \tag{2a}$$

$$\begin{aligned}&Rule^2:\ \text{IF } z(k) \text{ is about } -\pi \text{ or } \pi\\&\qquad \text{THEN } x(k+1)=\mathbf{A}_2(\theta)x(k)+\mathbf{B}_2(\theta)u(k)\end{aligned} \tag{2b}$$

where $x(\mathrm{k})=[x_1(\mathrm{k})\ \ x_2(\mathrm{k})\ \ x_3(\mathrm{k})]^{\mathrm{T}}$ is the state vector; $u(\mathrm{k})$ is the control input and

$$\mathbf{A}_1(\theta)=\begin{bmatrix}1-\frac{\theta(k)\Delta t}{L_2} & 0 & 0\\ \frac{\theta(k)\Delta t}{L_2} & 1 & 0\\ \frac{\theta(k)^2\Delta t^2}{2L_2} & \theta(k)\Delta t & 1\end{bmatrix},\ \mathbf{A}_2(\theta)=\begin{bmatrix}1-\frac{\theta(k)\Delta t}{L_2} & 0 & 0\\ \frac{\theta(k)\Delta t}{L_2} & 1 & 0\\ \frac{\varepsilon\theta(k)^2\Delta t^2}{2L_2} & \varepsilon\theta(k)\Delta t & 1\end{bmatrix},$$

$$\mathbf{B}_1(\theta)=\mathbf{B}_2(\theta)=\begin{bmatrix}\frac{\theta(t)\cdot\Delta t}{L_1} & 0 & 0\end{bmatrix}^{\mathrm{T}},\ \varepsilon=10^{-2}/\pi.$$

Given a pair of $(x(k),\ u(k))$, the final outputs of the T-S fuzzy model (2) can be inferred as follows.

$$x(k+1)=\sum_{i=1}^{2}h_i(z(k))\{\mathbf{A}_i(\theta)x(k)+\mathbf{B}_i(\theta)u(k)\} \tag{3}$$

where $h_i(z(k))=\omega_i(z(k))/\sum_{i=1}^{2}\omega_i(z(k))$, $h_i(z(k))\geq 0$, $\sum_{i=1}^{2}h_i(z(k))=1$, $\omega_i(z(k))=\prod_{j=1}^{2}\mathbf{M}_{ij}(z(k))$ and $\mathbf{M}_{ij}$ is the fuzzy set of the i-th fuzzy rule.

According to the PDC concept [1-3, 8], the fuzzy controller is given as follows.

$$u(k)=-\sum_{i=1}^{2}h_i(z(k))\{\mathbf{F}_i x(k)\} \tag{4}$$

Substituting (4) into (3), the closed-loop system can be thus described as

$$x(k+1)=\sum_{i=1}^{2}\sum_{j=1}^{2}h_i(z(k))h_j(z(k))(\mathbf{A}_i(\theta)-\mathbf{B}_i(\theta)\mathbf{F}_j)x(k) \tag{5}$$

The purpose of this paper is to find the feedback gains $\mathbf{F}_i$ such that the above closed-loop system (5) is asymptotically stable.

3 PDC-Based Fuzzy Controller Design

Considering the discrete LPV T-S fuzzy model (5), the sufficient stability conditions are derived in the following theorem.

Theorem 1

If there exists a parameter-dependent Lyapunov matrix variable $\mathbf{P}(\theta)$ and feedback gain $\mathbf{F}_i$ satisfying the following sufficient constraints, then the closed-loop system (5) is asymptotically stable.

$$\mathbf{G}_{ii}(\theta)^T\mathbf{P}(\theta)\mathbf{G}_{ii}(\theta)-\mathbf{P}(\theta)<0 \tag{6}$$

$$\left(\frac{\mathbf{G}_{ij}(\theta)+\mathbf{G}_{ji}(\theta)}{2}\right)^T\mathbf{P}(\theta)\left(\frac{\mathbf{G}_{ij}(\theta)+\mathbf{G}_{ji}(\theta)}{2}\right)-\mathbf{P}(\theta)<0,\ \text{for } i<j \tag{7}$$

Proof

Select a discrete parameter-dependet Lyapunov function as

$$V(x(\mathrm{k})) = x^{\mathrm{T}}(\mathrm{k})\mathbf{P}(\theta)x(\mathrm{k}), \quad \mathbf{P}(\theta) = \mathbf{P}_0 + \mathbf{P}_1\theta_1 + \cdots + \mathbf{P}_k\theta_k > 0 \tag{8}$$

Let $\mathbf{G}_{\mathrm{ij}}(\theta) = \mathbf{A}_{\mathrm{i}}(\theta) - \mathbf{B}_{\mathrm{i}}(\theta)\mathbf{F}_{\mathrm{j}}$ and $\mathbf{G}_{\mathrm{kl}}(\theta) = \mathbf{A}_{\mathrm{k}}(\theta) - \mathbf{B}_{\mathrm{k}}(\theta)\mathbf{F}_{\mathrm{l}}$. By evaluating the first-forward difference of the Lyapunov function $V(x(\mathrm{k}))$ along the trajectories of (8), one has

$$\begin{aligned}
\Delta V(x(\mathrm{k})) &= V(x(\mathrm{k}+1)) - V(x(\mathrm{k})) \\
&= x^{\mathrm{T}}(\mathrm{k}+1)\mathbf{P}(\theta)x(\mathrm{k}+1) - x^{\mathrm{T}}(\mathrm{k})\mathbf{P}(\theta)x(\mathrm{k}) \\
&= \frac{1}{4}\sum_{\mathrm{i}=1}^{2}\sum_{\mathrm{j}=1}^{2}\sum_{\mathrm{k}=1}^{2}\sum_{\mathrm{l}=1}^{2}\mathrm{h}_{\mathrm{i}}(z(\mathrm{k}))\mathrm{h}_{\mathrm{j}}(z(\mathrm{k}))\mathrm{h}_{\mathrm{k}}(z(\mathrm{k}))\mathrm{h}_{\mathrm{l}}(z(\mathrm{k})) \\
&\qquad \times x^{\mathrm{T}}(\mathrm{k})\left\{(\mathbf{G}_{\mathrm{ij}}(\theta) + \mathbf{G}_{\mathrm{ji}}(\theta))^{\mathrm{T}}\mathbf{P}(\theta)(\mathbf{G}_{\mathrm{kl}}(\theta) + \mathbf{G}_{\mathrm{lk}}(\theta)) - 4\mathbf{P}(\theta)\right\}x(\mathrm{k}) \\
&\le \frac{1}{4}\sum_{\mathrm{i}=1}^{2}\sum_{\mathrm{j}=1}^{2}\mathrm{h}_{\mathrm{i}}(z(\mathrm{k}))\mathrm{h}_{\mathrm{j}}(z(\mathrm{k}))x^{\mathrm{T}}(\mathrm{k}) \\
&\qquad \times\left\{(\mathbf{G}_{\mathrm{ij}}(\theta) + \mathbf{G}_{\mathrm{ji}}(\theta))^{\mathrm{T}}\mathbf{P}(\theta)(\mathbf{G}_{\mathrm{ij}}(\theta) + \mathbf{G}_{\mathrm{ji}})(\theta) - 4\mathbf{P}(\theta)\right\}x(\mathrm{k}) \\
&= \sum_{\mathrm{i}=1}^{2}\mathrm{h}_{\mathrm{i}}(z(\mathrm{k}))x^{\mathrm{T}}(\mathrm{k})\left(\mathbf{G}_{\mathrm{ii}}(\theta)^{\mathrm{T}}\mathbf{P}(\theta)\mathbf{G}_{\mathrm{ii}}(\theta) - \mathbf{P}(\theta)\right)x(\mathrm{k}) \\
&\qquad + 2\sum_{\mathrm{i}<1}^{2}\mathrm{h}_{\mathrm{i}}(z(\mathrm{k}))\mathrm{h}_{\mathrm{j}}(z(\mathrm{k}))x^{\mathrm{T}}(\mathrm{k}) \\
&\qquad \times\left\{\left(\frac{\mathbf{G}_{\mathrm{ij}}(\theta) + \mathbf{G}_{\mathrm{ji}}(\theta)}{2}\right)^{\mathrm{T}}\mathbf{P}(\theta)\left(\frac{\mathbf{G}_{\mathrm{ij}}(\theta) + \mathbf{G}_{\mathrm{ji}}(\theta)}{2}\right) - \mathbf{P}(\theta)\right\}x(\mathrm{k})
\end{aligned}$$

If the conditions (6) and (7) hold, the closed-loop system (5) is asymptotically stable due to $\Delta V(x(\mathrm{k})) < 0$.

The stability conditions derived in Theorem 1 are not LMI forms; hence, it is difficult to find their solutions via convex optimization technique [9]. In order to apply LMI technique to solve the proposed fuzzy controller design problem, it is necessary to transform the conditions of Theorem 1 into LMI formula. In the following theorem, the Schur complement [9] is used to accomplish the purpose desired.

Theorem 2

If there exists a parameter-dependent matrix variables $\mathbf{Q}(\theta)$ and $\mathbf{Y}_i(\theta)$ satisfying the following sufficient constraints, then the closed-loop system (5) is asymptotically stable.

$$\begin{bmatrix} \mathbf{Q}(\theta) & \{\mathbf{A}_i(\theta)\mathbf{Q}(\theta)-\mathbf{B}_i(\theta)\mathbf{Y}_i(\theta)\}^T \\ \{\mathbf{A}_i(\theta)\mathbf{Q}(\theta)-\mathbf{B}_i(\theta)\mathbf{Y}_i(\theta)\} & \mathbf{Q}(\theta) \end{bmatrix} > 0 \tag{9}$$

$$\begin{bmatrix} \mathbf{Q}(\theta) & \left\{\dfrac{\mathbf{A}_i(\theta)\mathbf{Q}(\theta)+\mathbf{A}_j(\theta)\mathbf{Q}(\theta)-\mathbf{B}_i(\theta)\mathbf{Y}_j(\theta)-\mathbf{B}_j(\theta)\mathbf{Y}_i(\theta)}{2}\right\}^T \\ \left\{\dfrac{\mathbf{A}_i(\theta)\mathbf{Q}(\theta)+\mathbf{A}_j(\theta)\mathbf{Q}(\theta)-\mathbf{B}_i(\theta)\mathbf{Y}_j(\theta)-\mathbf{B}_j(\theta)\mathbf{Y}_i(\theta)}{2}\right\} & \mathbf{Q}(\theta) \end{bmatrix} > 0, \text{ for } i<j \tag{10}$$

Proof

Let $\mathbf{P}(\theta)=\mathbf{Q}(\theta)^{-1}$ and $\mathbf{F}_i=\mathbf{Y}_i(\theta)\mathbf{Q}(\theta)^{-1}$, the stability conditions (6) and (7) can be rewritten as

$$\mathbf{Q}(\theta)-\{\mathbf{A}_i(\theta)\mathbf{Q}(\theta)-\mathbf{B}_i(\theta)\mathbf{Y}_i(\theta)\}^T\mathbf{Q}(\theta)^{-1}\{\mathbf{A}_i(\theta)\mathbf{Q}(\theta)-\mathbf{B}_i(\theta)\mathbf{Y}_i(\theta)\}>0 \tag{11}$$

$$\begin{aligned} &\mathbf{Q}(\theta)-\mathbf{Q}(\theta)\left\{\frac{\mathbf{A}_i(\theta)\mathbf{Q}(\theta)+\mathbf{A}_j(\theta)\mathbf{Q}(\theta)-\mathbf{B}_i(\theta)\mathbf{Y}_j(\theta)-\mathbf{B}_j(\theta)\mathbf{Y}_i(\theta)}{2}\right\}^T \\ &\times\mathbf{Q}(\theta)^{-1}\left\{\frac{\mathbf{A}_i(\theta)\mathbf{Q}(\theta)+\mathbf{A}_j(\theta)\mathbf{Q}(\theta)-\mathbf{B}_i(\theta)\mathbf{Y}_j(\theta)-\mathbf{B}_j(\theta)\mathbf{Y}_i(\theta)}{2}\right\}\mathbf{Q}(\theta)>0 \end{aligned} \tag{12}$$

Applying the Schur complement [9], the conditions (11) and (12) can be transformed into (9) and (10), respectively.

The conditions (9) and (10) are LMI formula with respect to variables $\mathbf{Q}(\theta)$ and $\mathbf{Y}_i(\theta)$. By solving the variables $\mathbf{Q}(\theta)$ and $\mathbf{Y}_i(\theta)$ from (9) and (10), the parameter-dependent Lyapunov matrix variable $\mathbf{P}(\theta)$ and feedback gain $\mathbf{F}_i$ can be obtained as

$$\mathbf{P}(\theta)=\mathbf{Q}(\theta)^{-1} \text{ and } \mathbf{F}_i=\mathbf{Y}_i(\theta)\mathbf{Q}(\theta)^{-1} \tag{13}$$

In order to demonstrate the usefulness and applicability of the proposed fuzzy controller design approach, some numerical simulations are provided in the following section.

4 Simulations

In this section, the simulation results for the control of TTMR system are provided. According to the parameters stated in [3], it is assumed that the length of truck $L_1 = 2.8\ (m)$, the length of trailer $L_2 = 5.5\ (m)$, the sampling time $\Delta t = 2\ (sec.)$ and the speed of the backward movement $\theta(k) \in [-1.2,\ -0.8](m/sec.)$. Applying the LMI-toolbox of MATLAB, one can find the following solutions to satisfy the stability conditions of Theorem 2.

$$\mathbf{P}(\boldsymbol{\theta}) = \begin{bmatrix} 0.8647 & -1.9790 & 0.3879 \\ -1.9790 & 7.2693 & -1.4170 \\ 0.3879 & -1.4170 & 0.5523 \end{bmatrix},$$

$$\mathbf{F}_1 = [-3.0904 \quad 4.4009 \quad -0.6257] \text{ and } \mathbf{F}_2 = [-3.1091 \quad 3.2882 \quad -0.6461]$$

Employing the above feedback gains, the PDC-based fuzzy controller can be obtained from (4). With the initial condition $x(0) = \begin{bmatrix} 70^\circ & 80^\circ & 8 \end{bmatrix}^T$, the simulation results can be shown in Figures 2-4. From these simulation results, one can find that the TTMR system driven by the designed PDC-based fuzzy controller is asymptotically stable.

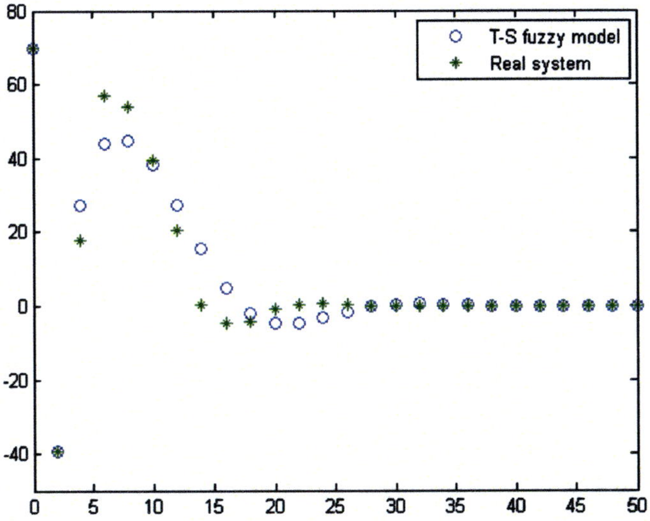

Fig. 2. Responses of state $x_1(k)$

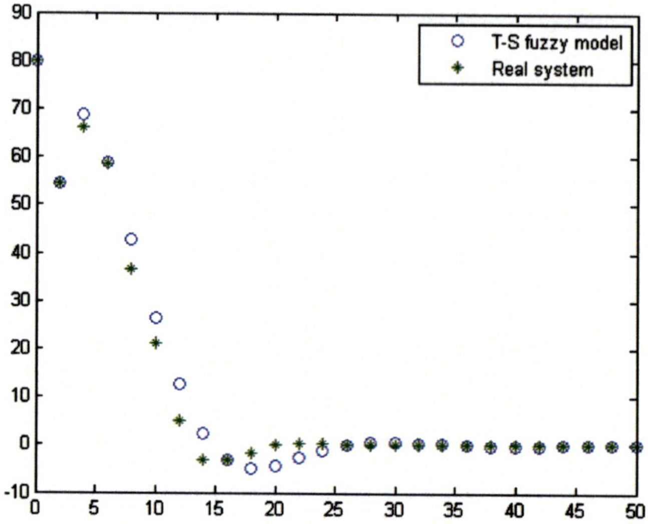

Fig. 3. Responses of state $x_2(k)$

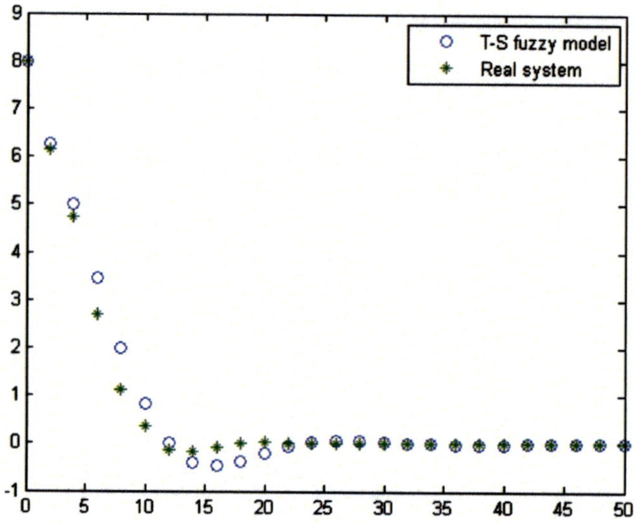

Fig. 4. Responses of state $x_3(k)$

5 Conclusions

The fuzzy controller design approach for the discrete LPV T-S fuzzy models has been developed in this paper. Based on the PDC concept, the stability conditions were derived via Lyapunov stability criteria. From the simulations for the control of TTMR system, it can be found that the proposed design methodology is applicable and effective.

Acknowledgements. This work was supported by the National Science Council of the Republic of China under Contract NSC100-2221-E-019-023.

References

1. Tanaka, K., Sano, M.: A robust stabilization problem of fuzzy control systems and its application to backing up control of a truck-trailer. IEEE Trans. Fuzzy Syst. 2(2), 119–134 (1994)
2. Chang, W.J., Shing, C.C., Chang, W.: Constrained discrete fuzzy control for a perturbed trailer type mobile robot stochastic system. In: Proc. of 2003 IEEE/ASME International Conference on Advanced Intelligent Mechatronics, Kobe, Japan, pp. 175–180 (2003)
3. Chang, W.J., Chang, W.: Synthesis of nonlinear discrete control systems via time-delay affine Takagi-Sugeno fuzzy models. ISA Transactions 44(2), 243–257 (2005)
4. Gahinet, P., Apkarian, P., Chilali, M.: Affine parameter-dependent Lyapunov functions and real parametric uncertainty. IEEE Trans. Automatic Control 41(3), 436–442 (1996)
5. Kajiwara, H., Apkarian, P., Gahinet, P.: LPV techniques for control of an inverted pendulum. Control Systems 19(1), 44–54 (1999)
6. Xie, W.: H_2 gain scheduled state feedback LPV system with new LMI formulation. In: IEE Proc. Control Theory and Applications, vol. 152(6), pp. 693–697 (2005)
7. Ge, D.M., Huang, X.L.: Fuzzy logic based gain-scheduled techniques. In: Proc. of the 7 World Congress on Intelligent Control and Automation, Chongqing, China, pp. 2170–2175 (2008)
8. Chang, W.J., Ku, C.C., Huang, P.H.: Robust fuzzy control of uncertain stochastic time-delay Takagi-Sugeno fuzzy models for achieving passivity. Fuzzy Sets and Systems 161(15), 2012–2032 (2010)
9. Boyd, S., et al.: Linear Matrix Inequalities in Systems and Control Theory. SIAM, Philadelphia (1994)

Discovering New Motor Primitives in Transition Graphs

Miha Deniša and Aleš Ude

Jožef Stefan Institute, Jamova 39, Ljubljana, Slovenia
{miha.denisa,ales.ude}@ijs.si

Abstract. In this paper we propose a methodology for discovering new movement primitives in a database of example trajectories. The initial trajectory data, which is usually acquired from human demonstrations or by kinesthetic guiding, is clustered and organized into a binary tree, from which transition graphs at different levels of granularity are constructed. We show that new movements can be discovered by searching the transition graph, exploiting the interdependencies between the movements encoded by the graph. By connecting the results of the graph search with optimized interpolation and statistical generalization techniques, we can construct a complete representation for new movement primitives, which were not explicitly present in the original database of example trajectories.

1 Introduction

Programming by demonstration, also called imitation learning, is a widely used approach for the acquisition of new behaviours [1,3]. Data suitable for learning by imitation is usually acquired from human demonstrations using optical marker systems or standard computer vision techniques, or by kinesthetic guiding. Due to the variability of problems that an autonomous robot might encounter in natural environments, it is not feasible to obtain all necessary movements from the demonstrated behaviors. To reduce the burden of teaching, a database of example motions must allow for new, not directly demonstrated motions to be discovered and new behaviors to be generated. In this paper we propose an approach that uses the concept of motion graphs and statistical generalization to create new movement primitives.

To organize the available example movements in a graph-like structure, we follow (with modifications) the approach of Yamane et al. [17]. While he proposed to cluster the state vectors from the example trajectories using principal component analysis (PCA) and minimum-error thresholding technique [5], we cluster our data with a k-mean algorithm [10]. Note that unlike Yamane et al., whose interest was primarily in full-body movements, we focus on arm trajectories and manipulation tasks, which often require fine precision. The results of clustering are used to construct a binary tree, which represents the captured data at different granularities. Every level of the binary tree is associated with a transition graph that describes transitions between the nodes at that level. The

S. Lee et al. (Eds.): Intelligent Autonomous Systems 12, AISC 193, pp. 219–230.
springerlink.com

nodes contain state vectors from all example trajectories. In contrast to Yamane et al. we do not assume connections between desired start and end nodes at the desired level. We employ a graph search algorithm and optimized interpolation to find best suited paths from desired start and end state vectors at the desired level of granularity. Statistical methods [16] are then utilized to generalize the newly discovered trajectories and to create a complete representation for a newly discovered motor primitive.

The rest of the paper is organized as follows. In Section 1.1 we describe previous research related to this paper. Section 2 presents the process of constructing the database. The following section deals with the discovery and generalization of motor primitives using the transition graphs and optimized interpolation. Experimental evaluation of our approach is presented in Section 4, while the conclusion and summary are given in Section 5. At the end a brief review of dynamic movement primitive representation is given in the appendix.

1.1 Related Work

A lot of research on imitation is focused on learning from a single demonstration, like for example in the case of dynamic movement primitives [4], or learning from multiple demonstrations, where each demonstration is a variant of the same type of movement, like in [2] when learning table tennis strokes. Other research includes segmentation and sequencing of motor primitives [7], storing movement primitives as basis functions [9] and the generation of mixtures of motor primitives [11]. However, compared to the work in computer graphics community, who has always assumed that a large database of diverse motion capture data is available for the generation of computer animations, the number of example movements considered in imitation research has usually been more limited and the example trajectories has been less diverse. The work of Kulić et al. [7,8] is a notable exception. On the other hand, the computer graphics community have shown that by exploiting the structured nature of motion capture data in large databases, which is made evident in motion graphs, smooth transitions between interconnected body movements can be found [13]. We therefore base the discovery of new movement primitives on a graph search process, which enables us to find subtler transitions between different example movements in the database.

Motion graphs have been proposed and investigated by the computer graphics community to encapsulate connections in the available motion capture data [6]. Kovar et al. [6] applied them to generate different styles of locomotion along arbitrary paths. A graph search algorithm was employed to find nodes that represent possible transitions between parts of the captured movements. Safonova and Hodgins [14] combined motion graphs and interpolation techniques to increase the number of paths through the graph. While the motion graph literature is vast, our work is most related to the approach of Yamane et al. [17], who used a binary tree to organize the data and the resulting transition graphs to generate human body locomotion on a desired path. A binary tree structure was also used by Sidenbladh et al. [15] for an efficient sampling of human poses.

2 Building the Database

In our approach, a database of example trajectories is organized using an approach similar to the one proposed by Yamane et al. [17]. It will be pointed out where our approach differs. In the developed system, example trajectories, which need to be organized into a graph-like data structure, can be given either in task or in joint space. For the purpose of building the database, we concatenate the acquired trajectories in a sample motion matrix:

$$\boldsymbol{X} = [\boldsymbol{x}_1 \ \boldsymbol{x}_2 \ \cdots \ \boldsymbol{x}_{N_f}], \tag{1}$$

where $\boldsymbol{x}_i$ denotes the state vectors sampled at a given discrete time interval and N_f is the total number of all postures belonging to the example trajectories incorporated into the database. State vectors for the end-effector trajectories specified in Cartesian space are given as

$$\boldsymbol{x}_i = [p_{xi} \ \dot{p}_{xi} \ p_{yi} \ \dot{p}_{yi} \ p_{zi} \ \dot{p}_{zi}]^T, \tag{2}$$

where p_{ji} and $\dot{p}_{ji}$, $j = x, y, z$, denote the position and velocity at time t_i. If the trajectories are given in joint space, then state vectors are defined by

$$\boldsymbol{x}_i = [q_{1i} \ \dot{q}_{1i} \ q_{2i} \ \dot{q}_{2i} \ \cdots \ q_{di} \ \dot{q}_{di}]^T, \tag{3}$$

where j-th joint angle and its velocity at time t_i are denoted by q_{ji} and $\dot{q}_{ji}$, and d is the number of the robot degrees of freedom (DOF).

Once example trajectories are stored and arranged in a sample matrix, they can be utilized to build a binary tree. The complete sample data matrix $\boldsymbol{X}$ represents the root node, which we split using the k-mean algorithm and thus gain its two child nodes. Those nodes are then split again to gain nodes at the next depth of our binary tree. Yamane el al. [17] based their criterion for when to stop splitting the tree nodes solely on the number of state vectors that remain in the two following child nodes. Instead, we use a criterion based on the variability of data contained in the node. We define the mean distance d_k of node k as

$$d_k = \frac{\sum_{i=1}^{n_k} d(\boldsymbol{x}_{ki}, \boldsymbol{c}_k)}{n_k}, \tag{4}$$

where n_k denotes the number of state vectors clustered at node k. For $d(\boldsymbol{x}_{ki}, \boldsymbol{c}_k)$ euclidean distance between state vectors $\boldsymbol{x}_k$ associated with this node and the nodes centroid $\boldsymbol{c}_k$, gained with k-mean algorithm, is used. If d_k is lower than a predefined threshold, then state vectors contained in the node are similar. By not splitting the nodes in this case, we avoid new nodes that would make the binary tree unnecessarily deep. On the other hand, we split the nodes even if they include a small number of state vectors if these state vectors are sufficiently diverse. In this way we gain on precision of our representation while preventing the binary tree and the resulting transition graphs from becoming unnecessarily large. With these criteria we cluster the data into nodes until we do not have any nodes left to split. We extend every branch to the last level by copying the leaf nodes and so ensuring that all state vectors are represented at all levels of the binary tree.

For further processing it is not necessary to store all state vectors at each node of the binary tree. Storing the mean of all state vectors associated with the node is sufficient. Only if the node contains exactly one final configuration on a trajectory, we store this final configuration instead of the mean of all state vectors. In this way we ensure that movements generated by graph search end in the same end points as the original movements. At each depth of the binary tree, we build a transition graph that represents all possible transitions between the nodes at the current depth. The edge weights in the transition graph represent the probability of transition from one node to another. Transition probability from node k to node l is estimated by

$$p_{kl} = \frac{t_{kl}}{n_k} \tag{5}$$

where $t_{k,l}$ denotes the number of transitions observed in all trajectories of the original data, i. e. the number of all state vectors clustered in node k that have a successor in node l. Given the binary tree, average state vectors (or end points) at nodes of the binary tree, and transition graphs for every depth, we can now search for new movements in the transition graphs.

3 Discovering New Movement Primitives

We start the process of discovering new discrete movement primitives, i. e. point-to-point movements such as reaching and grasping, by selecting the desired start and end points on two different trajectories. If the desired start and end point do not belong to one of the original example movements, we first establish the closest start and end node in the transition graph. At this point we also need to decide for a level in the binary tree, which determines the fidelity of reproduction compared to the original trajectories. We then try to find a path between these two nodes at the chosen level by using the A* algorithm. If the desired start and end nodes belong to the same connected component in the transition graph at this layer, the shortest path is found. If they do not, we can use the binary tree like structure of our database to find a level at which the connection exists. However, this higher level does not have the desired granularity. To achieve the proper granularity we need to move down the layers to the desired depth (see Fig.1). We do that by finding the node which division caused the break in the connection. This is the node that has only one child accessible from the start node at the next layer. We then store both parts of the path, i. e. the path connecting start node to that child and the path connecting the second child to the desired end node. If another break occurs while we move down the layers, we repeat this process and store the associated path parts. Once we return to the desired level, we have all parts of the path we are searching for.

Once the shortest path or its parts have been found, we obtain a series of nodes, i. e. mean values of state vectors. New example trajectories (or its parts)

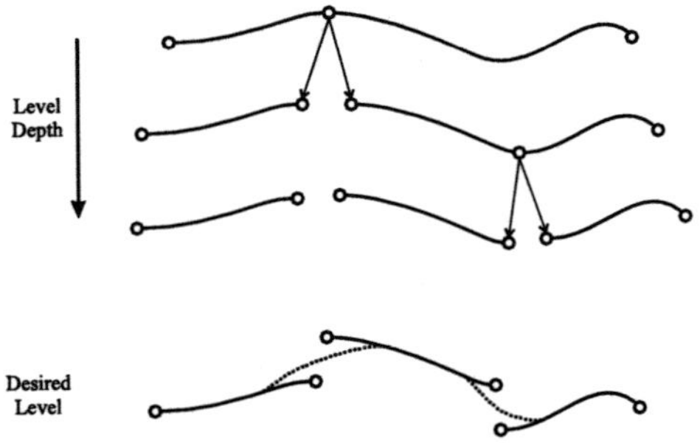

Fig. 1. Finding non-connected path parts at the desired level. Nodes are denoted by circles, while the connections are shown with lines. Dashed lines denote interpolated transitions.

are defined by these state vectors, which need to be enhanced by the time evolution of the trajectory. For this purpose we define

$$\tau_{i,j} = \frac{\tau_i}{2} + \frac{\tau_j}{2}, \tag{6}$$

where $\tau_{i,j}$ denotes the time between two successive nodes and τ_i is the duration of a single node, which is defined by

$$\tau_i = \frac{n_i}{p_i} \Delta t, \tag{7}$$

where $1/\Delta t$ is the sampling frequency of example movements. In Eq.(7), n_i denotes the number of state vectors clustered in node i and p_i the number of trajectories passing through node i.

If our process found non-connected path parts at the desired level, we use optimized interpolation to generate a complete movement. We interpolate a

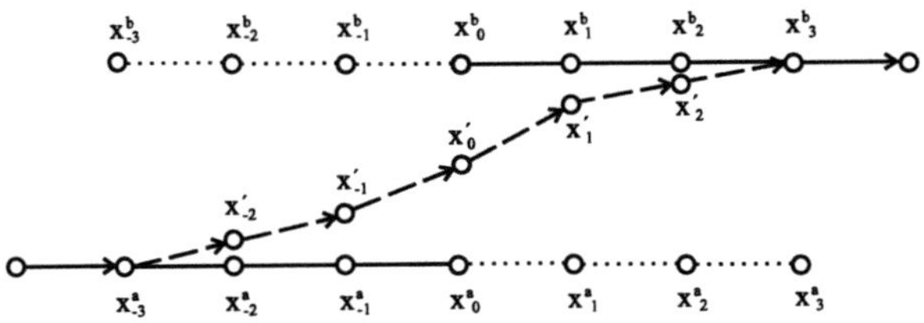

Fig. 2. Transition by interpolation. Nodes are represented by circles and connections by lines. Solid line denotes found connections i.e. parts of the path at the desired level. Dashed lines connect interpolated nodes i.e. state vectors. Nodes that are needed for interpolation but are not on our path parts are connected by dotted lines.

certain number of state vectors belonging to the nodes around the two break nodes (see Fig. 2)

$$\mathbf{x}_i' = \lambda_i \mathbf{x}_i^a + (1 - \lambda_i)\mathbf{x}_i^b. \tag{8}$$

The number of nodes we choose to interpolate is a compromise between the smoothness of the transition and the preservation of original paths. Non-linear local optimization algorithm BOBYQA [12] is used to find weights λ and time durations of individual nodes in such a way as to minimize the jerk of the transition trajectory.

After specifying a number of such start and end points, we end up with a set of new example trajectories $\boldsymbol{M}_k$. They are encoded by state vectors associated with the graph nodes and the start and end configurations on the trajectory:

$$\boldsymbol{M}_k = \left\{ \{\boldsymbol{x}_i^k, t_i^k\}_{i=1}^{n_k}, \boldsymbol{p}_{start}^k, \boldsymbol{p}_{end}^k \right\}. \tag{9}$$

The start and end configurations are always specified in task space and are the same as the first and last state vectors on the path if examples are given in task space, whereas they are related to these configurations via forward kinematics if the example trajectories (and consequently state vectors) are given in joint space. Even if a large database of example movements is available, it is highly unlikely – at least in the case of manipulation tasks – that an exact desired movement can be found via direct graph search. To accomplish a task such as reaching towards an object, we need to generalize the discovered example movements to

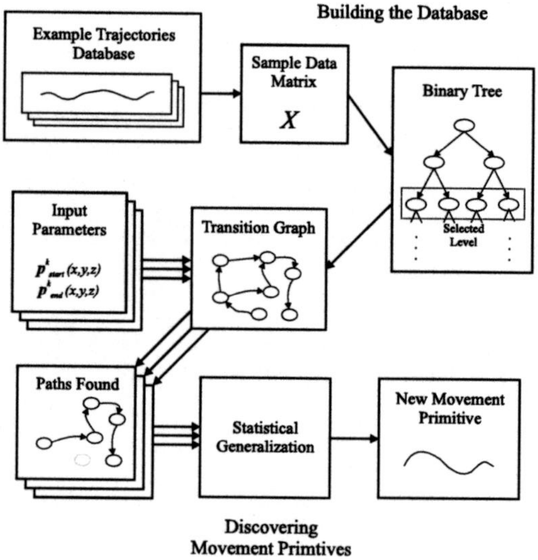

Fig. 3. Process of discovering new movement primitives. Block diagram represents our process of building the database and then using it to generate new movement primitives that were not present in the example database.

new reaching configurations. This can be accomplished by means of statistical generalization, where a movement suitable for the current object configuration is synthesized from a number of example trajectories. The complete discovery process is illustrated in Fig. 3.

For execution of a discrete movement on a robot we represent the trajectory of every degree of freedom by a dynamic movement primitive (DMP) [4]. Given the DMP representation (see appendix for a brief review of DMP representation) and trajectory data (9), Ude et al. [16] showed how to generalize the example trajectories to any combination of start and end configurations $\{\boldsymbol{p}_{start}, \boldsymbol{p}_{end}\}$ within the trajectory training space. Essentially, they proposed a methodology to learn a function

$$\boldsymbol{F}(\{\boldsymbol{M}_k\}) : [\boldsymbol{p}_{start}^T, \boldsymbol{p}_{end}^T]^T \longmapsto [\boldsymbol{w}^T, \boldsymbol{g}^T, \tau]^T, \tag{10}$$

where $\boldsymbol{w}$, $\boldsymbol{g}$ and τ are the parameters that define a DMP that starts in $\boldsymbol{p}_{start}$ and ends in $\boldsymbol{p}_{end}$. Function $\boldsymbol{F}$ is learned by a combination of locally weighted regression and Gaussian process regression. We omit the details. What is important for us here is that for this method to work, the discovered trajectories $\boldsymbol{M}_k$ need to transition smoothly between each other as a function of start and end configuration $\boldsymbol{p}_{start}$ and $\boldsymbol{p}_{end}$. It is the task of the graph search process to find paths through the transition graph that fulfill this condition.

4 Evaluation

Our process was evaluated with two experiments. First experiment was done in a simulation with a set of constructed movements with position end-effector trajectories. The second used sets of movements captured through kinesthetic guidance of a robot arm.

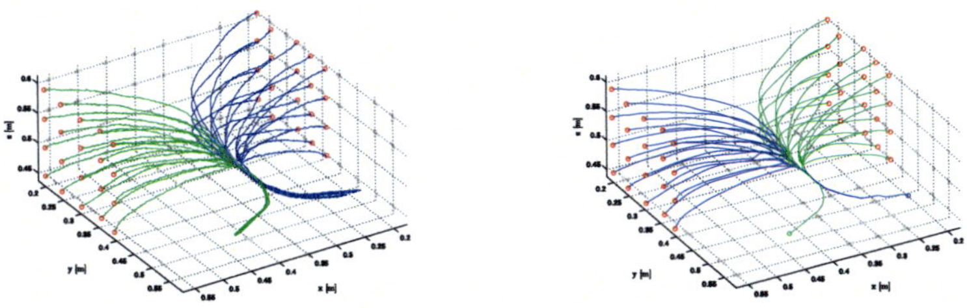

Fig. 4. Two sets of reaching movements used to construct the database. Every set of reaching movement has roughly the same starting point but 20 different end points, as shown on the left figure. Trajectories from one set are in green, from the other in blue. For better visualization, the end points are marked with circles.Two new sets of reaching trajectories that were discovered by graph search, but were not present in the original example database. They are shown on the right figure. One set of movements is shown in green, the other in blue. End points are again marked in red.

In a simulation experiment we constructed a database using two sets of 20 reaching movements. In both sets all movements have approximately the same starting point, but different ending points. Ending points cover a plane at roughly 5 cm intervals. The end-effector trajectories in task space can be seen in Fig. 4.

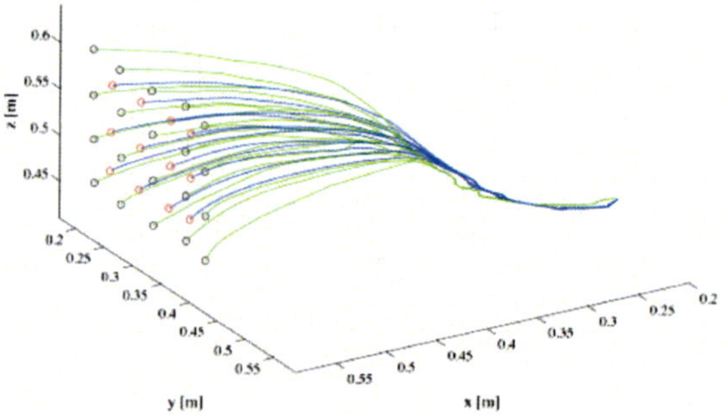

Fig. 5. Generalization of new reaching movements. The newly found trajectories, which were discovered by graph search, are shown in green with end points shown as black circles. Trajectories towards in-between reaching configurations (red circles) computed by statistical generalization (see Eq. (10)) are shown in blue. Note that the generalized trajectories preserve the shape of the initial reaching movements. A new movement primitive has been discovered and learned.

From these two sets of movements (altogether 40 trajectories) we constructed a database as described in Section 2. The resulting binary tree had 20 levels and 1862 nodes at the deepest level. Using the transition graph associated with the deepest level of the binary tree, we generated new movements by selecting a starting point in the first set of reaching movements and the ending point in the second set. Due to the structure of the training data, there always existed a path between these two points. Our graph search procedure was able to generate two new sets of movement, where each of the new movements starts in one set of reaching movements and ends in the seconds set (see Fig. 4). In this way we could generate two new sets of reaching trajectories that were not part of the initial database. It is especially important that the newly generated movements smoothly transition from one to another because this is a precondition for successful generalization. The results of the generalization process are shown in Fig. 5. Only the end points were used as input parameters for generalization because the starting points were same over the complete set of example movements. Note that the generalized trajectories have a similar shape as the ones discovered by graph search. In our second experiment we constructed a database using two sets of 15 reaching movements acquired by kinesthetic guiding on anthropomorphic Kuka Light Weight Robot arm. Again all movements have approximately the same starting point, but different ending points in each set. Ending points

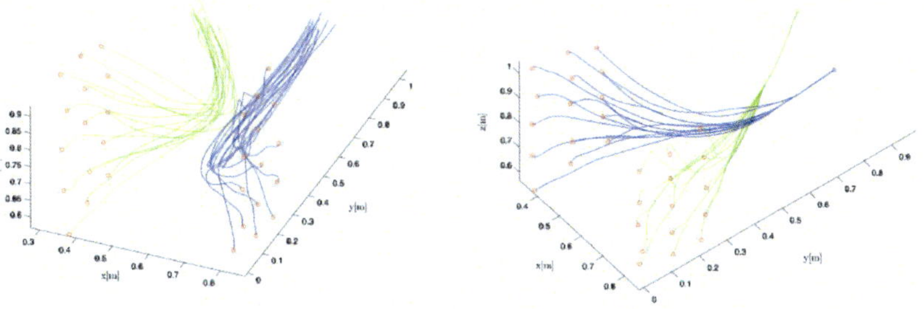

Fig. 6. Two sets of reaching movements, which were acquired by kinesthetic guiding of the Kuka Lightweight Robot arm. Every set of reaching movement has roughly the same starting point but 15 different end points, as shown on the left figure. Trajectories from one set are in green, from the other in blue.Two new sets of reaching trajectories that were discovered by our procedure, but were not present in the original example database. They are shown in the right figure. One set of movements is shown in green, the other in blue. End points are again marked in red.

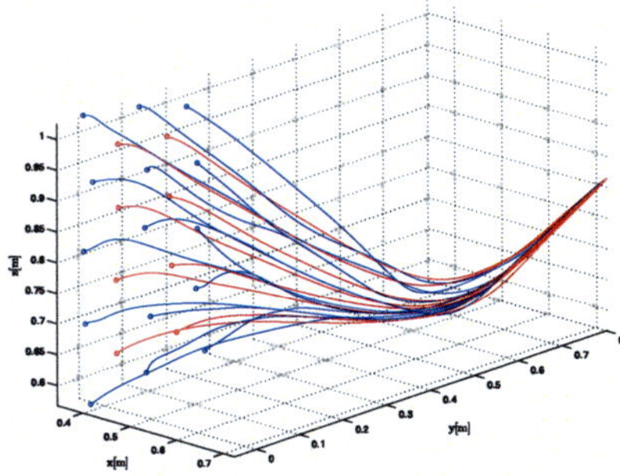

Fig. 7. Generalization of new reaching movements. The newly found trajectories, which were discovered by our procedure, are shown in blue with end points shown as blue circles. Trajectories towards in-between reaching configurations (red circles) computed by statistical generalization (see Eq. (10)) are shown in red. Note that the generalized trajectories preserve the shape of the new reaching movements, which are themselves obtained by concatenating the original trajectories acquired by kinesthetic guiding.

cover a plane at roughly 10 cm intervals. The end-effector trajectories in task space can be seen in Fig. 6.

From these two sets of movements (altogether 30 trajectories) we constructed a database as described in Section 2. The resulting binary tree had 17 levels and 2954 nodes at the deepest level. As in previous experiment, we generated new movements by selecting a starting point in the first set of reaching movements and the ending point in the second set. Since we chose the deepest level of the binary tree, new movements could not be generated only through path search in transition graph as the two sets did not share any nodes at that level. That can

Table 1. Properties of the constructed database

	Database 1	Database 2
No. of example trajectories	40	30
No. of state vectors	23536	8877
No. of levels	20	17
No. of nodes at the deepest level	1862	2954

also be observed in Fig. 6, where it is seen that our sets do not have any common parts. Despite of that, two new sets of movement were generated by our procedure (see Fig. 6). As in the first experiment every new movement starts in one set of reaching movements and ends in the seconds set. Although the initial sets of movements were not as carefully constructed as in the previous experiment, newly generated movements smoothly transition from one to another, which is a precondition for a successful generalization. The results of the generalization process are shown in Fig. 7. Note the similar shape of the generalized trajectories in regards to those discovered by graph search. At the end we executed new sets of smooth movements on our LWR Kuka robot arm.

5 Conclusion

While more experiments need to be conducted to fully evaluate the proposed approach, the first experiments presented here demonstrate that new movement primitives can be discovered and learned from a database of training movements using a combination of graph search, optimized interpolation and statistical generalization. Unlike many other approaches in imitation learning, we assume that the training database consists of different types of movements, which we combine to new movements by exploiting the interconnections between them. This approach does not need a direct connection between desired points at the selected level of granularity to generate new sets of example movements that can be used for generalization. In this way a complete representation for the new movement primitive can be calculated.

Acknowledgements. The research leading to these results has received funding from the European Communitys Seventh Framework Programme FP7/20072013 (Specific Programme Cooperation, Theme 3, Information and Communication Technologies) under grant agreement no. 270273, Xperience.

Appendix - Dynamic Movement Primitives

DMPs are defined by the following nonlinear system of differential equations

$$\tau\dot{v} = K(g - y) - Dv + f(x), \quad (11)$$

$$\tau\dot{y} = v. \quad (12)$$

The linear part of Eq. (11) – (12) ensures that y converges to the desired final configuration, denoted as g. The nonlinear part $f(x)$ modifies the shape of a movement and is defined by a linear combination of radial basis functions

$$f(x) = \frac{\sum_{i=1}^{N} w_i \psi_i}{\sum_{i=1}^{N} \psi_i} \quad (13)$$

$$\psi_i = \exp(-h_i(x - c_i)^2), \quad (14)$$

where ψ_i defines the basis functions, with centres at c_i and widths $h_i > 0$. As seen in Eq. (13), $f(x)$ is not directly time dependent. Instead, phase variable x defined in Eq. (15), with initial value $x(0) = 1$, is used to make the dependency more implicit:

$$\tau\dot{x} = -\alpha_x x \quad (15)$$

The phase is common across all DOF. By specifying the time evolution through phase, it becomes easier to stop the clock in case of external perturbations which cause the robot to deviate from the desired trajectory. It can be shown that – given the properly defined constants K, D, τ, $\alpha_x > 0$ – the above system is guaranteed to converge to the desired final configuration g.

References

1. Breazeal, C., Scassellati, B.: Robots that imitate humans. Trends Cognitive Sci. 6(11), 481–487 (2002)
2. Calinon, S., D'halluin, F., Sauser, E.L., Caldwell, D.G., Billard, A.G.: Learning and reproduction of gestures by imitation: An approach based on Hidden Markov Model and Gaussian Mixture Regression. IEEE Robot. Autom. Mag. 17(2), 44–54 (2010)
3. Dillmann, R.: Teaching and learning of robot tasks via observation of human performance. Robotics and Autonomous Systems 47(2-3), 109–116 (2004)
4. Ijspeert, A.J., Nakanishi, J., Schaal, S.: Movement imitation with nonlinear dynamical systems in humanoid robots. In: Proc. IEEE Int. Conf. Robot. Autom., Washington, DC, pp. 1398–1403 (2002)
5. Kittler, J., Illingworth, J.: Minimum error thresholding. Pattern Recognition 19(1), 41–47 (1986)
6. Kovar, L., Gleicher, M., Pighin, F.: Motion graphs. ACM Transactions on Graphics 21(3), 473–482 (2002)
7. Kulić, D., Takano, W., Nakamura, Y.: Online segmentation and clustering from continuous observation of whole body motions. IEEE Trans. Robot. 25(5), 1158–1166 (2009)

8. Kulić, D., Takano, W., Nakamura, Y.: Incremental learning, clustering and hierarchy formation of whole body motion patterns using adaptive hidden Markov chains. Int. J. Robot. Res. 27(7), 761–784 (2008)
9. Lim, B., Ra, S., Park, F.C.: Movement Primitives, Principal Component Analysis, and the Efficient Generation of Natural Motions. In: Proceedings of the 2005 IEEE International Conference on Robotics and Automation, ICRA 2005, pp. 4630–4635 (2005)
10. MacQuuen, J.B.: Some methods for classification and analysis of multivariate observation. In: Proc. of the 5th Berkley Symposium on Mathematical Statistics and Probability, pp. 281–297 (1967)
11. Peters, J., Mülling, K., Kober, J., Nguyen-Tuong, D., Krömer, O.: Towards Motor Skill Learning for Robotics. In: Pradalier, C., Siegwart, R., Hirzinger, G. (eds.) Robotics Research. STAR, vol. 70, pp. 469–482. Springer, Heidelberg (2011)
12. Powell, M.J.D.: The BOBYQA algorithm for bound constrained optimization without derivatives, Department of Applied Mathematics and Theoretical Physics, Cambridge England. Technical report NA2009/06 (2009)
13. Rose, C., Cohen, M.F., Bodenheimer, B.: Verbs and adverbs: multidimensional motion interpolation. IEEE Computer Graphics and Applications 18(5), 32–40 (1998)
14. Safonova, A., Hodgins, J.K.: Construction and optimal search of interpolated motion graphs. ACM Transactions on Graphics 26(3) (2007)
15. Sidenbladh, H., Black, M.J., Sigal, L.: Implicit Probabilistic Models of Human Motion for Synthesis and Tracking. In: Heyden, A., Sparr, G., Nielsen, M., Johansen, P. (eds.) ECCV 2002, Part I. LNCS, vol. 2350, pp. 784–800. Springer, Heidelberg (2002)
16. Ude, A., Gams, A., Asfour, T., Morimoto, J.: Task-specific generalization of discrete and periodic dynamic movement primitives. IEEE Trans. Robotics and Autom. (5), 800–815 (2010)
17. Yamane, K., Yamaguchi, Y., Nakamura, Y.: Human motion database with a binary tree and node transition graphs. Presented at the Robot. Sci. Syst. Conf., Seattle, Washington (2009)

Design of a Space Robot System to Simulate Climbing of Astronaut Based on Binocular Vision System

Que Dong, Yu He, Hongjie Li, Bo Wei, Xiaopeng Chen, Hui Li, Zhihong Jiang*, and Qiang Huang

Intelligent Robotics Institute, Key Laboratory of Biomimetic Robots and Systems, Ministry of Education
School of Mechatronical Engineering, Beijing Institute of Technology, 5 Nandajie, Zhongguancun, Haidian, Beijing, 100081, China

Abstract. The exploration and development of space science nowadays is explosive, the tasks which astronauts need to achieve in space are also more dangerous and complex. Therefore, space robot is needed to implement the EVA of astronauts to do these space missions (EVA—Extravehicular activity). Firstly, we designed a kind of space robot with vision system and also we discussed a vision information processing method based on a binocular stereo vision system which is used on our space robot. Secondly, we designed a passive experimental platform which can simulate the weightless environment in space for space robot to simulate climbing of astronaut, and a method is adopted to improve the stability of the passive mechanism by analyzing force and vibration condition. Lastly, based on this robot system, space robot could simulate the extravehicular movement of astronaut based on its binocular vision system, and this validated presented methods in this paper.

Keywords: space robot, visual system, trajectory planning.

1 Introduction

With the increase of space exploration, there is an increasing need for space robots. Space robots can not only replace the astronauts for space operations, reduce the burden and reduce operational risk for the astronauts, but also expand the ability of astronauts for space operations, complete the tasks which the astronauts can't do. It can improve the efficiency and achieve unmanned and automatic space operation. At present, the main tasks of space robots are assisting the astronauts or ground operators to complete space station docking, equipment installation, space science experiments, space cabin inspection or maintenance and other tasks.

In this paper, the experimental platform is a kind of space robot we designed, According to the movement of the robot's task, we analyzed the method by which the vision system process the vision information and control system use the information. Also we improved the passive mechanism to avoid vibration. The whole system is composed of the robot with its vision system, surveillance and control system, and

* Corresponding author.

S. Lee et al. (Eds.): Intelligent Autonomous Systems 12, AISC 193, pp. 231–243.
springerlink.com

passive mechanism with climbing rods to counteract gravity. Movement which the robot needs to do is climbing by rods in horizontal plane to simulate the movement which astronauts do outside space station.

The organization of this paper is as follows. In section 2, we introduced our experimental system, including the mechanical configuration of robot, passive mechanism, setting function of visual targets and hardware of the vision system. In section 3, we introduced the method and steps of vision information processing. We have done the experiment to validate the method and the experiment results are given in section 4. Finally, our conclusion is given in section 5.

2 Background Experimental System of Space Robot

When astronauts need to move outside of the space station, and have to climb along the rods which are set aforehand. Motion of an astronaut outside space station is shown in Fig.1.

Fig. 1. Motion of astronauts outside space station

Based on this motion, we designed a system to simulate an astronaut climbing outside space station. So the object we researched is a space robot system. The space climbing robot system is shown in Fig.2. It is composed of space climbing robot, passive mechanism, climbing rods and support mechanism.

Fig. 2. Space climbing robot system

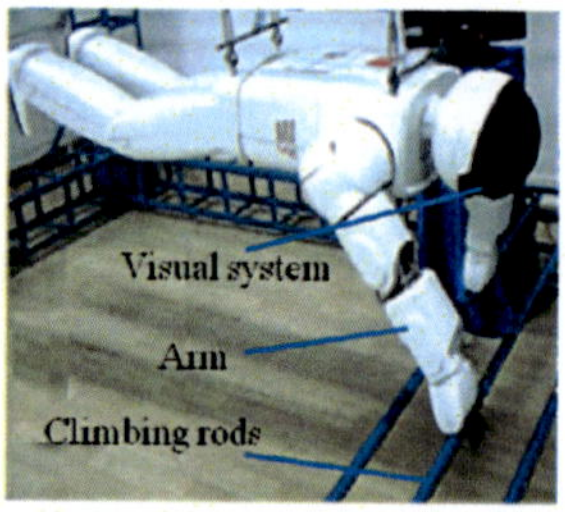

Fig. 3. Space climbing robot

2.1 Space Robot

The space robot is composed of head, trunk, and two arms, and expected to replace the astronauts to do the extravehicular activity such as space station maintenance and construction. As shown in Fig 3, it has 16 degrees of freedom. The head has two degrees of freedom, each arm has six degrees of freedom and the hand has one degree of freedom. The binocular stereo vision system is set in the head of the robot, and the position of the climbing rod can be calculated by the binocular stereo vision system. With the vision system guidance, the hand of the space climbing robot can grab the climbing rod, two hands of the robot alternately grab the climbing rod, and the robot can move forwards in the horizontal plane.

Space climbing robot control system consists of industrial computer and joint-controller. The industrial computer and joint controller are connected by CAN bus. RT-Linux is operating system of the industry computer, and RT-Linux is used to meet the real-time and deadline requirements [1-4].

2.2 Passive Mechanism and Climbing Rods

To conduct earthbound research on robotic applications for outer space environments, compensation for the gravitational field must be considered. Several schemes have been employed to compensate for the earth's gravity, including free falling motion and underwater experiments. Free falling motion method is that the robot does flat projectile motion on a plane to simulate the weightless environment, but the free falling motion method is high cost, and testing time is short. In underwater experiments, the robot is immersed in water, and the floating buoyancy is precisely adjusted to make the water buoyancy be equal to the gravity. However, the underwater experiment is expensive, and sealing requirements of the robot are very high. Therefore, in our system, the space climbing robot is suspended by the passive mechanism to simulate the weightless environment [5].

Passive mechanism is designed to simulate the weightless environment. As shown in Fig.2, the passive mechanism is composed of support mechanism, 4 guides, 8 slides, and the support board. Two guides are placed horizontally, and the other two guides are placed vertically. Two horizontal guides and four slides ensure the robot to achieve lateral movement; two vertical guides ensure the robot to achieve vertical movement. Therefore, four guides and eight slides can guarantee the robot to achieve any movement in the horizontal plane.

2.3 Vibration Restraint

The robot is combined with the passive mechanism by four chains which are same length. The chains are flexible. When the robot is moving, the payloads of motors change in a large range in different motion process, and the robot is easily vibrated. There are two kind of most serious vibration of this robot, one is fore-and-aft vibration when two arms grasping and the other is deflection vibration when only one arm grasping the rod.

- Vibration when two arms grasping

When the robot is moving forward by two arms, as shown in Fig.4, there is a force F on the robot from the arm grasping rod. Chains are pulled by this force, and a

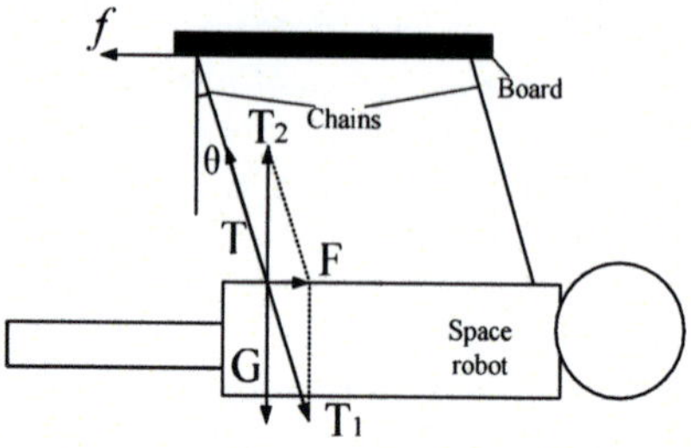

Fig. 4. Forces of robot during motion

deflection angle θ is generated. Chains pulled the board which is on the top of the passive mechanism to overcome the friction and move forward with the robot.

Because sliding friction is smaller than static friction, when the board of passive mechanism begin moving, the friction would reduce suddenly, the board would move forward accelerated, and the velocities of the robot and board are different. Thus, the angel θ would reduce also, it even become a negative value. In a short time, the chains will be flabby, which means the force T in Fig.4 is disappeared, so the robot will move down by its gravity, until the chains are pulled tight again. The process all above will generate swing of robot, and vibration is also generated. The flexible chains can magnify the vibration and even generate sympathetic vibration.

■ Deflection when one arm grasping

Because the pull from the support mechanism is not vertical, there's component of forces backwards. When the robot grasps a rod by one hand, the pull is taken on one arm which is set on one side of the trunk, however it was taken on two arms before, at this time, the grasping force might be limited because of the motor's ability, so the deflection will occur on the trunk of the robot. As shown in Fig 4, the robot can keep balance when it grasps by two hands (Fig.5 (a)). But if one hand releases, the force gained by it will disappeared, the robot will deflect by the force shown in Fig.5(b), then deflection happens (Fig.5(c)).

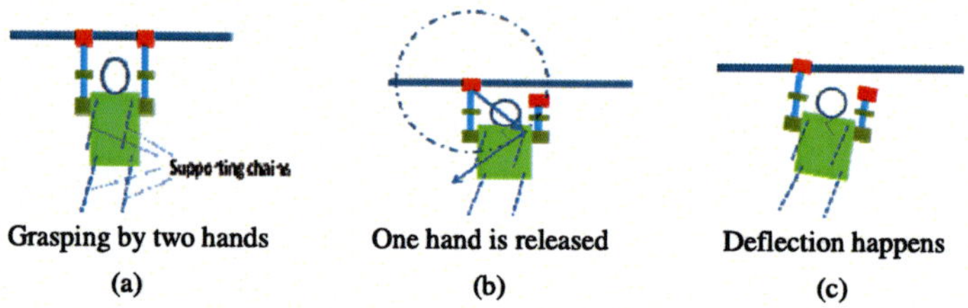

Fig. 5. Deflection caused by the pull from the chains

■ Vibration and deflection restraint

From all above we can know that the reason of vibration is flexibility of chains, if we combine rigid structure with the chains, the vibration will be restraint. So we designed a structure as shown in Fig.6. We designed an "X" structure which is composed of two crossed girder to combine the robot with the top board of passive mechanism.

Thus the force on the board which is given by robot is transferred by "X" structure, but not the flexible chains. At the same time, we put four rigid short rods to combine chains, the rods and "X" structure make a rigid and steady structure to combine the robot and top board. All the reasons which can generate vibration and deflection can be avoided by improving this structure. This structure is validated by experiment. The "X" structure can reduce for-and-aft vibration, and short rods can reduce deflection vibration.

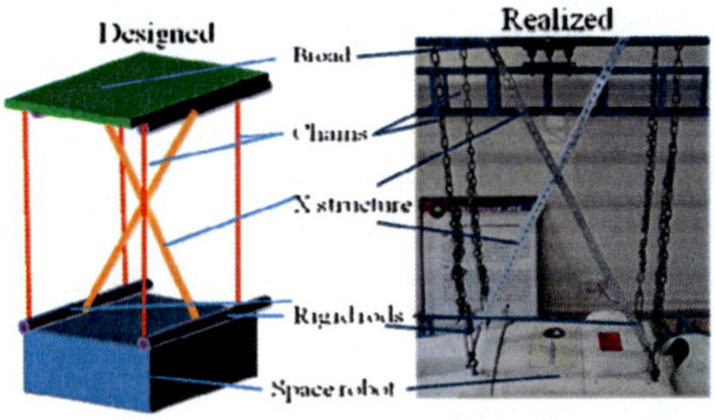

Fig. 6. Structure to restraint vibration and deflection

2.4 Bayesian Network

The binocular stereo vision system is composed of two parts, the vision processing circle board and two cameras. The two cameras are set parallel on the tripod head which is in the robot's head. The tripod head has two degrees of freedom, to achieve movement of pitching and rotation, its configuration is shown as Fig.7.

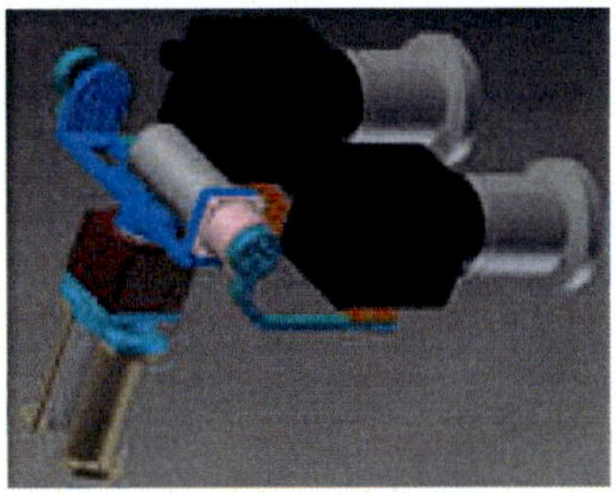

Fig. 7. Tripod head

Fig. 8. Vision processing circle board

The vision processing circle board is shown as Fig.8. Images got from two cameras can be translated to analogue video signal by the board. This circle board is a basic hardware to capture images, process and transfer video information, it is composed of double ADC units, FPGA chip, DSP, network and communication unit, serial communication unit and memory unit.

The structure of binocular stereo vision system is shown as Fig.9, images are captured by cameras and DSP chip transform the target from the reference frame of camera to the reference frame of robot. If the target is out of visual field, DSP will control the tripod head to search the grasping target.

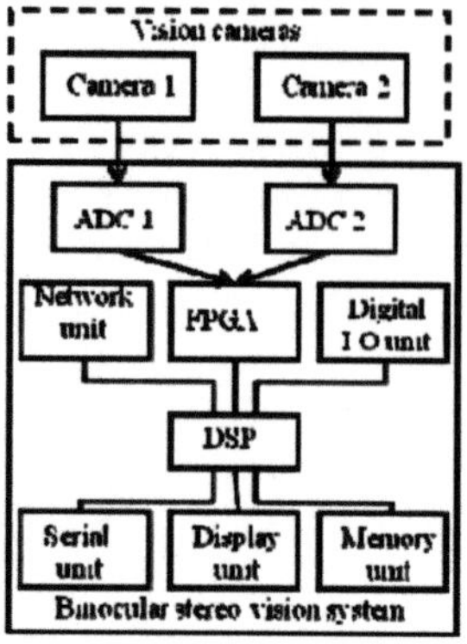

Fig. 9. Structure of binocular stereo vision system

3 Vision Signal Processing Method

Through the vision processing system, positions of visual targets can be calculated on time, and the process that visual system gets the three-dimensional position of the climbing rod is as follows.

1. The intrinsic parameters of camera can be obtained by the monocular calibration of the left camera and right camera. The extrinsic parameters describing the relative position of two cameras can be obtained through the binocular calibration;
2. The left camera and right camera get the image at the same time respectively, and get the information of analog images;
3. The analog images are converted into digital images by two ways of analog conversion modules;
4. FPGA module filters the digital image;
5. According to the left and right camera matrix, by the three-dimensional measurement of binocular vision algorithms, the digital signal processing module obtains the image's feature points of the two images from the filtered digital image. The target's coordinate in the real world's coordinate can be got from the optimal corresponding points of left and right images by the matching algorithm. Then the three-dimensional coordinate of the climbing rod is got.

The process of dual visual system programming is shown as Fig.10. This visual system can get two images by two cameras on time, if the target grasping position target is out of its visual field, it needs to make global searching to the target rod. When the target grasping position is appeared in both the two cameras' visual fields, the system can confirm position and direction of grasping by the method of dual visual position confirming. Thus, the robot can get both the position of target rod and itself, then the trajectory planning can be progressed, and the robot can achieve its climbing motion by itself.

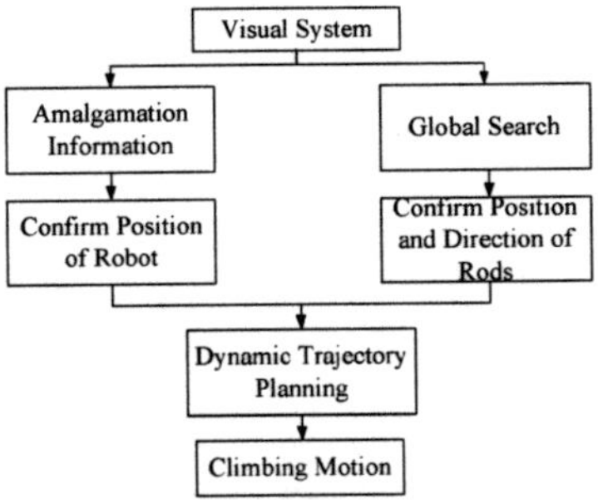

Fig. 10. Process of visual system programming

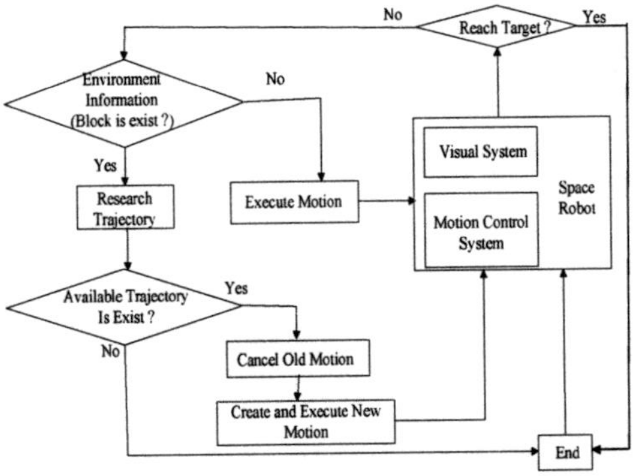

Fig. 11. Process of motion planning

The trajectory planning and motion control system complete the motion planning and obstacle escaping, and drive the joint motors by progressing the environment information including the robot, target and obstacle which are gotten by visual system. The process is shown in Fig.11.

3.1 Vision Processing System

Camera calibration is necessary in vision information process. It includes two parts, which are cameras external calibration and cameras internal calibration. We need to complete cameras internal calibration first because when light is through the lens, it might refract to bring errors, so we need to adapt the images by external calibration , also we need to complete cameras external calibration to ensure the relative position of two cameras, so the positions of visual targets can be captured more accurately.

First, we calibrate internal parameters for the two cameras by Zhang Zhengyou calibration method [8].By this method the internal and external parameters can be calculated as (1).

$$\min \sum_{i=1}^{n} \sum_{j=1}^{m} \left\| m_{ij} - \hat{m}(A, R_i, t_i, M_j) \right\|^2 \tag{1}$$

By this method we can calibrate parameters of the two cameras and the steps are shown as below:

- Single camera parameters calibration.

First we know the scale parameter of the calibration board, so we can ensure the coordinate of any angel point in the real world reference frame, and then the camera captured some images of the board from different angel. Complete angel points can be scanned from these images, so their coordinates in image reference frame can be calculated, and the internal parameters can be got from formula 5 when n images were captured.

- Double cameras external calibration

After the internal parameters calibration, we also need the relative position between two cameras. According to (1), by inputting the coordinate of angel points, real world coordinate and internal parameters of the left camera to this formula, we can get the external parameters of two cameras of any image.

So the transform matrix from left camera to right camera is:

$$\begin{bmatrix} {}^{r}\boldsymbol{R}_l & {}^{r}\boldsymbol{O}_l \end{bmatrix} = \begin{bmatrix} \boldsymbol{R}_{ir}\boldsymbol{R}_{il}^{-1} & \boldsymbol{t}_{ir} - \boldsymbol{R}_{ir}\boldsymbol{R}_{il}^{-1}\boldsymbol{t}_{il} \end{bmatrix} \quad (2)$$

So double cameras calibration can be achieved as below, according to the scale parameter of the calibration board, and the coordinate of any angel point in the real world reference frame, opposite angel points in real world reference frame should be the same in left and right images. We can capture several couples of images, scan complete angel points from them, calculate the coordinate of angel points in left and right image reference frame, and then get the external parameters of left and right cameras by formula 1 so external parameters of the couple of cameras can be calculated by (2).

■ Measure in three-dimension

The internal and external parameters are calculated by the method above. We assume the reference frame of the left camera is real world reference frame, then the transform matrix of two cameras is

$$\begin{cases} \boldsymbol{P}_l = \boldsymbol{A}_l \begin{bmatrix} \boldsymbol{I} & 0 \end{bmatrix} \\ \boldsymbol{P}_r = \boldsymbol{A}_r \begin{bmatrix} {}^{r}\boldsymbol{R}_l & {}^{r}\boldsymbol{t}_l \end{bmatrix} \end{cases} \quad (3)$$

If there is one point in workspace its coordinate is

$$\tilde{\boldsymbol{M}} = [X_w \quad Y_w \quad Z_w \quad 1]^{\mathrm{T}}$$

Then its coordinate in left and right images are

$$\begin{cases} \tilde{\boldsymbol{m}}_l = \boldsymbol{P}_l\tilde{\boldsymbol{M}} \\ \tilde{\boldsymbol{m}}_r = \boldsymbol{P}_r\tilde{\boldsymbol{M}} \end{cases} \quad (4)$$

So we can get left and right images and pick-up characteristic point, get the optimal opposite characteristic point, and calculate the solution of this point in real world reference frame

$$[X_w \quad Y_w \quad Z_w]$$

Then it can be output into control system, and the three-dimensional measurement is completed.

3.2 Ensuring Position of Target

The visual targets are set on the climbing rods, and two visual targets with same color are set on each rod, different rod is signed by different color.

When the control system requires the position where the robot need to grasp at next step, first the vision system ensures the color information of the next rod, then process the image by binaryzation and get the monochrome image, lastly, searching for the central position of maximum connected region in the image can make sure the central position of visual target. Also we can filter interferer by limiting the least points through maximum connected region.

3.3 Coordinate Transform

The reference frames of tripod head is shown in Fig.12, reference frame O_0 is immovable. Reference frame O_1 is driven by motor 1 and it can rotate around axes Y_1. Reference frame O_2 is driven by motor 2 and it can rotate around axes X_2. Reference frame O_3 and O_4 are reference frame of left camera and right camera. The Position shown is the original position, and we need to know the coordinate transform matrix when reference frame O_1 rotates θ_1 around axes Y_1, and reference frame O_2 rotates θ_2 around axes X_2.

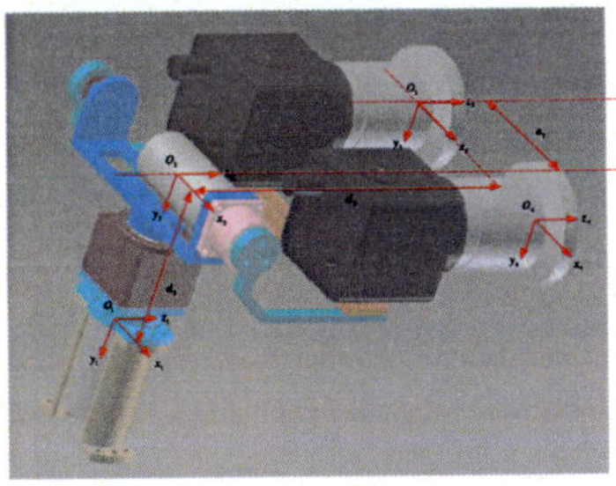

Fig. 12. Reference frames of tripod head

According to Fig.12, the transform matrix from one reference frame to its next reference frame can be calculated. So the total transform matrix of left camera is

$$ {}^{w}_{4}T = {}^{w}_{0}T\,{}^{0}_{1}T\,{}^{1}_{2}T\,{}^{2}_{3}T\,{}^{3}_{4}T = \begin{bmatrix} {}^{w}_{0}R & {}^{w}_{0}t \\ 0 & 1 \end{bmatrix} \cdot \begin{bmatrix} c\theta_1 & s\theta_1 s\theta_2 & c\theta_2 s\theta_1 & a_3 c\theta_1 + d_3 c\theta_2 s\theta_1 \\ 0 & c\theta_2 & -s\theta_2 & -d_2 - d_3 s\theta_2 \\ -s\theta_1 & c\theta_1 s\theta_2 & c\theta_1 c\theta_2 & -a_3 s\theta_1 + d_3 c\theta_1 c\theta_2 \\ 0 & 0 & 0 & 1 \end{bmatrix} $$

From all above, the transform from reference frames of camera to reference frame of the robot is

$$ {}^{w}p = {}^{w}_{4}T\,{}^{4}p \tag{5} $$

4 Experiment

4.1 Visual System Error Experiment

Before motion experiment, we need to test the visual system and get the error range to judge that if the system can give acceptable information of the robot.

First part is calibration error experiment. The absolute error of position between visual target and camera can be compensated by an immovable value, so we only need to analyze the image information of a same target that captured by the two cameras. The two images in Fig.13 are captured by two cameras to a same target, left is captured by the left camera, and right is captured by the right.

According to the method in section 3, we calculated that the excursion matrix is

$$T_{lr} = [-79.4785 \quad -1.9765 \quad -1.1319]^T$$

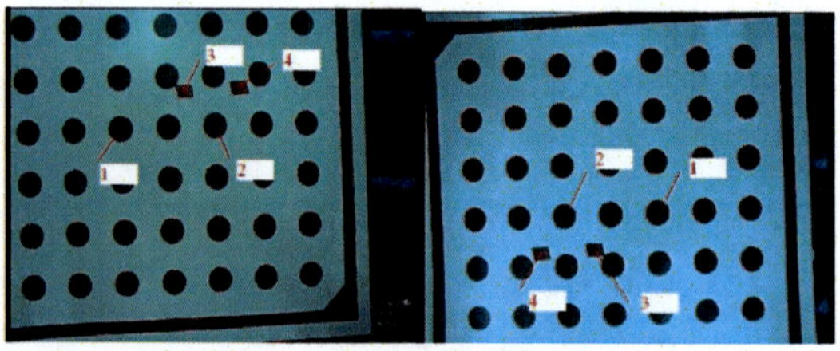

Fig. 13. Image capturing results of two cameras

We validated the calibration results by calculating the distance between point 1 and point 2, and distance between point 3 and point 4. These two distances can be measured from the calibration board directly. The experimental result is shown in table 1. The error is less than 2mm, so we consider that the calibration results are correct.

Table 1. Experiment results of calibration (mm)

	Camera left	Camera right	Measured value
Point 1	(200, 208)	(411, 348)	(-120, -49, 420)
Point 2	(363, 202)	(249, 355)	(3, -53, 427)
Point 3	(310, 142)	(303, 414)	(-37, -96, 428)
Point 4	(405, 135)	(211, 420)	(35, -103, 437)
distance	Calculated value	Measured value	error
Point1 and 2	123.5	125	1.5
Point 3 and 4	72.3	71	1.3

Second part is central point error experiment. The central point error is a direct standard to estimate the performance of a vision system. We captured images of a same target by the two cameras, and they are shown in Fig.14. The left is from left camera, and right is from the right.

The coordinates which are measured directly and calculated by vision system are compared in Table 2, we can know the precision of the vision system is according with requirements, furthermore, we calculated the distance between the two red visual targets, it is 208 mm, and the real distance is 213mm, so the error is less than 5mm.

Fig. 14. Images of same targets captured by two cameras

Table 2. Experiment results of central point error (mm)

Visual target	Camera	Real coordinates	Calculated coordinates
Red 1	left	x=408 y= 46	x=409 y= 47
Red 1	right	x=210 y=510	x=210 y=509
Red 2	left	x=134 y= 45	x=135 y= 46
Red 2	right	x=492 y=509	x=495 y=509
Orange	left	x=240 y=303	x=238 y=304
Orange	right	x=376 y=252	x=375 y=252

4.2 Movement Experiment of Robot

The robot is hanged at an immovable height, so the coordinate of climbing rod is also immovable on axes Z opposing to the camera's reference frame, the vision system only need to calculate the coordinate of visual target on axes X and axes Y. According to the position of visual target, the control system calculated the position of next rod and its angel of deflection. After scanning the visual target, the trajectory plan is progressed in time.

According to the experiment result of simulating climbing of astronaut, we found that without the improved structure, the robot usually starts vibrating when it moves its body by two arms and starts reflecting when one of its hands is just released, then it hold its body only by one hand. It is very hard to adjust control system parameters to restrain vibration and sometimes even sympathetic vibration occurs.

So we improved the structure which combined the robot and passive mechanism. The structure we proposed is introduced in section 2. By the experiment result, we found that the vibration and reflections which appeared previously was obviously reduced and the robot can keep steady during moving. The sympathetic vibration never appeared again. The experiment showed that the robot could move forwards and backwards by grasping rods on the passive mechanism based on getting visual information from the visual system.

The snapshots of the robot movement experiment are shown in Fig.15. As it is shown, the robot can recognize the climbing rods correctly and achieve the movement forward.

From all the experiments above, it is validated that the vision system can work effectively. The error is tiny and acceptable. So the vision information process method is correct, it can transform correct data to control system, and make the robot achieve the missions.

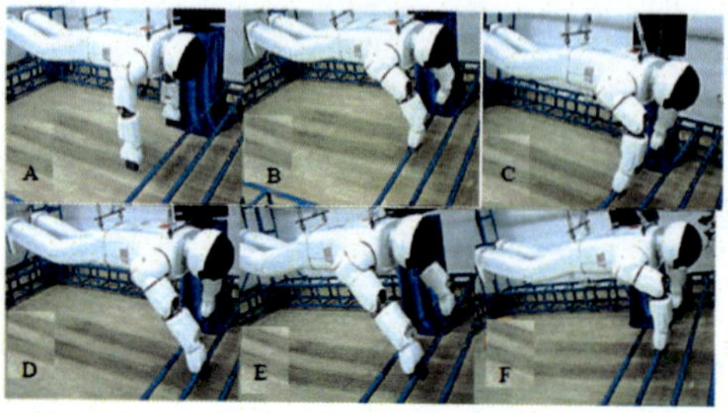

Fig. 15. Snapshots of robot movement experiment

5 Conclusion

This paper is focus on a space robot system based on binocular stereo vision system, and a kind of passive experimental platform is designed for space robot to stably simulate extravehicular climbing of astronauts. The main results of this paper are as follows.

1) This paper designed a space robot system including the space robot, passive mechanism. Especially, we proposed a method to reduce vibration by ameliorating the combination between the robot and passive mechanism.

2) The binocular stereo vision system of space robot could recognize the visual targets which are set on climbing rods.

3) The error of this vision system is acceptable to navigate space robot to move on the passive mechanism.

4) We have achieved that the robot can climb forward on the rods correctly by designed binocular stereo vision system while the vibration and reflection can be reduced well.

Acknowledgements. This work is supported by Beijing Science Foundation (Grant No.4122065), and National High Technology Research of China (863 Project) (Grant No.2011AA040202 and Grant No.2009AA04Z207), National Science Foundation for Distinguished Young Scholar (Grant No.60925014), "111 Project" (Grant No.B08043).

References

1. Yu, Z., Huang, Q., Li, J., et al.: Distributed Control System for a Humanoid Robot. In: The Proceedings of IEEE International Conference on Mechatronics and Automation, Harbin, China, August 5-8, pp. 1166–1171 (2007)
2. Hirai, K., Hirose, M.: The Development of Honda Humanoid Robot. IEEE Transaction on Robotics and Automation, 1321–1326 (1998)
3. Sakagami, Y., Watanabe, R.: The intelligent ASIMO: System Overview and Integration Robots and System. In: IEEE/RSJ Int. Conf. on Intelligent Robots and System, pp. 2478–2483 (2002)

4. Kaneko, K., Kanehiro, F.: Design of prototype humanoid robotics platform for HRP. In: IEEE/RSJ Int. Conf. on Intelligent Robots and System, pp. 2431–2436 (2002)
5. White, G.C., Xu, Y.: An Active Z Gravity Compensation System. In: The Proceedings of the 1993 IEEE/RSJ International Conference on Intelligent Robots and Systems, Yokohama, Japan, July 26-30, pp. 1181–1187 (1993)
6. Yasushi, S., Yoshihiro, K., Takashi, Y.: 3D Object Recognition in Cluttered Environments by Segment-based stereo vision. International Journal of Computer Vision 46(1), 5–23 (2002)
7. Khaleghi, B., Ahuja, S., Wu, Q.: An Improved Real-time Miniaturized Embedded Stereo Vision System (MESVS-I). In: Fifth Canadian Conference on Computer and Robot Vision, 26–33 (2008)
8. Zhang, Z.: Flexible Camera Calibration by Viewing a Plant from Unknown Orientations. In: The Seventh IEEE International Conference on Computer Vision, vol. 1, pp. 666–673 (1999)
9. Zhang, Z.: A Flexible New Technique for Camera Calibration. IEEE Transactions on Pattern Analysis and Machine Intelligence, 1330–1334 (2000)
10. Tian, J., Huang, Y., Yan, F.: A Variable-step Detecting Algorithm for Interested Boundary. In: Sixth World Congress on Intelligent Control and Automation (2006)
11. Mehling, J.S., Rovekamp, R., et al.: Centaur: NASA's Mobile Humanoid Designed for Field Work. In: The Proceedings of IEEE International Conference on Robotics and Automation (ICRA 2007), Roma, Italy, April 10-14, pp. 2928–2933 (2007)
12. Sakagami, Y., Watanabe, R.: The intelligent ASIMO: System Overview and Integration Robots and System. In: IEEE/RSJ Int. Conf. on Intelligent Robots and System, pp. 2478–2483 (2002)

Fast 6D Odometry Based on Visual Features and Depth

Salvador Domínguez Quijada[1], Eduardo Zalama[2], Jaime Gómez García-Bermejo[2], Rainer Worst[3], and Sven Behnke[3,4]

[1] Cartif Foundation, Valladolid, Spain
[2] ITAP, University of Valladolid, Spain
[3] Fraunhofer IAIS, Sankt Augustin, Germany
[4] University of Bonn, Germany

Abstract. The availability of affordable RGB-D cameras which provide color and depth data at high data rates, such as Microsoft MS Kinect, poses a challenge to the limited resources of the computers onboard autonomous robots. Estimating the sensor trajectory, for example, is a key ingredient for robot localization and SLAM (Simultaneous Localization And Mapping), but current computers can hardly handle the stream of measurements. In this paper, we propose an efficient and reliable method to estimate the 6D movement of an RGB-D camera (3 linear translations and 3 rotation angles) of a moving RGB-D camera. Our approach is based on visual features that are mapped to the three Cartesian coordinates (3D) using measured depth. The features of consecutive frames are associated in 3D and the sensor pose increments are obtained by solving the resulting linear least square minimization system. The main contribution of our approach is the definition of a filter setup that produces the most reliable features that allows for keeping track of the sensor pose with a limited number of feature points. We systematically evaluate our approach using ground truth from an external measurement systems.

1 Introduction

Optimizing the use of the available resources in mobile robotics like energy or processing capabilities is important, because some robots do not have any external physical connection and have to work autonomously. One of the most important tasks a mobile robot must perform is self-localization with respect to a map. This task is, however, also one of the most resource-consuming when no global positioning system is used. When the robot pose is known, it can be used for navigation as well as other purposes such as mapping the environment. The measurements needed for localization are usually provided by expensive and heavy laser range sensors. The availability of affordable lightweight RGB-D cameras offers the possibility to acquire color images and depth maps at high frame rates which must be processed in real time in order to be used for robot control.

S. Lee et al. (Eds.): Intelligent Autonomous Systems 12, AISC 193, pp. 245–256.
springerlink.com

In this paper, we propose a reliable method to obtain 6D odometry from color and depth images. Our method can be applied to estimate the sensor pose in real time, requiring limited computational resources.

We use a Kinect RGBD sensor as the only capturing device. The sensor driver provides color and depth images in a standard format. At each frame, we extract a set of visual features from the RGB image. They are filtered and mapped to 3D space using the associated depth image. We associate the features of consecutive frames in 3D and finally obtain the sensor pose increment. Key to our method is a carefully designed filtering process, which takes the most reliable features to estimate odometry. Finally, we calculate the transformation that minimizes the mean square error between two sets of associated features. Because our method processes only a limited number of features, it can be easily applied for pose estimation on mobile robots with limited computational resources, like e.g. a 1.6 GHz CPU with 1 GB RAM that we used for our experiments. Even with such a small computer, we achieve odometry rates near 20 Hz in low resolution, which makes it possible to use our method on a robot that moves at full speed (1 m/sec linear velocity and / or 45°/sec angular velocity in our case). Our method does not rely on any assumption about the kinematics of the movement. The odometry obtained can be used later as an input data stream for 6D-SLAM.

Raw detected features are intrinsically noisy and would produce low-quality odometry estimates if they were applied directly. The filtering reduces the number of features and consequently the time consumption of the overall process significantly. Instead of trying to match the features extracted from consecutive frames, we first stabilize the consistency of the current feature set with respect to a reference feature set. As explained below, such stabilization is performed by dynamically adapting the current set of features to a reference set.

The remainder of the paper is structured as follows. In Section 2, we discuss related work. In Section 3, we present our approach to the estimation of 6D odometry based on visual features. Finally, we demonstrate the advantages of our method using systematic experiments in Section 4.

2 Related Work

Visual odometry is the process of calculating the increments of pose based on the analysis of images captured by a camera. One of the first works in visual odometry is attributed to Moravec [1]. In 1976 he managed to implement a visual odometry system on the Stanford Cart to perform corrections on the control of the vehicle autonomously. In that technical report, the basic concepts of tracking visual features and depth estimation were introduced.

We can identify two main types of visual odometry using cameras as the only capturing device: monocular and stereo. In monocular odometry depth information must be estimated from motion parallax as the camera moves through the environment. Interesting examples of monocular odometry implementation include Nister et al. [2] and Davison [3]. In stereo odometry, the camera does not have to move to calculate depth information. Some implementations of stereo odometry are described by Moravec [1], Matties [4] and Nister et al. [2]. On the other hand, it is also possible to use additional devices for the acquisition of the depth like laser range scanners, time-of-flight cameras or infrared pattern projectors. The resulting input data stream for visual odometry estimation is in most cases a sequence of images with depth information

The tracking of the features plays an important role in visual odometry, because it is a process that consumes less time than extracting the features in every frame, and can be applied to several consecutive frames while as long as enough features are within the image boundaries. Barron [5] compiled a survey of optical flow algorithms for tracking, like the gradient-based technique proposed by Lucas and Kanade [6] (LK). LK is implemented in OpenCV and used in our approach. The selection of good features to track is also important in order to reduce resource consumption and increase accuracy. An interesting attention-driven feature selection is described by Einhorn [7].

3 Method

Figure 1 summarizes our approach for the estimatation of the odometry. No prior knowledge about the environment is assumed. We can distinguish two main loops, the extraction loop and the tracking loop. The extraction loop is only executed when a new set of features is needed (generated features GF), i.e. when there are not enough features to obtain a good odometry estimate. The tracking loop is executed in every frame and performs, among other things, tracking, filtering, and matching of the features to get the transformation. The key steps that make this approach suitable for real time application and less sensitive to noise are the filtering of the features and the consistency adaptation (marked with stars in Figure 1) which will be explained in detail below.

3.1 Visual Features Extraction

Visual features are interesting parts of the image with properties that make them suitable for detection. Different steps to detect robust, distinctive invariant features are explained in Lowe [8]. There are different kinds of such features, e.g. corners and blobs. Because their extraction is computationally less demanding, we focus in this paper on the so called corners. Although corner features are not completely robust and stable, they can be stabilized by the 3D information, which in our case is available.

Corners are points which have two dominant directions in the luminosity gradient, in a local neighborhood. Some of the most important techniques for extracting corners are presented by Harris and Stephens [9] as well as Shi and Tomasi [10].

An evaluation of three features detectors (Harris, SIFT (Scale Invariant Feature Transform) and LKT (Lucas-Kanade-Tomasi)) is described by Klippenstein and Zhang [11]. They found LKT to have a better performance in general. We test our approach with two feature extractors: Harris corner detector and LKT

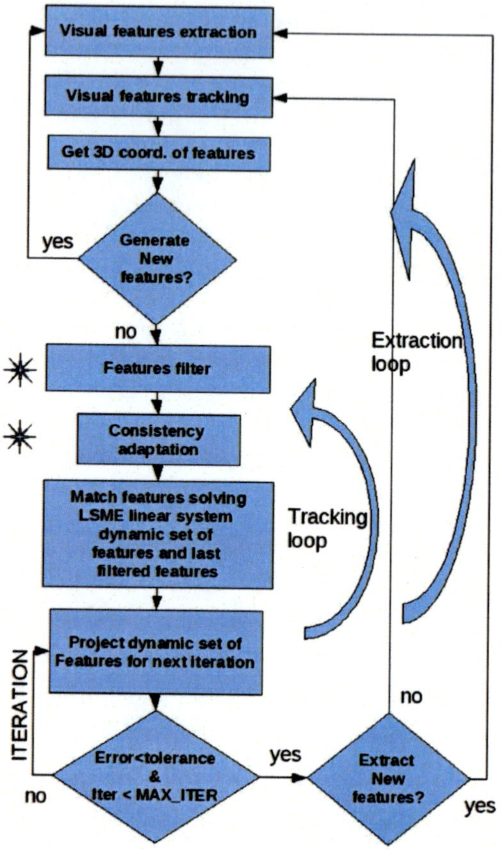

Fig. 1. Flow diagram of the overall process of odometry estimation

corner detector which are both included in the OpenCV library getting lightly better results with LKT.

3.2 Tracking of Visual Features

The tracking of features is a crucial step of visual odometry. Shi and Tomasi [10] propose a matching algorithm for tracking visual features based on a model of affine image changes. This technique has been adopted in our approach. For the tracking process, a pair of images is used: the most recently captured image and the previous one (Figure 2). The visual features corresponding to the previous image can be obtained either by feature extraction or as a result of the previous tracking iteration. We apply Shi-Tomasi's method to track the features in the

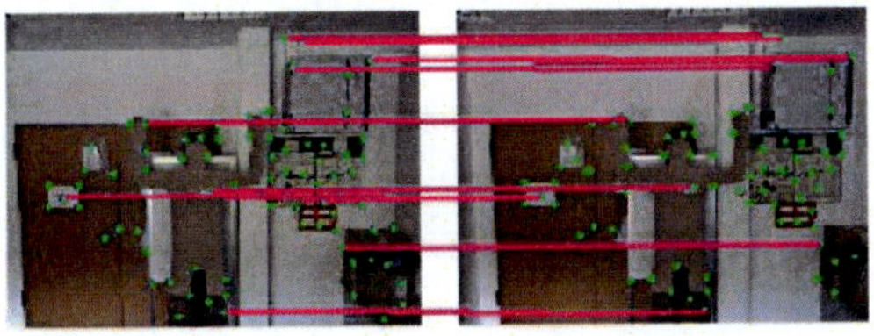

Fig. 2. Features in one frame (previous) are tracked during the tracking process to the next frame (tracked features TF). Some of the correspondences are shown in magenta.

first image, so we know where to expect them in the second image. If we cannot establish correspondence for a feature in the second image, it is marked as "non-paired" feature.

3.3 Generation of New Features

Because we are looking for a disambiguous solution to the linear error minimization system (see paragraph III-E), a minimum of 4 features are needed so the number of equations is larger or equal to the number of unknowns. When no such solution can be found we seek to generate a new set. There are several reasons for this to happen:

- the scene has no interest points to be easily distinguished like flat colored surfaces with few corners.
- when moving the sensor, the features will likely change their positions in the image and some of them can disappear on the image borders.
- the distribution of the features can lead to clustering in some parts of the image, for example when the camera moves backwards and distance between tracked features reduces and some of the features fuse together.

Moreover, the grouping of features decreases the precision of the movement estimate. For better performance, the visual features should be distributed roughly uniformly in the image. In order to deal with this problem, we measure the distance traveled and the rotated angles of the sensor from the last feature extraction. When either the distance or the angle, exceeds a given threshold, we trigger the generation of a new set of features.

3.4 Projection of Visual Features in 3D

RGB-D sensors like MS Kinect provide the depth information at most pixels in the color image. However, sometimes the physical properties of the surface at some positions are not suitable for the measurement of depth. Many of the visual features lie on object boundaries where depth information is often missing.

Moreover, visual features might jump from the front edge of an object to the background side.

If the entire neighborhood of the feature has similar depth values, then the depth of the feature point can be obtained by averaging these values. It could happen that the feature does not have depth information or that there is a large difference among the neighbors depths, probably because the feature is lying at the jump edge caused by a border. In this case, we utilize border ownership, which means that we average the depth of the closest neighboring points.

Averaging has also the effect of reducing the noise of the depth measurements. It can happen that no points in the neighborhood have depth information and it will not be possible to calculate the 3D depth of the feature. Those features will be discarded in the filtering process.

Preferring closer points is also advantageous for the measurement precision of triangulation sensors like Kinect because the error increase quadratically with the distance. The optimal radius of the neighborhood depends on the image resolution. In our experiments with a 640x480 resolution, a radius of 3 pixels has been proven to work suitably. Because our neighborhood is defined as a square, this implies a neighborhood of 49 pixels.

3.5 Filtering and Consistency Adaptation

In order to retain only the most reliable features for pose estimation, we designed a set of rules for discarding features. Firstly, we filter out the features with missing depth, then the consistency of the distribution is checked. We analyze the relative 3D positions of the last tracked features TF compared to the respective positions of the generated features GF, as is shown in Figure 3.

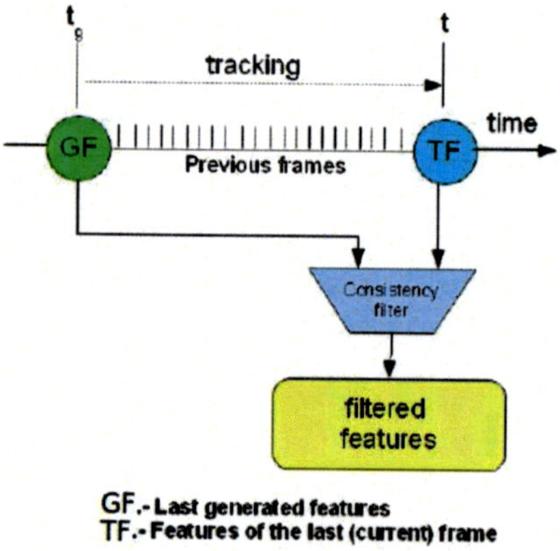

Fig. 3. Filtering process. The last extracted features (GF) and the last tracked features (TF) are used to filter the features.

This filter is based on the idea that every feature has to correspond to a fixed point in the real world. This way, the position of the TF in global coordinates must not differ too much from the global position of GF during the tracking process. In our approach, we model the set of features as a cloud of 3D points where each point is connected to the rest by elastic links.

To move a point from its equilibrium point in such elastic net, it is necessary to apply a force (see Figure 4). If we consider the GF as the balanced state where the deformation energy is zero, then the TF has an energy accumulated because the relative position of the features is not exactly the same due to non-perfect tracking. The force required at every node to move it from its equilibrium point to its current position, can be calculated through a vectorial sum of the forces in every link connecting that node to the rest. Every link can only apply a force in its longitudinal direction and can be calculated because we know the 3D coordinates of every feature in both sets of features. The condition to pass the filter is that the displacement force must stay below a threshold value.

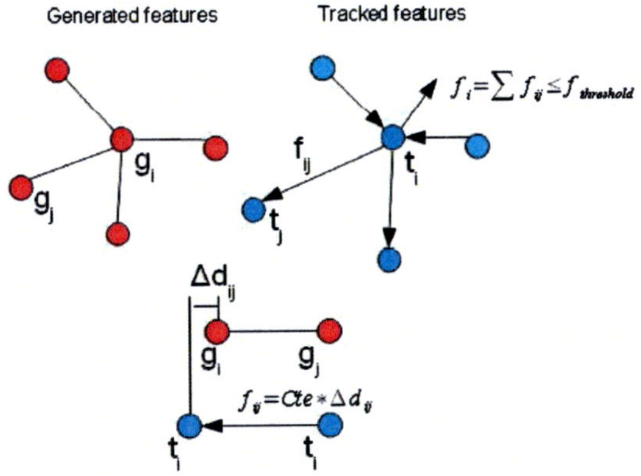

Fig. 4. The force in each feature is determined as a vectorial sum of the deformation forces in each link. The force in every link (i,j) is proportional to the difference of distance between that link in TF and GF.

The consistency filter discards features that moved more than a threshold distance from their equilibrium position. However, probably most of the rest of the features are still noisy in depth because the sensor accuracy decreases on the square of the distance to the point in the real world. Noisy depth values lead to decreased odometry accuracy, especially at distances above 2 meters using the MS Kinect sensor. To solve this problem we apply a filter to adapt and stabilize the consistency of the TF. This is what we call consistency adaptation in Figure 1. This filter works on so-called adaptive features TF^*. This new set of features is the result of the gradual consistency adaptation from set TF to GF. It can be seen as a stable version of the TF that we use for calculating the odometry transformation.

Figure 5 represents the process of consistency adaptation of the features. TF^* have the properties of having consistency similar to GF and a global pose similar to TF. Little changes in the relative average positions of the tracked features in 3D will be adapted progressively to the equilibrium position. TF^* and GF are finally used to calculate the transformation of the odometry leading to stable odometry.

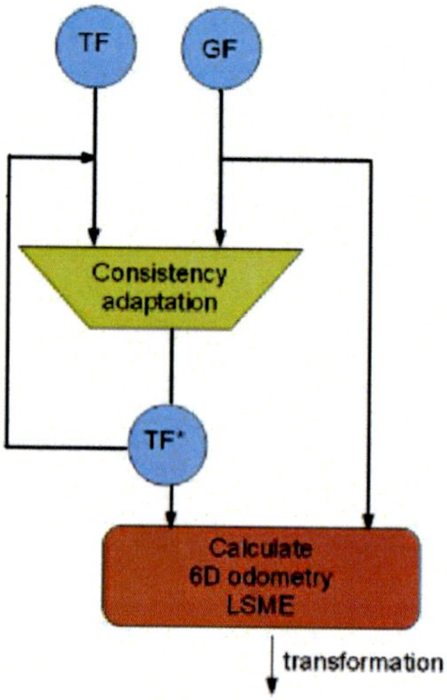

Fig. 5. Calculation and use of the adaptive features

The adaptation of the consistency has positive effects over the noise in the estimated position. Figure 6 shows the difference without and with adaptation, respectively. The ground truth in this experiment was composed by perpendicular straight lines following X, Y and Z axes. When no adaptation is applied, the noise increases with the distance to the features, on the contrary with adaptation the noise is less dependent on that distance.

3.6 Estimation of Pose Increments

Let us consider that we have tracked a set of N visual features p from the previous frame to the last one producing the set q whose features are paired one by one with those in p. These features have passed the filtering process and consistency adaptation so they are stable.

We are looking for a rigid transformation $(\Re, t)$ that minimizes the value of the following expression

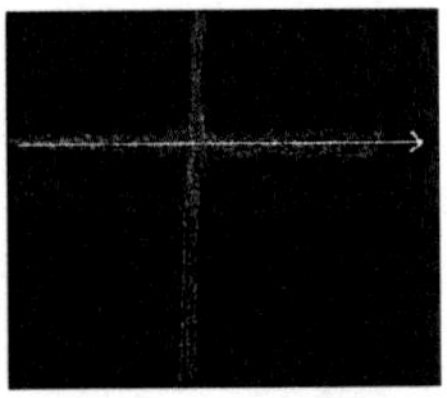

Without adaptation

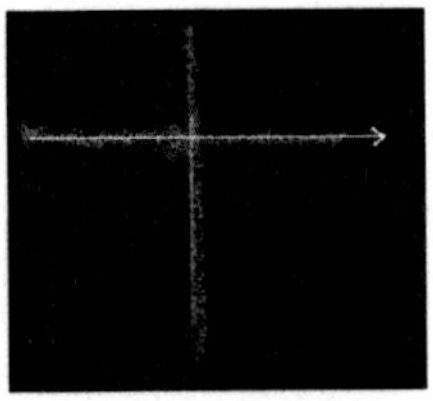

With adaptation

Fig. 6. Effect in the adaptation of the consistency. The distance from the sensor to the features increases in the direction of the arrow. Notice that the dispersion of the position increases with the distance to the sensor when no adaptative consistency is used.

$$J(\Re, t) = \sum_N \|p_i - \Re q_i - t\|^2 \tag{1}$$

where p_i and q_i represent the 3D position of the paired feature i in the previous frame and the current frame, respectively.

This is a typical problem of Linear Least Square Minimization. The linear system is represented by 3×N equations (N features × 3 coordinates) with 12 unknowns (9 for the rotation matrix $\Re$ and 3 for the translation vector t). Therefore, to have a unique solution we need at least 4 features.

Equation (1) can be represented in a different way, being equivalent to minimize the following expression

$$\begin{aligned} &\sum_N \|p_i - A(q_i) * \wp\|^2 \rightarrow p_i \approx A(q_i) * \wp \\ &A(q_i)^T * p_i = A(q_i)^T * A(q_i) * \wp \\ &\wp = (A(q_i)^T * A(q_i))^{-1} * A(q_i)^T * p_i = R(q_i) * p_i \end{aligned} \tag{2}$$

In this expression, $\wp$ is the 12×1 vector of unknowns which is composed by elements of the rigid transformation $(\Re, t)$ and $R(q_i)$ is a 12×3N matrix which depends on the coordinates of the features in the last frame q_i, and can be determined by classical methods such as QR or Cholesky factorization [12].

The error of the transformation is checked by projecting every feature in the last frame to the previous frame using the transformation $(\Re, t)$. If the error for the feature i ϵ_i is less or equal to a threshold δ the feature will be used for the next iteration (3).

$$\epsilon_i = \|p_i - \Re q_i - t\| \leq \delta \tag{3}$$

The process of pose estimation stops when

$$\sum_N \|p_i - \Re q_i - t\| \leq threshold \tag{4}$$

4 Experiments and Results

To evaluate the quality of our method, we have used datasets with a known ground truth. The datasets were recorded at Gelsenkirchen University of Applied Sciences using an industrial Kuka KR60HA robot (Figure 7). One example of the scene seen by the sensor is shown in Figure 8.

Fig. 7. Robot used for ground truth experiments

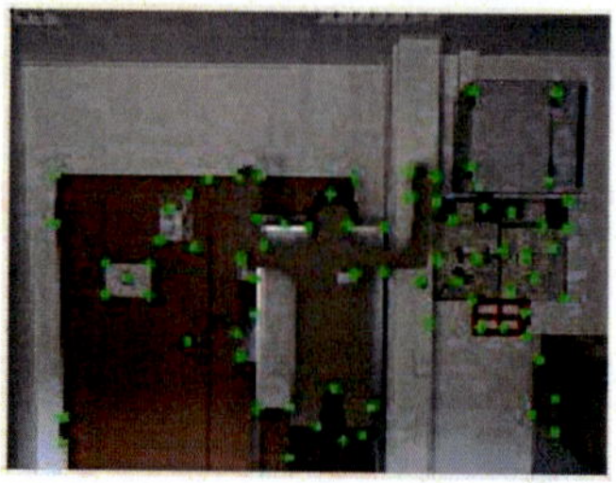

Fig. 8. Scene seen by the camera for experiments 1 and 2

The sensor trajectories used were geometric paths such as lines and arcs (see Figure 9) traced at about 15 cm/sec. The number of visual features always was bigger than 10 and the distance to the features covered a range between 2 and 5 m. Every test starts and finishes in the same position so that the difference between the initial position and the final one will measure the total error. We also pay attention to the maximum distance of the estimated position to the theoretical trajectory.

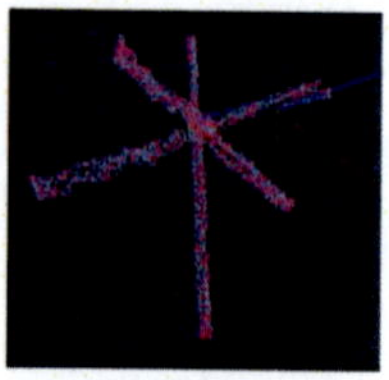

(a) Experiment 1

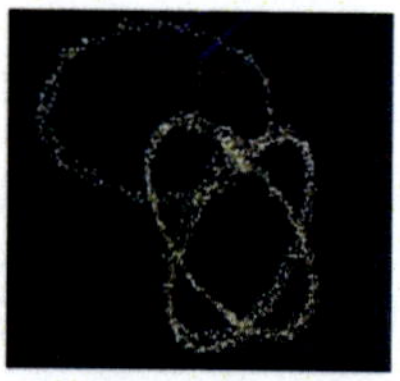

(b) Experiment 2

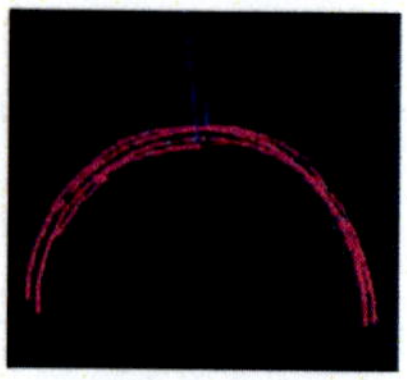

(c) Experiment 3

Fig. 9. Trajectories obtained on the experiments

Experiment 1: Linear translation along X, Y, and Z directions. The traveled distance along each direction was 1 m, and was repeated three times (see Figure 9a). This way, the total traveled length was 18 m. The final error found in the position was 11 cm in translation and 1.5° in inclination, which correspond to a 0.6% accumulated position error. The maximum deviation from the theoretical trajectory was only 6 cm.

Experiment 2: Circles traced on the XY, XZ, and YZ planes (see Figure 9b). Each circle was 50 cm in diameter. Every circle was repeated three times. The total length of the trajectory was 14.13 m and the final error was 4.5 cm and 1.2° in inclination, that is less than 0.3 % of the total length of the trajectory. The maximum deviation from the theoretical curve was 7 cm.

Experiment 3: Translation and rotation along an arc of 180° and 1 m radius, with the camera pointing along the radial direction (see Figure 9c). Each movement was repeated three times, so the total traveled distance was 18.85 m and the total angle rotated was 1080°. The final pose of the camera had an error of 36 cm in position and 10° in orientation, which corresponds to a 1.9 % linear error and 0.9 % rotation error.

5 Conclusion

We have presented a method for estimating the trajectory of a moving RGB-D sensor using 6D odometry based on visual features obtained from the RGB images, which are mapped to 3D using their measured depth. Key to the success of our method is a filtering process that removes the features that are not reliable, thus making our solution robust. The consistency adaptation of the features improves the stability and precision of the result. Our algorithm is computationally very efficient, because only a limited number of points is processed every frame.

We evaluated our method by moving the camera with an industrial arm, which provides ground truth for the trajectories. The obtained results are satisfactory for visual odometry considering the limitations of the sensor with regards to depth noise.

Our 6D visual odometry is in use as one of the key parts of the 3D SLAM system for the ground robot used in NIFTi project which is equipped with a small computer with a processor of 1.6 GHz. Working at 640×480 and 15-20 fps the CPU usage stays near 35 % on that computer.

Acknowledgments. We would like to thank to both Cartif Foundation and Fraunhofer IAIS for having made this work possible. This work has been funded by the European Commission under contract number EU FP7 NIFTi / ICT-247870 and also supported by Junta de Castilla y León (CCTT/10/VA/0001) and by the Science and Innovation Spanish ministry (Pr.Nb.DPI2008-06738-C02-01/DPI). Thanks also to the NIFTi team of IAIS, especially to Thorsten Linder, Viatcheslav Tretyakov and Nenad Biresev for their support in the experiments and helpful suggestions.

References

1. Moravec, H.: Obstacle avoidance and navigation in the real world by a seeing robot rover. Tech. report and doctoral dissertation, Robotics Institute, Carnegie Mellon University, Stanford University, CMU-RI-TR-80-03 (1980)

2. Nister, D., Naroditsky, O., Bergen, J.: Visual odometry for ground vehicle applications. Journal of Field Robotics 23(1) (2006)
3. Davison, A.J.: Real-time simultaneous localization and mapping with a single camera. In: Proceedings of the International Conference on Computer Vision, Nice (2003)
4. Matties, L.: Dynamic Stereo Vision. PhD thesis, Dept. of Computer Science, Carnegie Mellon University, CMU-CS-89-195 (1989)
5. Barron, J.L., Fleet, D.J., Beauchemin, S.S.: Performance of Optical Flow Techniques. The International Journal of Computer Vision 12(1), 43–77 (1994)
6. Lucas, B.D., Kanade, T.: An Iterative Image Registration Technique with an Application to Stereo Vision. In: Proceedings of 7th International Joint Conference on Artificial Intelligence, pp. 674–679 (1981)
7. Einhorn, E., Schrter, C., Gross, H.-M.: Can't Take my Eye on You: Attention-Driven Monocular Obstacle Detection and 3D Mapping. In: IROS 2010
8. Lowe, D.G.: Distinctive image features from scale invariant keypoints. International Journal of Computer Vision 60(2), 91110 (2004)
9. Harris, C., Stephens, M.: A Combined Corner and Edge Detector. In: Proc. of The Fourth Alvey Vision Conference, Manchester, pp. 147–151 (1988)
10. Shi, J., Tomasi, C.: Good features to track. In: IEEE Conference on Computer Vision and Pattern Recognition (1993)
11. Klippenstein, J., Zhang, H.: Quantitative Evaluation of Feature Extractors for Visual SLAM. J. Department of Computing Science University of Alberta Edmonton, AB, Canada T6G 2E8 (2007)
12. Golub, G. H., Van Loan, C. F.: Matrix Computations, 3rd edn. Johns Hopkins (1996) ISBN 978-0-8018-5414-9

Moving Region Segmentation Using Sparse Motion Cue from a Moving Camera

Jungwon Kang, Sijong Kim, Taek Jun Oh, and Myung Jin Chung

KAIST, 291 Daehak-ro, Yuseong-gu, Daejeon 305-701, Republic of Korea
{kctown,terre,buljaga}@rr.kaist.ac.kr, mjchung@ee.kaist.ac.kr

Abstract. This paper presents a method for pixel-wise segmentation of moving regions using sparse motion cues on an image from a freely moving camera. The main idea is to utilize residual motion, i.e., motion relative to a background, on sparse grid points. Our algorithm consists of three parts: global motion estimation, characterization of points based on sparse motion cue, and pixel-wise labeling of moving regions. Experimental results on real image sequences are presented, showing the effectiveness of the proposed method.

1 Introduction

Recognizing moving regions in an image is an important problem in a variety of applications for surveillance [1], driver assistance [2], and mobile robot navigation [3].

For a fixed and static camera, the background modeling and subtraction is popular for moving region detection [4]. This method learns the background by modeling appearances of each pixel, commonly using Gaussian mixture models [5]. Moving regions in the current frame are detected by comparison between the current frame and the background. However, in many applications, a camera needs to move freely. For a freely moving camera, the background modeling and subtraction method can not be easily adopted as the image changes by the camera motion. One feasible solution for a moving camera is to utilize temporal correspondences, i.e., the displacement of a point or a pixel between two frames of image sequences. The dense optical flow [6] provides motion of whole pixels in an image, and has been utilized as a motion cue in [7][8]. However, calculating the dense optical flow is computationally expensive. Considering the online operation, instead of computationally burdensome dense optical flow, sparse motion information from corner points [9] or points on a lattice [10] or edge [11] has been utilized as a motion cue. However, sparse points are not enough to describe a whole moving region. For the dense representation of a moving region, pixels need to form a cluster from sparse motion and dense appearances.

In this paper, we aim at doing pixel-wise segmentation of moving regions using sparse motion cues on an image from a moving camera. For the online operation with a freely moving camera, we utilize sparse motion obtained at points on a lattice, motivated by the work in [10]. With the sparse motion, we get dense pixel-wise moving regions by solving the pixel-wise labeling problem.

The rest of this paper is organized as follows. We present an overview in the upcoming section. In the section 3, we describe our method for moving region segmentation. We show experimental results in the section 4, and conclude this work in the section 5.

S. Lee et al. (Eds.): Intelligent Autonomous Systems 12, AISC 193, pp. 257–264.
springerlink.com

2 Overview

An overview of our algorithm is shown in Fig. 1. A moving region is defined as a set of pixels that move relative to the background. Our algorithm consists of three parts: global motion estimation, characterization of points based on sparse motion cues, and pixel-wise labeling of a moving region. The global motion indicates motion of background structures in an image, caused by camera movement. We estimate the global motion using the robust m-estimation. With the estimated global motion, we compute residual motion, i.e., motion relative to global motion, at the sparse grid points on the current frame. After characterization of each pixel node by its motion and appearance, each node is labelled moving or background on a CRF segmentation framework.

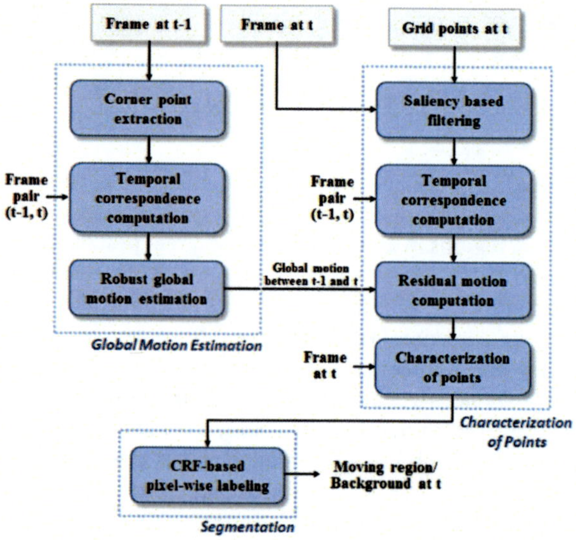

Fig. 1. An overview of our algorithm

3 Moving Region Detection and Segmentation

3.1 Global Motion Estimation

The global motion can be represented by 2D motion models such as translational, geometric, affine, and perspective model [12]. Among the motion models, we used the perspective model as it is the most general for modeling global motion. The perspective model is parameterized by an 8 parameter vector $\mathbf{m} = [m_0, m_1, \cdots, m_7]^T$. Let (x, y) and (x', y') be the coordinates in the current and the reference frame respectively. Then, the transformation by the perspective model is given by

$$x' = \frac{m_0 x + m_1 y + m_2}{m_6 x + m_7 y + 1}, \; y' = \frac{m_3 x + m_4 y + m_5}{m_6 x + m_7 y + 1} \tag{1}$$

The corner points extracted at the current frame are tracked by KLT algorithm [13], giving the corresponding points at the reference frame. The temporal correspondences contain outliers caused by not capturing real motion and deviated from the global motion model.

In order to estimate **m** from the temporal correspondences with outliers, we adopted the M-estimator [14] as it has shown superior performance on motion estimation [15]. The parameter estimation is formulated as the least square problem minimizing $\|r_i\|^2$ where r_i is the residue between the ith measured flow vector and the estimated flow vector by the motion model. Using the algebraic distance to arrange r_i makes the least square problem linear, providing the over-determined linear system $\mathbf{Am} = \mathbf{b}$ where N is the total number of correspondences, $\mathbf{A} = [\mathrm{A}_1; \cdots; \mathrm{A}_N]$ and $\mathbf{b} = [\mathrm{b}_1, \cdots, \mathrm{b}_N]^T$ with

$$\mathrm{A}_i = \begin{bmatrix} x_i \; y_i \; 1 \; 0 \; 0 \; 0 \; -x'_i x_i \; -x'_i y_i \\ 0 \; 0 \; 0 \; x_i \; y_i \; 1 \; -y'_i x_i \; -y'_i y_i \end{bmatrix} \tag{2}$$

$$\mathrm{b}_i = \begin{bmatrix} x'_i \; y'_i \end{bmatrix} \tag{3}$$

The M-estimation is implemented using a simple iterative re-weighted least squares algorithm, solving the following system $\mathbf{DAm} = \mathbf{Db}$ where $\mathbf{D} = diag(w_1, \cdots, w_N)$. Each w_i reflects the confidence of each point and is calculated by the weight function proposed by Beaton-Tukey [16].

3.2 Characterization of Points for Moving Region Segmentation

After estimating the global motion, all locations in a frame are characterized by their position, motion, and color. Let S be a set of sites of all the pixels in a frame of size $H \times W$, $S = \{(u,v) | 1 \leq u \leq W, 1 \leq v \leq H\}$. Since we utilize sparse motion cue, $S = S_A \cup S_B$ where S_A is a set with motion and S_B is the other set without motion. Note that we made no assumption of the shape and texture of moving regions. Thus it is desirable to keep uniform spatial density of points with motion over a whole image region, leading us to use regular grid points for obtaining motion.

The grid points on a textureless region are likely to give inaccurate motion. Such grid points are filtered out by checking eigenvalues of the coefficient matrix introduced in [17]. The motion of the remaining points at the current frame is obtained by finding the corresponding points at the previous frame using the KLT algorithm [13]. The locations of points with motion are described as $S_A = \{(u,v) \in S | u \equiv k \pmod{k}, v \equiv k \pmod{k}, \min(\lambda_1, \lambda_2) > \lambda_t, R(u,v) = 1\}$ where λ_1, λ_2 are eigenvalues of the coefficient matrix at (u,v), λ_t is a threshold, and $R(u,v) = 1$ indicates the success of tracking by the KLT. On the other hand, the positions of points without motion are given as $S_B = S - S_A$.

Each point in S_A, indexed by i, is characterized by $z_{A,i} = [c_i, m_i]$ consisting of the RGB color feature $c_i = [r_i, g_i, b_i]$ and the motion feature $m_i = [\Delta u_i, \Delta v_i]$ while a point in S_B is characterized by $z_{B,i} = [c_i]$, the color feature only. Here $m_i = [\Delta u_i, \Delta v_i]$ is a residual motion, i.e., motion relative to the global motion. The residual motion is defined as $\Delta u_i = \Delta u_{i,m} - \Delta u_{i,g}$, $\Delta v_i = \Delta v_{i,m} - \Delta v_{i,g}$ where $(\Delta u_{i,m}, \Delta v_{i,m})$ is a measured motion, $(\Delta u_{i,g}, \Delta v_{i,g})$ is motion estimated by the global motion model. The measured motion is $\Delta u_{i,m} = u_{i,m}^{t-1} - u_i^t$, $\Delta v_{i,m} = v_{i,m}^{t-1} - v_i^t$ where $(u_{i,m}^{t-1}, v_{i,m}^{t-1})$ is a point at frame $t-1$ corresponding to the point (u_i^t, v_i^t) at frame t, given by the KLT. The estimated global motion is $\Delta u_{i,g} = u_{i,g}^{t-1} - u_i^t$, $\Delta v_{i,g} = v_{i,g}^{t-1} - v_i^t$ where $(u_{i,g}^{t-1}, v_{i,g}^{t-1})$ is a point at frame $t-1$, estimated from the the point (u_i^t, v_i^t) at frame t with the global motion model.

3.3 Moving Region Segmentation

The pixel-wise segmentation of moving regions is formulated as a 'Foreground (moving region)/Background (non-moving region)' binary labeling problem. That is, a label $l \in \{1(F), -1(B)\}$ is assigned to each pixel. For the segmentation, the random field modeling such as MRF, CRF is popular as it incorporates contextual information in the form of spatial dependencies in an image. Let $\mathbf{l}$ be a set of output labels and $\mathbf{z}$ a set of observation data. Then, the labeling problem is to find $\mathbf{l}^*$ that maximizes $p(\mathbf{l}^*|\mathbf{z})$, the posterior over the labels given the observation data. Compared to the MRF which learns $p(\mathbf{z}|\mathbf{l})$ and $p(\mathbf{l})$ with conditional independence assumption of $p(\mathbf{z}|\mathbf{l})$, the CRF [18] directly models $p(\mathbf{l}|\mathbf{z})$, giving more accurate models than those of MRF. Thus, we adopted a CRF-based segmentation framework.

By the Hammersley-Clifford theorem, $p(\mathbf{l}|\mathbf{z})$ can be defined by a set of clique potential functions: $p(\mathbf{l}|\mathbf{z}) \propto \exp(-E(\mathbf{l}|\mathbf{z}))$. Then, the labeling problem is expressed as finding $\mathbf{l}^*$ that minimizes $E(\mathbf{l}^*|\mathbf{z})$. $E(\mathbf{l}|\mathbf{z}) = E_d(\mathbf{l}|\mathbf{z}) + \lambda_s E_s(\mathbf{l}|\mathbf{z})$ where $E_d(\mathbf{l}|\mathbf{z})$ is the data term, $E_s(\mathbf{l}|\mathbf{z})$ is the smoothness term, and λ_s is a regularization constant. The data term $E_d(\mathbf{l}|\mathbf{z})$ evaluates an association of the assigned label l_p to the observation z at each pixel node p. The $V_p(l_p,z)$ in formula (5) depends on $\|m_p\|$, the magnitude of a motion feature at each pixel node.

$$E_d(\mathbf{l}|\mathbf{z}) = \sum_{p \in S} V_p(l_p,z) \tag{4}$$

$$V_p(l_p,z) = \begin{cases} \frac{\lambda_d}{1+e^{-\alpha \cdot l_p \cdot (\|m_p\| - \|m_0\|)}} & \text{if } p \in S_A \\ 0 & \text{otherwise} \end{cases} \tag{5}$$

Here α, λ_d, $\|m_0\|$ are constants.

The smoothness term $E_s(\mathbf{l}|\mathbf{z})$ describes the interaction between neighbor nodes, and evaluates the assigned labels depending on the similarity of the color features when the two labels are different, as in formula (7).

$$E_s(\mathbf{l}|\mathbf{z}) = \sum_{p \in S} \sum_{q \in N_p} V_{pq}(l_p,l_q,z) \tag{6}$$

$$V_{pq}(l_p,l_q,z) = \begin{cases} \lambda_{s1} \exp(-\beta \|c_p - c_q\|^2) + \lambda_{s2} & \text{if } l_p \neq l_q \\ 0 & \text{otherwise} \end{cases} \tag{7}$$

Here β, λ_{s1}, λ_{s2} are constants, and N_p is the neighbor nodes of the node p.

With the models defined above, the inference algorithm such as GC (Graph Cut), TRW (Tree-Reweighted Message Passing), LBP (Loopy Belief Propagation) can be used to assign the labels $\mathbf{l}^*$ that minimize $E(\mathbf{l}^*|\mathbf{z})$ [19].

4 Experimental Result

4.1 Datasets and Implementation

The proposed method has been tested on the real image sequences. The datasets from [20] provide image sequences as well as ground-truth segmentation mask images. For the implementation, we have used the KLT software [21] and the modified version of inference algorithm packages [19]. In the test, we used $k = 4$ for setting grid points, and the following parameters to define segmentation models: $\alpha = 2.5$, $\lambda_d = 10000$, $\|m_0\| = 2$, $\beta = 5$, $\lambda_s = 1$, $\lambda_{s1} = 1000$, $\lambda_{s2} = 0$. For the inference, four algorithms including GC-Expansion, GC-Swap, TRW, LBP are tested.

4.2 Segmentation Results

Fig. 2 shows segmented moving regions on the image sequences from a moving camera for (a), (b) and a static camera for (c). Fig. 3 shows precision and recall curves, where

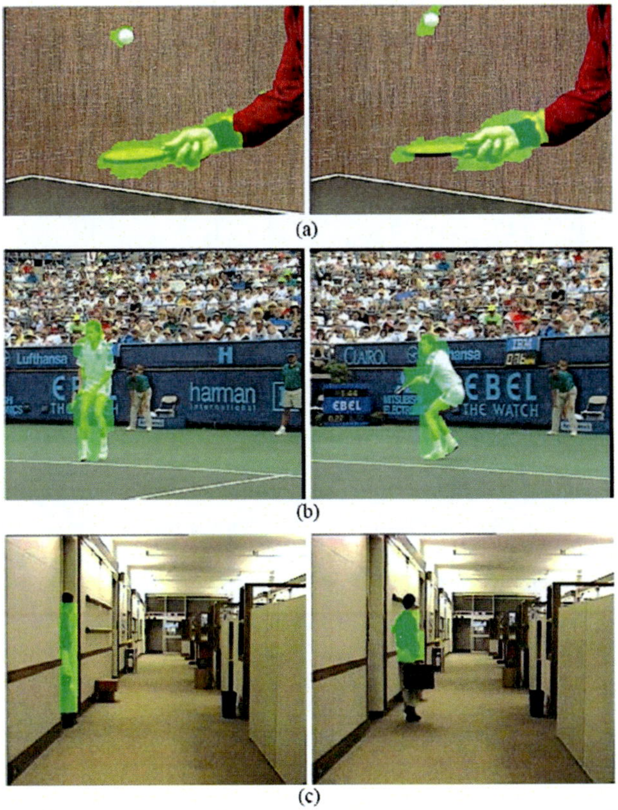

Fig. 2. Segmented (green colored) moving region on the image sequences (a) table tennis, (b) stefan, and (c) hall monitor

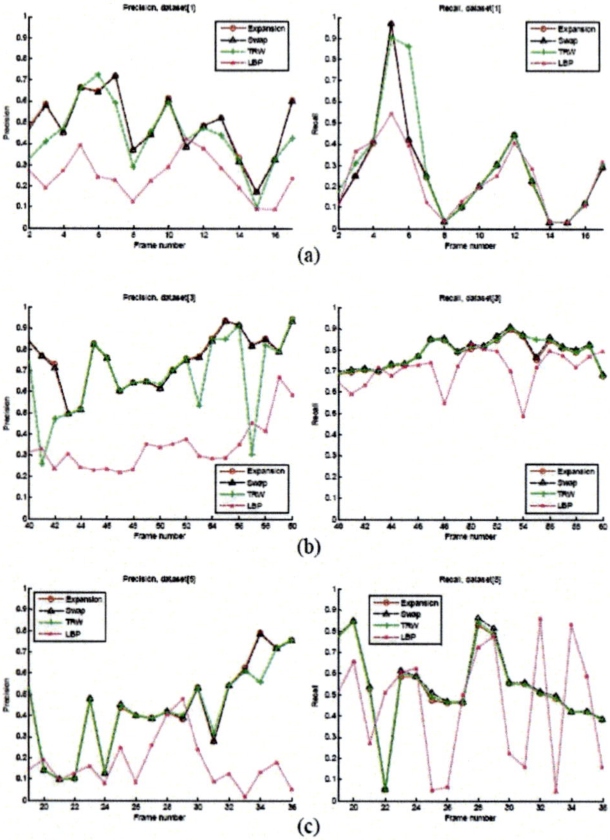

Fig. 3. Precision, recall curve on the image sequences (a) table tennis, (b) stefan, and (c) hall monitor

the precision is the percentage of the correctly segmented moving pixels to the total number of segmented moving pixels, and the recall is the percentage of the correctly segmented moving pixels to the total number of actual moving pixels.

The proposed method works although it only utilizes sparse motion cues. However, the result is sensitive to the error in temporal correspondences on grid points. This leads to poor precision and recall rates. Furthermore, since our method uses motion between only two consecutive frames, moving regions can not be detected if the regions move slowly. For example, the upper part of the arm in Fig. 2 (a) is not recognized as a moving region, as the part moves slightly between consecutive frames. One solution for this problem is to use selected key-frames with large motion discrepancy, as in [7].

5 Conclusion

In this paper, we have presented a method for pixel-wise segmentation of moving regions using sparse motion cues on an image from a freely moving camera. The proposed method utilizes residual motion on sparse grid points, and densely segments moving regions on a CRF segmentation framework.

The future work includes the integration of an appearance model into the segmentation framework, as in [22]. The use of motion from multiple frames tracking [23] is also needed to obtain discriminative motion feature.

Acknowledgement. This research was supported by the MKE(The Ministry of Knowledge Economy), Korea, under the Human Resources Development Program for Convergence Robot Specialists support program supervised by the NIPA(National IT Industry Promotion Agency) (NIPA-2011-C7000-1001-0007).

References

1. Poppe, C., Bruyne, S.D., Paridaens, T., Lambert, P., Walle, R.V.D.: Moving object detection in the H.264/AVC compressed domain for video surveillance applications. Journal of Visual Communication and Image Representation 20(6), 428–437 (2009)
2. Lenz, P., Ziegler, J., Geiger, A., Roser, M.: Sparse scene flow segmentation for moving object detection in urban environments. In: IEEE Intelligent Vehicles Symposium, pp. 926–932 (2011)
3. Kundu, A., Krishna, K.M., Sivaswamy, J.: Moving object detection by multi-view geometric techniques from a single camera mounted robot. In: IEEE/RSJ International Conference on Intelligent Robots and Systems, pp. 4306–4312 (2009)
4. Bouwmans, T., El Baf, F., Vachon, B.: Background modeling using mixture of gaussians for foreground detection - A Survey. Recent Patents on Computer Science 1(3), 219–237 (2008)
5. Stauffer, C., Grimson, W.: Learning patterns of activity using real-time tracking. IEEE Transactions on Pattern Analysis and Machine Intelligence 22(8), 747–757 (2000)
6. Baker, S., Scharstein, D., Lewis, J.P., Roth, S., Black, M.J., Szeliski, R.: A Database and Evaluation Methodology for Optical Flow. International Journal of Computer Vision 92, 1–31 (2011)
7. Liu, F., Gleicher, M.: Learning color and locality cues for moving object detection and segmentation. In: IEEE Conference on Computer Vision and Pattern Recognition, pp. 320–327 (2009)
8. Wang, Y., Ji, Q.: A dynamic conditional random field model for object segmentation in image sequences. In: IEEE Computer Society Conference on Computer Vision and Pattern Recognition (CVPR), pp. 264–270 (2005)
9. Klappstein, J., Vaudrey, T., Rabe, C., Wedel, A., Klette, R.: Moving object segmentation using optical flow and depth information. In: Proc. Pacific-Rim Symposium on Image and Video Technology, Tokyo, Japan, pp. 611–623 (2009)
10. Bugeau, A., Perez, P.: Detection and segmentation of moving objects in complex scenes. Computer Vision and Image Understanding 113(4), 459–476 (2009)
11. Han, M., Xu, W., Gong, Y.: Video object segmentation by motion-based sequential feature clustering. In: 14th Annual ACM International Conference on Multimedia (2006)

12. Su, Y., Sun, M.-T., Hsu, V.: Global motion estimation from coarsely sampled motion vector field and the applications. IEEE Transactions on Circuits and Systems for Video Technology 15(2), 232–242 (2005)
13. Lucas, B.D., Kanade, T.: An iterative image registration technique with an application to stereo vision. In: International Joint Conference on Artificial Intelligence, pp. 674–679 (1981)
14. Huber, P.J.: Robust statistics. John Wiler, New York (1981)
15. Haller, M., Krutz, A., Sikora, T.: Evaluation of pixel- and motion vector-based global motion estimation for camera motion characterization. In: Proc. WIAMIS, pp. 49–52 (2009)
16. Meer, P., Mintz, D., Rosenfeld, A., Kim, D.Y.: Robust Regression Methods for Computer Vision: A Review. International Journal of Computer Vision 6(1), 59–70 (1991)
17. Shi, J., Tomasi, C.: Good Features to Track. In: IEEE Conference on Computer Vision and Pattern Recognition(CVPR), pp. 593–600 (1994)
18. Lafferty, J., McCallum, A., Pereira, F.: Conditional random fields: Probabilistic models for segmenting and labeling sequence data. In: ICML, pp. 282–289 (2001)
19. Szeliski, R., et al.: A Comparative Study of Energy Minimization Methods for Markov Random Fields with Smoothness-Based Priors. IEEE Transactions on Pattern Analysis and Machine Intelligence 30(6), 1068–1080 (2008)
20. Datasets - Segmented foreground objects, `http://www.sfu.ca/~ibajic/datasets.html`
21. KLT: An Implementation of the Kanade-Lucas-Tomasi Feature Tracker, `http://www.ces.clemson.edu/~stb/klt/`
22. Wu, M., Peng, X., Zhang, Q.: Segmenting moving objects from a freely moving camera with an effective segmentation cue. Measurement Science & Technology 22(2) (2011) (article Number 025108)
23. Sand, P., Teller, S.: Particle Video: Long-Range Motion Estimation using Point Trajectories. In: IEEE Computer Society Conference on Computer Vision and Pattern Recognition (CVPR), pp. 2195–2202 (2006)

Fast and Robust Multi-people Tracking from RGB-D Data for a Mobile Robot

Filippo Basso*, Matteo Munaro*, Stefano Michieletto, Enrico Pagello, and Emanuele Menegatti

Department of Information Engineering of the University of Padova, via Gradenigo 6B, 35131 - Padova, Italy
{bassofil,munaro,michieletto,epv,emg}@dei.unipd.it

Abstract. This paper proposes a fast and robust multi-people tracking algorithm for mobile platforms equipped with a RGB-D sensor. Our approach features an efficient point cloud depth-based clustering, an HOG-like classification to robustly initialize a person tracking and a person classifier with online learning to manage the person ID matching even after a full occlusion. For people detection, we make the assumption that people move on a ground plane. Tests are presented on a challenging real-world indoor environment and results have been evaluated with the CLEAR MOT metrics. Our algorithm proved to correctly track 96% of people with very limited ID switches and few false positives, with an average frame rate of 25 fps. Moreover, its applicability to robot-people following tasks have been tested and discussed.

Keywords: People tracking, real-time, RGB-D data, mobile robots.

1 Introduction and Related Work

People detection and tracking are key abilities for a mobile robot acting in populated environments. Such a robot must be able to distinguish people from other obstacles, predict their future positions and plan its motion in a human-aware fashion, according to its tasks.

Many works exist about people detection and tracking by using RGB cameras only ([6], [21], [4]) or 3D sensors only ([14], [19], [20], [5], [15]). However, when dealing with mobile robots, the need for robustness and real time capabilities usually led researchers to tackle these problems by combining appearance and depth information. In [2], both a PTZ camera and a laser range finder are used in order to combine the observations coming from a face detector and a leg detector, while in [13] the authors propose a probabilistic aggregation scheme for fusing data coming from an omnidirectional camera, a laser range finder and a sonar system. Ess *et al.* [7],[8] describe a tracking-by-detection approach based on a multi-hypothesis framework for tracking multiple people in busy environments from data coming by a synchronized camera pair. The depth estimation provided

* These authors contributed equally to this work.

S. Lee et al. (Eds.): Intelligent Autonomous Systems 12, AISC 193, pp. 265–276.
springerlink.com

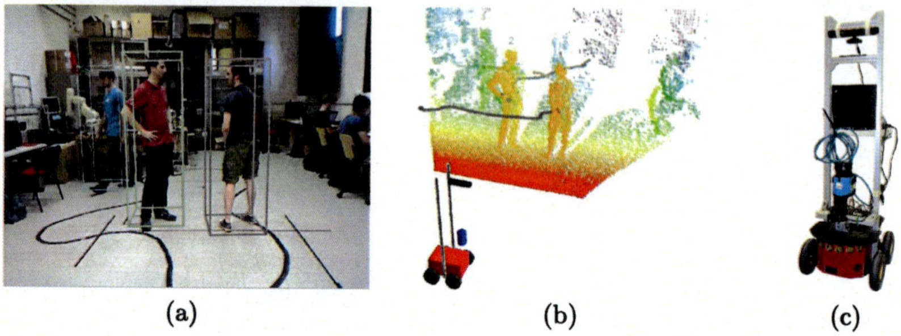

Fig. 1. Example of our system output: (a) a 3D bounding box is drawn for every tracked person on the RGB image, (b) the corresponding 3D point cloud is reported, together with people trajectories. In (c) the mobile platform we used for the experiments is shown. Note that in this paper we only employ RGB-D data from a Microsoft Kinect sensor and do not make use of other sensors such as Laser Range Finder or sonars.

by the stereo pair allowed them to reach good results in challenging scenarios, but their algorithm reached real time performance if one does not take into account the time needed by their people detection algorithm which needs 30s to process each image. Stereo cameras continue to be widely used in the robotics community ([1],[17]), but the computations needed for creating the disparity map always impose limitations to the maximum frame rate achievable, thus leaving less room for further algorithms operating in series with the tracking one.

With the advent of reliable and affordable RGB-D sensors a rapid boosting of robots capabilities can be envisioned. For example, the Microsoft Kinect sensor allows to natively capture RGB and depth information at good resolution (640x480 pixels) and frame rate (30 frames per second). Even though the depth estimation becomes very poor over eight meters of distance and this technology cannot be used outdoors because the sunlight can change the infrared pattern projected by the sensor, it constitutes a very rich source of information that can be simply used on a mobile platform. In [12] a tracking algorithm on RGB-D data is proposed that exploits the multi-cue people detection approach described in [18]. It adopts an on-line detector that learns individual target models and a multi-hypothesis decisional framework. No information is given about the computational time needed by the algorithm and results are reported for some sequences acquired from a static platform equipped with three RGB-D sensors.

In this work, we propose a multi-people tracking algorithm with RGB-D data specifically designed for mobile platforms. By assuming that people are moving on a ground plane, our method is able to robustly track them with a medium frame rate of 25 frames per second by relying only on CPU computation. We tested our approach on sequences of increasing complexity, reaching very good results even when dealing with complex people trajectories and strong occlusions. The track initialization procedure, which relies on a HOG people detector, allows to minimize the number of false positives and the online learning person classifier is used every time a person is lost, in order to recover the correct person ID even after a full occlusion.

The remainder of the paper is organized as follows: in Section 2 the various parts of our approach are introduced. Section 3 describes the downsampling scheme that we adopt for reaching real time performances, while the detection phase is described in Section 4. Section 5 details the tracking procedure and in Section 6 we describe the tests performed and we report the results evaluated with the CLEAR MOT metrics. Conclusions and future works are contained in Section 7.

2 System Overview

In this section, we outline the different software modules of our people tracking system. As reported in Fig. 2, the RGB-D data are processed by three main software blocks: filtering, detection and tracking. The filtering block consists in a smart down-sampling of the 3D point cloud. This is a crucial step for reducing the size of the big amount of data we deal with. The detection block performs the necessary operations for clustering the remaining points and selecting the clusters of points containing people, that are subsequently passed to the tracking module. In the last block, the clusters are associated to the correct track according to the predictions of an Unscented Kalman Filter and, whether the filter fails because of full occlusions, an online-learned person classifier is able to recover the correct association between the person and the track when the person return visible.

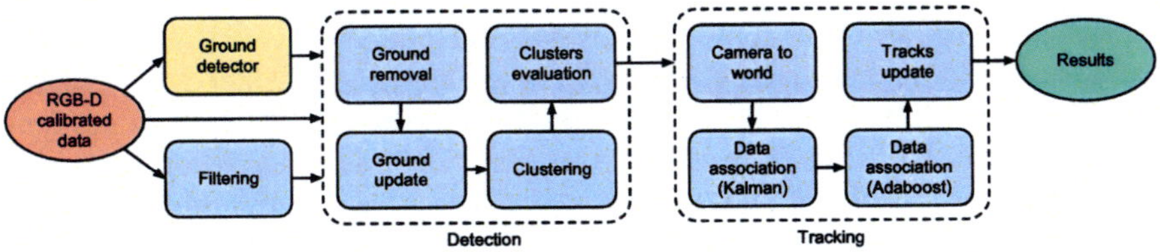

Fig. 2. System overview highlighting the main software blocks

3 Filtering

At each frame, the data provided by the RGB-D sensor are processed by a voxel grid filter, i.e. the space is subdivided in a set of tiny 3D boxes and all points inside each box are approximated with the coordinates of their centroid. By default, we chose the voxel size to be of 0.06m. This value allowed us to downsize the point cloud of our RGB-D sensor by an order of magnitude, thus reaching high real time performance and having enough data for performing the tracking procedure. Moreover, after this operation, points density no longer depends on their distances from the sensor.

4 Detection

4.1 Clustering

In order to divide the scene into different clusters, our algorithm remove all 3D points belonging to the floor from the output of the previous software block. In the first frame, the ground plane equation is estimated from an initial input from the user (i.e. click on three points of the ground plane in the image). Then, at every frame, this equation is updated by considering as initial condition the estimation at the previous frame. All points of the cloud within a threshold distance from the hypothetical plane are selected. Then, using a least squared method on these points, the plane coefficients are refined. This procedure permits to adapt to small changes in the floor slope and to the camera oscillations during robot movements. Once this operation has been performed, the different clusters are no longer connected through the floor, so they can be easily calculated by labeling neighboring 3D points on the basis of their Euclidean distances.

4.2 Clusters Evaluation

The objective of clusters evaluation is to preserve only those clusters containing a person and extract a series of features necessary for the subsequent tracking phase. For each cluster, we estimate: the height from the ground plane, the centroid, the distance from the sensor and the corresponding blob in the RGB image. Clusters with height out of a plausible range for an adult person[1] are discarded before computing the subsequent features and do not pass to the tracking phase. Moreover, we determine if a cluster is occluded by any other cluster or it is partially out of the camera view. When a cluster is not occluded, we compute also its HOG confidence. This consists in the confidence obtained by applying a HOG people detector [6] to the part of the RGB image corresponding to the cluster theoretical bounding box, that is the bounding box that should contain the whole person, from the head to the ground. For the people detector, we used Dollár's implementation of HOG[2] and the same procedure and parameters described by Dalal and Triggs [6] for training the detector.

5 Tracking

In this section we present the overall tracking approach we have designed. From now on, the positions relative to the detections (clusters) found in the previous phase are considered in world coordinates with respect to the starting position of the platform[3].

---

[1] In this work we considered clusters with height in the range 1.4 - 2.3m.

[2] Contained in his Matlab toolbox `http://vision.ucsd.edu/~pdollar/toolbox`

[3] Our platform estimates its position by means of odometry and a self localization system not described here.

5.1 Motion Model

Since we use a tracking-by-detection approach and our detector has been designed to minimize false negatives, we just choose to use an Unscented Kalman Filter to maintain people positions and velocities along the two world axes (x, y).

There are many methods to model human motion in tracking. However, a good way to manage full occlusions is to consider a constant velocity model, as described in [2]. For what concerns the observation model, and in particular the covariance matrix of the measurement noise, we expressed it as a sum of two different components, σ_v^2 and σ_d^2. σ_v^2 models the quantization error introduced by the voxel filter[4] while σ_d^2 models the measurement error introduced by the depth sensor. This error, for our particular sensor, is proportional to the distance squared[5]. Among the different extensions to the basic Kalman filter, we decided to use the Unscented version because it has prediction capabilities near those of a particle filter, but it is only slightly more computationally expensive than an Extended Kalman Filter [2].

5.2 Online Classifier

To overcome the limits imposed by the constant velocity model when dealing with full occlusions, we maintain for each track an online classifier based on Adaboost, like the one used in [10] or [12]. But, unlike these two approaches, that make use of features directly computed on the RGB (or depth) image, we calculate our features in the color histogram of the target, as following:

1. we select the image pixels belonging to the person by exploiting the blob mask given by the detector;
2. we compute the 3D color histogram of these points on the three RGB channels;
3. we select a set of randomized parallelepipeds (one for each weak classifier) inside the histogram. The feature value is given by the sum of histogram elements that fall inside a given parallelepiped.

For the training phase, since the approach introduced by [10] gave us poor performances, we use as positive sample the color histogram of the target, while as negative samples we consider the histograms calculated on the detections not associated to the current track. This approach has the advantage to select only the colors that really characterize the target and distinguish it from all the others.

5.3 Data Association

Once a new set of detections is available, we perform the detection-track association with two subsequent steps: at first only the prediction given by the Kalman

[4] This error is considered to be uniformly distributed between 0 and the voxel size multiplied by $\sqrt{3}$, namely the voxel diagonal.

[5] `http://www.ros.org/wiki/openni_kinect/kinect_accuracy`

filter is considered, while, as a second step, we consider only the results given by the classifier. We perform the data association based on the Kalman filter prediction with the Global Nearest Neighbor method, described in [11] and [2]. For each detection, we calculate the Mahalanobis distance from the predicted state of all the active tracks, we create the cost matrix based on this distance and then perform the association with the Munkres algorithm. Only at this point the remaining detections and tracks are considered for the association based on the classifier. If, after this step, there are some detections that are still unassociated, they are taken into account for the creation of new tracks.

We have not yet implemented more sophisticated methods, such as multiple hypothesis tracking or probabilistic data association filters, because in the considered scenarios the potential increase in performance would not compensate for the considerably higher computational load.

5.4 Tracks Management

The policies of creation/update/deletion of the tracks are really important to get good results from the whole tracking process. They can be summarized as following.

5.4.1 Initialization

A new track is created from an unassociated detection if the confidence value given by the HOG classifier is over a defined threshold for N frames. In our case N has been set to three. If this happen, the track is promoted to "validated", otherwise it is discarded.

5.4.2 Update

After the detection-track association, the Kalman filter is updated with the measurement of the theoretical centroid of the cluster. Then, if the cluster is not occluded also the classifier is updated. This is given by the fact that, when a person is occluded, some colors can be missing, updating the classifier would decrease the importance of these color features, thus distorting the results of successive classifications.

5.4.3 Recovery

After a full occlusion, a person could be wrongly assigned to a new track instead of the old one. This happens if neither the Kalman filter, nor the classifier correctly associate the new detection to the previous track. But, while the trajectory error of the Kalman filter cannot be corrected, the classifier can manage to recover from this mistake. In particular, during the first frames after its creation, the new track histograms are evaluated by the classifiers of the missing tracks and, if the result is above a determined threshold, the new track is recognized to refer to an old one and associated to it. An example of this behavior is depicted in Fig. 3. From left to right it can be seen that the person associated to track #2 goes out of the field of view of the camera and, when it comes back in, it is

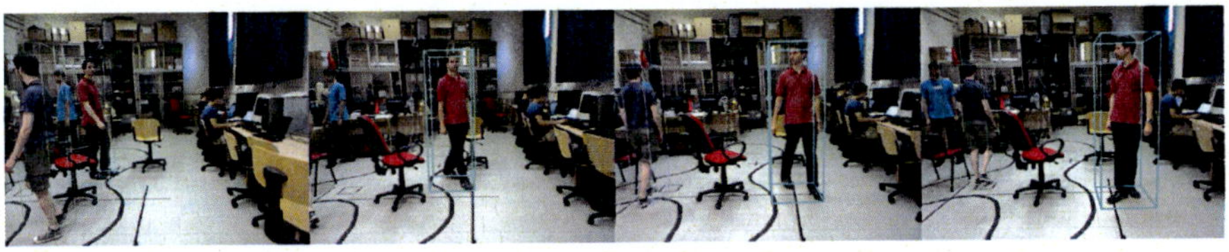

Fig. 3. From left to right, an example of track recovery. Track #2 becomes #4 after the full occlusion, but, in a few frames, it returns to its original ID.

assigned to a new track (#4), but, in a few frames, it returns to its original ID (#2).

5.4.4 Removal

After a person becomes occluded or goes out of the scene, the correspondent track is marked as missing. A track is deleted and no more considered if it remains in that state for a certain number of consecutive frames or, as described above, if it is not validated before time runs out.

6 Experiments

6.1 Experimental Setup

In this section we show the performance of our system on RGB-D video sequences collected in an indoor environment with the mobile robot shown in Fig. 1(c). It consists of a Pioneer P3-AT platform equipped with a Microsoft Kinect sensor, that is endowed with a standard RGB camera, an infrared camera and an infrared projector. This low cost hardware can provide RGB-D data with 640 x 480 pixel resolution at 30 frames per second. In these tests we acquired depth data at a reduced resolution, namely 160 x 120 pixel, for speed.

We performed tests both while keeping our platform as static and while moving it on a predefined path. For both cases we acquired videos in three different scenarios of increasing difficulty:

1. no obstacle is present, people move with simple (linear) trajectories;
2. no obstacle is present, people move with complex trajectories and interact with each other;
3. obstacles are present, people move with complex trajectories and interact with each other.

Every video sequence extends over about 750 frames, thus the total test set includes 4671 frames, 12272 instances of people and 26 tracks that have been manually annotated on the RGB image and that constitute the ground truth. The minimum distance between people is 0.2m while the minimum people-object distance is 0.05m.

Table 1. Frame rates for two test videos with our moving platform

	Kinect Data	Detector	Detector+Tracker
Complex traj.	29.971	26.268	26.215
With obstacles	29.967	25.243	25.227

6.2 Runtime Performance

The entire system is implemented in C++ within ROS, the Robot Operating System[6], making use of highly optimized libraries for 2D computer vision[7], 3D point cloud processing[8] [16] and bayesian estimation[9]. On a notebook with an Intel i5-520M 2.40GHz processor and 4GB of memory, we measured the frame rates reported in Table 1. They indicate the average rates reached by the Kinect data publication, the detection phase and our complete system (detection and tracking). These tests refer to two videos captured with our moving platform. A small frame rate reduction can be noticed when some obstacles are present in the scene. This happen because more clusters have to be evaluated by the detector. As it can be inferred from the data, the bottleneck of the system is the detector, while the tracking algorithm is very fast. Overall, the whole system can track multiple people in the scene at a medium frame rate of 25 frames per second.

6.3 Tracking Evaluation

For the purpose of evaluating the tracking performance we adopted the CLEAR MOT metrics [3], that consists of two indexes: MOTP and MOTA. The MOTP indicator measures how well exact positions of people are estimated, while the MOTA index gives an idea of the amount of errors that are made by the tracking algorithm in terms of false negatives, false positives and mismatches. In particular, given that our ground truth does not consist of the metric positions of all persons, but of their positions inside the image, we computed the MOTP index as the average PASCAL index [9] (intersection over union of bounding boxes) of the associations between ground truth and tracker results by setting the validation threshold to 0.5. Instead, we computed the MOTA index with the following formula

$$MOTA = 1 - \frac{\sum_t (fn_t + fp_t + ID_t^{sw})}{\sum_t g_t} \tag{1}$$

where fn_t is the number of ground truth people instances (for every frame) not found by the tracker, fp_t is the number of output tracks instances that do not have correspondences with the ground truth, ID_t^{sw} represents the number of times a track corresponding to the same person changes ID over time and g_t is the total number of ground truth instances present in all frames.

[6] `http://ros.org`

[7] OpenCV - `http://opencv.willowgarage.com`

[8] PCL - `http://pointclouds.org`

[9] Bayes++ - `http://bayesclasses.sourceforge.net`

In Table 2 and 3 we report, for every test sequence, the MOTP and MOTA indexes, the percentage of false positives and false negatives, the number of ID switches and the number of recoveries. With recovery we mean when a person that was lost by the Kalman-based tracking is re-identified by means of the classifier. The average precision of our tracker (MOTP) is always around 80%, while its accuracy (MOTA) slightly decreases from 93% to 89% as we increase the scenario complexity. The minimum accuracy is reached for the video that we captured while moving our platform inside an environment with obstacles that can significantly occlude people. These occlusions mainly have the effect to delay tracks initialization because we compute the HOG confidence only for not occluded clusters. Thus the false negative percentage increases and the MOTA index decreases. Nevertheless, the track initialization based on the HOG confidence makes our algorithm to totally avoid false tracks. The percentage of false positives shown in the tables are due to bounding boxes generated by real persons that are not perfectly centered on the ground truth data, thus they do not pass the PASCAL test.

Table 2. Tracking evaluation results for static tests

	MOTP	MOTA	FP	FN	ID Sw.	Rec.
Simple traj.	79.0%	93.2%	3.8%	2.9%	0	0
Complex traj.	79.2%	96.1%	3.1%	0.7%	0	0
With obstacles	82.6%	89.7%	2.9%	7.5%	1	0

Table 3. Tracking evaluation results for moving tests

	MOTP	MOTA	FP	FN	ID Sw.	Rec.
Simple traj.	81.9%	92.8%	2.8%	4.4%	0	1
Complex traj.	83.9%	89.2%	4.8%	6.1%	0	1
With obstacles	84.1%	89.2%	5.3%	5.5%	0	0

As we said in Section 5, the method we use for selecting which clusters are passed to the data association algorithm is based on a simple check of clusters height from the ground plane. This approach proved to be very effective for people tracking. In fact, if only the clusters that have an HOG confidence above a threshold (e.g. -3) were used to update the tracks, the MOTA index would drop of about 30%. This is due to the fact that the HOG people detector score decreases very quickly in case of occlusion.

In general, our algorithm showed to be very robust even for tracking partially occluded people moving with complex trajectories. In fact, we correctly track 96% of every ground truth track. The only ID switch happened when a person was lost due to a sudden movement and after then the Kinect automatically adjusted its image brightness. This episode led the appearance classifier to fail, thus a new track was initialized. Apart from this case, the on-line person-specific classifier correctly managed the most part of track recoveries after a full person occlusion. In particular, for the most complicated scenario we tested (moving platform, obstacles and complex trajectories), it allowed to track 20% more of ground truth tracks without ID switches.

In Fig. 4 we report some examples of correctly tracked frames from our test set. Different IDs are represented by different colors and the bounding box is drawn with a thick line if the algorithm estimates a person to be completely visible, while a thin line is used if a person is considered occluded. It can be noticed that our tracking system obtains very good performance even in presence of very strong occlusion (>70%). Furthermore, people are correctly classified as visible or occluded.

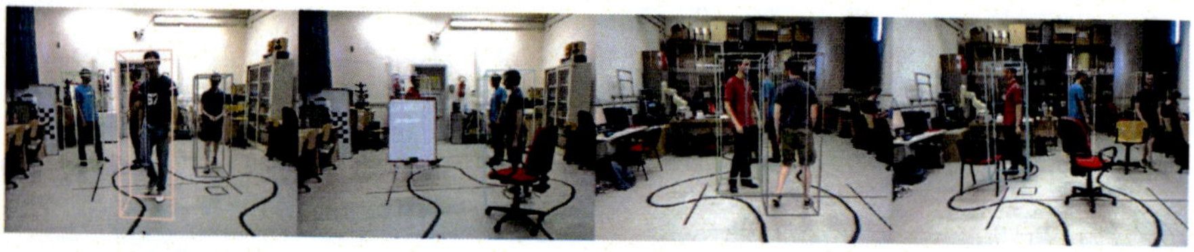

Fig. 4. Tracking output on some frames extracted from the test set

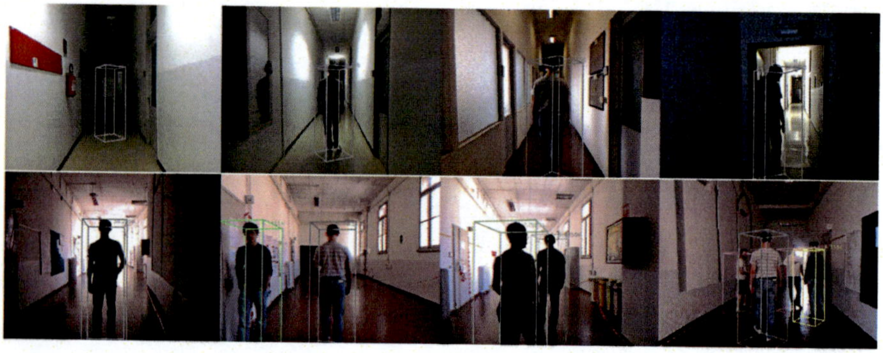

Fig. 5. People following test. First row: examples of tracked frames while a person is robustly followed along a narrow corridor with many lighting changes. Second row: other examples of correctly tracked frames when other people are present in the scene.

As a further test for proving the robustness and the real time capabilities of our tracking method we asked the robot to follow a particular person along a narrow corridor with many lighting changes and in a hall where many people are present. In Fig. 5 the output of the tracking algorithm is visualized for some frames collected during this experiment. In the first row it can be seen that the person is correctly tracked and followed by the platform along the corridor, while the lighting conditions considerably change. This is allowed by a good prediction of people position in ground plane coordinates, obtained with the use of the Unscented Kalman Filter that can also benefit of many available measures thanks to the high frame rate. In the second row the tracking (and following) robustness is shown when other people walk next to the followed person or between him and the robot.

7 Conclusions and Future Works

In this paper we have presented a very fast algorithm for multi-people tracking from mobile platforms equipped with a RGB-D sensor. A robust track initialization is obtained by checking the confidence consistence of a HOG people detector and tracks are recovered after a full occlusion thanks to an online learned classifier based on features extracted from RGB histograms. A proper downsampling is applied to the raw sensor data in order to make them tractable in real time.

Tests were performed in scenarios of increasing complexity and our tracking system proved to be very effective even when dealing with strong occlusions and complex trajectories. Some limits of our system are due to the limited field of view of the RGB-D sensor we are using and the poor depth estimates over eight meters of distance.

As future works, we plan to improve our clustering method in order to correctly handle situations when people are too close or touching each other and mount more Kinect sensors for extending the field of view of our platform. Moreover, given the high real time potential of our tracking algorithm, we envision to study motion planning algorithms that can allow robots to move in a human-aware fashion and fully integrate in a populated environment.

References

1. Bajracharya, M., Moghaddam, B., Howard, A., Brennan, S., Matthies, L.H.: A fast stereo-based system for detecting and tracking pedestrians from a moving vehicle. International Journal of Robotics Research 28, 1466–1485 (2009)
2. Bellotto, N., Hu, H.: Computationally efficient solutions for tracking people with a mobile robot: an experimental evaluation of bayesian filters. Auton. Robots 28, 425–438 (2010)
3. Bernardin, K., Stiefelhagen, R.: Evaluating multiple object tracking performance: the clear mot metrics. J. Image Video Process. 2008, 1:1–1:10 (2008)
4. Breitenstein, M.D., Reichlin, F., Leibe, B., Koller-Meier, E., Gool, L.V.: Robust tracking-by-detection using a detector confidence particle filter. In: IEEE International Conference on Computer Vision (October 2009)
5. Carballo, A., Ohya, A., Yuta, S.: People detection using range and intensity data from multi-layered laser range finders. In: IEEE/RSJ International Conference on Intelligent Robots and Systems (IROS), pp. 5849–5854 (2010)
6. Dalal, N., Triggs, B.: Histograms of oriented gradients for human detection. In: Computer Vision and Pattern Recognition, vol. 1, pp. 886–893 (June 2005)
7. Ess, A., Leibe, B., Schindler, K., Van Gool, L.: A mobile vision system for robust multi-person tracking. In: IEEE Conference on Computer Vision and Pattern Recognition 2008, pp. 1–8 (2008)

8. Ess, A., Leibe, B., Schindler, K., Van Gool, L.: Moving obstacle detection in highly dynamic scenes. In: Proceedings of the 2009 IEEE International Conference on Robotics and Automation, ICRA 2009, Piscataway, NJ, USA, pp. 4451–4458 (2009)
9. Everingham, M., Gool, L., Williams, C.K., Winn, J., Zisserman, A.: The pascal visual object classes (voc) challenge. Int. J. Comput. Vision 88, 303–338 (2010)
10. Grabner, H., Bischof, H.: On-line boosting and vision. In: Proceedings of the 2006 IEEE Computer Society Conference on Computer Vision and Pattern Recognition - Volume 1, Washington, DC, USA, pp. 260–267 (2006)
11. Konstantinova, P., Udvarev, A., Semerdjiev, T.: A study of a target tracking algorithm using global nearest neighbor approach. In: Proceedings of the 4th International Conference Conference on Computer Systems and Technologies: e-Learning, New York, NY, USA, pp. 290–295 (2003)
12. Luber, M., Spinello, L., Arras, K.O.: People tracking in rgb-d data with online boosted target models. In: IEEE/RSJ International Conference on Intelligent Robots and Systems, IROS 2011 (2011)
13. Martin, C., Schaffernicht, E., Scheidig, A., Gross, H.-M.: Multi-modal sensor fusion using a probabilistic aggregation scheme for people detection and tracking. Robotics and Autonomous Systems 54(9), 721–728 (2006)
14. Mozos, O., Kurazume, R., Hasegawa, T.: Multi-part people detection using 2d range data. International Journal of Social Robotics 2, 31–40 (2010)
15. Navarro-Serment, L.E., Mertz, C., Hebert, M.: Pedestrian detection and tracking using three-dimensional ladar data. In: FSR, pp. 103–112 (2009)
16. Rusu, R.B., Cousins, S.: 3D is here: Point Cloud Library (PCL). In: IEEE International Conference on Robotics and Automation (ICRA), Shanghai, China, May 9-13 (2011)
17. Satake, J., Miura, J.: Robust stereo-based person detection and tracking for a person following robot. In: Workshop on People Detection and Tracking IEEE ICRA (2009)
18. Spinello, L., Arras, K.O.: People detection in rgb-d data. In: IEEE/RSJ International Conference on Intelligent Robots and Systems, IROS 2011 (2011)
19. Spinello, L., Arras, K.O., Triebel, R., Siegwart, R.: A layered approach to people detection in 3d range data. In: Proc. 24th AAAI Conference on Artificial Intelligence, PGAI Track (AAAI 2010), Atlanta, USA (2010)
20. Spinello, L., Luber, M., Arras, K.O.: Tracking people in 3d using a bottom-up top-down people detector. In: IEEE International Conference on Robotics and Automation (ICRA 2011), Shanghai (2011)
21. Viola, P., Jones, M.: Rapid object detection using a boosted cascade of simple features. In: CVPR 2001, vol. 1, pp. 511–518 (2001)

Multi-agent Based Optic Flow

Kiwon Sohn[1], Paul Oh[1], and M. Anthony Lewis[2]

[1] Drexel University, Philadelphia, Pennsylvania 19104
skw1125@gmail.com, paul@coe.drexel.edu
[2] Iguana Robotics Inc, Tucson, Arizona 85715
m.anthony.lewis@gmail.com

Abstract. In this article, the authors present a novel algorithm for computing optic flow using a multi-agent based feature point tracking method. In this multi-agent based optic flow method, feature points which are invariant to scale, orientation and illumination changes are extracted and tracked in parallel using independent agents. Each agent is run by a separate light-weight thread which can be implemented using parallel processes on a multicore processor. The agents use a Kalman filter to predict the frame to frame position of the feature points in the image, producing position and velocity data for each feature point, which can then be used to perform optic flow, while simultaneously producing feature descriptors that can be used for object recognition and stereopsis. We show that in a parallel implementation, this algorithm provides significant performance advantages over other feature point tracking object recognition methods. It therefore may provide a plausible basis for a unified computer vision architecture including optic flow, object recognition, and stereopsis.

1 Introduction

We would like to develop robots that can navigate in the environment while implementing multiple aspects of computer vision, such as optic flow and object recognition. This motivates the desire to develop a unified underlying framework for image processing that combines aspects of object recognition with optic flow. Object recognition has already been demonstrated by David Lowe in his SIFT algorithm [1]. Many approaches to optic flow have been widely used. For example, both the Lucas-Kanade method [2] and Horn-Schunck method [3] provide estimation of the velocity of points in a scene. While Lucas-Kanade and Horn-Schunck provided a flow of information at high speed over various image sizes, unlike SIFT they are not able to match specific points between ordered images. With SIFT, Lowe showed that highly distinctive keypoints can be matched between ordered or stereo images with high probability [1]. This keypoint matching can be used to track objects between frames in video processing. However, SIFT is primarily used for single image object recognition and is not fast enough to be used for real-time video processing as would be needed for robot navigation.

In Lowe's implementation of SIFT using 2 GHz Pentium 4 processor, total recognition time to identify 3 objects in a test image (640 by 315 pixels) took 0.3

S. Lee et al. (Eds.): Intelligent Autonomous Systems 12, AISC 193, pp. 277–287.
springerlink.com

second. That is, Lowe's algorithm, running on a modest computer, can roughly duplicate the image processing time of a human [4] [5]. However, further speedup is possible, and in fact is needed for functions such as optic flow and stereopsis–particularly when such algorithms will be used to guide robots.

We propose the following key changes to SIFT for our application:

1. Agent Based Implementation : we make each keypoint intelligent by adding knowledge of its state: position and velocity. This allows the use of a Kalman filter to *predict* the position of keypoints in subsequent images. Note that the Kalman equation has two parts: Prediction and correction. So, velocity has meaning because it is part of the prediction part of the Kalman equation. We note that velocity *emerges* as an side-effect of tracking.
2. Parallelization: By converting each keypoint into an agent, the tracking of each keypoint can be processed in a separate thread. There is no communication between agents at this time. This makes the algorithm inherently parallelizable. Thus, we can expect significant speedup as processor continue to add additional cores, and the algorithm automatically adjusts itself to the presence of new processor cores.
3. Limited use of scale space: Lowe's algorithm is meant to recognize objects over a range scale changes. In applications for optic flow and stereomatchings, large scale changes are unlikely.

2 Agent Based Optic Flow

We propose an optic flow algorithm that uses intelligent agents to track keypoint descriptors within an image from frame to frame. Each agent predicts the future position of the keypoint, given knowledge about the past motion of the keypoint [1]. Given a prediction of the keypoint's position, the agent searches within a small window centered around the predicted position, rather than the whole frame.

The agent based algorithm consists of the following steps for each agent:

1. Extracting Keypoints using DOG filter
2. Orientation assignment
3. Building descriptor
4. Estimate Keypoint position in current frame using Kalman filter
5. Matching Keypoints between previous frame and window around estimated keypoint

2.1 Extracting Keypoints

The first step in finding potential keypoints is extracting stable keypoints over all possible scales of image to make this algorithm insensitive to image scale changes. To implement this, a scale space representation which represents an

[1] Movement of the camera, measured by a sensor, e.g. gyro, could easily be incorporated as well.

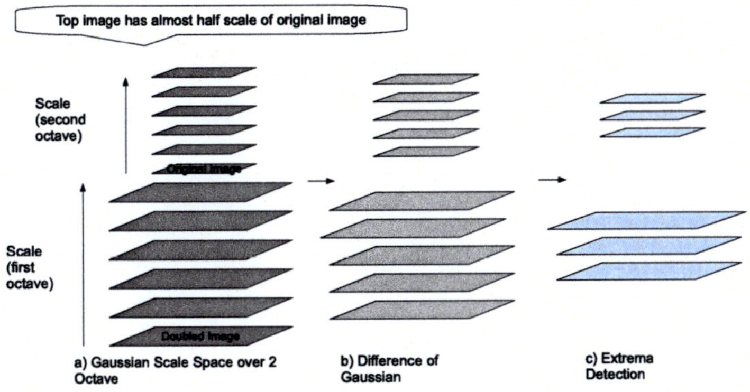

Fig. 1. Extraction of Potential Keypoints. To make the keypoint descriptors invariant to scale, the images are smoothed over a Gaussian kernel with a variety of different scales. These images are then subtracted to create Difference of Gaussian Images, and extrema are extracted by localized pixel comparisons. This process produces six DOG images, for various scales.

image with a set of smoothed images is produced using a Gaussian kernel. Since it can be assumed that the scale of any object in the frame will not increase or decrease by more than a factor of two between consecutive frames, only 2 octaves were used, compared to Lowe's SIFT algorithm, which searches over as many octaves as possible. After building the scale space, Difference of Gaussian images were created by subtracting Gaussian smoothed images. The extrema of Difference of Gaussian images were found by comparison with neighboring pixels. Lowe proved that Difference of Gaussian can be approximated to scale normalized Laplacian of Gaussian [1]. Mikolajczyk found that the extrema of scale normalized Laplacian of Gaussian generates more stable keypoints than the gradient function or Harris corner function [6] [7]. This process results produces several images that each represent the scale-features of the original image at different scales.

All scales of each octave are precomputed and Gaussian smoothing on each octave is implemented in parallel. Since there are six images per interval of each octave, speed can be improved on a multicore processer by processing in parallel.

Although SIFT implements a keypoint localization process after scale space extrema detection, this process is omitted in our simplified agent-based version. Though localization and keypoint selection might be helpful for object recognition tasks in which objects in the new image could be deformed or degraded, these processes did not greatly improve matching of keypoints between ordered images in optic flow. Keypoint localization also serves to reduce the number of keypoints by filtering out the less useful ones, but in our algorithm the number of keypoints is directly controlled by adjusting the number of agents.

2.2 Orientation Assignment and Building Descriptor

Following scale space extrema detection, a process to make each keypoint insensitive to orientation changes is implemented. Like SIFT, local image gradients around each keypoint are calculated by comparing pixels to the central keypoint pixel. These gradients are then weighted by a Gaussian centered at the keypoint, and divided into eight histogramming bins which are uniformly divided over 360 degrees (i.e. 8 cardinal directions). The dominant orientation angle is extracted by averaging over each bin, then the keypoint descriptor is rotated relative to this angle so that all keypoints have their dominant gradient angle pointing along the X axis [1]. This process allows feature descriptors to be easily compared across differing orientations.

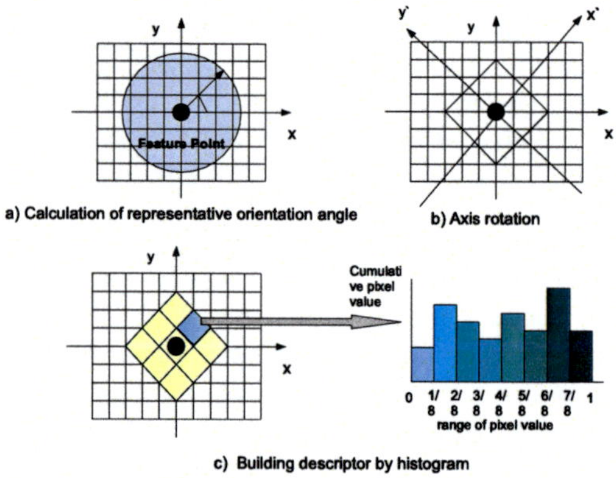

Fig. 2. Orientation Assignment and Building Descriptor. First (a), gradients around the keypoint are computed and weighted by a Gaussian at the scale of the image. Second (b) the image is rotated so that the dominant gradient is oriented along the X axis. Third (c), pixels are normalized to correct for intensity changes and divided into 9 regions. Finally (d), pixels are divided into 9 histograms of 8 bins each, so that 72 numbers represent a description of the image around the keypoint. This is the keypoint descriptor.

The next step is producing a feature descriptor which makes keypoints reasonably distinctive and able to be matched with points in another frame with a high probability. In our approach, pixel values at the selected scale around keypoint are normalized first to prevent variation due to illumination change. Normalized values are put into histogram bins which have a segmented range of pixel values. The region around the keypoint is divided into nine squares and each divided region is histogramed individually into eight intensity bins, making a total of dimension for the feature descriptor of 72. Fig 2 shows the orientation assignment and building of the feature descriptor around keypoints at the

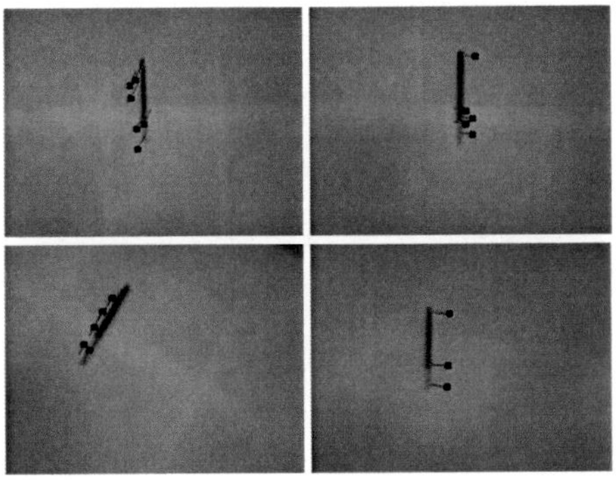

Fig. 3. Matching in Aperture Case

selected scale. As with the scale space process, the building of feature descriptors is implemented in parallel. The set of keypoints is divided into six subsets and descriptor is built in parallel. This dividing task might be faster on a machine which has more than six cores. The number six is assigned according to our machine performance, so it could be adjusted to the ability of the machine.

Motion perception usually encounters the problem that moving contours with different orientations can cause identical responses if they are seen through the small hole. Likewise, number of unknowns is larger than number of equations in a two frame differential method like Lukas-Kanade. This is known as the aperture problem in optical flow. However, the aperture problem could be solved in Lowe's SIFT and our approach by using highly distinctive descriptor around each keypoint. This high dimensional descriptor results in accurate matching of feature points between consecutive frames. Fig 3 shows the performance of our simplified flow model in the aperture case. It shows that even keypoints on edges can match with keypoints in the next frame with a high probability.

2.3 Agent Based Matching

Unlike SIFT or other typical stereo matching techniques, our approach assigns an individual agent to each feature point. Each agent attempts to track a feature point continuously until the point goes outside of the image or becomes unstable. Agents treat feature points as particle in a swarm and each particle is a separate light-weight thread. The agents use a Kalman filter to track position and estimate velocity. Each agent calculates and saves translational movement between previous (i-1th) and (i-2th) frame and send those data to the filter to estimate translational movement between previous (i-1th) and current (ith) frame. The estimated velocity is added to position of feature point in previous frame to predict the position in the current frame.

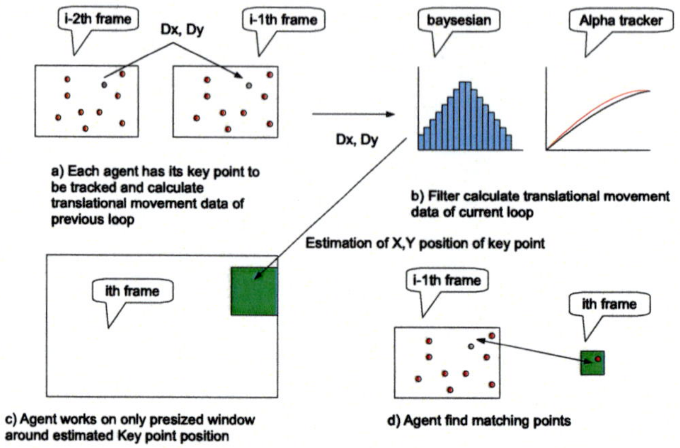

Fig. 4. Agent Based Matching. First (a) the agent estimates it's velocity from previous data. Second (b), a prediction of the position in the next frame is made. Third (c) a small window around the estimated keypoint position is extracted. Fourth (d), the agent performs keypoint processing and matching on the limited window within the image.

$$v_i = v_{i-1} + K * (p_{i-1} - p_{i-2})$$

$$p_i = p_{i-1} + v_i * \Delta T$$

Second, keypoint extraction and building descriptor processes are executed in a window around the estimated position of the keypoint. If there are not any keypoints found in the window, the agent attempts to take a new frame and repeats all process in a small sized window of the new frame. After extracting keypoints in the window, the agent attempts to match it's previous keypoint with a keypoint in the new frame. Keypoints in window are first organized as a k-d tree and searched in order of Euclidean distance between feature descriptors by using a Best Bin First search similar to SIFT [8] [9]. Using the Best Bin First search, two nearby feature points in the window are selected and the ratio between distances is computed (shorter over longer). If the ratio is smaller than 0.5, the closer feature is assumed to be a stable match and is selected. If the number of keypoints in selected window image is less than two or the best comparative ratio of distances is greater than 0.5, the agent attempts to find a new keypoint. These criterion are based on previous experiments with SIFT which showed that they are more effective at tracking stable feature points than other methods [1] [8].

Since each agent in this approach occupies its own thread, this algorithm can provide its best performance in a multicore processor such as a GPU (Graphics Processing Unit). However, the number of agents (and hence keypoints) to be tracked can be adjusted to the number of cores in the machine. Therefore, it can be compatible with a multicore PC as well as GPU. Moreover, the size of window around the estimated feature point position is relatively small, so it is expected that this algorithm can provide good performance on a single core machine if the number of agents is limited to less than N^2/n^2 as shown above.

2.4 Time Complexity of Agent Based Architecture

Traditional SIFT performs each operation on every pixel in the image, making it's time complixty scale as $O(N^2)$, where N is the width of the image. By using a smaller window, the complexity is reduced to $O(Kn^2)$, where K is the number of agents and n is the width of the search window. Since n can be fixed, this algorithm scales linearly with the number of agents (keypoints), rather than quadratically with the size of the image. Hence, to get an improvement in speed, K must be less than $(N/n)^2$.

In addition, the reduced search window size imposes a limit on the speed at which the camera may move. If the keypoint moves outside of the search window between consecutive frames, then the agent will fail to track the correct keypoint on the next frame. If f is the number of frames per second we can impose an upper limit on the velocity, $V_{max} < nf/2$ (factor of 2 since the keypoint is in the middle of the search window). If the image is moving faster than V_{max}, keypoint tracking accuracy will be degraded. It should be noted that a faster frame rate from the camera will improve tracking accuracy (and hence camera speed) within the same size window. We will discuss these tradeoffs later in the Discussion section.

3 Experiments

For all experiments, we used an eight core 2.3 GHZ computer with 1 GB memory and a QuickCam Communicate Deluxe was for image capture which provided up to 30 frames per second. For accurate comparison, a fixed image size (240 by 320 pixels) was used and the number of feature points was restricted to 25, under the rule described earlier. During the experiments, the camera was rotated with a consistent speed of 45 degrees per second. Four different approaches were implemented in our experiment:

1. Lowe's standard SIFT implementation. All SIFT processes are implemented in cascade using a single core.
2. Partially parallelized SIFT implementation. Scale space and descriptor building processes are implemented in parallel using six cores.
3. Simplified Flow model. The number of octaves used for the DoG filter is reduced to 2 and the feature descriptor is reduced to just the 72 pixel intensity bins, eliminating many SIFT processes. All feature points in image were tracked using one single agent and feature point matching was executed using the whole frame as in SIFT. Position and velocity of feature points were not tracked continuously.
4. Multi-agent implementation. Simplified process is implemented in parallel using 25 agents to track feature points and a search window size of 24x36. Each feature point was assigned to an independent agent and was tracked continuously until it moves outside of image or became unstable.

4 Results

Table 1 shows the processing time per frame used by the four algorithms. In the cascade SIFT implementation each frame took from 0.38 to 0.39 seconds. This result is comparable with David Lowe's results [1], described earlier. When scale space and descriptor building processes were parallelized, some execution time (0.03 to 0.06 seconds) was saved, without losing accuracy of feature point matching between ordered frames.

Table 1. Comparison of Speed

Step	Range of Total Process Time
SIFT Flow	0.38 - 0.39 (second/frame)
Parallelized SIFT Flow	0.32 - 0.36 (second/frame)
Simplified Flow Model (without Agent Based Matching)	0.29 - 0.32 (second/frame)
Multi Agent Based Optic Flow Model	0.05 - 0.25 (second/frame)

Third, the simpified algorithm described in Section 2 was implemented. This simplified flow algorithm took 0.29 to 0.32 second per frame, generating 0.1 second of time savings over SIFT. Also, accuracy of matching feature points between consecutive frames was preserved.

Last, the multi-agent based method was implemented. As was described in table 1, this algorithm showed much better performance in our experiment than other approaches which are mentioned above. It ranged from 0.05 second in minimum to 0.25 second in maximum. The wide range of variation in processing times was due to a degradation in feature point tracking accuracy. Because of the reduced search window size, the Kalman filter in the agent based approach is more likely to lose track of a feature point between frame. At this point, extracting keypoint and building descriptor process were repeated until agent found a new feature point to be tracked, resulting in significant time delays.

5 Discussion

Though the multi-agent based optic flow model is significantly faster, there is a tradeoff with respect to the ability to track features consistently, as well as the speed at which the camera may move. As described earlier, we have an upper limit on the camera speed of $V_{max} < nf/2$. It should be noted, however, that consistency in feature point tracking is not necessarily needed for all vision tasks, and does not necessarily translate to degradation in performance when navigating by optic flow. Compared to Multi-Agent Optic Flow, the simplified flow model consistently tracks the same points at higher camera speeds, since it uses the entire frame to track keypoints, though this comes at the cost of slower processing. These conflicts introduce a number of tradeoffs. A larger search window is slower in terms of processing speed, but allows the camera to move faster, while more consistently track the same points in the environment. Consistent feature point tracking may improve performance of tasks such as object recognition and hence navigation by way landmark recognition. However, faster processing allows us to process more frames per second, which feeds back into the maxmimum camera speed.

As noted earliet, keypoint tracking accuracy and camera motion limits implicitly depend on the frame rate at which we are able to process images. Thus, as camera frame rates and CPU processing speeds improve, the multi-agent based approach will become more robust to camera movement. We can restate the computational complexity as $O(K(V_{max}/f)^2)$, showing that the computational complexity of this approach will decline with increasing frame rate (by allowing the search window size to shrink). Alternatively, we can trade off camera speed for computational performance by increasing the size of the search window. Effectively, there is a three-way tradeoff between computational speed, camera speed, and feature point tracking consistency.

In principle, the processing can be automatically varied to maximize processing rate, minimize error on the fly and potentially adapt dynamically to the immediate computational ability of the host system. If the system is loaded down with processing, agents can be deleted to compensate. As more computation becomes availible, additional agents can be added. Thus we anticipate a graceful degradaton in performance. These tradeoffs may also be useful for task switching. For instance, while optic flow performance is less strictly dependent on tracking of specific feature points consistently, object recognition rquires more feature points and more consistent tracking of them. hence, an agent could, for instance, slow down in order to accomodate more agents and more stable feature point tracking, in order to improve object recognition performance. When it wants to navigate quickly, it could reduce the number of agents and search window size to accomdate smoother optic flow processing.

The combined optic-flow and object recognition architecture introduces a number of relevant consideratiosn with respect to biological vision systems. The Cortex is a highly regular, six layer structure in human beings, yet it can decompose a visual stream into seemingly very different visual modalities: Form (leading to object recognition), depth from stereopsis, spectral sensitivity (color) and motion. What is interesting is that these seemingly very different types of processing are performed in the cortex with a very regular structure. On the other hand, computer vision has largely treated topics such as optic flow, stereo, and object recognition as separate functions supported by separate families of algorithms. What is the common element between these computations?

Superficially, they can be tied together via correlation based methods. For example, the basic Reichart detector in insect vision is a spatiotemporal correlation detector. It matches simultaneously through space and time. Stereopsis can be implemented using a spatial correlator between two eyes. Object recognition can be implemented as correlation through time (with memorized templates). To all of these methods spectral information (color) can be used (or not used) to improve matching. Or image color can be correlated with remembered prototypes, or across the image.

Correlation is typically thought of as block matching between image areas. However, we point out that a recently very successful algorithm for object recognition, David Lowe's Scale Invariant Feature Transform (SIFT)[1], is essentially based on intelligent correlation. This point may not be completely obvious.

SIFT works as follows: Scaled Difference of Gaussian filters (DOG) are applied at every point in an image. At each point in the transformed image, a certain multi-resolution signature exists based on these DOG filters. Stable feature points can be identified by examining each image point and its surrounding area. By stable, Lowe means that it is unlikely to change as the target, an object,undergoes scale changes and rotations, and lighting changes. These points have unique signatures which makes the matching with stored templates much easier and more robust.

We can now restate Lowe's algorithm by saying that (1) We operate in a frequency domain instead of the luminosity domain; (2) The features points are points in the image that are given a much higher weight or credibility than surrounding points, i.e. we differentially weight points in the frequency space; (3) We effectively push the description of each image point into a much higher dimensional space making matching more robust.

6 Conclusion

In this paper, we introduced a new computer vision algorithm, using feature descriptors from object recognition techniques as the basis for computing optic flow. Each feature point is tracked by an individual agent, which independently predicts the feature's position and velocity, and can be implemented in parallel with other agents on a multicore processor. Experiments showed that agent based flow algorithm is significantly faster per frame than previous object recognition methods that rely on feature point tracking. We discuss the tradeoff in speed with respect to stability of feature point tracking and speed limits on the camera. Although multi-agent based feature point tracking can be less accurate at higher camera speeds, the control of the number of agents and the search window size allows us to dynamically adjust to different situations and computer vision tasks. Because this method combines aspects of both object recognition and optic flow, it can potentially be used as the basis of a more unified computer vision system.

Acknowledgement. The authors would like to thank Theresa Klein for helping calculation of computational complexity of multi agent based optic flow.

References

1. Lowe, D.G.: Distinctive Image Features from Scale-Invariant Keypoints. Journal of Computer Vision (2004)
2. Lucas, B.D., Kanade, T.: An iterative image registration technique with an application to stereo vision. In: Imaging Understanding Workshop, pp. 121–130 (1981)
3. Horn, B.K.P., Schunck, B.G.: Determining optical flow. AI 17, 185–204 (1981)
4. Subramaniam, S., Biederman, I., Kalocsai, P., Madigan, S.R.: Accurate identification, but chance forced-choice recognition for rsvp pictures.In: Annual Meeting of the Association for Research in Vision and Ophthalmology, Ft. Lauderdale, FL (1995)
5. DiGirolamo, G., Kanwisher, N.: Accessing stored representations begins within 155 ms in object recognition. In: 36th Annual Meeting of the Psychonomics Society, Los Angeles (1995)
6. Harris, C., Stephens, M.: A combined corner and edge detector. In: Fourth Alvey Vision Conference, Manchester, UK, pp. 147–151 (1988)
7. Mikolajczyk, K., Schmid, C.: An Affine Invariant Interest Point Detector. In: Heyden, A., Sparr, G., Nielsen, M., Johansen, P. (eds.) ECCV 2002, Part I. LNCS, vol. 2350, pp. 128–142. Springer, Heidelberg (2002)
8. Beis, J., Lowe, D.G.: Shape indexing using approximate nearest-neighbour search in highdimensional spaces. In: Conference on Computer Vision and Pattern Recognition, Puerto Rico, pp. 1000–1006 (1997)
9. Beis, J.S.: Indexing without invariants in model based object recognition. PhD thesis, University of British Columbia (1997)

3D Face Recognition Based on Curvature Feature Matching, with Expression Variation

Shu-Wei Lin, Shu-Shen Hao, Jui-Lun Chang, and Sheng-Yi Li

Department of Electronic and Electrical Engineering Chung Cheng Institute of Technology
National Defense University
Taoyuan Country, 33551, Taiwan
haoshu@ndu.edu.tw

Abstract. In this paper, we try to improve face recognition system by taking better advantage of the inherent 3D nature of the face. Face recognition can be greatly improved because the abundant 3D face features can be obtained from different angles. During the simulation, we try to extract the curvatures of the eyes, nose and mouth, which can be used as features for face recognition. The Gaussian curvature is an important component of our work. The distribution of this curvature is used to construct the feature vectors. In order to raise the recognition rate, the projection method is used to intensify the edge information. The mesh modification method is also applied to the 3D mesh models. Finally, the distance between the 3D normalized curvatures of the features is compared between the query and database images for recognition. Even when the facial expression of the query image has changed, we can still achieve a 92% recognition rate with our 3D face recognition algorithms.

Keywords: 3d face recognition, Gaussian curvature, 3d mesh model, curvature feature matching.

1 Introduction

Face recognition using 3D face shape is known as an attractive field in the biometric system. Recent research [1-3] shows that the application of the 3D face model in face recognition achieves better performance in recognizing the pose variation of the face than intensity imaging does. We capture a 3D face image based on structure light with the LT 3D Face Camera [4] produced by the LOGISTIC TECHNOLOGY CORPORATION. The LT 3D Face Camera can produce a great amount of native vertices from the 3D face image. The challenge is organizing the massive amount of data and then using it to identify the face effectively and precisely.

We present a three-step framework in order to increase the speed and performance of 3D face recognition. First, we compute the Gaussian curvature [5,6] values for every point on the 3D face image. Within is first step we detect landmarks (nose tip/base, left/right nose border point, etc.) on the face according to the distribution of Gaussian curvature values on the 3D face image. After detecting landmarks, we extract the nose region from the 3D face image as a face feature based on the location

S. Lee et al. (Eds.): Intelligent Autonomous Systems 12, AISC 193, pp. 289–299.
springerlink.com

of landmarks. Second, we use the mesh modification method [7] to improve the quality of the mesh of nose region from the 3D face image so that we can achieve better face recognition rates during experiments. Third, we use the iterative closed point (ICP) algorithm [8, 9] to compute the difference between two nose regions, one from the probe set and the other one from the gallery set. In the face recognition system, the gallery is the set of 3D face images that shows all individuals known in the database and the probe is the 3D face image on which recognition is performed.

We perform two face recognition tasks in this paper. The first task is identification. The system reports the person who is shown on the probe as recognized or not recognized. The second task is verification, in which the system verifies a person as a particular subject in the gallery. All the 3D face images are collected using the LT 3D Face Camera with the resolution per pixel up to 1 mm. We build three gallery images for each individual with different expressions (neutral and non-neutral). There are 65 individuals in the gallery. In the experiment section, we compare the results of our algorithm with LVC [14]and PCA[15]. LVC is a method which chooses K-means clustering to identify basic facial elements, and it is an effective 3D face recognition method. PCA is a mathematical procedure that uses an orthogonal transformation to convert a set of observations of possibly correlated variables into a set of values of uncorrelated variables called principal components, and it is also a good method for 3D face recognition. The experiment section shows that our framework has a better and more stable recognition rate with different expressions.

2 Curvature Feature Matching

2.1 Detecting Landmarks and Extracting Face Features

The core step to detecting landmarks is computing the Gaussian curvature K_g, the curvature value of a point on a surface. It is the product of the principal curvatures [10, 11]. The formula is defined as following

$$K_g = k_1 * k_2 \tag{1}$$

where k_1 and k_2 are the maximum and minimum curvature values of a point on the surface. Then, we let S be a surface equation defined as

$$S(u,v) = au^2 + buv + cv^2 + \dots \tag{2}$$

where a , b and c are the coefficients of the surface equation S. We can define the first fundamental form of S as the following

$$I = \begin{bmatrix} E & F \\ F & G \end{bmatrix} \tag{3}$$

where $E = \frac{\partial S}{\partial u}\frac{\partial S}{\partial u}$, $F = \frac{\partial S}{\partial u}\frac{\partial S}{\partial v}$ and $G = \frac{\partial S}{\partial v}\frac{\partial S}{\partial v}$

We also define the second fundamental form of S as the following

$$II = \begin{bmatrix} L & M \\ M & N \end{bmatrix} \tag{4}$$

where $L = \frac{\partial S^2}{\partial^2 u} * n$, $M = \frac{\partial S^2}{\partial u \partial v} * n$, $N = \frac{\partial S^2}{\partial^2 v} * n$ and $n = \frac{\frac{\partial S^2}{\partial u} \times \frac{\partial S^2}{\partial v}}{\left| \frac{\partial S^2}{\partial u} \times \frac{\partial S^2}{\partial v} \right|}$

Here * is the product operator, × is the outer product operator, | | is the absolute value operator and n is the unit normal vector. We can rewrite the Gaussian curvature equation (1), which is based on (2)-(4), as the following

$$K_g = k_1 * k_2 = \frac{LN - M^2}{EG - F^2} \tag{5}$$

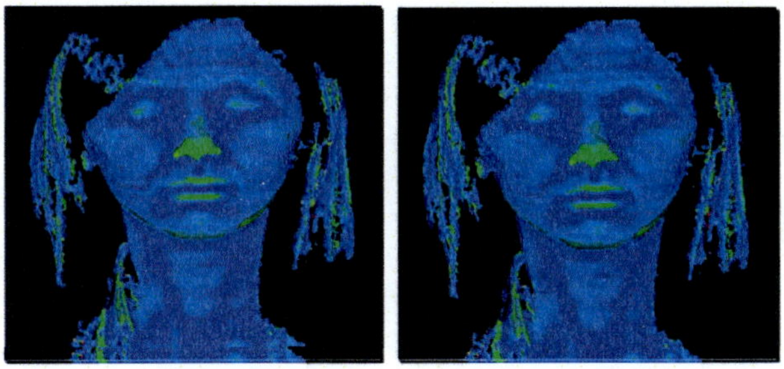

(a) k-mean = 5 (b) k-mean = 5

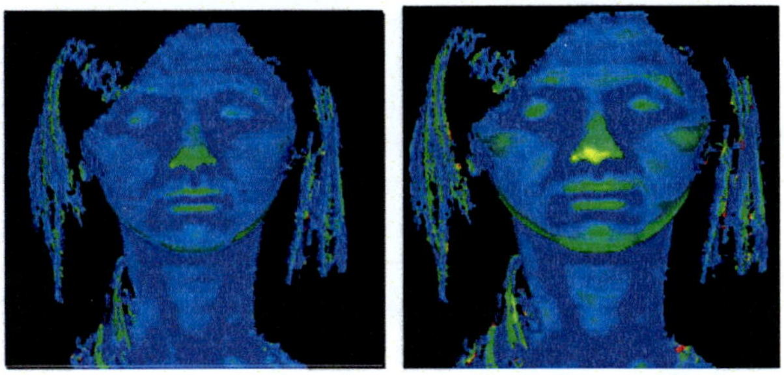

(c) k-mean = 15 (d) k-mean = 20

Fig. 1. The location of the nose region is observed clearly by using k-means, where k = 20 heuristically

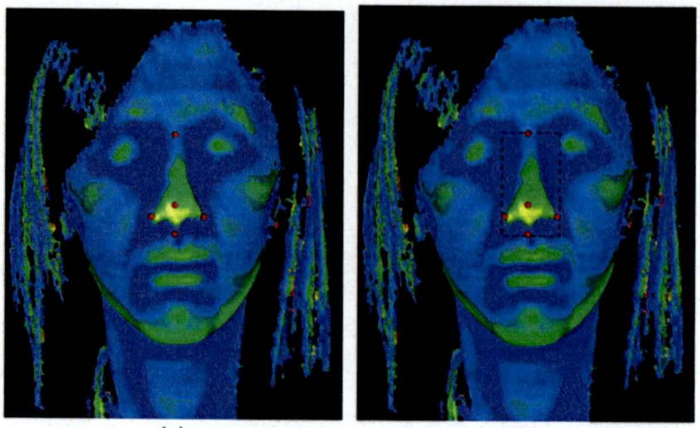

(a) (b)

Fig. 2. We determine the landmarks on the 3D face image as the nose tip, left/right nose border, nose base and nasion in (a). The nose region on the face image can be located based on landmarks in (b).

After computing the Gaussian curvature value for every point on the 3D face image, we have a distribution of the Gaussian curvature on the 3D face image. Then we run a clustering procedure using the distribution of the Gaussian curvature in order to locate the nose region, which we are interested in. In Fig. 1, we aim to partition the distribution of the Gaussian curvature into k clusters using the k-means [12] algorithm for locating the nose region. Using this heuristic process, we determine that if k is equal to 20, the cluster procedure is most successful for locating the nose region.

In Fig. 1(b), after using the cluster procedure, the distribution of the Gaussian curvature on the 3D face image is partitioned into twenty clusters, allowing us to identify landmarks in the next step. First we find the nose tip, the smallest z value (the depth value) on the 3D face image because the nose tip is closest to the camera. Then in Fig. 2(a), the left/right nose border is determined by searching the left/right boundary of the cluster to which the nose tip belongs. Next, we determine the location of the nose ridge by looking at the distance between the nose tip and the top boundary of the cluster that the nose tip and left/right nose border belong to. The nasion can be determined based on the smallest Gaussian curvature value along the direction of the nose ridge. Finally, we landmark the nose base depending on the smallest Gaussian curvature between the bottom of the cluster, which is where the nose region is and the cluster where the lip is. After the location of all the landmarks is determined, we obtain the region in Fig. 2(b).

2.2 Mesh Modification for Face Features

The mesh type we use for 3D face imaging is the triangular mesh. It is comprised of a set of three dimensional triangles that are connected by their common edges or corners. Because of environmental conditions, an irregular mesh may appear when taking the face image from the 3D camera and degrade the quality of the triangular

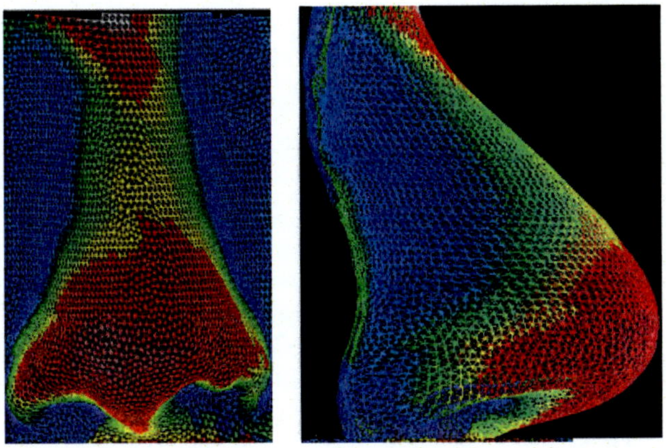

Fig. 3. Shown is the mesh of the nose region with the distribution of the Gaussian curvature. It is the view on X-Y plane in (a) and it is the view on Y-Z plane in (b).

mesh. To reduce the influence of the irregular mesh, we use the mesh modification [7] operation to improve the mesh quality by replacing poor elements with better ones. In Fig. 4, we show that the concept of the mesh modification operation transfers the irregular tetrahedral to the regular tetrahedron. The four triangles are equilateral in a tetrahedral. The pseudo code [7] of the mesh modification operation is shown as following:

```
initialize a dynamic list
insert all elements with quality below a given threshold to the dynamic list
while the dynamic list is not empty
    pop an element from the list
    analyze the popped element
    determine the mesh modification improving element quality the most
    if such a mesh modification exists
        apply the mesh modification
            update the dynamic list
    end if
end while
```

Having a nose region with irregular mesh could increase the error in face feature matching and reduce the performance of face recognition rate. Before face feature matching, we apply the mesh modification pseudo code to the nose region extracted from the native 3D face image. Then we can smooth the shape of the nose region and reduce the error caused by the irregular mesh. After applying the mesh modification operation, we find that the purple region (portion of the nose region) with irregular mesh in Fig. 5(b) is smoother than the one in Fig. 5(a).

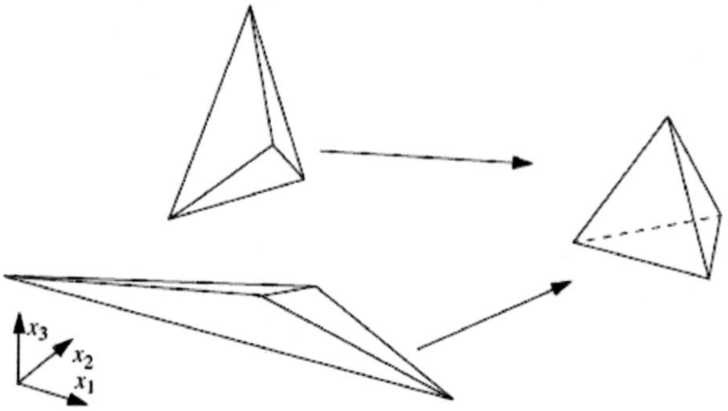

Fig. 4. A poor quality mesh is modified to a better quality mesh by applying the mesh modification operation

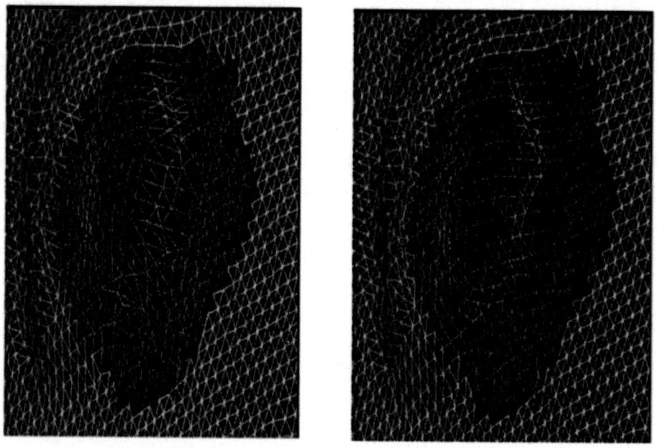

(a) (b)

Fig. 5. (a) is a portion of the nose region with irregular mesh, and (b) is the result after applying the mesh modification operation on (a)

2.3 Face Feature Matching

Face feature matching is based on ICP, which is widely used for geometric alignment of three-dimensional models when the initial relative pose is known. In this paper, we consider the Chen-Medioni (point-to-plane) framework [13] for ICP. We assume that we have a collection of points (p_i, q_i) with normals n_i. The first collection of points p_i is applied to align with the second collection of points q_i. Thus, the alignment error we use is defined as

$$E = \sum_{i=1}^{N_a} [(Rp_i + t - q_i) \cdot n_i]^2 \qquad (6)$$

where p and are both a collection of N_a points in the range image (3D image), R is the rotation parameter, t is the translation parameter and n_i is the normal. When determining the optimal rotation, we already know that the rotation is a nonlinear function. If we assume that every incremental rotation is small, we can approximate $\cos\theta \approx 1$ and $\sin\theta \approx \theta$. An example of a rotation in x coordinate is as in equation (7).

$$R_x = \begin{bmatrix} 1 & 0 & 0 \\ 0 & \cos\alpha & -\sin\alpha \\ 0 & \sin\alpha & \cos\alpha \end{bmatrix} \approx \begin{bmatrix} 1 & 0 & 0 \\ 0 & 1 & -\alpha \\ 0 & \alpha & 1 \end{bmatrix} \tag{7}$$

Thus, we can define the rotation as the following

$$R \approx \begin{bmatrix} 1 & -\gamma & -\beta \\ \gamma & 1 & -\alpha \\ \beta & \alpha & 1 \end{bmatrix} \tag{8}$$

where α, β and γ are the rotation parameters corresponding to the X,Y and Z coordinates. If we substitute (6) into (8), we obtain

$$E = \sum_{i=1}^{N_a} \begin{matrix} [(p_{i,x} - \gamma p_{i,y} + \beta p_{i,z} + t_x - q_{i,x})n_{i,x} + \\ (\gamma p_{i,x} + p_{i,y} - \alpha p_{i,z} + t_y - q_{i,y})n_{i,y} + \\ (-\beta p_{i,x} + \alpha p_{i,y} + p_{i,z} + t_z - q_{i,z})n_{i,z}]^2 \end{matrix} \tag{9}$$

which we rewrite alignment error as

$$E = \sum_{i=1}^{N_a} \begin{matrix} [(p_i - q_i)\cdot n_i + t\cdot n_i + \alpha(p_{i,y}n_{i,z} - p_{i,z}n_{i,y}) \\ + \beta(p_{i,z}n_{i,x} - p_{i,x}n_{i,z}) + \gamma(p_{i,x}n_{i,y} - p_{i,y}n_{i,x})]^2 \end{matrix} \tag{10}$$

From (10), we define $c = p \times n$ and $\gamma = \begin{pmatrix} \alpha \\ \beta \\ \gamma \end{pmatrix}$

We rewrite the alignment error as

$$E = \sum_{i=1}^{N_a} [(p_i - q_i)\cdot n_i + t\cdot n_i + r\cdot c_i]^2 \tag{11}$$

Now, we can determine the minimal rotation and translation parameter based on equation (10) using ICP. In Fig. 6, we compute the alignment error between the meshes in (a) and in (b) of the nose region. The final alignment with minimal alignment error between (a) and (b) is shown in (c).

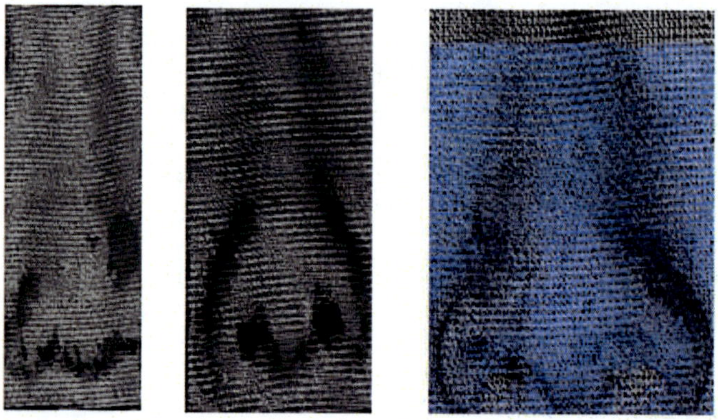

(a) (b) (c)

Fig. 6. (a) and (b) show the mesh of the nose region and (c) is the final result of the alignment with minimal error between (a) and (b)

3 Experiment

In this paper, the database that we use to evaluate our algorithm contains 65 individuals, each showing 4 neutral and 3 non-neutral (happy or sad) expressions. We have 455 3D face images in total. We use the images with non-neutral expressions, such as Fig. 7(a) and (c), as the probe, and the images with neutral expressions, such as in Fig. 7(b) and (d), as the gallery.

We compare the nose region with mesh modification and without mesh modification. We use 3 neutral faces of each person for the gallery set. Table 1 shows the different recognition rates for the nose region with the mesh modification operator and without the mesh modification operator.

Next, we compare our method, which is face recognition based on the nose region, with two other base methods, which are LVC [14] and PCA [15]. We use different test images with non-neutral expressions as the probe. For the gallery set, we first use three neutral faces of each person. We then use three different non-neutral expressions of the same person. The recognition results are presented in Table 2. From the table we can see that our method is the most successful one.

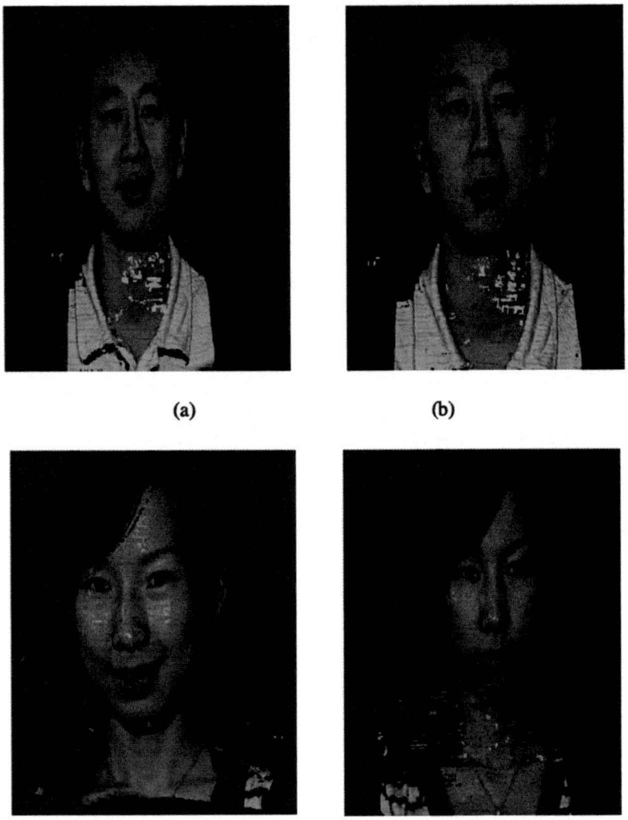

(a) (b)

(c) (d)

Fig. 7. (a) and (b) show the mesh of the nose region and (c) is the final result of the alignment with minimal error between (a) and (b)

Table 1. Comparison of the nose region in the gallery set with mesh modification operation in fallery ser and the mesh without mesh modification

Method	3 images with neutral expression as gallery set without mesh modification operator	3 images with neutral expression as gallery set with mesh modification operator
Our method	89.03%	92.71%

Table 2. Comparsions with different methods

Method	3 images with a neutral expression as the gallery set	3 images with different expressions as the gallery set
LVC	89.15%	83.50%
PCA	86.69%	68%
Our method	92.71%	92.61%

4 Conclusions

In this paper, we propose a method to extract the nose region from a face image and use it to perform 3D face recognition, thus avoiding the influence of the face expression on recognition. In order to increase the system performance, we also propose the use of the mesh modification operation, which improves the mesh quality of the nose region. One of our experiments shows that the mesh modification operation also reduces the alignment error when the probe and the gallery are from the same individual. Another experiment shows that when the gallery set only features neutral expressions, the recognition rates of both LVC and PCA are low. When the gallery set features different expressions, the performance of LVC and PCA is enhanced.

However, most importantly, the result of the experiment shows that our method is stable and is independent of the expressions in the gallery set. The total system performance of the recognition rate reaches 92%. While the experiment results show that the nose region is robust, we still plan to develop an effective algorithm to extract the eye, eyebrow and forehead regions as face features for building a real-time 3D face recognition system.

References

1. Mahoor, M.H., Abdel-Mottaleb, M.: Face recognition based on 3D ridge images obtained from range data. Pattern Recognition 42, 445–451 (2009)
2. Xu, C., Li, S., Tan, T., Quan, L.: Automatic 3D face recognition from depth and intensity Gabor features. Pattern Recognition 42, 1895–1905 (2009)
3. Li, L., Liu, F., Li, C., Chen, G.: Realistic wrinkle generation for 3D face modeling based on automatically extracted curves and improved shape control functions. Computers & Graphics (2010) (in press, corrected proof)
4. LT 3D Face Camera, `http://www.ltech.com.tw/`
5. Jagannathan, A., Miller, E.L.: Three-Dimensional Surface Mesh Segmentation Using Curvedness-Based Region Growing Approach. IEEE Transactions Pattern Analysis and Machine Intelligence 29, 2195–2204 (2007)

6. Ceron, A., Salazar, A., Prieto, F.: Relevance analysis of 3D curvature-based shape descriptors on interest points of the face. In: 2010 2nd International Conference Image Processing Theory Tools and Applications (IPTA), Paris, Frence, July 7-10 (2010)
7. Li, X., Shephard, M.S., Beall, M.W.: 3D anisotropic mesh adaptation by mesh modification. Computer Methods in Applied Mechanics and Engineering 194, 4915–4950 (2005)
8. Besl, P.J., McKay, H.D.: A method for registration of 3-D shapes. IEEE Transactions on Pattern Analysis and Machine Intelligence 14, 239–256 (1992)
9. Rusinkiewicz, S., Levoy, M.: Efficient variants of the ICP algorithm. In: 2001 Proceedings of the Third International Conference on 3-D Digital Imaging and Modeling, pp. 145–152 (2001)
10. Jagannathan, A., Miller, E.L.: Three-Dimensional Surface Mesh Segmentation Using Curvedness-Based Region Growing Approach. IEEE Transactions Pattern Analysis and Machine Intelligence 29, 2195–2204 (2007)
11. Peng, J., Li, Q., Kuo, C.-C.J., Zhou, M.: Estimating Gaussian curvatures from 3D meshes. In: Human Vision and Electronic Imaging VIII, vol. 5007, pp. 270–280 (2003)
12. Verevka, O., Buchanan, J.W.: Local K-means Algorithm for Colour Image Quantization. M.Sc. Dissertation, Department of Computing Science. University of Alberta, Canada (1995)
13. Chen, J., Medioni, G.: Detection, Localization, and Estimation of Edges. IEEE Trans. PAMI PAMI-11(2), 191–198 (1989)
14. Zhong, C., Sun, Z., Tan, T.: Robust 3D face recognition using learned visual codebook. Pattern Recognition, 1–6 (2007)
15. Turk, M., Pentland, A.: Eigenfaces for recognition. Journal of Cognitive Neuroscience 3, 72–86 (1991)

Advances in 3D Camera: Time-of-Flight vs. Active Triangulation

Daesik Kim[1] and Sukhan Lee[2]

[1] Department of Electrical and Computer Engineering, Sungkyunkwan University, Suwon, Korea 440-746
daesik80@skku.edu

[2] Department of Electrical and Computer Engineering and Department of Interaction Science, Sungkyunkwan University, Suwon, Korea 440-746
lsh@ece.skku.ac.kr

Abstract. Over the last decade, numerous 3D camera techniques have been proposed and advanced dramatically. One main approach is time-of-flight (TOF) and the other is active triangulation. Each has its own strengths and weaknesses. In this paper, we overview the principle of each method and compare the advantages and disadvantages in detail, and introduce several commercially available 3D cameras and their characteristics.

1 Introduction

A demand for real-time range images has increased explosively in the computer vision and robotic community. The advances in 3D camera have created a new research area in many applications, such as gesture recognition, human/object tracking, 3D SLAM, 3D scene reconstruction, and mixed/augmented reality etc. The principle of the 3D camera can be largely classified into time-of-flight (TOF) and active triangulation. Although TOF and active triangulation had become obsolete technologies, they have come in the limelight again recently, because of their potential ability to measure the 3D scene with a high frame rate and high spatial and depth resolution. In this paper, we overview the principles of 3D camera techniques, and analyze the advantages and disadvantages of each method.

This paper is organized as follows. The principles and characteristics of the time-of-flight and active triangulation technologies are described in section 2 and 3, respectively. The specifications of the commercially available sensors and their pros and cons are summarized in section 4. Finally section 5 concludes the work.

2 Time-of-Flight

2.1 Pulse Runtime Measurement

The basic principle of the TOF camera is to measure the travel time when the pulsed light is emitted, reflected by an object, and received as shown in Fig. 1 (left). Since the speed of light is known, the distance can be computed as

$$d = \frac{t}{2} \cdot c, \tag{1}$$

S. Lee et al. (Eds.): Intelligent Autonomous Systems 12, AISC 193, pp. 301–309.
springerlink.com

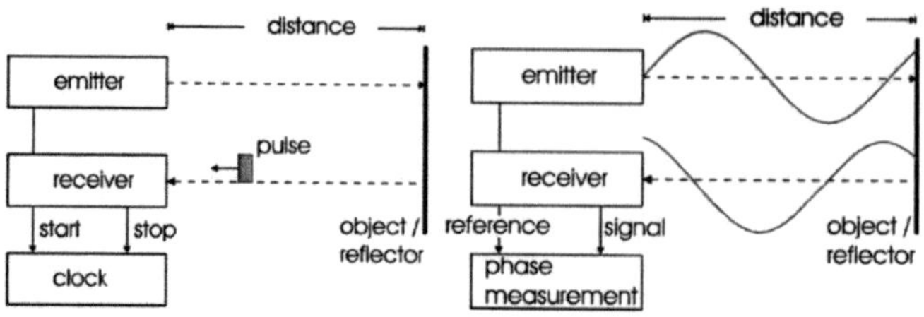

Fig. 1. Time-of-flight distance measurement principle. Pulse Runtime (left) and Phase Shift (right). [1]

where d is the distance between the sensor and the object; t is the travel time between emitting and receiving; c is the speed of light ($c \approx 3 \cdot 10^8 m/s$).

Although the concept looks simple, there are several challenges to achieve higher accuracy; high-accuracy clock time measurement and the short light pulse generation with high repetition rates should be overcome.

2.2 Phase Shift Measurement

In order to avoid high precision clocks and short light pulse, an alternative approach is to measure the phase shift between the emitted and received sinusoidal modulated light as shown in Fig. 1 (right) [1–6].

Let the emitted sinusoidal signal $g(t)$ and received signal $s(t)$ be as follows:

$$\begin{aligned} g(t) &= cos(\omega t), \\ s(t) &= b + acos(\omega t + \phi), \end{aligned} \tag{2}$$

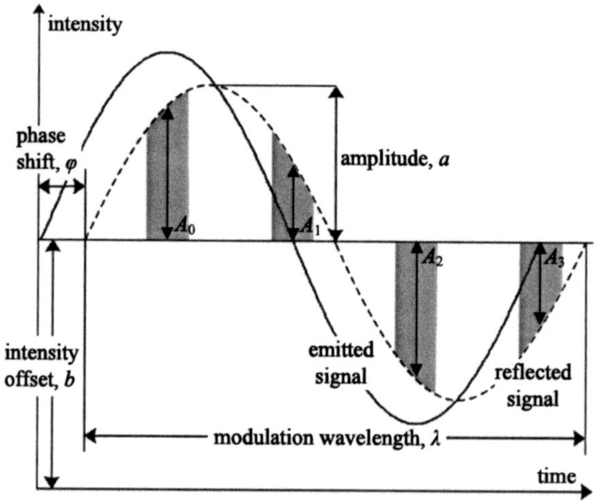

Fig. 2. Phase shift distance measurement principle

where ω is the modulation frequency; a is an amplitude; b is an intensity offset; φ is a phase shift as shown in Fig. 2. In order to retrieve the phase shift φ, it should be demodulated by cross-correlation of received signal with emitted signal:

$$c(\tau) = s * g = \int_{\infty}^{\infty} s(t) \cdot g(t+\tau)dt, \tag{3}$$

where τ is the offset. This function can be simplified as

$$c(\tau) = \frac{a}{2}cos(\omega\tau + \varphi) + b. \tag{4}$$

Let the four measurements be as follows:

$$A_i = c(i \cdot \frac{\pi}{2}), \qquad i = 0, \cdots, 3 \tag{5}$$

then the unknown a, b, and φ can be obtained as follows:

$$\begin{aligned} a &= \frac{\sqrt{(A_3 - A_1)^2 + (A_0 - A_2)^2}}{2}, \\ b &= \frac{A_0 + A_1 + A_2 + A_3}{4}, \\ \varphi &= arctan2(A_3 - A_1, A_0 - A_2). \end{aligned} \tag{6}$$

Finally, the distance between the sensor and the object can be computed as follows:

$$d = \frac{c}{4\pi\omega}\varphi, \tag{7}$$

where c is the speed of light.

The drawback of this approach is the light should be integrated over time to improve the measurement accuracy and to reduce the noise, so this limits the frame rate of the sensor. Moreover, it is hard to obtain a high-resolution depth image.

2.3 Intensity Measurement by a Shuttered Sensor

Another approach is to measure the time of flight with a fast image shutter [7–11]. The concept is to measure the accumulated light intensity instead of the pulsed light runtime. The emitter generates an IR laser pulse (*light wall*) and the receiver accumulates the reflected light carrying an imprint of the objects. The depth information can now be extracted from the reflected *deformed light wall*, by deploying a fast image shutter in front of the CCD chip, and blocking the incoming light, as shown in Fig 3.

In practice, objects have various reflectivity coefficients, so this should be compensated. By dividing the front portion of intensity $I_{shutter}$ (see Fig. 3 (d)) by the corresponding portion of the total intensity I_{total} (see 3 (c)), the ratio, which indicates the normalized depth, can be obtained as follows:

$$\alpha = I_{shutter}/I_{total}. \tag{8}$$

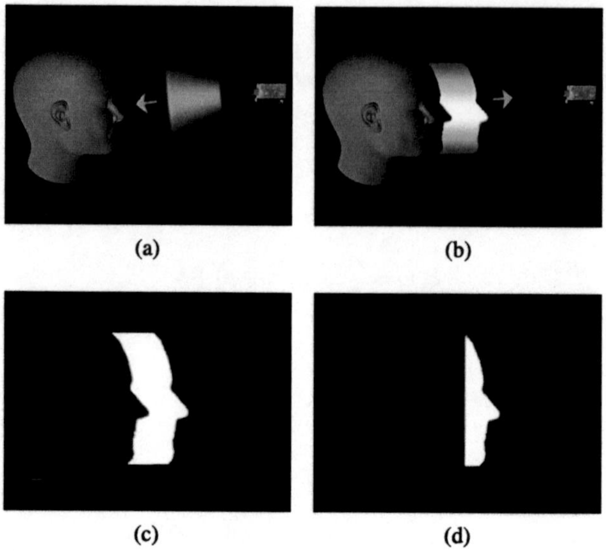

Fig. 3. The principle of the intensity measurement approach by the shutter sensor: (a) A *light wall* is emitted from the sensor. (b) It is reflected by the object. (c) The total reflected light from the object. (d) The front portion of the light is cut by a shuttered sensor. If the object distance is close, the front portion of the reflected light is big, otherwise small [7].

The measurable minimum and maximum distances are as follows:

$$d_{min} = \frac{t_{min} \cdot c}{2},$$
$$d_{max} = \frac{t_{max} \cdot c}{2}, \tag{9}$$

where c is the speed of light; t_{min} and t_{max} is the time when the shutter opens and closes, respectively. The absolute distance is

$$d = d_{max} - (d_{max} - d_{min})\alpha. \tag{10}$$

Ideally, α is equal when the same depth is measured regardless of the objects' reflectivity, but in practice, this is not true [11]. Therefore, this should be taken care of.

Theoretically, this overcomes the drawback of the pulse run-time and phase shift measurement approach, and can achieve the high-resolution depth image; however, not many research results are available.

3 Triangulation

3.1 Point or Line Light Measurement by a Camera

The basic geometry for an active triangulation system is shown in Fig. 4a. Here, it is assumed that the origin of the reference frame is attached to the center of projection of

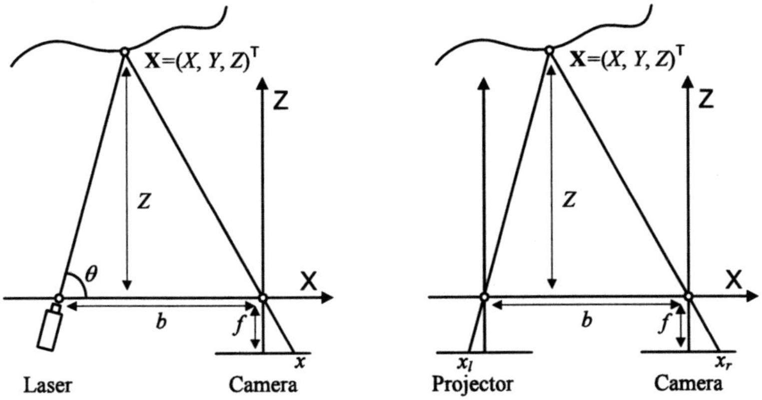

Fig. 4. The principle of active triangulation

(a) (b)

Fig. 5. Coded Pattern: (a) Kinect's pattern. (b) Willow Garage's pattern [12].

a camera. Let f be the focal length and b be the baseline, which is a distance between the line (or point) light source (e.g., laser) and the origin. In this setup, the coordinates of a point $\mathbf{X} = (X,Y,Z)^{\mathsf{T}}$ are given by

$$\begin{bmatrix} X \\ Y \\ Z \end{bmatrix} = \frac{b}{f cot\theta - x} \begin{bmatrix} x \\ y \\ f \end{bmatrix}, \tag{11}$$

where x and y is the corresponding image point.

Since this is a line (or point) measurement system, in order to measure the area, the angle of the light source should be scanned (rotated). Therefore, this approach is not suitable for moving objects.

3.2 Coded Pattern Measurement by a Camera

Fig. 4b and Fig. 5a show the principle of another simple active triangulation, which is composed of an imaging device (e.g., camera) and a projecting device (e.g., projector) [13, 14].

The difference between the projector and the camera is the direction of the projected light. A projector is a device to project the light to outside, but a camera is a device to acquire the light from outside. Since the projector model is the same as the camera model, the principle of the active triangulation becomes the same as the conventional passive stereo system. Here, it is assumed that there is no rotation between two devices. Let the focal length, f, the baseline, b, which is the distance between two devices, and the disparity $d = x_l - x_r$, then the distance Z can be estimated by the triangle similarity:

$$\frac{b}{Z} = \frac{b-d}{z-f}, \tag{12}$$

$$Z = \frac{fb}{d}. \tag{13}$$

X and Y are also computed as follows:

$$X = \frac{xZ}{f} \quad \text{and} \quad Y = \frac{yZ}{f}. \tag{14}$$

The distance Z is proportional to the focal length, f, and the baseline, b, but inversely proportional to the disparity.

3.3 Coded Pattern Measurement by Stereo Camera

The system composed of two cameras and one projector is another approach [12]. In order to robustly solve the correspondence problem between two cameras, the pattern is coded so that every block is well discriminated from every other block. The projector is for generating the texture to the objects and the conventional block matching algorithm is used for a stereo camera as shown in Fig. 5b.

3.4 Depth Resolution

The depth resolution of the triangulation can be computed by differentiating the equation of the triangulation (13):

$$\frac{\partial Z}{\partial d} = -f\frac{b}{d^2}. \tag{15}$$

The disparity is derived from the equation (13) as follows:

$$d = \frac{fb}{Z}. \tag{16}$$

By plugging equation (16) into (15),

$$\frac{\partial Z}{\partial d} = -\frac{Z^2}{fb}. \tag{17}$$

Since the equation (17) means the rate of change of Z with respect to d, if we denote $\partial Z = \Delta Z$ and $\partial d = \Delta d$, and take the absolute value to make the resolution positive, the depth resolution is

$$\Delta Z = \frac{Z^2}{fb}\Delta d. \tag{18}$$

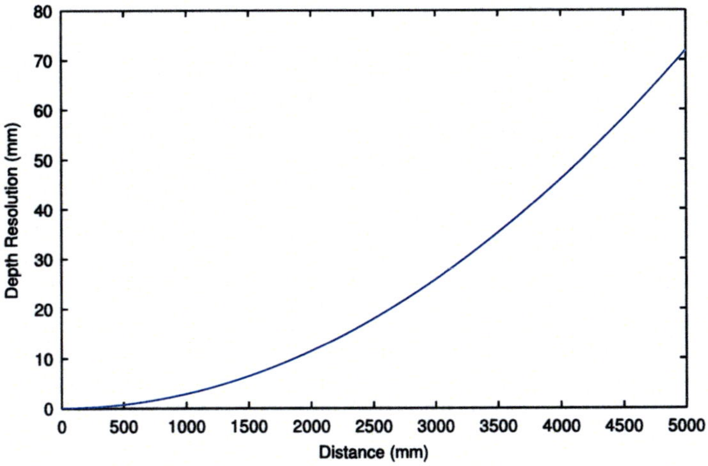

Fig. 6. The depth resolution of the Kinect

From equation (18), we can infer that the depth resolution becomes higher (i.e., ΔZ becomes smaller.) if the focal length f and baseline b become bigger and the distance Z becomes shorter. Fig. 6 shows the depth resolution of the Kinect.

Table 1. TOF-Based 3D Camera Specification

	SR4000 (Mesa)	CamCube (PMD)
Light Source	LED (NIR)	LED (NIR)
Image Resolution	176×144	200×200
Measurement Range	0.8–5.0m	0.3–7.0m
Field of View	43°×34°(1) 69°×56°(2)	40°×40°
Frame Rate	50fps	40(5), 60(6), 80(7)fps
Size	65×65×68mm(3) 65×65×76mm(4)	194×60×60mm(8)
Repeatability (1σ)	4mm @ 2m(1) 6mm @ 2m(2)	3mm @ 4m

(1)Standard Field of View, (2)Wide Field of View
(3)USB Camera, (3)Ethernet Camera
(5)200×200, (6)176×144, (7)160×120 pixels
(8)One camera and two illumination units

4 Specification Comparison

In this section, we compared the tow commercially available TOF-based 3D cameras and two active triangulation-based 3D camera (see Table 1 and 2).

Table 2. Active Triangulation-Based 3D Camera Spec

	Kinect (X-tion)	Artec 3D Scanner
Light Source	LED (NIR)	Flash Bulb
Image Resolution or X-Y Resolution*	640×480	0.5mm*(5) 1.0mm*(6)
Measurement Range	0.8–3.5m	0.4–1.0m*(5) 0.8–1.6m*(6)
Field of View	58×45°	30×21°(5) 41×32°(6)
Frame Rate	30fps(1) 60fps(2)	15fps
Size	283×39×62mm(3) 180×35×50mm(4)	72×222×120mm(5) 70×353×114mm(6)
Repeatability(1σ) or Accuracy**	1.5mm @ 1m 6mm @ 2m	0.1mm @ 1m**(5) 0.2mm @ 1m**(6)

(1)640×480, (2)320×240
(3)Kinect, (4)X-tion, (5)Artec M, (6)Artec L
*3D X-Y Resolution (up to)
**3D Point Accuracy (up to)

In summary, the advantages of the TOF-based 3D Camera are that it is compact and accurate at a relatively long distance. The disadvantage is that the image resolution is very low. On the other hand, the active triangulation approach is more accurate in the near range measurement and it is able to provide a high-resolution image. The weakness is that in order to measure the relatively long distance, the baseline should be longer to preserve the accuracy. This makes the sensor size big.

5 Conclusion

In this paper, we overview the principles of the 3D camera techniques, and analyze the advantages and disadvantage of each method.

Acknowledgment. This research was performed for the Intelligent Robotics Development Program, one of the 21st Century Frontier R&D Programs (F0005000-2010-32), and in part for the KORUS-Tech Program (KT-2008-SW-AP-FSO-0004) funded by the MKE, Korea. This work was also partially supported by the MEST, Korea, under the WCU Program supervised by the KOSEF (R31-2008-000-10062-0), and by MKE, Korea under ITRC NIPA-2010-(C1090-1021-0008)(NTIS-2010-(1415109527)).

References

1. Kahlmann, T., Remondino, F., Ingensand, H.: Proceedings of the ISPRS Commission V Symposium 'Image Engineering and Vision Metrology', pp. 136–141 (2006)
2. Lange, R.: 3D time-of-flight distance measurement with custom solid-state image sensors in CMOS/CCD-technology. Ph.D. thesis, University of Siegen (2000)
3. Oggier, T., Lehmann, M., Kaufmann, R., Schweizer, M., Richter, M., Metzler, P., Lang, G., Lustenberger, F., Blanc, N.: Proceedings of SPIE: Specific Applications: Sensors and Medical Optics
4. Blanc, N., Oggier, T., Gruener, G., Weingarten, J., Codourey, A., Seitz, P.: Proceedings of IEEE Sensors, pp. 471–474 (2004)
5. Oggier, T., Lustenberger, F., Blanc, N.: Miniature 3D TOF Camera for Real-Time Imaging. In: André, E., Dybkjær, L., Minker, W., Neumann, H., Weber, M. (eds.) PIT 2006. LNCS (LNAI), vol. 4021, pp. 212–216. Springer, Heidelberg (2006)
6. Gokturk, S., Yalcin, H., Bamji, C.: Conference on Computer Vision and Pattern Recognition Workshop, p. 35 (2004)
7. Iddan, G., Yahav, G.: Proceedings of SPIE: Videometrics and Optical Methods for 3D Shape Measurements, pp. 48–55 (2001)
8. Gvili, R., Kaplan, A., Ofek, E., Yahav, G.: Proceedings of SPIE Video-Based Image Techniques and Emerging Work, pp. 564–574 (2003)
9. Medina, A., Gaya, F., del Pozo, F.: Journal of the Optical Society of America A 23(4), 800 (2006)
10. Yahav, G., Iddan, G., Mandelboum, D.: Digest of Technical Papers of International Conference on Consumer Electronics, pp. 1–2 (2007)
11. Davis, J., Gonzalez-banos, H.: IEEE International Workshop on Projector-Camera Systems (2003)
12. Konolige, K.: Proceedings of IEEE International Conference on Robotics and Automation, pp. 148–155 (2010)
13. Sazbon, D., Zalevsky, Z., Rivlin, E.: Pattern Recognition Letters. 26(11), 1772 (2005)
14. García, J., Zalevsky, Z., García-Martínez, P., Ferreira, C., Teicher, M., Beiderman, Y.: Applied Optics 47(16), 3032 (2008)

A Novel 3D Registration Algorithm Using Parallel-Light Association

Han Wang and Ying Ying

School of Electrical and Electronic Engineering
Nanyang Technological University, Singapore
50 Nanyang Avenue, Singapore, 639798
hw@ntu.edu.sg, yi0002ng@e.ntu.edu.sg

Abstract. This paper presents a novel method for free-form registration of multiple point clouds. The method adopts a parallel-light data association design inspired from torchlight structure which improves the correctness of point correspondence. When two sets of point clouds are placed together, assume a set of parallel light beams are passing through them. Each light beam will pass the point clouds twice, one on each data set. The Euclidean distance on each light beam between the two sets are taken as measurement of the separation. The fitness is the reciprocal of the mean distance of all light beams. When the two sets are optimally aligned, the fitness is maximized. Hence, the registration problem is reduced to a six degree of freedom search. Preprocessing and acceleration modules such as Genetic Algorithm (GA) are introduced to reduce the exploration space and execution time. Unlike the Iterative Closest Point (ICP) algorithm, the proposed algorithm does not require pre-alignment information. Secondly, ICP does not perform well when the overlapped area between two sets are not sufficiently large. And the proposed algorithm does not suffer from this partial overlapping problem. Based on various experiments with real data, the proposed method has superior performance compared to ICP.

Index Terms: 3D registration, parallel light, Iterative Closest Point, Genetic Algorithm.

1 Introduction

Surface registration is an intermediate and crucial step inside the whole computer vision complexity, which benefits by previous steps such as object detection and subject tracking, and is the input for further steps such as object modeling. Consequently, lots of various applications are resulted, such as reverse engineering, manufacturing process, virtual reality, and map building in robotics. Devices involved have their own pros and cons, but all of them provide only partial surface due to the limited field of views. To solve this issue, multiple point clouds taken from different positions are acquired and proceeded for data fusion. Although it will be more precise if surface registration is done by high-overlapped data, meanwhile, more data capturing work needs to be done and more execution

S. Lee et al. (Eds.): Intelligent Autonomous Systems 12, AISC 193, pp. 311–319.
springerlink.com

time will be costed. In practice, robust registration approach which can handle low-overlapped point clouds with no prior information is strongly needed.

This paper presents a novel approach using parallel-light structure for data association to tackle surface registration problem. Given two surface measurements of an object from different viewing locations, transformation between these point clouds can be optimized utilizing GA with a robust fitness function. We focus on obtaining precise alignment among low-overlapped views without initial information given, and modified structure accelerates computational speed of GA mechanism. The developed registration algorithm is found to be fast, accurate, and robust, comparing to other well known algorithms. Especially, our approach has the capability of dealing low quality data acquired from Kinect sensor.

The remaining of this paper is organized as follows. First, work related to the registration topic is presented in Section 2. Second, proposed free-form registration method using parallel-light structure is detail described in Section 3. Next, Section 4 provides experimental results in the presence of data acquired from Kinect sensor to demonstrate the effectiveness and robustness. The article ends with conclusion in Section 5.

2 Related Work

The goal of registration is to find the spatial relationships between a set of data of a given object taken from different positions in order to present them all with respect to a reference frame.

The best known methods for surface registration are the variations on ICP algorithm [2]. ICP is an iterative procedure minimizing the alignment error between points in one view and the respective closest points in another view. The proper convergence of ICP cannot be guaranteed if low-overlapped data is given. Another drawback is that it requires good pre-alignment of views to converge to global optimum. Although variant ICPs [3] have been well studied and experimental performance has been improved, the limitations of ICP-based algorithms cannot be totally overcomed.

Mostly, two-stage registration strategy [4][9][10][11] is preferred to complement the shortcomings of ICP. According to the classification in [1], differing as whether initial information is required, work can be usually separated into coarse and fine stage. Coarse stage is to compute an initial estimation, while fine one is to obtain the most accurate solution as possible. In order to calculate the rough transformation in the coarse step, Brunnstrom et al. [4] calculated the approximate transformation by searching for the correspondences between range images, Liu et al. [9] applied dynamic matching tree structure to find the correspondence points, Wang et al. [10] selected crease and corner points as features based on the curvature estimation, and Dai et al. [11] utilized point normal information within small neighborhood as local invariant feature to establish correspondence. Afterwards, ICP is chose for accurate tuning in the fine step.

Another class of approaches is to find the geometric transformation through space searching, in which contains the global optimum can be directly used to precisely align views. These registration methods [5][6][7][8] are simple and robust, because good performance can be achieved with no initial information necessarily. Among them, GA is a popular choice providing optimization searching. However, the drawback is generally time consuming due to GA's stochastic characteristic and evolution optimization. Silva et al. [5] and Chow et al. [6] proposed similar work which optimized transformation parameters using dynamic structure or robust fitness function. Robertson et al. [8] focused on computational time reduction using parallel GA strategy, while the performance was not improved.

3 Proposed Registration Method

3.1 Parallel-Light Data Association

Given two sets of point cloud, input data $P = \{p_i\}$ and target data $Q = \{q_j\}$, translation T and rotation R are involved in transformation Tr, thus points can be related by the following mapping

$$q_j = Tr(p_i) = Rp_i + T \tag{1}$$

If Q is the data of P under transformation $Tr(T, R)$, in theory, the Euclidean distance between $Tr(p_i)$ and its correspondence q_j on Q should be zero for all i. However, various depth sensors provide more or less accurate data. Thus, the Euclidean distance between $Tr(p_i)$ and q_j will not be zero in real situation, which can be treated as registration error. Therefore, it is natural to choose minimizing registration error among all correspondences be the measurement, in which finding correspondence between P and Q is essential.

Although different registration error functions were defined, "closest point" (CP) distance [2] is the most popular correspondence matching measurement. In spite of being widely used in ICP-based or other registration algorithms, CP matching strategy potentially generates large numbers of incorrect pairings in the presence of noise and outlier (Fig. 1(a)), or with low-overlapped data (Fig. 1(c)), and hence slows down the rate of convergence [3].

Supposing $Tr(P)$ is well merged into Q, using closest point matching, correspondences are built by selecting minimum Euclidean distance among data. However, multiple input points have correspondences with the same target point, which is unreasonable in practice situation. In order to get robust pairing, a parallel-light strategy is developed to improve pair association which is inspired from the work utilizing light beam structure [12]. Invisible parallel lights which are composed by lots of light beams with small radius r are projected originated from the depth sensor which acquires target data Q. As long as the matching pair $(Tr(p_i), q_j)$ is located inside the torchlight range, the correspondence is built, as is shown in Fig.1(b) and (d). The process of correspondence selection boosts the robustness of registration error computation, not only handling the disturbance of noise and outliers which cannot be avoided since low quality consumer depth camera is used, but also workable with low-overlapped data with which reduces acquiring data while keeping the completeness of object.

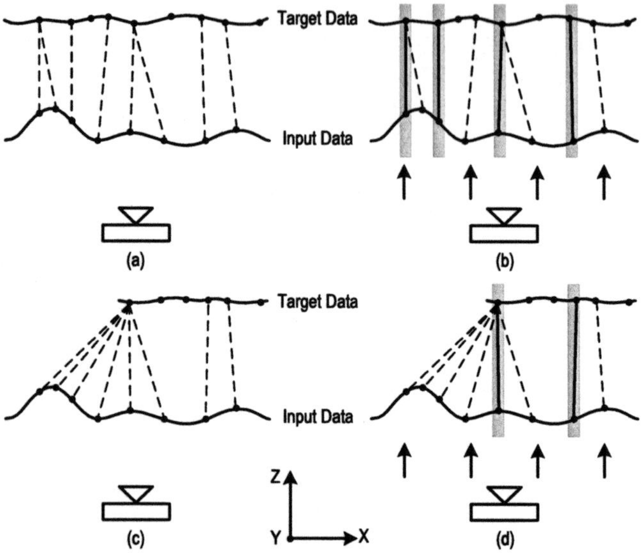

Fig. 1. Comparisons of data association strategies: (a) closest point; (b) parallel-light; (c) closest point with low-overlapped data; (d) parallel-light with low-overlapped data.

After data association is built, the registration error function can be defined as

$$F = \frac{L}{\sum Dist[q_j - Tr(p_i)]} \tag{2}$$

which only utilizes those correspondences passing through the parallel-light structure, and L is the pairing number. With the proposed cost function F, improved registration results are achieved and details will be explained in Section 4.

3.2 Optimization Mechanism for Registration

As a powerful optimization tool, GA is employed in our algorithm. The general principle underlying GA is to maintain a population of possible solutions encoded in form of chromosomes and submit to an evolutionary process until convergence is achieved, which means the strongest individual is to be the winner in the competitive environment. During evolution, each solution calculates a fitness value by a user-defined fitness function and evolution is executed based on the fitness result. Hence, registration error function Eq. 2 is measured as the fitness function, and at the end of optimization, the final optimum solution should be the one with smallest registration error.

Referring to the time consuming issue, various efforts [5][6] have been done to accelerate convergence speed. Heavily subsampling the input data is straightforward technique since point population for pair matching is largely decreased.

Multi-layer structure is another way which gives coarse-to-fine results in different layers. Considering the low quality data captured by Kinect sensor, reasonable acceleration strategies are proposed here.

Preprocessing. Before any processing, due to the existing of noise and outliers in the raw data, smoothing module is applied to relief interference. Here, median filter is applied to only z-axis not the whole coordinate avoiding data distortion.

Data registration can be treated as coordinate transformation rather than point transformation, and 3D data illustrates information referring to respective depth camera. Hence, mean value should be removed from data and unbiased surface can be sent out for global searching.

Grid Structure. In short, parallel-light data association is to restrict distance between $Tr(p_i)$ and the light ray through q_j to be less than radius r. At a speed consideration, this idea can be simplified as grid structure which has the similar function. Surface Q can be separated into grids along x and y-axis with a big step size d relative to light ray radius r. Before being involved in optimization, all points of data Q should be labeled according to x and y values. After being classified into corresponding grid according to the transformed x and y values, each transformed point $Tr(p_i)$ applied proposed data association method only with target points which locate in the same grid. Thus for each $Tr(p_i)$ point, the computational complexity reduces $O(M)$ to $O(M_{grid})$, where M_{grid} and M are the point numbers in one grid and in one surface set, respectively. Fig. 2(b) demonstrates the converted design.

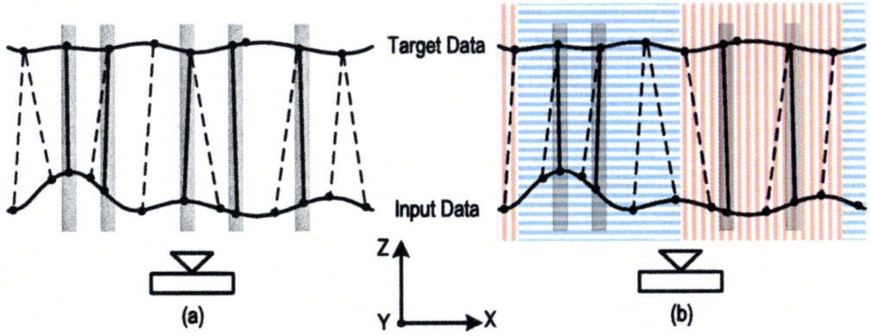

Fig. 2. Correspondence strategies (a) without, and (b) with acceleration structure

4 Experiments

For evaluation, we use data captured by Kinect sensor and target at multi-view face registration application. Registration experiments in this paper only utilize depth information (range images). Illustration in this section are all 2D figures projected from 3D data, and from different degrees, same figure with and without facial texture is shown for interpenetration visualization purpose. Red color represents target data, while blue color represents input data. Before approaching to registration, face data needs to be processed as illustrated in Fig. 3. The face is detected using [13] and followed by the segmentation to the face center. Smoothing and bias removal are used to adapt a generic face data. The average depth and dimension of generic face data are around 1m and 0.25m by 0.25m, respectively. Usually each set data contains five to seven thousand points.

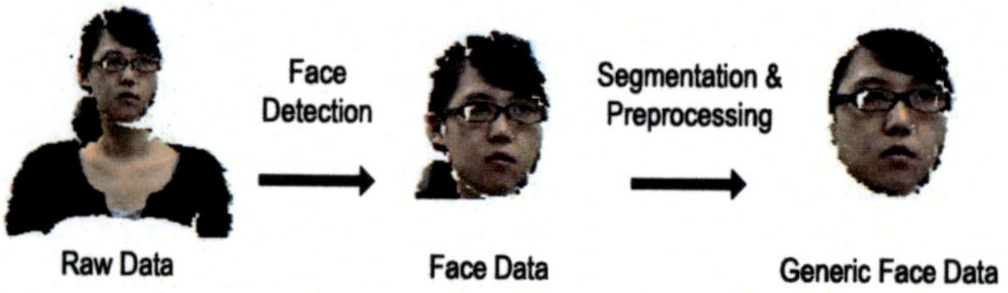

Fig. 3. Illustration for data captured by Kinect sensor and processing before registration

ICP and another approach, which uses "closest point" as data association strategy and GA as optimization tool (CP-GA), are implemented here for experimental comparison. In order to accelerate the optimization, CP-GA is aided with Kd-tree.

4.1 Qualitative Experiment

In experiments Fig. 4 and Fig. 5, data are acquired by one Kinect device, and another set data with different pose is artificially generated by self-defined transformation. Since self-transformed data is used for registration, the ideal result is two data are perfectly matching with zero registration error. Graphic results are illustrated in the corresponding experiment figures and processing time is summarized in Table 1 with two sets self-defined transformation values.

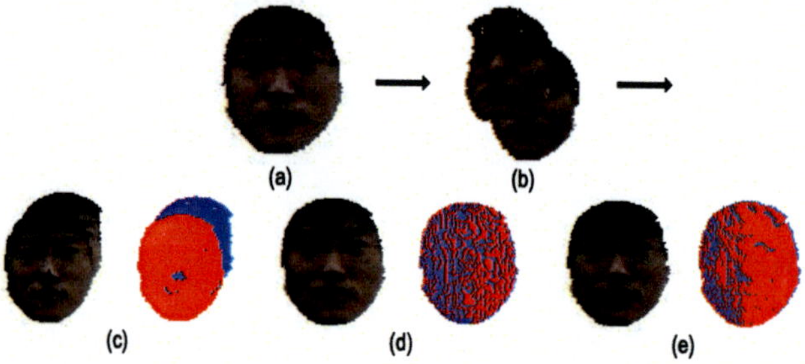

Fig. 4. (a) Generic face data; (b) artificial generated data from (a); (c) ICP; (d) CP-GA with Kd-tree; and (e) proposed method

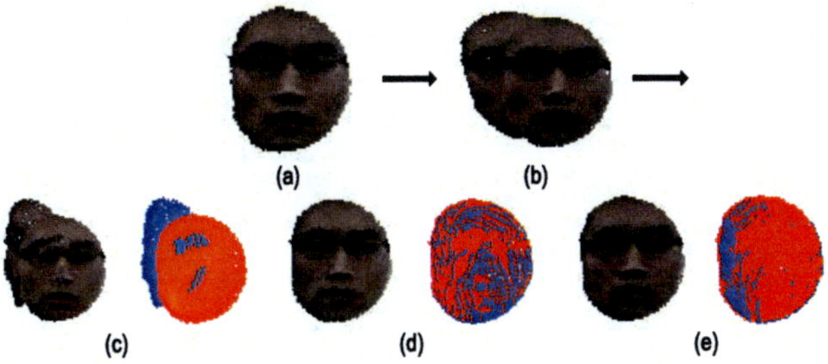

Fig. 5. (a) Generic face data; (b) artificial generated data from (a); (c) ICP; (d) CP-GA with Kd-tree; and (e) proposed method

4.2 Quantitative Experiment

In experiments Fig. 6 and Fig. 7, point clouds are simultaneously captured by different Kinect devices with various view poses. Considering the requirement of ICP, Matlab calibration toolbox is applied to calculate the initial transformation information. Graphic results are still shown in the corresponding experiment figures, and last two rows of Table. 1 contains calibration results and computational costs for these experiments.

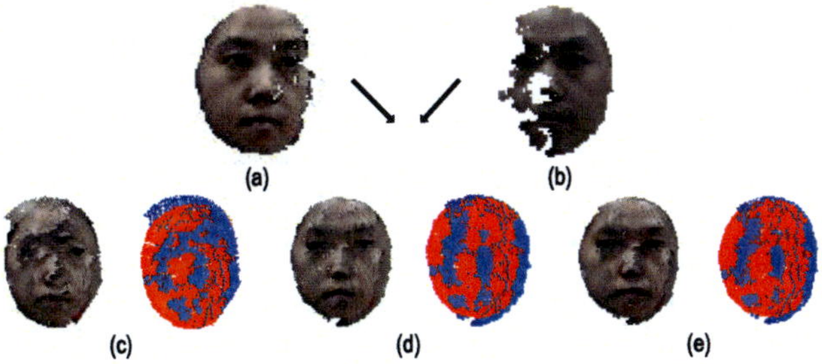

Fig. 6. (a)(b) Multiple generic face data; (c) ICP; (d) CP-GA with Kd-tree; and (e) proposed method

4.3 Discussion

From both registration graphic results and processing speeds of different approaches, it is obviously that performances of both ICP and CP-GA are far from satisfactory and proposed work well merges multiple point clouds with reasonable execution speed. In qualitative experiments, ICP already shows its weakness in the need of prior information, and only converges to the local optimums. Its performance is even worse in quantitative experiments. Regarding to the CP-GA method, when facing large rotation angle (or called low-overlapped data)

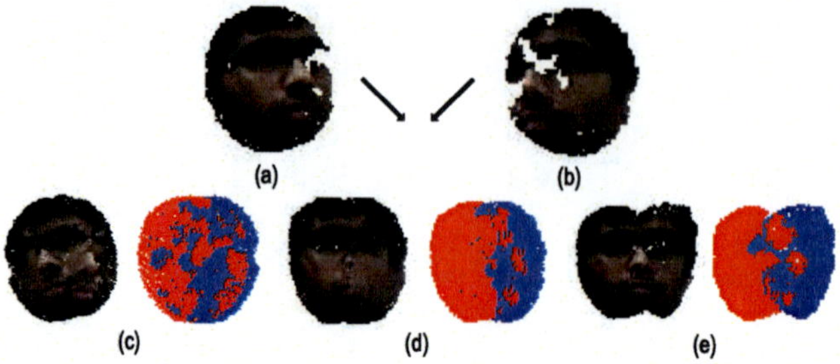

Fig. 7. (a)(b) Multiple generic face data; (c) ICP; (d) CP-GA with Kd-tree; and (e) proposed method

Table 1. Computational cost comparisons of experiments. (Note: iteration=20 for all experiments).

	Transformation		Time (sec)		
	Translation (m)	Rotation (deg)	ICP	GA	Our
Fig.4	[0.03,0.05,-0.05]	[20.0,-10.0,-5.0]	18.0	13144.0	39.7
Fig.5	[-0.05,0.01,-0.03]	[-10.0,5.0,15.0]	14.7	8228.9	28.7
Fig.6	[-0.70,0.025,0.44]	[3.95,20.62,-4.15]	11.3	9337.7	31.0
Fig.7	[-0.93,0.052,0.36]	[3.06,50.19,-6.52]	8.61	7581.3	18.6

like Fig. 7, it loses the robustness. Therefore, proposed method is competent for various cases and outperforms than other two.

5 Conclusion

This paper introduces a robust approach for free-form registration of multiple point clouds, which does not need prior information and also is insensitive to noise and outliers. The major contribution of our work is the proposed data association with parallel-light structure to evaluate the registration error, which highly improves the pair matching correction. Besides that, acceleration module is introduced to relieve the computational cost for global searching process. The target data is separated into grids and the registration error function is calculated based on the grid structure. As experimental results illustrate, the proposed method provides robust performance in finding accurate registration, especially is capable of dealing low-overlapped surfaces, and completes all integration process in a reasonable duration. As future work, color information may be adopted in fitness calculation which is helpful for correspondence creation. Also real-time or near real-time registration can be explored utilizing multi-layer or hybrid method.

References

[1] Salvi, J., Matabosch, C., Fofi, D., Forest, J.: A review of recent range image registration methods with accuracy evaluation. Image and Vision Computing 25(5), 578–596 (2007)
[2] Besl, P.J., McKay, H.D.: A method for registration of 3-D shapes. IEEE Transactions on Pattern Analysis and Machine Intelligence 14(2), 239–256 (1992)
[3] Rusinkiewicz, S., Levoy, M.: Efficient variants of the ICP algorithm. In: Proceedings of the 3rd International Conference on 3-D Digital Imaging and Modeling, pp. 145–152 (2001)
[4] Brunnstrom, K., Stoddart, A.J.: Genetic algorithms for free-form surface matching. In: Proceedings of the 13th International Conference on Pattern Recognition, vol. 4, pp. 689–693 (1996)
[5] Silva, L., Bellon, O.R.P., Boyer, K.L.: Enhanced, robust genetic algorithms for multiview range image registration. In: Proceedings of the 4th International Conference on 3-D Digital Imaging and Modeling, pp. 268–275 (2003)
[6] Chow, C.K., Tsui, H.T., Lee, T.: Surface registration using a dynamic genetic algorithm. Pattern Recognition 37(1), 105–117 (2004)
[7] Lomonosov, E., Chetverikov, D., Ekárt, A.: Pre-registration of arbitrarily oriented 3D surfaces using a genetic algorithm. Pattern Recognition Letters 27(11), 1201–1208 (2006)
[8] Robertson, C., Fisher, R.B.: Parallel evolutionary registration of range data. Computer Vision and Image Understanding 87(13), 39–50 (2002)
[9] Liu, R., Burschka, D., Hirzinger, G.: A novel approach to automatic registration of point clouds. In: IEEE International Geoscience and Remote Sensing Symposium, pp. 401–404 (2007)
[10] Wang, L.R., Xu, F., Hagiwara, I.: An efficient registration algorithm of multi-view three-dimensional images. World Congress on Computer Science and Information Engineering, 703–706 (2009)
[11] Dai, J.J., Yang, J.: A novel two-stage algorithm for accurate registration of 3-D point clouds. In: International Conference on Multimedia Technology, pp. 6187–6191 (2011)
[12] Felsberg, M., Larsson, F., Wang, H., Ynnerman, A., Schon, T.B.: Torchlight Navigation. In: Proceedings of the 20th International Conference on Pattern Recognition, pp. 302–306 (2010)
[13] Ying, Y., Wang, H., Xu, J.: An automatic system for multi-view face detection and pose estimation. In: Proceedings of the 11th International Conference on Control Automation Robotics and Vision, pp. 1101–1108 (2010)

Real-Time Model Based Visual Servoing Tasks on a Humanoid Robot

Amine Abou Moughlbay[1], Enric Cervera[2], and Philippe Martinet[3,*]

[1] Institut de Recherche en Communications et Cybernétique de Nantes, Ecole Centrale de Nantes, 1 rue de la noë, 44321 Nantes, France
Amine.Abou-Moughlbay@ec-nantes.fr

[2] Robotic Intelligence Lab, Jaume-I University, 12071 Castello, Spain
Ecervera@icc.uji.es

[3] Institut de Recherche en Communications et Cybernétique de Nantes, Ecole Centrale de Nantes and Institut Pascal in Clermont Ferrand, France
Philippe.Martinet@irccyn.ec-nantes.fr

Abstract. Several model based techniques have been used to apply various domestic service tasks on humanoid robots (through teleoperation, learning, ...). But for many reasons, it is more suitable to study the interaction between the robot and its environment using the Sensor Based Control in these cases. In this paper we present a work of integration of real-time visual servoing techniques in performing self localization and different manipulation tasks on a humanoid robot in closed loop.

Real-time model based tracking techniques are used to apply 3D visual servoing tasks on the Nao humanoid robot. Elementary tasks used by the robot to perform a concrete scenario are detailed with their corresponding control laws. Experimental results are presented for the following tasks: self-localization of the robot while walking, head servoing for the visibility task, detection, tracking and manipulation of environment's objects.

1 Introduction

To perform manipulation tasks, the robot interacts with its environment through contact points, such as applying forces and moments on objects. By controlling the position and velocity of these points, as well as the forces acting on them, the robot performs the desired tasks. In complicated and uncertain environments, end-effector motion may be subject to online modifications in order to accommodate unexpected events or to respond to sensor inputs. When controlling the interaction between manipulator and environment, the desired task is often specified in the Operational Space and requires precise control of the end-effector motion. Joint Space Control schemes are not suitable in these situations. Thus programming of service and manipulation tasks is most conveniently accomplished by directly specifying data at the contact points, rather than specifying the joint positions and velocities required to achieve them [1].

* This work has been funded by the French National Agency of Research under the reference ANR-10-SEGI-002.

S. Lee et al. (Eds.): Intelligent Autonomous Systems 12, AISC 193, pp. 321–333.
springerlink.com

To apply service tasks, many approaches have been used: different works have been carried out in the area of teleoperation, by controlling a robotic system to perform tasks at a distance using a multi-modal human-system interface which provides sensory feedback to the operator and allows him to interact with the remote environment by mapping his actions. It was applied to bi-manual manipulation and walking [2], haptic interface for mobile teleoperators [3] and recently on a teleoperation system with haptic device for micro-manipulation [4].

Furthermore, many practical learning control systems are used to control complex robots involving multiple feedback sensors and multiple command variables during both repetitive and nonrepetitive operations [5]. The issue of teaching a robot to manipulate everyday objects through human demonstration has been studied by [6] who proposed a method that enables a robot to decompose a demonstrated task into sequential manipulation primitives, series of sequential rotations and translations [7]. Other earlier works used also the Model Based Control strategies, executed by automatically generating a control sequence that moves the robot to the states specified by the program to develop executives that emphasize model based approaches and deep integration of automated planning [8].

Self-localization is one of the fundamental problems in robotics, as in many applications a robot needs to know its location in order to perform its tasks. Most of the indoor localization algorithms use particle based filters or Kalman type filters to solve the problem of noisy sensors and controls. Particle filters inherently help solve the problem of ambiguous landmarks, whereas Kalman filters must track multiple hypotheses to work in ambiguous environments [9]. In some divisions of RoboCup, algorithms are very well established, given the rich sensor data provided by laser scanners, omnidirectional cameras etc. However, in more general cases, there are substantial sensor limitations particularly with the rapid motion of the camera, and the need for active perception [10]. In addition to odometry sensors, some recent alternatives use also the ambient magnetic field to control the heading of a robot in case of one-dimensional localization problem [11].

As previously presented, most actuated systems use sensors to obtain information about their environment. These can be a camera, ranging devices, or temperature and force sensors. Among all these feedbacks, the visual information provides the most important and instant cues for perception of the interaction with the working environment. Compared to already mentioned methods, visual servoing provides very efficient solutions to control robot motions. It supplies high positioning accuracy, good robustness to sensor noise and calibration uncertainties, and reactivity to environment changes [12].

In this paper, only Sensor Based Control formalism is used to perform the desired tasks. More especially, 3D visual feedback data and Model Based Tracking (MBT) techniques are used to execute, in real-time and closed loop, many tasks on the humanoid mobile robot Nao in a semi structured environment. The robot's camera is calibrated, and a rough geometric model of the objects is available (doors, tables, pieces to grasp ...). The envisioned scenario consists of a Nao robot which can carry out service tasks, moving around the room and manipulating objects. One of the missions that can be requested from such robot would be to "Pick up the orange and green pieces and deposit them on the table nearby the door" (see Fig. 1). But to execute this mission, many problems should be addressed and resolved:

Fig. 1. Nao picks up the orange and green pieces and deposits them on the table near the door

- How can the robot localize itself wrt. its environment?
- How to decompose the robot's mission into elementary tasks?
- What information should be given to the robot to execute each task?
- What are the more useful techniques the robot can use to perform these tasks?

In the second section of this paper, we present the architecture of the Nao robot and the Model Based Tracking technique. In section 3, we define the common elementary tasks and frames employed during the application of the desired tasks. Implemented tasks are presented in section 4 and the experimental results in section 5. The first part of this latter includes the localization of the robot in its environment when walking. The second one includes the object tracking and manipulation by visual servoing. The final section draws some conclusions and outlines future works.

2 System Architecture

2.1 Nao Architecture

Nao Robot [13], developed by Aldebaran robotics, is a biped robot with 25 Degrees of Freedom (DOF). It has 3-fingered robotic hands used for grasping and holding small objects (it can carry up to 300g using both hands). It is equipped with: 2 ultrasound devices situated in the chest that provide space information in a range of 1 meter, 2 cameras situated on the top and bottom of the head, 2 bumpers (contact sensors on the robot's feet), a gyrometer and an accelerometer (to determine whether the robot is in a stable or unstable position).

2.2 Visual Servoing and Tracking Techniques

A large variety of positioning or target tracking tasks can be implemented by controlling from one to all DOF of the system. For whatever the sensor configuration, which can vary from one camera mounted on the robot end-effector to several free-standing cameras, a set of visual features $\mathbf{s}$ has to be designed from the visual measurements obtained from the system configuration $\mathbf{x}(t)$ $(\mathbf{s} = \mathbf{s}\,[\mathbf{x}(t)])$, allowing control of the desired DOF. A control law is thus designed so that these features $\mathbf{s}$ reach a desired value $\mathbf{s}^*$, defining a correct realization of the task. Indeed, if the camera velocity is considered as input of the robot controller, the control law which performs the desired exponential decoupled decrease of the error $\mathbf{e} = (\mathbf{s} - \mathbf{s}^*)$ is:

$$\mathbf{V}_c = -\lambda \mathbf{L}_s^+ (\mathbf{s} - \mathbf{s}^*) \tag{1}$$

where λ is a proportional gain that has to be tuned to minimize the time-to-convergence, and $\mathbf{L}_s^+$ is the pseudo-inverse of the interaction matrix (for more details refer to [14]).

Many tracking tools have been implemented in several visual servoing toolboxes [15]. On the Visual Servoing Platform (ViSP) [16], we find a dot tracker, a moving edges tracker, and a 3D model based tracker. The last one [17] tracks a 3D model thanks to the moving edges method, using a virtual visual servoing technique. It requires a 3D model and needs to compute the initial pose which is used to project the model on the image. The tracking method assumes that the pose corresponding to the previous image is known, the new lines are tracked, and the goal is to move the pose to match the object in the new image with the projection of the model.

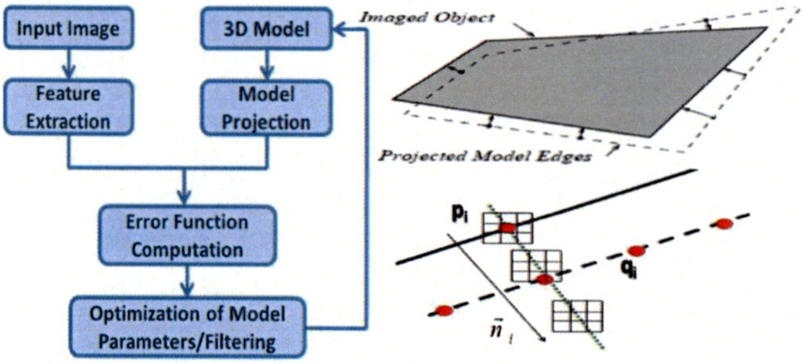

Fig. 2. Model Based tracking system (left) using the Moving edge detection technique (right)

The error function (err) between image features p_i and model projection q_i is thus minimized along the normal direction n (see Fig. 2):

$$err = \sum_i \Delta(p_i, q_i) = \sum_i |(q_i - p_i) \cdot (n_i)| \tag{2}$$

3 Scenario and Task Identification

3.1 Frames and Control Law

To execute the different tasks, we should define frames on the robot's body and environment's items (see Fig. 3). Using the inverse kinematic model of the robot and the 3D visual servoing technique, the general control law used to define a task is given by:

$$\dot{\mathbf{q}} = -\lambda (\mathbf{L_s J})^{+} (\mathbf{s} - \mathbf{s}^*) \tag{3}$$

where $\dot{\mathbf{q}}$ is the robot's joint velocity and $\mathbf{J}$ is the kinematic matrix which relates camera velocity with the robot's joint velocity ($\mathbf{V}_c = \mathbf{J}\,\dot{\mathbf{q}}$).

3.2 Generic Tasks

While Nao is a capable platform, due to the complexity of the problem to be solved, the capabilities of the robot can be enhanced and the complexity of the problem can be reduced by decomposing it into simpler generic tasks. Thus, the desired scenario is executed by answering the following questions:

1. **Where am I ? → Self-Localization Task**
   Depending on the desired mission to be executed, the robot searches for the corresponding set of models in its environment to be localized wrt. them (table, door,

Fig. 3. Useful frames in Nao's environment: In Nao's body we consider the following frames: Nao's space frame $\mathscr{F}_n$ (between robot's feet), $\mathscr{F}_t$ on robot's torso, $\mathscr{F}_c$ a camera attached frame, robot's hand frame $\mathscr{F}_h$, robot's gripper frame $\mathscr{F}_g$ and a pre-grasping frame $\mathscr{F}_{pg}$. The following frames are defined in the robot's environment: object's frame $\mathscr{F}_o$, desired object's pose frame $\mathscr{F}_d$ and environment's frame $\mathscr{F}_e$ for the localization task.

light switch, corner, ...). After detecting the items in the field of view of the robot's camera, and choosing the nearest one to the object to manipulate ($\mathscr{F}_e$), the robot calculates, using the MBT technique, its position/orientation wrt. the item in form of a homogeneous transformation matrix (${}^n\mathbf{M}_e$).

2. **What must I do and How to do it ? → Task Scheduler**
   The desired scenario is interpreted to the robot's language and decomposed into elementary tasks [18]:
   - Localize the desired objects to manipulate
   - Move in the appropriated direction
   - Detect and track the desired objects
   - Keep these pieces in the robot's field of view
   - Move the robot's arms and grasp these pieces
   - Go to the desired table and deposit objects
3. **Which data will I use ? → Definition of used data**
   To execute the defined elementary tasks, the scheduler chooses the appropriate models for each one: models of the objects to manipulate, model of the environment's items (for the self-localization task), data of the robot's camera, parameters of the robot's gripper and the used items...

4. **Where to move ? → Robot Locomotion Task**
   This task is used to move to the region where the task is to be executed. Knowing the position of at least one item of the environment (${}^n\mathbf{M}_e$), the robot walks in the appropriate direction of ($\mathscr{F}_e$) until enterring the range of a defined distance from the object. During this task, an obstacle avoidance technique is used to adapt the robot's trajectory [19].

5. **How to perceive ? → Detection and Tracking Tasks**
   Using the MBT technique of ViSP, the tracker is manually initialized and the pose of the desired item is determined. Hence it allows us to track in real-time the pose of the object to manipulate. An automatic re-initialization of the tracker is implemented: it uses the last poses of the tracked object to be reinitialized in case of a failure due to an occlusion or the fast motions of the robot's camera (especially when walking). The automatic transition between the tracking of different objects is also implemented using the knowledge of the rough relative position between environment's objects.

6. **How to keep the concerned points in the robot's field of view ? → Visibility Task**
   This task consists of controlling the robot's head position/ orientation to focus a (fixed/mobile) point of the environment (item's center, gripper, virtual point...) in the center of the camera's image. This task can be used in hand-eye coordination for dynamic grasping of objects by focusing on the gripper-item midpoint, to keep the robot's hand and object in the robot's field of view. The used DOF in this task depends on the geometry of the robot's head and the desired complexity of the task.

The head's Yaw/Pitch can be controlled to focus on the object's center, and a more complex task can also control the distance between the robot's head and the object.

7. **How to perform the manipulation ? → Grasping Task**
   This task uses the hand's control point and allows the robot to move it to a desired static/mobile pose. This task can be used to perform pre-grasping, grasping, and displacing objects tasks. In case of pre-grasping task, the goal position (${}^{g}\mathbf{M}_{pg}$) is determined using the grasping strategies. These strategies depend on the geometry of the object to manipulate and the geometry of the robot's gripper. The grasping strategy controls the relative position and/or the angle between the gripper and the item to grasp. For a grasping task, the same technique is considered: the robot's arm moves to a desired pose by minimizing the relative distance/orientation between the object and the robot's hand (${}^{g}\mathbf{M}_{o}$) in the Gripper's Frame [20].

4 Tasks Definition

In this part, we present preliminary results of the general scenario detailed in the previous section applied on the humanoid robot Nao. The experimented tasks are: the self-localization of the robot when walking, and the head and arm's visual servoing to track and grasp an object. During the execution of the first task, only a rough model of the door and a part of the room is used, and for the second one only the approximate model of the item to grasp is given. Note that no other exteroceptive data is given from the robot's environment.

4.1 Self-localization Task during Locomotion

During this task, the robot tracks the door and thus it is localized wrt. the environment while walking. We use the model of the door and the lines of the room around it to initialize the MBT which gives then the pose of the door in the camera's frame (${}^{c}\mathbf{M}_{e}$) . From the given pose, the robot's pose can be calculated: ${}^{e}\mathbf{M}_{n} = ({}^{c}\mathbf{M}_{e})^{-1}\,({}^{n}\mathbf{M}_{c})^{-1}$.

Note that the tracker was automatically reinitialized each time the tracking failed due to large camera displacements during walking; it used the last found pose of the object to reinitialize the tracking. In our experiments, the robot walks in open loop in the direction of the door for a distance of 1 meter.

4.2 Detection and Tracking Task

The MBT is used to track simple item models; it's initialized manually at the beginning of the application. Afterwards, the model is automatically detected and tracked; this tool allows us to determine instantly the pose of the desired item frame in the robot camera's frame (${}^{c}\mathbf{M}_{o}$).

4.3 Visibility Task

Throughout this application, the visibility task is used for controlling the robot's head orientation to focus the item's center in the center of the camera's image. Two DOF are used by this task to control the head's Yaw and Pitch. The task's goal is thus to regulate exponentially ($\dot{\mathbf{e}} = -\lambda \mathbf{e}$) the horizontal and vertical position of the center of the object projection $\mathbf{s}_{x,y} = {}^c\mathbf{T}_{o(x,y)}$ to zero ($\mathbf{s}^* = (0,0)$).

Using the object 3D pose ${}^c\mathbf{T}_o = (X,Y,Z)^T$, and the 2D pose $(x,y)^T$ of the tracked point (projection of 3D point in the normal image plane), we apply the control law defined in (3) using the visual primitive $\mathbf{s} = (x,y)$ and its corresponding interaction matrix $\mathbf{L_s}$ given by:

$$\mathbf{L_s} = \begin{bmatrix} -\frac{1}{Z} & 0 & \frac{x}{Z} & xy & -(1+y^2) & y \\ 0 & -\frac{1}{Z} & \frac{y}{Z} & 1+x^2 & -xy & -x \end{bmatrix} \tag{4}$$

Note that for this task the $\mathbf{J}$ matrix in (3) uses the Jacobian of the robot's head control point calculated from the robot's geometric model.

4.4 Pre-grasping Task

Through this task the robot's arm is supposed to move close to the item to grasp. According to Nao's gripper's geometry (of one DOF) and the item's shape (rectangular model), the number of constrained DOF and the pre-grasping position is predefined in the robot's manipulation parameters. This position is defined wrt. the object's position in Nao's frame ^{n} (${}^o\mathbf{M}_{pg}$).

In our case, regarding the gripper's and item's shapes, 4 DOF are enough to execute this task: 3 DOF constraints the gripper's pose and 1 DOF (Yaw angle) for the gripper's orientation. Furthermore, the pre-grasping distance is fixed to 5 cm, but we should not forget that the tracker gives the item's pose in real-time, thus the desired pre-grasping pose is calculated in closed loop. The task's target is then to move the robot's arm to the pre-grasping pose. The task's error is extracted from the relative pose between the gripper and pre-grasping point (${}^g\mathbf{M}_{pg}$) which is regulated to zero.

Note that ${}^g\mathbf{M}_{pg} = ({}^n\mathbf{M}_g)^{-1}\, {}^n\mathbf{M}_c\, {}^c\mathbf{M}_o\, {}^o\mathbf{M}_{pg}$ where ${}^n\mathbf{M}_g$ and ${}^n\mathbf{M}_c$ are given by the robot's proprioceptive sensors.

Considering that the visual primitive is parameterized by $\mathbf{s} = (\mathbf{t}, \mathbf{u}\theta)$ where $\mathbf{t}$ is the position error between the current and desired frame, while $\mathbf{u}\theta$ is the orientation error, decomposed as the axis $\mathbf{u}$ and angle θ of the rotation between these two frames. The control law (3) is then applied using the Jacobian at the robot's gripper and the corresponding interaction matrix $\mathbf{L_s}$ given by:

$$\mathbf{L_s} = \begin{bmatrix} -\mathbf{I}_3 & [\mathbf{t}]_\times \\ \mathbf{0}_3 & \mathbf{L}_\omega \end{bmatrix} \tag{5}$$

where $\mathbf{I}_3$ and $\mathbf{0}_3$ are the 3×3 identity and zero matrices respectively, the $\mathbf{L}_\omega$ matrix is given by $\mathbf{L}_\omega = \mathbf{I}_3 - \frac{\theta}{2}[\mathbf{u}]_\times + \left(1 - \frac{sinc(\theta)}{sinc^2(\theta/2)}\right)[\mathbf{u}]_\times{}^2$, and $[\mathbf{t}]_\times$ is the skew symmetric matrix associated with vector $\mathbf{t}$.

4.5 Grasping Task

Along this task the robot's arm moves to grasp the desired item. The same number of DOF is constrained and the same technique is used to define the grasping task as in the previous case of pre-grasping, but the desired gripper pose is changed to the object's pose. Thus the task's error will be extracted from the relative pose between the gripper and the item ${}^{g}\mathbf{M}_o$ which is also regulated to zero (this pose is calculated using the relation: ${}^{g}\mathbf{M}_o = ({}^{n}\mathbf{M}_g)^{-1}\,{}^{n}\mathbf{M}_c\,{}^{c}\mathbf{M}_o$.

After arriving to the desired pose, the robot's gripper closes to catch the item. We should note that the tasks are automatically sequenced and a task is completed once the error norm reaches a predefined threshold value. This threshold varies wrt. the executed task: during the grasping task of small objects, a high precision is necessary unlike in the case of pre-grasping task. In our experiments, the threshold is predefined to 0.8 mm in the camera's image for the visibility task. In case of pre-grasping and grasping tasks, the precision is predefined to 3 mm and 0.2 mm respectively and 2 deg for the orientation.

5 Experimental Results

The presented tasks in the previous section have been implemented and tested several times on the Humanoid Nao robot to ensure the efficiency of this method. We used the 2 cameras embedded on the robot's head: the top one for localization and the bottom one for manipulation tasks. The control rate of the robot is equal to that of the camera (20 Hz) [1].

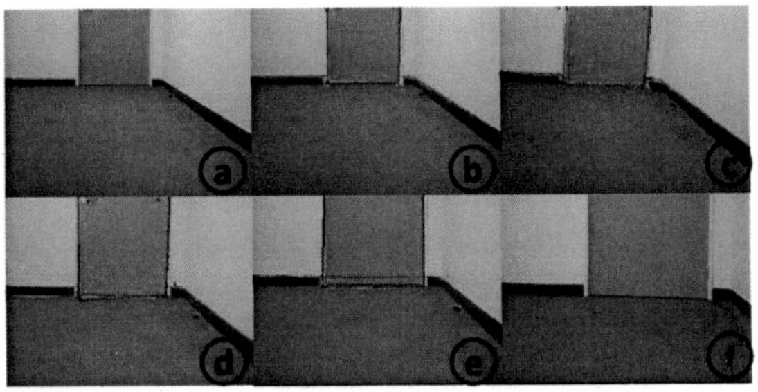

Fig. 4. Experiment photos of Localization Task during robot's locomotion

5.1 Self-localization Task Results

This part corresponds to the task defined in section 4.1. Fig. 4-a shows the robot's environment before launching the tracking and locomotion tasks. Fig. 4-b...e show the tracking of the door during the task's execution, and when the task is completed (Fig. 4-f). In Fig. 5, the distance between the robot's frame $\mathscr{F}_n$ and the origin of the door's frame is plotted, in addition to the X, Y and Z components of this pose in $\mathscr{F}_n$. The distance decreases from 3.94 m to 2.88 m. These results show then that the robot successfully tracks the door while walking the desired distance (1 m) with an error of 6 cm, which allow us to localize the robot.

[1] See a video of the applied tasks on `www.youtu.be/WflA_jRBMkM`

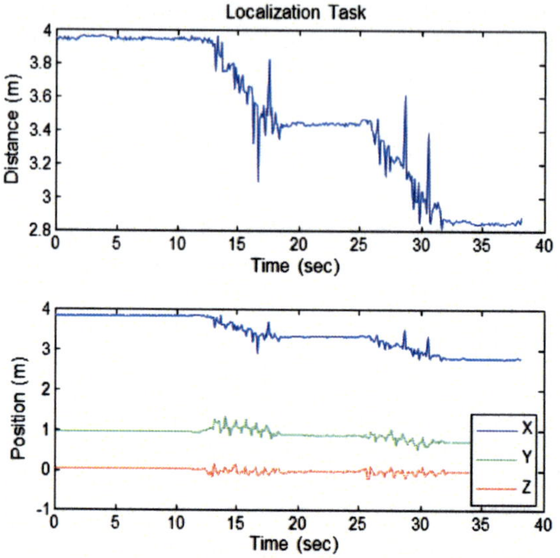

Fig. 5. Experimental results of the localization task in $\mathcal{F}_n$ frame

5.2 Visibility and Grasping Task Results

Experiment photos of the grasping tasks executed by Nao robot are presented in Fig. 6 and correspond to the tasks presented in sections 4.2...4.5 above. Fig. 6-a, shows the item to grasp before launching the MBT to detect and track it (Fig. 6-b). The visibility task is used to center the object on the camera's image, and the pre-grasping task is executed in Fig. 6-c, where we can identify the different frames on the robot's arm and gripper in addition to the object's frame. Afterwards, the gripper's frame approaches the object's when executing the grasping task (Fig. 6-d). Finally the gripper closes and the manipulation task is completed (Fig. 6-e-f).

The variation of the error in each task is presented in Fig. 7: the first graph represents the horizontal and vertical position error during the head servoing task (visibility task), initially the object is at a distance of approximately 30 cm from the center of the camera's image, we remark that this error is successfully regulated to zero during 18 sec with a precision of 0.8 mm. For the 2nd and 3rd graphs, we present the pre-grasping and grasping tasks errors on X, Y and Z components (in Nao's frame), and the Yaw angle of the gripper orientation: the robot's hand is initially at an approximate distance of 15 cm from the predefined pre-grasping position, and the gripper is rotated of 80 deg wrt. the object. During this task, the position error and the Yaw angle are regulated exponentially to zero during 30 sec. Finally, we recall that these (pre) grasping tasks are successfully executed with a precision of 3 mm and 0.2 mm respectively and 2 deg for the orientation.

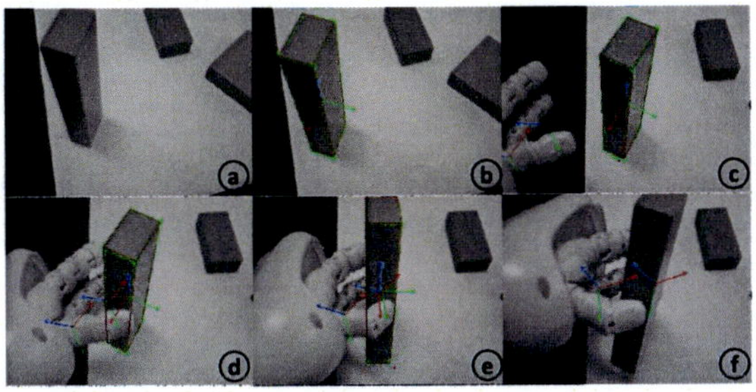

Fig. 6. Experiment photos of tracking and grasping of an item showing the arm's, gripper's and object's frames

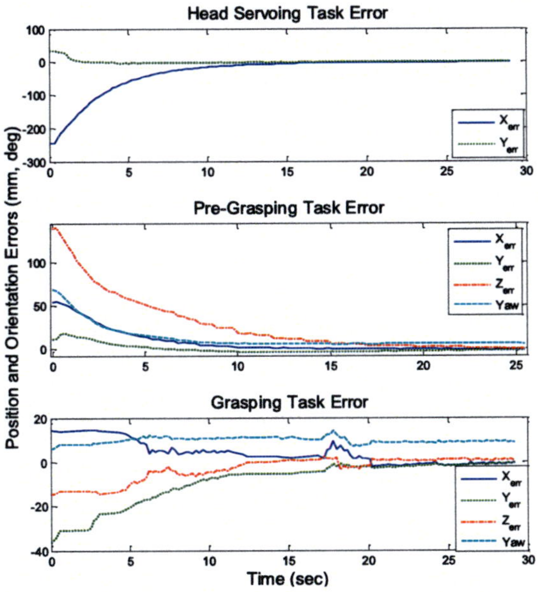

Fig. 7. Experimental results: Visibility task, Pre-Grasping task and Grasping task errors

6 Discussions and Conclusions

In this paper, we presented a concrete scenario of a humanoid mobile robot executing self-localization task while walking and manipulation tasks in an everyday life environment, and we detailed the elementary tasks used by the robot to perform this scenario. The experimental results point out the possibility and efficiency of using the MBT techniques to apply real-time 3D visual servoing on a simple humanoid robot (Nao) in case of localization and manipulation tasks.

For the self-localization task, the error may be relatively large wrt. other localization methods, but it is acceptable and sufficient when dealing with indoor locomotion for manipulation tasks, because of the high robustness of the MBT technique and the implemented automatic re-initialization of the tracking process. In Fig. 5, between the 10th and 32nd sec, we can identify some disturbances which are caused by the displacement of the camera during the robot's locomotion. During this period, the automatic re-initialization module of the tracking is used to prevent localization task from failure. To improve this method, an additional module could be used to dynamically compensate the camera's motions during robot's walking (using data from robot's accelerometer/gyrometer), or by fusion with other sensors data; for example using an external camera on the ceiling of the room to provide a wider view and more visual data, or the Nao's odometry module.

For the manipulation part, the results of the tracking and grasping tasks, which were tested several times, show that this method is robust to camera occlusion by the robot's hand, and robust to slight object movement due to hand-object collision. These tasks should be improved and tested on other robots. Indeed, the mechanic and software architecture of Nao is very constraining, especially for the small camera field of view, the inability of simultaneous use of the two cameras and the constrained operational space of the hands. For this, future works will concentrate on the improvement and implementation of this method on other platforms with different objects of complex shapes with mobile or articulated elements. Other sensors' feedbacks can also be used to improve the manipulation robustness and reactivity against dynamic or unusual changes in the environment.

References

1. Brock, O., Kuffner, J., Xiao, J.: Motion for manipulation tasks. In: Siciliano, B., Khatib, O. (eds.) Springer Handbook of Robotics, pp. 615–645. Springer, Heidelberg (2008)
2. Evrard, P., Mansard, N., Stasse, O., Kheddar, A., Schau, T., Weber, C., Peer, A., Buss, M.: Intercontinental, multimodal, wide-range telecooperation using a humanoid robot. In: IEEE/RSJ Int. Conf. on Intelligent Robots and Systems, IROS 2009, pp. 5635–5640 (2009)
3. Nitzsche, N., Schmidt, G.: A mobile haptic interface mastering a mobile teleoperator. In: IEEE/RSJ Int. Conf. on Intelligent Robots and Systems, IROS 2004, vol. 4, pp. 3912–3917 (September-October 2004)
4. Houston, K., Sieber, A., Eder, C., Vittorio, O., Menciassi, A., Dario, P.: A teleoperation system with novel haptic device for micromanipulation. Int. Journal of Robotics and Automation 26(3) (2011)
5. Miller, W.T.: Sensor-based control of robotic manipulators using a general learning algorithm. IEEE Journal of Robotics and Automation 3(2), 157–165 (1987)
6. Dang, H., Allen, P.: Robot learning of everyday object manipulations via human demonstration. In: IEEE/RSJ Int. Conf. on Intelligent Robots and Systems, IROS 2010, pp. 1284–1289 (October 2010)
7. Jain, A., Kemp, C.: Pulling open novel doors and drawers with equilibrium point control. In: IEEE-RAS Int. Conf. on Humanoid Robots, pp. 498–505 (December 2009)
8. Mansard, N., Stasse, O., Chaumette, F., Yokoi, K.: Visually-guided grasping while walking on a humanoid robot. In: IEEE Int. Conf. on Robotics and Automation, pp. 3041–3047 (April 2007)

9. Quinlan, M.J., Middleton, R.H.: Multiple Model Kalman Filters: A Localization Technique for RoboCup Soccer. In: Baltes, J., Lagoudakis, M.G., Naruse, T., Ghidary, S.S. (eds.) RoboCup 2009. LNCS, vol. 5949, pp. 276–287. Springer, Heidelberg (2010)
10. Billington, D., Estivill-Castro, V., Hexel, R., Rock, A.: Using Temporal Consistency to Improve Robot Localisation. In: Lakemeyer, G., Sklar, E., Sorrenti, D.G., Takahashi, T. (eds.) RoboCup 2006: Robot Soccer World Cup X. LNCS (LNAI), vol. 4434, pp. 232–244. Springer, Heidelberg (2007)
11. Haverinen, J., Kemppainen, A.: Global indoor self-localization based on the ambient magnetic field. Int. Journal of Robotics and Autonomous Systems 57(10), 1028–1035 (2009)
12. Thuilot, B., Martinet, P., Cordesses, L., Gallice, J.: Position based visual servoing: keeping the object in the field of vision. In: IEEE Int. Conf. on Robotics and Automation, vol. 2 (2002)
13. Gouaillier, D., Hugel, V., Blazevic, P., Kilner, C., Monceaux, J., Lafourcade, P., Marnier, B., et al.: Mechatronic design of nao humanoid. In: IEEE Int. Conf. on Robotics and Automation, pp. 769–774 (May 2009)
14. Chaumette, F., Hutchinson, S.: Visual servo control. i. basic approaches. IEEE Robotics Automation Magazine 13(4), 82–90 (2006)
15. Cervera, E.: A Cross-Platform Network-Ready Visual Servo Simulator. In: IEEE Int. Conf. on Intelligent Robots and Systems (IROS 2006), Beijing (China), pp. 2314–2319 (2006)
16. Marchand, E., Spindler, F., Chaumette, F.: Visp for visual servoing: a generic software platform with a wide class of robot control skills. IEEE Robotics Automation Magazine 12, 40–52 (2005)
17. Comport, A., Marchand, E., Pressigout, M., Chaumette, F.: Real-time markerless tracking for augmented reality: the virtual visual servoing framework. IEEE Transactions on Visualization and Computer Graphics 12(4), 615–628 (2006)
18. Galindo, C., Fernandez-Madrigal, J.-A., Gonzalez, J., Saffiotti, A.: Robot task planning using semantic maps. Robotics and Autonomous Systems 56(11), 955–966 (2008)
19. Stasse, O., Verrelst, B., Vanderborght, B., Yokoi, K.: Strategies for humanoid robots to dynamically walk over large obstacles. IEEE Transactions on Robotics 25(4), 960–967 (2009)
20. Sorribes, J., Prats, M., Morales, A.: Visual tracking of a jaw gripper based on articulated 3d models for grasping. In: IEEE Int. Conf. on Robotics and Automation, pp. 2302–2307 (May 2007)

Visual Gyroscope for Omnidirectional Cameras

Nicola Carlon and Emanuele Menegatti

Department of Information Engineering (DEI),
Faculty of Engineering,
University of Padua,
Via Gradenigo 6/a, I-35131 Padova, Italy
nicola.carlon@gmail.com, emg@dei.unipd.it

Abstract. At present, algorithms for attitude estimation with omnidirectional cameras are predominantly environment-dependent. This constitutes a significant limitation to the applicability of such techniques. This study introduces an approach aimed at general mobile camera attitude estimation. The approach extracts features to directly estimate three-dimensional movements of a humanoid robot from its head-mounted camera. By doing so, it is not subject to the constraints of Structure from Motion with epipolar geometry, which are currently unattainable in real-time. The central idea is: movements between consecutive frames can be reliably estimated from the identity on the unit sphere between external parallel lines and projected great circles. After calibration, parallel lines match optical flow tracks. The point of infinity corresponds to the expansion focus of the movement. Simulations and experiments validate the ability to distinguish between translation, pure rotation, and roto-translation.

1 Introduction

Omnidirectional vision is often used to adding robustness to inertial sensor, which are extremely sensitive to measurement errors since the estimate of rotation angles is made through double integration of sensed accelerations.

Omnidirectional cameras have many applications in robotics: tracking of objects, S.L.A.M., attitude estimation of U.A.V., etc...

In estimating egomotion, the use of a omnidirectional camera can solve the ambiguity between rotations and translations because it allows the analysis of traces in all directions [2]. Most of the attitude estimation algorithms used in UAVs [4] are based on the extraction and the tracking of the horizon: this is relatively easy to extract, but it only works outdoors and only at rather high altitude.

For indoor egomotion estimation, algorithms based on central catadioptric line detection have been propoesed in [1], [3], [11], and [12] where the vanishing points of the image are located and use to make the tracking. These are not particularly robust because groups of objects positioned inconsistently with the principal directions (such as inclined books on the shelves or beams) may compromise the resulting attitude estimation.

S. Lee et al. (Eds.): Intelligent Autonomous Systems 12, AISC 193, pp. 335–344.
springerlink.com

2 Outline

To broaden the applicability of attitude estimation techniques, we introduce a temporal approach which tracks non-specific features (such as corners). Hence, the need to find specific ones (such as lines and segments) is avoided.

> The introduced method tracks general features, considering time as a spatial dimension to calculate a Motion Vanishing Point (MVP). The MVP is the point which all tracks excerpts tend to.
>
> This is in contrast to most other algorithms, which locate vanishing points of lines projected through the 3D world on the image plane. Previous methods employ combine methods for time tracking and subsequent attitude calculation.

In humans, movement is perceived in an environment with static objects as the projected flow on the retina. At any specific time point, the environment seems to come from a fixed point on the retina, the Focus of Espansion (FOE), and to enter at the opposite point, namely the Focus of Contraction (FOC)).

The FOE corresponds to the point where the 3D vector that specifies the instantaneous direction of motion intersects the image plane.

The motion vector is the one tangent to the path between the instant baseline and the current one. The FOE coincides with the MVP calculated by the intersections between the Great Circles (GCs), as described in detail in the next Section.

3 Method

As shown in [5] and [8], the projection of each line of the real world on the sphere forms a circle with center and radius coinciding with those of the sphere. These circles are called Great Circles. In this case, instead of having a straight line in space, we consider a straight temporal line in passing through the ends of the track. Following the same approach as in [1], consider $P_i = (X_i, Y_i, Z_i)$ as the position of the i-th feature at time $t = 0$ and $P_i' = (X_i', Y_i', Z_i')$ at time $t = 1$.

These two points on the sphere define a single GC $\mathscr{C}$ on plane $\mathscr{P}$ passing through the center of the sphere O, whose normal is $n = \overrightarrow{OP_i} \times \overrightarrow{OP_i'}$. Consider $\overrightarrow{n_i}$ and $\overrightarrow{n_j}$ as the normals of two GCs. Vector $\overrightarrow{u_{ij}}$ passing through the intersections I_{ij} and I_{ij}' between the two CGs is computed by the cross product $\overrightarrow{u_{ij}} = \overrightarrow{n_i} \times \overrightarrow{n_j}$ Considering all the tracks we get a sheaf of circles that intersect at two opposite points. In this way we obtain the FOE and the FOC (Figure 1).

As analyzed in [9] and in [13], if we have a pure translation FOE and FOC will be antipodal.

If, on the other hand, we have a rototranslation FOE and FOC will be separated by an angle due to the rotation (different than 180°).

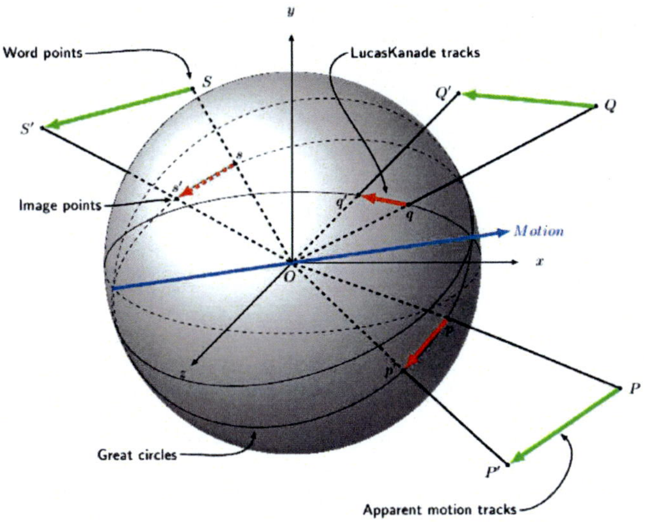

Fig. 1. The direction of two parallel lines is the same as that of the line through the points of intersection and the corresponding GCs

In case of pure rotation, the points identified by the intersections of GCs are neglected since they play no role in the estimation. Some groups of tracks can produce misaligned sheaves of GCs due to image noise, close approximation, different resolution between numerical points and distant points, moving objects in the scene and errors in the calculation of the correspondence between features. For this reason, the antipodal points of intersection of each pair of GCs are grouped according to their position.

We have used the same clustering algorithm of [1] where the new centroid of the cluster is estimated when the distance from the previous centroid and the new point is less than a certain threshold. Otherwise, an additional list with a new centroid is generated. The centroid which refers to the cluster with more elements will be the candidate MVP. If there are two not-antipodal cluster with similar number of elements it will be a roto-translation.

3.1 Algorithm

3.1.1 Region of Interest (ROI) Definition and Image Enanchement

The selection of features is critical for the robustness of the method, so the omnidirectional image was divided into 8 symmetric sections. We select a fixed number of features for every sector. This is very important to ensure that features are extracted in all directions. Without this step, it would be not be possible to take advantage of omnidirectionality.

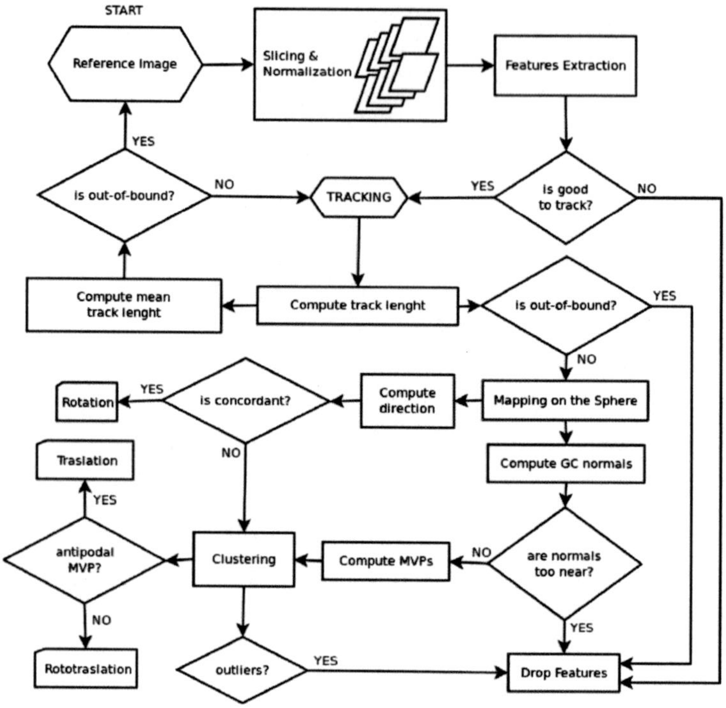

Fig. 2. System overview

3.1.2 Features Extraction and Tracking

The selection is performed by the method cvGoodFeaturesToTrack from OpenCV. It selects the corners by computing minimal eigenvalue of gradient matrix [10]. The ROIs are used only during the initialization. Hence, tracking occurs even if the features are out of their fields. This is very important to be sure to extract features in all directions, otherwise you couldn't use the advantages of using an omnidirectional camera.

For each extracted feature cvFindCornerSubPix is applied to yield accurate estimates of feature positions at sub-pixel resolution.

Taking advantage of the calibration, points on the image plane can be projected on the unit sphere and we can calculate the direction of each track. Finally we use the piramidal Lucas-Kanade method to track the selected features [14].

3.1.3 Attitude Estimation

Attitude estimation: FOE and FOC are not useful in the case of pure rotations. Thus, pitch, roll and yaw (based on tracks) are calculated from the beginning.

- we calculate pitch, roll e yaw from projection of a single track on planes XY, YZ and XZ respectively.

- we calculate the direction of a single track (as angle sign). If there is a predominant direction (based on a threshold) on all tracks, then the camera is rotating in that direction
- we calculate the angle (Section 4) by averaging the angle produced by the tracks with consistent directions.

When no predominant direction is detected, the camera must be translating or roto-translating. In all cases, the calculation of centroid proceeds by checking whether two not-antipodal MVPs are detected. If this is the case, pitch, roll and yaw are calculated from the angle between the unit vectors pointing to the two MVPs.

3.1.4 Removing Outliers

Removing outliers: there exists a significant turnover of selected features during dynamic execution. Some features are discarded by the tracking algorithm becaure they end up being out of sight or being considered unreliable by the cvCalcOpticalFlow-PyrLK method of OpenCV.

Another skimming occurs as a function of the length of the tracks. This is to remove the too long one, which are due to matching errors. The same applies to the too short ones, which are predominantly due to the presence of noise or of distant objects.

Finally, features producing outlier MPVs are eliminated. Outlier detection is performed with respect to atypical distance from the respective cluster centroid. At each iteration, the identified outliers are replaced by new features. The total number of features in each sector is constant.

Before looking for new features, all the previous ones are deleted together with their respective 8-Neighbours. This creates circular deleted areas on the image and avoids reintroducing outliers or selecting the same features twice. Statistical estimates of the tracks are performed to evaluate and update the reference image based on their average length.

In other words, when most of the tracks exceed a certain threshold, it means that the robot has moved from the old location. At this point, the reference image can be updated.

4 Results

The experimental configuration consists of the humanoid robot Robovie-X produced by V-Stone, size 343x180x71mm with a weight of 1.3kg and 17 degree of freedom [16].

On the head of the robot a webcam USB Philips with a hyperbolic mirror [17] provided with CMOS sensor (with resolution up to 1280 x 1024 and 30fps) is placed horizontally. The mirror allows a 360° vision around the robot with the vertical field of view of 75°. The calibration has been performed with OcamClib Toolbox for Matlab [18] Software has been developed with C++ for Linux using OpenCV and QT for the GUI (Figure 3).

Fig. 3. Robovie-X with omnidirectional camera and GUI

4.1 Ground-Truth Validation

To study the configurations of the cluster MPVs, we have introduced the possibility of replacing the automatic extraction of features with a supervised selection tool that looks for features in the neighborhood of the point clicked by the user. The testbed consists in a series of bright points in space at different distances and heights as shown in Figure 4.

To maximize the contrast we used passive fluorescent markers excited by a UV lamp placed over the worktop.

Instead of translating the camera we have choosen to translate the worktop on rails (keeping the camera fixed on a bracket) in order to reduce the measurement error.

Fig. 4. UV photography of the worktop

To rotate the camera we used a servomotor *Hitec HS422* driven by the card *Adafruit Motor/Stepper/Servo Shield* for Arduino.

Using this software we can send to the servomotor the value of the angle (in degrees) we want the camera to rotate of.

4.2 Translation

To study the attitude estimator we used 8 markers placed at the center of the sections in which the image is divided. Working on a sphere the translation of the camera in any direction will produce the same effects on the optical flow, thus the translation is

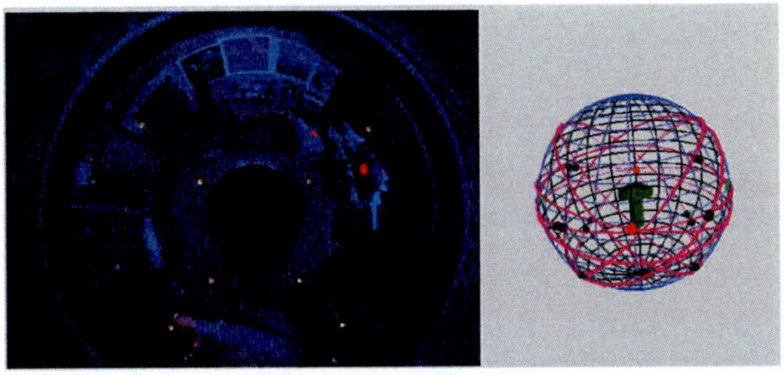

(a) Tracks and GCs displacement

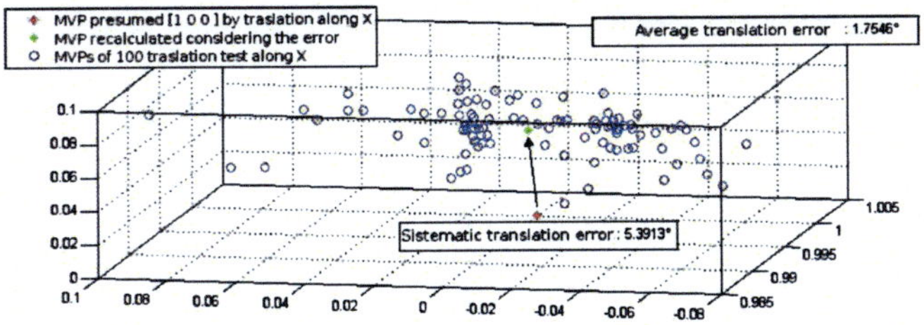

(b) Error

Fig. 5. Cluster of FOE in traslation along X axis

considered for simplicity along the axis x which coincides with the sagittal axis of the robot on which the camera is placed. After 100 tests (translating the camera of $10cm$ along the x axis) it is observable that the x axis is actually slightly shifted from than what indicated after calibration (Figure 5).

We calculated the centroid Q of the cluster of the intersection between the unit vector movement, which should coincide with the x axis, and the unit sphere (the FOE). The difference between this centroid and the point $P = (1,0,0)$ is a systematic error due to the mapping of the image in the unit sphere (5.39°).

4.3 Rotation

Also when we study the pure rotation, we can take advantage of the spherical mapping and consider the rotation along a single axis.

We consider the structured environment constituted by 4 markers lying on the plane orthogonal to the axis of rotation in order to evaluate the drift due to the update of the image reference (Figure 6). We must consider that, every time the average length of the tracks exceeds the threshold, the reference image is updated.

This update during continuous rotation produces a systematic error because we start to measure the rotation after the tracks have a value greater than a threshold.

We measured this systematic error in a series of tests in a structured environment, obtaining an error value $< 1°$. With its correction (verified in unstructured environment) we obtain a rotation error about $3°$.

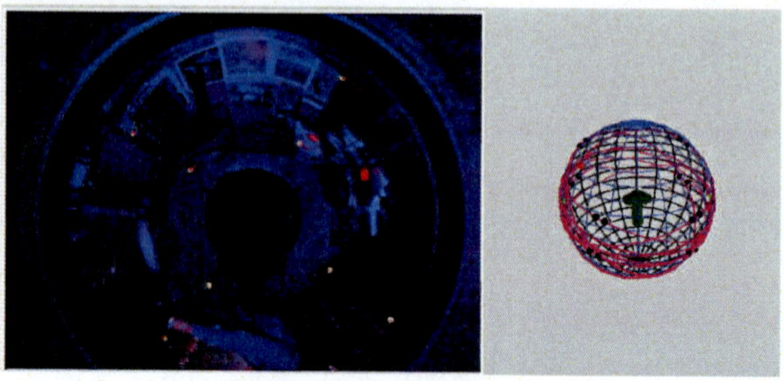

(a) Tracks and GCs displacement

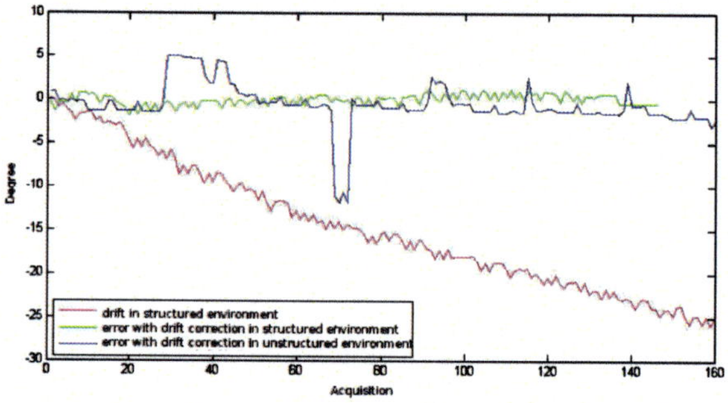

(b) Error

Fig. 6. Errors during a continuous rotation

4.4 Rototranslation

We have considered only a subset of possible rototranslation i.e. that can be produced due to a tilting of the robot. To get an estimate of the error in the calculation of rototranslation we used the same worktop used for the rotation, but decentralizing the camera from the actuator. The calculation of error was made possible by the fact that the angle of rotation of the actuator corresponds to the rotation angle of the camera.

In this case the error is greater (about 7°) because the calculation is based on the corner of the angle between the two not-antipodal MVPs, detected by clustering and not calculated directly from the tracks.

5 Conclusion

The following errors have been estimated in the experimental unstructured environment (Figures 7, 8):

- approx. 3 degrees in case of pure rotation,
- less than 2 degrees in case of pure translation,
- less than 7 degrees in case of roto-translation.

Ongoing testing aims at addressing the dependency of these values upon the type of environment. Computationally light methods have been employed to implement the software in an embedded system. As expected, substantial performance degradation has been detected in the case of insufficient light. This is due to the excessive shutter requirements, which are necessary to compensate exposure. Shutter limitation is due to hardware constraints and is expected to play a marginal role in future experiments. In conclusion, current approach and hardware enable the detection of the correct direction. Yet the lack of depth information prevents quantitative estimates of the path of movement. Future work will focus on robust estimation and parameter optimization. Indeed, significant improvements are expected due to the incorporation of environment-specific configurations. Furthermore, consistency with the Bingham distribution [15] indicates the possibility of incrementing the predictive power of the spherical clustering process.

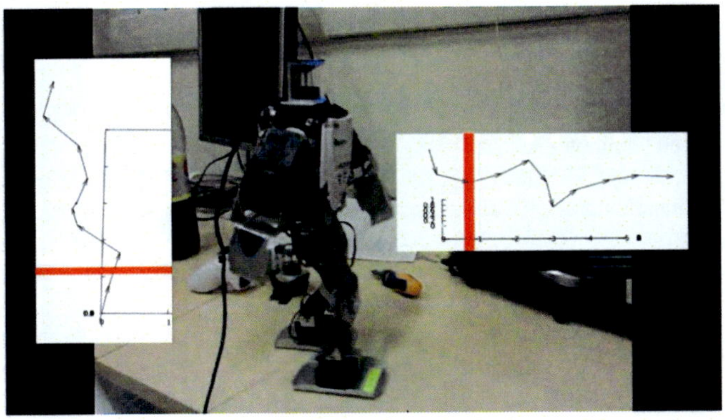

Fig. 7.

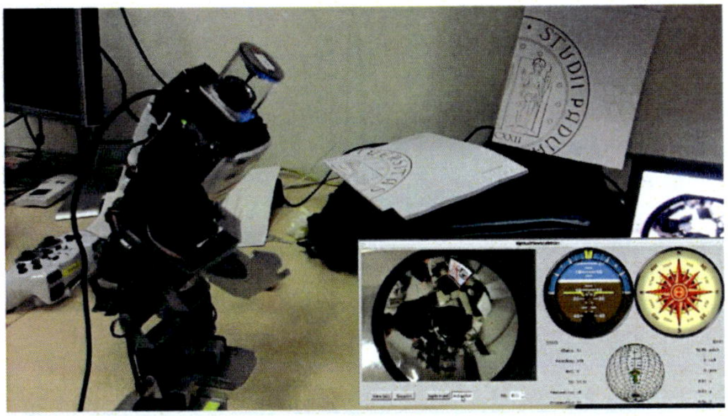

Fig. 8.

Acknowledgment. The authors would like to thank Alberto Pretto for his advice and his review work.

References

1. Finotto, M., Menegatti, E.: Humanoid Gait Stabilization based on Omnidirectional Visual Gyroscope. Department of Information Engineering (DEI), Faculty of Engineering. University of Padua, Padova, via Gradenigo 6/a, I-35131 (2009)
2. Corke, P., Mahony, R.: Sensing and control on the sphere. In: 14th International Symposium on Robotics Research, ISRR 2009 (2009)
3. Bazin, J.C., Kweon, I., Demonceaux, C., Vasseur, P.: Rectangle Extraction in Catadioptric Images. In: IEEE 11th International Conference on Computer Vision, ICCV 2007 (2007)
4. Demonceaux, C., Vasseur, P., Pegard, C.: UAV Attitude Computation by Omnidirectional Vision in Urban Environment. In: IEEE International Conference on Robotics and Automation (2007)
5. Baker, S., Nayar, S.K.: Single Viewpoint Catadioptric Cameras. In: Panoramic Imaging: Sensors, Theory, and Applications (2001)
6. Baker, S., Nayar, S.K.: A theory of catadioptric image formation. In: Sixth International Conference on Computer Vision (1998)
7. Barreto, J.P., Araujo, H.: Geometric properties of central catadioptric line images and their application in calibration. IEEE Transactions on Pattern Analysis and Machine Intelligence (2005)
8. Geyer, C., Daniilidis, K.: Catadioptric Projective Geometry. International Journal of Computer Vision (2001)
9. Nelson, R.C., Aloimonos, J.: Finding motion parameters from spherical motion fields (or the advantages of having eyes in the back of your head). Biological Cybernetics (1988), doi:10.1007/BF00364131
10. Shi, J., Tomasi, C.: Good features to track. In: Proceedings CVPR 1994, IEEE Computer Society Conference on Computer Vision and Pattern Recognition (1994), doi:10.1109/CVPR.1994.323794
11. Ly, S., Demonceaux, C., Vasseur, P.: Vasseur, Translation estimation for single viewpoint cameras using lines. In: IEEE International Conference on Robotics and Automation, ICRA (2010), doi:10.1109/ROBOT.2010.5509555
12. Vasseur, P., Mustapha Mouaddib, E.: Central Catadioptric Line Detection. C.R.E.A. (Centre de Robotique d'Electrotechnique et d'Automatique), University of Picardie Jules Verne (U.P.J.V.), France (1999)
13. Kim, J., Suga, Y.: An Omnidirectional Vision-Based Moving Obstacle Detection in Mobile Robot. International Journal of Control, Automation, and Systems (2007)
14. Bouguet, J.-Y.: Pyramidal Implementation of the Lucas Kanade Feature Tracker Description of the algorithm. Intel Corporation Microprocessor Research Labs (2000)
15. Kent, J.T.: Asymptotic Expansion for the Bingham distribution. Appl. Statist. University of Leeds, UK Online Document (1987)
16. Robovie-X, V-Stone, `http://www.vstone.co.jp/english/products/robovie_x/`
17. Hyperbolic Mirror, V-Stone, `http://www.vstone.co.jp/english/products/sensor_camera/`
18. oCamCalib, tutorial e download, `http://asl.epfl.ch/scaramuz/research/DavideScaramuzzafiles/Research/OcamCalibTutorial.htm`

Panoramic 3D Reconstruction with Three Catadioptric Cameras

Miriam Schönbein, Holger Rapp, and Martin Lauer

Institute of Measurement and Control Systems, Karlsruhe Institute of Technology (KIT),
D-76131 Karlsruhe, Germany
{miriam.schoenbein,holger.rapp,martin.lauer}@kit.edu

Abstract. In this paper, we present a new system setup combining three catadioptric cameras. Our application is in autonomous vehicles. This camera setup allows for panoramic stereo vision of the environment and therefore proves to be useful for ego motion estimation and localization by 3D feature points all around the vehicle. The three catadioptric cameras are arranged in a triangle on top of the vehicle and are horizontally aligned. In the paper we discuss two methods for 3D scene point reconstruction with the system. We perform experiments for the 3D reconstruction in a simulated environment and evaluate the accuracy by means of a Monte-Carlo-Simulation. The proposed system improves the mean accuracy of the 3D reconstruction significantly compared to a system with two cameras.

Keywords: catadioptric camera, 3D reconstruction, stereo vision.

1 Introduction

For intelligent robotic applications environment perception and 3D scene reconstruction are current fields of research. Especially for autonomous driving stereo vision and camera based reconstruction of a large area of the environment are very important tasks. Therefore it is desirable for camera systems utilized for autonomous vehicles to have a large monoscopic as well as stereoscopic field of view (FOV) of the environment. However, cameras typically mounted on vehicles have a limited FOV and are looking only in driving direction. But for many autonomous vehicle applications like blind spot detection or lane tracking it is necessary to perceive also the areas beside and behind the vehicle. Moreover, landmark detection on the sides of the vehicle could improve self-localization.

One common way to enhance the FOV with only one camera is using catadioptric camera systems. Catadioptric cameras are a combination of a convex mirror above a lens and provide a panoramic image of the environment at the cost of image resolution. They have been extensively studied in the past [3,1] and are often used in indoor robotic applications for navigation and surveillance tasks. There are also some approaches using single catadioptric cameras for driver assistance applications like lane tracking [4], blind spot detection [5] or ego motion estimation [11]. However, up to now the potential of stereo vision with catadioptric cameras for outdoor vehicles has not been explored in depth.

S. Lee et al. (Eds.): Intelligent Autonomous Systems 12, AISC 193, pp. 345–353.
springerlink.com

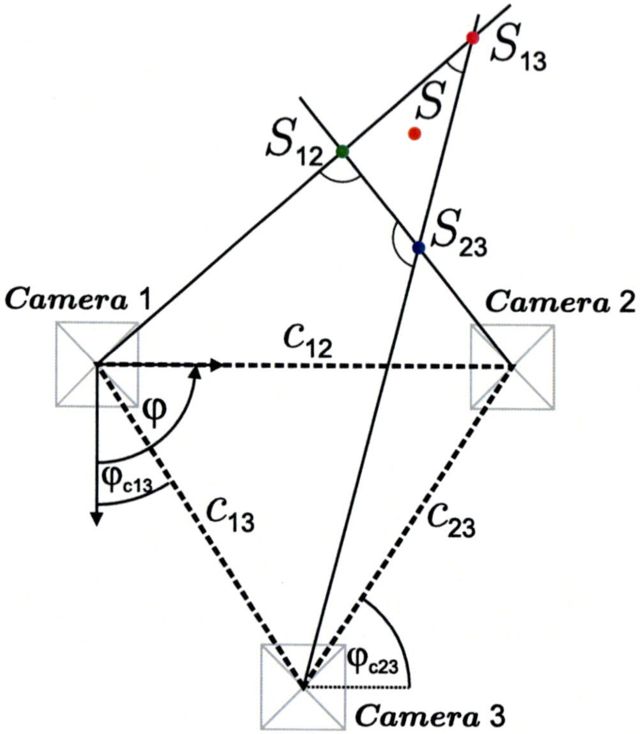

Fig. 1. Camera Setup with three catadioptric cameras. S_{12}, S_{13} and S_{23} show the 3D intersection points in each case by two cameras. S is the 3D scene point calculated by all three cameras.

In recent years different types of omnidirectional stereo camera systems combining catadioptric cameras have been proposed. A large number of approaches uses stereo vision with a vertical camera baseline. This can be achieved by two catadioptric cameras mounted on top of each other [8], by two mirrors mounted on top of each other with only one lens [14] or a special designed mirror with two slopes and only one lens [2]. However, vertically aligned stereo camera systems are not very useful for autonomous outdoor vehicles due to the small baseline of the camera configuration and the height of the setup. A large baseline can only be archived by a high pole mounted on the roof of the vehicle which is not suitable.

Gandhi and Trivedi [6] present a catadioptric camera system for visualizing and analyzing the nearby surrounding with two horizontally aligned cameras mounted on the left and the right side mirror of a vehicle. In [12] we discuss a similar catadioptric stereo camera setup with two catadioptric cameras mounted on the left and the right side on top of the vehicle for 3D reconstruction. This camera configuration allows for accurate stereo vision in the area in front and behind the vehicle. A drawback of such a system with horizontally aligned cameras is the dependency of the depth error from the magnitude of the intersection angle between the two rays. Thus, the advantage of a panoramic stereo view exists only in front and behind the vehicle.

In this paper we propose a new panoramic camera setup consisting of three planar aligned catadioptric cameras arranged in a triangle. With this configuration we increase the monoscopic FOV of the nearby environment by the improved position on the vehicle as well as the stereoscopic FOV. The camera configuration allows for 3D reconstruction in the whole environment around the vehicle and therefore improves many stereo based downstream application for autonomous driving. We show that three cameras overcome the problem of stereo blind spots along the baseline connecting two cameras.

The paper is divided as follows. In Section 2 we present the theoretical model of our catadioptric camera and the 3D reconstruction. Section 3 describes our new camera setup and the reconstruction with three cameras. In Section 4 we show results for enhancement of the 3D reconstruction with cameras on simulated data. Finally, Section 5 concludes this paper and shows future work.

2 Catadioptric Scene Reconstruction

2.1 Catadioptric Camera Model

For our trifocal panoramic system we use catadioptric cameras consisting of an hyperbolic mirror and a perspective lens. These catadioptric cameras fulfill the single view point condition [1]. Different projection models for single view point catadioptric cameras exist to define the relationship between a 3D world point $\mathbf{S}$ and a 2D point $\mathbf{u}$ on the catadioptric image plane ([7,13]). In our approach we use the projection model presented by Scaramuzza [10] which takes into account natural misalignments between the camera and the mirror. This model describes the projection function as a polynomial. The dependency between an image point $\mathbf{u} = [u,v]^T$ and a vector $\mathbf{p}$, which presents a ray from the focal point of the mirror F_C to the scene point $\mathbf{S}$, is encoded by the function g.

$$\mathbf{p} = g(\mathbf{u}') = g(A\mathbf{u}+t) = P\mathbf{S} \tag{1}$$

$\mathbf{S} \in \mathbb{R}^4$ is expressed in homogeneous coordinates, $P \in \mathbb{R}^{3\times4}$ is the perspective projection matrix and $A \in \mathbb{R}^{2\times2}$ and $t \in \mathbb{R}^{1\times2}$ are affine transformation parameters for misalignment errors and digitizing artefacts.

The function g has the following expression

$$g(\mathbf{u}) = (u', v', f(u', v'))^T \tag{2}$$

and the polynomial f is defined as

$$f = a_0 + a_1 \cdot \rho + a_2 \cdot \rho^2 + a_3 \cdot \rho^3 + a_4 \cdot \rho^4 \tag{3}$$

with

$$\rho = \sqrt{u'^2 + v'^2}. \tag{4}$$

2.2 3D Reconstruction

Out of two corresponding image points in a catadioptric image pair the according 3D world points $\mathbf{S}$ can be reconstructed by calculating the points on the mirror surfaces

$\mathbf{X}_{M_i}$ corresponding to the image points $\mathbf{u}_i$ for both catadioptric cameras ($i = 1,2$). Afterwards, we construct the rays through the focal point of the mirrors F_{C_i} and the points on the mirror surfaces $\mathbf{X}_{M_i}$. The 3D scene point $\mathbf{S}$ is given as the midpoint of the smallest distance [9] between the two rays.

In [12] we discuss the accuracy of the 3D reconstruction with two horizontally aligned cameras i.e. Camera 1 and 2 in Fig. 1. We demonstrate that the accuracy depends on the position of the scene point $\mathbf{S}(r, \varphi, \theta)$ given in polar coordinates. The accuracy decreases from $\varphi = 0°$ to $\varphi = 90°$ with 0° being behind the two cameras and 90° being on the right side of the cameras as shown in Fig. 1. In Fig. 2 the dependency of the 3D reconstruction from the angle φ and the accuracy are displayed. The red function shows the normalized mean Euclidean error μ_d for the reconstruction of a circular simulated point cloud. The points are located in a circle with constant height around the two aligned catadioptric cameras with a radius of 8 m. Consequently, the reconstruction error is maximal at $\varphi = 90°$ and $\varphi = 270°$ which is along the baseline of the camera combination.

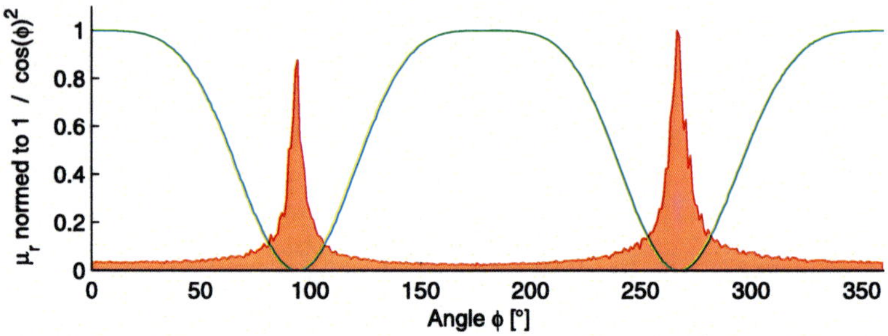

Fig. 2. Normalized mean Euclidean error μ_d as a function of angle φ of the 3D scene point (red) and the weight function $cos^2(\varphi)$ (green)

3 Trifocal Vision

3.1 Camera Setup

In our approach we use three catadioptric cameras arranged in a plane parallel to the ground in a triangle as shown in Fig. 1. The cameras are mounted on top of the vehicle, two in the front on the left and right side and the third one centered in the rear. With this camera configuration all points around the vehicle are seen by at least two cameras. Moreover, points which lie on the baseline from two cameras and can not be reconstructed with the two cameras are visible by the third camera, if not covered by another object. Thus, every 3D point around the vehicle can be calculated out of the image points according to the best positioned cameras with a sufficient accuracy. In addition this camera configuration with the cameras mounted at the edges of the vehicle has the advantage that also small objects next to the vehicle can be seen with at least one camera.

3.2 3D Reconstruction

First of all, for the 3D reconstruction out of three single catadioptric cameras we calculate the corresponding 3D intersection points S_{12}, S_{23} and S_{31} out of two cameras between all camera combinations as shown in Fig. 1. The straight forward approach is to determine $\mathbf{S} = \frac{1}{3}\sum \mathbf{S}_n$ with $n \in \{12,13,23\} =: N$ as the average of all three intersection points. Obviously, this computation is not advantageous due to the varying accuracy of the pairwise 3D reonstruction as shown in Fig. 2.

To establish the best 3D point out of the intersection points $\mathbf{S_n}$ we evaluate two different methods. Our first method calculates the best intersection point $\mathbf{S}$ by a sum over all intersection points each multiplied by a window function $w_n(\varphi)$ depending on the angle φ.

$$\mathbf{S} = \sum_{n \in N} w_n(\varphi) \cdot \mathbf{S}_n \tag{5}$$

with the following window functions $w_n(\varphi)$ for all three camera combinations.

$$w_{12}(\varphi) = \begin{cases} 1, & \text{for } \varphi \in [0^\circ, 30^\circ] \cup [150^\circ, 210^\circ] \cup [330^\circ, 360^\circ] \\ 0, & \text{else} \end{cases}$$

$$w_{13}(\varphi) = \begin{cases} 1, & \text{for } \varphi \in [90^\circ, 150^\circ] \cup [270^\circ, 330^\circ] \\ 0, & \text{else} \end{cases}$$

$$w_{23}(\varphi) = \begin{cases} 1, & \text{for } \varphi \in [30^\circ, 90^\circ] \cup [210^\circ, 270^\circ] \\ 0, & \text{else} \end{cases}$$

This method uses only the best intersection points depending on the angle of the 3D point.

To use all three intersection points our second method calculates the 3D points with a weighted sum over all three points.

$$\mathbf{S} = \frac{\sum g_n \cdot \mathbf{S}_n}{\sum g_n} \tag{6}$$

The weight functions g_n depend on the angle φ_{c_n} between the pairwise translation vector connecting the cameras $\mathbf{c_n} = [c_{x_n}, c_{y_n}, c_{z_n}]$ and the angle φ of the 3D point, shown in Fig. 1.

$$\varphi_{c_n} = \arccos \frac{c_{y_n}}{\sqrt{c_{x_n}^2 + c_{y_n}^2}} \tag{7}$$

For the weight function $g_n(\|\varphi_n - \varphi_{c_n}\|)$ we use

$$g_n(\|\varphi_n - \varphi_{c_n}\|) = \cos^2(\|\varphi_n - \varphi_{c_n}\|) \tag{8}$$

as depicted in green in Fig. 2. The weight function penalizes intersection points with a small spatial angle to the baseline between the cameras and rewards points perpendicular to the baseline.

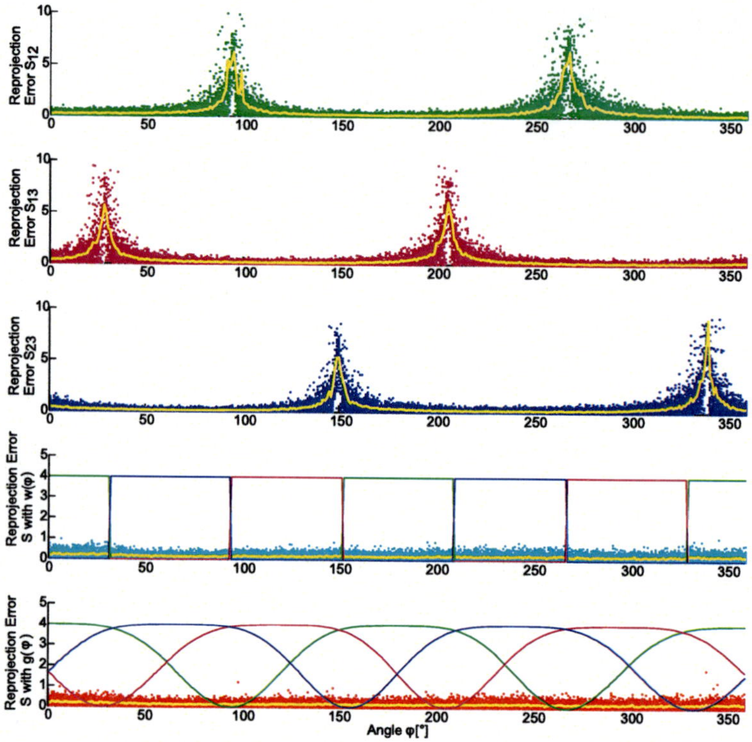

Fig. 3. Euclidean reprojection error and mean μ of the three camera combinations (S_{12} - green, S_{13} - magenta, S_{23} - blue). The fourth figure depicts the error for the 3D reconstruction calculated with Eq. 5 (cyan) and the last one the error for the 3D points calculated with Eq. 6 (red). The weight functions are shown as colored lines.

4 Results

In this section the results of our approach on a simulated 3D environment are presented. We created 3D scene points in a circle ($\varphi = 0°, 1°, ..., 360°$) around the three cameras with radius 8 m and a constant height. The cameras are located at position $[0\ 0\ 0]^T$, $[0\ 1\ 0]^T$ and $[1\ 0.5\ 0]^T$ as shown in Fig. 1. The accuracy of the 3D reconstruction is evaluated by means of a Monte-Carlo-Simulation. For this purpose the circular scene points are mapped onto the image planes of the catadioptric cameras. For each of the 360 mapped image points we created 100 samples and disturbed them with Gaussian noise with a standard deviation of 1.5 pixels. Using these disturbed image points we calculated the corresponding points on the mirror surfaces and determined the 3D intersection points by reprojection with the presented methods using Eq. 5 and Eq. 6.

The 3D reconstruction can be judged by the reprojection error which is the Euclidean distance $\|\mathbf{X} - \mathbf{S}\|$ between the reconstructed point $\mathbf{S}$ and the ground truth scene point $\mathbf{X}$. Fig. 3 shows the error for the reconstruction with two cameras for Camera 1 and 2

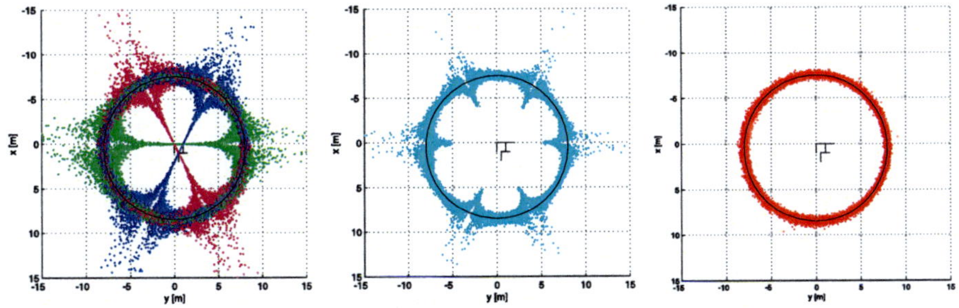

(a) Distribution for the 3D reconstruction of Camera 1 & 2 (green), Camera 1 & 3 (magenta) and Camera 2 & 3 (blue).

(b) Distribution for the 3D reconstruction calculated as average of all three intersection points.

(c) Distribution for the 3D reconstruction calculated as weighted sum of all three intersection points.

Fig. 4. 3D reconstruction for a simulated noisy circular point cloud with radius 8 m around the cameras and a constant height

(green), for Camera 1 and 3 (magenta) and for Camera 2 and 3 (blue). The dependency on the angle φ can clearly be seen. The same effect is visible in Fig. 4(a) which shows the reconstructed 3D point cloud from each camera pair in the same color. Obviously, the reconstructed points along the baselines deviate from the true scene points.

The reconstruction error for the estimated 3D point by the two computation methods are also depicted in Fig. 3. The fourth figure shows the reprojection error for the reconstruction method using Eq. 5 and the last figure displays the reprojection error for the reconstruction using Eq. 6. The weight functions $w(\varphi)$ and $g(\varphi)$ are also shown as colored lines in the figures. It can be seen that the reprojection error with both methods is smaller than for the 3D points calculated by only two cameras. Most importantly, it is independent of the angle φ.

Fig. 4 shows the reconstructed point clouds with the black circle as ground truth. In Fig. 4(b) the reprojected point cloud by the straight forward mean combination is displayed. The reconstructed point cloud by the combination of the three cameras with the weight functions (Eq. 6) is illustrated in Fig. 4(c). The reprojected point cloud with Eq. 5 looks similar to this one. It became clear that the reconstructed scene points with the weighted combinations (Eq. 5, 6) are closer to the ground truth than the mean combination of the three intersection points and are independent from the angle φ. Hence, points on all sides of the vehicle can be reconstructed with these methods with similar accuracy.

For a quantitative result we analyze the mean μ_d and standard deviation σ_d of the error shown in Tab. 1. The mean and standard deviation for the proposed methods are considerable smaller than for the intersection points calculated with two cameras. The values for the reconstruction with Eq. 6 are slightly smaller. The reconstruction with Eq. 6 also copes better with outliers because it always uses three cameras.

Table 1. Mean μ_d and standard deviation σ_d of the error

Setup	S_{12}	$\frac{1}{3}\Sigma(S_n)$	$\Sigma w_n S_n$	$\frac{\Sigma(w_n S_n)}{\Sigma g_n}$
μ_d [m]	0.5785	0.4273	0.1792	0.1738
σ_d [m]	0.8951	0.4282	0.0282	0.0247

5 Conclusion and Future Work

In this work, we have presented a new catadioptric camera system for autonomous vehicles. The proposed setup consists of three cameras aligned on a plane and arranged in a triangle on top of a vehicle. We presented two methods for the 3D reconstruction with three catadioptric cameras which achieved good results. Both methods calculate the 3D point by a weighted average, one with a window function and one with a weight function based on the cosine. We have shown that the accuracy of the 3D reconstruction with three cameras is by far better as the reprojection with two cameras and independent of the spatial angle of the expected 3D point.

Our next steps will include a validation of the 3D reconstruction with the camera setup on our experimental vehicle with real world data. Furthermore, we want to use the accurate 3D points all around the vehicle to improve the ego motion estimation and localization with catadioptric cameras. Moreover, the extrinisic calibration between the cameras can be enhanced with this 3D reconstruction.

Acknowledgement. The authors would also like to thank the Karlsruhe School of Optics and Photonics (KSOP).

References

1. Baker, S., Nayar, S.K.: A theory of single-viewpoint catadioptric image formation. International Journal of Computer Vision 35(1), 1–22 (1999)
2. Baldwin, J., Basu, A.: 3d estimation using panoramic stereo. In: 2000 Proceedings of the 15th International Conference on Pattern Recognition, vol. 1, pp. 97–100 (2000), doi:10.1109/ICPR.2000.905283
3. Benosman, R., Kang, S.E.: Panoramic vision: sensors, theory, and applications. Springer-Verlag New York, Inc., Secaucus (2001)
4. Cheng, S.Y., Trivedi, M.M.: Toward a comparative study of lane tracking using omni-directional and rectilinear images for driver assistance systems. In: ICRA 2007 Workshop: Planning, Perception and Navigation for Intelligent Vehicles (2007)
5. Ehlgen, T., Pajdla, T., Ammon, D.: Eliminating blind spots for assisted driving 9(4), 657–665 (2008)
6. Gandhi, T., Trivedi, M.: Vehicle surround capture: Survey of techniques and a novel omni-video-based approach for dynamic panoramic surround maps 7(3), 293–308 (2006)
7. Geyer, C., Daniilidis, K.: A Unifying Theory for Central Panoramic Systems and Practical Implications. In: Vernon, D. (ed.) ECCV 2000, Part II. LNCS, vol. 1843, pp. 445–461. Springer, Heidelberg (2000)

8. Gluckman, J., Nayar, S.K., Thoresz, K.J.: Real-time omnidirectional and panoramic stereo. In: Proceedings of the 1998 DARPA Image Understanding Workshop, pp. 299–303. Morgan Kaufmann (1998)
9. Hartley, R.I., Zisserman, A.: Multiple View Geometry in Computer Vision, 2nd edn. Cambridge University Press (2004) ISBN: 0521540518
10. Scaramuzza, D., Martinelli, A., Siegwart, R.: A flexible technique for accurate omnidirectional camera calibration and structure from motion. In: IEEE International Conference on Computer Vision Systems, ICVS 2006, p. 45 (2006), doi:10.1109/ICVS.2006.3
11. Scaramuzza, D., Siegwart, R.: Monocular Omnidirectional Visual Odometry for Outdoor Ground Vehicles. In: Gasteratos, A., Vincze, M., Tsotsos, J.K. (eds.) ICVS 2008. LNCS, vol. 5008, pp. 206–215. Springer, Heidelberg (2008)
12. Schönbein, M., Kitt, B., Lauer, M.: Environmental perception for intelligent vehicles using catadioptric stereo vision systems. In: Proceedings of the 5th European Conference on Mobile Robots (ECMR), Örebro, Schweden (2011)
13. Svoboda, T., Pajdla, T.: Epipolar geometry for central catadioptric cameras. Int. J. Comput. Vision 49(1), 23–37 (2002), `http://dx.doi.org/10.1023/A:1019869530073`
14. Yi, S., Ahuja, N.: An omnidirectional stereo vision system using a single camera. In: ICPR 2006: Proceedings of the 18th International Conference on Pattern Recognition, pp. 861–865. IEEE Computer Society, Washington, DC (2006), `http://dx.doi.org/10.1109/ICPR.2006.263`

Omnidirectional Vision for Indoor Spatial Layout Recovery*

J. Omedes, G. López-Nicolás, and J. J. Guerrero

Instituto de Investigación en Ingeniería de Aragón. Universidad de Zaragoza, Spain
jason.omedes@gmail.com, {gonlopez, jguerrer}@unizar.es

Abstract. In this work, we study the problem of recovering the spatial layout of a scene from a collection of lines extracted from a single indoor image. Equivalent methods for conventional cameras have been proposed in the literature, but not much work has been done about this topic using omnidirectional vision, particulary powerful to obtain the spatial layout due to its wide field of view. As the geometry of omnidirectional and conventional images is different, most of the proposed methods for standard cameras do not work and new algorithms with specific considerations are required. We first propose a new method for vanishing points (VPs) estimation and line classification for omnidirectional images. Our main contribution is a new approach for spatial layout recovery based on these extracted lines and vanishing points, combined with a set of geometrical constraints, which allow us to detect floor-wall boundaries regardless of the number of walls. In our proposal, we first make a 4 walls room hypothesis and subsequently we expand this room in order to find the best fitting. We demonstrate how we can find the floor-wall boundary of the interior of a building, even when this boundary is partially occluded by objects and show several examples of these interpretations.

1 Introduction

Indoor structure recovery from images is an easy task for humans but not that easy for computers. At the same time, it is a very useful task since knowing floor-wall boundaries can give us valuable information for navigation, motion planning, obstacle detection or 3D reconstruction.

This problem has been studied several times and still attracts the effort of many researchers to implement each time better algorithms. Most of these contributions work under the Manhattan-World assumption [1], which assumes the scene is composed of 3 main directions orthogonal to each other. Indoor environment usually satisfies this condition so is understandable this hypothesis is extensively used. Some examples are [2], that uses extracted lines and geometric reasoning to generate hypothesis and select the best fit, or [3] which represents the room as a 3D box and tries to recognize floor-wall boundary in cluttered rooms. There are also other works as [4] that uses Bayesian filtering over a set of floor-wall boundary hypotheses without the restriction

* This work was supported by the Spanish project VISPA DPI2009-14664-C02-01 and FEDER funds.

S. Lee et al. (Eds.): Intelligent Autonomous Systems 12, AISC 193, pp. 355–364.
springerlink.com

of Manhattan-World assumption, but there still being 3 main directions without the imposition of orthogonality between them.

Lately, research on omnidirectional vision is taking more importance due to the wide range of vision of these images, which helps in the detection of VPs and makes visible lines much longer. However, in central catadioptric images, straight lines from the real world become conics adding the issue of geometrical complexity, implying that many of existing algorithms for conventional images are not applicable. So it is needed to come up with new methods that take in account characteristics of this kind of images.

Here, we present our work for structure recovery from images. Starting from a single omnidirectional indoor image, we extract lines from it, classify them depending on their orientation. From this classification, we select a set of points which will lead us to generate possible wall-floor boundaries, and imposing geometrical constraints we generate a first 4 walls-room hypothesis to later on expand or not this room according on how the data is distributed.

Inspired by [5], we propose a new method. This approach is more robust since it does not rely in finding corners which often are not easy to detect, also we do not need to specify the number of walls we are looking for. In addition, it is much faster, as trying every possible combination of normals vector to classify the extracted lines or combination of corners to find the room hypothesis was high time-consuming. With our new approach we avoid all these long iterations making viable its use in a sequence of images at real-time.

2 Vanishing Point Estimation through Line Detection

The first step of our proposal begins with extraction of lines from the image. Regarding to line extraction for catadioptric systems two methods are [6] [7]. Both start using Canny edge detector and linking edge pixels. The main difference is that [7] works on the catadioptric image, where lines and VPs are extracted by RANSAC. Whereas [6] uses the unitary sphere model proposed in [8] where points from the image $\mathbf{p_I} = (X_o, Y_o, 1)$ are projected as $\mathbf{p_S} = (X_S, Y_S, Z_S)$. By doing this, each chain of pixels from a line in the image defines a *great circle* on the sphere which can be represented by its normal vector $\mathbf{n} = (n_x, n_y, n_z)$.

We use Bazin's Matlab toolbox[1], but adapting the equations from para-catadioptric $(\xi = 1)$ to hyper-catadioptric $(0 < \xi < 1)$ system in order to generalize the method for a more general mirror shape. From this point, [6] proposes to test every possible combination between pairs of normal vectors to identify main directions (1 vertical, 2 horizontal) at the same time as the 3 corresponding VPs. However, this method is time-consuming and sometimes comes up with misclassifications.

We propose a new robust and fast method to classify lines parallel to the 3 dominant directions, taking in consideration two hypotheses: a) Manhattan-world assumption [1] which states the scene is build on a cartesian grid, b) $\mathbf{Z}$ camera's axis is aligned with $\mathbf{Z}$ reference's axis of the world, since catadioptrical systems are mainly used in wheel-based robots, so planar-motion is assumed. It is easily demonstrable that under these assumptions and with perfect data, normal vectors $\mathbf{n} = (n_x, n_y, n_z)$ corresponding to

[1] `http://graphics.ethz.ch/~jebazin`

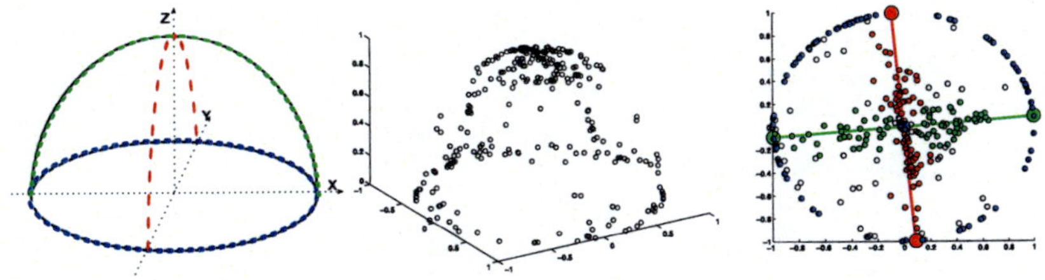

Fig. 1. In the sphere model, every line from the image is represented by its normal on the sphere. The figure represents the sphere where each point corresponds to a normal vector (Colorcode: X=Red, Y=Green, Z=Blue). From left to right: Sphere with perfect data; Sphere of a real image; classification of the previous data using our algorithm in the horizontal plane. Big dots represent VPs.

the three different directions $\mathbf{X}, \mathbf{Y}, \mathbf{Z}$ does not has its own component (i.e. line from the real world belonging to direction $\mathbf{X}$, has a normal vector $\mathbf{n}$ whose component $n_x = 0$), Fig. 1. However, often data is not perfect so it will suffer deviations from this configuration. The classification process for these normals is done as follows:

1) Under assumption $\mathbf{b}$, image lines whose normal vectors has n_z component below a threshold (experimentally we find 0.2 is a good value) are automatically classified as vertical lines, and removed for next steps.

2) Suppress n_z component of remaining normals so every $\mathbf{n} = (n_x, n_y, 0)$ will fall in a 2D plane, and using RANSAC we seek two orthogonal lines which minimize $\frac{error}{inliers}$ (with number of $inliers$ greater than a minimum). These lines will define the two horizontal main directions.

3) Image lines are labeled depending on the distance between its normal vector and one of the two main directions. It is remarkable that normal vectors whose component $n_z \simeq 1$ are conflictive as they are conics which degenerate into circles and can not be properly classified, so it is better remove them to avoid errors.

4) Finally, VPs are estimated as the points where the lines defining main directions cut the sphere at the hemisphere ($Z = 0$), see Fig. 1.

3 Hierarchical Layout Hypothesis Method

Due to noise and imperfections of real images often there is not enough information to clearly define where the floor-wall boundary is, so with the extracted lines and a set of geometric constraints we must seek the best approach to find where these boundaries are. In order to do this we generate conics (possible boundaries) from a set of points belonging to the lines previously classified.

3.1 Selection of Set of Points

The first factor to notice is that information obtained by vertical lines is more robust and less susceptible to noise than horizontal lines, which sometimes are difficult to classify

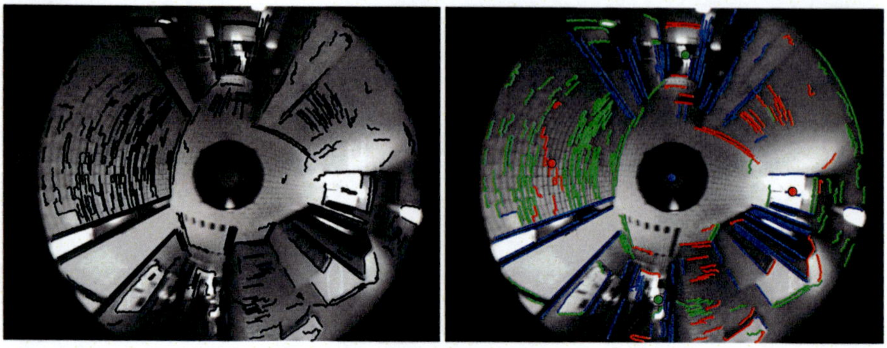

Fig. 2. Left: Lines extracted by Canny Edge detector after pruning step. Right: Same lines grouped in the 3 dominant directions according to our classification. Big dots represent VPs.

or are not well detected. Furthermore, studying typical images (see examples Fig. 6), we have noticed that in absence of objects most of vertical lines have their origin around the region that define floor-wall boundary, and if objects are present, they are standing over the floor and use to be close to the wall. So, unless these objects are placed all around the room, the origin of the lines that define them still being close to the desired boundary.

With these points of vertical lines we have to generate conics which will define possible floor-wall boundaries. In [9], it has been demonstrated how with a calibrated camera it is possible to define a conic in the image from only two points. If we apply the condition that every line in the image must pass through a vanishing point, just one point belonging to the floor-wall intersection is needed to define a conic in the image being a boundary.

Due to these 2 facts, let us denote as group G_Z the set of points composed by the closest point from each vertical to the center of the image. Additionally, we select a homogeneously distributed set of points from horizontal lines situated at the same height as the points in G_Z. Carrying out this selection for lines in X and Y direction, we obtain two more groups, G_X and G_Y, respectively. This is done in order to remove noisy horizontal segments, such as those found in objects, windows, doors,... and prevent them voting, Fig. 3(left).

3.2 Generation of Conics

Since we do not know which points of these groups are situated in the floor area, we apply RANSAC to identify the most voted conic, candidate to represent our desired boundary.

As we mentioned before, only two points are needed, one Vanishing Point and one point from the previous sets. Cross product between VP and each of these points $\mathbf{p_i}$ generates a normal vector $\mathbf{n_i}$, which defines a conic $\overline{\Omega}$ finally obtaining $\widehat{\Omega}$ after a projective transformation $\mathsf{H_C}$[10].

$$\mathbf{n_i} = \begin{pmatrix} n_{i_x} \\ n_{i_y} \\ n_{i_z} \end{pmatrix} = \begin{pmatrix} VP_x^S \\ VP_y^S \\ VP_z^S \end{pmatrix} \times \begin{pmatrix} P_{i_x}^S \\ P_{i_y}^S \\ P_{i_z}^S \end{pmatrix} \tag{1}$$

$$\overline{\boldsymbol{\Omega}_i} = \begin{bmatrix} n_{i_x}^2(1-\xi^2) - n_{i_z}^2\xi^2 & n_{i_x}n_{i_y}(1-\xi^2) & n_{i_x}n_{i_z} \\ n_{i_x}n_{i_y}(1-\xi^2) & n_{i_y}^2(1-\xi^2) - n_{i_z}^2\xi^2 & n_{i_y}n_{i_z} \\ n_{i_x}n_{i_z} & n_{i_y}n_{i_z} & n_{i_z}^2 \end{bmatrix} \tag{2}$$

$$\widehat{\boldsymbol{\Omega}_i} = \mathsf{H_C}^{-t}\overline{\boldsymbol{\Omega}_i}\mathsf{H_C}^{-1} \tag{3}$$

Now, distance between conic and every point $\mathbf{p_j}$ is computed using an approximation [9] to [11]. We compute the polar line of a point $\mathbf{p_j}$ in the conic $\widehat{\boldsymbol{\Omega}}$, calculate the perpendicular specifying it lies over $\mathbf{p_j}$. This perpendicular line intersects the conic in two points $\mathbf{q}^+$ and $\mathbf{q}^-$, the minimum Euclidean distance between $\mathbf{p_j}$ and $\mathbf{q}^+$ or $\mathbf{q}^-$ corresponds to distance from point to conic, Fig. 3.

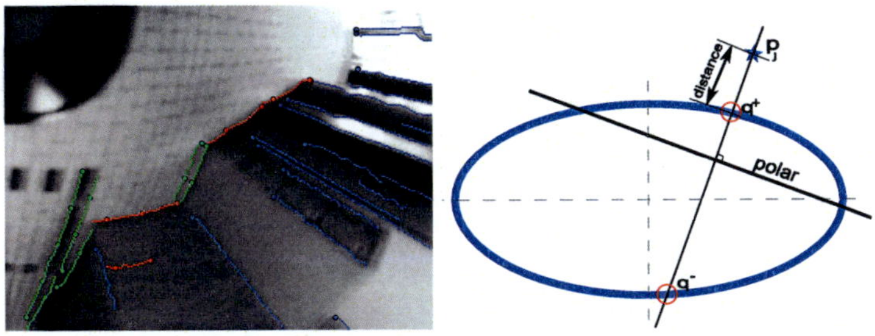

Fig. 3. Left: Selection of points as explained in Section 3.1. Right: Graphic explanation for distance measurement between point and conic in Section 3.2.

A new normal vector is estimated from the average of all points with minor distance than a threshold, and we iterate the whole process until its convergence (no more points are added). Points voting for this conic are removed from the list, and one of the remaining is chosen to generate an new conic, repeating the procedure and stopping when every point has been assigned.

3.3 Initial Boundaries Hypothesis

Computers cannot tell from a bunch of raw data how many walls a room is made of, but it is known that the most common indoor places are halls and rooms with similar shape to those shown in Fig. 4. All these geometrical shapes can be depicted by a central square with branches arising from all or some of its faces which at the same time must meet a geometric constraint: Parallel faces have to be one at each side of the imaginary line formed by joining their two corresponding VPs. This comes from the definition of vanishing point as the geometric place where parallel lines appear to converge.

Due to this constraint, the searching algorithm to extract conics (Section 3.2) is executed for four different cases, in order to find the first four boundaries (better seen in Fig. 5):

- Boundaries 1 and 3: Are sought using points of G_Z and G_X at each sides of the imaginary line defined by VPs in direction X (Fig. 5 (left)) .
- Boundaries 2 and 4: Are sought using points of G_Z and G_Y at both sides of the imaginary line defined by VPs in direction Y (Fig. 5 (center)) .

Another property is that the four vanishing points define a conic which corresponds to points situated at the same height as the camera, so every point falling within the conic might be on the floor and will be a possible candidate, while the ones out of the conic are automatically eliminated.

Remaining points are projected onto the sphere for each of the four cases previously defined, and we proceed to generate conics with these points.

Conics more voted are now selected as possible candidates and the nearest to the center of the image is chosen. We rather chose this closest line to other which might be more voted because it will have chance to be selected in the expansion process (Section 3.4), and if chosen now, it would be possible the loss of information. Once the four boundaries have been found, they are combined to conform walls and floor of our first hypothesis (Fig. 5).

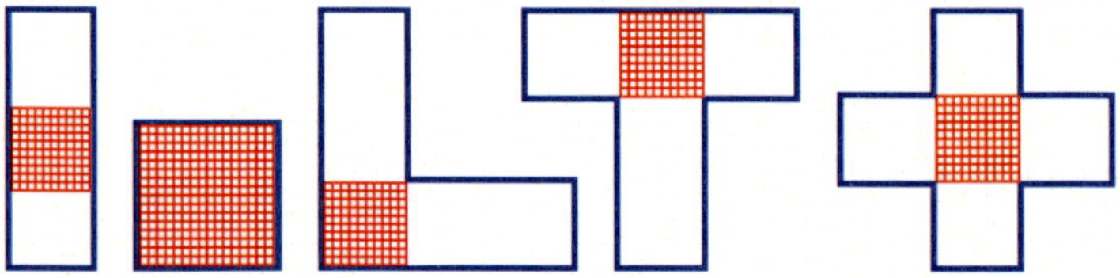

Fig. 4. Most common room/hall shapes (top view). Red grid represents the basic square we are seeking in section 3.3.

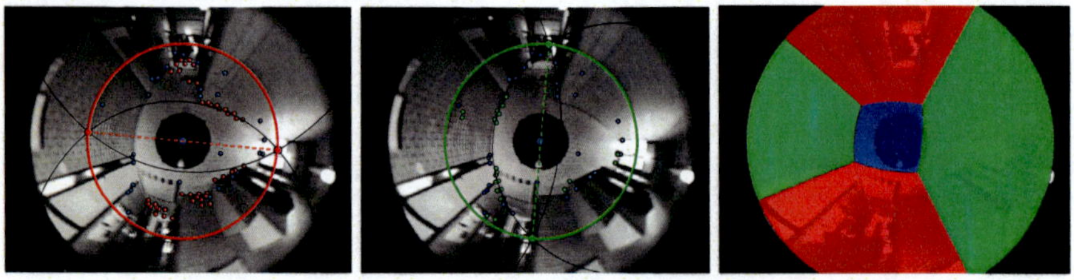

Fig. 5. First two images show points from groups G_Z, G_X and G_Y under constraints exposed in section. 3.3, where blue, red and green dots correspond to G_Z, G_X and G_Y points respectively. Dashed red and green lines are the imaginary lines, going through the VPs, which divide the image in 2 parts. Finally, black conics represent the most voted boundaries for each case. Right image shows the result of combining those boundaries to generate the first hypothesis.

3.4 Hierarchical Expansion Process

Let us denote as B_1, B_2, B_3 and B_4 the four boundaries defined in previous section 3.3. The area between those and the end of the image defines four sectors. These sectors may correspond to actual walls or may exist the possibility they can be expanded, understanding *expand* as replacing the boundary B_i for others which enlarge the area of the first-hypothesis room layout. For each of these sectors we repeat the same method described in Section 3.3, obtaining a maximum of 3 new boundaries. Let them be B_i^L, B_i^M and B_i^R in clockwise order as shown in Fig. 7.

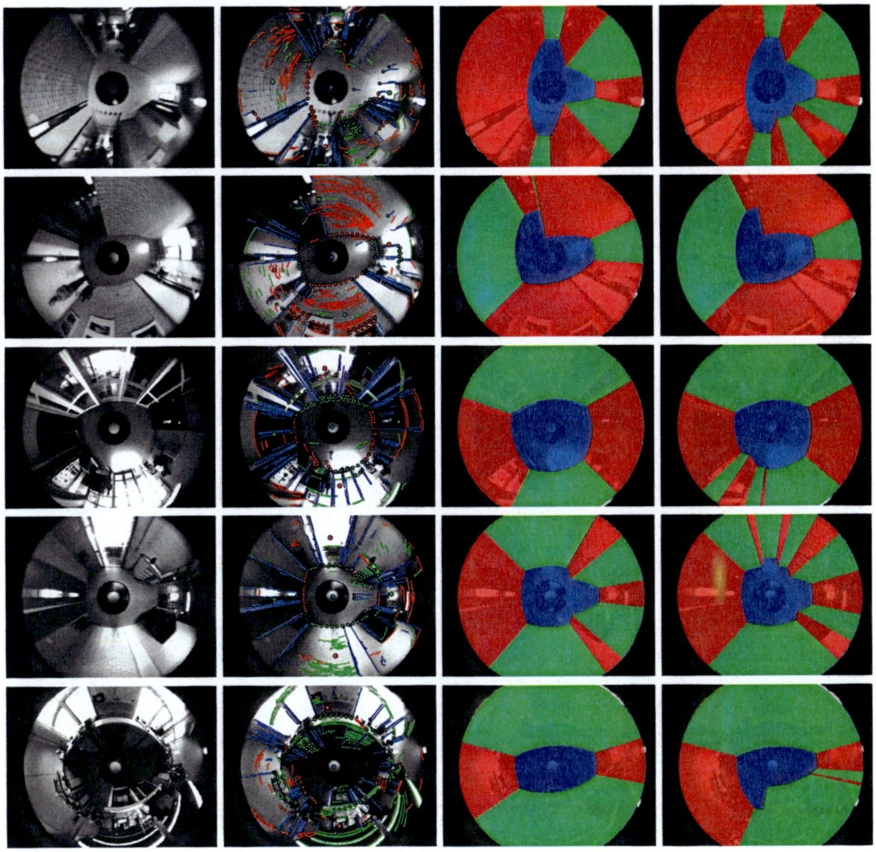

(a) (b) (c) (d)

Fig. 6. Examples of experimental results obtained for five different images. (a)Input images. (b) Line classification and extracted points which vote for boundary selection. (c) Output images by our algorithm. (d) Ground truth, manually labeled.

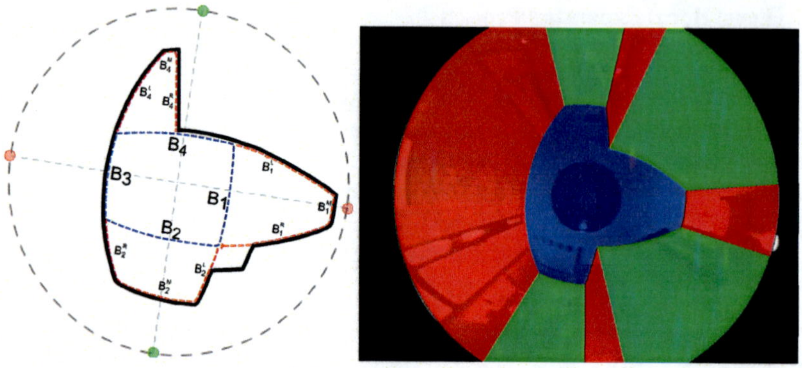

Fig. 7. Left: Synthetic example depicting the possible cases (B_1 and B_2 are expandable regions, B_3 will not be expanded, and B_4 corresponds to an occluded corner). Black line represents the actual room boundaries, first hypothesis in dashed blue, and final expansions in dashed red. Right: final result of a real example.

When looking for expansion three cases can happen:

- Enough data is available to define the 3 boundaries B_i^M, B_i^L and B_i^R; so there will be expansion in the current sector.
- B_i^M is very close to B_i, this means the most voted wall still being the same and will not be expansion.
- Data only allow us to find 1 or 2 boundaries. This last case can be originated for different situations and should be studied.

Third case is present when we lack of data, caused by lines not detected or by an occluded corner (Fig. 7(left)). Both cases imply expanding floor area but results are completely different, therefore care must be taken.

If missing boundary is a lateral (B_i^L, B_i^R) or lateral plus middle (B_i^M), and points from well-detected boundaries only fall at one side from the VP within the current sector, this is due to an occluded corner. Thus the missing border is defined as a radial line through the center of the image and the point, belonging to the well-detected boundary, whose angle is the closest to the angle defined by the VP.

On the other hand, if previous conditions are not satisfied, we assume some line was not detected, hence if the missing boundary is any of B_i^L or B_i^R, it will be defined as the resulting conic passing through its corresponding VP and the last point belonging to B_i^M. By contrast, if the missing edge is B_i^M, we consider it should be no expansion except if there are a relevant number of voting points in the lateral boundaries B_i^L and B_i^R.

4 Results

Our experiments have been performed using Matlab, running at 3 sec per frame, and with the dataset COGNIRON composed of indoor images (768×1024) with a wide variety of rooms. These images were taken by a camera with a hyperbolic mirror spotted on a mobile robot. The calibration of the camera is also available online [12].

We show some of our results in different kind of indoor situations Fig. 6. First two examples correspond to T and L shape halls (like the ones shown in Fig. 4), walls are not too saturated with objects so the result is accurate. In this second example we also observe an occluded corner at the superior part of the image. Third picture is taken in a room where walls are made of glass (top and bottom of the image); due to these walls very bright areas appear in the scene, but we still achieve a good approximation of its structure.

Forth case shows a hall with a desk and a shelf, where our algorithm is able to recognize these obstacles. However, it does not detect the open door situated at the top part of the image, probably due to all the light going through it.

Last scene correspond to a room with many objects, colors are very dark, which makes difficult line extraction at some areas. At the same time, most of the longest detected lines fall over the objects, what might lead to misclassify wall-floor boundaries, but as we can see, we are still achieving good results.

Comparing results from our algorithm with their respective ground truth, we define as true positives (tp) the number of pixels both have in common, false positives (fp) the number of pixels identified as floor by our method but do not correspond to the floor in the ground truth, and false negatives (fn) the number of pixels identified as not floor when they result to be floor in the ground truth. With these values we compute **precision** ($\frac{tp}{tp+fp}$), **recall** ($\frac{tp}{tp+fn}$) and $\mathbf{F_1}$ ($\frac{2\,precision\,recall}{precision+recall}$) for several images, Table 1.

Table 1. Performance values obtained for images of Fig. 6

	Image1	Image2	Image3	Image4	Image5
Precision	0.973	0.984	0.896	0.964	0.904
Recall	0.887	0.969	0.992	0.937	0.878
F_1	0.928	0.977	0.942	0.950	0.891

5 Conclusion and Future Work

We have proposed a new method to extract vanishing points from an omnidirectional image and to perform line classification. Since for the database of images the Z axis from camera and world reference were aligned, our proposal have been designed with this constraint, but we believe this can be extended for the general case where the alignment of axis is not known. We also have proposed a new simple and robust method for scene layout recovery and we have shown its performance in experimental results. This is useful in many applications, since knowing where floor-wall boundary are and the height of the camera, we can known exactly where every point of the scene is. Currently we are working into spread out this method for a whole sequence of images in order to improve accuracy in those images with possible misclassifications.

References

[1] Coughlan, J.M., Yuille, A.L.: Manhattan world: Compass direction from a single image by bayesian inference. In: Int. Conf. on Computer Vision, pp. 941–947 (1999)
[2] Lee, D., Hebert, M., Kanade, T.: Geometric reasoning for single image structure recovery. In: IEEE Conference on Computer Vision and Pattern Recognition, pp. 2136–2143 (June 2009)
[3] Hedau, V., Hoiem, D., Forsyth, D.: Recovering the spatial layout of cluttered rooms. In: IEEE International Conference on Computer Vision, pp. 1849–1856 (2009)
[4] Tsai, G., Xu, C., Liu, J., Kuipers, B.: Real-time indoor scene understanding using bayesian filtering with motion cues. In: ICCV, pp. 121–128 (2011)
[5] Ozisik, N.D., López-Nicolás, G., Guerrero, J.J.: Scene structure recovery from a single omnidirectional image. In: ICCV Workshops, pp. 359–366 (2011)
[6] Bazin, J.C., Kweon, I., Demonceaux, C., Vasseur, P.: A robust top-down approach for rotation estimation and vanishing points extraction by catadioptric vision in urban environment. In: IEEE/RSJ International Conference on Intelligent Robots and Systems, pp. 346–353 (2008)
[7] Bermudez, J., Puig, L., Guerrero, J.J.: Line extraction in central hyper-catadioptric systems. In: OMNIVIS (2010)
[8] Geyer, C., Daniilidis, K.: A Unifying Theory for Central Panoramic Systems and Practical Implications. In: Vernon, D. (ed.) ECCV 2000, Part II. LNCS, vol. 1843, pp. 445–461. Springer, Heidelberg (2000)
[9] Bermudez-Cameo, J., Puig, L., Guerrero, J.J.: Hypercatadioptric line images for 3d orientation and image rectification. Robotics and Autonomous Systems 60(6), 755–768 (2012)
[10] Barreto, J.: General central projection systems: Modeling, calibration and visual servoing. Ph.D. dissertation (2003)
[11] Sturm, P., Gargallo, P.: Conic fitting using the geometric distance. In: Proceedings of the Asian Conference on Computer Vision, Tokyo, Japan, pp. 784–795 (2007)
[12] Zivkovic, Z., Booij, O., Krose, B.: From images to rooms. Robotics and Autonomous Systems 55(5), 411–418 (2007)

EPI Analysis of Fish-Eye Images

Kenji Terabayashi[1], Toru Morita[2], Hiroya Okamoto[3], Takaaki Oiwa[1], and Kazunori Umeda[3]

[1] Shizuoka University, 3-5-1 Johoku, Naka-ku, Hamamatsu-shi, Shizuoka 432-8561, Japan
tera@eng.shizuoka.ac.jp
[2] Sony Corporation, 2-15-3 Konan, Minato-ku, Tokyo 108-6201, Japan
Toru.Morita@jp.sony.com
[3] Chuo University, 1-13-27 Kasuga, Bunkyo-ku, Tokyo 112-8551, Japan
umeda@mech.chuo-u.ac.jp

Abstract. This paper proposes a method to get depth information from image sequences obtained from a moving fish-eye camera. The motion is assumed to be along the optical axis of the fish-eye camera. In this case, a point on a fish-eye image moves to radial direction. This is one of the epipolar constraints that are very effective to find corresponding points in an image sequences. Epipolar-plane image (EPI) is an image having epipolar constraint in a volume of accumulating an image sequence. The gradient of a curve on EPI has depth information of the corresponding measuring point. Measurement accuracy of the proposed method is examined over a wide-angle range.

1 Introduction

Wide-area sensing at one time is important for intelligent mobile systems such as mobile robots and intelligent cars. These systems often use wide-field cameras to sense surrounding environments while keeping them simple and cheap. One typical example of these is a visual support system for a driver. This has been developed using fish-eye cameras with large field of view to provide birds-eye view over a car to the driver [1,2].

In addition to visual support use, fish-eye cameras are used for taking 3D information. Stereo systems using two fish-eye cameras have been discussed about calibration and image rectification for stereo matching [3–5]. Stereo matching strategy of fish-eye images has been also discussed in a particular situation of actual forest environments [6].

Omnidirectional camera similarly has wide field, and is used for acquiring 3D structure of surrounding environments [7–10]. The wide-field feature is typically given by two ways; (i) swiveling a camera around the vertical axis [7,8], (ii) using a catadioptric system [9–12]. The first way is not suitable for 3D measurement from a mobile platform because the camera position changes before capturing entire omnidirectional images. Omnidirectional images captured from the second way include the camera lens at the image center. This implies that the installation position of an omnidirectional camera is limited to a particular area such as over a mobile system. In contrast to omnidirectional cameras, fish-eye cameras can capture a wide-field image at a time without an inclusion of the camera lens on the image. Consequently, this study focuses on a fish-eye camera to measure the 3D structure around an intelligent mobile system.

S. Lee et al. (Eds.): Intelligent Autonomous Systems 12, AISC 193, pp. 365–373.
springerlink.com

To obtain 3D information from image sequences of video data captured by a moving camera, there are two typical approaches [13]; (i) epipolar-plane image (EPI) analysis [14–17], (ii) factorization method [18]. The EPI analysis recovers depth information from known motion of a camera, in contrast to the factorization method estimating both of scene geometry and camera motion. It is usually difficult to extract stable feature points in an image sequence for the factorization method due to occlusion, noise, and complicated texture. In the EPI analysis, assumptions of camera motion such as constant speed and straight movement simplify the correspondence problem, and bring robustness of the matching between sequential images.

We focus on the EPI analysis instead of the factorization method because motion information of intelligent mobile systems is often available from odometry or other sensors. The EPI analysis using wide-field cameras has not been investigated except omnidirectional camera [16, 17].

In this paper, we propose a method to get depth information of static environment from fish-eye images based on the EPI analysis. The proposed method is on the assumption that an image sequence for the EPI analysis is captured from a fish-eye camera moving along the optical axis at a constant velocity. In this case, objects passing through the side of an intelligent mobile system can be seen for longer time by a fish-eye camera on the system than a pinhole camera of perspective projection. This leads to wide-area sensing of depth information of a static environment around the mobile system.

2 EPI Analysis of Fish-Eye Images for Depth Sensing

2.1 Fish-Eye Camera Model

Fish-eye lenses are designed to be radially symmetric, in particular, usually obeying one of the following projections [19]:

$$r = 2f\tan(\theta/2) \quad \text{(stereographic projection)}, \tag{1}$$
$$r = f\theta \quad \text{(equidistance projection)}, \tag{2}$$
$$r = 2f\sin(\theta/2) \quad \text{(equisolid angle projection)}, \tag{3}$$
$$r = f\sin\theta \quad \text{(orthogonal projection)}, \tag{4}$$

where θ is the angle between the optical axis and the incoming ray, r is the distance on the image plane between the image point and the focal one, and f is the focal length (Fig.1).

Taylor expansion of these projection model can be expressed by the following equation:

$$r = k_1\theta + k_3\theta^3 + k_5\theta^5 + k_7\theta^7 + k_9\theta^9 + \cdots, \tag{5}$$

where parameters k_i $(i = 1, 3, 5, \cdots)$ are appropriate constant values for fitting (5) into an actual projection almost obeying (1) – (4). The equidistance projection (2) is equivalent to the first-order approximation of (5), and is the most common model in these projections.

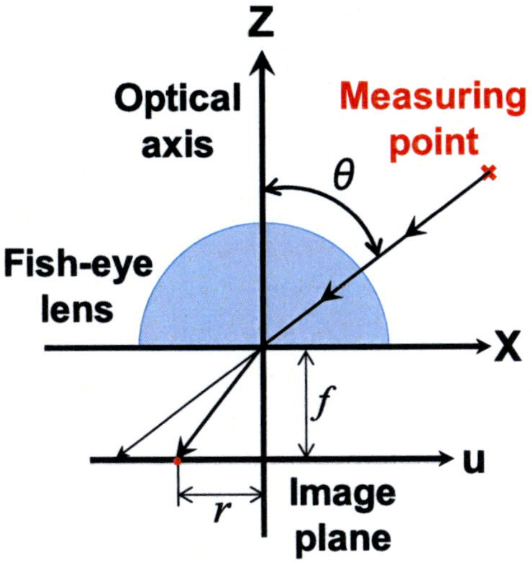

Fig. 1. Image projection of fish-eye camera on the cross-section view including the optical axis

2.2 EPI Analysis of Fish-Eye Images

If a fish-eye camera moves along the optical axis Z with constant velocity v_c and captures an image sequence with constant frame rate f_c as shown in Fig.2, the geometric relationship between the camera and a measuring point (x_m, z_m) is

$$\theta_n = \tan^{-1} \frac{x_m}{z_m - \frac{v_c}{f_c} n} \tag{6}$$

where θ_n is the angle between the optical axis and the direction of the measuring point and n is the number of steps of the camera motion. The accumulation of captured image sequences is called spatio-temporal image (STI). That motion assumption makes epipolar-plane constraint in STI. This epipolar plane includes the optical axis.

In case of the equidistance projection (2), a measuring point with direction angle θ_n is imaged on the image plane with the distance r_n from the focal point:

$$r_n = f\theta_n. \tag{7}$$

The pixel position of the imaged point (Fig.1) is

$$u_n = \frac{r_n}{w} \tag{8}$$

where w is the size of a pixel.

Epipolar-plane image (EPI) $I(u,n)$ is an image on the epipolar plane in STI. On the above camera condition, a measuring point makes a curve on EPI:

$$n = \frac{f_c}{v_c}\left(z_m - \frac{x_m}{\tan \frac{wu_n}{f}}\right) \tag{9}$$

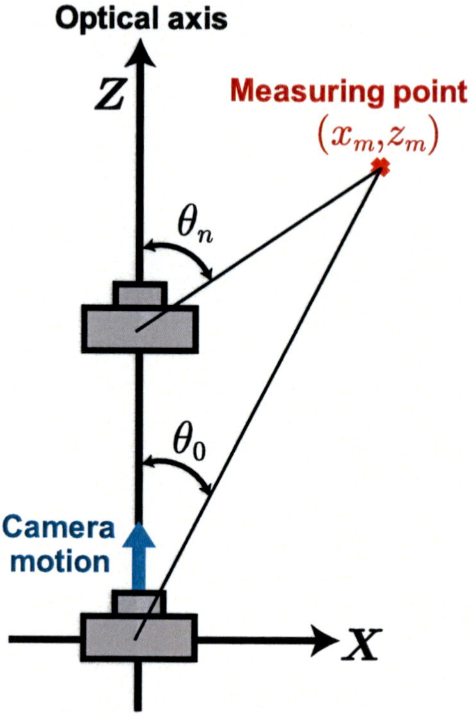

Fig. 2. Motion direction of fish-eye camera

which is from (6), (7), and (8). The differential of n with respect to u_n is

$$\frac{dn}{du_n} = \frac{f_c w}{v_c f} \frac{x_m}{\sin^2 \frac{wu_n}{f}}. \tag{10}$$

A point (x_m, z_m) can be measured by using gradient information of curves on EPI:

$$x_m = \frac{v_c f}{f_c w} \sin^2 \frac{wu_n}{f} \frac{dn}{du_n}, \tag{11}$$

$$z_m = \frac{v_c}{f_c}(n + \frac{f}{2w} \sin \frac{2wu_n}{f} \frac{dn}{du_n}). \tag{12}$$

2.3 Implementation of EPI Analysis of Fish-Eye Images for Depth Sensing

EPI is generated from an image sequence captured by a fish-eye camera. On the EPI, edges are detected by Canny operator [20]. Then, detected edges are categorized as connected components that are used to estimate gradient information of curves on EPI. Using this gradient, depth information can be calculated based on (11) and (12).

3 Experiment

Measurement accuracy of the proposed method is examined for car applications that require depth sensing of surrounding environments.

3.1 Conditions

The layout of this experiment is shown in Fig.3. The conditions θ of measuring direction are 15, 30, 45, 60, and 75 deg. The conditions d of distance between the optical axis and measuring points are 200, 250, 300, 350, 400, 450, and 500 mm. The image-capturing pitch length $\delta_z (= v_c/f_c)$ is $d/50$ mm, for example, 10 mm in condition of $d = 500$ mm.

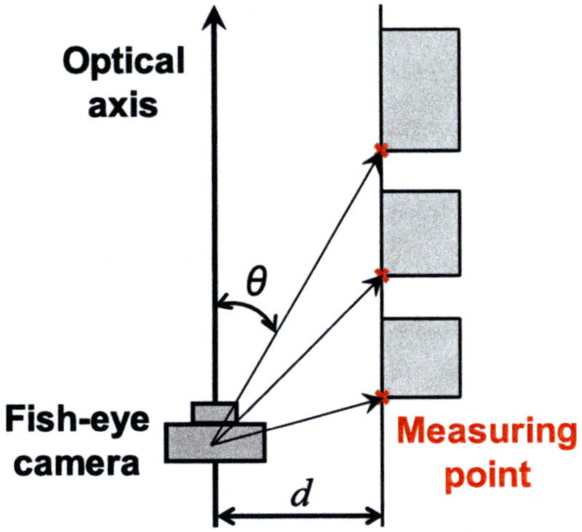

Fig. 3. Layout of a fish-eye camera and measuring points for evaluation of measurement accuracy

Fig. 4. Experiment system

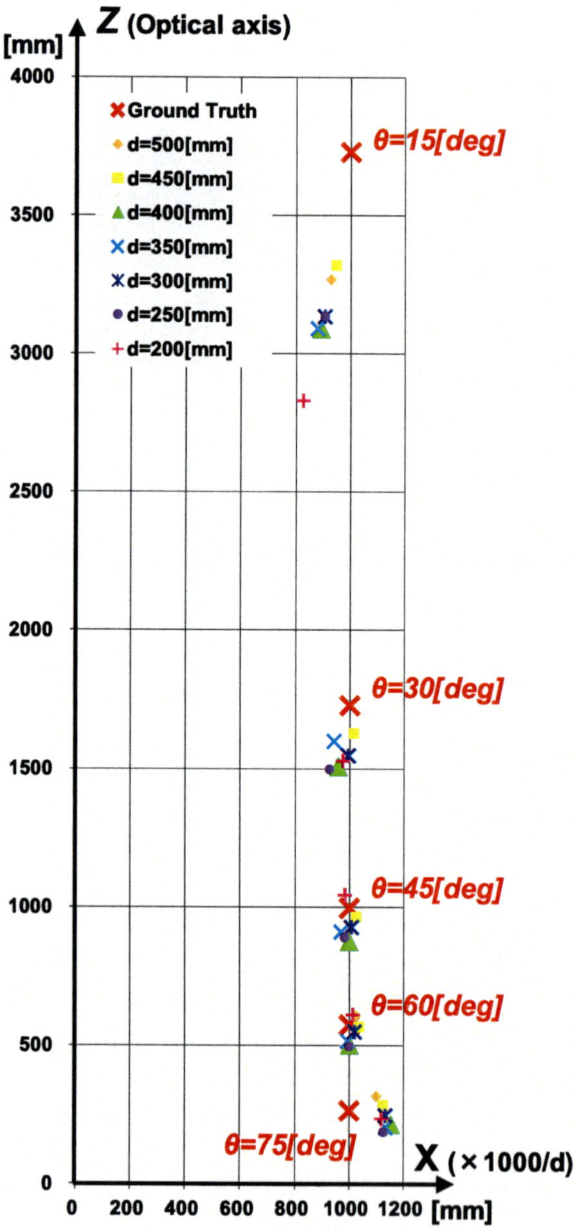

Fig. 5. Measured points and ground truth in overhead view

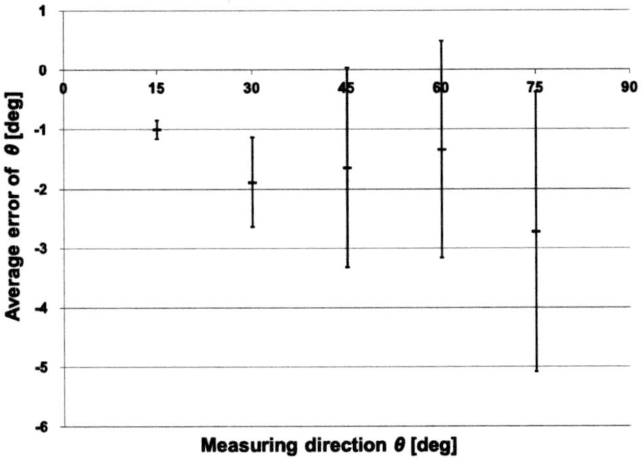

(a) Measurement accuracy of θ

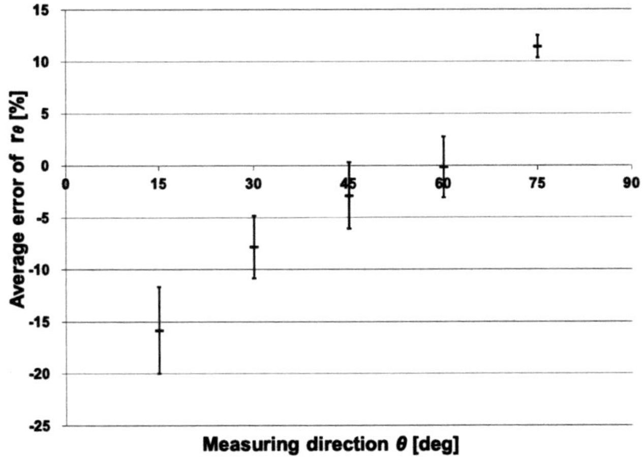

(b) Measurement accuracy of r_θ

Fig. 6. Measurement accuracy of average error with standard deviation

The experiment system is shown in Fig.4. The fish-eye lens is TV1634 (SPACE Inc.). The nominal focal length is $f = 1.6$ mm. The camera is Dragonfly2 (Point Grey Research Inc.). The size of a image pixel is $w = 4.65$ μm. The image size is XGA (1024 $\times$ 768 pixel).

3.2 Results

The comparison between measured points and ground truth is shown in Fig.5. Each measured point are normalized to be $d = 1000$ mm. In this figure, red crosses represent ground truth. The proposed method can measure wide-area points with directional angle $\theta = 75$ deg.

Measurement errors are summarized in Fig.6: Fig.6 (a) is directional error, Fig.6 (b) is radial error.

Measurement accuracy in conditions of $\theta = 45$, 60 deg is higher than one in other conditions. In Fig.6 (a), standard deviation of directional error increases with increase in directional angle θ. This is mainly because spatial resolution is lower in wider-angle area. In Fig.6 (b), standard deviation of radial error increases with decrease in directional angle θ. This is mainly because a measuring point with small directional angle is far from the camera and is close to motion direction of the optical axis. In addition, gradient estimation at this area is not good due to pixel-accuracy edge detection.

4 Conclusion

This paper proposes EPI analysis of fish-eye image sequences for depth acquisition. This method can measure wide-area points with directional angle $\theta = 75$ deg.

In this method, there are three assumptions: (i) straight camera motion along the optical axis, (ii) constant-interval image capture, (iii) equidistance projection. These are reasonable assumption in car applications.

Measurement accuracy will be improved by modifying gradient estimation and fish-eye projection model.

References

1. Around view monitor, `http://www.nissan-global.com/EN/TECHNOLOGY/OVERVIEW/avm.html`
2. Shigang, L.: IEEE Transactions on Intelligent Transportation Systems 7(4), 541 (2006)
3. Abraham, S., Forstner, W.: ISPRS Journal of Photogrammetry and Remote Sensing 59(5), 278 (2005)
4. Shigang, L.: IEEE Transactions on Intelligent Transportation Systems 9(4), 589 (2008)
5. Heller, J., Pajdla, T.: IEEE Conference on Computer Vision and Pattern Recognition, pp. 1414–1421 (2009)
6. Herrera, P.J., Pajares, G., Guijarro, M., Ruz, J.J., Cruz, J.M.: Sensors 11(2), 1756 (2011)
7. Ishiguro, H., Yamamoto, M., Tsuji, S.: IEEE Transactions on Pattern Analysis and Machine Intelligence 14(2), 257 (1992)
8. Kang, S.B., Szeliski, R.: International Journal of Computer Vision 25(2), 167 (1997)
9. Liancheng, S., Chuanjiang, L., Feng, Z.: 2006 IEEE International Conference on Information Acquisition, pp. 48–52 (2006)
10. Lhuillier, M.: Computer Vision and Image Understanding 109(2), 186 (2008)
11. Yagi, Y., Kawato, S., Tsuji, S.: IEEE Transactions on Robotics and Automation 10(1), 11 (1994)
12. Nayar, S.K.: 1997 IEEE Computer Society Conference on Computer Vision and Pattern Recognition, pp. 482–488 (1997)
13. Yokoya, N., Shakunaga, T., Kanbara, M.: IEICE Transactions on Information and Systems E82D(3), 523 (1999)
14. Bolles, R.C., Baker, H.H., Marimont, D.H.: International Journal of Computer Vision 1(1), 7 (1987)

15. Baker, H.H., Bolles, R.C.: International Journal of Computer Vision. 3(1), 33 (1989)
16. Kawasaki, H., Ikeuchi, K., Sakauchi, M.: 15th International Conference on Pattern Recognition, vol. 1, pp. 379–383 (2000)
17. Kawasaki, H., Ikeuchi, K., Sakauchi, M.: IEEE Conference on Computer Vision and Pattern Recognition, vol. 2, pp. 577–584 (2000)
18. Tomasi, C., Kanade, T.: International Journal of Computer Vision 9(2), 137 (1992)
19. Kannala, J., Brandt, S.S.: IEEE Transactions on Pattern Analysis and Machine Intelligence 28(8), 1335 (2006)
20. Canny, J.: IEEE Transactions on Pattern Analysis and Machine Intelligence 8(6), 679 (1986)

An Adaptive Fuzzy Based System for Time Critical Real World Applications

Christoph Kattmann[1], Oliver Zweigle[2], Kai Häussermann[2], and Paul Levi[2]

[1] IEH - University of Stuttgart, Pfaffenwaldring 47, 70569 Stuttgart
christoph.kattmann@ieh.uni-stuttgart.de
[2] IPVS - University of Stuttgart, Universitätsstr. 38, 70569 Stuttgart
zweigle@ipvs.uni-stuttgart.de

Abstract. From a control theory point of view, robotics and artificial intelligence offer exceptionally complex problems. Often the examined control systems have a high number of inputs and outputs, show a black-box behaviour and can not be systematically analyzed due to a vague output and long reaction times. Adaptive behaviour control and enhancement during runtime of robots moving with high speed is such a problem, with the added requirement of real-time capability. In this paper, an adaptive pre-calculated fuzzy system is proposed as a possible solution for this task. The basic structure, construction process and adaption mechanism are described, furthermore the runtime for various dimensions is benchmarked as efficiency aspects are a major contribution of the approach.

1 Introduction

Robotics and Artificial Intelligence have been vibrant fields of research in the past years. Humans are still very much superior to robots concerning the ability to master complex, high dimensional control situations with an uncertain and fuzzy input, as well as on the fields of autonomous adaptation and learning. These are necessary capabilities for an algorithm, which controls the behavior of a robot that interacts in a highly dynamical real world environment and is navigating with a high speed. In this paper we introduce a method for the automatic efficient adaption of robot behaviors using a pre-calculated fuzzy system in combination with an automatic parameter adaption during runtime in a real-time based system. It includes the consideration of experiences of the robot, based on data collected during runtime. The method can later also be applied to adapt strategic team decisions in a multi-robot system.

For evaluation purposes, the RoboCup Middle Size soccer scenario is used. In this scenario teams of five autonomous robots play soccer against each other on a field of 18x12m. The robots reach top speeds of up to 5,5m/s. This demands for fast reactions and adaption capabilities to the environment. During a RoboCup match human intervention is not allowed. The only possibility to get external information, e.g. about the game state is to interpret the commands given by the referee and fuse this information with the sensor information of the robots. As there is very little time to process data (e.g. because of the high-speed of the robots) the efficiency of the proposed framework is a major issue and demands for a very specialized solution.

S. Lee et al. (Eds.): Intelligent Autonomous Systems 12, AISC 193, pp. 375–386.
springerlink.com

Based on two examples from the RoboCup we clearify the problem statement:

- *Shoot adaption:* the robot is shooting toward the opposing goal using parameters for speed, shoot strength and shoot angle. Every time the robt shoots a goal the referee event "Goal" is triggered. If the robot misses the goal there will be an event "Goal Kick". In combination with the position information of the ball this knowledge can be interpreted as "Left of the goal" or "Right of the goal" and can be used as an input parameter for the online adaption of the shoot parameters.
- *Strategy adaption:* a team of robots uses a random strategy to attack the opponent goal (e.g. by searching the shortest path without obstacles). During this attack strategy a goal is scored every time the team attacks over the left side. Consequently the strategy is adapted to preferentially use the left side.

For the desired system that is modeled as a controller consisting of a pre-calculated fuzzy system and an adaption algorithm, the input is provided by the sensors of the robot, mainly an omnidirectional camera on top of the robot and by virtual sensors that interpret the referee events during a game. The output is a set of parameters which control the robots behavior, e.g. the shooting mechanism or the team strategy.

Generally the proposed control system is usable for all kinds of different behavior control tasks and is not limited to the RoboCup scenario. We only use this as a concrete example.

2 Related Work

Up to now mostly algorithms that are exactly tailored to the scenario and that don't allow for any autonomous adaption processes are used for the behavior control of robots, e.g. [1], [2], [3], [4]. In those cases the robots have certain pre-parameterized behaviors that adapt to a set of environmental changes (e.g. obstacles etc.) using different methods. This does not include the enhancement of behaviors based on experiences the robot gained during the execution of the behavior.

There are established theoretic concepts for robust, non-linear controllers with adaptation and learning algorithms for highly complex tasks, mainly soft-computing techniques like artificial neural nets [5], fuzzy systems [6] or bayesian networks [7], but unmodified, none of them meets all the prerequisites described before. A fuzzy system based on a Mamdani controller [6] is able to master non-deterministic control tasks with uncertain inputs. It can be easily set up with the definition of fuzzification and defuzzification maps and a set of rules. However, there are no clear, indisputable adaption algorithms for fuzzy controllers, and the execution time is usually too high to be used unmodified in a real-time environment except for simpler systems.

An artificial neural net is a tool for system identification for highly complex tasks and usually very fast to execute. However, it is hard to setup when the control system (plant) is not completely available and real-life samples for the essential backtracking learning algorithms are hard to come by and are not very numerous. The weakness of artificial neural nets, the elaborate construction process via training using input-output pairs is critical in a highly dynamical and time critical environment when it comes to the control of general behavior with few of these samples available.

Bellas et al. [8] introduce an approach for creating and adapting a model of an unknown control system (plant), but their "Evolutionary Multilevel Darwinist Brain" certainly does not meet the runtime requirements. Jian et al. [9] have developed a fuzzy controller implementation for VHDL with extensive parallelization, thus speeding it up substantially, but without specialized hardware their approach is not applicable in the proposed scenario of this work.

3 Robot Behavior Adaption as a Control Challenge

Mapping a set of sensor input parameters to a set of output parameters that control the robot's behavior is a control problem. However, the methods of classical control theory fail in a highly dynamic scenario that has to deal with uncertain sensor data and uncertain situation knowledge. For the design of a controller for this kind of process, different methods and even new controller structures have to be applied.

The system that has to be controlled contains not only the robot, but in a sense the entire environmental scenario, as the input parameters are translated to a certain behavior, which influences the environment and causes a response from other robots in the system, e.g. the opponents and the own team in the RoboCup scenario. Figure 1 shows the typical elements of the controlled process as subsystems. The environmental situation detected by the sensors of the robot results in the output of the plant after certain processing steps (e.g. data fusion). Those output variables are inflicted with uncertainty. All sensor imprecisions and simplifications during the processing are carried over to the input of the controller.

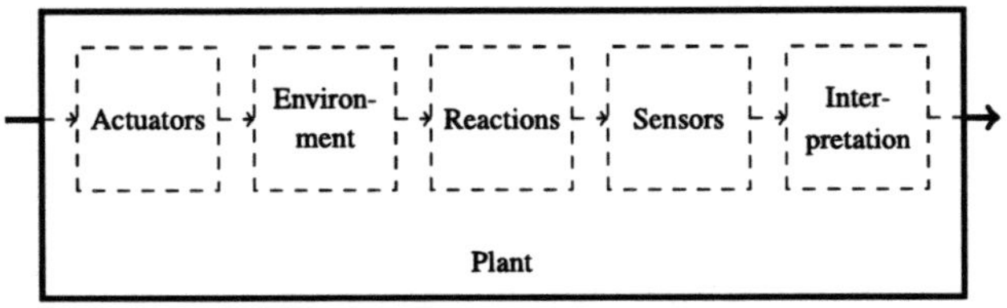

Fig. 1. Elements of the controlled process as subsystems

The internal processes of the control system (plant) are almost completely hidden and show black-box-behavior. It is practically impossible to model and simulate the system in order to use methods like PID or model-predictive control or to test out the controller in a lab environment as the scenario can never be completely predicted and is influenced by highly dynamic parameters. Furthermore it is time variant. Considering the RoboCup scenario as an example the dynamic changes of the environment caused by the opponent influences and changes the system at least to some extent.

Considering the precondition of applying the controller in a time critical application, the runtime needs to be short enough so that the system can be integrated in robots software and executed in the normal cycle of operations, which is in the case of the

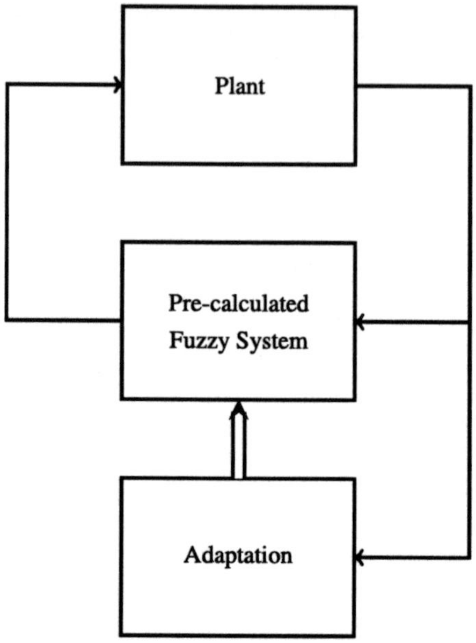

Fig. 2. The control look with the adaption mechanism. The adaption module does not influence the plant, but only changes the controller based on the outputs of the plant, which allow an evaluation of the controllers performance. Generally, this holds true for the referee events, but other measured values may provide insight on flawed controller behavior.

RoboCup team of the University of Stuttgart 30ms. This demands for a high performance algorithm, for the adaption process.

To summarize, all the mentioned difficulties call for a controller, which can deal with such a non-deterministic plant, has adaption and learning capabilities, is exceptionally robust to uncertainties both in the inputs and the system itself and is fast enough to execute its task nearly in real-time.

For this purpose we are introducing an extended pre-calculated fuzzy system (see section 4) tailored to the requirements of the scenario with a super ordinated adaption loop (see section 5) that grades the systems success and accordingly performs an optimization on the parameter set that defines the pre-calculated fuzzy system. This way, the advantages of a fuzzy system, the easy initialization of a system with complex non-linear behavior and the robustness, is combined with the computation speed and the comprehensive optimization algorithms of a mapping system that is essentially a adaptive MIMO P-controller. The plant, the pre-calculated system and the adaption mechanism and its relationship is shown in figure 2. The system is designed in order to allow the fuzzy controller to correct existing parameters used for controlling robot behaviors. In the proposed RoboCup scenario there already exist parameter sets for behaviors that can be reused. The obtained values of the fuzzy controller and the predefined values are added up. Consequently those predefined parameter sets become part of the new control system (plant).

4 A Pre-calculated Fuzzy System

One method to achieve a shorter runtime for fuzzy systems is to pre-calculate them. This approach has already been described in [10] and implemented in [11], [12]. In these papers part of the fuzzy controller is pre-calculated, explicitely the inference step and the dufuzzification step, respectively, to speed up evaluation. By computing the output for every possible input in advance, the complex procedure of evaluating a fuzzy controller, which consists of the fuzzification of the inputs, the inference step using the rules in the rule base and the defuzzification to obtain crisp output values, is reduced to the simple evaluation of a one-dimensional function or map K for each input-output relation. To simplify the pre-calculation, instead of constructing a complete fuzzy controller and then calculating a map for every relation it is also possible to use the rules and fuzzification information to compute base points for the maps and linearly connecting them. Either way, the pre-calculated fuzzy system here takes the form outlined in figure 3.

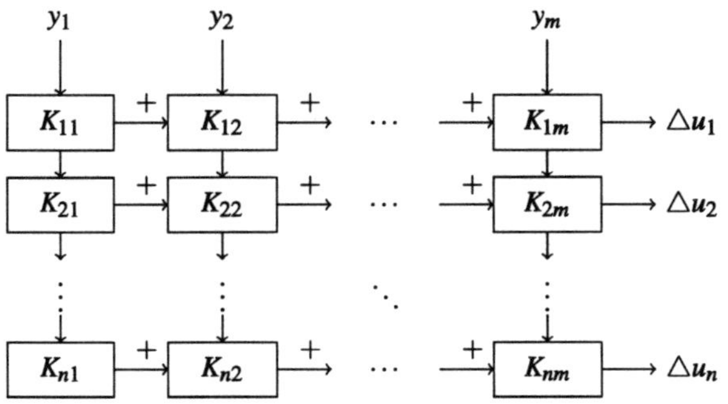

Fig. 3. Basic layout of the pre-calculated fuzzy system

The inputs are fed into all the maps in a vertical line, and the output is formed by summing up the results of all maps in a horizontal line. So, the output is

$$\triangle u_i = \sum_{j=1}^{m} K_{ij}(y_j).$$

Of course, this simplification impairs the capability of the original fuzzy system to represent non-linear, non-smooth systems. The maps define linear, smooth dependencies between input and output variables, which is sufficient for most of the desired behavior. There are mechanisms like internal feedback to implement more complex functions, but as the primary goal of the system is a minimized runtime, this basic structure has to suffice. Some trade-off between complexity of the system and runtime has to be made.

The translation from fuzzy-type rules and fuzzification info into base points and values for the maps can happen separately from the machine that executes the system. The information given by a human to initialize the system can be identical to what would be necessary to configure a Mamdani-type fuzzy controller, or it can be even simpler by using singleton variables for defuzzification, practically assigning each linguistic defuzzification value a numerical value. This information, given in a single file, is translated by the construction system into a configuration file, which then contains blueprints for the elements.

In order to construct the pre-calculated system, the following information is necessary:

- the rulebase,
- the range of values for the inputs for normalization
- fuzzy sets for defuzzification or directly provided crisp values for linguistic output values

To give an example of the construction process from the RoboCup scenario, a simple rule which affects the shoot strength of the robot depending on the distance from the goal is

$$\text{IF}\ \texttt{distance} = \texttt{high}\ \text{THEN}\ \texttt{shootstrength} = \texttt{higher}.$$

This information is processed to form the base point of an element in the map that describes the relation between distance and shoot strength. The normalization to a range of $0 \ldots 99$ allows us to directly assign the linguistic input value `high` the value 99. The outputs use linguistic values like `higher` and `lower` to reflect the corrective nature of the system. To quantify the linguistic output value `higher`, either a defuzzification map can be evaluated using standard methods, or a crisp value can be directly assigned. Using the latter method, the corresponding crisp value could be $+20$. The base point $(99, 20)$ is then added to the map. If this rule is the only information on the map, the point $(0, 0)$ is automatically included and a linear connection between those two points is added as an element. The map has the form outlined in figure 4.

It is possible to include information on more than one map into a rule by mentioning more inputs and outputs. By defining more linguistic values for both input and output variables it is possible to define a complex behaviour like one would for a regular fuzzy controller and translate it into a structure that is fast to evaluate and easy to adapt.

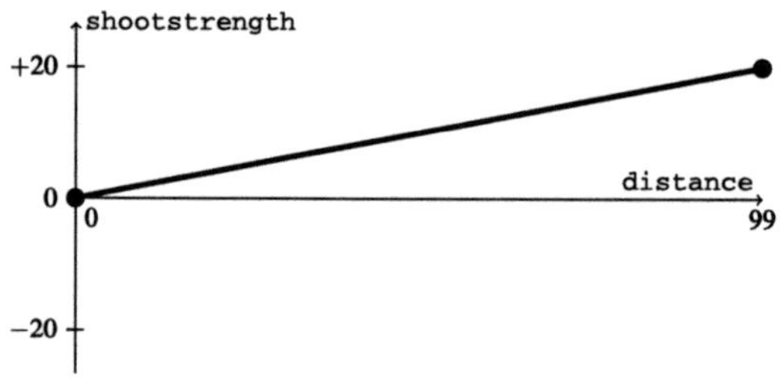

Fig. 4. Example of a map constructed from a simple rule, with the base points $(0, 0)$ and $(99, 20)$

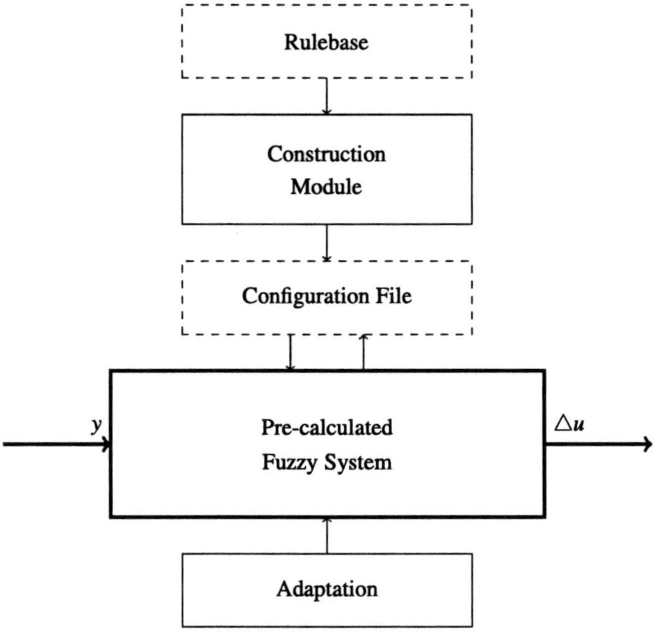

Fig. 5. Structure of the implemented system including communication via configuration files. The construction module using the rule base creates the configuration files. The files are read by the constructor of the fuzzy system class to initialize all the rules and maps and connect them to the right inputs and outputs. It is also possible to store the current configuration of the pre-calculated fuzzy system including the conducted optimizations back to the configuration file.

The structure of the entire system including the construction of the system and the adaption is outlined in figure 5.

The system is implemented as a vector of *rules*, which represent one horizontal line of maps and defines all dependencies for one output variable. The rule contains a vector of maps. The maps are, for performance reasons, not implemented directly as such, but as a list of *elements*, which are linear pieces of the function K_{ij} that forms the map. An evaluation of the map is hence accomplished by searching for the matching element for the given input in the list. The list arrangement can be altered when it proves inefficient by moving regularly used elements to the front. This implementation was benchmarked with a varying number of inputs and outputs on a regular Intel Quadcore machine with 2.6 GHz CPU clock and 4GB of RAM (this machine is used on the RoboCup robots). The resulting runtimes are outlined in figure 6.

The results show that the system is easily useable in a high performance environment (e.g. RoboCup) with up to 100 inputs and outputs, where it only takes 1ms. The plot in figure 7 shows that for square systems with an equal number of inputs and outputs an execution time of 30ms is reached by a system with approximately 500 inputs and outputs. The runtime is comprehensibly linearly dependent on the number of maps. Figure 7 shows a graphical representation of the measured runtimes.

i/o	1	10	100	1000	10000
1	0	0	0	0	4
10	0	0	0	1	23
100	0	0	1	12	121
1000	0	1	11	117	68k
10000	1	12	115	1154	dnf

Fig. 6. Runtimes of the pre-calculated fuzzy system with a varying number of outputs in milliseconds. From left to right, a number of outputs in logarithmic scale. From top to bottom, the number of inputs. A 0 signifies a runtime below the minimal detection range of around 1ms. For the numbers of inputs and outputs that occur in the RoboCup scenario, the runtime is sufficient. The unusually long time of more than 68s is most likely caused by memory swapping, as the time recorded by `clock()` is only 1960ms. The run with 10000 inputs and outputs was interrupted after 15 minutes after thoroughly negating any real-time capabilities.

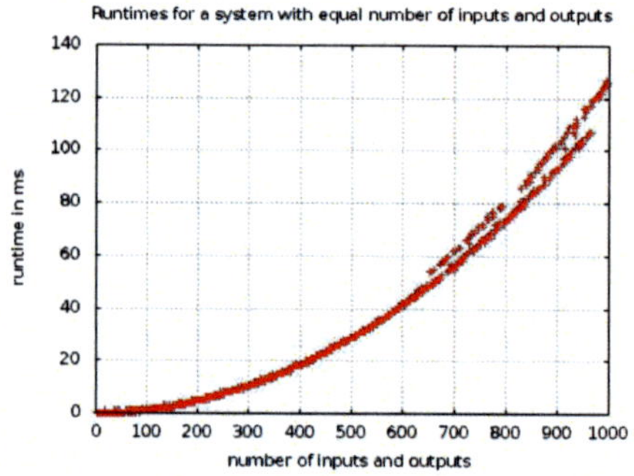

Fig. 7. Runtimes of the pre-calculated fuzzy system with an equal number of inputs and outputs

This pre-calculated system is the base for the adaption process that is described in the next section.

5 Autonomous Adaption Mechanism

There are multiple theoretic concepts on how to add adaption and learning capabilities to a classical fuzzy controller [13], [14], [15], [16]. This is a nontrivial task because there is no clearly defined parameter set which can be altered to observe the results. Instead, special methods to access and alter the rulebase and the fuzzification and defuzzification maps need to be developed. This is much easier for a pre-calculated fuzzy system as its core information is available through the values of the elements.

The usage of an adaption mechanism for a pre-calculated fuzzy system is recommended in order to correct the imprecise nature of the construction process.

An element of a map is defined by four parameters, the x- and y-coordinate of its left and right base point, and two flags that provide shortcuts, if the element is constant or constantly zero. If one of those flags is set, the calculation of the exact value inside of the element, which is necessary for linearly increasing or decreasing elements, is bypassed and the constant value or zero is directly returned. These values can directly be accessed and have a clear influence on the system. For their optimization, the usual tools of parameter optimization, like hillclimbing and evolutionary algorithms can be used [17], [18]. The effort that has to go into calibration of the optimization algorithm is much higher when using evolutionary algorithms, so greedy hillclimbing was used.

When the system grows larger and more inputs and outputs are added, it makes sense to make some maps, for which the input-output relationship is unknown, more responsive to the adaptation than others. To achieve this, every map is assigned a *certainty value* which steers the extensiveness of the adaption, especially when it comes to changing the two flags which qualitatively change the map.

With a certainty value c and a random integer r, both ranging from 1 to 100, the event of an adaptation of the four possible parameters (1) `isZero`, (2) `isConstant`, (3) the positions of the left and right base points and (4) their values can be evaluated by

$$c + r < p_i, i = 1 \ldots 4,$$

with for instance $p_1 = 10$, $p_2 = 30$, $p_3 = 60$, $p_4 = 100$. This particular values prevent any change for a map with certainty value 100, and gradually allow a more grave change only for elements with lower certainty value.

The used greedy hillclimbing approach basically choses a value in the neighbourhood as candidate. The entire optimization space that contains every possible solution is denoted by S. A solution is a set $s = e^n$ that consists of n quadruples $e \in E$, for instance

$$e_1 = \{0, 0, (0,0), (99,20)\},$$

in the case of the example element constructed in section 4. A neighbor of e_i in E is formed by alternating these four values according to the random event described in equation 5 and depending on the certainty c_i, so a new solution candidate element

$$e_i^{\mathtt{next}} = N(e_i, c_i)$$

is generated by randomly choosing one candidate out of the neighborhood N, which is greater if c_i is low. The optimization then is the search for the solution $s \in S$ that maximizes the performance value r which is obtained by evaluating the target function

$$T : s \rightarrow r$$

The choice of T is not trivial. To evaluate the performance of the robot and thus of the solution candidate s, events like referee broadcasts are graded and an average is taken over a certain time.

For a really large system, which should control hundreds of inputs and outputs, it might be promising to apply an evolutionary algorithm. By defining the entire parameter set of one specific system as an individuum and by adapting it through recombination and mutation, patterns can be preserved. Consequently the optimization can be effectively separated into exploration and exploitation. This is very helpful in the RoboCup scenario, because during competition matches, a radical change of parameters with potentially bad performance through recombination is undesired, while a slight adaptation through mutations to adapt to the opponents properties and strategy is still welcome.

As the plant is only very rarely entirely available, the calibration of the optimization algorithm relies mostly on experience and intuition. The lack of testing possibilities influences the adaption, too.

The adaption mechanism is separated from the regular system in order to not slow it down. Optimization steps can occur when the robot is idle and there are enough resources to calculate the performance of the robot and the next solution candidate.

6 Test Results

The system was evaluated using the robots of the 1. RFC Stuttgart RoboCup team of the University of Stuttgart. To try out the effectiveness of the system and the adaption algorithm, a test problem involving a shot on the goal was devised. The pre-calculated system was initialized using 9 rules which defined relationships between the 5 input variables position (horizontal and vertical), rotation velocity, translation velocity and angle and the 3 output variables shoot strength and shoot point (horizontal and vertical). All maps that were not defined by the rules were set to zero and left to the adaption by assigning to them a certainty value of 0. The robot performed 30 shots. After each 3 shots the performance was evaluated by manually grading the outcomes. The resulting optimization path is shown in figure 8.

As expected, the untouched maps with low certainty values were heavily altered during the optimization, but none of the maps not predefined by the rules proved to have a positive influence. The maps, which were affected by the rules and subsequently had a higher certainty value, were only slightly changed into a configuration, which produced better results. The four-stage optimization process thus proved functional.

The runtimes of this system had no influence on the runtimes of the overall control cycle of the system. Furthermore another test with a random pre-calculated system with

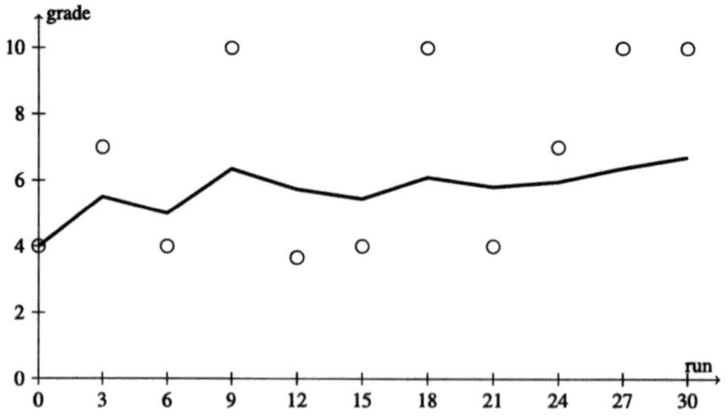

Fig. 8. Optimization values for 30 runs. The line shows the average grade.

400 inputs and outputs was executed. It proved that the cycle was not slowed down. Though as measured above, a system with 500 inputs and outputs can be executed in 30ms, additional time has to be taken into account to read in the inputs and write the outputs.

Nevertheless the test results show the efficiency of the system. With the presented approach it is possible to run large fuzzy rule systems in a very short cycle time what is especially for real-time robotic applications a major contribution.

7 Conclusion

The tests show that the devised system works as intended, and that through efficient implementation, the runtime is sufficiently short for up to around 400 inputs and outputs under realistic conditions. This is enough for a comprehensive control of the robots behaviour.

A trade-off between complexity and runtime has to be made, and so the precalculated system loses some abilities to model non-smooth relationships at the benefit of a much lower runtime. In the case of the RoboCup, the system is still capable of approximating the desired behaviour. The simple structure of the final precalculated fuzzy sytem also simplifies the development of an adaption mechanism.

The implemented system shows the ability to influence the robot based on verbally given rules and the adaption steers the behaviour in a desired direction.

As future work, it would be useful to implement an evolutionary algorithm for the optimization, especially for a more comprehensive system according to the second scenario, strategy adaption outlined in section 1. The current implementation of greedy hillclimbing could in that case be used as a mutation method.

References

1. Pineau, J., Thrun, S.: High-level robot behavior control using pomdps. In: AAAI Workshop Notes (2002)
2. Tunstel, E.: Mobile robot autonomy via hierarchical fuzzy behavior control. In: 6th Intl. Symp. on Robotics and Manufacturing, WAC 1996, pp. 837–842. ASME Press (1996)
3. Mota, L., Reis, L.P.: Setplays: achieving coordination by the appropriate use of arbitrary predefined flexible plans and inter-robot communication. In: Proceedings of the 1st International Conference on Robot Communication and Coordination, RoboComm 2007, p. 13:1–13:7. IEEE Press, Piscataway (2007)
4. Mota, L., Reis, L.P., Lau, N.: Multi-robot coordination using setplays in the middle-size and simulation leagues. Mechatronics (July 2010)
5. McCulloch, W., Pitts, W.: A logical calculus of the ideas immanent in nervous activity. Bulletin of Mathematical Biology 5, 1–13 (1943)
6. Mamdani, E.H., Assilian, S.: An experiment in linguistic synthesis with a fuzzy logic controller. International Journal of Man-Machine Studies 7, 115–133 (1975)
7. Lee, S.H., Suh, I.H.: Bayesian network-based behavior control for skilligent robots. In: IEEE International Conference on Robotics and Automation, ICRA 2009, pp. 2910–2916 (May 2009)

8. Bellas, F., Duro, R.J.: Introducing long term memory in an ann based multilevel darwinist brain. In: Proceedings of the Artificial and Natural Neural Networks 7th International Conference on Computational Methods in Neural Modeling - Volume 1, IWANN 2003, pp. 590–597. Springer, Heidelberg (2003)
9. Chen, J., Patyra, M.J.: Vhdl modeling of a multivariable fuzzy logic controller hardware system. In: Proceedings of the Third IEEE Conference on IEEE World Congress on Computational Intelligence Fuzzy Systems, vol. 1, pp. 129–132 (1994)
10. Chiueh, T.: Optimization of fuzzy logic inference architecture. Computer 25(5), 67–71 (1992)
11. Aja-Fernandez, S., Alberola-Lopez, C.: Fast inference in sam fuzzy systems using transition matrices. IEEE Transactions on Fuzzy Systems, 170–182 (April 2004)
12. Ostrowski, P.Y.K., Cheung, D.J., Roubaud, K.: An outline of the intuitive design of fuzzy logic and its efficient implementation. In: Proceedings of the Second IEEE International Conference on Fuzzy Systems, vol. 1, pp. 184–189 (1993)
13. Wang, L.-X.: Stable adaptive fuzzy control of nonlinear systems. IEEE Transactions on Fuzzy Systems 1(2), 146–155 (1993)
14. Li, H.-X., Tong, S.: A hybrid adaptive fuzzy control for a class of nonlinear mimo systems. IEEE Transactions on Fuzzy Systems 11(1), 24–34 (2003)
15. Chen, B., Liu, X., Tong, S.: Adaptive fuzzy output tracking control of mimo nonlinear uncertain systems. IEEE Transactions on Fuzzy Systems 15(2), 287–300 (2007)
16. Golea, N., Golea, A., Barra, K., Bouktir, T.: Observer-based adaptive control of robot manipulators: Fuzzy systems approach. Applied Soft Computing 8(1), 778–787 (2008)
17. Coello, C.A., Lamont, G.B., Veldhuizen, D.: Evolutionary Algorithms for Solving Multi-Objective Problems, XXII, 2nd edn. Springer-Verlag New York, Inc. (2007)
18. Cao, Y., Chen, Y., Zhao, Y.: Stochastic System Identification by Evolutionary Algorithms. In: Huang, D.-S., Gan, Y., Premaratne, P., Han, K. (eds.) ICIC 2011. LNCS, vol. 6840, pp. 247–252. Springer, Heidelberg (2012)

Ontology Representation and Instantiation for Semantic Map Building by a Mobile Robot

Gi Hyun Lim[1], Chuho Yi[1], Il Hong Suh[2], Dong Wook Ko[3], and Seung Woo Hong[3]

[1] Department of Electronics and Computer Engineering, Hanyang University, Seoul, Korea 133–791
{hmetal,d1uck}@hanyang.ac.kr

[2] Department of Computer Science and Engineering, Hanyang University, Seoul, Korea 133–791
ihsuh@hanyang.ac.kr

[3] Department of Intelligent Robot Engineering, Hanyang University, Seoul, Korea 133–791
{pumpblack, swcoke01}@gmail.com

Abstract. To offer sustainable robotic services, service robots must accumulate knowledge by using recognition results and choose a action for services intelligently. Robust knowledge instantiation and update by using imperfect sensing data such as misidentification of perception is a main issue to implement semantic robot intelligence. In this paper, robust knowledge acquisition method is proposed to enable robots to detect falsity of object recognition for robust knowledge instantiation, where spatial reasoning, temporal reasoning, movable properties and data confidences are considered.

1 Introduction

The robotic service is up-and-coming application to receive attention in various research fields of robotics [1]. Various robots is expected to be one of the most important issues to occur in the residential space to offer services. However, service environments are said to be partially observable and uncertain with respect to service robots. Even though a robot may often fail to perform service tasks, given service tasks should be sustained to satisfy the user's requests.

Performing service tasks successfully, such as taking orders semantically from a person, moving to various type of residential space, and finding an object which meets person's needs, autonomous robots are required to have substantial knowledge [2]. Service robots are designed to complete service tasks semi or fully automatically in a service environment [3]. Service robots will need to understand semantic relationships of objects, spaces and contexts in order to assist humans in their everyday lives, and then the robot must carry out its service tasks with its primitive behaviors. For this, a robot needs many kinds of data from low level sensor data to high level symbolic data. High level perceptual tasks such as context awareness, object recognition and navigation are essential for intelligent robots. Also a robot must combine its atomic behaviors.

S. Lee et al. (Eds.): Intelligent Autonomous Systems 12, AISC 193, pp. 387–395.
springerlink.com

Robots must interact semantically with humans as well as understand user intentions to enable socially collaboration [4], [5]. When interacting with humans, it will be needed for robots to be able to understand semantic and contextual information regarding objects. For example, if a robot needs to find a cup from a table that contains other objects such as kettle and the cup are occluded by the kettle, some additional information can give more chances to find the cup. This might take the form of a spatial relationship, such as "kettle is left of the cup." If a service robot recognized the kettle, the cup might be right of the kettle. Semantic knowledge also evaluates candidate objects based on spatial semantic properties such as the door handle should be found in one of a few areas once the door is detected [6]. Semantic navigation is another example to show the validity of semantic robot knowledge. Humans do not necessarily use accurate quantitative information to perceive space where one is located or to move to another place. Instead, they remember a few landmarks that constitute the space such as a specific structure or distinct objects, they restructure the knowledge based on a spatial contexts and then utilize the knowledge again [7]. Moreover, humans interact others with symbolic information, which are not just simple data but semantic knowledge which is represented by network or graph which represents semantic relations between concepts. Through the relationships, humans can acquire additional information beyond given information in interaction.

Service tasks always creates some social situation which is framed to pay attention to restricted and related domains [8]. Moreover, service tasks have successive interrelation of number of subtasks as service flow. From the service provider's vies, there are obstacles to look for recurrent sequence of subtasks from many various kinds of service tasks. Service tasks tend to be repeating pattern of task sequence. Each service captures specific tasks to be carried out (forming a very flexible sort of subtasks) and their respective precondition and post conditions. These service tasks can be modeled with each subtasks being further linked to a rich description as a scenario. However, a situation cannot be understood using sensory information such as object recognition, robot localization and human recognition, it requires contextual interpretation of the scene using not only robot-embodied sensors but also exogenous sensors [9]. Understanding situation provides clues for the appropriate and efficient action selection for sustainable service tasks.

In knowledge management literature it has often been pointed out that the relation between knowledge, information and data is important, and often misunderstood. It has also been argued that this misunderstanding leads to problems in information system design [10]. Data has commonly been seen as simple facts that can be structured to become information. Information, in turn, becomes knowledge when it is interpreted, put into context, or when meaning is added to it. There are several variations of this widely adopted theme. The common idea is that data is something less than information, and information is less than knowledge. Moreover, it is assumed that we first need to have data before information can be created, and only when we have information, knowledge can emerge.

2 Robot-Centered Ontology

OWL ontology [11] is used for representation of robot knowledge. Here, ontology that expresses semantics with concept hierarchy and their relationships is mainly formulated in the knowledge representation formalism and is necessary to enable the assimilation of information from diverse sources [12]. Semantics for robot knowledge refers to five knowledge classes including feature, object, space, context, and action, and how these classes are related to each other. Also, there is production system with two types of rules, uni-directional rules and bi-directional rules. Robot knowledge can be integrated from the different sources of knowledge including encyclopedic knowledge designed manually from expert knowledge or from Internet, action knowledge derived from observations of human motion, and robot knowledge about self-model and its surrounding world model [13], [14]. These robot knowledge should reflect actual environments as practice knowledge [15]. A robot perceives objects with its own sensors, models a world where it exists, plans some sequences of tasks, performs tasks with its own behaviors, and then perceives again [2], [16]. To comply with such cognitive capabilities, robot-centered ontology is designed to be composed of knowledge boards (KBoards) and rules, where KBoards is composed of five classes of knowledge (KClass): feature, object, space, action, and context as shown in Fig. 1. Here, context is a characteristic environmental situation around robots which

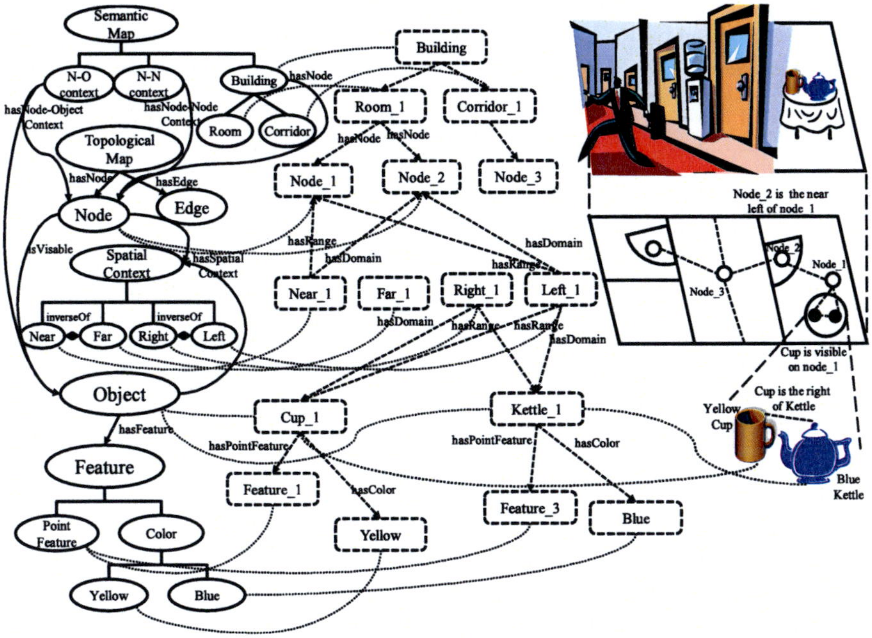

Fig. 1. An example of robot-centered ontology including schematic knowledge, an experimental environment and their knowledge instances

can provide clues of the proper action selection mechanism for a robot. A world model represents a robot's internal states reflecting perceived environments and consists of context, object, and space (COS) classes.

Each knowledge class has three knowledge levels (KLevel) such as high level, middle level and low level knowledge. Low level knowledge of feature class and action class (numerical descriptor, and behavior) are robot-specific knowledge for its own sensory motor capabilities, and COS classes of low-level knowledge consists of spatial context, part object, and metric map. Middle level knowledge visual feature, object, topological map, temporal context, and task are common robot knowledge at the abstract level and hide details of a particular set of low-level sensor data and motor commands. High-level knowledge of compound object, semantic map, situation, and service classes are common knowledge for robots as well as humans. Each knowledge level has three ontology layers (OLayer), such as the meta ontology layer for generic knowledge, ontology schema layer for domain knowledge, and ontology instance layer for knowledge instance. Meta ontology can be a template of the ontology schema layer, and the ontology schema layer is instantiated to instance layer.

The feature KClass has three KLevels. The low level knowledge of feature is the numerical descriptor level that includes a set of numerical descriptors of image processing algorithms, which are produced by robots' own sensors and data processing algorithms. The middle level knowledge of feature is the visual feature level that includes visual features, which are extracted by numerical descriptors in the low level. The high level knowledge of feature is the visual concept level that is grounded with visual features in feature knowledge class and object feature in the object knowledge class or metric map in the space knowledge class. The following is a DL representation of *blue.*

$$
\begin{aligned}
Blue := ColorFeature \wedge \exists hasColor.HueValue \\
\wedge \exists hasAlgorithm.extractColor
\end{aligned}
$$

Each COS class consists of three knowledge levels for the representation of the robot world model. Low levels such as an object part, metric map, and spatial context are used for matching with perceptual feature. Middle levels include object, topological map, and temporal context that contain its name and functionality. High levels such as situation, compound object, and semantic map are abstract level, which easily describe relationships between other knowledge classes. In the Description Logic (DL) example below, a cup has a *colorFeature.*

$$
\begin{aligned}
Cup := Object \wedge \exists isObject.Container \\
\wedge \exists hasColor.ColorFeature
\end{aligned}
$$

Context is not a list of objects and their locations, but implies abstract and characteristic situations that can be represented by relationships between objects and object properties. The low level knowledge of context denotes the spatial context level that has a spatial concept such as *on*, *in*, *left* and *right*, which are inferred by using the object level and space level instances in the model class.

The middle level knowledge of context represents the temporal context level that has temporal concepts defined by Allen [17]. Finally, the high level knowledge of context is the high-level context. The following is a DL representation of *left.*

$$Left := SpatialContext \wedge \forall hasSubjective.Object$$
$$\wedge \forall hasObjective.Object$$

3 Robust Knowledge Acqusition

Successfully accomplishing everyday service tasks requires autonomous robots to have substantial knowledge with their own sensors. These robot knowledge should be instantiated and updated by using imperfect perception data, such as misidentification of object recognition [15]. The feature knowledge of visual observation can be instantiated as ontology, which ensures that only sound and complete data are asserted and propagated with ontology inference. Noisy sensor data, such as false positives and false negatives, should be filtered for robust robot knowledge acquisition. For instance, a misidentified object may make erroneous spatial relationships. Moreover, inferred erroneous facts will result in false consequences for reasoning; this generates a vicious cycle, and errors are difficult to correct, even with additional true negative results.

To address the failure of knowledge instantiation, a reasoning mechanism with knowledge acquisition rules are proposed to instantiate and update knowledge by estimating confidence of the perception results. There are four properties to build robust knowledge acquisition rules: temporal reasoning to check the validity of relationships between time intervals, statistical reasoning to determine the confidence level of the perception results, ontological reasoning to check if a perceived feature satisfies object properties or space properties.

3.1 Likelihood Confidence Interval (LCI)

Our way to estimate the confidence of likelihood for perception is to extend some form to formalize the *common sense law of inertia*, whereby an event is assumed to persist unless there is reason to believe otherwise, and to be perceived perfected without recognition failure or misidentification [18]. The extend form considers the uncertainty of perception on the basis of the *confidence law of inertia*, whereby knowledge instance is assumed to persist unless there is confidence to believe otherwise as shown in Fig. 2.

Confidence of recognition is determined by an likelihood interval-counter (γ) from the measurement likelihood for each object recognition result. If the measurement likelihood of object A is x_A, then (1-x_A) is the probability that the recognition data for A can be false. From that, (1-x_A)$^{\gamma_A}$ can be calculated to define probability when the values of γ_A consecutive data are all false. If the result of (1-x_A)$^{\gamma_A}$ is less than 5% (0.05), then it can be said that the data have been obtained within a confidence interval (1.96σ, $P = 0.05$) of the 95% confidence level. For example, if the measurement likelihood of object A is 80%successively,

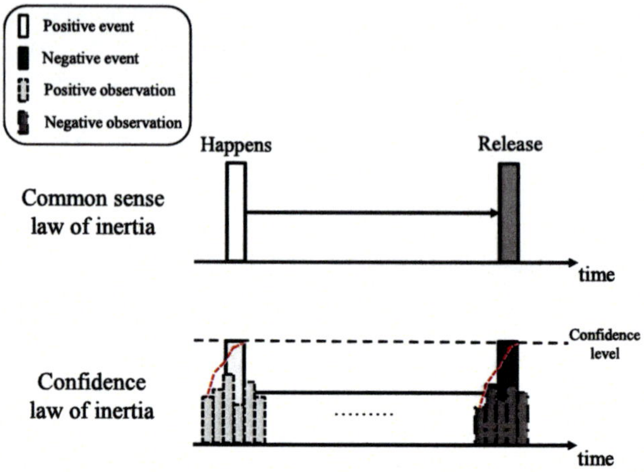

Fig. 2. The *confidence law of inertia* as an extended form of the *common sense law of inertia*

the recognition failure rate of object A might be 20% (0.2). The result rate of recognition failure of two consecutive observations is 4% (0.04) and 4% is beyond the 95% confidence interval($P = 0.05$), so γ of object A is 2. At that time, the instance of object A is created and vice versa. The likelihood interval-counter using β likelihood distribution can be represented as follows:

$$\gamma_\beta = \min\{\gamma \in I | \prod_{i=1}^{n} (1 - x_{\mathtt{obj}}) \leq P\}, \tag{1}$$

where $P = 0.05 = 1 - 95\%$ confidence level.

3.2 Temporal and Statistical Reasoning

According to continuous observations from robot movement, object instances might be created or deleted whether certain number of consecutive observation likelihoods exceed the likelihood confidence interval. Time intervals of object instances which exist or not is determined by the durations between the changes of confidence. Temporal relations between intervals are inferred using temporal reasoning. The temporal relation was first proposed by Allen [17] and represents temporal relations using before, after, meets, met-by, overlaps, overlapped-by, and so on. If two intervals meet or overlap, then they are merged into one interval. The merged interval begins at the start point of the former and ends at the end point of the latter. Temporal confidence reasoning (TCR) is based on the assumption that recognized objects cannot go away and come back within a single time interval.

When an object instance of A is registered, if other objects are also considered to be true positive instances and to have a temporal relation of *overlapped* with

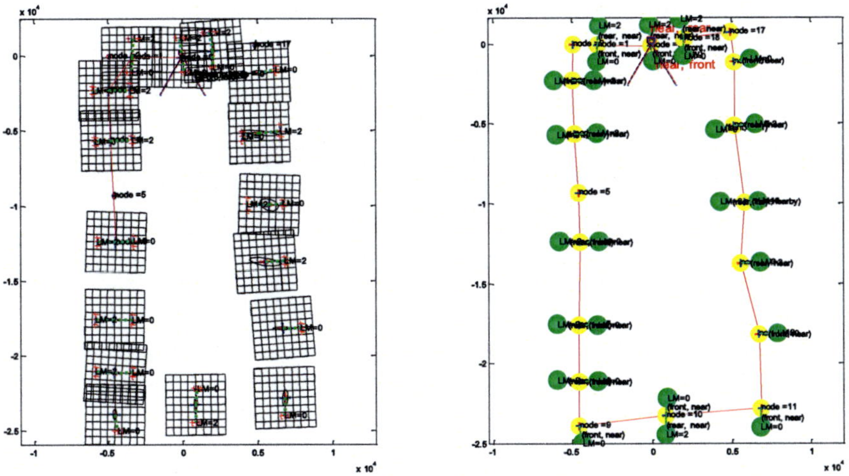

(a) Metric-Topology Map (b) Semantic-Topology Map

Fig. 3. An example of space knowledge instance

object `A`, then spatial relations among the objects can be inferred. Then, the spatial relation between them can be reasoned and set using spatial reasoning. All object instances and their spatial relations can be registered in the instance database.

3.3 Space Knowledge Class

Figure 3 show an experimental result of space knowledge class instance including metric-topology map and semantic-topology map. For this experiment, a pioneer 3 AT robot carrying a single consumer-grade camera was driven around the fourth floor of IT&BT building in Hanyang university. The TCR rules are applied to check the validity of the relationships between intervals and statistical reasoning to determine the LCI of visual perception from commercial vision system. As a result, there are 19 nodes and 27 objects.

4 Concluding Remarks

In this paper, a robust knowledge acquisition method that makes imperfect information of epistemic results toward robust and consistent knowledge is proposed as one of knowledge management methods. The method uses temporal reasoning to check the validity of relationships between intervals and statistical reasoning to determine the confidence interval (CI) of object recognition. The experimental results indicate that false positives in recognition results were corrected. The proposed method had difficulty registering some objects with a recognition rate less than 52.9% ($\gamma = 4$). In spite of this, the method can determine misidentification well, and thus dependable semantic knowledge for service robots can be instantiated.

Acknowledgement. This work was supported for the Intelligent Robotics Development Program, one of the 21st Century Frontier R&D Programs funded by the Korea Ministry of Knowledge Economy.

References

[1] Stefanov, D., Bien, Z., Bang, W.: The smart house for older persons and persons with physical disabilities: structure, technology arrangements, and perspectives. IEEE Transactions on Neural Systems and Rehabilitation Engineering 12(2), 228–250 (2004)
[2] Lim, G.H., Suh, I.H., Suh, H.: Ontology-based unified robot knowledge for service robots in indoor environments. IEEE Transactions on Systems, Man and Cybernetics, Part A: Systems and Humans 41(3), 492–509 (2011)
[3] Kawamura, K., Pack, R., Iskarous, M.: Design philosophy for service robots. In: IEEE International Conference on Systems, Man and Cybernetics, Intelligent Systems for the 21st Century, vol. 4 (1995)
[4] Severinson-Eklundh, K., Green, A., Hüttenrauch, H.: Social and collaborative aspects of interaction with a service robot. Robotics and Autonomous Systems 42(3-4), 223–234 (2003)
[5] Choi, C., Huckaby, J., Rogers III, J., Trevor, A., Case, J.P., Christensen, H.I.: Towards Semantic Perception Towards Semantic Perception for Mobile Manipulation. In: International Conference on Intelligent Robots and Systems (IROS) 2009 Workshop on Semantic Perception for Mobile Manipulation, Citeseer (2009)
[6] Klingbeil, E., Saxena, A., Ng, A.: Learning to open new doors. In: Robotics Science and Systems (RSS) Workshop on Robot Manipulation, Citeseer (2008)
[7] Yi, C., Suh, I.H., Lim, G.H., Jeong, S., Choi, B.-U.: Cognitive Representation and Bayeisan Model of Spatial Object Contexts for Robot Localization. In: Köppen, M., Kasabov, N., Coghill, G. (eds.) ICONIP 2008, Part I. LNCS, vol. 5506, pp. 747–754. Springer, Heidelberg (2009)
[8] Wetzel, I., Klischewski, R.: Serviceflow Beyond Workflow? Concepts and Architectures for Supporting Inter-organizational Service Processes. In: Pidduck, A.B., Mylopoulos, J., Woo, C.C., Ozsu, M.T. (eds.) CAiSE 2002. LNCS, vol. 2348, pp. 500–515. Springer, Heidelberg (2002)
[9] Lim, G.H., Chung, J.L., Ryu, G.G., Kim, J.B., Lee, S.H., Lee, S., Suh, I.H., Choi, J.H., Park, Y.T.: Ontological Representation of Vision-based 3D Spatio-Temporal Context for Mobile Robot Applications. Artificial Life and Robotics, AROB (2007)
[10] Tuomi, I.: Data is more than knowledge: implications of the reversed knowledge-hierarchy for knowledge management and organizational memory. In: Proceedings of the 32nd Annual Hawaii International Conference on System Sciences, HICSS-32 (1999)
[11] Smith, M., Welty, C., McGuinness, D.: Owl web ontology language guide. W3C recommendation, vol. 10 (2004)
[12] Doerr, M., Hunter, J., Lagoze, C.: Towards a core ontology for information integration. Journal of Digital information 4(1) (2011)

[13] Tenorth, M., Beetz, M.: Knowrob - knowledge processing for autonomous personal robots. In: IEEE/RSJ International Conference on Intelligent Robots and Systems, IROS 2009, pp. 4261–4266. IEEE (2009)
[14] Waibel, M., Beetz, M., Civera, J., D'Andrea, R., Elfring, J., Galvez-Lopez, D., Haussermann, K., Janssen, R., Montiel, J., Perzylo, A., Schiessle, B., Tenorth, M., Zweigle, O., van de Molengraft, R.: Roboearth. IEEE Robotics Automation Magazine 18(2), 69–82 (2011)
[15] Lim, G.H., Suh, I.H.: Robust robot knowledge instantiation for intelligent service robots. Intelligent Service Robotics, 115–123 (2010)
[16] Suh, I.H., Lim, G.H., Hwang, W., Suh, H., Choi, J.H., Park, Y.T.: Ontology-based multi-layered robot knowledge framework (OMRKF) for robot intelligence. In: IEEE/RSJ International Conference on Intelligent Robots and Systems, IROS 2007, pp. 429–436 (2007)
[17] Allen, J.: Planning as temporal reasoning. In: Proceedings of the Second International Conference on Principles of Knowledge Representation and Reasoning, vol. 13, pp. 25–45. Citeseer (1991)
[18] Shanahan, M.: The event calculus explained. Artificial Intelligence Today, 409–430 (1999)

Recognizing Hardware Faults on Mobile Robots Using Situation Analysis Techniques

Oliver Zweigle, Benjamin Keil, Markus Wittlinger, Kai Häussermann, and Paul Levi

IPVS - University of Stuttgart, Universitätsstr. 38, 70569 Stuttgart, Germany
zweigle@ipvs.uni-stuttgart.de

Abstract. Current research in the field of robotics deals with planning and executing of increasingly complex tasks as well as applications in which autonomous mobile robot systems interact with highly dynamic environments. In order to operate successfully in a highly dynamic physical world a mobile robot needs awareness to adapt its behavior according to the environment. This awareness also has to cope with uncertainty due to imperfect hardware sensors and actuators or network delays of the robot itself. Thus another important ability of a mobile robot is to "understand" its internal state to detect any kind of hardware faults in the system. In this work a highly customizable framework for situation awareness is presented. For the evaluation of this framework we take advantage of its generality and assume fault diagnosis as a more specific subclass of the general situation awareness process to extend the situation detection framework for fault detection and diagnosis in a multi robot system.

1 Introduction

In order to solve complex tasks mobile robots usually need some kind of awareness of their internal state and the physical environment. Endsley defined situation awareness [1] as the perception of elements in the environment within a volume of time and space, the comprehension of their meaning, and projection of their status in the near future. Recent works in the domain of situation awareness usually refer to a well-defined scenario, e.g. the field of autonomous vehicles or the domain of driver assistance systems [2], [3], [4]. Although there already exists some work of a general architecture for the recognition of situations, which base on temporal aspects [5], these approaches are not well suited to be used in multi-robot-systems. The architecture of an awareness-framework in multi robot systems should be able to cope with uncertainties, e.g. based on inconsistent team negotiations, delays and faults in the communication network or imperfect hardware sensors and actuators. Thus the main focus of attention in this work will be to develop a highly customized framework for situation awareness, which is able to handle uncertainties. Consequently we use fault detection as a more specific subclass of the general situation awareness process. Both situation awareness and fault detection base on a given stream of observations to determine the state of a system, its behavior and the environment over time. In order of this assumption, fault detection is defined as the recognition of all anomalies from common behavior, which can be interpreted as symptoms of a disorder or a fault in the system. In section 2 a short overview of related work in the domain of situation awareness and fault detection is given. In section 3 the main

S. Lee et al. (Eds.): Intelligent Autonomous Systems 12, AISC 193, pp. 397–409.
springerlink.com

idea and the principles of the developed situation recognition framework is presented. In section 4, the benefits of the framework are used to create a fault-detection system. In order to allow not only the detection of some anomalies based on the comparison of the common behavior with the current system, but also to allow the recognition and identification of the current fault, this framework is able to detect, classify and label a hardware fault. This fault detection and classification process is done by defining a fault as a special kind of (fault) situation and using the situation recognition framework. This enables the system to fault awareness, which can be used further to adjust the behavior planning. Consequently a hardware fault can be described as a set of constraints gained e.g. from the collected sensor data over a certain time t and the uncertainty of the observed data. The results of the evaluation of two exemplary fault scenarios are presented in section 5.

2 Related Work

A common approach of situation and fault detection is model-based ([6], [7], [8], [9], [10]). The main idea of this approach is to construct an appropriate model of the system, which describes how the system is supposed to behave. During the recognition process the actual behavior will be compared to the predicted behavior and variances or anomalies will be interpreted as symptoms of a fault.

According to model based situation recognition, the Unified Modeling Language (UML) is used by [11] for modeling contextual aspects. Another modeling approach in this domain is presented in [12], where the Object-Role Modeling (ORM) is extended by context information. In the work of Hofer et. al. [13] context is modeled by using object-oriented (OO) techniques, where objects represent different context types. This approach also includes all typical benefits of object-orientation, e.g. encapsulation, reusability, inheritance. A much higher degree of formality uses logic-based models for situation recognition [14]. In this approach a set of logical facts, expressions and rules are used to define a context model. A logical reasoning system is used to remove, update and derive new facts based on existing rules. A similar approach to detect multiple faults is used by de Kleer [15]. In this work a general diagnosis engine (GDE) was described, which considers not the full multiple fault combination, but a minimal subset of broken components that can explain the observations.

Another approach of model-based fault detection is the use of artificial neural networks for fault detection and diagnosis [16]. Here the input-output relationship for the components in a system is trained in multiple local neural networks to learn which input leads to a fault. Supervised learning is used to train the local model neural networks. To evaluate some neural network models a mobile robot has been used. Here the models were trained to detect anomalies in the motion of the robot, where the voltage given to the motors (input) was compared with the predicated speed of the wheels (output). Another biologically inspired approach is the use of artificial immune systems (AIS) [17], [18]. Similar to the biologically immune system which distinguish between foreign objects and the bodys own self, the artificial AIS works as a classifier which distinguishes between two states ("self" and "non-self") [19]. Typically in this domain self corresponds to fault-free operation while non-self-refers to observations resulting from a faulty behavior. This idea has been successfully applied to robotics by [20]. In this work AIS is used as a fault detector for a Khepera robot. The work of Jakimovski et. al. [21] introduces RADE (Robot Anomaly Detection Engine), which applies principles of fuzzy logic for representing and processing the information within the artificial immune system. In [22] a framework for a software architecture for modeling sensor-based control systems is introduced. It is only focused on modeling sensor systems. The presented approach is applicable to a wider range of scenarios, including abstracted knowledge derived e.g. from reasoning processes.

Most model-based approaches mentioned above usually refer to a well-defined scenario and are, if anything, only less able to handle uncertainties adequately. These uncertainties typically arise in the field of robotics due to noisy sensor data, a highly dynamical environment and the non-accurate hardware actuators of a mobile robot system. To cope with these uncertainties this work uses a Bayesian model to detect hardware faults.

3 Situation Recognition Framework

In this section the principle of operation of the developed situation recognition framework will be discussed. Fig. 1 shows the basic idea and the schema of the framework.

The most important parts of the framework are so-called situation templates, which are located in a situation library. A situation template consists of a pre-defined situation schema, and a situation id. The situation id is a unique identifier used for referencing and identifying a situation template. A situation schema is an abstract description of a situation based on a proper concatenation of attributes, which includes a set of sensor- and context information, combined with corresponding constraints and coherences. These constraints and the corresponding concatenations are used for the recognition process of (particular) situations.

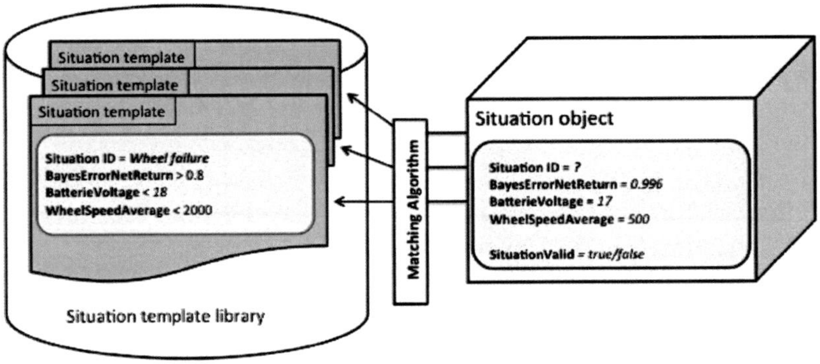

Fig. 1. Schematic representation of situation recognition process using situation templates, a situation library and situation objects. A matching algorithm is used to decide if a situation defined in a situation template is valid when it is instantiated with current sensor and context data from an application.

Each attribute contains a description of its corresponding range of values including the possible uncertainty of those data sets or alternatively a logical comparison. All defined schema attributes can be furthermore linked with temporal or logical operators.

Furthermore the situation schema includes a description how to concatenate, combine and calculate the result of the situation analysis. During runtime a situation template is instantiated as a situation object. The situation object contains the current sensor- and context data provided by the robot. During a simple recognition process these data sets are compared to the well-defined value ranges in the situation template.

The quality of a situation template is defined by the quality of the sensor-/context data and the corresponding value ranges, which are used as a reference-value for recognizing a situation. In the case that the selected sensors and context data or even the value ranges are not properly defined the situation recognition process can be aborted.

The actual recognition process, this means the analysis of a situation object in correspondence to the defined situation attributes, is done by an exchangeable matching algorithm. Several methods for the matching algorithm can be used, e.g. logical reasoning, Bayesian Networks or Neuronal Networks. The recognition process is executed by transforming the situation object into a situation graph defined in [23]. The result of the recognition process defines the probability of the occurrence of the correspondent situation. A second value called confidential metric defines the accuracy of the recognition process. This confidence value is based on the reliability of the sensor (e.g. noise, tolerance, variance) and the concatenation of the attributes of the underlying template. Especially for situational awareness the differentiation between those values is

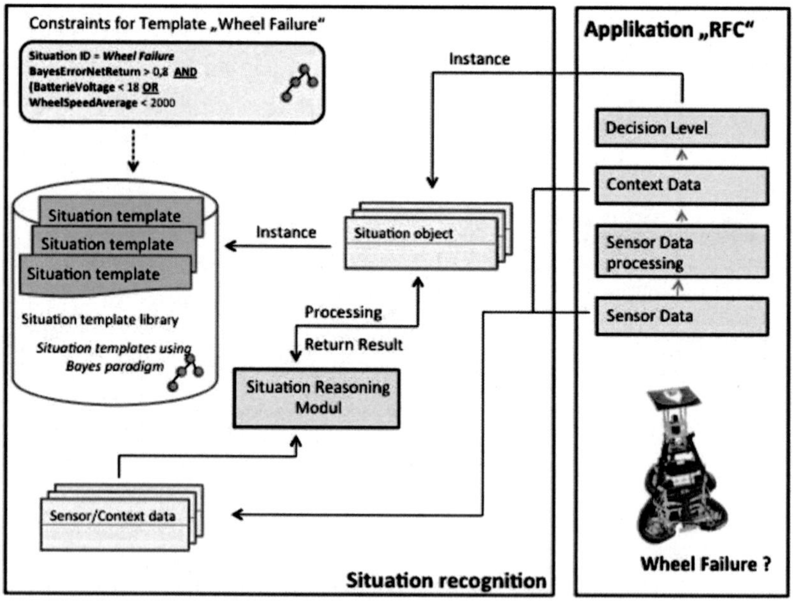

Fig. 2. Data flow of the situation recognition process

essential in order to interpret the result of the recognition process correctly. If the general situation A on a mobile robot is recognized with a probabilistic value of 99%, but the confidential metric value is only 8% due to e.g. bad sensors, the probability that the situation A really occurs is relatively low. Consequently the robot should not directly adapt its behavior, as the situation cannot be clearly affirmed, as maybe some sensor measurements are highly disturbed.

Fig. 2 shows the flow of a situation recognition process. In this scenario the situation is defined in a situation template. This template is part of the situation library. During runtime a situation object is instantiated from the situation template [24]. Every situation that should be recognized corresponds to one situation template. In contrast every situation template can be instantiated several times with different data. The depicted situation-reasoning module contains the described matching algorithm. Here the real sensor and context data is collected. Using these values a situation object is instantiated and a situation graph is created. This situation graph is used to process the reasoning process. The probabilistic and confidential metrics are calculated with Bayesian probabilistic methods. In this paper the complete process of the calculation of those metrics is not in the center of interest. It is described in [25]. In contrast the focus in this work is on the integration of an external algorithm (in this case also Bayesian Networks) in the situation recognition framework to detect fault situations.

4 Integration of External Algorithms

The integration of any external algorithms like Neural Networks or Bayesian networks can be done easily within this framework. A node in a situation graph cannot only contain logical conjunctions and constraints. A node can also be used as a trigger for an external algorithm that provides certain return values that can be afterwards used as input values for a logical conjunction or a comparison in a node. In this work we want to show the integration of Bayesian networks. Bayesian Networks can return a probability value that can be directly integrated in a situation schema attribute. If a Bayesian network is used its return state is represented by a node in the situation graph. The Bayesian Network can be seen as a kind of virtual sensor [25] that can contribute further information to the situation recognition process. Consequently the return value of the Bayesian network represented in a situation schema attribute can be brought into conjunction with other situation schema attributes using logical or temporal relations. As a consequence it is not important for the system from which data source the values are coming. If an external algorithm is included into a situation schema attribute the algorithm is triggered during runtime and the return value is directly used.

5 Fault Analysis and Evaluation

In the following section the presented recognition framework will be evaluated. Fault detection is assumed to be a more specific subclass of the general situation awareness process.

In the following a fault will be modeled as an ordinary situation template. The requirement of adequate fault detection in multi robot systems is due to the fact that hardware fault detection can improve the system performance significantly. The tasks in a multi robot system are usually distributed over the available agents, e.g. via roles. If one of the robots has any kind of hardware fault it may still execute an important task, as the complete system has no awareness of the fault. Consequently the system performance rapidly decreases. In case of an automatic fault detection the possibility of a fault can be taken into account for the distribution of tasks or roles. Consequently the system performance can be enhanced, as the erroneous agent does not get an important role or task anymore.

In the following scenarios several typical fault situations for a mobile robot will be discussed. The robot is equipped with several sensors (e.g. camera, compass, odometry) and is omni-wheel-driven. In order to evaluate the framework these scenarios will be used as a proof of concept that the presented situation recognition architecture in conjunction with Bayesian networks is able to detect different kinds of hardware faults on mobile robots. The following scenarios were chosen:

- Fault in the compass sensor
- Fault of one or several wheels of the robot (defined waypoints are not reachable)

As mentioned above all scenarios will be defined as situations. The reasoning process does not only require pure sensor or context data but also an interpretation step considering past events. In order to use the described situation recognition framework the corresponding situation templates have to be defined.

Fig. 3 shows a simple example of a situation template that can be used for integrating general fault detection in a situation template. For these scenarios the corresponding attributes as well as environmental knowledge have to be integrated and adapted.

SituationID = FaultSituation
BayesFaultNetReturn $< | > | = | <= | >=$ *value*

SituationValid = true/false

Fig. 3. Example of a simple general situation template integrating the return values of a Bayesian network

In this example a simple dynamic Bayesian network for the fault recognition process is used. The values for this network can be easily computed without using a lot of hardware resources. Consequently the system can also run on low-end systems often used on mobile robots.

The purpose of the (dynamic) Bayesian network is to recognize hardware faults by observing the change of sensor or context data over a time slot t. Three nodes are used: *Evidence*, *Fault* and *Value*. The *Evidence* node is used to integrate a temporal episode of discretized senor and context values.

The discrete co-domain has to be defined individually for every fault situation. The *Fault* node is responsible for the calculation of the probability of a fault situation considering the evidences. The calculation is done by a summation of the evidences. The conditional probability table of the node is defined according to the corresponding situation that a sequence of evidences that may indicate a fault automatically raises the overall probability of the whole Bayesian network and consequently of the fault situation. Using this coherence the inherent noise of sensor data can be automatically pre-filtered. The *Value* node is a non-dynamic node. It provides the probability of a fault after using the evidences and the calculation method in the *Fault* node. In order to calculate the probabilities time slices are used. They have to be defined according to the fault situation that should be detected. It is important that a Bayesian network always integrates the complete amount of evidences. A network with 100 time slices that only gets 5 evidences cannot return reasonable values.

5.1 Scenario A (Compass)

In this scenario we assume a hardware fault in the compass-module of the robot. For this fault recognition process the orientation of the robot, the angle provided by the compass and the internal state of the compass will be used as input. In the next step the deviation calculated by the angle and the orientation of the robot is used as input for the Bayesian network.

For the calculation the least 10 corresponding time slices are considered. Fig. 4 shows a graphical representation of the measured data in case of a faulty compass and in case of a properly working compass. It can be seen that the compass sensor has variations in the normal operation of up to 35°. Fault cases, which have a critical influence on the success of a complex task, will start from variations of 65°.

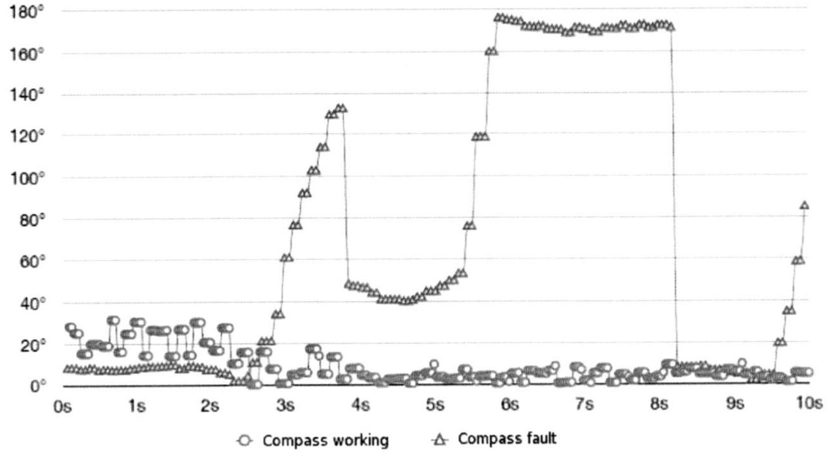

Fig. 4. Deviation of the compass sensor. It is obvious that the compass sensor has variations in the normal operation of up to 35°. Fault cases, which have critical influence for the success of a complex task, will start from variations of about 65°.

Table 1. Probabilities of compass deviation for t and t -1

	Small deviation	Big deviation
No Fault	0,7	0,3
Fault	0,4	0,6

	Deviation (t)			
	Small deviation		Big deviation	
	No Fault (t-1)	*Fault (t-1)*	*No Fault (t-1)*	*Fault (t-1)*
No Fault	0,9	0,6	0,6	0,2
Fault	0,1	0,4	0,4	0,8

To decrease the input-space the angles are discretized as:

- Small deviation $\rightarrow$ angle deviations $< 70°$
- Big deviation $\rightarrow$ angle deviations $>= 70°$

In the next step the discretized deviation is interpreted as evidence and set as input to the evidence node of the Bayesian net and the calculation of the probability is executed. According to the probability table given in Table 1, the probability value of a fault based on 10 sequent observations of evidence e = *Big deviation* will be $P(fault|e) \approx 0,6$. This value will be used further as an input condition of the template in form of a threshold for the recognition process to infer the fault situation compass fault.

5.2 Scenario B (Waypoints)

For executing a complex task, a planner typically splits this task into several ordered primitive-behaviors, which have to be achieved first. Each one of these primitive-behaviors sets a specific intermediate target position to the robot, which has to be achieved. In the case that the robot cannot reach one of these intermediate target positions, we assume that a fault-case occurred, e.g. due to a defect of the wheels. As input data for the recognition of the fault situation "target position not reached" the position calculated from a Monte-Carlo method, as well as the robots intention to reach the target point will be used.

To recognize a fault, the position and intention data during a time interval of 5s (i.e. 10 data points each 500ms) is examined and compared. The following equation gives the relation of the position and the intention:

$$\Delta t = (intention_t, position_t) - d(intention_t, position_{t-1})$$

$d(p1, p2)$ calculates the Euclidean distance between two points in the Cartesian 2-dimensional space.

Fig. 5 shows the values for t in the fault-case and in the normal-case. In Fig. 5 positive values indicate that the distance between the target and current position has increased.

The values in the time interval of 7.5s till 10s represent the movement of the robot in the direction to the target. Because the robot swings across the target position it

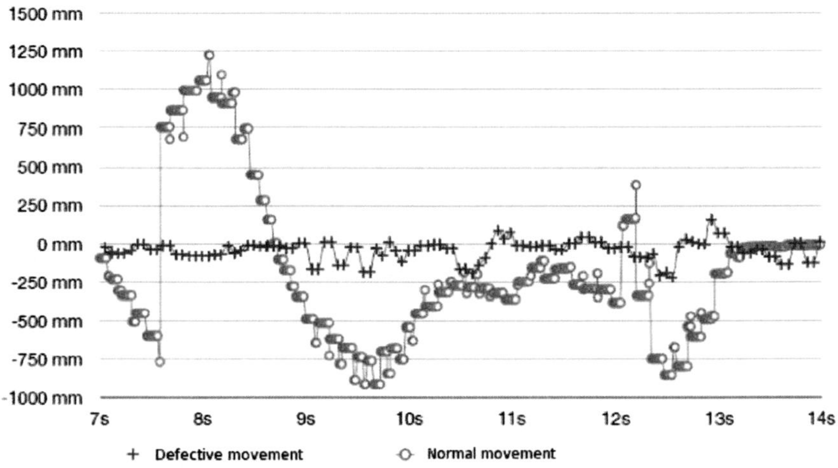

Fig. 5. The graph shows the derivation of the target position over time. In fault-case (blue) it can be seen that the robot does not move and its target position does not change over time.

receives in the next step the new intention in the opposite direction. To achieve this, the movement of the robot has to slow down and accelerate in the opposite direction. This means, even if the robot is in a normal navigate behavior (non-fault-case), the robot does not reach the target position exactly. As you can see in Fig. 5, the robot is neither getting closer nor getting further away from its target in the fault-situation. This implies the definition of the fault - situation: Fault-case means in this scenario and for this robot, that either the robot moves away from its target (minimum distance 0.5 m), or the robot does not approach the target during observation time.

According to this data the values will be discretized into two classes: Positive (Distance) Change and Negative (Distance) Change. Negative Change is true if the distance the robot moves away from its target position is bigger than a defined threshold. This can be formulated according to the following criteria:

$$(|\Delta_t| > \text{MinTraveledDist} \wedge \Delta_t > 0 \vee (|\Delta_t| < \text{SmallDist} \wedge d(\mathit{Intention}_t, \mathit{Pos}_{(t-1)}) > \mathit{PosMargin})$$

Positive Change is defined as opposite condition of the above-formulated criteria. The discretized distance changes are used as input observations for the *Evidence* node of the (dynamic) Bayesian Network. The *Fault* node consists of four $state_i$ variables with $i \in 0...3$. The definition of the probabilities can be found in Table 2. If the evidence *Negative Change* is occurs several times in a row the probibilty for the $state_i$ variables is increasing consecutively. Conseqeuntly the probability $P(state_1)$ increases as soon as $P(state_0)$ as reached a certain threshold. Only if $P(state_3)$ has reached a certain threshold the fault case is true. The occurance of 10 *Negative Change* states is defined as the minimum requirement for a fault case.

Table 2. Definition of probabilities for the Waypoint fault sceanrio ($t = 0$ and $t > 0$)

d(t)	Positive Change	Negative Change
$state_0$	0,99	0,2
$state_1$	0,01	0,8
$state_2$	0	0
$state_3$	0	0

d(t)	Positive Change				Negative Change			
Self[t-1]	$state_0$	$state_1$	$state_2$	$state_3$	$state_0$	$state_1$	$state_2$	$state_3$
$state_0$	1	0,6	0	0	0,25	0	0	0
$state_1$	0	0,4	0,6	0	0,75	0,2	0	0
$state_2$	0	0	0,4	0,6	0	0,8	0,35	0,01
$state_3$	0	0	0	0,4	0	0	0,65	0,99

5.3 Evaluation

For the evaluation of the fault recognition process a set of templates were defined and stored in the situation library, each of them representing a corresponding fault-situation. For fault-recognition each template has been tested and inferred. In the case that a situation occurred and the recognition process result was recognized the process was rated as true if not as false. In the case that a recognition result was incorrect because of another fault in the system, the recognition result was rated as conditional. Furthermore if the case that the situation did not occur even though the recognition process result was true all incorrect recognition results were rated as Situation not valid - Fault.

The evaluation is based on data recorded during the final game of the RoboCup Middle-Size soccer World Championship in Graz 2009. Based on the complete log of every robot the fault recognition process was executed. Furthermore a video of this game was analyzed to find out if the recognized faults really occurred when they were indicated by the system. The data used during the evaluation, ranges over a time period of 30 minutes corresponding to the length of RoboCup game. Table 3 shows the results. According to those measurements it is feasible that the proposed methods bring very good results for the recognition of faults in a real world environment.

Furthermore several different sampling algorithms were examined and their runtimes were compared. For this test a standard PC (Core2 2.2 GHz processor, 2 GB RAM, Linux OS) that is mounted on the robot and also executes all other algorithms, was used. The results of the time measurements are listed in Table 4. Corresponding to those measurements the Clustering algorithm is the best choice for the fault detection scenario. Of course the proposed architecture is not limited to the presented examples, but can be used with any kind of fault detection scenario.

The results of the fault recognition process are used to adapt the behavior of a single robot system or of a complete multi-agent system consiting of several robots. In case of a fault according to *Scenario A* the compass data is not used e.g. for the localisation process. For *Scenario B* the consequences of a fault can be more complex. In the introduced RoboCup scenario robots get certain tasks that differ in their importance. If a robot can

Table 3. Fault analysis of the recognition process rating all correct and incorrect recognized results

Scenario	Situation Valid				Situation not valid
	True	*False*	*Conditional*	*Sum*	
Way Points not Reached					
	15	2	0	17	0
Wrong Localization					
	7	2	1	10	0
Kicker Fault					
	1	0	0	1	0
Compass Fault					
	32	1	4	37	0

Table 4. Overview of the different sampling algorithms and the corresponding calculation times. This demonstrates that the clustering algorithm is the fastest method for calculating the probabilities in this scenario.

Inference Algorithms	**Compass**	Kicker	Local.	**Waypoints**	**ø(ms)**
Clustering	139,40	67,31	70,36	153,57	107,66
Probabilistic Logic Sampling	258,82	130,37	134,87	279,62	200,92
Liklihood Sampling	297,40	151,40	156,44	314,35	229,15
Self Importance Sampling	351,71	179,69	182,83	352,29	266,63
Backward Sampling	276,79	139,59	144,42	285,07	211,46
AIS Sampling	3774,36	2032,35	2086,12	3755,23	2912,02
EPIS Sampling	252,06	126,30	129,21	275,61	195,80

not reach the defined waypoints anymore a wheel failure is obvious. Consequently this robot will not be used for important tasks like getting the ball. It can be still used for taks of minor imprtance. This is handled by a specific task allocation algorithm. The results of the fault recognition process have direct influence on the task allocation in the system. In the worst case the robot has to be removed and repaired. In this case the source for the fault can be easily located and the repair process can be done more fast.

6 Conclusion

To operate successfully in the physical world, a mobile robot needs awareness to adapt its behavior according to the environment, its internal state and in multi robot system even the state of its teammates. In this paper a general and highly customized framework for situation awareness has been presented. The significance of the requirement of adequate fault detection in multi robot systems to improve the system performance has been discussed. In order to evaluate the universal applicability of the framework two typical fault cases were modeled and detected. The interchangeability of the recognition algorithm has been demonstrated by the simple integration of the Bayesian network approach.

References

1. Endsley, M.R.: Toward a theory of situation awareness in dynamic systems. Human Factors: The Journal of the Human Factors and Ergonomics Society 37, 32–64 (1995)
2. Rombaut, M., Nigro, J.M.: Idres: A rule-based system for driving situation recognition with uncertainty management. Information Fusion 4 (December 2003)
3. Regele, R.: Using ontology-based traffic models for more efficient decision making of autonomous vehicles. In: Proceedings of the Fourth International Conference on Autonomic and Autonomous Systems, ICAS 2008, pp. 94–99. IEEE Computer Society, Washington, DC (2008)
4. Breuel, G., Tamke, A., Dang, T.: Integrierte simulations- und entwicklungsumgebung fr die ausweichassistenz zum fugngerschutz. In: Proceedings of the 11. Braunschweiger Symposium AAET 2010 (2010)
5. Dousson, C., Gaborit, P., Ghallab, M.: Situation recognition: representation and algorithms. In: Proceedings of the 13th International Joint Conference on Artifical Intelligence - Volume 1, IJCAI 1993, pp. 166–172. Morgan Kaufmann Publishers Inc., San Francisco (1993)
6. Gertler, J.J.: Survey of model-based failure detection and isolation in complex plants. IEEE Control Systems Magazine 8 (2002)
7. Isermann, R., Ball, P.: Trends in the application of model based fault detection and diagnosis of technical processes. Control Engineering Practice 5, 709–719 (1997)
8. Ferrara, A., Brambilla, D., Capisani, L.M., Pisu, P.: Fault detection for robot manipulators via second-order sliding modes. IEEE Transactions on Industrial Electronics 55, 3954–3963 (2008)
9. Savas, S., Eski, I., Erkaya, S., Yildirim, S.: Fault detection on robot manipulators using artificial neural networks. Robotics and Computer Integrated Manufacturing (2010)
10. O'Grady, R., Christensen, A.L., Dorigo, M.: Synchronization and fault detection in autonomous robots. In: IEEE/RSJ International Conference on Intelligent Robots and Systems, IROS 2008, pp. 4139–4140 (2008)
11. Sheng, B., Benatallah, Q.Z.: Contextuml: a uml-based modeling language for model-driven development of context-aware web services. In: Proceedings of the International Conference on Mobile Business (ICMB 2005), pp. 206–212 (2005)
12. Henricksen, K., Indulska, J., Rakotonirainy, A.: Generating context management infrastructure from high-level context models. In: 4th International Conference on Mobile Data Management (MDM) - Industrial Track, pp. 1–6 (2003)
13. Hofer, T., Schwinger, W., Pichler, M., Leonhartsberger, G., Altmann, J., Retschitzegger, W.: Context-awareness on mobile devices - the hydrogen approach. In: Proceedings of the 36th Annual Hawaii International Conference on System Sciences (HICSS 2003) - Track 9 - Volume 9. IEEE Computer Society, Washington, DC (2003)
14. McCarthy, J., Buvac, S., Costello, T., Fikes, R., Genesereth, M., Giunchiglia, F.: Formalizing context (expanded notes). In: AAAI Fall Symposium on Context in Knowledge Representation and Natural Language, pp. 99–135 (1995)
15. de Kleer, J., Williams, B.C.: Diagnosing multiple faults. Artif. Intell. 32(1), 97–130 (1987)
16. Skoundrianos, E.N., Tzafestas, S.G.: Finding fault - fault diagnosis on the wheels of a mobile robot using local model neural networks. IEEE Robotics & Automation Magazine 11, 83–90 (2004)

17. Forrest, S., Hofmeyr, S.A., Somayaji, A.: Computer immunology. Commun. ACM 40(10), 88–96 (1997)
18. Das, S., Gui, M., Pahwa, A.: Artificial immune systems for self-nonself discrimination: Application to anomaly detection. Advances of Computational Intelligence in Industrial Systems, 231–248 (2008)
19. Cohn, M.: On the opposing views of the self-nonself discrimination by the immune system. Immunol. Cell Biol. 87(2), 113–119 (2009)
20. Canham, R., Jackson, A.H., Tyrrell, A.: Robot error detection using an artificial immune system. In: Proceedings of the 2003 NASA/DoD Conference on Evolvable Hardware, EH 2003, IEEE Computer Society, Washington, DC (2003)
21. Jakimovski, B., Maehle, E.: Artificial Immune System Based Robot Anomaly Detection Engine for Fault Tolerant Robots. In: Rong, C., Jaatun, M.G., Sandnes, F.E., Yang, L.T., Ma, J. (eds.) ATC 2008. LNCS, vol. 5060, pp. 177–190. Springer, Heidelberg (2008)
22. Dekhil, M., Henderson, T.C.: Instrumented sensor system architecture. International Journal of Robotic Research 17, 402–417 (1997)
23. Zweigle, O., Häussermann, K., Käppeler, U.-P., Levi, P.: Supervised learning algorithm for automatic adaption of situation templates using uncertain data. In: Proceedings of the 2nd International Conference on Interaction Sciences: Information Technology, Culture and Human, pp. 197–200. ACM, New York (2009)
24. Zweigle, O., Käppeler, U.-P., Lafrenz, R., Levi, P.: Situation recognition for reactive agent behavior. In: Artificial Intelligence and Soft Computing, pp. 1–8. IASTED, Palma de Mallorca (2006)
25. Zweigle, O., Häussermann, K., Käppeler, U.-P., Levi, P.: Extended TA Algorithm for Adapting a Situation Ontology. In: Kim, J.-H., Ge, S.S., Vadakkepat, P., Jesse, N., Al Manum, A., Puthusserypady K, S., Rückert, U., Sitte, J., Witkowski, U., Nakatsu, R., Braunl, T., Baltes, J., Anderson, J., Wong, C.-C., Verner, I., Ahlgren, D. (eds.) Progress in Robotics. CCIS, vol. 44, pp. 364–371. Springer, Heidelberg (2009)

A Decision System for Aircraft Faults Diagnosis Based on Classification Trees and PCA

ZeFeng Wang, Jean-Luc Zarader, Sylvain Argentieri, and Karim Youssef

Institute for Intelligent Systems and Robotics
University Pierre and Marie Curie (Paris VI)
4 Place Jussieu, Paris 75005, France
{Zwang,zarader,argentieri,youssef}@isir.upmc.fr

Abstract. Aircrafts are complex systems that require permanent and precise monitoring and troubleshooting. The automation of these tasks is thus of a high importance. This paper presents an intelligent decision system for faults diagnosis of aircrafts. The system relies on decision trees, being easier to interpret, quicker to learn than other data-driven methods, and able to work even with missing pieces of information. The used C4.5 algorithm automatically "learns" the best decision tree by performing a search through the set of possible trees according to the available training data. And Principal Component Analysis (PCA) is used to decrease the input data's dimension. Compared to other methods, the proposed one is more advantageous and some presented evaluations demonstrate its abilities. High correct faults detection rates and low missed detection and false alarm rates are obtained. Such a decision system is highly useful for engineering consulting services, accumulating the knowledge for the operational rules of diagnosis, and the design of new aircrafts.

Keywords: Faults Diagnosis, Aircrafts, Decision System, Decision Tree, PCA.

1 Introduction

Faults diagnosis is of a high importance for modern aircrafts, with structures and systems becoming more and more complex. Hundreds of sensors are being used to supervise an aircraft, but abnormal information and faults are still difficult to find. In order to deal with this problem, many faults diagnosis systems have been invented, with intensive studies of data mining. One can cite: the expert system [1] that needs to establish the knowledge base and the rules by experts and has to face an intractable problem - the conflict of rules [2]. Also, neural networks [3] and Support Vector Machines (SVMs) [4] which require much more time for the learning process.

In this paper, a novel intelligent decision system for faults diagnosis of aircrafts is proposed. We decided to use a decision tree learning algorithm since it provides much easier data to interpret than other algorithms, such as neural networks and SVMs [5]. Decision tree learning is a method for approximating discrete-valued target functions [6]. A decision tree algorithm automatically "learns" a decision tree by performing a search through the space of possible trees to find the one that best fits the training data.

S. Lee et al. (Eds.): Intelligent Autonomous Systems 12, AISC 193, pp. 411–422.
springerlink.com

The particular algorithm used in this paper is known as C4.5 (see Section 2). As Figure 1 shows, the decision system uses the output information of sensors, and its diagnosis results are delivered to the fault tolerance system and the decision system which will demonstrate the faults and suggest decisions to the pilot and the ACC (airport command center) (emergency landing, return to base or mission continuation if the fault tolerance system can deal with the fault, etc.). In accordance with the results of the decision system, the pilot or ACC can decide their next action.

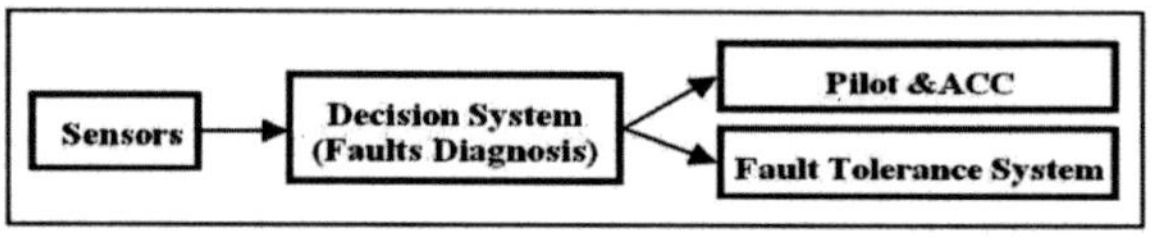

Fig. 1. Overall Diagnosis System

Aside the normal working conditions of the aircraft and thus of the fault diagnosis system, the real risk events in the system should be taken into consideration during a flight. Such cases can be seen for example in the loss of some sensors' information in special situations, especially for the fighting aircrafts. And in these cases, the diagnosis system should be ensured of working continuously. The presented system is able to deal with these cases. This paper studies the case where 3 faults of the anti-icing system of the aircraft X (Dassault Aviation) need to be diagnosed. Usually the hot air exhausted by the aircraft's engine is used in the anti-icing system to de-ice the aircraft wings. 33 sensors supervise the anti-icing system. They mainly provide values of the engine and air wings' temperatures, pressures and so on. So our diagnosis system has input vectors of a dimension 33 used to analyze and build the decision tree. Note that the natures of the faults cannot be revealed in this paper because of confidentiality issues.

The rest of this paper is organized as follows. Section 2 demonstrates the classical decision tree fundamentals, and shows a study of the tree's over fitting and the pruning of unnecessary branches. Section 3 demonstrates the usage of PCA in decreasing the initially high input data dimension. Both the procedure of that application and its effect on the system are included in this section. Section 4 reveals the robustness and the fault-tolerance of the system even in the case where some sensors are lost. Section 5 shows the results of evaluations studying the cases presented before; and the last section concludes the paper and presents a preview of future works.

2 Decision Tree

2.1 Decision Tree Fundamentals

Decision tree learning is a method commonly used in data mining. Its goal is to create a model that predicts a target value (that can be the class of an input example), based on several input values that describe the example. This method uses a hierarchical tree structure that progressively checks input variables and makes corresponding decisions like classification. Like the system of expert, learned trees can also be re-represented as sets of if-then rules to improve human readability, but they don't need human experts to establish a knowledge and rules database, they build rules by themselves.

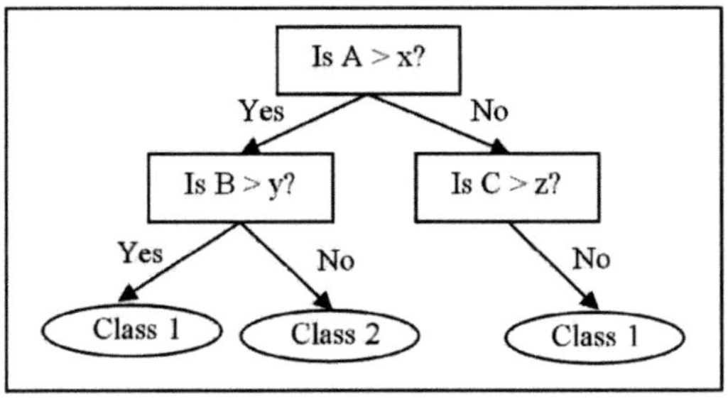

Fig. 2. A decision tree used to discriminate 2 classes with inputs consisting of 3 values (A, B and C)

An example is shown in Figure 2. Data represented by three input variables can be classified into two classes. Each node corresponds to one of the input variables – A or B or C, and inside each node, the variable is compared to a corresponding edge. The result of this comparison is translated by a leaf that leads to another node, or to the final classification decision. But in the presented example, one can ask several questions, like: why is it that A is in the root, not B or C? How was the threshold x found for A? So, in a decision tree algorithm, and in a more general manner, two questions are to be answered, in order to find the best possible classifier: 1) How to decide the sequencing of variables and corresponding nodes in the tree? 2) How to determine the used thresholds values?

In order to answer the first question, and thus to determine the best sequencing of the variables, we resort to a measurement that is commonly used in information theory, called entropy. Entropy characterizes the impurity of an arbitrary collection of examples [7]. For example, given a collection S containing positive and negative examples of a target concept, the entropy of S relative to this Boolean classification is:

$$Entropy(S) = -P_{\oplus}\log_2 P_{\oplus} - P_{\ominus}\log_2 P_{\ominus} \quad (1)$$

Where $P_{\oplus}$ and $P_{\ominus}$ are respectively the proportions of positive and negative examples in S. Note that in calculations involving entropy we consider 0* log (0) to be equal to 0. In our case, a collection will be the set of values provided by a sensor, that is to say, a variable. Such a variable can be Boolean, like in the example, or can take multiple values. So at first, we calculate the entropies of all the variables. Then, a measurement of the effectiveness of an attribute or value that a variable can take is needed. We use a measure called information gain that is simply the expected reduction in entropy caused by partitioning the according to the attributes and values that a variable can take. More precisely, the information gain, Gain(S, A), of an attribute A, relative to a collection of examples S, is defined as:

$$Gain(S,A) \equiv Entropy(S) - \sum_{v \in Values(A)} \frac{|S_v|}{|S|} Entropy(S_v) \quad (2)$$

where values(A) is the set of all possible values for the variable A, and S_v is the subset of S for which the variable A has the value v. So, as said before, at first the entropies and the information gains of all the variables are calculated. The variable with the highest information gain is the root node from which leaves go out, each leaf corresponds to a value that its variable can take. Following this, entropies and information gains corresponding to the other variables and to each of the leaves going out of the root nodes are computed, and new nodes are taken corresponding to the highest gains. And so on, until all variables are used and the final tree is formed. Classification is done by using the values of the variables constituting an example and following the resulting tree leaves and nodes from the root to the final leaf which gives the class.

In this system, all the input variables are not Boolean but can take multiple values. Therefore, the values of each attribute are sequently arrayed firstly, and then all the expected pair values which are neighboring but sorted in different classes are picked out. After that, all possible threshold values are calculated, as the halves of the sums of the pairs mentioned before. Finally, the threshold value with maximum information gain is taken.

2.2 Tree Pruning to Avoid over Fitting

Cross Validation (CV) is used to determine how deep the decision tree will grow. This is known as the post-pruning rule. Being used by the C4.5 algorithm [8], it involves the following steps:

1. Use the CV to infer the decision tree from the training set, grow the tree until the training data is as fit as possible and allow over fitting to occur.
2. Convert the learned tree into an equivalent set of rules by creating one rule for each path from the root node to a leaf node.
3. Prune (generalize) each rule by removing any preconditions that result in improving its estimated accuracy.
4. After sorting the pruned rules based on their estimated accuracy, involve them into classification of subsequent instances.

So in summary, CV performs a search through multiple pruned tree configurations, and keeps the configuration that gives the best validation results. In this paper, the decision system is built with 33 sensors' information. As shown in Figure 3, CV operated during a learning procedure allowed us to find the best size of the tree based on these 33 sensors, with 11 terminal nodes. This is obtained while the number of nodes without CV can actually grow much higher than 11. Figure 4 shows the pruned tree, originally 59 nodes are obtained without pruning, the dotted lines are the leaves

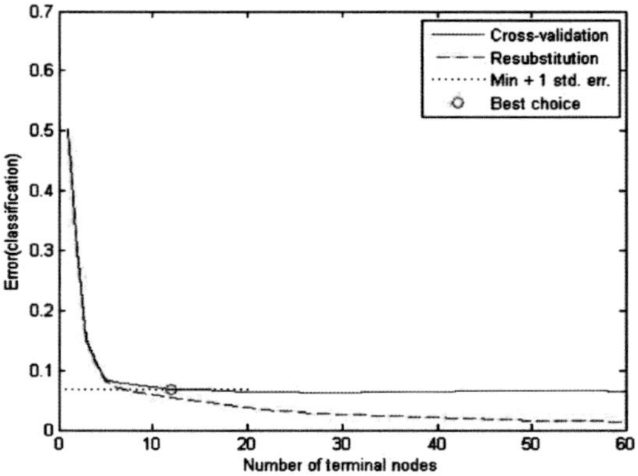

Fig. 3. Using Cross-Validation to find the optimal tree size

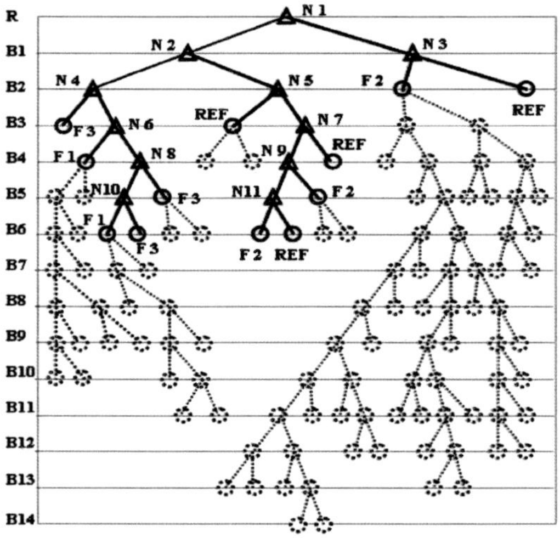

Fig. 4. Diagnosis tree and leaves to prune, R: Root; B: Branch (B1-B14); N: Node (N1 – N11); F1: Fault 1; F2: Fault 2; F3: Fault 3; REF: No-Fault

to prune, and the remaining nodes contain the rules that the system finally used. It is clearly that there is one root with 14 branches in the original, and only 6 branches left after optimization, which highly enhances the working efficiency.

3 Principal Component Analysis

For the sake of safety, plenty of the sensors installed in the aircraft are redundant. Multiple sensors can do the same role, in order not to lose the information when one of them malfunctions. Indeed, the sensors' redundancy is indispensable for the aircraft, especially with sensitive and important parts. But for the diagnosis system used here, using all the sensors' data directly in the input space at once is not very necessary and it affects the arithmetic speed of the diagnosis system. Principal Component Analysis (PCA) is employed, as being a tool that effectively reduces the input data dimension and improves the computational loads of the diagnosis process, and thus of the overall system. PCA is mathematically defined [9] as an orthogonal linear transformation that converts a set of observations of possibly correlated variables into a set of values of uncorrelated variables called principal components. The data is projected to a new coordinate system where the greatest variance of the data obtained by any projection comes to lie on the first coordinate (principal component). The second greatest variance lies on the second coordinate, and so on. The number of principal components is generally less than the number of original variables. So PCA reduces the dimension of our aircraft diagnosis system input data. For example, in our case, Figure 5 shows that the first principal component's variance constitutes more than 50% of the total sum of all variances over all components. This variance decreases as the order of the corresponding component increases. Finally, 6 principal components were found to add up to 95% of the total sum of the variances in our case. Note that these principal components are computed based on the 33 used anti-icing system sensors and thus a big dimension reduction is performed, while keeping most of the beneficial information.

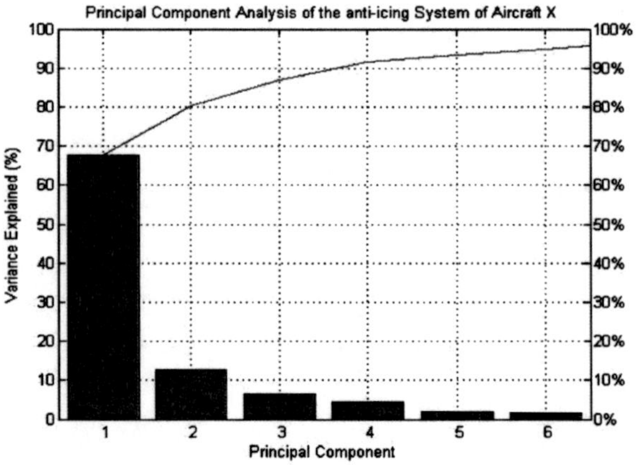

Fig. 5. PCA of the anti-icing system of the aircraft X: 6 first principal components, their corresponding variances and the cumulative variance

4 Information Loss and Rebuilding of the Tree

In the real conditions of an aircraft's flight, especially with the fighting aircrafts, there is a risk of losing sensors or sensors' information. This loss of information is harmful for the diagnosis system's operation as it can lead to inadequate responses. The problems of such cases especially arise with diagnosis systems using neural networks or SVMs, where the absence of signals or the zero-valued inputs lead to inactivity of certain parts of the networks. However, while the diagnosis tree cannot resolve this problem, it can perform a quick relearning based on the initial learning database and using only the data of the currently active sensors. For example, with a computer – Intel Pentium Dual-Core T4400 and 2G memory, it only needs 46 seconds for relearning with a 33x160000 dataset while with an SVM or a neural network it takes more than one hour.

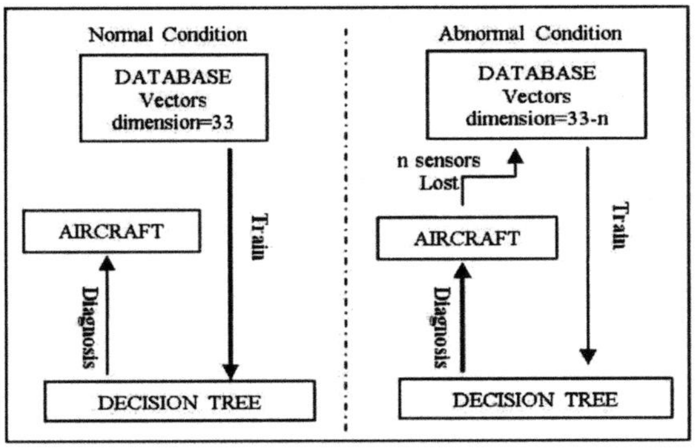

Fig. 6. Relearning of the diagnosis tree

Figure 6 shows data provided by 33 sensors and m learning examples. If the aircraft loses n sensors or sensors' information, the diagnosis tree will automatically detect the loss and get relearned using new vectors of dimension 33-n. This ensures an appropriate online functioning of the system during a flight.

5 Results

5.1 Experimental Procedure

A database consisting of 160000 examples (160000 records, one record per second) is used to build the diagnosis system and test it. It is constituted of data vectors obtained with the pre-described 33 sensors of the anti-icing system of the aircraft X. Data is measured during a flight and describes the faults and the normal aircraft state without faults. 75% of the data were chosen randomly for the training and the remaining 25% are used to test the tree.

During the training, CV is used to avoid the over fitting problem. Subsection B shows the PCA's utility in reducing the data dimension, subsection C simulates a dangerous situation of sensors loss and presents corresponding results. In order to get a valid scientific and compelling evaluation scheme, a trail of one hundred repetitions of all the processes is studied as shown in Figure 7. The evaluations study the ratios of the following criteria: Correct Decision, Missed Detection, and False Alarms as shown in the last two subsections.

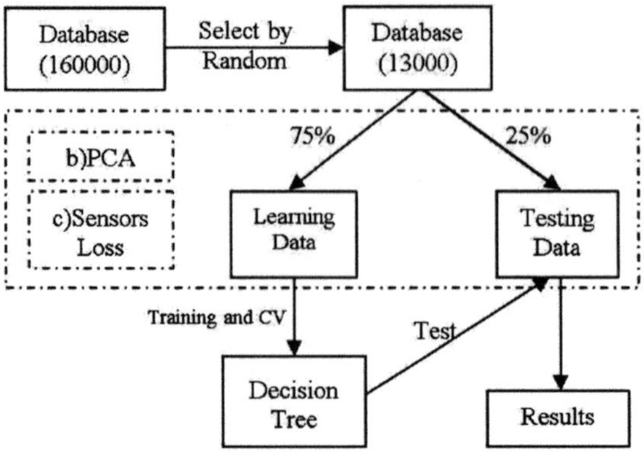

Fig. 7. Experiments Drawing

5.2 Results in the Normal Condition

The test results will be presented in two forms corresponding to two different conditions: the condition of maintenance and the condition of mission on-line.

1) In the condition of maintenance, the system must distinguish clearly each fault and show the diagnosis results to an engineer. With the Confusion Matrices which are shown in the Tables 1 - 4, three observations can be reached:

1-a) Performance of system with PCA: Although PCA reduces the dimension of the data; the performance of the decision system is not reduced. The score of the system based on PCA is nearly the same as that of the system without PCA, especially for the systems with a pruned tree. For example, the score of Table 2 is 95.16% and the score of Table 4 is 95.17%.

1-b) Performance of system based on Pruned Tree: In comparison with Table 1 and Table 2 (or Table 3 and Table 4), as mentioned in Section 2, the results confirm that a system with a pruned tree works better than a system with a tree without pruning. The pruned trees add about two to four percentage points to the scores of diagnosis.

1-c) Research of faults with confusion matrix: Some words about data mining of Aircraft X's anti-icing system: as shown in these tables, the fault 2 and fault 3 are close to each other and Ref and Fault 1 are similar. In these tables, the system never confused Ref or Fault 1 with Fault 2 or Fault 3, and vice versa. In addition, Fault 2 can be perfectly diagnosed by a system with a pruned tree.

Table 1. Confusion matrix, testing with a whole tree (without pruning), Score = 93.64%

Target/Output	REF	Fault 1	Fault 2	Fault 3
REF	93.67%	6.33%	0.00%	0.00%
Fault 1	2.61%	97.39%	0.00%	0.00%
Fault 2	0.00%	0.00%	94.59%	5.41%
Fault 3	0.00%	0.00%	11.11%	88.89%

Table 2. Confusion matrix, testing with a pruned tree, Score = 95.16%

Target/Output	REF	Fault 1	Fault 2	Fault 3
REF	98.73%	1.27%	0.00%	0.00%
Fault 1	6.96%	93.04%	0.00%	0.00%
Fault 2	0.00%	0.00%	100.00%	0.00%
Fault 3	0.00%	0.00%	11.11%	88.89%

Table 3. Confusion matrix, testing with a whole tree (without pruning), and PCA = 95% (with 6 vectors), Score = 91.65%

Target/Output	REF	Fault 1	Fault 2	Fault 3
REF	94.50%	5.50%	0.00%	0.00%
Fault 1	8.14%	91.86%	0.00%	0.00%
Fault 2	0.00%	0.00%	92.11%	7.89%
Fault 3	0.00%	0.00%	11.76%	88.24%

Table 4. Confusion matrix, testing with a pruned tree, and PCA = 95% (with 6 vectors). Score = 95.17%.

Target/Output	REF	Fault 1	Fault 2	Fault 3
REF	93.58%	6.42%	0.00%	0.00%
Fault 1	1.16%	98.84%	0.00%	0.00%
Fault 2	0.00%	0.00%	100.00%	0.00%
Fault 3	0.00%	0.00%	11.76%	88.24%

2) In the condition of mission on-line, the pilot doesn't need to know exactly which type of fault of anti-icing system appeared, he only needs to know that the anti-icing system is broken or not. If it doesn't work, he must go back to ACC or make an emergency landing. Therefore, all the types of faults are grouped as one fault in this condition, the correct detection ratio; missed detection ratio and false alarm ratio will be studied in this part.

Table 5. Testing results with a tree without pruning

CASE	Amount of vectors	Correct Detection	Missed Detection	False Alarm
Without PCA	33	96.49%	3.51%	4.42%
PCA 99%	12	96.04%	3.96%	5.54%
PCA 95%	6	95.49%	4.51%	5.52%
PCA 90%	4	90.73%	9.27%	7.84%
PCA 80%	2	85.06%	14.94%	8.96%

Table 6. Testing results with a pruned tree

CASE	Amount of vectors	Correct Detection	Missed Detection	False Alarm
Without PCA	33	97.43%	2.57%	4.16%
PCA 99%	12	97.08%	2.92%	4.76%
PCA 95%	6	95.62%	4.38%	4.76%
PCA 90%	4	92.95%	7.55%	5.45%
PCA 80%	2	88.83%	11.17%	5.96%

As can be seen in Table 5 and Table 6, the system offers satisfying performances with high correct detection rates, and low missed detection and false alarm rates. The performances slightly decrease with decreasing input data dimensions, but the results remain highly accurate. This may be due to one of two hypotheses:

a) The faults detection and discrimination remains easy for the system, even with a change of input conditions and tree structure.
b) PCA finds the principal components and the system can get well adapted to them and classify the aircraft's situations.

But, there still exists a big problem: all the data is recorded per second during the mission of the aircraft, if there is 5% missed detection ratio or false alarm ratio, the pilot will receive the missed detection or false alarm 3 times per minute, which is unacceptable in the real applications. To solve this problem, statistics of the results of the diagnosis system during a period of time, as 15 or 30 seconds, are studied. That is to say, the system records the diagnosis results from a time; then compares the fault's ratio diagnosed in the results with pre-set threshold; at last the system will judge it as fault occurred if its value is above the threshold. For example, the pre-set threshold is 15%, and only two results are diagnosed as fault by the system during 15 seconds. Since the frequency is below the threshold, the diagnosis of these 15 seconds is considered as "no-fault". This criterion is used to test the continuous data of 18000 seconds: the Correct Detection ratio equals 100%, the Missed Detection ratio equals 0, and the False Alarm ratio equals 0.0013% (using 6 vectors of input data whose PCA is about 95%).

5.3 Results in the Abnormal Condition

In this subsection, a state of emergency is described: a number of sensors are lost. To simulate this, at first the number of the sensors is randomly reduced, without PCA. And then PCA is used to pre-treat the data if a few sensors like 1 ~ 5 are lost. In this case, PCA wasn't able to reduce the dimension of the input data if there were already many sensors lost. We run the system one hundred times, and at each time sensors are lost by random; at last we calculate the mean of the correct detection, the missed detection and the false alarm of the tests series.

Table 7. Results with a tree without pruning in an abnormal condition

Lost Sensors	Correct Detection	Missed Detection	False Alarm
3	95.46%	4.54%	4.51%
5	93.06%	6.94%	8.10%
10	90.71%	9.29%	13.56%
15	83.84%	16.16%	18.25%

Table 8. Results with a pruned tree in an abnormal condition

Lost Sensors	Correct Detection	Missed Detection	False Alarm
3	97.25%	2.75%	2.77%
5	95.85%	4.15%	5.44%
10	93.57%	6.43%	8.67%
15	87.94%	12.06%	11.03%

Table 7 and Table 8 show the stability of the system within several lost captors. While the missed detection and false alarm ratios rise with the loss of sensors information, the performances of the system remain sufficiently accurate.

6 Conclusion and Future Works

In this paper, a faults diagnosis system of aircrafts using decision trees is presented. The learning speed of this intelligent system is much higher than the speed of other diagnosis systems, like neural networks or SVMs, which allows it to perform a quick relearning in the cases of data loss. PCA is used to accurately reduce the system's input dimension. Tests results of normal and abnormal flight conditions demonstrate the high abilities of the system. For the future work, we will focus deeply on how to prognosis the situations (failure prediction) of an aircraft with a decision system.

Acknowledgements. All the data are supplied by Project MODIPRO "MOdelisation, DIagnosis and PROnostic".

Financiers : DGCIS, Conseil régional d'Ile de France, Conseils généraux 77 et 78.
Partners: DASSAULT AVIATION, ISIR/UPMC, IT4 CONTROL, INRIA, SupElec, SNECMA, BAYESIA, KBS.

References

1. Chen, W., Liu, X., He, C., Liu, Y.: Knowledge Base Design for Fault Diagnosis Expert System Based on Production Rule. In: 2009 Asia-Pacific Conference on Information Processing (2009) 978-0-7695-3699-6/09 IEEE
2. Chen, W., Liu, X., Fang, Y., Zhang, J.: Inference Engine Design of Expert System Based on Blackboard Model and Fault Tree. In: 2009 Asia-Pacific Conference on Information Processing (2009) 978-0-7695-3699-6/09 IEEE

3. Wei, L., Hua, W., Pu, H.: Neural Network Modeling of Aircraft Power Plant and Fault Diagnosis Method Using Time Frequency Analysis. In: 2009 Control and Decision Conference (2009) 978-1-4244-2723-9/09 IEEE
4. Li, B., Zhang, W.-G., Ning, D.-F., Yin, W.: Fault Prediction System Based on Neural Network Model. In: Innovative Computing, Information and Control Conference (2007) 0-7695-2882-1/07 IEEE
5. Schwabacher, M., Aguilar, R., Figueroa, F.: Using Decision Trees to Detect and Isolate Simulated Leaks in the J-2X Rocket Engine. IEEEAC 1408, Version 1 (2008)
6. Jamehbozorg, A., Mohammad Shahrtash, S.: A Decision-Tree-Based Method for Fault Classification in Single-Circuit. IEEE Transactions on Power Delivery 25(4) (October 2010)
7. Mitchell, T.M.: Machine Learning. McGraw-Hill Science/Engineering/Math. (1997)
8. Quinlan, Ross, J.: C4.5: programs for machine learning. Morgan Kaufmann Publishers Inc. (1993)
9. Jolliffe, I.T.: Principal Component Analysis, 2nd edn. Springer Series in Statistics. Springer, NY (2002)

A Sociology of Intelligent, Autonomous Cothinkers and Coagents

Jean-Daniel Dessimoz[1], Jean-François Gauthey[1], and Hayato Omori[2]

[1] HEIG-VD, School of Business and Engineering
HES-SO, West Switzerland University of Applied Sciences
CH-1400 Yverdon-les-Bains, Switzerland
[2] Department of Precision Engineering, Chuo University, Tokyo, Japan
Jean-Daniel.Dessimoz@heig-vd.ch,
h_omori@bio.mech.chuo-u.ac.jp

Abstract. Scientific and technological progress has brought robots where machine-based cognition and cooperation abilities start to emerge; not only between robots and humans but also among multiple robots themselves. In order to technically improve performances in latter case, as well as, by analogy, to better understand how humans can interact with one another and grow communities, time as come to further, scientifically and technically develop sociology-related knowledge and ontologies. Critical theoretical bases for cognition have been built and demonstrated, both in the human and machine-based cases, providing valuable contributions in this regard. Now sociable competences are considered, allowing for incrementally binding individuals and small groups into holistic units of increasing scope. Ultimately, what is also considered here is a kind of common, meta-human, secular framework where robots and humans can best co-think and co-act. Concepts have now been complemented and validated by real size implementation and experimentation in the domain of homes, as well as industrial and public environments. This should motivate the reader to get familiar with the proposed formal, quantitative MCS framework, thereby getting better insight in judgment and better ability to quantify requirements.

Keywords: Sociology of Robots, Sociology of Automated Systems, Meta-human Secular Sociology, AI, Robotics, Robocup-at-Home, Cooperating Robots, Service Robots, Sociable Robots.

1 Introduction

International initiatives, and in particular "RoboCup" [1, 2], which began in the late 1990s, have given a boost to progress in the fields of Robotics and Artificial Intelligence (AI) and, more generally, machine-based cognition. In particular, in the cognitive context, it has become more and more apparent that a solid formal basis and a quantitative approach are necessary to be successful. Consider the analogy for a human to jump over a wall: success is strongly related to the height of the wall; 30 cm is easy, 3 m impossible. Similarly in the cognitive domain, appropriate metrics are necessary.

S. Lee et al. (Eds.): Intelligent Autonomous Systems 12, AISC 193, pp. 423–434.
springerlink.com

When an attempt is made to estimate quantitatively the complexity of any domain of reality, it appears that the domain is not within reach. Some kind of modeling is always required, and this modeling can make sense only when very specific goals have been defined [3].

History shows that humans generally have a particularly strong sense of curiosity and strive to make their lives easier. The research and development of intelligent robots that will be capable of helping humans in all kinds of situations has thus become a desirable goal. Such a general goal however may be too broad; therefore, it was initially decided that the goal would be to design artificial soccer players. Then later some researchers found it preferable to specify a goal that seemed to call for more complex cognitive situations and moreover would be likely to deliver more obviously useful benefits for humans: domestic helpers [4, 5]. Concrete examples will be shown in this context, where several levels of multi-agent systems can be found: 1. about 20 of them in the core processing of supervisory Piaget environment; 2. about 20 of them at robot-level where specialized resources act as specialized agents (computers, servocontrollers, cameras, arms, audio units, PLC, etc.); and 3. at group level where individual robots (RH-Y, OP-Y, Nono-Y) and humans cooperate.

Thus now the interesting shift in paradigm goes from individual bodies or systems (humans, robots) towards collective entities, including robot-human associations, or groups of multiple robots. This is the essence of social systems; the Latin origin of the word "social" refers to the notions of united, allied, living or liking to live with others (re. Online Etymology Dictionary). Some researchers have started to address the topic of a sociology of robots (re. in particular the International Journal of Social Robotics, Springer; and Interaction Studies, the journal of John Benjamins Publishing Comp that explores Social Behavior and Communication in Biological and Artificial Systems). This paper has a similar purpose, yet it differs in its efforts to connect to two fundamental areas; the first one is the classical (human-related) sociology corpus of knowledge (ref. follow in the paper), and the second one hosts the formal and quantitative definitions of the Model for Cognitive Sciences (MCS, [6]). In the latter case, we focus here on new material relating to sociologies, but more can be found on closely related material in our other publications and online videos, in particular in terms of emotions and validation of concepts in world-level contests [7].

The paper is organized as follows. Section 2 summarizes the general concepts of machine-based cognition and cooperation; Section 3 focuses on human and robotic arts of cooperation, with some grounding examples relating to the "RoboCup@Home" league; Section 4 discusses the principles of sociology from a cognitic and robotic perspective; Section 5 introduces a meta-sociology where humans and machines are associated as co-thinkers and co-agents, and where soccer players and domestic helpers find their place.

2 Machine-Based Cognition and Cooperation

Intelligent robots are a kind of "thinking machine", as defined in the middle of 20th century [8]. Machines have many capabilities, in particular to physically sense and act in the world. Cognition is a special capability though, the only one to be explicitly referred to with the specific attributes of "intelligent" or "thinking" or "with A.I." added to the general notions of machines and robots.

When machines look like humans, the boundary between these two groups is blurred. Statues have been built for ages, and machines and automata have been commonly used for centuries. Robots, then humanoids (functionally similar to humans), and androids (humanoids moreover looking like humans) even more so, bring the boundary of machine domain even closer to the one of humans. Yet, cognition is a very controversial area. Some people consider it to be a faculty continuously extending toward human capabilities (and possibly beyond) from some of the capabilities of very basic existing machines; for others, cognition does not follow from the current capabilities of existing machines and is definitely restricted exclusively to the human domain (notice that for the latter group, AI is essentially impossible). This controversy, or the related fuzzy state of the situation, may call for explicit and separate mention of the cognitive properties ("intelligent", "thinking"), the only specific terms used to complement the general words "robot" or "machine", that implicitly represent all of the other, widely accepted components in very complex systems.

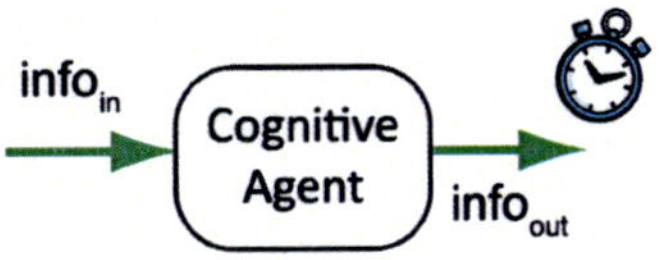

Information:	$n = \sum p_i \log_2(1/p_i)$ [bit]
Knowledge:	$K = \log_2(n_{out}\, 2^{n_{in}} + 1)$ [lin]
Fluency:	$F = 1/\Delta t$ [s^{-1}]
Expertise:	$E = K \cdot F$ [lin/s]
Learning:	$\Delta E = E(t_1) - E(t_0)$; >0 [lin/s]
Experience:	$R = r(n_{in} + n_{out})$ [bit]
Intelligence:	$I = \Delta E / \Delta R$ [lin/s/bit]
relative Agility:	$A_r = \tau / T$

T: Fluency^{-1} and communication delays
τ: Reaction time of target system, to be controlled

Fig. 1. The MCS framework provides essential definitions and metrics to evaluate complexity, information flows (re. Shannon), knowledge, expertise and other cognitive and dynamic properties, quantitatively and conveniently. This is in particular useful for RoboCup@Home competition task specifications; and/or for rating demonstrated performance level of participating robots. Just as metric units conveniently allow for describing the height of a wall. Some illustrated elements: behavioral description of cognitive agents (*left*), and core definitions and metrical equations in MCS (*right*).

Length and time can be quantified with their specific units, meter and second, no matter whether the context relates to a machine or to a human. It is our conviction that, in the same way, cognitive properties can be similarly quantified both for the case of humans and for robots. Inheriting from the advantages as well as from the essential limits of the concepts of model and information, cognitive concepts such as complexity, knowledge, expertise and intelligence have been rigorously defined [6]. It is therefore demonstrated, in the case of RoboCup as well as in many other daily applications that machine-based cognition has not only emerged but has even become pervasive.

The "Model for Cognitive Sciences" framework, MCS, provides mathematical definitions and measurement units for the core entities in cognition – natural or automated. As concisely shown in Fig.1 and Fig.2, the core properties in cognition domains can be conveniently and quantitatively characterized by a cognition calculus, ultimately based on information calculus and time measurements. MCS being firmly rooted in information theory domain, similar specific properties are present in terms of sensitivity to time, subjectivity, and of course, like always since humans exist, on some more or less implicit goal-oriented world modeling.

Life implies more than cognition. It also involves materials, energy, mechanical structures, in one word "operation" or as implied from etymologic root, some kind of work. Furthermore, it is evident that not much can be done alone and consequently cooperative approaches, i.e. some kind of social organization are required.

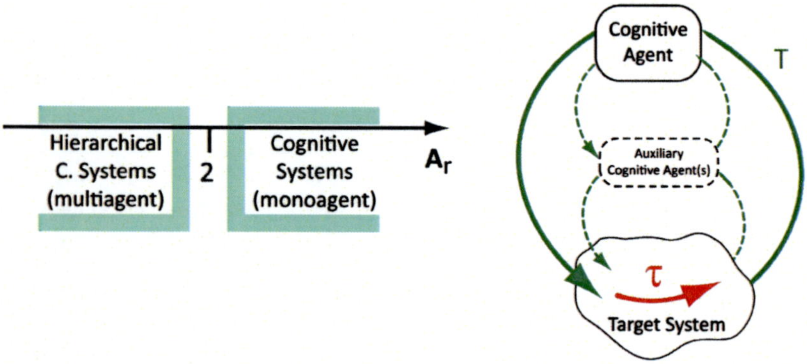

Fig. 2. Other illustrated elements of MCS framework : Agility criterion for selecting a structure adapted to system dynamics (*left*); structure and critical dynamic aspects of controlled systems (*right*)

The role of machines and technologies in the production and processing of materials, energy and mechanical structures is well identified and keeps progressing as it has done so for centuries. And as just mentioned in previous paragraph, cognition has gained a sound theoretical basis, and machine-based cognition is already very significantly operative.

So a critical factor that remains to be addressed now is a scientific and technical delineation of (human) sociology, then itself, if required, possibly enlarged to accommodate artificial cognitive agents, robots and other systems.

3 Human and Robotic Arts of Cooperation

Life requires cooperation. This section presents three complementary arts of cooperation. It starts with the human-robot kind of cooperation, a need that is made evident e.g. in the practical and representative case of domestic service robots; with some grounding examples relating to the "RoboCup@Home" league. Then cooperation among robots is reviewed. And finally the complex case of human sociology is addressed.

3.1 Human - Robot Cooperation

From an historical and conceptual perspective, robots have been invented as systems essentially capable of cooperating with humans. Robots were sorts of slaves serving humans in the original art and literature context for which the word was coined. Then Alan Turing, without fully reaching to cooperation was nevertheless defining possible intelligence in machines essentially as a capacity of effectively (and with anthropomorphic style) chatting with humans. Asimov introduced his famous three laws of robotics with a critical reference to humans.

The topic of Human-Robot Cooperation has also been concretely addressed for several years (e.g. survey in [9]), even though such robots are not found to be ready yet for large-scale implementations in practice.

Various degrees of socially relevant interaction have been defined, ranging from relatively passive solutions to more proactive, sociable systems.

In RoboCup@Home context, vocal and visual communication, ad hoc body presence and motions, as well as close or remote gestures, possibly detected by machine-specific sensors and interfaces, are often practiced, with a certain success (like jumping over the wall). Humanoids have also been used as mediators between humans and other machines, as shown for example in Singapore in 2010 (team RH-5, re. Fig. 3), where our robot group, PRG-Y, has performed; it can been developed, programmed, and controlled in real-time with our Piaget environment and language.

Fig. 3. Human and robot share their representations, the robot following first the human, and then both vocally synchronizing their respective English names for describing specific locations, such as the plant in the living room (*left*). Nono-Y, our Nao-typed humanoid, on the lower right, ensures the mediation between human and other machines (OP-Y platform where Nono-Y sits; and RH-Y robot, which has brought drink and snacks) (*right*).

From a scientific and technical point of view, the main simplification for machines is that the limits of human interaction apply; human communication capacity is relatively small, in particular in terms of perceptive and expressive, physical channels, as well as in terms of signal to noise ratio and bandwidth, i.e. information flows (30 bit/s). On the opposite, the main challenge is probably related to the lack of shared experience between machines and humans (e.g. known objects, names of robots and people, etc.), thereby preventing a full communication capability. Notice that this latter handicap similarly applies to mutual communication among humans in their huge (life-styled and cultural) diversity (e.g. language mismatch). Robots have a limited capability to share human-like emotions at group level, but on the other side, they offer novel specific channels, such as graphic displays and tactile screens, very information-rich and robust against disturbances, for coordination with human partners [7].

3.2 Cooperation among Robots

Cases of cooperation of multiple robots have also been documented, including more extreme systems like swarm robots, where many units cooperate.

Technically, the boom of communication hardware, networks and procedural techniques has brought a practically unlimited quantity of physical resource for inter robot exchange of information. On the operational side, servo-ed or compliant techniques allow for the effective cooperation of multiple agents.

In the so-called field of "sociology of robots" a niche has been considered for cases where only robots interact and cooperate. Freedom is much larger than in the case of human-robot interaction, for considering alternative physical channels and modes, even if a similar paradigm remains in terms of conventions and shared values. Section 4 will develop this common paradigm.

Robots are machines and as such they do benefit from numerous developments in terms of technologies and standards, which have been developed for other applications and now are useful as a common culture (e.g. Ethernet, TCP/IP protocol).

3.3 Cooperation among Humans

Humans are social being living in communities of various sizes, ranging from a minimal scope of couples or families, the small scale of prehistoric tribes, to the pan-telluric collective body of modern times.

Socialization implies mutual help and also mutual control of individuals. Concepts tightly bounded to the notion of society usually include the ones of common culture, recurrently, religions, stratification of classes and networks of influence.

Complex social structures have especially developed in the context of cities, leading to the notion of political organization, which along with secularization has spread and finally lead to the dominant, current, world-level, global order. At large-scale, human societies are organized as nations all coordinated in UNO.

Sociology is the core discipline for studying cooperation among humans. Established to apply scientific rationalism to human behavior (Comte, Durkheim), it integrated the progressing power of conflicts (Marx), and soon attempted to interpret the causal and goal-oriented properties (meaning) of social action (Weber). Language has been considered as a rich repository of social references and constraints (de Saussure). Nowadays, multiple dimensions are considered although the tension between individuals (agency) and collective norms (structure) remains the most significant one in sociology (e.g. Giddens).

As history shows, social structures have their own life cycle and from time to time human communities adapt themselves to new situations, sometimes with small, gradual changes, and sometimes with brisk discontinuities (revolutions), always in search for a better world.

4 Principles of Sociology from a Cognitic and Robotic Perspective

Science has developed with constant, balanced and coherent endeavor to acquire from reality experimental, qualitative and quantitative evidence of phenomena, and on this basis, to create coherent and concise theories.

Success is particularly evident for physical processes and properties. In "soft" sciences, which include human sciences and in particular sociology, progress has probably been as significant, even though much doubt is often expressed, in numerous domains. It is our goal to (contribute to) bridge the gap between these two classes of sciences. This goal is naturally of top priority in an era where machines get closer to humans.

Cognitive sciences provide a path of high interest on the way to a scientific, well-grounded sociology. It is even more so for the case of sociology applied to automated, non-human individuals and collectivities (The Free Dictionary - collectivity: the people considered as a body or whole).

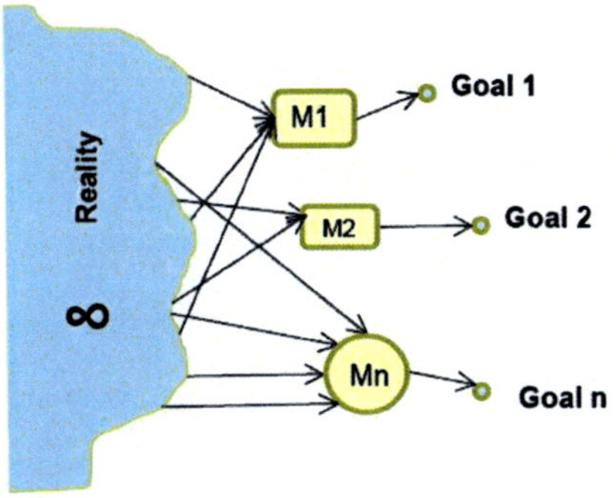

Fig. 4. As developed in [6], the MCS theory relies with great benefits on the concept of information, for anchoring the original definition of core cognition properties. But this is not a free lunch. There are also extremely strong limits, associated with the concept of information (time-variation, subjectivity). And there is still especially another limit, evidenced in the context of information, and inherited from a more fundamental notion yet, the one of modeling (domain-dependant, infinitesimal, goal-oriented representation of reality; re. Fig. 4).

The cognition theory and further down the road, sociology as a well-founded conceptual framework, are bound to inherit both from the essential merits of their precursors and from their limits as well..

Notice that in classical sociology quantitative, scientific methods have always been practiced. They typically appear in terms of collected statistics. Similarly, information is also quantitatively based on a probability calculus. It would therefore not be a total revolution to complement ad hoc traditional statistical approaches with corresponding, compatible, more standard information-theoretic techniques; values measured in [bit].

In principle, for MCS theory, a group can globally be considered just like any other cognitive system (CS). The behavioral model adopted for CS can be applied at any granularity level: at the global level of an entity (re. in particular Fig. 1.a), but also at lower levels (re. internal information flows while thinking, among multiple differentiated subsystems), or reciprocally at higher levels where integration prevails (notion of a group; re. Fig. 5).

The sequel of this paragraph mostly borrows from [6].

Yet a group is, in principle, more than an individual. Additional concepts are useful to handle this more complex form of organization. After a general definition, three views follow where we then adjust the focus on the holistic aspects, the individual properties, and, finally, the way the structure can be coordinated.

Essential properties of a group organization can typically be viewed as including the following elements:

- The holistic behavior of the group as a body with collective properties
- The individual properties of group members
- The way the structure and means can support coordination (culture, communication, spirit).

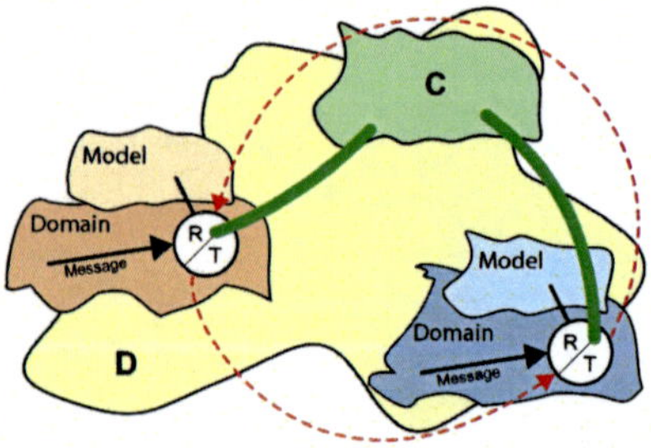

Fig. 5. Group. Individual cognitive agents (blue, brown) may coordinate each other, and thus may collectively form a group. For this purpose, a common culture (C, green), in reference to some common domain of interest (D, yellow) and some communication media are required among agents (R: receive; T: transmit). At a meta-level, the individual members may be considered as merging, to yield a new individual (the group) with its own collective model (C). (From [6]).

In the context of MCS core theory, the behavior at global level, as an integrated CS, characterizes a group. Quantitatively, all the metrics defined for the properties of CS apply. With the behavioral approach, there is, in principle, no necessity to explicitly describe how groups are internally organized. Nevertheless, in the same way as cognition may address subsystems for analytical or design purpose, groups may be considered at the level of individual group members as well.

In the context of MCS theory, group members appear themselves as ordinary, elementary CS with their respective cognitive domains mostly defining their specificity. Surprisingly, even though a group is essentially made of individual members, the observation of those members alone may not say much of overall group behavior. In the context of sociology, an attribute of particular interest for a group member would be its sociable character, its ability to relate with others in order to foster building-up of the group.

The communication channel is typically a physical medium that supports the transmission of information. For example in @Home groups, interaction between robots and humans may notably rely on vision, gestures, voice, distance estimation and forces. Additionally, robots in a group may typically communicate with each other using wifi. All these channels can be quantitatively characterized.

The socialization process operationally binds members together (coordinates them) in order to yield overall group properties. Prerequisites include the availability of a communication channel as well as some common culture and spirit.

We can view spirit and culture as a set of intangible underlying factors that ensure the coordination of individuals in order to achieve a specific collective identity and behavior. "Spirit" and "culture" consist in a system of common, shared references, values, and objectives, in reference to some common domain of interest, which may dynamically evolve. For example in @Home competitions a list of 20 names is typically defined as the common standard for persons involved in Who's who tests. Always necessary, culture and spirit, like the group itself, may not exist actually per se (that is, they are not implemented out of the members themselves; even though books, bylaws, charts or rulebooks for example may attempt to describe them). Consider an orchestra playing without conductor. Here the group is the orchestra, members are musicians, and the spirit is the name retained for what makes it possible for the musicians to play together in a coherent way, even when there is no additional conductor nor outside regulating factor. For @Home competitions for example, the rules typically define a short-term, ad hoc, standard set of objects, names, processes (re. checkpoints) or locations in rooms (mirror, showcase, kitchen etc.).

For humans, through ages numerous kinds of collective structures have been explored and defined in addition to the notion of group (re. sociology). Nevertheless for MCS theory, only the cognitive aspects are retained; moreover, all cognitive concepts denoted in sociology-jargon should just be considered as domain-specific synonyms of corresponding standard MCS entities.

From a cognitic perspective, the strong modeling limit illustrated in Fig. 4, calls for a pragmatic process whereby first, as a prerequisite, goals are chosen, and proper modeling is ensured. Symmetrically, the necessity of results requires in a last phase, that cognitive results be put to work with energy, for example by robots. These stages are illustrated in Fig. 6, where two critical, additional components appear, complementing purely cognitive processes: selection (re. ethics) and action (operability, agency).

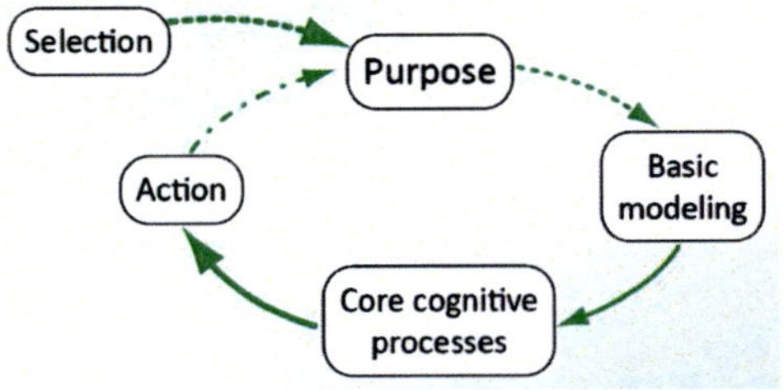

Fig. 6. From a pragmatic point of view, the complexity of reality requires first the selection of a goal (purpose); only then can modeling and cognition proceed, finally leading to action, operation possibly of robotic nature, that can ensure closing of the loop (iteration, and ultimately, in general, survival); from [3]

The first new domain to consider for a full solution in life refers to the selection of purpose, of goal. Notions closely related to this issue include egocentric aspects (self-oriented needs and desires, in particular for survival and freedom), and constraints set by the environment (realism, ethics, possibly including altruistic and group-oriented aspects). Interestingly, in the classical domain of sociology, the respected work by John Child [10] also addresses the critical role of goal selection, "the role of strategic choice" in the context of organizational structure, environment and performance. In @Home experimental validation tests, notice that goals are precisely defined for all, humans and machines, in very technical and detailed way, reaching to the concrete value points to be collected.

The second required domain, to complement cognition for effective changes in life, is the one of action, operation (re. the end of section 2), possibly of robotic nature. Notice that the etymologies of the word robot, as well as of operation given above, consistently refer to the concept of work.

5 Meta-Human and Secular Sociology of Cothinkers and Coagents

Section 4 has presented the concepts of CS and group, adding that they are applicable at multiple granularity levels, or scopes. Humans and robots can similarly enter collective structures of various scales.

Humans commonly belong to small social structures such as families or friendship circles. The most extreme case however, in theory and in terms of small scale, could be the self; this is the zero level of sociability and typically corresponds in humans to pathologies, such as autism. At the other extreme, reaching to large scales, collectivities appear as nations, religious communities, to name the most common, or globally, mankind.

Members of a group in principle share values and goals, and are ready to give up some of their own judgment and freedom for the sake of group success and collective well-being; in short they cooperate. By contrast, competition typically applies with respect to possible entities outside the group under consideration, especially when resources are scarce.

It is therefore very important well to identify the limits and priorities of group memberships, which in time may vary and simultaneously apply to numerous different groups and corresponding scopes, for example as follows: companion (shares the bread), comrade (shares the room), compatriot (shares the state ancestors), co-religionists.

From an historic perspective, innovation in specific focus, here (planet Earth) and now (this century), has allowed to define a common cooperative framework for multiple and otherwise possibly conflicting religions (re. secularization process).

Today some people extend beyond mankind the scope of collectivity they adhere to, to possibly encompass, in addition to other humans, animals, insects, even plants (vegetable kingdom) and all forms of life (it is consequently not surprising in such a context that radical limits to the rights of humans may follow). Domestication is a mild example in that direction.

Getting back to the topic addressed in title, an effective progress can be expected if the scope of thinking people and human agents is extended to accommodate machine-based resources capable of cognition and action, yielding collectively a meta-human, secular sociology of co-thinkers and co-agents. In such a context, it might be rational e.g. for a robot to stop operation before hurting a human, or for a human, to give some of one's time to relieve a machine resource perceived as critical for collective success at group level.

In RoboCup@Home for example, great care and numerous measures have been especially taken in order to avoid human casualties. Furthermore, guidelines have been proposed for safe operation in the cardinal task for a robot to follow humans [11]. Similarly a modest move in the direction of technical standards might allow for better results in medium term, even though some experts still prefer to keep providing a constraint-free context, perceived by them to be more prone to foster ideal, universal robotic and cognitive solutions.

Concretely, in the future, the international community should develop the responsibilities of helping to provide a summary of the progress achieved, and of defining in a rigorous way an ontology of the objects and operations typical in a domestic context. Standards might also be recommended standards to simplify robotic automation and make it more reliable and cost-effective, thus achieving useful and widespread automation. In human societies, a culture seems always to have developed, which allows for members to behave and to communicate efficiently with each other, about issues relating to their common life and environment. The natural language is an evident example of such an approach in all societies through history. More recently, car driving conventions or Internet protocols are other instances of such an approach. In the view of improving the quality of life of humans at home, by the means of designing intelligent machines and providing novel solutions, it should be natural for the international community to help in providing also a common culture for the field; for humans and also for robots.

6 Conclusion

Advances in sciences and technologies are such that today machine-based cognition and cooperation abilities start to emerge in robots, as e.g. RoboCup and more especially its @Home league can prove it.

Primarily, cooperation is understood here between robots and humans; it can however also be considered in other fields: among multiple robots; and, either by analogy, or as epistemologically founded, making use of millennia of experience, among humans.

Consequently, time has come to develop knowledge in sociology, in order to improve robotic performance and possibly better explain how humans can cooperate and build-up collectivities.

Sociable behavior is crucial for gathering individuals and subgroups into larger structures.

As machine-based agents get capable of cognitive performance and robots contribute to get things done, time as come to expand the secular social framework traditionally exclusively reserved for humans, in order to possibly integrate machine-based co-thinkers and co-agents.

Key concepts have now been made and exposed. They have started to be validated by real size implementation and experimentation in the domain of homes (re. e.g.

RoboCup@Home initiative), as well as in industrial or public environments. This is worthe to be further developed.

Acknowledgements. The authors acknowledge numerous contributions from colleagues and students at HEIG-VD, and especially from Michel Cornioley, Sudarat "Amy" Tangnimitchok, and Promphan "Kim" Ounchanum, recently.

Thanks also for several helpful suggestions from the reviewers.

References

1. Asada, M., et al.: Overview of RoboCup-98. In: Asada, M., Kitano, H. (eds.) RoboCup 1998. LNCS (LNAI), vol. 1604, pp. 1–21. Springer, Heidelberg (1999)
2. International RoboCup Federation, http://wiki.robocup.org/
3. Dessimoz, J.-D.: Cognition for a Purpose - Cognitics for Control. In: 4th International Conference on Cognitive Systems, CogSys 2010, ETH Zurich, Switzerland, January 27&28 (2010)
4. van der Zant, T., Wisspeintner, T.: RoboCup X: A Proposal for a New League Where RoboCup Goes Real World. In: Bredenfeld, A., et al. (eds.) RoboCup 2005. LNCS (LNAI), vol. 4020, pp. 166–172. Springer, Heidelberg (2006)
5. Nardi, D., Dessimoz, J.-D., Dominey, P.F., Iocchi, L., Ruiz-del-Solar, J., Rybski, P.E., Savage, J., Schiffer, S., Sugiura, K., Wisspeintner, T., van der Zant, T., Yazdani, A., Holz, D., Kraetzschmar, G., Gossow, D., et al.: "Rulebook" for the League RoboCup@Home (2006.2011), http://www.ai.rug.nl/robocupathome/
6. Dessimoz, J.-D.: Cognitics - Definitions and metrics for cognitive sciences and thinking machines, Roboptics Editions, Cheseaux-Noréaz, Switzerland, p. 169 (January 2011) ISBN 978-2-9700629-1-2
7. Dessimoz, J.-D., Gauthey, P.-F.: What Role for Emotions in Cooperating Robots? – The Case of RH3-Y. In: Gottscheber, A., Obdržálek, D., Schmidt, C. (eds.) EUROBOT 2009. CCIS, vol. 82, pp. 38–46. Springer, Heidelberg (2010)
8. John McCarthy, M.: What is Artificial Intelligence? (January 21, 2010), http://www.ormal.stanford.edu/jmc/whatisai/whatisai.html (rev. November 12, 2007)
9. Fong, T., Nourbakhsh, I., Dautenhahn, K.: A survey of socially interactive robots? In: Robotics and Autonomous Systems. Elsevier (2003)
10. Child, J.: Organizational structure, environment and performance: the role of strategic choice. Sociology (1972)
11. Dessimoz, J.-D., Gauthey, P.-F.: Domestic Service Robots in the Real World: the Case of Robots Following Humans. In: Proc. Domestic Service Robots in the Real World Workshop, SIMPAR 2010, 2nd International Conference on Simulation, Modeling and Programming for Autonomous Robots, Darmstadt, Germany, November 15-18 (2010)

An Algorithm Model for Gross Cognitive Reappraisal Strategy

Xiaolan Peng[1], Lun Xie[1], Xin Liu[2], and Zhiliang Wang[1]

[1] School of Computer and Communication Engineering,
University of Science and Technology Beijing,
Beijing, China, 100083
pengxiaolanlvlv@sina.com, {xielun,wzl}@ustb.edu.cn
[2] School of Automation and Electrical Engineering,
University of Science and Technology Beijing,
Beijing, China, 100083
liuxin87@yeah.net

Abstract. In this paper, we use the mathematical model of Finite State Machine to describe the conversion process of individual's emotional state based on Gross's process model of emotion regulation strategies. In the model, we abstract Gross cognitive reappraisal strategy into a quantitative parameter and propose a kind of preliminary algorithm description for the influence cognitive reappraisal strategy has on the emotional conversion process. At last, we make simulation experiment for the algorithm model, and the experimental results show that the proposed algorithm proposed can effectively describe the relationship between reappraisal strategy and emotion-generative process.

1 Introduction

Emotion is a state constantly being aroused and experienced by the individual. Sometimes, emotions are triggered virtually automatically, At other times, emotions arise only after considerable meaning analysis. Emotion regulation refers to the processes by which we influence which emotions we have, when we have them, and how we experience and express them [1,2]. Since emotion regulation has a significant influence on the cognition, behavior and health in our daily life, in recent years, emotion regulation has become a frontier scientific issue in emotional psychology.

In psychology, cognition refers to the acquisition process of knowledge, it contains all the mental activities which make the whole process of translating a physical stimulus into the expression of the real world and thus guiding the behavior can be achieved [3]. In the process of emotion regulation, cognition refers to individual's possible explanation for the emotional events [4]. To illustrate the process of emotion regulation, Gross present a groundbreaking process model of emotion regulation that provides an overarching conceptual framework. He thinks that emotion regulation involves changes in emotion dynamics, or the latency, rise time, magnitude, duration, and offset of responses in behavioral,

S. Lee et al. (Eds.): Intelligent Autonomous Systems 12, AISC 193, pp. 435–446.
springerlink.com

experiential, or physiological domains [1]. He has also done a lot of experiments and measurement studies which effectively describe the important influence reappraisal and suppression strategies has on emotion regulation, Gross's theory provides a theoretical basis for the research on the relationship between cognition and emotion.

Up to now, there is still not a specific emotional mathematical method of the abstract emotional theory which can be implemented by computer because of the complexity and multiple correlation of human emotion, which leads to the relationship between cognition and emotion can not be reflected. This relationship is interdependent, not separate from each other, we should establish a new algorithm framework to describe the relationship between the cognition and emotion based on the behavioral and neuroscience research [4]. The mathematical description of this relationship will be expected to make machine to own the ability of natural emotion regulation process, thus to achieve a more harmonious and natural human-computer interaction. This paper begins with a brief overview of the Gross's process model of emotion regulation strategies, and builds a basic mathematical model for the conversion process of individual's emotional state described by the Finite State Machine, and then quantifies the reappraisal parameter, proposed a algorithm description of the influence reappraisal had on the emotion regulation process, lastly, we simulate the algorithm and the experimental results show the effectiveness of the algorithm proposed.

2 Gross's Process Model of Emotion Regulation Strategies

2.1 Gross's Process Model of Emotion Regulation

Gross believe that emotion regulation starts with the emotion-generative process, different emotion regulations will be generated at different emotional stages. According to this, he proposed the model of emotion regulation (see Figure 1) [5,6]. Gross divided the emotional regulation process into five stages, those are situation selection, situation modification, attentional deployment, cognitive change, response modulation. In this paper, we focus on the influence cognitive change (reappraisal strategy) has on the emotion-generative process.

Situation selection refers to approaching or avoiding certain people, places, or things so as to regulate emotion. Situation modification has also been referred to as problem focused coping or as primary control. Attentional deployment is used to select which of the many aspects of the situation you focus on. Cognitive change refers to selecting which of the many possible meanings, you will attach to that aspect. Cognitive change is often used to decrease the emotional response. However, it also may be used to magnify the emotional response, and even to change the emotion itself. The personal meaning that is assigned to the situation is crucial because it powerfully influences which experiential, behavioral, and physiological response tendencies will be generated in that particular situation. Response modulation refers to attempts to influence emotion response tendencies once they already have been elicited [1].

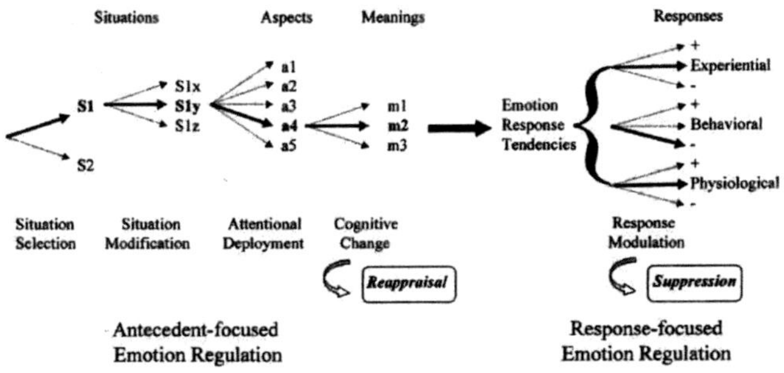

Fig. 1. A process model of emotion regulation. According to this model, emotion may be regulated at five points in the emotion generative process: (1) selection of the situation; (2) modification of the situation; (3) deployment of attention; (4) change of cognitions; and (5) modulation of experiential, behavioral, or physiological responses. The first four of these are antecedent focused, the fifth is response focused. The number of response options shown at each of these five points is arbitrary, and the heavy line indicates the particular option selected in the example given in the text. Two specific emotion regulation strategiesreappraisal and suppressionare the primary focus of this review. Adapted from Gross(2001)[1].

2.2 Gross's Emotion Regulation Strategies

According to Gross's model, the emotion regulation is divided into antecedent-focused emotion regulation and response-focused emotion regulation, the first four of these are antecedent focused, the fifth is response focused. Two specific emotion regulation strategies (reappraisal and suppression) are the primary focus of this review. Large number of experiments and measurement studies conducted by Gross show that reappraisal is a very important and effective emotion regulation strategy, it is also an important method and technique to regulate state-based, procedural, and situational emotion [1].

Gross's review focuses on two commonly used strategies for down-regulating emotion. The first, reappraisal, comes early in the emotion-generative process. It consists of changing the way a situation is construed so as to decrease its emotional influence. The second, suppression, comes later in the emotion generative process. It consists of inhibiting the outward signs of inner feelings.

The cognitive reappraisal strategy in the Gross emotional regulation has an influence on the offset of responses in behavioral, experiential, or physiological domains before emotions are stimulated. A lot of results from measurements and experiments by Gross and his group also show that cognitive reappraisal will reduce or increase the emotional reactions which are originally about to be triggered, and even change the nature of emotions and to alleviate negative mood occurring[1].For example, someone steps on our feet, if you stubbornly interpret that he is deliberate, you must be very angry, if some explanation make you interpret that he is unintentional, you may not be so angry and forgive it. The theoretical analysis of state-based emotion regulation strategies provides an important theoretical support for the followed study.

3 The Algorithm Model for Gross Cognitive Reappraisal Strategy Based on FSM

3.1 Finite State Machine and Emotional State Space

DFA(Deterministic Finite Automaton) is the combination constituted by $M-(I,O,S,\delta,\lambda)$ of which matrix model uses matrix theory and Boolean algebra as its tools, by introducing the Boolean of I (input set), O (output set) and S (status set),with δ (state transition mapping) and λ (output mapping) described by the mapping matrix equations. Based on DFA, for the case of none output of finite automaton, paper [7] proposed a FSM(Finite State Machine), which is a mathematical model with discrete input and output of the system, only contains the I (input set), O (output set) and δ (state transition mapping),that is $M-(I,S,\delta)$,the corresponding matrix model is expressed as $Q=B(x)\times Q$, the input variables is $x=(x_1,x_2,\cdots,x_L)\in X$, among which $X=\{(a_1,a_2,\cdots,a_L)\,|a_i=0,1\}$ and $Q=\{Q_1,Q_2,\cdots,Q_n\}$ are the Boolean express of the input status set and the output status set.

Q^{t-1} \ $A(x)$ \ Q	Q_1	Q_2	...	Q_l	...	Q_n
Q_1^{t-1}	$a_{11}(x)$	$a_{12}(x)$	...	$a_{1l}(x)$	...	$a_{1n}(x)$
Q_2^{t-1}	$a_{21}(x)$	$a_{22}(x)$	...	$a_{2l}(x)$	...	$a_{2n}(x)$
...	...	...	...	...	...	...
Q_l^{t-1}	$a_{l1}(x)$	$a_{l2}(x)$	...	$a_{ll}(x)$	...	$a_{ln}(x)$
...	...	...	...	...	...	...
Q_n^{t-1}	$a_{n1}(x)$	$a_{n2}(x)$	...	$a_{nl}(x)$	...	$a_{nn}(x)$

Fig. 2. The tabular form of the matrix model

The basic and congenital emotions of human form initially only when being stimulated by a series of scenes or events clearly defined and the activation of individual's emotion relates not only with the input stimuli at the time, but also with the individual's own emotional state (mood). It is easy to find that this inherent nature of human's emotions is in line with the function described by FSM. Since the correspondence between emotions and external input stimuli is discrete, like the theory of the discrete relationship between the input and output described by FSM mathematical model, therefore, based on this theory of FSM, we can use the matrix equation to describe the emotional states transition process [8].

Define the individual's own emotional state space as $Q=\{Q_1,Q_2,\cdots,Q_n\}$, individual's emotion states convert among the state space defined above which contains n kinds of emotion states. Taking into account the influence which the external input stimulus has on individual emotional states, define the external input stimuli emotional space as $X = \{\omega_1, \omega_2, \cdots, \omega_m\}$, which consists of m kinds of input stimuli. When the input stimulus is ω_i, the matrix of emotion states is A:

$$A = \begin{pmatrix} a_{11} & a_{12} & \cdots & a_{1n} \\ a_{21} & a_{22} & \cdots & a_{2n} \\ \vdots & \vdots & \ddots & \vdots \\ a_{n1} & a_{n2} & \cdots & a_{nn} \end{pmatrix}$$

a_{jk} refers to that the probability of individual's emotional state convert from Q_j to Q_k,

$$a_{jk} = p(\omega = \omega_i) \times p(\omega_k/\omega_j)/p(\omega_j)$$

As this FSM is the whole state transition machine, given one input stimuli, the summation of all the probability of the state space Q that convert from one emotion state is 1, that means the constraint of A is: given an arbitrary ω_i ,there is $\sum_{k=1}^{k=n} a_{jk} = 1$ [9].

3.2 Emotion Strategies Parameter

In order to use mathematical methods to study the influence of cognitive reappraisal strategies on the output state of emotions and get Gross theory of emotional strategy to become an algorithm which can be understood and implemented by computer. In our emotion model, the individual understanding of the significance of emotional events is abstracted as a cognitive reappraisal parameter α to reflect the positive and negative level of individuals understanding of current emotional event and to study how it influences the emotional state transition probabilities under certain incentive to stimulate the emotions. In order to facilitate the theoretical analysis and experimental simulation, define $\alpha \in (-10, +10)$. the bigger the parameter α, the more positive individuals understanding of emotional event, and the better individuals negative emotions about to be triggered such as frustration, anger, disgust, etc. can be relieved, and may even lead negative emotions to positive emotions; the smaller the parameter α, the more negative individuals understanding of emotional event, which exacerbate the negative emotions about to be triggered, increase the probability of emotions converted to negative and intensify contradictions; and $\alpha = 0$ means that individual has little understanding of the significance of current emotional event.

Based on reference [10], the special output of individual emotional states are divided into the following three categories in this paper:(1) positive emotions such as happiness, joy and cheer, etc., which will bring positive emotional psychological experience and have a positive influence on the social communication and interaction in general;(2)neutral emotions, which usually wont cause strong emotional experience generally;(3)negative emotions, such as pain, anger, sadness, etc., which will bring negative emotional psychological experience and have a negative influence on the social communication and interaction in general.

3.3 Emotion Regulation Strategies Model

In order to make further research and analysis, we divide the individual's emotion state space $Q = \{Q_1, Q_2, \cdots, Q_n\}$ into $Q = \{Q_1, Q_2, \cdots, Q_{n_1}, Q_{n_1+1}, \cdots, Q_{n_1+n_2}, Q_{n_1+n_2+1}, \cdots, Q_{n_1+n_2+n_3}\}$, $n_1 + n_2 + n_3 = n$, Q_1 Q_{n_1} is for n_1 kinds of positive emotional states, Q_{n_1+1} $Q_{n_1+n_2}$ is for n_2 kinds of neutral emotional states, $Q_{n_1+n_2+1}$ $Q_{n_1+n_2+n_3}$ is for n_3 kinds of negative emotional states. The corresponding emotional state transition probability matrix is extended to be:

$$A = \begin{pmatrix} a_{11} & a_{12} \cdots a_{1n_1} & \cdots & a_{1(n_1+n_2+1)} & \cdots a_{1(n_1+n_2+n_3)} \\ a_{21} & a_{22} \cdots a_{2n_1} & \cdots & a_{2(n_1+n_2+1)} & \cdots a_{2(n_1+n_2+n_3)} \\ \vdots & \vdots \quad \ddots \quad \vdots & \ddots & \vdots & \ddots \quad \vdots \\ a_{n1} & a_{n2} \cdots a_{nn_1} & \cdots & a_{n(n_1+n_2+1)} & \cdots a_{n(n_1+n_2+n_3)} \end{pmatrix}$$

a_{jk} means that the probability of individual's emotion state convert from Q_j to Q_k.

The model described in this paper is about the process with fixed and single input stimulus (that means our work focus on the whole emotional regulation process of one emotional event, not the interaction among different stimuli, so if another stimulus occur, that will be seen as another new regulation process), we argues that the emotional state won't jump from one kind to another, that is to say, the individual's emotional change from negative emotions to positive emotions must go through the neutral emotions. In the probability matrix model, for one certain input emotional stimulus, the individual emotional states transit according to A in the case of none cognitive reappraisal (cognitive reappraisal parameter $\alpha = 0.0$), while after taking the reappraisal parameter into account, the way of individual's emotional convert probabilities will change as below:

In the figure, the upward arrow indicates that the probability of the emotional state will reduce and the down arrow indicates that the probability of the emotional state will increase. To better describe the regulation process via the mathematical parameter α, the affect level is specifically defined as τ.

When $\alpha > 0$, negative emotions will transfer to neutral emotions and neutral emotions will transfer to positive emotions. That means each of the probability of negative emotion will drop $\tau \times \alpha_{jk}$, and each of the probability of neutral emotion will rise according to the proportion of their original probability as $\tau \times \sum_{k=n_1+n_2+1}^{k=n_1+n_2+n_3} a_{jk} \times \frac{a_{jk}}{\sum_{k=n_1+1}^{n_1+n_2} a_{jk}}$ (perhaps you can think this increased part comes from the probability dropped from negative emotions) and drop its probability of $\tau \times \alpha_{jk}$ to positive emotions just as negative emotions at the same time, and each of the positive emotion will rise according to the proportion of their original probability of $\tau \times \frac{a_{jk}}{\sum_{k=1}^{n_1} a_{jk}} \times \sum_{k=n_1+1}^{n_1+n_2} a_{jk}$ which comes from the probability dropped from negative emotions.

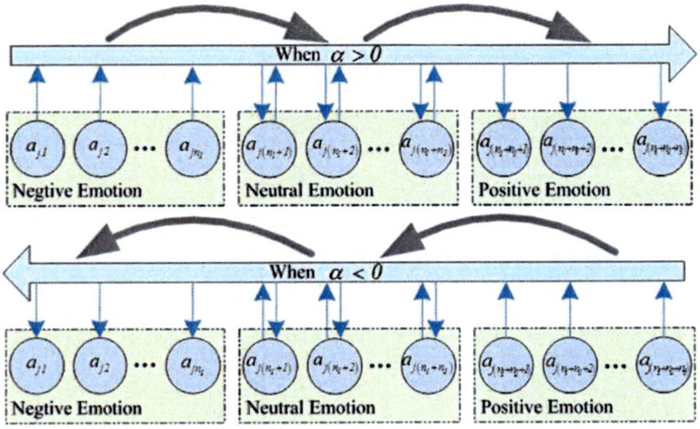

Fig. 3. The change process of emotional convert probabilities

When $\alpha < 0$, positive emotions will transfer to neutral emotions and neutral emotions will transfer to negative emotions just like the reversed process stated above.

In both the case, the change level of will depends on $|\alpha|$. Through the analysis above, we derive the equations as below:

When $\alpha > 0, \tau = \frac{2}{\pi} \arctan \frac{1}{2}\alpha$,

$$a_{jk}{}' = \begin{cases} a_{jk} + \tau \times \frac{a_{jk}}{\sum_{k=1}^{n_1} a_{jk}} \times \sum_{k=n_1+1}^{n_1+n_2} a_{jk}, k \in [1, n_1] \\ a_{jk} + \tau \times \frac{a_{jk}}{\sum_{k=n_1+1}^{n_1+n_2} a_{jk}} \times \sum_{k=n_1+n_2+1}^{n_1+n_2+n_3} a_{jk} - \tau \times a_{jk}\ , \\ \qquad k \in [n_1+1, n_1+n_2] \\ a_{jk} - \tau \times a_{jk}, \quad k \in [n_1+n_2+1, n_1+n_2+n_3] \end{cases}$$

And when $\alpha < 0, \tau = \frac{2}{\pi} \arctan \frac{1}{2}\alpha$,

$$a_{jk} = \begin{cases} a_{jk} + \tau \times a_{jk}, \quad k \in [1, n_1] \\ a_{jk} - \tau \times \frac{a_{jk}}{\sum_{k=n_1+1}^{n_1+n_2} a_{jk}} \times \sum_{k=1}^{n_1} a_{jk} + \tau \times a_{jk}\ , \\ \qquad k \in [n_1+1, n_1+n_2] \\ a_{jk} - \tau \times \frac{a_{jk}}{\sum_{k=n_1+n_2+1}^{n_1+n_2+n_3} a_{jk}} \times \sum_{k=n_1+1}^{n_1+n_2} a_{jk}\ , \\ \qquad k \in [n_1+n_2+1, n_1+n_2+n_3] \end{cases}$$

When the input stimulus is ω_i , a_{jk} means that the probability of individual's emotion state convert from Q_j to Q_k. after taking the reappraisal parameter α into account. We can verify that when $\alpha > 0, \tau = \frac{2}{\pi}\arctan\alpha$.

$$\sum a_{jk} = \sum_{k=1}^{n_1} (\mathrm{a_{jk}}+\tau \times \frac{\mathrm{a_{jk}}}{\sum_{k=1}^{n_1} \mathrm{a_{jk}}} \times \sum_{k=n_1+1}^{n_1+n_2} \mathrm{a_{jk}})+$$
$$\sum_{k=n_1+1}^{n_1+n_2} (a_{jk} + \tau \times \frac{\mathrm{a_{jk}}}{\sum_{k=n_1+1}^{n_1+n_2} \mathrm{a_{jk}}} \times \sum_{k=n_1+n_2+1}^{n_1+n_2+n_3} \mathrm{a_{jk}} - \tau \times a_{jk})$$
$$+ \sum_{k=n_1+n+1}^{n_1+n_2+n_3} (a_{jk} - \tau \times a_{jk})$$
$$= \sum_{k=1}^{n_1} a_{jk} + \sum_{k=n_1+1}^{n_1+n_2} a_{jk} + \sum_{k=n_1+n_2+1}^{n_1+n_2+n_3} a_{jk}+$$
$$\tau \times \sum_{k=n_1+1}^{n_1+n_2} \mathrm{a_{jk}} + \tau \times \sum_{k=n_1+n_2+1}^{n_1+n_2+n_3} a_{jk} - \tau \times \sum_{k=n_1+1}^{n_1+n_2} \mathrm{a_{jk}}$$
$$-\tau \times \sum_{k=n_1+n_2+1}^{n_1+n_2+n_3} a_{jk}$$
$$= \sum_{k=1}^{n} a_{jk} = 1$$

And when $\alpha < 0, \tau = \frac{2}{\pi}\arctan\alpha$,

$$\sum a_{jk} = \sum_{k=1}^{n_1} (\mathrm{a_{jk}}+\tau \times a_{jk}) + \sum_{k=n_1+1}^{n_1+n_2} (a_{jk} - \tau \times \frac{\mathrm{a_{jk}}}{\sum_{k=n_1+1}^{n_1+n_2} \mathrm{a_{jk}}}$$
$$\times \sum_{k=1}^{n_1} \mathrm{a_{jk}} + \tau \times a_{jk}) + \sum_{k=n_1+n+1}^{n_1+n_2+n_3} (a_{jk}-$$
$$\tau \times \frac{\mathrm{a_{jk}}}{\sum_{k=n_1+n_2+1}^{n_1+n_2+n_3} \mathrm{a_{jk}}} \times \sum_{k=n_1+1}^{n_1+n_2} \mathrm{a_{jk}})$$
$$= \sum_{k=1}^{n_1} a_{jk} + \sum_{k=n_1+1}^{n_1+n_2} a_{jk} + \sum_{k=n_1+n_2+1}^{n_1+n_2+n_3} a_{jk} + \tau \times \sum_{k=1}^{n_1} \mathrm{a_{jk}}-$$
$$\tau \times \sum_{k=1}^{n_1} a_{jk} + \tau \times \sum_{k=n_1+1}^{n_1+n_2} \mathrm{a_{jk}} - \tau \times \sum_{k=n_1+1}^{n_1+n_2} a_{jk}$$
$$= \sum_{k=1}^{n} a_{jk} = 1$$

We can see the new definition of the transition matrix is deduced to meet the constraint of the emotional state transition matrix, that is $\sum_{k=1}^{k=n} a_{jk} = 1$.

It is noteworthy that the mathematical model is set to study the influence of cognitive reappraisal parameter α has on the conversion probability of emotional state. It is in the context that the initial A is a probability matrix which has considered other factors in the process of emotion regulation, actually, cognitive reappraisal parameter is not the only factor affecting the conversion probability. In addition, although this model is a study under the premise that emotions are divided into three categories, it is equally applicable to other cases of the emotional conversion process whose emotion state space is classified by the degree of positive of emotion.

4 Simulation Results

Ekman proposed six basic emotions in his theory of emotion division: anger, disgust, fear, happiness, sadness, and surprise [11]. In order to better perform simulation experiment, we add neutral emotions "joy" "like" and "calm" to this six basic emotional states space to form a new individuals emotional states space $Q = \{Q_1, Q_2, Q_3, Q_4, Q_5, Q_6, Q_7, Q_8, Q_9\}$, Q_1 refers to happiness, Q_2 refers to joy, and Q_3 refers to like, happiness, joy and like are three kinds of positive emotions; Q_4 refers to surprise and Q_5 refers to calm, surprise and calm are neutral emotions; Q_6 refers to angry, Q_7 refers to sadness, Q_8 refers to disgust and Q_9 refers to fear, angry, disgust, fear and sadness are four kinds of negative emotions. The corresponding probabilities of each state converted from other states are a_{j1}, a_{j2}, a_{j3}, a_{j4}, a_{j5}, a_{j6}, a_{j7}, a_{j8} and a_{j9}, and $\sum_{k=1}^{k=9} a_{jk} = 1$. We first assume that given one single input stimulus, the probability of each emotion state has been already obtained (the initial probability a_{jk}). Then we discuss two kinds of the change process of transition probability, one of negative initial state and another one of positive initial state. Lastly, we use Matlab software packages to perform simulation analysis, and observe the intervention process of the parameter on the transition probability in the model by changing the level of cognitive reappraisal parameter α . The simulation results are shown as figure 4 and figure 5.

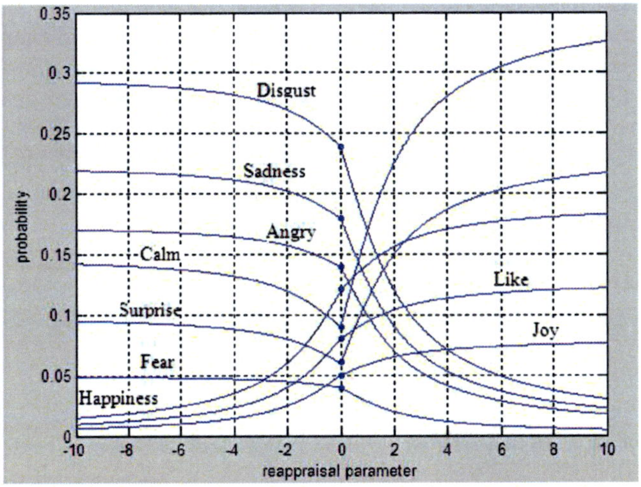

Fig. 4. The change process of transition probability (a_{jk}) when initial state is negative(disgust,sadness and angry)

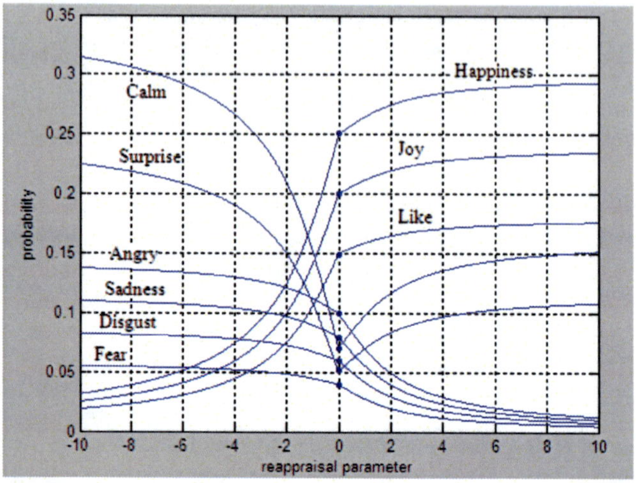

Fig. 5. The change process of transition probability (a_{jk}) when initial state is positive(happiness)

In figure 4, the probability of the states is set as: $a_{j8} > a_{j7} > a_{j6} > a_{j1} > a_{j5} > a_{j3} > a_{j4} > a_{j2} > a_{j9}$, so the initial state is disgust(Q_8) whose probability is the largest. When $\alpha > 0$, we can see that both the probabilities of positive emotions and neutral emotions rise with the increasing level of the reappraisal parameter, and the probability of happiness becomes larger than disgust's when the reappraisal parameter gets up about to 1.65. The probability of two kinds of neutral emotions rise rapidly partly because of the fewer dimensions, while the most important reason is that it is easier to transfer from negative emotions to neutral emotions. The main emotion states have been transferred from negative to neutral and positive when the reappraisal parameter gets up about to 4.00.When $\alpha < 0$, both negative and neutral emotions drop their probabilities to positive emotions. In figure 5, the probability of the states is set as: $a_{j1} > a_{j2} > a_{j3} > a_{j6} > a_{j7} > a_{j5} > a_{j8} > a_{j4} > a_{j9}$, so the initial state is happiness(Q_1) whose probability is the largest. We can see that When $\alpha < 0$, both the probabilities of negative emotions and neutral emotions drop with the increasing level of the reappraisal parameter, and the probability of happiness becomes larger than disgust's when the reappraisal parameter gets up about to -1.80. The probability of two kinds of neutral emotions rise rapidly partly because of the fewer dimensions, while the most important reason is that it is easier to transfer from neutral emotions to positive emotions. The main emotion states have been transferred from positive to neutral and negative emotions when the reappraisal parameter gets down about to -5.90.When $\alpha > 0$, positive emotions drop their probabilities to negative and neutral emotions. From the simulation results above, we can see that whether the initial emotion is biased to positive or negative, the algorithm can effectively display that the reappraisal parameter will reduce or increase the emotional reactions which are originally about to be triggered, and even change the nature of emotions and to alleviate negative mood occurring which Grosss theory of cognitive reappraisal strategy.

5 Conclusion

In this paper, we first briefly described Gross's process model of emotion regulation and focus on the influence cognitive reappraisal strategy has on the emotion-generative process. Then, we established a emotional state space matrix model on the basis of the theory of cognitive reappraisal strategy and Finite State Machine, in the model, we abstract Gross cognitive reappraisal strategy into a quantitative parameter and get the algorithm description for the influence cognitive reappraisal strategy has on the conversion process of emotional states. At last, with the application of classic Ekman's six emotions, we made the simulation for the algorithm we proposed above, and the experimental results verified that the algorithm can effectively describe the relationship between reappraisal strategy and emotional states in Gross emotional theory.

As the complexity and multiple correlation of human emotion regulation, a lot of factors will affect this process, it is very difficult to consider all the factors. In this paper, we only selected one emotion regulation strategy that would exert great influence on emotion states for the establishment of the mathematical model and the algorithm's implement on computer, and multiple dimensions of emotional status considered can undoubtedly make a clearer simulation results. In a word, giving a specific algorithm which can be implemented by computer for emotion regulation process still requires for further research and effort.

Acknowledgement. This work is supported by National Natural Science Foundation of China (No.61170115 and No.61105120) and the Open Research Project Funds for Key Laboratory of Complex Systems and Intelligence Science (No. 20110108).

References

1. Gross, J.J.: Emotion regulation: affective, cognitive, and social consequences. Psychophysiology 39, 281–291 (2002)
2. Gross, J.J.: Emotion regulation in adulthood: Timing is everything. Current Directions in Psychological Science 10, 214–219 (2001)
3. Wang, H., He, X.: Knowledge Representation by Naive Mental Image. Chinese Journal of Computer 24(8), 891–896 (2001)
4. Liu, Y., Fu, Q., Fu, X.: Cognitive and emotional interaction. Chinese Science Bulletin 18(54), 2783–2796 (2009)
5. Gross, J.J.: The emerging field of emotion regulation: integrative review. Review of General Psychology 2, 271–299 (1998)
6. Gross, J.J.: Antecedent-and response-focused emotion regulation: divergent consequences for experience, expression, and physiology. Journal of Personality and Social Psychology 74, 224–237 (1998)

7. Zhu, Z., Zhu, Q.: Algebraic properties of state automaton matrix model. Computer Engineering and Applications (04) (2003)
8. Liu, X., Xie, L., Yang, W., Wang, Z., Fu, S.: Dynamic regulation process of facial expression robot. Journal of Control Theory and Applications 28(7), 936–946 (2011)
9. Johnston, V.S.: Why we feel the science of human emotions. Perseus Books Group Massachus 8 (April 1999)
10. Wang, L., Markham, R.: The development of a series of photographs of Chinese facial expressions of emotion. Journal of Cross Cultural Psychology 31(4), 397–410 (1999)
11. Ekman, P.: Universals and cultural differences in facial expressions of emotion. In: Cole, J. (ed.) Nebraska Symposium on Motivation, vol. 19, pp. 207–282. University of Nebraska Press, Lincoln (1971)

Actor Studio: Development of User Friendly Action Editing System for Cultural Performance Robots

Ho Seok Ahn, Dong-Wook Lee, Dongwoon Choi, Duk-Yeon Lee, Manhong Hur, and Hogil Lee

Department of Applied Robot Technology, Korea Institute of Industrial Technology
Ansan 426-171, Republic of Korea
hoseoka@gmail.com, {dwlee,cdw,proldy,hmh426,hglee}@kitech.re.kr

Abstract. Social robots are developed for social interacting between humans and robots by various ways. One of the ways is sharing cultural performances such as singing, dancing, and acting. Various technologies are required for developing cultural robots, and one of the most important things is generating contents. We develop a user friendly action editing system with graphical user interface. It is comfort method to generate long contents of performance according to musical notes or scenarios. Using this method, we generate action data for some performances with real robot system in the theater.

Keywords: cultural robotics, performance robots, action editing program, action generation, android, EveR.

1 Introduction

Social robots interact with us by various ways in our daily life, for example, assisting and healing humans, showing performance in the theater, landscaping, etc. Service robots are assisting purposed robot systems and some robots express their emotions [1-6]. On the other hand, cultural robotics is a novel approach to apply robots to our cultural life. Some cultural robots are used for entertainment purposed toy systems. AIBO, a dog type robotic pet, is one of the most popular robots in the world [7]. PARO is modeled of sea seal and listed in the Guinness Book of World Records as the most efficient therapeutic robot [8]. Some cultural robots are used for landscaping. The robotic flower was firstly proposed by Carnegie Mellon University researches and has seven degree of freedom [9]. ROHINI is a humanoid type flower robot that mediates interactions between occupants and various devices and equipments embedded in a smart home environment [10].

Some cultural robots perform such as singing, dancing, acting, and playing musical instruments in the theater and theme parks. Especially, these kinds of cultural robots are called as cultural performance robots. Wakamaru from Mitsubishi Heavy Industries, the robot actor in the play, was programmed to speak his bit and move around the stage alongside human actors [11]. Janet and Thomas, both biped humanoid robots, are the protagonists of the robot theater [12]. Cybernetic human HRP-4C form AIST acted in a one man show in Digital Contents Expo 2009 and sang

S. Lee et al. (Eds.): Intelligent Autonomous Systems 12, AISC 193, pp. 447–455.
springerlink.com

a song in CEATEC JAPAN 2009 [13]. KITECH (Korea Institute of Industrial Technology) has developed the android robot EveR series for various uses. For example, EveR-1 is an announcer, EveR-2 is a singer, EveR-3 is an actress [14-18].

In this paper, we present an android robot 'Ever-3' and action editing method. EveR-3 is developed for an actress robot system. It is human scale robot system and has mobile lower body instead of bipedal to make active performances. The skin of the head and limbs is made of special silicon. The motion data is made by capturing human motion and converted to performance data by action editing method according to scenario. For this, we have developed a scenario editing tool [19]. It is helpful to make robot's motion according to the scenario, but it is difficult to use for the people who is unfamiliar with this tool. For easy editing of performance, we develop an 'Acting Studio' that is an action editing system based on formal tool with graphical user interface. This paper is organized as follows. In Section 2, we introduce the overview of the android EveR-3. We explain the acting studio in Section 3 and present some performances in Section 4. Finally, we conclude this paper in Section 5.

2 The Android 'EveR-3'

EveR-3 is an actress robot that has human-like appearance including face and body as well as shows human-like performance. For human-like appearance, EveR-3 is designed base on three-dimensional data of a young Korean female, as shown in Fig. 1. EveR-3 is covered with soft material as well. For human-like performance, EveR-3 speaks and sings a song with lip synchronization, facial expression, and gestures. To imitate human motion, EveR-3 uses the motion data captured from a real actress. To go around easily with stability, EveR-3 has mobile lower body composed of two controllable wheels and two auxiliary wheels. EveR-3 recognizes speech and responds with voice synthesis function for basic interaction with human. EveR-3 also recognizes human faces and tracks the detected faces. Table 1 shows the whole specification of EveR-3.

Fig. 1. The actress android, EveR-3, developed in KITECH (Korea Institute of Industrial Technology). The appearance of EveR-3 is based on Korean female. EveR-3's name derives from the Biblical 'Eve', plus the letter 'R' for robot. She is the third version of the EveR series.

Table 1. Whole Specification of EveR-3

Appearance		Young Korean female singer
Height		167 cm
Weight		65 Kg
DoFs	Face	23 DoFs
	Neck	3 DoFs
	Arms	6 DoFs * 2
	Hands	5 DoFs * 2
	Torso	3 DoFs
	Wheels	2 DoFs
	Total	53 DoFs
Eyes		2 small CCD cameras
Skin		Synthetic pliable silicone jelly (face & hand: soft material, body: hard material)
Control time		20 ms

3 The Action Editing System 'Acting Studio'

The Acting Studio, an action editing system for performance robots, is composed of three windows: a scene window, an action window, and a simulation window as shown in Fig. 2. The Acting Studio has graphical user interface for easy editing environments. Scripts are displayed with pictures or movies by scene. Users select music, speech, basic pose, facial expression, and body gesture according to scenario and put it on timeline. Finally, users can test the edited action data through the robot simulator. The Acting Studio can reduce the time and efforts of users.

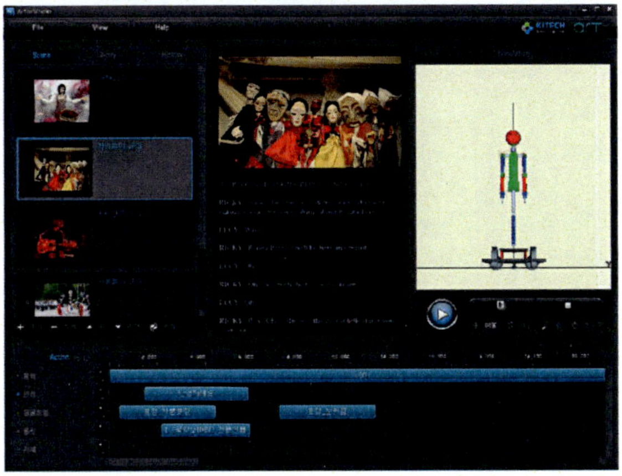

Fig. 2. The Actor Studio, which is an action editing system for performance robots, consists of three windows: a scene window (the left part), an action window (the bottom part), and a simulation window (the right part). It is useful and easy tool to generate performance data of cultural robots.

Fig. 3 shows some functions of the Acting Studio. The Acting Studio can load and save the edited scene data as well as the edited scenario data. In the view option, users can select the format of scenario viewer and the type of the simulator. The Acting Studio has user convenience functions such as library and history function. The library data is a bundled data of music, speech, basic pose, facial expression, and body gesture. It is useful when same music and its motions are used repeatedly. The history data is the working record in order and can be used when users want to cancer the previous work.

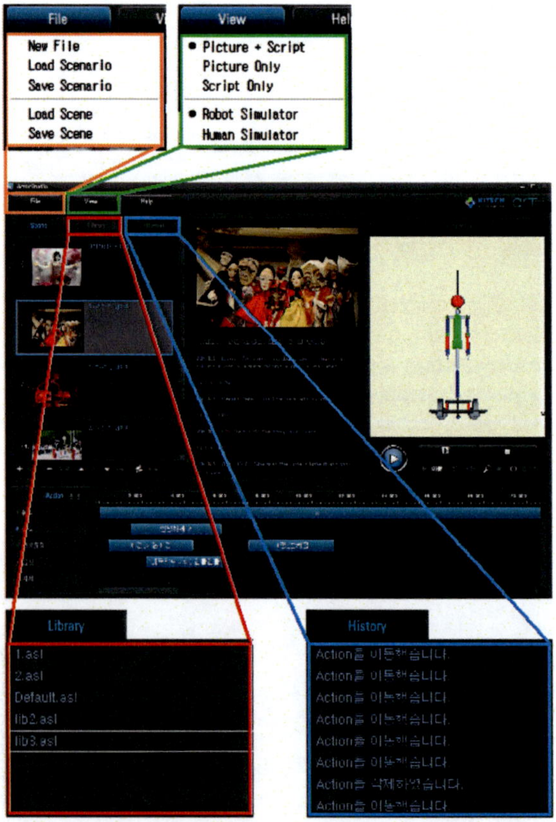

Fig. 3. Functions of the Actor Studio

3.1 The Scene Window

The scene window as shown in Fig. 4, the left part of the Acting Studio, is composed of two parts: a scene list part and a scene script part. The order of scenes is changeable and new scene can be added among them. Representative image and explanation of each scene is displayed on the scene list. The scene script is displayed by three types: a picture and script mode, a picture only mode, and a script only mode.

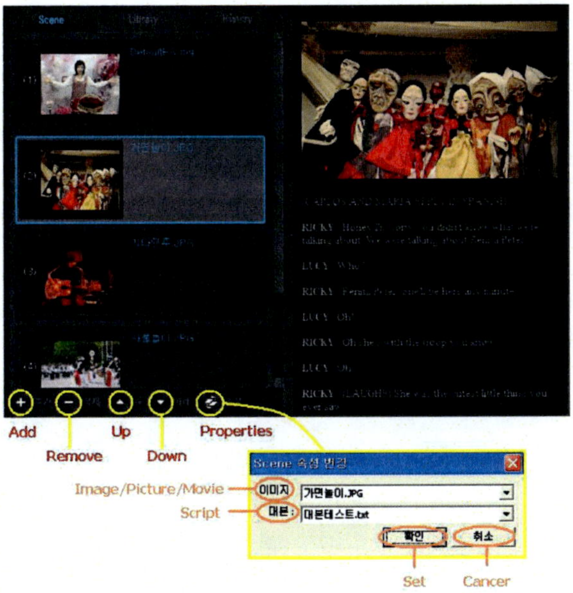

Fig. 4. Functions of the scene window

3.2 The Action Window

The action window as shown in Fig. 5, the bottom part of the Acting Studio, is composed of five lines: a music line, a speech line, a facial expression line, a body gesture line, and a basic pose line. Each line can be added or removed and has action

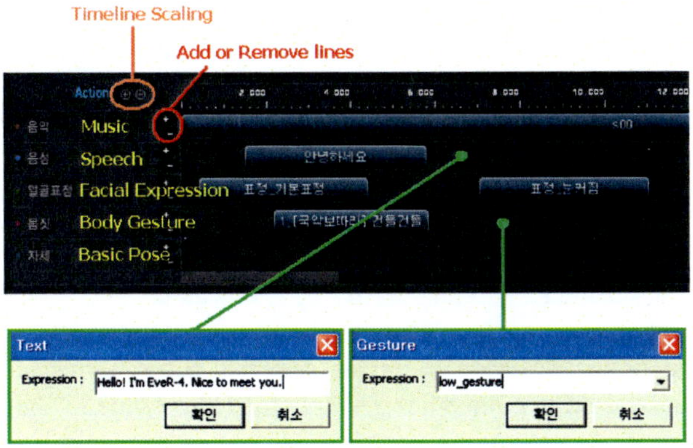

Fig. 5. Functions of the action window

items on timeline. In the case of the speech line, users can type a text for speaking. In the case of the other lines, users can select an action item among the lists, which can be added.

3.3 The Simulation Window

The simulation window as shown in Fig. 6, the right part of the Acting Studio, is a simulator of the edited action data on the timeline. The Acting Studio has two type of the simulator: a robot simulator as shown in the left of Fig. 6 and a human simulator as shown in the right of Fig. 6. Users can test the motion of the performance robot simply before applying the edited action data to the real robot system.

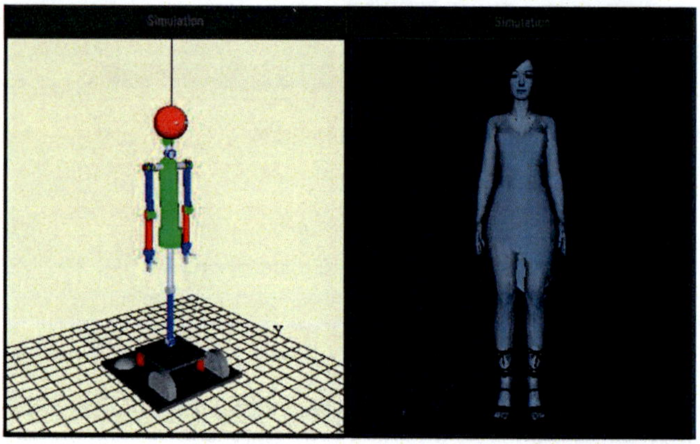

Fig. 6. Two types of simulator on the simulation window

4 Performances

We have applied the edited action data of the Acting Studio to the actress android EveR-3. Fig. 7 shows the experimental process of performance. Firstly, we get the motion data from human. Secondly, action items for EveR-3 are generated by motion retargeting. Thirdly, action data for performance is edited using the Acting Studio. Fourthly, the edited action data is applied to EveR-3. For modifying the motion of EveR-3, from the second step to the fourth step is repeated.

We have showed some performances as shown in Fig. 8. EveR-3 acts her role that is edited by the Acting Studio with human actors in many performances. We could reduce our time and efforts for editing actions of Ever-3 due to the Acting Studio. Also motor trouble of EveR-3 was also reduced because we could simulate the edited action data using the Acting Studio.

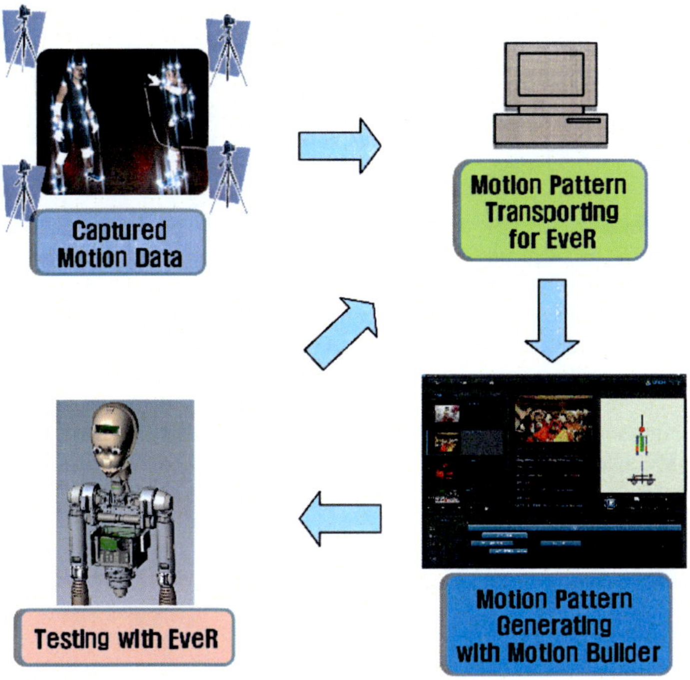

Fig. 7. The experimental process of EveR-3

Fig. 8. The performances of EveR-3

5 Conclusions

Cultural robotics is one of novel approaches of social robotics. Cultural robots are used for performances as well as assisting humans. Especially, cultural performance robots are used in various areas, such as acting, singing, dancing, playing musical instruments, playing games, teaching dancing, landscaping, and theme parks. We have developed the actress android EveR-3 for performing with humans in the theater. For this, action data of EveR-3 according to scenario is required. It takes very long time and efforts because performance takes so long time. We have developed the Acting Studio that is an action data editing system with graphical user interface. Using this system, we could reduce our time and efforts to editing action data of EveR-3. From various performances of EveR-3, we verify the efficiency of the Acting Studio for editing action data of cultural performance robots.

Acknowledgements. This research is supported by Ministry of Culture, Sports and Tourism (MCST) of Korea and Korea Creative Content Agency (KOCCA) in the Culture Technology (CT) Research & Development Program 2010.

References

1. Ahn, H.S., Beak, Y.M., Sa, I.-K., Kang, W.S., Na, J.H., Choi, J.Y.: Design of Reconfigurable Heterogeneous Modular Architecture for Service Robots. In: Proceedings of the IEEE International Conference on Intelligent Robots and Systems 2008 (IROS 2008), pp. 1313–1318 (2008)
2. Ahn, H.S., Sa, I.-K., Choi, J.Y.: PDA Based Mobile Robot System with Remote Monitoring for Home Environment. IEEE Transactions on Consumer Electronics 55(3), 1487–1495 (2009)
3. Ahn, H.S., Choi, J.Y.: Emotional Behavior Decision Model Based on Linear Dynamic Systems for Intelligent Service Robots. In: Proceedings of the IEEE International Symposium on Robot and Human Interactive Communication 2007 (RO-MAN 2007), pp. 786–791 (2007)
4. Lee, D., Ahn, H.S., Choi, J.Y.: A General Behavior Generation Module for Emotional Robots Using Unit Behavior Combination Method. In: Proceedings of the IEEE International Symposium on Robot and Human Interactive Communication 2009 (RO-MAN 2009), pp. 375–380 (2009)
5. Ahn, H.S., Choi, J.Y., Lee, D.-W., Shon, W.H.: Behavior Combination Generating Method for Reflecting Emotional Probabilities using Simulated Annealing Algorithm. In: Proceedings of the IEEE International Symposium on Robot and Human Interactive Communication 2011 (RO-MAN 2011), pp. 192–197 (2011)
6. Ahn, H.S., Sa, I.-K., Lee, D.-W., Choi, D.: A Playmate Robot System for Playing the Rock-Paper-Scissors Game with Humans. Artificial Life and Robotics 16(2), 142–146 (2011)
7. Fujita, M.: On activating human communications with pet-type robot AIBO. Proceedings of the IEEE 92(11), 1804–1813 (2004)
8. Shibata, T., Ohkawa, K., Tanie, K.: Spontaneous Behavior of Robots for Cooperation - Emotionally Intelligent Robot System. In: Proceedings of the 1996 IEEE International Conference on Robotics & Automation (ICRA 1996), pp. 2426–2431 (1996)

9. http://www.sciencedaily.ocm/releases/2007/04/070425103027.htm
10. Chang, S., Ham, S., Suh, D.: ROHINI: A robotic flower system for intuitive smart home interface. In: Proceedings of the 2010 International Conference on Control, Automation and Systems (ICCAS 2010), pp. 1773–1776 (2010)
11. Nameral, K., Takasugil, S., Takanol, K., Yamamoto, T., Miyake, Y.: Timing Control of Utterance and Body Motion in Human-Robot Interaction. In: Proceedings of the IEEE International Symposium on Robot and Human Interactive Communication 2008 (RO-MAN 2008), pp. 119–123 (2008)
12. Lin, C.-Y., Tseng, C.-K., Teng, W.-C., Lee, W.-C., Kuo, C.-H., Gu, H.-Y., Chung, K.-L., Fahn, C.-S.: The Realization of Robot Theater: Humanoid Robots and Theatric Performance. In: Proceedings of the IEEE International Conference on Advanced Robotics 2009 (ICAR 2009), pp. 1–6 (2009)
13. Miura, K., Nakaoka, S., Kajita, S., Kaneko, K., Kanehiro, F., Morisawa, M., Yokoi, K.: Trials of Cybernetic Human HRP-4C toward Humanoid Business. In: Proceedings of the 2010 IEEE International Workshop on Advanced Robotics and its Social Impacts (ARSO 2010), pp. 165–169 (2010)
14. Lee, H.-G., Baeg, M.-H., Lee, D.-W., Lee, T.-G., Park, H.-S.: Development of an Android for Emotional Communication between Human and Machine. In: Proceedings of the International Symposium on Advanced Robotics and Machine Intelligence, pp. 41–47 (2006)
15. Lee, D.W., Lee, T.G., So, B.R., et al.: Development of an Android for Emotional Expression and Human Interaction. In: Proceedings of the World Congress on International Federation of Automatic Control (IFAC 2008), pp. 4336–4337 (2008)
16. Lee, D.W., Che, D.W., Lee, D.Y., Jung, J.Y., Park, H.S., Lee, H.G.: Development of an Android for a Performance. In: Proceedings of the International Conference on Ubiquitous Robot and Ambient Intelligence (URAI 2010), p. 664 (2010)
17. Ahn, H.S., Lee, D.-W., Choi, D., Lee, D.Y., Hur, M.H., Jung, J.Y., Lee, H., Shon, W.H.: Development of an Android for Singing with Facial Expression. In: Proceedings of the Annual Conference on IEEE Industrial Electronics Society (IECON 2011), pp. 104–109 (2011)
18. Choi, D., Lee, D.-W., Lee, D.Y., Ahn, H.S., Lee, H.: Design of an android robot head for stage performances. Artificial Life and Robotics 16(3), 315–317 (2011)
19. Lee, D.-W., Choi, D., Jung, J.Y., Ahn, H.S., Park, H.S., Lee, H.: Remote control and conversation system between human and android robot vis internet. In: Proceedings of the International Symposium on Artificial Life and Robotics 2011 (AROB 2011), pp. 1021–1022 (2011)

RoboEarth Action Recipe Execution

Daniel Di Marco[1], Moritz Tenorth[2], Kai Häussermann[1], Oliver Zweigle[1], and Paul Levi[1]

[1] Department Image Understanding, Universität Stuttgart, Universitätsstr. 38, 70569 Stuttgart, Germany
{dimarco,haeussermann,zweigle,levi}@ipvs.uni-stuttgart.de
[2] Intelligent Autonomous Systems Group, Technische Universität München, Karlsstr. 45, 80333 München, Germany
tenorth@cs.tum.edu

Abstract. The ability of reusing existing task execution plans is an important step towards autonomous behavior. Today, the reuse of sophisticated services allowing robots to act autonomous is usually limited to identical robot platforms and to very similar application scenarios. The approach presented in this paper proposes a way to mitigate this limitation by storing and reusing task plans on a global accessible database. We describe the task execution engine for the RoboEarth project to demonstrate its ability to execute tasks in a flexible and reliable way.

1 Introduction

Robot control and execution planning is a challenging task, particularly in complex indoor household environments. Autonomous service robots have to achieve high-level task goals while performing in a reactive and flexible way. To allow a robust and reliable high-level task execution, autonomous robots have to detect unexpected situations, handle exceptions and manage their limited resources by adapting their behavior accordingly to cope with issues of dynamics and uncertainty.

Therefore high-level capabilities, such as knowledge-processing and reasoning, are essential for task planning and decision making processes. Pre-programming a plan for a given robot to solve a task is usually a complex and tedious act and most of the time, plans once written cannot be used again under different premises. Even if a plan is programmed for one robot in a robust manner it is usually impossible to copy this plan on another robot type and achieve the same success. To cope with these typical issues, RoboEarth was established. The prime objective of RoboEarth [23] is to enable robots to reuse previously generated data; i.e. maps of the environment, object models or task execution plans. The most important component in the RoboEarth project thus is a distributed database [18] for storing and processing this data. The data can be generated from participating robots, humans or processor programs running on the database itself. For the interpretation and execution of abstract task descriptions we developed an execution engine which is able to serialize and trigger the execution of all commands, supervise them, cover failure states and handle incoming events.

The focus of this work is to describe the task execution engine of the RoboEarth project and to demonstrate its ability to execute tasks in a flexible and reliable way.

S. Lee et al. (Eds.): Intelligent Autonomous Systems 12, AISC 193, pp. 457–466.
springerlink.com

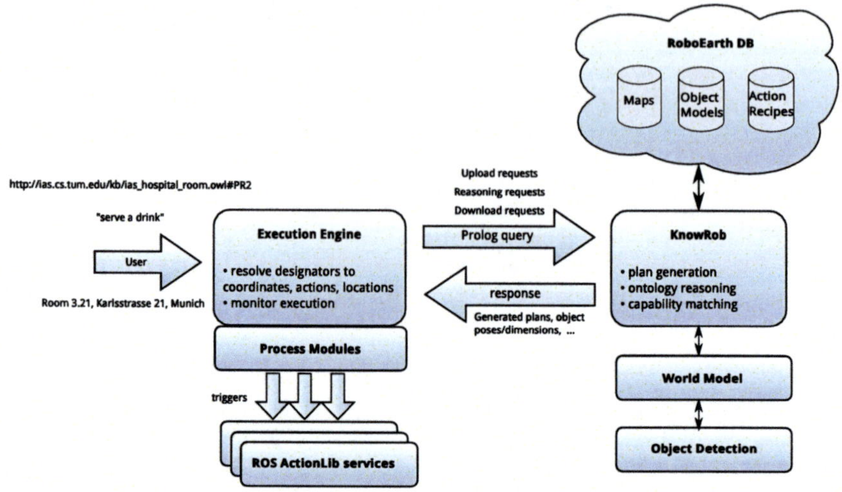

Fig. 1. Interaction between RoboEarth components, as seen from the execution engine's point of view

The remainder of the paper is organized as follows: first, a brief overview of the related work in the domain of plan-based robot control and decision-making is given in section 2. The current architecture for the RoboEarth demonstrator implementation is presented briefly in section 3. In the following section 4.1 the notion of action recipes is described in detail. The process for selecting an action recipe is detailed in section 5. Afterwards we describe the chosen language for implementing execution plans along with the necessary modifications in Section 6. Section 7 presents some experimental results. A discussion in Section 8 analyzes and sums up the contributions of this paper.

2 Related Work

Within the last few years several applications of autonomous service robots and demonstrations of plan-based control have been developed and successfully evaluated in real world environments. With respect to the implementation of a reliable plan-based high-level controller on autonomous mobile robots the work of [19, 2, 15] should be mentioned. All mentioned approaches so far have in common that the robot system will not be given the detailed sequence of commands to execute, but a list of goals which have to be achieved instead. During the run, the robot will generate a plan to accomplish these goals, handle unexpected situations, manage failures and thus execute the plan in a robust manner. The approach developed in [1] has the objective to control and manage a group of autonomous mobile robots for transshipment tasks in a decentralized and cooperative way. Further AI-based cooperative systems were proposed by [7, 6, 5].

In the plan-language and -execution domain several sophisticated systems have been developed. One of the first situation-driven interpretations of plans was proposed in RAP [9] to enable reactive planning and execution of *sketchy* plans. A descendant of

RAP is the language Reactive Plan Language (RPL) which was designed by [14] to incorporate a richer set of control constructs and to allow the creation of reactive plans for robotic agents. Further plan execution systems used by autonomous robots are the Reactive Model-based Programming Language (RMPL)[24] and the Executive Support Language (ESL) [10].

The execution system of the RoboEarth project builds upon the CRAM planning language [3], a sophisticated plan language that shares several concepts with RPL, but its focus lies on the plan execution on real physical robots, e.g. providing fast feedback-loops. In [4], the authors demonstrate the system by using a robot which downloads and parses a set of natural language instructions from a web page and translates these into a sketchy CRAM plan.

In contrast to the works mentioned above, this paper focuses on describing an approach to sharing task execution plans among different robot types. The instructions to be downloaded (called *action recipes*) along with object models, maps and other requirements are stored in a well-defined, computer-readable way.

3 System Overview

The overall architecture is shown in Figure 1. All data for RoboEarth is stored by a cloud computing database based on Apache Hadoop[1]. The implementation is described in [18]. The data stored in the database includes action recipes, object detection models, navigation maps, etc. Every file of this data is annotated with an OWL description to allow semantic and taxonomic queries. The execution engine takes textual commands and a description of the robot and environment from the user and poses requests for information like plans and object positions to the knowledge processing framework KnowRob. KnowRob is responsible for the knowledge processing (section 4). It can access both the world model (for answering object positions queries) and the global database. To generate a plan, it queries the database and retrieves a matching, symbolic task plan to the execution engine. During execution, the execution engine resolves symbolic plan annotations (*designators*) by querying the KnowRob component. Process modules are basically an interface for basic action primitives, in the given example implemented as ROS actionlib services.

Assume a scenario where a robot is commanded to find a mug in order to provide a human with a drink. In this case it is possible to query the RoboEarth database for all object models that represent an object of type mug and its specializations (e.g. coffee mug). Should this inquiry fail or not give the desired results, it is also possible to query for all models representing an object that can store liquids.

4 Representations for Exchanging Knowledge

The knowledge processing framework KnowRob [20] is used as interlingua to exchange knowledge between different components of the RoboEarth system. When the execution engine queries for a recipe, KnowRob searches the RoboEarth database, downloads

[1] `http://hadoop.apache.org`

recipes, checks for missing components (such as environment maps or object models that are required for the task), and triggers their download. During runtime, KnowRob serves as a semantic interface to the object positions provided by the world model.

In the current implementation, the KnowRob knowledge base is running locally on each robot. This comes at the cost of increased processing power and memory requirements on the robot, but has the advantage of the RoboEarth framework still providing basic functionality without having access to the Internet.

4.1 Action Recipes

Action recipes describe in a hardware-agnostic way how tasks are to be executed by a robot and are thus a kind of generalized task specification. Recipes are stored in description logic in a language built upon the W3C-standardized Web Ontology Language [22] and the KnowRob knowledge base [20]. The language itself is described in [21]. An excerpt from an exemplary action recipe is given in Figure 2.

This OWL-based language is optimized for automated reasoning about the requirements that need to be fulfilled in order to execute the action on a given robot platform, and for the description of actions, objects, environment models and robot capabilities in a single, coherent format. The following two sections will introduce some of the key concepts of this language. Before execution, the recipes are translated into the CRAM Plan Language CPL, which supports better execution monitoring, failure handling, and grounding of abstract object descriptions. This language will be described in Section 6.

```
Class: MoveBaseToHandoverPose
  EquivalentTo:
    (knowrob:toLocation some
        robotPose-handover1)
  SubClassOf:
    roboearth:Translation-LocationChange
...

Individual: ServeADrinkOrder10
  Types:
    knowrob:PartialOrdering-Strict
  Facts:
    knowrob:occursBeforeInOrdering
             MoveBaseToGraspPose,
    knowrob:occursAfterInOrdering
             GraspBottle
```

Fig. 2. Action Recipe example: Subaction and partial ordering specification (Manchester Syntax, [11])

Figure 2 gives examples of action specifications in RoboEarth action recipes, describing the action to move to a position for handing an object to a patient (*MoveBaseToHandoverPose*) as a specialization of an intentional location change (*Translation-LocationChange*). Task specifications are divided in sub-tasks that are partially ordered, thus allowing descriptions of both sequential and parallel task execution order.

```
Individual: Bottle1
  Types: knowrob:DrinkingBottle
Individual: Bed1
  Types: knowrob:Bed-PieceOfFurniture

Individual: handPose-handover1
  Types: knowrob:Point3D
  Facts: knowrob:aboveOf Bed1
```

Fig. 3. Description of two objects and one location for the example use case

4.2 Objects and Locations

In order to fully specify a task, it is not only necessary to describe the actions to execute, but also the objects and locations that the actions will interact with.

Pictured in Figure 3 is the corresponding description for some of the objects and locations used in the previously given example plan. *DrinkingBottle*, *Bed-PieceOfFurniture* and *Point2D* are the names for OWL classes defined in the RoboEarth ontology.

5 Reasoning about Robot Capabilities

To show the essential need of capability matching, we start with a typical scenario. Assume the RoboEarth database stores task descriptions for very different robot platforms. Assume further that some of the stored recipes were originally intended for e.g. Willow Garage's PR2 service robot and others were provided for a Roomba vacuum cleaner robot (i.e. a robot without an arm and gripper). Obviously, in this case there are recipes in the database that the Roomba robot cannot use, no matter how generic and platform agnostic these are formulated.

Thus two key features are required:

1. a way for describing the capabilities of different robot hardware
2. a method for selecting recipes from the database which are usable for a robot platform

These two points will be addressed in the following subsections.

5.1 Specifying Capability Requirements of an Actions

The Unified Robot Description Format (URDF) [25] has been adopted by the Robot Operating System [16] for describing a robot platform by its joints and links. This format is used within the ROS framework for visualizing robots and for configuring motion planners.

To decide whether a given recipe can be executed by a given robot, enumerating the joints and links is not enough [13]. Additionally, a semantic description of sensors and actuators is required, which can be checked with requirement descriptions in the action recipes.

Recently, a robot capability language incorporating sensor and actuator semantics (called SRDL) has been proposed by [13]. The OWL classes it uses for designating sensors and actuators are extensions of the KnowRob ontology, and are thus easily integrable into the RoboEarth system. We chose the successor of the language, called SRDL2, for integration into RoboEarth [21].

5.2 Selecting Action Recipes

To decide whether a given robot is capable of executing a recipe, it is necessary to annotate said recipe with capability requirements. Thanks to formulating recipes in terms of classes in an OWL ontology, there are multiple ways to do that. For instance, in the KnowRob ontology there is an OWL class called *Reaching*, whose parent is *VoluntaryBodyMovement* and which has the children (among others) *ReachForObjectPose* and *HandoverObject*. Both children inherit the dependency on a *ReachingAbility* from their parent class, specified as OWL restriction *hasCapabilityDependency some ReachingCapability* for the class *Reaching*.

With this adaption, for every recipe containing a sub-action that is a descendant of *Reaching* (for example *HandoverObject*), we can infer by OWL reasoning that a robot needs to provide the *ReachingCapability* capability to execute the recipe.

6 CRAM Plan Language

The CRAM planning language (CPL) [3] is used as the base language for executing RoboEarth. CPL extends on concepts from McDermott's "Reactive Plan Language" [14]. It provides amongst other things several control structures for executing and synchronizing tasks in parallel and failure handling. As it is based on Common Lisp, the expressiveness of a proven, Turing-complete programming language is available for implementing plans. Together with ROSLisp, the Lisp client library for the Robot Operating System, a whole range of tested robotics implementations are readily available for use in an execution plan. An exemplary plan describing the task of fetching a bottle to a person in a hospital bed is shown in Figure 4.

Objects, locations and actions are described by *designators*. Designators are basically symbolic descriptions that are resolved at the latest possible instant during task execution. For example, *robot-pose-handover1* in Figure 4 is a location designator that gets resolved to an actual robot base position such that the robot can reach to *hand-pose-handover1*, while the latter designator gets resolved to a position above the bed.

Actions to be executed in CPL are specified by describing corresponding goals. These goals in turn can be specified by a set of other goals or by calling a process module. Process modules implement hardware-specific behavior, for example for navigating to a given pose or for controlling the manipulators. They can thus be seen as a classic hardware abstraction layer, abstracting manipulators as well as sensors.

```
(def-top-level-plan serve-a-drink ()
  (with-designators
      ((bed1 (object '((name bed1)
         (type bed_piece-of-furniture))))
      (location-bed1 (location
         '((of ,bed1))))
      (hand-pose-handover1 (location
         '((on ,location-bed1))))
      (bottle1 (object '((name bottle1)
         (type drinking-bottle))))
      (robot-pose-handover1 (location
         '((to reach)
           (loc ,hand-pose-handover1)))))
    (achieve '(object-in-hand ,bottle1 :right))
    (achieve '(loc Robot ,robot-pose-handover1))
    (achieve '(object-handed-over ,bottle1
      ,hand-pose-handover1))))
```

Fig. 4. CRAM example plan for serving a drink to a patient

6.1 Designator Resolution

There are three different designator types in the current implementation:

1. Object designators describe objects that the robot has to interact with. In the example plan in Figure 4 both the bed and the bottle are designators of this kind. These kinds of designators are resolved to objects defined in the semantic map of the environment by evaluating the given description.

   The bed object in the given example is resolved by the object type; i.e. all objects of the type *bed_piece-of-furniture* found in the semantic map of the environment are candidates for resolution. It is worthwhile to mention that specializations of object types are also considered: E.g. assume that the above plan is to be run in a hospital room with a semantic map containing an object described as *hospital_bed_piece-of-furniture*. Presuming this object type is defined as a subclass of *bed_piece-of-furniture*, the bed object designator gets still bound to the hospital bed object.
2. Location designators can be bound to objects to reference the pose of objects, as is the case with in the example in Figure 4. They can also describe poses relative to other locations and thus provide a basic form of geometric reasoning, as is the case with *hand-pose-handover1*.
3. Action designators provide symbolic annotations to a movement or perception action. They get resolved by the platform-dependent process modules responsible for the corresponding manipulator. In order to keep action designator descriptions consistent across different hardware platforms, we plan to extend the RoboEarth ontology as a basis for action designators.

6.2 Translating Action Recipes into CRAM Plans

The language for formulating action recipes described in [21] is designed to be a language for specifying and sharing hardware-independent concepts and enabling automated reasoning on these concepts, i.e. it is a language to describe the semantics of a task. Therefore, before execution it is necessary to translate these descriptions into a plan description language, containing calls to manipulator commands, sensor interpretation commands, etc. Older prototypical implementations of an execution engine for the RoboEarth project are translating recipes into a finite state machine[2].

As the sub-tasks in an action recipe are partially ordered, an action recipe can be represented as a directed, acyclic graph. To create a plan from this graph, our implementation first applies a topological sorting (using an implementation of the algorithm proposed by [12]). Our implementation thus currently does not account for parallel execution of sub-tasks.

Having found a valid topological ordering, we then proceed to generate a list of designators from the task descriptions in the recipe. The creation of designators for objects and locations in the recipe is straight forward: Object designators are created from the object type and name as specified in the recipe; e.g. the object *Bed1* from Figure 3 is translated to the designator *bed1 (object '((name bed1) (type bed_piece-of-furniture)))*. Similarly, location descriptions like *handPose-handover1* are translated to location designators as *hand-pose-handover1 (location '((above-of,location-bed1)))*

For translating tasks to goals, we have defined a set of mappings. E.g. the *MoveBaseToHandoverPose* from Figure 2 is mapped to the goal *(loc Robot pose)*, whereas *pose* is the location designator for the target pose. In the future, we plan to integrate these mappings into the action recipe language, to increase the applicability of the approach.

7 Experiments

The work described in this paper was used for the demonstrator for the third RoboEarth workshop at TU Munich in February 2012 on a PR2 and the TUE's AMIGO robot platform [8]. Both robots had the same task similar to the examples given above, i.e. to fetch a bottle from inside a cabinet to a hospital bed, but were situated in different rooms in different locations. Two recipes were used, one describing the task of grasping an object and one describing fetching an object to a hospital bed whereas the former was used by the latter. There were several resources that had to be downloaded prior to task execution; i.e. maps describing the environments and object detection models. Additionally, the robots were downloading an articulation model [17] to open the cabinet door or creating one, if none was available.

Both robots were able to successfully retrieve the required resources from the database, thus synthesize the plan and complete the task. We noticed that due to the static nature of the OWL to CRAM translation, we had to take care that the process modules for both platforms are equivalent in their effects.

8 Conclusions and Future Work

In this work, a plan-based execution engine for the RoboEarth-Project has been presented. The primary focus of this work was the tight integration of the concept in the

[2] A video of a public demonstration where one implementation of this kind was used is available at `http://youtu.be/r5ZRxbo0pSQ`

whole RoboEarth framework. The key aspect of RoboEarth is the use of action recipes, which act as symbolic, high-level task descriptions and which are described in a description language in a platform independent way to allow a platform independent sharing and reasoning.

In the current implementation, the plans describing how to achieve sub-goals (e.g. *(achieve '(object-in-hand ...))* in Figure 4) for picking up objects) are implemented on the robot itself. A more generic approach would be to have these sub-plans generated automatically from action recipes available in the database, as we do it with the top level plan (described in section 6.2).

Another aspect that we will investigate further is how to move the knowledge processing onto the database servers. This would enable robots with less computing power to retrieve and execute action recipes, if a reliable Internet connection is given.

Acknowledgement. The research leading to these results has received funding from the European Union Seventh Framework Programme FP7/2007-2013 under grant agreement number 248942 RoboEarth.

References

[1] Alami, R., Fleury, S., Herrb, M., Ingrand, F., Robert, F.: Multi-robot cooperation in the martha project. IEEE Robotics & Automation Magazine 5(1), 36–47 (1998)

[2] Beetz, M., Arbuckle, T., Belker, T., Cremers, A., Schulz, D., Bennewitz, M., Burgard, W., Hahnel, D., Fox, D., Grosskreutz, H.: Integrated, plan-based control of autonomous robot in human environments. IEEE Intelligent Systems 16(5), 56–65 (2001)

[3] Beetz, M., Mösenlechner, L., Tenorth, M.: CRAM – A Cognitive Robot Abstract Machine for everyday manipulation in human environments. In: 2010 IEEE/RSJ International Conference on Intelligent Robots and Systems (IROS), pp. 1012–1017 (2010)

[4] Beetz, M., Klank, U., Kresse, I., Maldonado, A., Mösenlechner, L., Pangercic, D., Rühr, T., Tenorth, M.: Robotic roommates making pancakes. In: 2011 11th IEEE-RAS International Conference on Humanoid Robots (Humanoids), pp. 529–536 (2011), doi:10.1109/Humanoids.2011.6100855

[5] Brumitt, B., Stentz, A.: GRAMMPS: A generalized mission planner for multiple mobile robots in unstructured environments. In: 1998 Proceedings of the IEEE International Conference on Robotics and Automation, vol. 2, pp. 1564–1571. IEEE (1998)

[6] Clement, B., Durfee, E.: Top-down search for coordinating the hierarchical plans of multiple agents. In: Proceedings of the Third Annual Conference on Autonomous Agents, pp. 252–259. ACM (1999)

[7] Collins, J., Tsvetovat, M., Mobasher, B., Gini, M.: Magnet: A multi-agent contracting system for plan execution. In: Proc. of SIGMAN, pp. 63–68 (1998)

[8] Control Systems Technology Group, Eindhoven University of Technology (2011) AMIGO specifications (2011), `http://www.roboticopenplatform.org/wiki/AMIGO` (accessed December 19, 2011)

[9] Firby, R.: An investigation into reactive planning in complex domains. In: Proceedings of the Sixth National Conference on Artificial Intelligence, vol. 1, pp. 202–206 (1987)

[10] Gat, E.: Esl: A language for supporting robust plan execution in embedded autonomous agents. In: 1997 Proceedings of the IEEE Aerospace Conference, vol. 1, pp. 319–324. IEEE (1997)

[11] Horridge, M., Patel-Schneider, P.: Manchester syntax for owl 1.1. In: OWL: Experiences and Directions, OWLED (2008)
[12] Kahn, A.B.: Topological sorting of large networks. Commun. ACM 5, 558–562 (1962)
[13] Kunze, L., Roehm, T., Beetz, M.: Towards semantic robot description languages. In: IEEE International Conference on Robotics and Automation (ICRA), Shanghai, China (2011)
[14] McDermott, D.: A reactive plan language. Tech. rep., Citeseer (1991)
[15] Muscettola, N., Dorais, G., Fry, C., Levinson, R., Plaunt, C.: Idea: Planning at the core of autonomous reactive agents. In: Proceedings of the Workshops at the AIPS-2002 Conference, Tolouse, France (2002)
[16] Quigley, M., Gerkey, B., Conley, K., Faust, J., Foote, T., Leibs, J., Berger, E., Wheeler, R., Ng, A.: Ros: an open-source robot operating system. In: ICRA Workshop on Open Source Software (2009)
[17] Rühr, T., Sturm, J., Pangercic, D., Cremers, D., Beetz, M.: A generalized framework for opening doors and drawers in kitchen environments. In: IEEE International Conference on Robotics and Automation (ICRA), St. Paul, MN, USA (to appear, 2012)
[18] Schießle, B., Häussermann, K., Zweigle, O.: Deliverable D6.1: Complete specification of the RoboEarth platform. Tech. rep., (December 1, 2010), `http://www.roboearth.org/wp-content/uploads/2011/03/D61.pdf`
[19] Simmons, R., Goodwin, R., Haigh, K., Koenig, S., O'Sullivan, J.: A layered architecture for office delivery robots. In: Proceedings of the First International Conference on Autonomous Agents, pp. 245–252. ACM (1997)
[20] Tenorth, M., Beetz, M.: KnowRob - knowledge processing for autonomous personal robots. In: 2009 IEEE/RSJ International Conference on Intelligent Robots and Systems (IROS), pp. 4261–4266. IEEE (2009)
[21] Tenorth, M., Perzylo, A., Lafrenz, R., Beetz, M.: The RoboEarth language: Representing and Exchanging Knowledge about Actions, Objects, and Environments. In: IEEE International Conference on Robotics and Automation (ICRA), Saint Paul, USA (accepted for publication, 2012)
[22] W3C OWL Working Group OWL 2 Web Ontology Language Document Overview. W3C recommendation, W3C (2009) (accessed December 1, 2011)
[23] Waibel, M., Beetz, M., Civera, J., D'Andrea, R., Elfring, J., Galvez-Lopez, D., Häussermann, K., Janssen, R., Montiel, J., Perzylo, A., et al.: RoboEarth. IEEE Robotics & Automation Magazine 18(2), 69–82 (2011)
[24] Williams, B., Ingham, M., Chung, S., Elliott, P.: Model-based programming of intelligent embedded systems and robotic space explorers. Proceedings of the IEEE 91(1), 212–237 (2003)
[25] Willow Garage: Xml robot description format (urdf) (2011), `http://www.ros.org/wiki/urdf/XML` (accessed December 1, 2011)
[26] Zweigle, O., van de Molengraft, R., D'Andrea, R., Häussermann, K.: RoboEarth: connecting robots worldwide. In: Proceedings of the International Conference on Interaction Sciences: Information Technology, Culture and Human, pp. 184–191. ACM (2009)

Exchanging Action-Related Information among Autonomous Robots

Moritz Tenorth and Michael Beetz

Intelligent Autonomous Systems Group
Department of Informatics, Technische Universität München
{tenorth,beetz}@cs.tum.edu

Abstract. In this paper, we describe representations and inference techniques that are used in the RoboEarth system for the web-based exchange of information between robots. We present novel representations for environment maps that combine expressive semantic environment models with techniques for selecting suitable maps from the web-based RoboEarth knowledge base. We further propose techniques for improving class-level object models with additional information as needed for distributed learning of object properties. In an integrated experiment, we show that the system enables robots to perform mobile manipulation tasks including the retrieval of suitable environment maps and the estimation and exchange of object property information.

1 Introduction

The RoboEarth system [1] is designed as a web community by robots for robots: Similar to community-driven web pages for humans, such as Wikipedia, which enable humans to exchange knowledge among each other, RoboEarth allows knowledge exchange between robots. When one robot has learned how to perform a task or has created an object model, it can upload this information to RoboEarth. Other robots can later download it and do not have to learn everything on their own again.

Since the platform is intended to serve for exchanging various kinds of information among heterogeneous robots with different capabilities, the development of appropriate techniques for finding relevant information becomes a challenging problem. When humans search for information on the Web, they can easily distinguish relevant from irrelevant information (e.g. search results caused by ambiguous meanings of the search terms) and filter out information they cannot make use of (e.g. text written in a language they do not understand). Realizing these capabilities on a robot requires powerful knowledge representation and reasoning techniques.

A central part of the RoboEarth project is thus the development of an expressive, formal representation language to encode the exchanged information. It helps the robot answer questions such as: Which descriptions exist for the task to be performed? Which of them can be used given the robot's hardware and software configuration? Are all required capabilities available? Which object models

S. Lee et al. (Eds.): Intelligent Autonomous Systems 12, AISC 193, pp. 467–476.
springerlink.com

are needed to perform the task? Are there maps that describe the environment the robot is to operate in, which kinds of information do they provide?

Autonomously taking these decisions requires access to meta-information that describes which kind of knowledge is encoded in which format, including for instance which coordinate systems are used for spatial information or which capabilities are needed to perform a task. This requires more comprehensive representations than commonly used for describing actions, objects, or environment structures in robotics. Related approaches usually focus on the representation of the information content, e.g. the spatial information in a map, but do not describe meta-information, e.g. *which* environment is actually described in that map. Hierarchical Task Networks (HTN [2]) and related plan languages are similar to the action representation used in RoboEarth but focus on the description of the task itself, i.e. its sub-actions, goals, and ordering constraints. XABSL [3], mainly used in the RoboCup soccer context, describes actions in terms of hierarchical finite state machines. AutomationML [4] is a standard for describing task information and spatial configurations, mainly used in industrial applications. The FIPA [5] standard primary deals with the definition of communication standards for software agents. Object description formats like the proprietary DXF [6] or the open Collada [7] standard describe objects using meshes and textures, but without further specifying semantic properties.

In [8], we gave a general overview of the representation language used in RoboEarth. In this paper, we will first describe the work-flow for downloading information from RoboEarth to show how the reasoning processes are integrated into the overall system. Then, we will focus on two specific reasoning problems, namely the representation and discovery of map information in RoboEarth, and the retrieval, improvement and sharing of object models.

The main contributions of this paper are the following: We present novel representations for environment maps that combine the expressive semantic environment models introduced in [9] with explicit descriptions of the environment that is described in the map, and with techniques for selecting suitable maps from a web-based database. We further present novel techniques for improving class-level object models with additional information as needed for distributed learning of object properties.

2 Retrieving Information Needed for a Task

A common scenario in RoboEarth is that a robot is given a task that it does not have a plan for. This means the robot needs to obtain an action plan and identify and retrieve missing components such that it, in the end, has all information it needs to successfully accomplish the task.

We assume that the robot has initial knowledge about the environment it is operating in, for example the address of the room (see Section 3.3), and a semantic model of itself, including its hardware and software components and abstract capabilities. This self-model needs to be created only once for each kind of robot and is described using the Semantic Robot Description Language (SRDL [10]). When the robot acquires additional capabilities, e.g. by downloading an object model from RoboEarth, it adds them to its self-model so that they are available for future queries.

Figure 1 outlines the main inference steps involved in the retrieval of information. These inferences are performed before the robot starts to execute the task and ensure the availability of all components that are required for this task. First, the robot checks whether it has a map of the environment at hand and, if not, downloads one from RoboEarth to obtain information about free space and obstacles as well as about objects in the environment. The next step is to search for task descriptions, called "action recipes", that describe on the one hand which actions need to be performed, and, on the other hand, list dependencies the task has in terms of components and capabilities. The robot matches this specification against its self-model and iterates over all missing components to check whether they can be obtained from RoboEarth. If all missing components can be provided, the system converts the action recipe into an executable action plan in the CRAM plan language (CPL, [11]) and sends it to the CRAM executive.

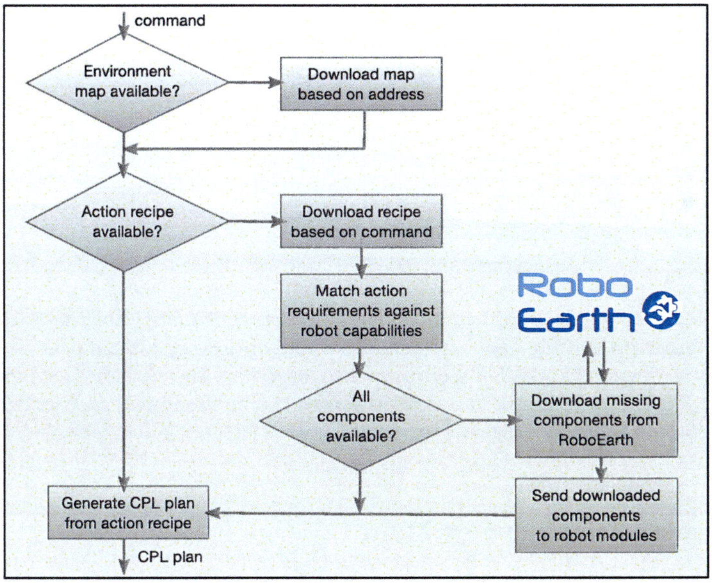

Fig. 1. Flowchart describing the process for downloading information from RoboEarth. Based on the command to perform a task, the robot determines which information it needs and downloads the required descriptions from the RoboEarth knowledge base.

The representation language [8] represents knowledge using the Web Ontology Language (OWL) as an extension of the KnowRob knowledge base [12]. The ontologies and software modules are available online as open source software.[1]

3 Representation of Environment Information

Figure 2 visualizes different kinds of environment maps that can be described and exchanged via RoboEarth. Their descriptions consist of two parts: the representation of the map's content, described in Section 3.1, and of the environment that is described in this map, which will be introduced in Section 3.3.

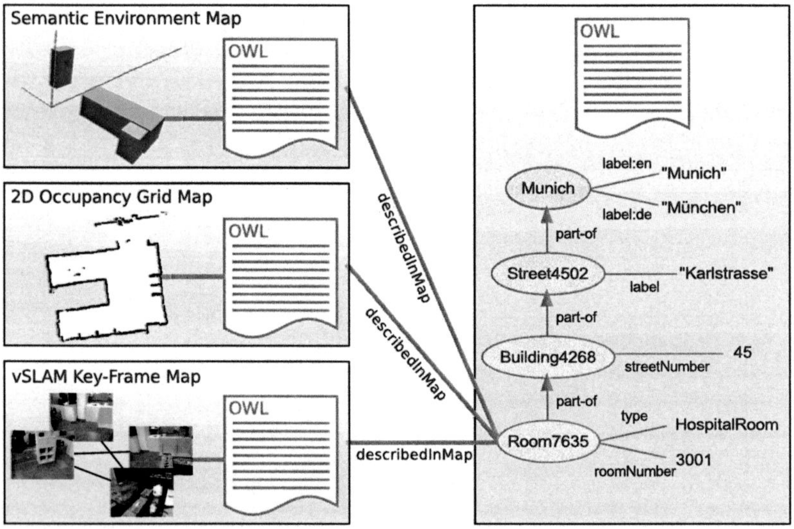

Fig. 2. Representations of different kinds of environment maps. Each map consists of an OWL description of its type and properties and optionally a binary file. In the case of semantic maps, the OWL description also describes the objects that are part of the map. The map is linked to a description of the environment it describes via the *describedInMap* relation.

3.1 Types of Environment Maps

For all tasks that include navigation or interaction with objects, a robot needs an environment model to plan its actions. Depending on the action to be performed, different kinds of maps are needed: Occupancy grid maps discriminate between free space and obstacles for localization, navigation and obstacle avoidance. Semantic environment maps contain localized object instances and can be used for grounding abstract object descriptions contained in an action recipe.

[1] Online: `http://ros.org/wiki/knowrob`, `http://ros.org/wiki/re_ontology`

Maps are described by an OWL specification of its type and properties that can optionally be linked to binary files. After download, the OWL description is parsed, which creates an instance of the respective type of map in the knowledge base such that the robot 'knows' that the map is available. Depending on the type of the map, the system can decide how to proceed after download: For example, a serialized occupancy grid is sent to the localization and navigation routines.

3.2 Spatio-temporal Object Pose Representation

Semantic environment maps consist of a set of typed, localized object instances. Based on the type, one can retrieve semantic information about the objects that is described at the class level, for example descriptions of physical parts, articulation properties, or CAD models to be used for recognition, spatial reasoning and visualization. Section 4 describes in detail how these properties are described for object classes.

To account for the dynamics in the environment and in order to be able to describe changes in the world, introduced by the robot itself or external effects, the representation of object instances can store and reason about changes in object poses. Multiple detections at different poses can be attached to an object instance; each of them is annotated with a time point and the perception method that has been used. This spatio-temporal object representation can be generated automatically from perceptual information and provides a flexible way of integrating novel information about objects into the knowledge base. It is described in more detail in [8].

3.3 Discovery of Suitable Maps for the Robot's Environment

In order to exchange maps with other agents, a robot needs to be able to describe which environment the maps belong to. This information is very important in the RoboEarth setup as it allows to specify which maps are desired when sending a request to the knowledge base. Since environments can be described in many different ways, the representation should support a spatial hierarchy (city – street – building), a semantic hierarchy (room – kitchen), and different labels for the same room.

We have developed a flexible approach that is similar to the human way of describing an address: Maps are annotated with data like the city, street, building, floor, room number or room type they describe, as shown in the right part of Figure 2. These components are linked by a transitive part-of relation. This allows to query for combinations of these levels, e.g. to search for all maps of a kitchen in Karlstrasse in Munich, or for all rooms on the third floor of Karlstrasse 45. It further allows to combine labels, such as room numbers, with types of rooms (private homes usually do not have room numbers) and to attach multiple labels to the same physical entity (Karlsplatz and Stachus are two names for the same square in Munich).

The representation can be combined with other techniques for searching for environment maps: Searching by GPS coordinate would be useful for field robotics, while there is normally no GPS signal available indoors. Alternatively, one can also assume that the robot is continuously localized (quantitatively or qualitatively) in a hierarchy of metric and topological maps, and that it can detect when it runs out of one map. In this case, it can search for an extension map describing the new area. This method, however, requires modification of the lower-level map management infrastructure.

4 Retrieving and Improving Object Models

While the information a robot perceives is at first environment-specific, there are often generic pieces of information that can be extracted from it. For example, if the robot estimates the articulation properties of an object [13], this information is a priori specific to the object instance the robot was interacting with. If the object, however, is an instance of a common type, it may make sense to extract the generic information and share it with other robots. For this reason, we have developed methods to generate class-level object models, taking a specific object instance as an example.

Figure 3 shows the life cycle of an object model: During the initialization phase, shown in Figure 1, the model is downloaded from RoboEarth. If the

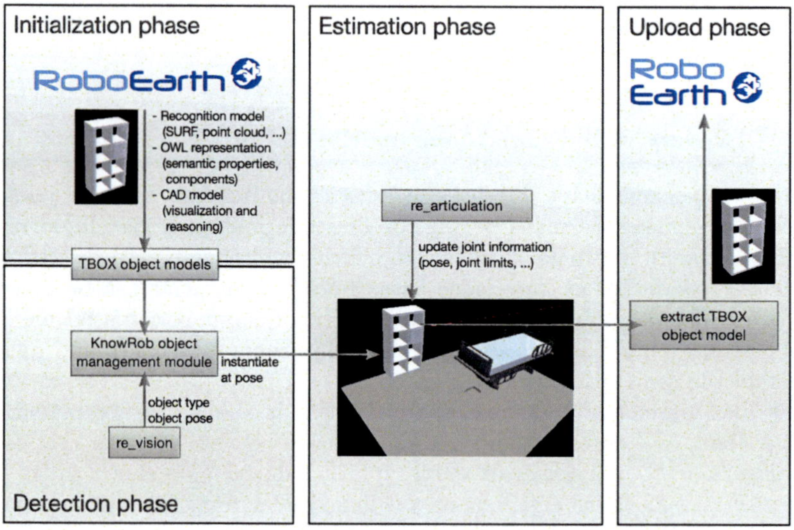

Fig. 3. Schematic overview of the object model up- and download. Robots can retrieve an object model from RoboEarth, instantiate the model if the object has been perceived in the environment, estimate additional properties, and upload a model generated from this object instance to share the new information with other robots.

robot detects the object in the environment, the abstract class-level model is instantiated at the respective position, recursively translating all object-intrinsic coordinates into global map coordinates. During interaction, the robot estimates object properties like the positions and types of joints and adds them to the object instance. After finishing the task, it checks which pieces of information have been updated and for which sharing makes sense.

For these objects, it extracts a class-level (TBOX) model and uploads it to the RoboEarth knowledge base. The RoboEarth language supports the explicit specification of coordinate frames (e.g. map-global, relative to other objects or other components of the same objects, or qualified using frames in the ROS tf system[2]), and provides methods for the conversion between coordinates in different frames. By recursively converting the coordinates and translating instance-level descriptions into class restrictions, the system can create an object model that includes the new information.

5 Experiments

In a recent experiment, we have investigated how these methods can enable robots to perform a drink serving task in a previously unknown environment. The experiment has been done in two locations with two different robot platforms, a PR2 and the Amigo robot. Both robots were only equipped with the command and the address of the environment they were operating in and successfully downloaded all information required for the task. Compared to prior experiments [8], the robots thus had to operate with less initial information and perform more inference to find, select and apply information from the RoboEarth knowledge base.

The upper part of Figure 4 shows the environment maps that have been downloaded from RoboEarth. RoboEarth offers a SerQL [14] query interface.[3] The following query has been used to download the map information:

```
select source from context source {R}
  kr:describedInMap {S} ;
  kr:roomNumber {N}
  where N like "3001"
  using namespace
  kr=<http://ias.cs.tum.edu/kb/knowrob.owl#>;
```

Based on this map, the robots (Figure 4 bottom) could navigate to the appropriate positions and locate the objects required for the task. The action recipe to be used was selected using the following query:

```
select source from context source {A}
  rdfs:label {"serve_a_drink"^^xsd:string}
  using namespace
  rdfs=<http://www.w3.org/2000/01/rdf-schema#>
```

The action recipe was then matched against the robots' capabilities with the result that all required capabilities were available, but some recognition models

[2] http://ros.org/wiki/tf

[3] http://api.roboearth.org

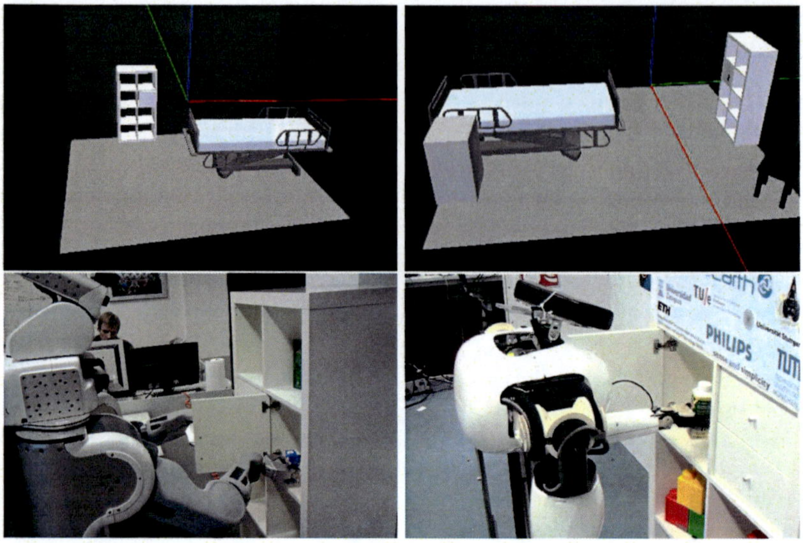

Fig. 4. Top: Semantic environment maps of the two hospital rooms, downloaded from RoboEarth based on the address and room number. Bottom: PR2 and Amigo robots opening the cabinet and picking up the drink to be served.

for some of the objects mentioned in the task were missing (namely the bottle and the bed), which have been downloaded, including the CAD models shown in Figure 4. The following CPL plan was then generated from the action recipe:

```
(def-top-level-plan serve-a-drink ()
(with-designators (
      (bottle1 (object '((name bottle1)
                         (type drinking-bottle))))
      (bed1 (object     '((name bed1)
                         (type bed_piece-of-furniture))))
      (hand-pose-handover1  (location '((on ,bed1))))
      (robot-pose-handover1 (location '((to reach)
                              (side :right)
                              (loc ,hand-pose-handover1))))
      (arms-at101 (action '((type trajectory)
                              (pose ,hand-pose-handover1)
                              (side :right))))
      (unhand-action102 (action '((type open-gripper)
                              (side :right))))
)

(achieve '(object-in-hand ,bottle1 :right))
(at-location (robot-pose-handover1))
(achieve '(arms-at ,arms-at101))
(achieve '(arms-at ,unhand-action102))))
```

This experiment shows that the representation and reasoning mechanisms in RoboEarth can be used to represent and select information about human environments and mobile manipulation tasks. They have successfully been used by two heterogeneous robots in different environments who both performed the same mobile manipulation task.

6 Discussion and Conclusions

In this paper, we presented novel methods for exchanging information between robots with a focus on two applications: the exchange of environment maps and of object models. Regarding maps, we discussed the representation of the different kinds of maps and the relation between the maps themselves and the environments they describe. These techniques allow easy discovery of information in the database using only information that can realistically be assumed to be available on the robot. Object models are described by a combination of a semantic description in OWL with (binary) recognition models and support the representation of a component hierarchy as well as articulation properties. We showed that object class descriptions can be downloaded and be instantiated once the corresponding object has been detected in the environment. Based on these instances, the robot can estimate additional information, attach this information to the object, and share the newly acquired information with other robots. The presented methods significantly raise the level of semantics and are a big step towards robots that autonomously exchange information. Challenges to be addressed in the future include the question of which pieces of information are actually worth being exchanged.

Acknowledgments. This work is supported in part by the EU FP7 Project *RoboEarth* (grant number 248942).

References

1. Waibel, M., Beetz, M., D'Andrea, R., Janssen, R., Tenorth, M., Civera, J., Elfring, J., Gálvez-López, D., Häussermann, K., Montiel, J., Perzylo, A., Schießle, B., Zweigle, O., van de Molengraft, R.: RoboEarth - A World Wide Web for Robots. Robotics & Automation Magazine 18(2), 69–82 (2011)
2. Erol, K., Hendler, J., Nau, D.: Htn planning: Complexity and expressivity. In: Proceedings of the National Conference on Artificial Intelligence, pp. 1123–1123. John Wiley & Sons LTD (1994)
3. Loetzsch, M., Risler, M., Jüngel, M.: XABSL-a pragmatic approach to behavior engineering. In: Proceedings of IEEE/RSJ International Conference of Intelligent Robots and Systems (IROS), Citeseer, pp. 5124–5129 (2006)
4. Drath, R., Luder, A., Peschke, J., Hundt, L.: AutomationML-the glue for seamless automation engineering. In: IEEE International Conference on Emerging Technologies and Factory Automation, ETFA 2008, pp. 616–623. IEEE (2008)
5. O'Brien, P., Nicol, R.: FIPA – towards a standard for software agents. BT Technology Journal 16(3), 51–59 (1998)
6. Rudolph, D., Stürznickel, T., Weissenberger, L.: Der DXF-Standard, Rossipaul (1993)
7. Arnaud, R., Barnes, M.: COLLADA: sailing the gulf of 3D digital content creation. AK Peters, Ltd. (2006)
8. Tenorth, M., Perzylo, A.C., Lafrenz, R., Beetz, M.: The roboearth language: Representing and exchanging knowledge about actions, objects, and environments. In: IEEE International Conference on Robotics and Automation (ICRA), St. Paul, MN, USA, May 14-18 (accepted for publication, 2012)

9. Tenorth, M., Kunze, L., Jain, D., Beetz, M.: KNOWROB-MAP – Knowledge-Linked Semantic Object Maps. In: 10th IEEE-RAS International Conference on Humanoid Robots, Nashville, TN, USA, December 6-8, pp. 430–435 (2010)
10. Kunze, L., Roehm, T., Beetz, M.: Towards semantic robot description languages. In: IEEE International Conference on Robotics and Automation (ICRA), Shanghai, China, pp. 9–13 (May 2011)
11. Beetz, M., Mösenlechner, L., Tenorth, M.: CRAM – A Cognitive Robot Abstract Machine for Everyday Manipulation in Human Environments. In: IEEE/RSJ International Conference on Intelligent Robots and Systems, Taipei, Taiwan, October 18-22 (2010)
12. Tenorth, M., Beetz, M.: KnowRob – Knowledge Processing for Autonomous Personal Robots. In: IEEE/RSJ International Conference on Intelligent Robots and Systems, pp. 4261–4266 (2009)
13. Sturm, J., Stachniss, C., Burgard, W.: Learning kinematic models for articulated objects. Journal on Artificial Intelligence Research, JAIR (2011)
14. Broekstra, J., Kampman, A.: Serql: A second generation rdf query language. In: Proc. SWAD-Europe Workshop on Semantic Web Storage and Retrieval, pp. 13–14 (2003)

Ubiquitous Network Robot Platform for Realizing Integrated Robotic Applications

Shuichi Nishio[1], Koji Kamei[2], and Norihiro Hagita[2]

[1] Hiroshi Ishiguro Laboratory, Advanced Telecommunications Research Institute International (ATR), Keihanna Science City, Kyoto, Japan
nishio@ieee.org

[2] Intelligent Robotics and Communication Laboratories, ATR, Keihanna Science City, Kyoto, Japan
{kamei,hagita}@atr.jp

Abstract. This paper introduces a common infrastructure for robotic applications to support our daily life: the *Ubiquitous Network Robot Platform* (UNR-PF). UNR-PF bridges the gaps among service application providers, customers and robotic devices such as robots, sensors and smartphones. For application programmers, UNR-PF provides a common API so that programmers can realize applications in a way independent to the actual variation in devices. UNR-PF also provides applications to span across multiple areas so that customers can receive seamless support in different scenes based on their abilities. Here, we first describe the motivation as well as requirements for UNR-PF based on past field studies. Based on these requirements, several real-world case studies has been performed which showed the effectiveness of the proposed concept of UNR-PF for continuously supporting our daily activities across multiple areas. We describe recent activities toward standardizing the key elements of UNR-PF and discuss future works.

1 Introduction

Demands for assistive robots are quickly emerging these days with the approach of aged society. In recent decades, researchers had been focusing on constructing robots that can support human daily activities by interactive means [3]. As the structure and functionality of robots grow large and complicated, constructing robots now requires greater time, cost and effort. Similar to traditional software engineering, people started to seek modularity and reusability of basic functional components, which has led to research and development of common libraries and middlewares [1, 2, 10, 11]. Such allows developers to utilize existing functional modules in combination with their own software and to rapidly develop working robots. This modularized development process has accelerated the development of stand-alone robots as well as individual functional components. At the same time, however, variations in both hardware and software has decreased the reusability of robotic applications.

Another approach to enhance capabilities of robots is the concept named *Network Robot System* (NRS) or *Networked Robots* [13]. The main goal of NRS was to integrate various types of devices such as robots, sensor networks and smartphones, so that the

S. Lee et al. (Eds.): Intelligent Autonomous Systems 12, AISC 193, pp. 477–484.
springerlink.com

whole system could act as an integrated system with enhanced capabilities that cannot be attained by a single robot or multiple uniform robots. Since its proposal in 2002, many research projects have been carried out accompanied with real-world field experiments, and they have successfully shown the concept to be effective [4, 12, 13, 14, 17].

However, people need robots to support us in a much wider range of daily activities. Robots for individual household tasks such as cleaning floors or folding laundry are not sufficient. Support for the elderly and disabled in a variety of daily activities is greatly demanded. Such support requires robots to accompany people to many different places (e.g. homes, shopping malls, hospitals) and to assist people for various activities (e.g. checking health, showing routes, carrying luggage) in ways appropriate for each place and for physical demands of each person (e.g. using wheelchairs). At the same time, these sequence of activities shall be well integrated to provide comfortable and seamless support throughout our daily lives.

Existing robotic systems cannot yet provide such continuous support in various aspects of our daily lives. Although they have had success in enriching capabilities of individual robotic applications, a general framework is missing for adapting behaviors of robots and composing them to form a integrated sequence of applications. Similar concepts can be seen in computer systems, such as *Service Oriented Architecture (SOA)*. SOA provides design principles for constructing large-scale information systems adopted to numbers of middlewares, especially in web service systems. In combination with the concepts of grid- or cloud-computing systems, many commercial services with high dependability are available today. However, these concepts are not sufficient for robotic systems.

One major element that makes information systems different from robotic systems is where the applications are performed. Information system applications are performed in cyberspace with information terminals such as personal computers or cellphones. This is especially clear for web-based systems where every application run through the web browsers. Based on the information shown on the screen, people performs actions such as business procedures, travel and cooking. Thus, the actual work in the physical world is left to the customers. And this is where the role of robotic systems starts. Robotic systems need to consider various factors in the real-world environments including physical abilities and limitations of both customers and devices. Moreover, as stated previously, a variety of different devices shall be integrated to realize continuous robotic support in our daily lives.

As such, different kinds of "ubiquity," not just for locations but also various applications and ways of providing each service on the basis of customer attributes, shall be focused on, as well as inter-application and inter-location continuity. As the complexity of robotic devices are much larger than traditional information systems, at the same time, we shall provide abstracted, common access methods for application programmers to make the development process easy and highly reusable. A new framework is now required that can bridge the gap between that improves and extends traditional robot systems that satisfy such demands. Based on such consideration, we have proposed the concept of *Ubiquitous Network Robot Platform (UNR-PF)* [5, 8, 16] and have been examining the concept through various field trials. In the following, we outline the requirements and related activities on UNR-PF.

2 Requirements for UNR-PF

For developing UNR-PF applications, flexible customizability must be provided for service developers. The development process of the services does not only depend on the pre-constructed design by service provides. While developing the services, the system should be modified for each environment and improved in accordance with the requests from service users through field experiments in the real environments. To provide the service at a different place, the system should be customized in accordance with the local environment. Also, robotic functional modules are assumed to be modified for performance improvement. If such improvements enforce modification in the whole system, they obstruct the whole development process. Therefore, the service application layer, which defines the process of the service content, shall be independent of robotic functions. Thus, the abstraction level of robotic functions shall be carefully designed to be controlled from service applications.

Services with UNR-PF will be provided in wide, continuous as well as discrete areas, such as in home, shops, hospitals, restaurants, etc., to support people's life ubiquitously. To coordinate such multi-locational services, a mechanism must be constructed for sharing information among these different areas. As robot development advances, more and more types of robots will be created to achieve a variety of tasks. The abilities of robots will also increase rapidly across the whole robotics field, and a kind of specialized robot will proliferate. For such a variety of robots to cooperate, the types of robot abilities must be classified and managed properly. As mentioned above, the design of a ubiquitous network robot system should address the following issues: appropriate abstraction and classification of robotic functions, information sharing and management between distributed locations, and common management platforms for robotic components with their capabilities and availabilities.

2.1 Multi-robot Management

The platform is required to classify the robot abilities and allocate appropriate robots in accordance with the service content and the execution environment. By implementing such a management function on the platform, each robot's functionalities can be covered so that the robot coordination can be improved in situations where various robots are available. Also, the compatibility has to be enhanced between similar functional modules on different robots.

2.2 Multi-area Management

To provide services over a wide area, it is important to improve coordination among multiple areas. To link several physical points, the platform needs a mechanism to share spatial information of each area such as map information. It is important to share not only static spatial information like map information but also dynamic location information (i.e., location of robots, user, obstruction and target object) that changes its relative and absolute location in accordance with circumstances. For the efficient multi-area coordination, this spatial and location information shall be managed on a multi-layered management structure that contains local area layers, which manage location information of each area, and global layers, which manage the relationships between the local areas.

2.3 User Management

To provide services suitable for service users, information related to the users has to be managed. Especially in the robotic service that does a physical task for the user, the robot system must select the proper equipment in accordance with the user attributes. The platform shall provide a mechanism for managing such user attributes commonly so that all services can refer to the user information efficiently.

2.4 Service Management

To execute a number of services across multiple areas, it is important to monitor the state of each area and determine the start of the services so that the service can be executed in appropriate situations. The platform requires a mechanism for managing the state of a service execution environment. Furthermore, the platform requires not only a mechanism to execute each service independently but also a mechanism to share information between each service.

3 The Ubiquitous Network Robot Platform

To solve these problems, UNR-PF consists of three layers: the functional module layer, the service application layer, and the common platform layer. The first layer consists of reusable functional modules that belong to individual robots, and the second layer

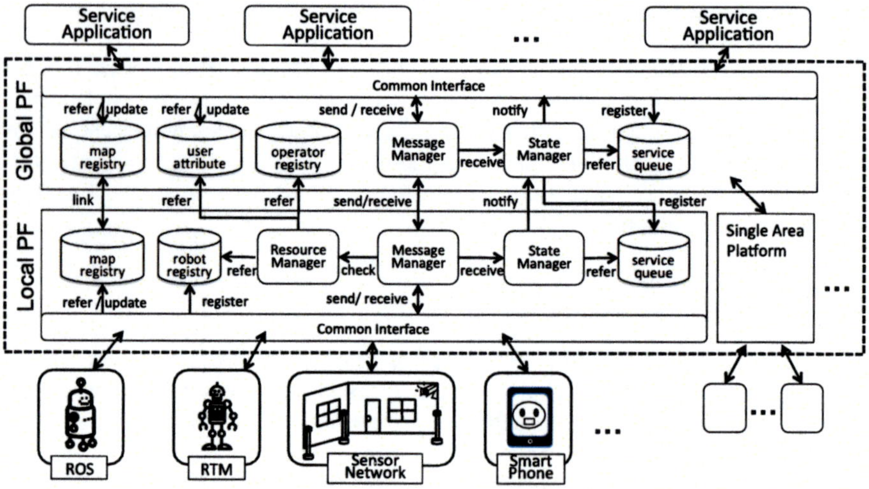

Fig. 1. Overview of the Ubiquitous Network Robot Platform

manages logic for service contents. The common platform layer should be independent of and serve common functions to the other two layers. Once such a common platform is developed and provided stably, it will separate robotic functions and services so that the whole robotic system will be able to be developed at lower cost and become more dependable in terms of reusability, scalability, and availability. The structure of UNR-PF is shown in Figure 1. Every interaction between applications and robots is established via UNR-PF.

As such, UNR-PF serves as a middleware between applications and robotic devices and is composed of two layers of platform: a local platform (LPF) and a global platform (GPF). The LPF is a platform for configuring the robotic system in a single area. The GPF is a platform for configuring the robotic system in a wider range of areas that includes a number of LPFs These platforms serve as a middle-layer between the service application layer and robotic function layer.

4 Field Experiments

We have evaluated the concept of UNR-PF through field experiments on six service cases: 1) remote listening support [21], 2) community formation, 3) healthcare [18], 4) shopping support, 5) attracting customers [6], and 6) navigation support. As an example, here we introduce the navigation support in the shopping mall as an example of UNR service that is provided across multiple areas. This service is provided through three areas: customer's home, shopping mall, and operator center.

First, the customer makes a reservation via the virtual-type robot on a mobile device at his/her home. The virtual-type robot is connected to a LPF in the home (LPF-h)) and LPF-h notifies the GPF of the reservation information. The touring support service application which is connected to the GPF receives the reservation and registers its trigger condition, i.e., the customer's arrival at the shopping mall, to the service queue in the GPF. The GPF refers to its map register to search an appropriate LPF that manages the shopping mall (LPF-s) and registers the trigger condition to LPF-s's service queue. When the customer approaches the shopping mall, the virtual-type robot on the customer's mobile device connects to LPF-s and notifies its arrival. This event fires the registered condition in the service queue and LPF-s notifies GPF, that lead to starting sequence of touring support. Then the service application requests the resource manager to reserve the robot in the shopping mall and the operator in the operator center via the GPF. To reserve the robot, the resource manager refers to the user attribute registry and selects the robot suitable for the customer. In the touring support service, the manager selects the wheelchair robot if the customer has difficulty walking. Otherwise, it selects a mobile robot available. After the allocation of the resources, the service is executed in accordance with the service flow defined in the service application. The service application refers to the map registry in the LPF and instructs the robot to move around the shopping mall.

In the case of a large shopping mall, the touring support service is provided across several LPFs. When the robot comes close to the boundary between the two areas, the robot alerts GPF's state manager via LPF's state manager of the current area. Then, GPF's state manager refers to its map register to find a LPF for the next area and then registers the service and its start condition, i.e., the customer's arrival at the next area. When the customer arrives at the next area, the robot disconnects from the LPF of the first area and connects to the next LPF if the robot can continue to serve in the next area. Otherwise, another robot in the next area is allocated and comes to pick the customer up. In this way, the customer can receive the service seamlessly in a wide area regardless of the area segmentation for the robotic system.

5 Standardization

To share the information between the robots and the service applications and achieve interoperability between different robots, the specifications of data structures and interfaces must be standardized. We have focused on four key elements of UNR-PF and have been working on standardizing these elements: architecture, map information, location information and common API. The architecture of UNR-PF is now under discussion at Q25/16 of ITU-T. The recommendation, F.USN-NRP, was accepted as a standardization work item in 2011 and is expected to be released in 2013 [7]. Standardization of map information has been processed in Open Geospatial Consortium (OGC). Update request for the *CityGML* specification has been made to allow maps to contain robot-specific information. This is scheduled to be published as the next revision, version 2.0 [9]. In the Object Management Group (OMG), two specifications related to UNR-PF has been proposed. One is for exchanging location / pose information among various robotic components and applications. This standard, the *Robotic Localization Service (RLS)* specification [8], has been published in February 2010. The other is the *Robotic Interaction Service (RoIS) Framework* specification [15] which defines a common API between robotic applications and components. A draft proposal was approved in June 2010 and is expected to be issued in 2012.

6 Conclusion and Future Works

In this paper, we introduced the Ubiquitous Network Robot Platform (UNR-PF), which is a common platform for implementing robotic applications over multiple areas. This is achieved by integrating various robotic devices such as robots, sensor networks and smartphones. Functional requirements for UNR-PF such as abstraction of robotic functions and managements of components with physical capabilities have been discussed through review of recent robot systems and from our past experiences. These requirements have been evaluated through field experiments in real-world environments.

To achieve a truly useful networked robot system, it is important to not only improve robotic functions and service contents but also to enhance the stability of the platform. Therefore, the UNR-PF and the development and experiment environment that uses the platform are planned to be made available to all robot developers and service providers to improve the capability and stability of the platform. We are currently re-implementing the prototype of UNR-PF based on the ongoing standard specifications. This prototype is scheduled to be released as open source software in Q2/2012.

In this prototype, various mechanisms that realizes the four major requirements of UNR-PF, multi-robot / multi-area / user / service managements, are implemented in a rather straightforward means. For example, service invocation is implemented by pattern matching of temporal-spatial features and multi-robot management by constraint satisfaction. While implementing, we found that there are some hard problems such as balancing optimal solution and real-time responses, and strategies for assigning mobile terminals and robots as well as sensors with limited spatio-temporal effective ranges. UNR-PF provides testbeds for such interesting issues that requires further studies.

Besides, future works on UNR-PF includes several items. Increasing in scalability and enhancing dependability of the system itself as well as robotic applications is one issue. Another major topic is the semantic extension of UNR-PF. Currently, UNR-PF decomposes application requirements into several sub-plans based on spatial nature (where the service is to be performed and what kind of place it is), customer attribute (whether the customer need special support), functional decomposition (what functionalities are required) and availability and ability of robotic components (what component is available, to what extent can it be used). This decomposition is basically structural and can be formulated as a constraint satisfaction problem. However, much semantic decomposition may be necessary to enhance the ability of UNR-PF and to optimize the usage of available resources. In the real human society, we often divide a single task into several tasks so that it can be performed quickly and exactly by several persons. When dividing, we often consider the nature of the task as well as the nature of the person who is going to perform it. Performing such decomposition requires semantic description of application tasks and various knowledge about real-world objects and phenomena. This is quite a difficult task that was once challenged by the AI researchers. Recently, several projects such as RoboEarth [19, 20] has been conducted to tackle this challenge. We will seek collaboration with these projects and aim to further enhance the UNR-PF framework to provide richer robotic support for our daily lives.

Acknowledgement. This work was supported in part by by the Ministry of Internal Affairs and Communications, Japan.

References

[1] Ando, N., Suehiro, T., Kitagaki, K., Kotoku, T., Yoon, W.K.: Rt-middleware: distributed component middleware for rt (robot technology). In: 2005 IEEE/RSJ International Conference on Intelligent Robots and Systems (IROS 2005), pp. 3933–3938 (2005), doi:10.1109/IROS.2005.1545521

[2] Bruyninckx, H.: Open robot control software: the orocos project. In: Proceedings 2001 ICRA. IEEE International Conference on Robotics and Automation, vol. 3, pp. 2523–2528 (2001), doi:10.1109/ROBOT.2001.933002

[3] Fong, T., Nourbakhsh, I., Dautenhahn, K.: A survey of socially interactive robots. Robotics and Autonomous Systems 42(3-4), 143–166 (2003), doi:10.1016/S0921-8890(02)00372-X

[4] Jung, H.S., Baek, J.K., Jeong, C.S., Lee, Y.W., Hong, P.D.: Unified ubiquitous middleware for u-city. In: Proceedings of the 2007 International Conference on Convergence Information Technology, ICCIT 2007, pp. 2374–2379. IEEE Computer Society, Washington, DC (2007), doi:10.1109/ICCIT.2007.387

[5] Kamei, K., Sato, M., Nishio, S., Hagita, N.: Cloud network robotics. IEEE Network Magazine (in press, 2012)

[6] Kanda, T., Shiomi, M., Miyashita, Z., Ishiguro, H., Hagita, N.: A communication robot in a shopping mall. Trans. Rob. 26(5), 897–913 (2010), doi:10.1109/TRO.2010.2062550
[7] Nishio, S.: F.UNR-NRP: Proposed modifications to the draft recommendaton. AVD-4200, ITU-T SG16 (2012)
[8] Nishio, S., Lee, J., Sakamoto, T., Noda, I., Tsubouchi, T., Doi, M.: Robotic Localization Service standard for ubiquitous network robots. In: Abdellatif, H. (ed.) Robotics 2010 Current and Future Challenges, ch. 20, pp. 381–400. In-Teh, Vukovar, Croatia (2010)
[9] Open Geospatial Consortium. OpenGIS City Geography Markup Language (CityGML) encoding standard version 2.0 (2012)
[10] Park, H.S., Han, S.: Development of an open software platform for robotics services. In: ICCAS-SICE, pp. 4773–4775 (2009)
[11] Quigley, M., Conley, K., Gerkey, B.P., Faust, J., Foote, T., Leibs, J., Wheeler, R., Ng, A.Y.: ROS: an open-source robot operating system. In: ICRA Workshop on Open Source Software (2009)
[12] Salvini, P., Laschi, C., Dario, P.: Do service robots need a driving license? IEEE Robotics Automation Magazine 18(2), 12–13 (2011), doi:10.1109/MRA.2011.941628
[13] Sanfeliu, A., Hagita, N., Saffiotti, A.: Network robot systems. Robotics and Autonomous Systems 56(10), 793–797 (2008), doi:10.1016/j.robot.2008.06.007
[14] Sanfeliu, A., Andrade-Cetto, J., Barbosa, M., Bowden, R., Capitn, J., Corominas, A., Gilbert, A., Illingworth, J., Merino, L., Mirats, J.M., Moreno, P., Ollero, A., Sequeira, J., Spaan, M.T.J.: Decentralized sensor fusion for ubiquitous networking robotics in urban areas. Sensors 10(3), 2274–2314 (2010), doi:10.3390/s100302274
[15] Sato, M., Hori, T., Nishio, S., Chi, S.Y.: Standardizing framework for robotic services and functions. In: Proc. ICRA 2011 Workshop on Robotics Modular Architecture Design and Standardization, Shaghai, China, pp. 1–6 (2011)
[16] Sato, M., Kamei, K., Nishio, S., Hagita, N.: The ubiquitous network robot platform: Common platform for continuous daily robotic services. In: Proc. 2011 IEEE/SICE International Symposium on System Integration (SII 2011), pp. 318–323 (2011), doi:10.1109/SII.2011.6147467
[17] Shiomi, M., Sakamoto, D., Kanda, T., Ishi, C., Ishiguro, H., Hagita, N.: Field trial of a networked robot at a train station. International Journal of Social Robotics 3, 27–40 (2011), doi:10.1007/s12369-010-0077-4
[18] Sugiyama, O., Shinozawa, K., Akimoto, T., Hagita, N.: Case Study of a Multi-robot Healthcare System: Effects of Docking and Metaphor on Persuasion. In: Ge, S.S., Li, H., Cabibihan, J.-J., Tan, Y.K. (eds.) ICSR 2010. LNCS, vol. 6414, pp. 90–99. Springer, Heidelberg (2010)
[19] Tenorth, M., Klank, U., Pangercic, D., Beetz, M.: Web-enabled robots. IEEE Robotics Automation Magazine 18(2), 58–68 (2011), doi:10.1109/MRA.2011.940993
[20] Waibel, M., Beetz, M., Civera, J., D'Andrea, R., Elfring, J., Galvez-Lopez, D., Haussermann, K., Janssen, R., Montiel, J., Perzylo, A., Schiessle, B., Tenorth, M., Zweigle, O., van de Molengraft, R.: Roboearth. IEEE Robotics Automation Magazine 18(2), 69–82 (2011), doi:10.1109/MRA.2011.941632
[21] Yomazoe, H., Koyama, Y., Yonezawa, T., Abe, S., Mase, K.: Estimation of user conversational states based on combination of user actions and feature normalization. In: Proceedings of the 5th ACM International Workshop on Context-Awareness for Self-Managing Systems, CASEMANS 2011, pp. 33–37 (2011), doi:10.1145/2036146.2036152

II: Unmanned Aerial and Underwater Vehicles and Bio-inspired Robotics

Kwang Joon Yoon

Autonomous Vehicles and Biologically-Inspired Robotics are representative robots doing real roles for human life with autonomous system.

A lot of research activities in aerial, ground and underwater vehicles have been done not only for developing efficient platforms, but also for making the platforms intelligent. Especially, intelligent autonomous systems in vehicles permit the vehicles to perform surveillance and reconnaissance missions and environmental monitoring without the human supervision. These autonomous technologies also help make the platforms have various applications in research and entertainment fields.

In recent decades, the development of unmanned autonomous vehicles have been of great interest, and different kinds of autonomous vehicles have been studied and developed all over the world. In particular, UAVs (unmanned aerial vehicles) have many applications in emergency situations; humans often cannot come close to a dangerous natural disaster such as an earthquake, a flood, an active volcano, or a nuclear disaster. Since the development of the first UAVs, research efforts have been focused on military applications. Recently, however, demand has arisen for UAVs such as aero-robots and flying robots that can be used in emergency situations and in industrial applications. Among the wide variety of UAVs that have been developed, small-scale HUAVs (helicopter-based UAVs) have the ability to take off and land vertically as well as the ability to cruise in flight, but their most important capability is hovering. Hovering at a point enables us to make more effective observations of a target. Furthermore, small-scale hovering UAVs offer the advantages of low cost and easy operation.

Biologically-inspired robotics has received much attention recently. It is an interdisciplinary field between biology and robotics. It is often referred as biomimetic robots or bio-robotics. One issue in biologically-inspired robotics is to apply biological motivations or ideas to engineering problems and another issue is to use robots as physical models to answer biological questions.

The several papers related with bio-inspired flight cover basic principles of specific natural species, strategy to mimic the features, fabrication and experimentation of flight systems, and their findings. Potential applications of flapping-wing systems are also discussed.

In the unstructured environment, such as, under-water, air, disastrous area, the localization/navigation technologies are essential for rescue and construction. The measuring devices and algorithms for the localization/navigation are discussed in the papers related with underwater robot. These papers introduce recent progress on underwater vehicles, robots including its control method, simulation, flow analysis, etc.

In this chapter, the contents of 39 papers presented in IAS-12 are introduced in the category of autonomous aerial/underwater vehicles, bio-inspired flight, biologically-inspired robotics, humanoids, intelligent transportation, mobile robot applications, special environment localization/navigation, and underwater robots.

Rapid Prototyping Framework for Visual Control of Autonomous Micro Aerial Vehicles

Ignacio Mellado-Bataller[1], Pascual Campoy[1], Miguel A. Olivares-Mendez[1], and Luis Mejias[2]

[1] Centre for Automation and Robotics (CAR), Universidad Politécnica de Madrid
C/ Jose Gutierrez Abascal, 2. 28006 Madrid, Spain
www.vision4uav.com

[2] Australian Research Centre for Aerospace Automation (ARCAA),
Queensland University of Technology
GPO Box 2434, Brisbane Queensland 4001
luis.mejias@qut.edu.au

Abstract. Rapid prototyping environments can speed up the research of visual control algorithms. We have designed and implemented a software framework for fast prototyping of visual control algorithms for Micro Aerial Vehicles (MAV). We have applied a combination of a proxy-based network communication architecture and a custom Application Programming Interface. This allows multiple experimental configurations, like drone swarms or distributed processing of a drone's video stream. Currently, the framework supports a low-cost MAV: the Parrot AR.Drone. Real tests have been performed on this platform and the results show comparatively low figures of the extra communication delay introduced by the framework, while adding new functionalities and flexibility to the selected drone. This implementation is open-source and can be downloaded from www.vision4uav.com/?q=VC4MAV-FW

Keywords: MAV, UAV, communications, software framework.

1 Introduction

"Fail early, fail often" is a wise mantra. The earlier you find the mistakes in your new idea, concept or system design, the sooner you can fix them and get your path to success. This is especially applicable in the field of visual control, where dynamic systems are controlled using images from one or more cameras as feedback. Visual control algorithms that work fine on the simulator may fail catastrophically in the real world. In this paper, we propose a flexible rapid prototyping software framework for visual control of Micro Aerial Vehicles (MAV).

In the last few years, personal Micro Aerial Vehicles and their components are hitting the consumer market like [1], [2] and [3]. For the work in this paper, we are using the Parrot AR.Drone as prototyping platform [1]. This drones is sold as a toy at amateur-affordable prices and can be remotely operated from a smart phone. It has

S. Lee et al. (Eds.): Intelligent Autonomous Systems 12, AISC 193, pp. 487–499.
springerlink.com

out-of-the-box onboard cameras and Inertial Measurement Units (IMU). Whereas the overall quality is low compared to a professional MAV, it can be 20-40 times cheaper. In addition, it can be bought at several toy stores and taken straight to the lab without worrying about delivery delays. Furthermore, because of its low cost, taking risks is acceptable: if you crash and break one, you can just buy a new unit. For these reasons, it is worth to be taken into account as prototyping platform, especially when developing algorithms for MAV swarms with many units, where the total cost might be prohibitive with more professional MAVs.

The Parrot AR.Drone has a Software Development Kit (SDK) that enables third-party developers to write applications for the drone. A third-party program that is run on an external workstation is able to send commands and receive information from the sensors, the camera and the IMU through a WiFi link. The drone has internal high frequency control loops that drive the rotors to maintain the demanded attitude and altitude. However, the SDK is too limited for our research requirements, as it only supports a single point-to-point link between the program and the drone, thus, a program can only communicate with a single drone. Besides that, we would like to work with networked communication schemes like those shown in Fig. 1.

The required new functionalities are provided by the proposed software framework, while increasing the isolation between the application and the hardware platform, and opening the possibility to easily port applications to other MAVs with either onboard or off-board computing.

In section 2, other related works are explored. In section 3, we introduce some general guidelines of the framework architecture, while a specific implementation for the Parrot AR.Drone and a C++ application is discussed in section 4. In section 5, some test results of this implementation are presented and, in section 6, they are discussed. Section 7 concludes the paper.

2 Related Work

The AR.Drone SDK already offers an API for developing third-party applications [10]. Examples of research works with the AR.Drone are [7], [8] and [9]. However, by the time this paper is written, it does not support communications with multiple drones across a network. With regards to the communications between the application and the drone, reference [4] points to an existing open project by ETH PIXHAWK. It offers a communication architecture for MAVs that is based on a library for message transmission over a network [5], but it does not have native support for the Parrot AR.Drone. On the other hand, there is a driver for AR.Drone by Brown University [6] for the Robot Operating System (ROS) that was used in [9]. Nevertheless, it does not implement either access control to the drone or parameter configuration. Moreover, we would like our framework to remain lightweight, without burdening the new developer with the installation of heavy and complex packages like ROS. The framework implementation presented in this paper tries to fill the gap left by the cons of the other alternatives. It is being used to carry out experiments of a see-and-avoid fuzzy controller, after tuning the system with the robot simulator Gazebo. More information can be found at [11].

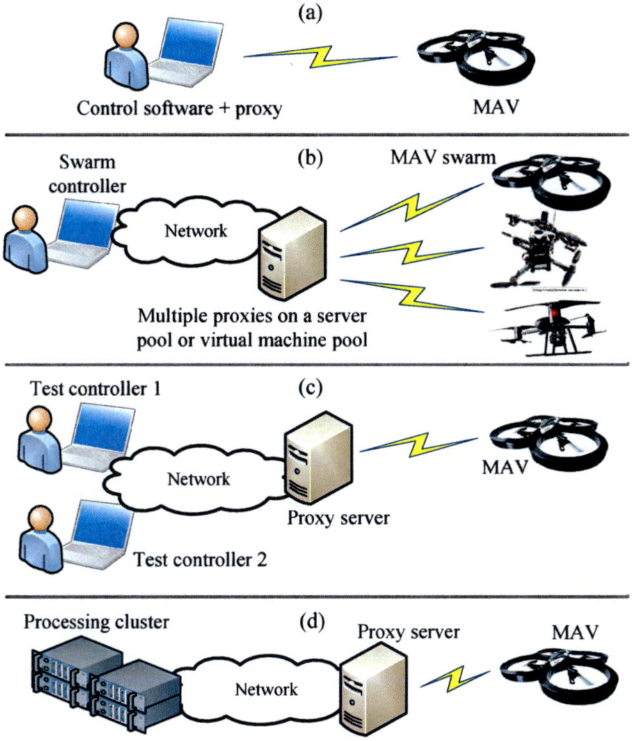

Fig. 1. New communication schemes provided by our framework. In (a), a point-to-point scheme, also allowed by the AR.Drone SDK; in (b), an application controls an MAV swarm; in (c), multiple researchers may share the same MAV resources -one at a time-; in (d) ,video is broadcasted over a cluster for parallel processing.

3 Framework Architecture

In this section, we define a general model for the implementation of our framework architecture. These are guidelines and requirements that are extensible to any MAV, (whose capabilities are similar to the AR.Drone's), any packet network and any application programming language. In the next section, the model will be applied to the AR.Drone and to specific network technologies and a C++ API.

To give network capabilities to the drone, a proxy-based architecture has been defined. The architecture is depicted in Fig. 2. The proxy is responsible for connecting a single drone with the network. With one proxy per drone, all drones can share the network as communication mean. At the application side, we define an Application Programming Interface (API). Thanks to this API, the application is able to communicate with the proxies of the different drones that it aims to control.

It is worth to notice that the framework components do not have fixed running locations. If the MAV processing platform has open access, the proxy can be run onboard. Then, the control application can reside either onboard, communicating locally, or off-board, through a wireless link.

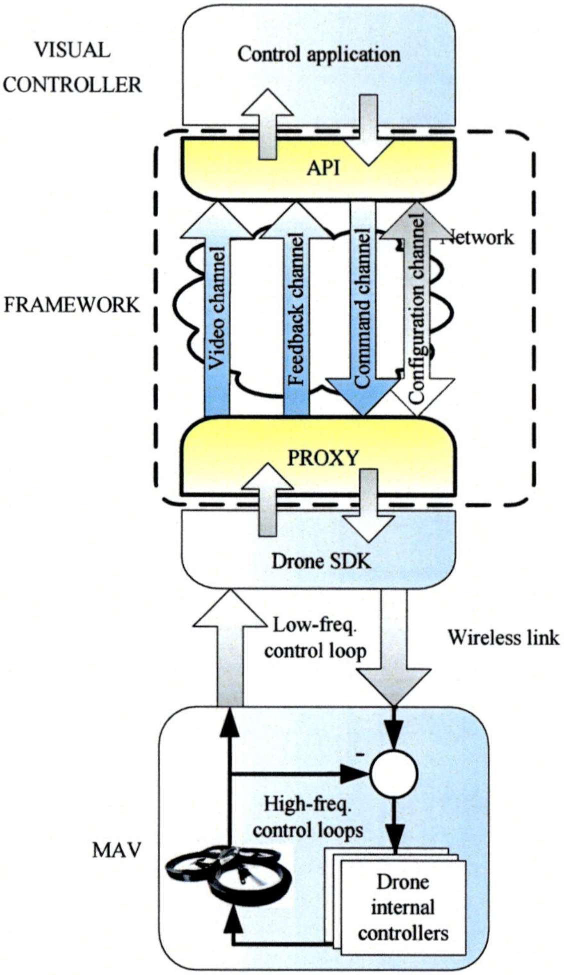

Fig. 2. System architecture. The framework interconnects the visual controller with each MAV through a network. The total system follows a cascade control structure. Usually, there are high-frequency controllers onboard the MAV, while the visual controller closes an outer loop with lower frequency. The network link for the control loop is formed by three independent channels. Configuration channel is used to read and change parameters and it is not part of the control loop.

Otherwise, if the MAV onboard processing is closed, which is the case for the AR.Drone, the proxy is run off-board, and the control application can be executed either on the same platform or on any other that is connected through a network, as seen in Fig. 1.

Regarding portability, while the proxy depends on the drone manufacturer, the Application Programming Interface (API) library is platform-independent. In other words, the proxy isolates the application from the drone specifics. In this way, there is no need to update the control applications every time the manufacturer releases a new SDK version. Most times, updating the proxy will be enough. Another advantage of this isolation, is the possibility of porting the API to programming environments or languages not supported yet by the manufacturer's SDK. For instance, a Matlab API could be programmed, despite not existing any specific software by the manufacturer.

3.1 Communications

The communication link between the proxy and the MAV depends on the manufacturer specification and it may vary between different models. It is the manufacturer who defines the communication protocol of the drone and it will not be discussed in this paper. Our framework is responsible of the link between the proxy and the application. This link is formed by four independent communication channels, named: command, feedback, video and configuration. These channels are logical, not necessarily physical, as they are established over the network. They just represent an information flow between both network nodes. To implement the channels, no specific communication protocols are defined as mandatory; there are only recommendations.

The network between the proxy and the application may fail. A cable might break, a router might stop or a WiFi link might lose the signal. Both ends of the link must be robust to these situations and implement self-recovery mechanisms, which must be transparent to the application. The application will be notified of a failure situation but will not have to perform any actions to fix it. In case of stateless protocols –those not requiring to establish a connection– there is no extra effort to be done, as packets will continue to be transmitted after the network is recovered. Nevertheless, the application must be notified if packets do not arrive at expected times. Oppositely, connected protocols must automatically try to reconnect until the network is restored, besides informing the application of the link state.

3.1.1 Command Channel

The command channel transports all the control actions from the application to the proxy: a signature, a sequence number and drone-specific commands (required flying mode, desired attitude, etc.). The signature identifies the packet as a command channel packet and may be used as a start token.

Through this channel, data packets are transmitted periodically. Low delays are favored by allowing packet dropping, because control loops depend on this channel and delays generally harm loop stability [12]. Datagram protocols, like UDP over an IP network, are suitable for implementing this mechanism because they do not have automatic retransmission of faulty packets. At the proxy side, if a packet is lost or it arrives after a previously sent one, it is discarded. Instead of asking for a retransmission or reordering packets after a sequence error, the proxy expects that a new packet with up-to-date command information will eventually arrive. The purpose of the sequence number is to determine if a packet has arrived out of sequence.

An MAV can only be commanded by one control application at a time. Therefore, no concurrent access is allowed on this channel. When the channel is in a free state, any control application can lock it by sending an initialization packet for write access. After that, no other packets from other applications are processed until the original application unlocks the channel or stays inactive for a time longer than a pre-configured threshold.

3.1.2 Feedback Channel

In the feedback channel, navigation information flows from the proxy to the application. The content of each feedback packet is: signature, timestamp and drone-specific information (proxy-drone link health, battery level, measured attitude, etc.). The signature identifies the packet as pertaining to this channel and may be used as start token. Like in the command channel, packet dropping at the receiver –based on the timestamp– is encouraged in order to minimize delays in control loops.

Through the feedback channel, the proxy can feed data to multiple applications simultaneously. In situations where network broadcast or multicast is not possible, applications can subscribe to a consumer list in the proxy by requesting a read-only access through the command channel. The multicasting is then emulated by the proxy by iteratively sending the information to its subscribers, at the expense of increasing processing time and delay.

3.1.3 Video Channel

In a video channel, video from a drone camera is transmitted to the control application. Like the feedback channel, multiple applications can request video channels from a proxy. While a feedback channel sample will usually fit in a network packet, a video channel sample, i.e. a frame, will need to be encoded, packetized and transmitted with some transport protocol. Like for the other channel types, the lower the transmission delay is, the higher the stability margin of a visual control loop will be. Hence, implementations with compression-ready encodings, low-delay protocols and frame-dropping mechanisms would be preferred.

The video channel transports periodic fragments with frame data that include a header with a signature (it may be used as a fragment start token), information about the video encoding and a timestamp, so the application knows how to decode the video stream and when each frame was captured. The timestamp must be as close as possible to the real capture time of a frame. If the MAV does not provide this information, the proxy will give an estimation. When possible, the timestamps of both video and feedback channels must use the same clock reference. Although this reference is unknown by the application, sample times of different channels can be compared and ordered if needed.

3.1.4 Configuration Channel

The configuration channel is used to read and write configuration parameters of the MAV from the application. It is intended for parameters that are not time-critical, such as allowed attitude ranges or video capture features, which are mainly changed at startup. In order not to disturb any other channels requiring a higher bandwidth and a lower delay, any fast changing parameters must be transferred through the command and feedback channels.

When the application writes a parameter through the configuration channel, it must have a confirmation that it has actually been changed in the MAV, as it might be safety-critical. Likewise, when reading a parameter, the application must know that it was actually read. Therefore, a connection-oriented transport protocol is required for this channel. For example, TCP on an IP network would suit these requirements.

3.2 Application Programming Interface

The API library enables the application to access the communication architecture programmatically. The control application processes the feedback information from the MAV and generates the commands to be sent in response, closing the loop. The API defines methods that are directly called to change these commands.

The application can gather the feedback information –video and navigation– in two ways. The first one consists in explicitly polling the data when needed. However, because of the asynchronous nature of the feedback channel, the data is not requested on demand to the drone, but periodically received. And, consequently, the request method returns the last sample that was received from the proxy. The second method for feedback retrieval is event-driven. The application registers a listener through the API and the listener gets a notification whenever the data is received from the proxy, so it can be processed immediately. Navigation data and video frame notifications are received independently, as they are transmitted through unrelated channels, due to their different bandwidth requirements.

4 Framework Implementation

The framework model has been implemented for a Parrot AR.Drone. The implementation is targeted to IP networks, and the API works on Linux as a C++ programming language library. A specific proxy has been built for the AR.Drone with the manufacturer's SDK examples [10], also for Linux. The proxy is a separate executable that runs off-board the MAV because the onboard computer is closed to third-party code. The manufacturer point-to-point communication with the drone is established via WiFi.

4.1 Channels

The command channel is implemented using a UDP socket. Each data packet carries the following information: sequence number, required flying mode, attitude desired values and desired altitude speed. The sequence number is required by the framework model, while the rest is payload information specific to the MAV.

The feedback channel uses a UDP socket, too. It transports the following information: timestamp, proxy-drone link health, drone state, battery level, measured attitude, measured altitude and measured velocities. The timestamp is defined as mandatory by the model. The other data is drone-specific.

As UDP is not a reliable protocol, command and feedback channels are provided with a retransmission mechanism. At the application side, as soon as commands are changed by the application, the API library transmits them to the drone through the proxy. When the application is not generating new commands, the API library keeps transmitting the last commands periodically to ensure that they eventually arrive to the other end. The proxy has the same mechanism: new sensor readings are sent immediately, but if they are not available at a predefined minimum frequency, the last readings are periodically sent through the feedback channel to ensure that they arrive to the other end. In this way, there is constant activity in the channels and both ends know that they are linked.

The video channel is implemented with a TCP socket. According to the model definition, this is not the most adequate protocol because it is not designed for real-time, but for reliability. Nonetheless, the transparent streaming capabilities of the protocol make the video channel implementation straight-forward. Each frame is transmitted with its own encoding. Currently, the supported encodings are JPEG and raw RGB with eight bits per plane.

To pass the received video frame to the application, the API library has a triple buffering mechanism: the reception buffer, the frame-ready buffer and the processing buffer. The first one is continuously retrieving the frames from the network, preventing the TCP buffers from overflowing, which would time out the transmission at the proxy side and would be interpreted as a connection failure. Right after a frame is received, it is copied to the frame-ready buffer, to keep it accessible by other program modules, while the reception buffer is free to receive the next frame from the network. However, the frame-ready buffer is overridden as soon as a new frame is received, therefore any operation on this buffer should last less than a frame period. For longer processing times, the processing buffer is provided. When the frame-ready buffer gets new contents, all the video channel listeners are notified. One of them is a video processor module that copies the frame-ready buffer contents to its own processing buffer only after the last processing operation has finished. Meanwhile, the frames are dropped for that video processor. Multiple video processors can be freely initiated by the application, allowing concurrent frame processing with independent frame dropping for each processor.

The configuration channel is implemented with a TCP socket, as low delays are not mandatory but reliability is. Each parameter operation is performed in a transaction consisting in a request and a response. Each request contains a signature, the request type, the parameter identifier and the parameter desired value. The desired value will only be interpreted by the proxy if it is a write request. The response is formed by a signature, a value indicating whether the last request was successful and the parameter value. The parameter value will only be meaningful if the last request was for reading.

4.2 Robustness

At both communication ends, there is code responsible for keeping communication channels synchronized. If faulty behavior occurs, the corresponding channel is restarted, so both ends are automatically synchronized back. The channel behavior can be understood as faulty when a malformed packet is received or when packets are not received as frequently as expected. Every time this happens, the application is notified so it can react accordingly. For example, it could display an alarm on a user interface. However, the channel recovery mechanism is completely transparent and all efforts for the channel restoration are performed by the framework.

At the application side, all the API errors are handled with C++ exceptions. This mechanism favors that errors show up during the development phase so they can be fixed early. In this API implementation, every thread has a last line of defense that catches all non-caught exceptions, writes the exception in a log file for debugging and prevents the thread from being terminated, so it can try to recover the normal state.

4.3 Extra Features

The API library is able to interface with a Vicon positioning system. With this system, position and attitude information of MAVs can be gathered inside a delimited space. This information can be very useful, for instance, to close control loops or as ground truth for visual pose estimation algorithms.

On the other hand, the API library provides data logging functionalities. The data logger can gather events generated by the channels and the Vicon interface. Hence, commands, navigation feedback, video feedback and Vicon data can be stored in a disk for later analysis. The data logger runs asynchronously, so the delays of the disk write operations do not bother other ongoing threads.

Finally, the API defines classes that help developing a controller by only overriding two methods. The methods are automatically called whenever navigation or visual feedback is received from the MAV. A controller may be implemented inside these methods. The received information is used as input to the controller and the controller's output is sent to the MAV directly calling the appropriate API methods. The images from the cameras are passed back with the encoding used by OpenCV. To help writing the controller code, the API also exposes matrix data types that perform common algebraic operations.

5 Experimental Results

The total communication delay between the application and the drone will be the sum of the delays introduced by the API library, the network, the proxy and the proxy-to-drone link. The drone manufacturer is accountable for the last one. The second one is given mainly by the physical network infrastructure. The first and third elements are responsibility of the framework implementation and must be measured.

In order to measure the framework contribution to the delay, the proxy is run in the same host where the application resides, so the API-proxy link is established through local sockets. Timestamps are added to channel packets at the sender and the time lapse is calculated at the receiver. As both processes run on the same computer, they share the same clock reference and time calculations can be performed without additional synchronization.

Regarding the proxy-to-application delay, the timestamps are obtained right after receiving the data from the drone, so all proxy processing time is also taken into account. The arrival time is acquired right after releasing the data to the application. For the application-to-proxy delay, the timestamps are taken right after issuing the commands to the API and the arrival time is calculated at the proxy, before sending the commands to the drone through the point-to-point link. The framework version 0.8 beta has been used in this test. The test was run on an Acer Aspire 5750G with a Intel Core i7-2630QM 2GHz processor and 8 Gbytes of DDR3 RAM. The Operating System was Linux Ubuntu 11.04. During the test, the data logging was disabled. The packet frequency for the command and feedback channels was set to 32 Hz. The video frame rate was 15 frames per second in average (this is determined by the

AR.Drone) and the video channel frames were encoded as raw RGB with eight bits per plane. The test application consists on a simple visual teleoperation interface with a waypoint-based path controller. The test duration is 5 minutes.

Figs. 3 and 4 show the distribution of delays introduced by the framework in the command and video channels (the feedback channel distribution is similar to the command channel one). Table 1 gives some numerical details about the delay distributions. Fig. 5 shows the time evolution of the delays. In all figures, delays are expressed as percentages of the channel period. The channel periods are 31.25 ms for command and feedback, and 66.7 ms for video.

Table 1. Characterization of channel delays

Channel	Delays (ms)			Num. samples	
	Mean [a]	Min.	Max.	Total [b]	Delay < 1% of channel period [c]
Command	0.091 (0.29%)	0.013	1.739	9,771	99.93%
Feedback	0.109 (0.35%)	0.025	1.670	9,828	99.73%
Video	1.038 (1.56%)	0.310	1.936	5,011	0.22%

a. Absolute delay in milliseconds and delay relative to channel period.

b. Total number of delay samples.

c. Delay samples lower than 1% of channel period (31.2 ms command and feedback; 66.7 ms video).

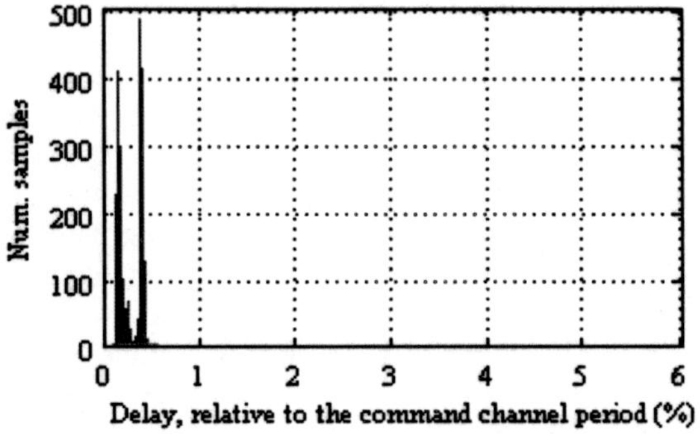

Fig. 3. Distribution of the delays introduced in the command channel by the framework. The relative delays are percentages of the command channel period, i.e. 31.25 ms. The highest sample is 5.56%.

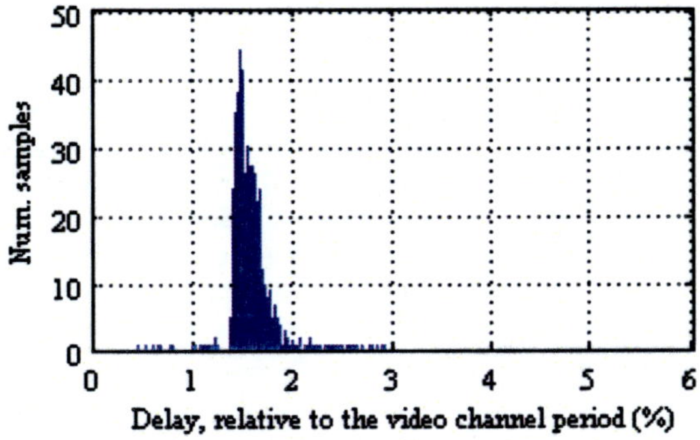

Fig. 4. Distribution of the delays introduced in the video channel by the framework. The relative delays are percentages of the average video channel period, i.e. 66.67 ms. The highest sample is 2.9%, but the horizontal scale is set as in Fig. 3 for easy comparison.

6 Discussion

As seen in table 1, the average delays introduced by the framework are considerably low, compared to frequencies of the channels. For a visual controller, the impact of the framework in the reaction time would be the result of adding the visual and command channel delays, i.e. the time it takes to see an event plus the time to react accordingly. In average, it is a contribution of 1.129 ms to the total loop delay. Assuming a visual control loop at 15 frames per seconds, this represents a 1.7% of the loop period.

In Fig. 5, there are spurious samples that might be caused by the fact that the implementation is not running on a real-time Operating System (OS). Instead, this OS has a preemptive scheduler that can interrupt a task anytime to yield some time for other tasks. Despite it might not cause problems during usual prototyping, it must be taken into account for high-frequency delay-sensitive applications.

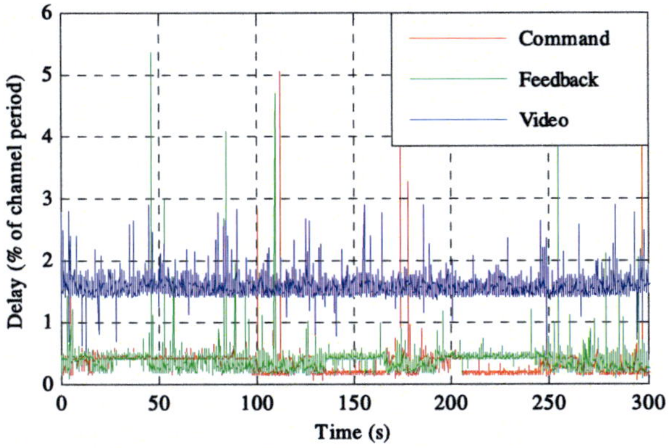

Fig. 5. Evolution of relative channel delays introduced by the framework. Most of the time, command and feedback delays are below 1% of the channel period. Highest peaks reach 5%, but are very unusual. The video channel delay is always under 3%.

7 Conclusion

We have introduced a framework for fast prototyping of visual control applications for Micro Aerial Vehicles (MAV). First, a framework model with general guidelines has been presented, without regarding specific technology details, in order to leave it open to other implementations. The framework architecture is able to transform cheap MAVs without onboard processing and networking capabilities into network nodes for off-board processing, opening the door to new prototyping configurations, like drone swarms, distributed vision processing or MAV sharing by multiple researchers. Moreover, the framework defines a common API that may be used to control MAVs with similar capabilities from different manufacturers, thus improving manufacturer independence with minimal code changes.

In order to show the framework applicability to cheap prototyping, an implementation on an amateur MAV –a Parrot AR.Drone– has been discussed. In the experimental results, the delays added by this implementation to the control loops are significantly low, compared to the loop periods. However, the timings are not deterministic because the implementation is not running on a real-time Operating System. Thus, the applicability to safety-critical controllers is disregarded. Anyhow, it does not affect most applications. On the contrary, the framework has proven to be a useful tool for rapid testing. This implementation is an open-source project. It is available at www.vision4uav.com/?q=VC4MAV-FW

Acknowledgements. The authors thank the Australian Research Centre for Aerospace Automation - Queensland University of Technologies (ARCAA-QUT), where a major part of the development and testing for this work was carried out. The authors would like to thank Caja Madrid for the mobility research grant of one of the authors. This work has been sponsored by the Spanish Science and Technology Ministry under the grant CICYT DPI2010-20751-C02-01 and by the IRSES Program Marie Curie FP7-PIRSES-GA-2009-230797 - ICPUAS (International Cooperation Program for Unmanned Aerial Systems Research and Development).

References

1. Parrot AR.Drone, `http://ardrone.parrot.com`
2. MikroKopter, `http://www.mikrokopter.us`
3. ArduCopter Quad, `http://www.udrones.com/product_p/acrtf1.htm`
4. MAVCONN Aerial Middleware, `http://pixhawk.ethz.ch/wiki/software/middleware/start`
5. MAVLINK protocol library, `http://www.qgroundcontrol.org/mavlink/start`
6. ROS driver for the Parrot AR.Drone, `http://code.google.com/p/brown-ros-pkg/wiki/ardrone_brown`
7. Visser, A., Dijkshoorn, N., van der Veen, M., Jurriaans, R.: Closing the gap between simulation and reality in the sensor and motion models of an autonomous AR.Drone. In: Proc. International Micro Air Vehicle Conference and Flight Competition (IMAV 2011), pp. 40–47 (September 2011)

8. Bills, C., Chen, J., Saxena, A.: Autonomous MAV flight in indoor environments using single image perspective cues. In: Int. Conf. Robotics and Automation (ICRA), Shanghai, China, pp. 5776–5783 (May 2011)
9. Koval, M.C., Mansley, C.R., Littman, M.L.: Autonomous quadrotor control with reinforcement learning, http://mkoval.org/projects/quadrotor/files/quadrotor-rl.pdf
10. ARDRONE open API platform, https://projects.ardrone.org
11. See and avoid with a fuzzy controller, http://vision4uav.eu/?q=researchline/SeeAndAvoidCE
12. Branicky, M.S., Phillips, S.M., Zhang, W.: Stability of networked control systems: explicit analysis of delay. In: Proc. American Control Conf, pp. 2352–2357 (June 2000)
13. Preemption (computing), http://en.wikipedia.org/wiki/Preemption_%28computing%29#User_mode_and_kernel_mo

UKF Applied for Position Estimation of Underwater-Beacon Precision

Ba-Da Yoon[1], Ha-Nul Yoon[2], Sung-He Choi[2], and Jang-Myung Lee[2]

[1] Dept. of Interdisciplinary Corporative Course Robot, Pusan National University
Jangjeon2-dong, Geumjeong-gu, Pusan 609-735, Republic of Korea
badayoon@pusan.ac.kr
[2] Dept. of Electrical Engineering, Pusan National University
Jangjeon2-dong, Geumjeong-gu, Pusan 609-735, Republic of Korea
{Haneul1696,dhmgbds,jmlee}@pusan.ac.kr

Abstract. A location estimation algorithm based on underwater beacon was proposed in this paper. Location estimation of beacon is prerequisite to the underwater localization and communication. Unscented Kalman Filter is verified through MATLAB simulations and experiments in real-time environments. And performance of Unscented Kalman Filter was evaluated through underwater-BEACON. In This experiments, MCU of Underwater-BEACON was used DSP(TMS320F28335). INS(Gyroscope Sensor, Accelerometer Sensor, Magnetic Compass) and GPS was Used for the location estimation of underwater BEACON. UKF combined GPS signals and INS signals to design Underwater vehicle's nonlinearity and non-normal distribution. Underwater-BEACON applies UKF. It can see that fewer errors were occurring.

Keywords: underwater Vehicles, Unscented Kalman Filter, Underwater Hovering System.

1 Introduction

70% of the earth surface takes possession of the sea. However, there is a limitation of human beings' accessing to it, the sea exists as the unknown world. Though the ocean is known to possess a lot of the underwater and seabed resources, it has not been developed due to the limitations of approach compared to the ground. Cases of the ocean exploration's application have been gradually increasing after the recent development of technology of unmanned submersibles. There is a disadvantage that the efficiency of ocean exploration is reducing because of the limitation of time and space constraints and the loss of mass of information data when exploring such a broadband of the ocean space. In order to compensate this, exploration technology should be developed to investigate the large area simultaneously using the various exploring equipment. The most important part in the ocean exploration is underwater wireless communications and location awareness area.

A transmission medium such as typical radio frequency, laser, or light waves is difficult to use due to attenuate sound waves at underwater. Thus, ultrasound can be understood as the only transmission means for long distance [1].

S. Lee et al. (Eds.): Intelligent Autonomous Systems 12, AISC 193, pp. 501–508.
springerlink.com

In order to the underwater location awareness and the underwater communication, a type of seabed-fixed and underwater-mobile BEACON is used. Development the BEACON and studies to make it practical are in progress, it is hard to recognize the precise location of the underwater BEACON because of the influence of waves, birds, and sea breeze. GPS and INS are used to recognize the location of BEACON. INS consists of the accelerometer sensor, the Gyroscope sensor, the magnetic compass and so forth to estimate the absolute location and the relative location at the same time. It provides the precise location information in a short time, but there occur accumulated errors when used for long time because of the errors that come from the characteristics of the sensor itself and the disturbance of the external environment. PS error is large in a short period of time. In the specific environment that isn't able to receive the GPS signals, those are cut off. However, it provides the long-time stable and absolute location information. It helps to correct the location by receiving the signals continuously from the satellite by real time and brings the result of not having a cumulative error. These two navigations complement each other's characteristics and make possible to estimate the precise location from outdoors [2-4].

This paper proposes precise localization and control of underwater BEACON that is preferentially required for underwater localization and underwater wireless communications using INS sensor and GPS. I will introduce outdoor precise localization estimation system applying KF to UKF for INS/GPS combination having the nonlinear and the irregular distribution noise in this paper.

2 Unscented KALMAN Filter

Many researchers study to solve estimation problem of states variables about the dynamic system. The method based on theory of probability is configured probability space consisting of state variables. It is based on the estimate about state variable using System's dynamic characteristics and the measurement value. Typically based on Bayesian estimation technique, Kalman Filter (KF: Kalman Filter), a particle filter and so forth have been actively studied in the field of localization.[5] The table 1 shows the characteristic of different Filters about the type of system models and noise.

Table 1. Filters for system model noise

Filter	System model	Noise
KF	Linear	Normal distribution
Extended KF	Non-linear	Normal distribution
Unscented KF	Non-linear	Non-normal distribution
PF	Non-linear	Non-normal distribution

Among filters that estimate non-linear system's state, PF has an advantage over EKF or KF which supposes noise as normal distribution. When there are enough samples, it approaches Bayesian optimum estimated figure, which makes it more precise than EKF or UKF. If PF is used on the Underwater BEACON in real time, amount of sample is increase, and when calculating the algorithm, DSP's performance is decrease. So in this paper, UKF is used for real-time filtering. Figure 1 shows the structure of Unscented Kalman Filter algorithm.

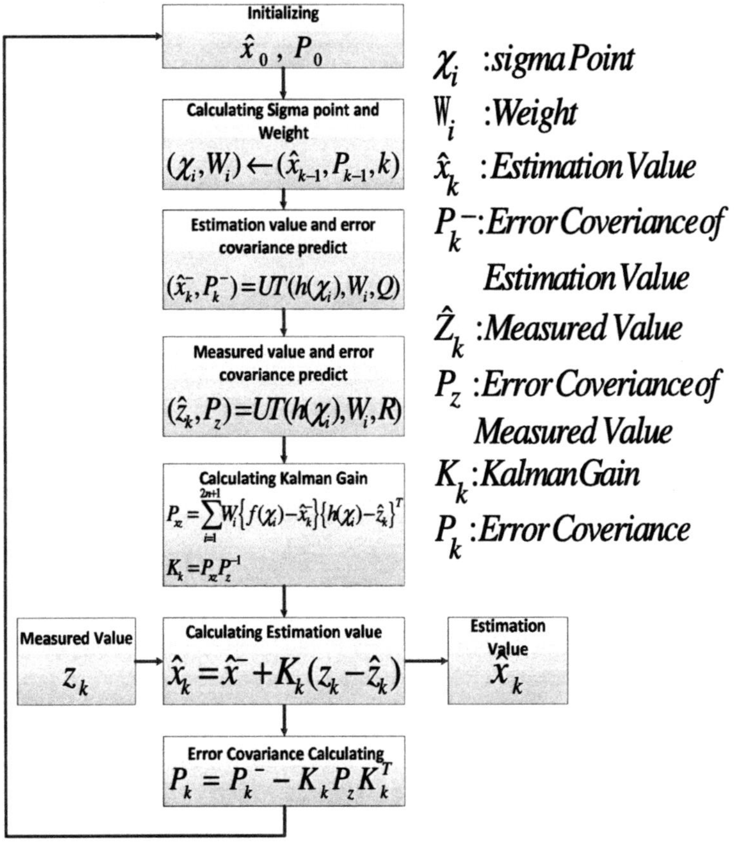

Fig. 1. Structure of Unscented Kalman Filter

Average is x_m, covariance is P_x. The state variables is x. $x \sim N(x_m, P_k)$ is normal distribution. When you convert arbitrary function $f(x)$ from x, approximation solution of the value is find using numerical analysis method.

An Obvious way to find approximate solution is to insert sufficient amount sample of x as function $f(x)$ using Monte Carlo simulation to calculate covariance and the standard in respect of result-value. However Monte Carlo simulation is not suitable real-time calculating because it processes huge amount of calculation.

Unlike Monte Carlo simulation which extracts samples randomly, Unscented Transformation selects samples and their weighted value elaborately. It makes UT calculate possible using much fewer samples than Monte Carlo simulation does. Moreover, Unscented Transformation has an advantage to calculate error covariance without Jacobian while Extended Kalman Filter uses it. Unscented Transformation calculates the average and covariance about x of arbitrary function $f(x)$. First, in (1), the sigma point x_i and weight W_i define about x.

$$\begin{aligned} x_1 &= x_m \\ x_{i+1} &= x_m + u_i \qquad i = 1, 2, \ldots, n \\ x_{i+n+1} &= x_m - u_i \\ W_1 &= \frac{k}{n+k} \\ W_{i+1} &= \frac{k}{2(n+k)} \qquad i = 1, 2, \ldots, n \\ W_{i+n+1} &= \frac{k}{2(n+k)} \end{aligned} \tag{1}$$

In (2), u_i is row vector of Matrix U and k is an arbitrary constant.

$$U^T U = (n+k)P_x \tag{2}$$

The average and covariance of function $y = f(x)$ can be obtained as shown in (3).

$$\begin{aligned} y_m &= \sum_{i=1}^{2n+1} W_i f(\chi_i) \\ P_y &= \sum_{i=1}^{2n+1} W_i \{f(\chi_i) - y_m\}\{f(\chi_i) - y_m\}^T \end{aligned} \tag{3}$$

Sigma-point x_i of unscented transform is sample values of Monte Carlo simulation. Weight W_i is the proportion of each sigma-point when calculating the determine average and covariance. Sigma points and weights which are defined in (1) are satisfied with the same characteristics in (4).

$$\begin{aligned} x_m &= \sum_{i=1}^{2n+1} W_i \chi_i \\ P_x &= \sum_{i=1}^{2n+1} W_i \{\chi_i - x_m\}\{\chi_i - x_m\}^T \end{aligned} \tag{4}$$

Unscented Transformation can represent statistical characteristic (average and covariance) of x if there are 2n+1EA of weight and sigma-points without using myriad samples. Conversely, the sigma points and weights have a role to represent the statistical characteristics of variable of state.

3 Experiment Result

INS Used in this experiment is composed of the 3-axis Accelerometer Sensor 1EA, 1-axis Gyroscope Sensor 3EA and Magnetic Compass Sensor 1EA. MCU use TI's TMS320F28335. 3-axis Accelerometer Sensor is composed of the VTI-Technologies's

SCA3000-D01($\pm 2g$), 1-axis Gyroscope Sensor is composed of the Analog-Devices's ADIS6100($\pm 300° / s$), Magnetic Compass Sensor is composed of Philips's KMZ10A. GPS and INS(Gyroscope Sensor, Accelerometer Sensor, Magnetic Compass) was modularity for the localization of underwater BEACON. It is made as one-board to be used on TMS320F28335. In Fig 2, INS sensor and TMS320F28335 used in the experiment are shown for.

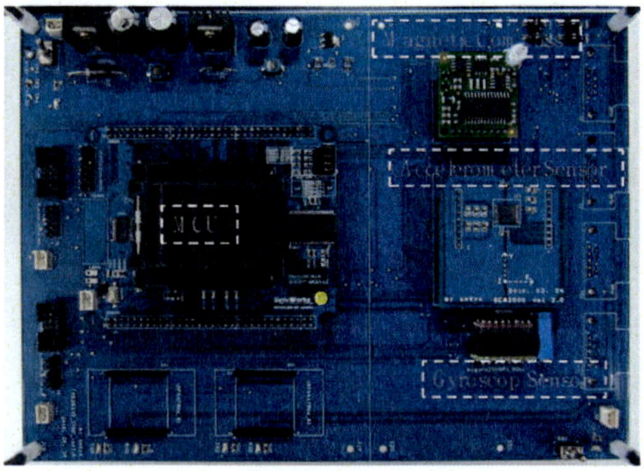

Fig. 2. Preciseoutdoor location estimation system INS/MCU

In this experiment, The GPS was uses UI_GOODS's UIGGUB02-R001's. Actuator (Thruster) of The Underwater-BEACON was uses TECHADYNE's TECHADYNE 300[6]. For the operation of the underwater-BEACON, 12-bit binary counter and D/A converter are used to compose actuator signals ($\pm 5V$). The Fig 3 shows underwater-BEACON used in experiment.

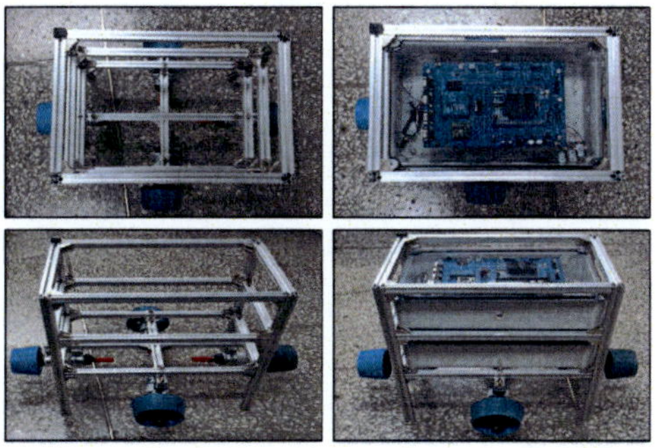

Fig. 3. Underwater-BEACON used in the experiment

Progression direction is in front of Underwater-BEACON. In INS, x-axis of the acceleration sensors is same as progression direction. Gyroscope sensor measures Yaw-axis and Pitch-axis. Magnetic compass sensor also measures Yaw-axis to compensate Yaw data of Gyroscope. Movement Characteristics of underwater-BEACON must be considered. So, We have experiment in near beach of Pusan to minimize the effect of environment[7-8].

For Performance evaluation of this system, Standard moving information of BEACON is obtained from BEACON's GPS. GPS can receive BEACON's Position information. And, Position error of BEACON was compensated through using INS values of UKF.

The UKF combined Acceleration signal, gyro signal and magnetic compass signals with NMEA (The National Electronic Association) GPRMC code of GPS.

INS data has many noise signals. So we used first order LPF(low pass filter). Fig. 4 shows measured INS data at uniform velocity and INS data used LPF.

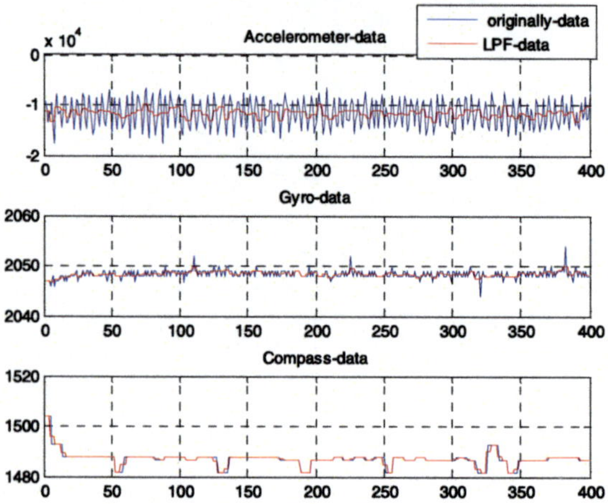

Fig. 4. Measured INS data at uniform velocity and INS data used LPF

Fig. 5 shows estimated driving-trajectory of underwater-BEACON. The blue line shows actual moving trajectory and the red line shows estimated moving trajectory. The jade line shows estimated moving trajectory using the INS data. The star line shows GPS data.

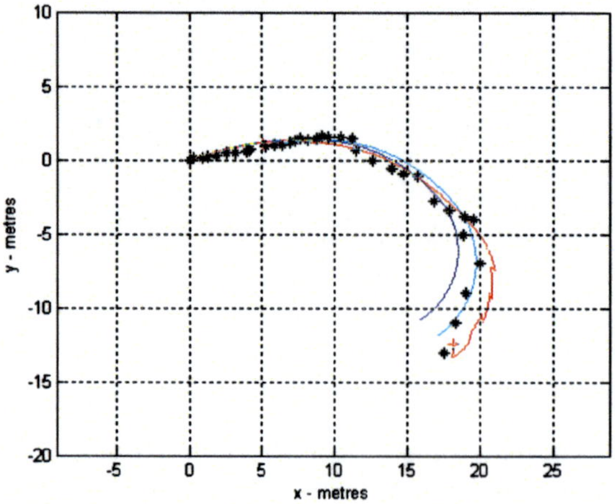

Fig. 5. Location estimation consequence of INS/GPS combined using UKF

The error of estimated trajectory increased over time for the reason that the error of yaw axis as the input of UKF increased. But estimation trajectory was combined GPS signals. It is convergence to the actual moving trajectory. Table 2 shows the errors in curve driving. This experiment is based on $20\,m^2$ driving.

Because the accumulated error of Yaw axis increases, the error of estimated trajectory also increases. In this paper, the estimated trajectory was combined with GPS signals to compensate increasing error. Consequently, we confirm that the estimated trajectory is convergence to the actual moving trajectory.

Table 2. Performance test of UKF

Error / Case	Mean error
Curvilinear	2m

4 Conclusion

The combination systems of INS and GPS compensate GPS errors in an outdoor. So, it generally uses in outdoor precision localization. However, there is a difficulty that it can't effectively remove natural noise used in INS. And, the motion of ship's body is assumed to be a linear. In addition, the noise of INS and GPS is assumed to be regular Distribution noise. Thus, Owing to System Configuration, it is many difficult in outdoor precision localization.

In addition, because of the influence of the disturbance(sea breeze, wave etc.). On the underwater environment, underwater beacon is affected when it moves along the trace.

This paper proposed the filtering technique using by the UKF which was coincided with non-linear motion properties of the underwater flight and non-normally distributed noises of INS and GPS. The performance was confirmed through test in the underwater. The effectiveness of the proposed filtering technique was confirmed through a driving test of BEACON. But effect of disturbance was not indemnified. Because effect of disturbance(sea breeze, wave) of underwater environment is not considered.

In addition to the results of this study, We are going to do control research considering characteristics of disturbance in the underwater[6-8]. We are also planning to apply most proper filter of particle to presume non-linear, irregular distribution by using probability distribution, among the filters for presuming non-linear system state.

Acknowledgements. This work was supported by the Brain Korea 21 project.

This research was supported by the MKE(The Ministry of Knowledge Economy), Korea, under Specialized Field Navigation/Localization technology research center support program supervised by the NIPA(National IT Industry Promotion Agency).

References

1. Im, G.-Y.: Develop of Underwater ultrasonic communication systems. Korea Ocean Research & Development Institute News 238, 8–9 (2003)
2. Kim, S.-H., Roh, C.-W., Kang, S.-C., Park, M.-Y.: Outdoor navigation of a mobile robot using differential GPS and curb detection. In: Proceedings of IEEE International Conference on Robotics and Automation, pp. 3414–3419 (April 2007)
3. Schmidt, G.T.: INS/GPS Technology Trends. NATO Research and Technology Organisation, 1–16 (May 2009)
4. Shaikh, K.N., Bin, A.R., Shariff, M., Jamaluddin, H., Mansoor, S.: GPS-Aided-INS for Mobile Mapping in Precision Agriculture. In: Map Asia Conference (2003)
5. Kim, K., Park, C.G., Yu, M.J., Park, Y.B.: A performance Comparison of Extended and Unscented Kalman Filters for INS/GPS Tightly Coupled Approach. Journal of Control, Automation, and Systems Engineering 12(8) (August 2007)
6. Byum, S.-W., Kim, J.-Y.: Development of Hovering AUV 'NOAH' Test-bed for Underwater Explorations. Journal of the Korea Academia-Industrial Cooperation Society 11(2), 414–419 (2010)
7. Choi, J.-W., Ha, T.-K., Binugroho, E.-H., Yu, C.-H., Seo, Y.-B.: Design of a Test bed and Performance Evaluation for a Hovering Type Autonomous Underwater Vehicle under Open Control Platform. Journal of Institute of Control, Robotics and Systems 16(5), 489–497 (2010)
8. Lee, C.-M., Choi, H.-T., Moon, I.-S., Lee, P.-M.: On Technique for the Thrust Performance of AUVs. The Korean Society of Ocean Engineers 23(4), 47–51 (2009)

Smart Filter Design for the Localization of Robotic Fish Using MEMS Accelerometer

Tae Suk Yoo[1], Sang Cheol Lee[2], Sung Kyung Hong[2], and Young Sun Ryuh[3]

[1] LIG Nex1 Co, Korea
[2] Department of Aerospace Engineering, Sejong University, Korea
[3] Korea Institute of Industrial Technology, Korea
taesuk.yoo@lignex1.com, tend@sju.ac.kr, skhong@sejong.ac.kr, ysryuh@kitech.re.kr

Abstract. This paper presents the design of smart filter for the 2D localization of robotic fish using low-cost MEMS (Micro-Electro Mechanical System) accelerometer. The main purpose of the paper is to minimize the drift error that is inevitable in the double integration process in accelerometer-only navigation system. The proposed approach relies on two parts: 1) an effective calibration method to remove the major part of the deterministic sensor errors and, 2) a novel smart filtering scheme based on fuzzy-logic in order to accurately estimate a 2D position with an accelerometer triad. In addition, we compare the results of the fuzzy logic based on 2D position estimation system with simulation result from a conventional Kalman Filter.

Keywords: robotic fish, 2D position estimation, MEMS inertial sensor, calibration, fuzzy filter.

1 Introduction

Performing reliable localization and navigation within highly unstructured underwater environments is a difficult task. A typical navigation sensor outfit for an underwater vehicle may consist of standard components such as compass, pressure sensor, and some class of INS (Inertial Navigation System) [1]. High precision (inertial grade) IMUs (Inertial Measurement Units) offer excellent strapdown navigation capabilities, but their size, power consumption, and cost have precluded their widespread use in small robotic fish. One possibility is to use MEMS based inertial sensors. In recent years, MEMSs have capability to fabricate cheap, lightweight, and low-power chip-based inertial sensors which have been adopted into various applications such as automobiles, consumer electronics, and medical electronics [2-3]. Although the MEMS inertial sensors have such advantages, they are still suffering from large errors including bias, scale factor error, non-orthogonality, random noise (drift), temperature effect, and so on. It is customary to distinguish between deterministic errors and stochastic errors of MEMS inertial sensors. Deterministic errors include bias, scale factor error and non-orthogonality. These error sources can be removed by calibration procedure. On the other hand, stochastic error sources are not easy to handle due to its random nature. For the navigation system employing MEMS inertial sensors, several

S. Lee et al. (Eds.): Intelligent Autonomous Systems 12, AISC 193, pp. 509–518.
springerlink.com

estimation techniques are chosen currently to eliminate noise problems. For example, the Kalman Filters use statistical error models to predict the behavior of sensor components. This approach, however, is not ideal for MEMS accelerometer based robotic fish system because the statistical models cannot represent well single "catastrophic" events [4-5]. Fuzzy systems have been rapidly emerging as a powerful resource in the field of instrumentation and measurement. Indeed, a variety of applications have been recently proposed encompassing control systems, multi-sensor data fusion, data analysis and digital signal processing. The main purpose of the paper is to minimize the drift error that is inevitable in the double integration process in accelerometer-only navigation system. We compare the results of our fuzzy logic system with simulation result obtained from a conventional Kalman Filter.

2 Experimental Setup for the 2D Position Estimate of the Robotic Fish

During the entire process of the paper, a set of data obtained in a water tank. As shown in Fig. 1, the test runs are performed in the water tank (700cm×400cm×70cm) with a small robotic fish prototype by moving along a rectangular trajectory (200cm×200cm). A MEMS triad IMU module, 3DM-GX3-35 of Microstrain Inc., was selected for this research [6]. The positioning estimation filters are implemented in a real-time DSP (Digital Signal Processor) (TMS320F28335) within the size of 6cm×4cm×3cm.

Fig. 1. Water tank and robotic fish prototype and DSP module with MEMS IMU (3DM-GX3 -35)

With acceleration measurement attached to the body of the robotic fish, the 2D position estimate process consists of three parts as shown in Fig. 2: 1) coordinate transformation of the acceleration in the vehicle's body frame to the navigation frame 2) butterworth bandpass filtering for the alleviation of both drift and noise, and 3) double integration process for the final position estimates.

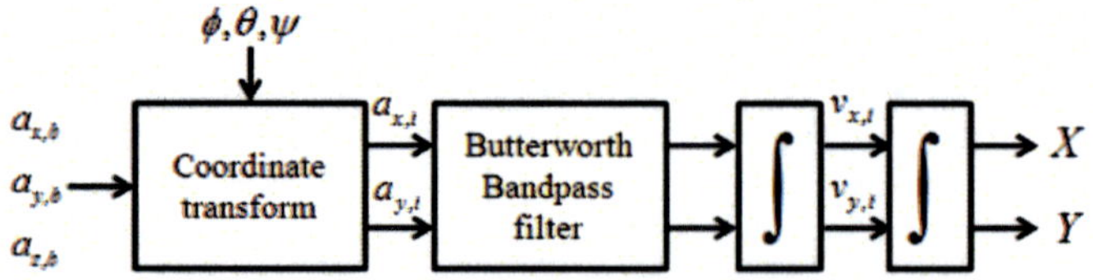

Fig. 2. 2D position estimation process

Clearly, an accelerometer will introduce an error proportional to t^2 in the position due to the drift by double integration. To alleviate drift in position estimates without employing other aided external sensors, one possible solution is to employ a linear bandpass filtering by choosing a cutoff frequency somewhere between the frequencies of low-frequency drift signal and that of the periodic motion which has relatively high frequency [7]. The butterworth filter is a type of signal processing filter designed to have as flat (no ripples) a frequency response as possible in the passband, so it is termed a maximally flat magnitude filter. In order to observe the nominal maneuvering frequency of robotic fish, the discrete Fourier transforms of the signal is found by taking the FFT (Fast Fourier Transform) algorithm, as shown in Fig. 3. Considering the nominal maneuvering frequency of robotic fish (≈2Hz), the 0.1 and 3 Hz cutoff frequencies are chosen for the butterworth filter. Bode plot is also shown Fig. 3.

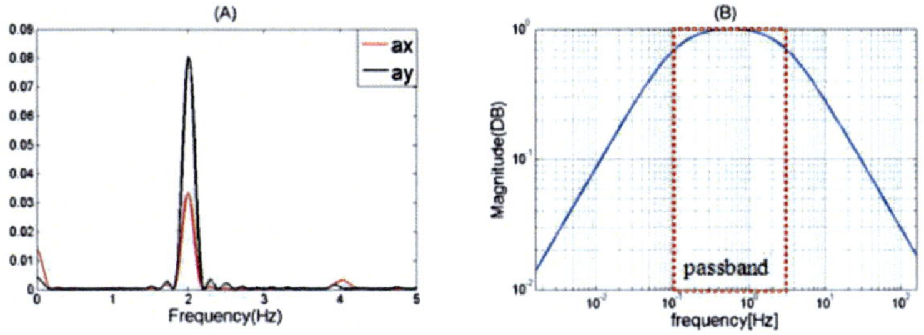

Fig. 3. Analyze (A) and bode plot of butterworth bandpass filter (B)

3 Calibration Method

3.1 Sensor Error Model

A measurement provided by an accelerometer may be expressed in terms of an applied acceleration and the sensor error coefficients. The following equation is typically used to describe the measured acceleration:

$$a_{m,i} = a_i + b_i + S_i a_i + m_{i,j} a_j + \varepsilon_i \tag{1}$$

Where $a_{m,i}$ represents the measured acceleration for i sensitivity axis, a_i is the true acceleration about the sensitivity axis, b_i is the accelerometer instrument bias, S_i is the scale factor error matrix, $m_{i,j}$ is the non-orthogonality matrix, and ε_i is the random noises. From Equation (1), it is evident that bias, scale factor, and non-orthogonality errors are the dominant deterministic elements. If these errors are minimized, their residual effects during navigation will minimize the position drift of the navigation system.

3.2 Least Squares Algorithm

Calibration is the procedure for comparing instrument output with known reference (true) information about the quantity to measure. Standard calibration method can be used to determine the bias and scale factors of the sensors, however, cannot estimate the non-orthogonality. An improved six-position test method which takes into account all three types of errors (bias, scale factor, and non-orthogonality) at the same time is provided.

$$\mathbf{G} = \mathbf{MA} + \mathbf{B}$$

$$\underbrace{\begin{bmatrix} a_{m,x} \\ a_{m,y} \\ a_{m,z} \end{bmatrix}}_{\mathbf{G}} = \underbrace{\begin{bmatrix} (1+S_x) & m_{xy} & m_{xz} \\ m_{yx} & (1+S_y) & m_{yz} \\ m_{zx} & m_{zy} & (1+S_z) \end{bmatrix}}_{\mathbf{M}} \underbrace{\begin{bmatrix} a_x \\ a_y \\ a_z \end{bmatrix}}_{\mathbf{A}} + \underbrace{\begin{bmatrix} b_x \\ b_y \\ b_z \end{bmatrix}}_{\mathbf{B}} \quad (2)$$

Equation (2) can be rearranged as:

$$\mathbf{G} = \tilde{\mathbf{M}}\tilde{\mathbf{A}}$$

$$\underbrace{\begin{bmatrix} a_{m,x} \\ a_{m,y} \\ a_{m,z} \end{bmatrix}}_{\mathbf{G}} = \underbrace{\begin{bmatrix} (1+S_x) & m_{xy} & m_{xz} & b_x \\ m_{yx} & (1+S_y) & m_{yz} & b_y \\ m_{zx} & m_{zy} & (1+S_z) & b_z \end{bmatrix}}_{\tilde{\mathbf{M}}} \underbrace{\begin{bmatrix} a_x \\ a_y \\ a_z \\ 1 \end{bmatrix}}_{\tilde{\mathbf{A}}} \quad (3)$$

By aligning the accelerometer using the six-position method, the ideal acceleration would be measured as:

Table 1. Ideal acceleration for the six-position method

	case1	case2	case3	case4	case5	case6
$a_x[g]$	1	-1	0	0	0	0
$a_y[g]$	0	0	1	-1	0	0
$a_z[g]$	0	0	0	0	1	-1

$\tilde{\mathbf{A}}$ is the matrix for the least squares adjustment. G is measured accelerometer matrix. The desire is to extract the components of the matrix $\tilde{M}$ in Equation (3). This can be done using the well-known solution of least squares:

$$\tilde{\mathbf{M}} = \mathbf{G} \cdot \tilde{\mathbf{A}}^{\mathbf{T}} \cdot (\tilde{\mathbf{A}}\tilde{\mathbf{A}}^{\mathbf{T}})^{-1} \quad (4)$$

4 Fuzzy Logic Signal-Processing Algorithm

Stochastic error source is not easy to handle due to its random nature. In addition, due to the high impact of perturbations within highly unstructured underwater environments, a simple linear filter is not sufficient for the elimination of noises. The random noise provides a large error which increases with time due to double integration process. Therefore, a smart signal-processing algorithm that is capable of the detection of corrupted data is required.

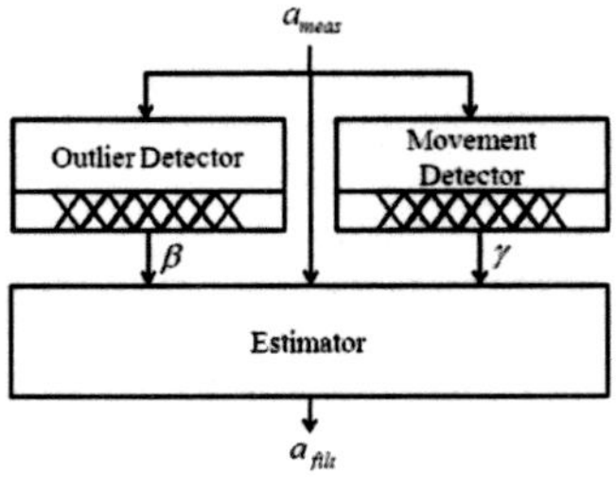

Fig. 4. Fuzzy filtering strategy

A fuzzy logic based on nonlinear filter is used to approximate the navigation trajectory as precisely as possible. The method to design fuzzy filtering for inertial sensors is explained in [9-9]. In the paper, the proposed fuzzy filtering strategy is composed of an outlier detector, a movement detector, and estimator processing. The structure of integrated fuzzy filtering is shown in Fig. 4.

4.1 Outlier Detector

The first step towards obtaining a valid data is the detection of outlaying observations. Detected outliers are candidates for aberrant data that may otherwise adversely lead to biased parameter estimation and incorrect results. In the paper, the outlier detector computes the last three data and the structure of outlier detector is shown in Fig. 5.

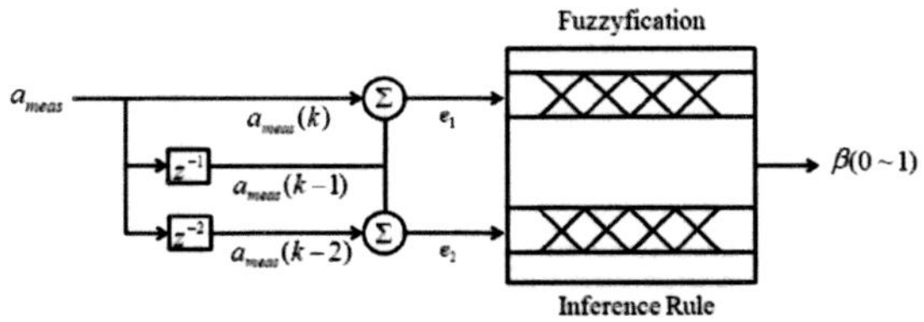

Fig. 5. Structure of outlier detector

In the proposed scheme, the outlier is identified based on the magnitudes e_1 and e_2, which are defined as:

$$e_1 = a_{meas}(k) - a_{meas}(k-1) \tag{5}$$

$$e_2 = a_{meas}(k-1) - a_{meas}(k-2) \tag{6}$$

Fuzzy rules have the following form:

$$\text{If } e_1 \text{ is } A_{1,i} \text{ and } e_2 \text{ is } A_{2,i}\text{, then } \beta = \beta_i \tag{7}$$

Where A 's is MFs (Membership Functions), β (the possibility of the outlier) is the output (β=0 means valid signal and β=1 means outlier), and the subscript i denotes the rule number. The MFs of the fuzzy sets for e_1 and e_2 are shown in Fig. 6. Note that β_i may be considered as a singleton MF as shown in Fig. 6. The truth value of the rule (μ_i) in Equation (7) is obtained by the product of the MF values

$$\mu_i = \mu_{A1,i}[e1] \cdot \mu_{A2,i}[e2] \tag{8}$$

Where $\mu_{A1,i}$ and $\mu_{A2,i}$ are the MF values of the fuzzy set given a value of e_1 and e_2, respectively. Based on μ_i, the defuzzification value of β yields the following:

$$\beta = \sum \mu_i \beta_i \tag{9}$$

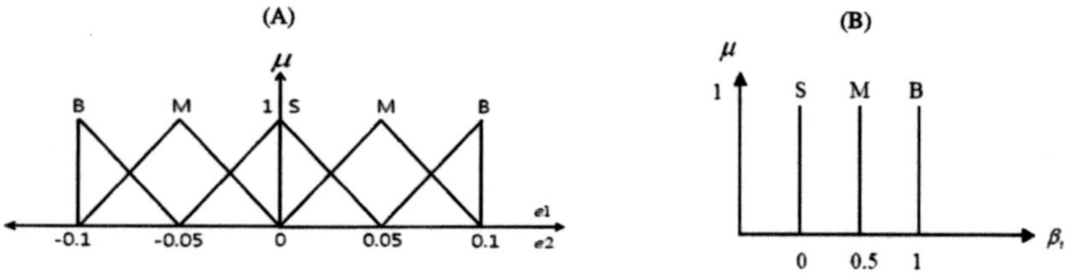

Fig. 6. Membership functions (A) and Singleton membership function (B)

4.2 Movement Detector

An easy way to avoid a position drift is to detect when the system is immobile. The concept of movement detector is shown in Fig 7.

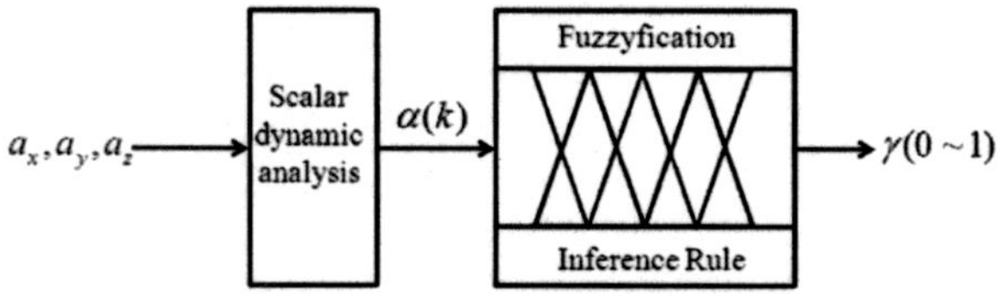

Fig. 7. Structure of movement detector

The scalar dynamic acceleration ($\alpha(k)$) is defined as:

$$\alpha(k) = \left| \sqrt{a_x(k)^2 + a_y(k)^2 + a_z(k)^2} - 1 \right| \tag{10}$$

Fuzzy rules have the following form:

$$\textit{If } \alpha(k) \textit{ is } A_i, \textit{ then } \gamma = \gamma_i \tag{11}$$

Where A's is MF, γ (the possibility of the movement) is the output (γ=0 means immobile and γ=1 means mobile), and the subscript i denotes the rule number. The MF's of the fuzzy sets for $\alpha(k)$ is shown in Fig. 8. Note that γ_i may be considered as a singleton MF as shown in Fig. 8.

$$\gamma = \sum \mu_i(\alpha) \cdot \gamma_i \tag{12}$$

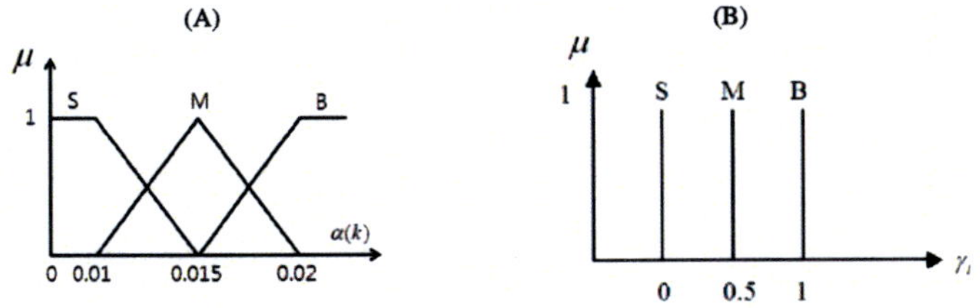

Fig. 8. Membership functions (A) and Singleton membership function (B)

4.3 Estimator

After these two parallel steps, a final decision has to be taken. At this point, all the needed values are known (outlier probability (β) and probability of movement (γ)). The resulting filtered acceleration ($a_{filt}(k)$) is determined by (13) with defuzzified values, β and γ

$$a_{filt}(k) = \gamma \cdot ((1-\beta) \cdot a_{meas}(k-1) + \beta \cdot a_{meas}(k-2)) \tag{13}$$

This outcome indicates that there is an improvement in the error reduction (as shown Table 2) with the aid of the proposed method. Fig. 9 is a comparison between the original and the total practical trajectory result.

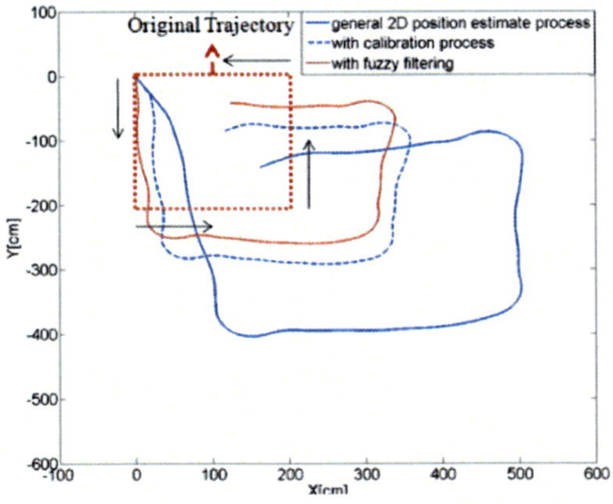

Fig. 9. Comparison of total trajectory test results

Table 2. Error reduction at the end point for proposed method

	General method	Calibration process	Fuzzy filtering
Error	≈210cm	≈150cm	≈110cm
Reduction	-	28%	48%

5 Comparison Simulation Result of Fuzzy Logic Based System with Kalman Filter

Fuzzy logic based position estimation system is fundamentally different from the well-known Kalman filter method. Kalman filters generally use statistical error models to predict the behavior of sensor components. A successful application of the Kalman filter is strongly dependent on whether the system dynamics is perfectly modeled in the mean sense. In practice, disturbances such as wind and current may have random variations (catastrophic events) superimposed on changing mean values (trend). This is very difficult task, if not impossible to model the system dynamics of a kinematics positioning system perfectly because the environment always changes. As far as measurement noise is concerned the system will require knowledge of the random errors, which will be assumed Gaussian. As a result, Kalman filter is not ideal for robotic fish system based on MEMS accelerometer. Fig. 10 shows a comparison simulation between setup Fuzzy logic based system and Kalman filter.

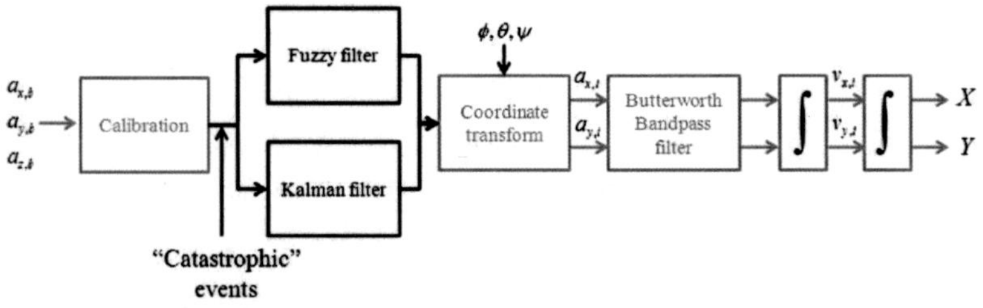

Fig. 10. Comparison simulation setup fuzzy logic based system with Kalman filters

Some random values into the data represent catastrophic events, such as collision or impact that can occur in natural underwater environment. Fig. 11(A) shows such data applying Kalman filter and Fuzzy filter. In case of Fuzzy filter, it removes the effect of the random values that describe "catastrophic" event. In case of Kalman filter, however, could not remove these effects. Fig. 11(B) is the simulation of 2D position estimation result by using the data from Fig. 11(A). This outcome indicates that Fuzzy filter is much robust in the "catastrophic" events compared with the famous Kalman filter method.

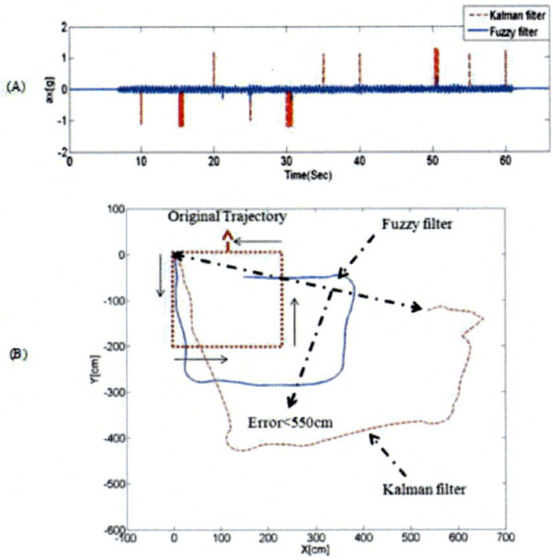

Fig. 11. Outlier detector effect(A) and Comparison of trajectory test-with "catastrophic" event (B)

Table 3. Position error at the end point for between Fuzzy filter and Kalman filter

	Without catastrophic events	**With catastrophic events**
Fuzzy filter	≈110cm	≈110cm
Kalman filter	≈150cm	≈550cm

6 Conclusion

The paper presents an effective calibration method and a fuzzy logic based smart signal-processing algorithm in order to accurately estimate a 2D position for small robotic fish with a low-cost MEMS accelerometer. The main purpose of the paper is to minimize the drift error that is inevitable in the double integration process in accelerometer-only navigation system. The performance of proposed method is evaluated by the trajectory test. The proposed scheme has shown good characteristics for the drift errors from the MEMS IMU measurement as well as the noise corruption from the fuzzy filter in all dynamic conditions. The proposed scheme gave an accuracy of better than 110cm compared to the end point of original trajectory (navigation for 70 seconds). The test results are very encouraging since the estimation process was performed with only MEMS accelerometer.

The prime purpose of the paper is to make low-cost MEMS accelerometer sensor as accurate as possible via proposed calibration method and fuzzy-based filtering scheme. As a further study, the accelerometer will be fused with multisensory data such as USBL (Ultra Short Baseline) for our miniaturized robotic fish. To this end, extend Kalman filter and their variants will be used.

Acknowledgements. This research was supported from Development of Biomimetic Robot System Research Project funded by Korea Institute of Industrial Technology of the Korean Government.

References

1. Hegrenaes, O., Berglund, E., Hsllingstad, O.: Model-aided Inertial navigation for underwater vehicles. In: Proceedings of the IEEE International Conference on Robotic and Automation, PASADENA, pp. 1069–1076 (2008)
2. Yoo, T.S., Hong, S.K., Yoon, H.M., Park, S.S.: Gain scheduled complementary filter design for a MEMS based attitude and heading reference system, vol. 11, pp. 3816–3830 (2011)
3. Hong, S.K.: Fuzzy logic based closed-loop strapdown attitude system for unmanned aerial vehicle (UAV). Sens. Actuat. A-Phys. 107, 109–118 (2003)
4. Grewel, M.S., Henderson, V.D., Miysako, R.S.: Application of Kalman Filtering to the Calibration and Alignment of Inertial Navigation Systems. IEEE Trans. Automatic Control 36, 3–13 (1991)
5. Lauro, O., Giuloi, R., Daniel, C., Johann, B.: The FLEXnav Precision Dead-reckoning System. International Journal of Vehicle Autonomous System 4, 173–195 (2006)
6. Miniature Attitude Heading Reference System (AHRS) with GPS, 3DM-GX3-35; Micro strain: 459 Hurricane Lane, Suite 102, Williston, VT 05495 USA (2011)
7. Win, T.L., Kalyana, C.V., Wei, T.A.: Drift-free position estimation of periodic or quasi-periodic motion using inertial sensors, vol. 11, pp. 5931–5951 (2011)
8. Park, M.H., Yang, G.: Error analysis and stochastic modeling of low-cost MEMS accelerometer. The Journal of Intelligent and Robotic Systems 46, 27–41 (2006)
9. Hong, S.K.: A Fuzzy Logic based Performance Augmentation of MEMS Gyroscope. Journal of Intelligent & Fuzzy Systems 19(6), 393–398 (2008)

View Planning of a Multi-rotor Unmanned Air Vehicle for Tree Modeling Using Silhouette-Based Shape Estimation

Dae-Yeon Won[1], Ali Haydar Göktoğan[2], Salah Sukkarieh[2], and Min-Jea Tahk[1]

[1] Division of Aerospace Engineering, KAIST, Daejeon, Republic of Korea
{dywon,mjtahk}@fdcl.kaist.ac.kr
[2] Australian Centre for Field Robotics, University of Sydney, Sydney, Australia
{a.goktogan,s.sukkarieh}@acfr.usyd.edu.au

Abstract. The use of a multi-rotor unmanned air vehicle (UAV) in image acquisition tasks is promising for three-dimensional (3D) object modeling. Such an autonomous data acquisition system can be useful to handle the geometric complexity of objects such as trees and the inherent difficulties of image capture. In this paper, we address the problem of view planning for a camera-equipped multi-rotor UAV to acquire an adequate set of images that leads to more detailed and complete knowledge of the 3D tree model. The proposed algorithm based on shape-from-silhouette methods incorporates both expected new visual information and vehicle movement. Occupancy estimation for volumetric object model serves as a baseline measure of new information. The outlined approach determines next best views across the viewpoint space bounded by the sensor coverage and the capability of the UAV with minimal a priori knowledge of the object. Simulation studies conducted with virtual reality environments show the effectiveness of the algorithm.

1 Introduction

Three-dimensional (3D) geometry of trees including outlines, shape, and height is widely required for scientific and commercial applications. In computer graphics, the use of realistic tree modeling provides a significant level of virtual scenes. In weed biological control, on the other hand, the geometric information of trees is important to improve detection, classification, and quantification between several species [1].

In this context, image-based tree modeling, the process of reconstructing 3D tree models from photographs, has been studied in computer vision area [2]-[6]. Image-based 3D modeling process typically involves taking multiple view images, reconstructing camera pose with sparse geometry, and registering geometric data in a reference frame [7,8]. These image-based modeling methods are based on structure from motion techniques that use correspondences between 2D image measurements to recover 3D information. Recently, digital image-based modeling software such as Autodesk's Photofly [9] or Microsoft's Photosynth [10] is getting popular and readily available online for the purpose of object modeling and exploring unstructured environments. Although automated reconstruction process reduced the amount of human work, multiple image acquisition from views is still done manually via trial-and-error approach to

S. Lee et al. (Eds.): Intelligent Autonomous Systems 12, AISC 193, pp. 519–531.
springerlink.com

achieve acceptable level of modeling results. Also, if the geometric complexity of trees increase, it becomes more difficult to get a proper set of images by hands. Concerning the data acquisition performance, employing a multi-rotor unmanned air vehicle (UAV) suits well for this purpose in terms of mobility and repeatability. An intelligent view planning mechanism designed to equip the UAV for the automated data acquisition is required to guarantee the fidelity of 3D tree reconstruction.

In this paper, we address the problem of view planning for 3D tree reconstruction using a camera-equipped multi-rotor UAV. To locate next viewpoints for improving reconstruction quality, shape-from-silhouette based approach [11] is employed. Since the silhouettes of a tree depend on both the tree itself and the region allowed to the viewpoints, backprojecting of these 2D information can be sources of volumetric information. As the UAV captures a number of views in different viewpoints, the reconstruction quality can be measured by shape estimation from multiple silhouettes. On the other hand, structure from motion approaches, which generates 3D point cloud, is challenging due to trees' immensely complex structure. The presented approach considers utilisation of a multi-rotor UAV platform capable of flying to a given way point and hovering there while holding the commanded heading. The platform will be carrying a light weight, low-cost, passive camera instead of expensive and relatively heavier active sensors such as laser scanners [12]. The performance of this approach is characterized by applying the algorithms to a simulated tree model in a virtual environment. Paper is organized as follows; Section 2 describes the problem statements, Section 3 discusses some related work. The proposed view planning algorithm for silhouette-based tree model reconstruction is addressed in Section 4. Section 5 provides the simulation model and the numerical results. The final section summarizes the overall discussion.

2 Problem Statement

The main aim of this paper is to develop a solution for the automated image acquisition of a single tree by using a small multi-rotor UAV equipped with a low-cost, camera. The proposed approach is illustrated in Fig. 1. It is assumed that the geolocation of the tree is known a priori and the desired distance between the centre of the tree and the

Fig. 1. Image acquisition task for 3D tree model reconstruction using a multi-rotor UAV

UAV is given. Because of on-board computational capabilities of conventional small-size UAVs, the live video stream from the on-board camera is transmitted to a ground control station for image processing. When the computation of next best viewpoints is completed, the desired viewpoint is uploaded to the flight control system. Also, we assume that the UAV always finds collision-free paths over desired viewpoints.

3 Related Work

Finding an optimal viewpoint plan necessary for the best possible reconstruction is an intractable problem and addressed by researchers as next-best-view (NBV) problem [13]. Following is a brief summary of view planning schemes that have been proposed for robotic applications.

Sujan et al. [14] conduct a study on view planning for unknown, unstructured environment mapping by cooperative sensing. To determine the agent's next best view, the authors considered optimizing a rating function over new information about the environment based on Shannon's information theory. Paul et al. [15] also propose a view planning algorithm for an industrial robot manipulator based on information theory. Their exploration approach incorporates multiple objectives to minimize manipulator movement and maximize its safety. Schmid et al. [16] address the problem of model-based view planning for 3D reconstruction of outdoor sites based on a 2.5D digital surface model. Applicable viewpoints from the list of possible viewpoints are chosen by heuristic constraints imposed by the multi-view stereo algorithm. This approach gives manual viewpoint planning guidelines resulted from reconstruction experiments.

Although the above mentioned algorithms claim good performance, most of them concerned with the use of range measurements. As opposed to those works, view planning algorithms using a number of images have been proposed with silhouette-based approaches [17,18,19]. While the resulting methods has been well established theoretically and applied to various class of objects, these work largely deal with the interactive object-specific reconstruction problems for convex polyhedra.

4 View Planning Algorithm

4.1 Overview

As described in Fig. 2, our automated data acquisition process consists of four stages. The first stage is to construct a volumetric hull and define boundaries of viewpoint space based on a priori knowledge of the tree of interest. The 3D tree model is considered as a probabilistic discretized occupancy grid. In the second stage, the silhouette of the tree is extracted by partitioning an input image into the tree and background. In the third stage, occupancy values of the volumetric model is updated up to the current viewpoint using silhouette-based shape estimation. If the end criteria of view planning is not satisfied, the final stage is to find the next best view by defining and optimizing a rating function over the possible viewpoint space and move the UAV to the desired pose.

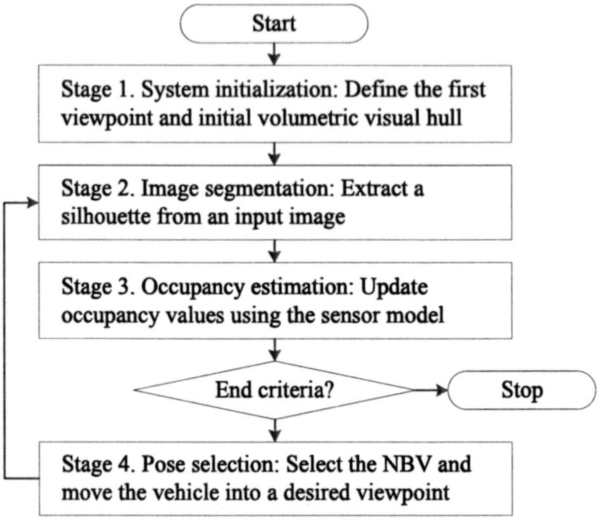

Fig. 2. Outline of the view planning algorithm

4.2 System Initialization

As a first stage, a circular path of the UAV is designed as a baseline for view planning. The path parameters including radius and height are bounded by the sensor's field of view to cover the volumetric object model entirely. These parameters are also limited by the UAV's operating range. The closest point on the circular path is taken as the first viewpoint. The dimension of the voxel model depends on both computational loads and the size of the object given as a priori knowledge. The volume of the object is stored in an octree data structure, which is a common method for 3D object modeling from a set of images [18,20,21]. In contrast to other modeling approaches in 3D such as a point cloud, elevation map or multi-level surface map, an octree-based approach is suitable for a multi-resolution representation in probabilistic occupancy modeling whereas it has the problem of memory consumption and over-confidence in the model [22]. During an image acquisition task, the octree is initialized as a cube bounding the object, and recursively subdivided all occupied voxels down to the maximum subdivision level as a number of views are added.

4.3 Image Segmentation

In the second stage, the input image from the on-board camera is segmented into the tree of interest and the background. The silhouette of the object, obtained as a perspective projection onto the image plane, provides 3D geometric shape understanding as the number of silhouettes increase [11]. To extract the object's silhouette, classifier based algorithms such as graph cuts or support vector machines have been applied to a wide

Fig. 3. Image segmentation. An input image (left) is segmented into the tree and the background (right) by using an image segmentation algorithm. The extracted silhouette is used to construct the volumetric model of the object.

range of problems in object segmentation [23,24,25]. Typically these algorithms can be carried out in polynomial time with minimal user interaction. Hence, these approaches are not considered in this paper. Fig. 3 shows an example of image segmentation for a tree object.

4.4 Occupancy Estimation

In the third stage, the 3D object model is computed by occupancy estimation up to the current viewpoint. If object's silhouettes and corresponding pose measurements are obtained with absolute confidence, this stage can be considered as shape modeling by volume intersection. In this case, intersection of the outline cones backprojected from all silhouettes carves free space lied outside of silhouettes and reveals the closest approximation of the tree model referred to as the visual hull [21]. The intersection process identifies whether each voxel lies entirely inside, entirely outside, or partially intersects the tree. Although the resulting visual hull cannot capture some of the details, it captures most of the shape appearance-defining elements [2]. Fig. 4 depicts an example of developing the visual hull for a simulated tree model using octree representation.

As the observations from the UAV system are not necessarily error-free, a statistical occupancy grid model is employed to carve the volumetric model by fusing multiple silhouette measurements. We calculate the posterior probability for each voxel of a discrete grid instead of intersecting volumetric cones directly. Each cube in octree representation has its occupancy probability based on whether its projection onto the image plane lies inside or outside of silhouettes from a view plan. This approach gives a constant-time access to grid cells and provide a proper way to represent uncertainties [26]. We consider that the model m consist of a discrete, three-dimensional grid of L voxels denoted as $m_1, ..., m_L$. The measurement z_k is backprojected space from the 2D silhouette of the kth image at the corresponding viewpoint x_k. Given independence assumptions for static environments with a non-changing object model, a posterior probability $p(m_l | x_{1:k}, z_{1:k})$ is estimated recursively as follows.

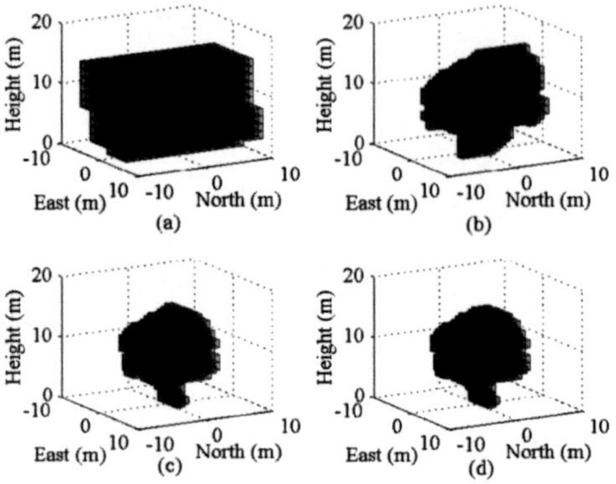

Fig. 4. Progression of the visual hull from (a) to (d) constructed by the intersections of volumes of four different silhouette cones

$$p(m_l|x_{1:k}, z_{1:k}) = \left(1 + \frac{1 - p(m_l|x_k, z_k)}{p(m_l|x_k, z_k)} \cdot \frac{p(m_l)}{1 - p(m_l)} \cdot \frac{1 - p(m_l|x_{1:k-1}, z_{1:k-1})}{p(m_l|x_{1:k-1}, z_{1:k-1})}\right)^{-1} \tag{1}$$

This updates voxels that fall into the sensor's field of view. To prevent overconfident state of the model, a clamping update policy is applied to the occupancy estimation [27]. The clamping update policy bounds the maximum and minimum probability so that it makes voxels remain updatable by adding new observations until the end of the task. The update equation of the clamping update policy is following:

$$p(m_l|x_k, z_k) = \max\left(\min\left(p(m_l|x_k, z_k), p_{\max}\right), p_{\min}\right). \tag{2}$$

The inverse sensor model $p(m_l|x_k, z_k)$ strongly depends on the type of sensor and its application environment. For 3D reconstruction from 2D image sensors, occupied voxels are classified by backprojecting a silhouette from the sensor, unlike ray-casting operations which use range sensors. If a voxel falls inside the silhouette's backprojection cone, the voxel is classified as occupied space in the octree representation. This inverse sensor model can be formulated as

$$p(m_l|x_k, z_k) = \begin{cases} p_{prior} & \text{, outside of the field of view} \\ p_{occ} & \text{, inside of backprojection} \\ p_{free} & \text{, outside of backprojection} \end{cases} \tag{3}$$

It holds $0 \leq p_{free} \leq p_{prior} \leq p_{occ} \leq 1$. Fig. 5 illustrates an example of voxel classification for 2D image sensors. This method assigns an occupancy value of p_{occ} to all voxels whose projection onto the viewpoint lies within the silhouette.

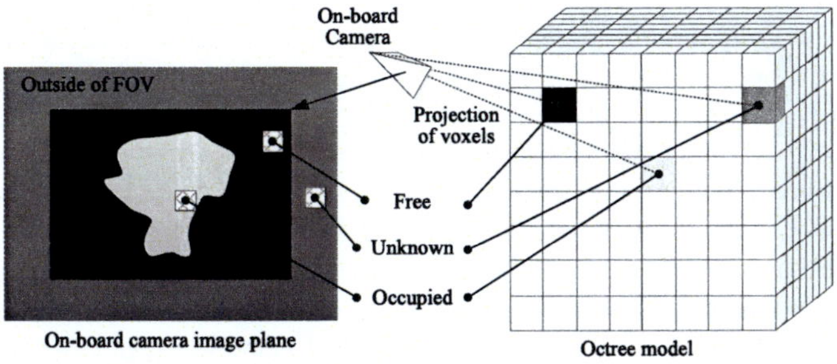

Fig. 5. Inverse sensor model for updating occupancy values of the octree model. Voxel-based shape estimation is updated by an occupancy classification of each volume element into one of the discrete states: free, unkonwn, and occupied.

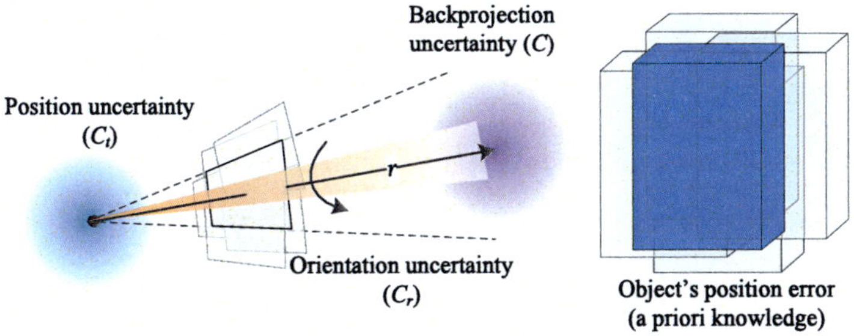

Fig. 6. Sources of uncertainty and its propagation to 3D points: pose measurement noise and unexpected position of an object

To take the uncertainty in a real environment into account, our sensor model incorporates three types of noise and uncertainty sources: pose measurement noise, silhouette extraction errors, and unexpected tree location. As depicted in Fig. 6, those uncertainty sources are essential to making the sensor model with silhouette-based shape estimation. Pose uncertainty makes a difference between a computed view plan and what will actually be executed by the UAV's flight control system. If the effects of pose uncertainty are not handled properly, data acquisition by the view plan can be compromised. Image segmentation, whether manual or automatic, can be achieved only to a certain finite accuracy level. Any silhouettes extracted from images, therefore, are subject to measurement errors. Uncertainty in tree location is due to the GPS errors in the system initialization stage. This locational error may be critical when a planning process has a pre-programmed view plan without an online correction manner.

Analytical expressions for the sensor uncertainty model should incorporate noise characteristics with the shape-from-silhouette technique. We consider how those errors in measurement systems propagate through the geometry in order to quantify the uncertainty on the final backprojected 3D point. The model can be simplified as

$$C = C_t + r^2 R(\theta) C_r R(\theta)^T \tag{4}$$

where C, C_t, and C_r are covariance matrices of the sensor model, position measurement, and orientation measurement noises, respectively. r is the distance to the environment point from the body frame, and $R(\theta)$ is the rotation matrix from the world frame to the body frame by the Euler angle θ. Errors due to image segmentations, which has erosion or dilation effects on segmented silhouettes, are modeled as a part of orientation errors. Since the orientation measurements are made in the camera coordinate frame, a transformation of covariance parameters from the observation frame to the world frame is required. From this probabilistic backprojection of silhouettes into the volumetric tree model, voxels inside of 1-standard deviation from the backprojected point are considered as occupied space. Fig. 7 shows a Monte Carlo simulation of this model for the backprojected 3D point placed at 25m distance from its corresponding viewpoint. The analytical and simulated distributions are almost perfectly overlapping.

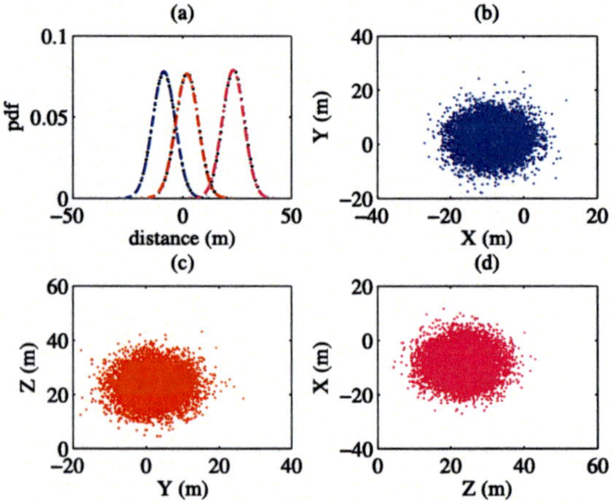

Fig. 7. Monte Carlo simulation for the uncertainty analysis: (a) the analytical (dashed) and simulated (dotted) distribution of the computed distance for each axis of the world coordinate. The two curves corresponding to each axis are almost perfectly overlapping. (b), (c) and (d) are distributions of the simulated data points backprojected from the viewpoint to the center of image plane.

4.5 Pose Selection

The final stage finds the next best view for the image-based 3D reconstruction until the end criteria are satisfied. View planning process can be terminated on two conditions: travelling around its circular path once or the low rate of carved volume by an added view. A quality viewpoint is selected from viewpoint candidates in a manner of optimization that incorporates both expected new visual information and distance travelled by the UAV. For the current occupancy probability model, the projected area S of occupied voxels onto the image plane of a possible viewpoint and the corresponding distance d to be moved by the UAV are combined into a single objective function formulated as

$$f = w_S \cdot S + w_d \cdot d^{-1} \quad (5)$$

where w_S and w_d are weighting coefficients. Maximizing the first term over a set of viewpoints promotes the pursuit of new geometric information to reveal remaining free space inside the current visual hull. The evaluation of S projects voxels which has the occupancy probability higher than p_{occ}. The second term of the objective function aims to minimize the movement of the UAV from the current then to the next pose. Fig. 8 depicts this evaluation of expected information from a viewpoint candidate.

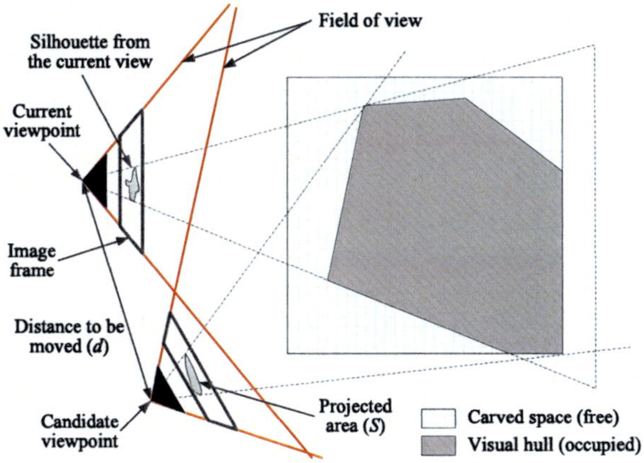

Fig. 8. Evaluation of the objective function for a viewpoint candidate

Since the objective function cannot be optimized analytically, finding an optimum next viewpoint requires searching algorithm over a set of uniformly distributed candidate viewpoints in the environment. While determining an appropriate viewpoint, the viewpoint candidates for the next pose are restricted by two constraints: a coverage constraint and a maximum angle constraint. The coverage constraint indicate that certain part of an object has to be seen in the field of view. The maximum angle constraint means that the angle between two consecutive views to the center of the volumetric model should not exceed the maximum angle. Fig 9 describes the geometry of the pose selection scheme. A number of geometric entities such the viewpoint region constraints and the maximum angle constraint are fixed by considering capabilities of both the UAV and the image sensor with a priori knowledge of the tree.

5 Numerical Results

Here we give the numerical results of the proposed view planning algorithm. Three viewpoint selection methods are compared. The view plan of the first method is selected randomly within viewpoint candidates for each step. The second method selects the next camera viewpoint sequentially from a set of uniformly gridding viewpoints on the circular path. In the third method, the next camera viewpoint is selected with maximum expected information by optimization.

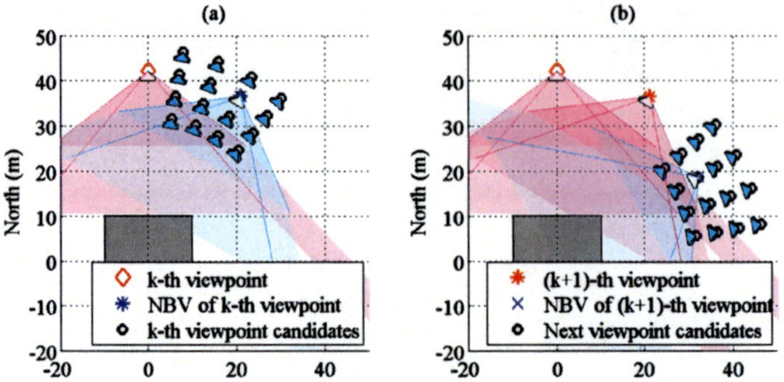

Fig. 9. Viewpoint selection scheme: (a) and (b) show top-down view with viewpoints picked out for next-best-view at each step. Viewpoint candidates are indicated by their centers (circle) and pyramids.

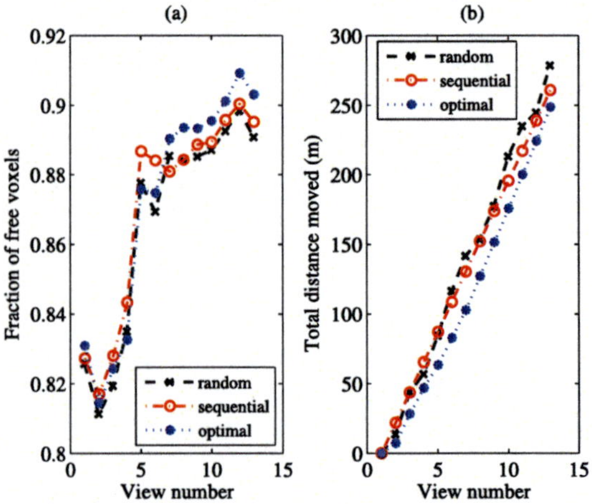

Fig. 10. Comparison of the three viewpoint selection methods for modeling an unknown tree: (a) fraction of free voxels as a function of view number. (b) total distance travelled by the multi-rotor UAV as a function of view number.

A 3D graphics scene including a tree is generated by using VRML model that is directly incorporated in MATLAB. The tree of interest, which is 15 m tall and a 10 m in diameter, is located at (7m, 5m, 0m) in a local East-North-Up (ENU) coordinate system whereas the expected tree location is the origin. The initial octree cube from a priori knowledge is set to 30 m $\times$ 30 m $\times$ 30 m with appropriate margins for unexpected tree location. Viewpoint candidates across the free space bounded by the size of the octree cube and design parameters. Since we employ the circular path as a baseline, the viewpoint space has a doughnut shape. The heading angle of the UAV and the tilting angle of the vision system frame are determined to point the center of the volumetric tree model. The occupancy probability of all voxels is initialized to the uniform prior of $p_{prior} = 0.5$ and we used values of $p_{occ} = 0.8$ and $p_{free} = 0.2$ for occupied and free

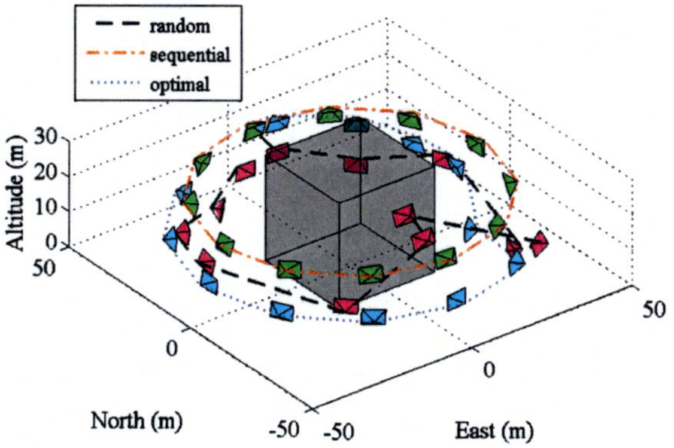

Fig. 11. Path of viewpoints planned by the three selection methods of view planning for modeling an unknown tree

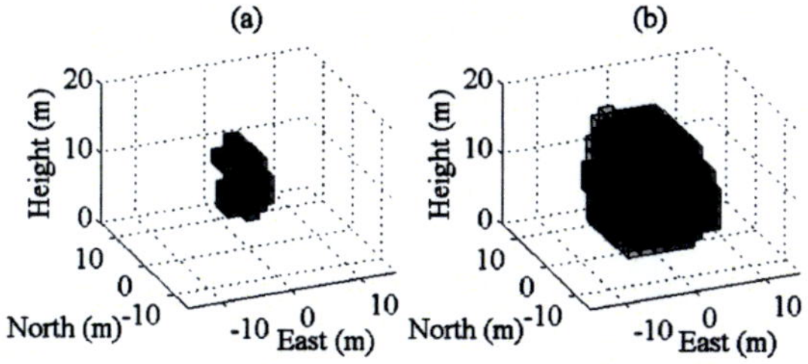

Fig. 12. The probability occupancy tree model from the overall viewpoints in the error-free situation (left) and the optimal selection method under measurement uncertainty (right). The probabilistic model of (b) is enlarged by uncertainty during occupancy estimation.

voxels, respectively. The clamping thresholds are set to $p_{max} = 0.9$ and $p_{min} = 0.1$. The uncertainties of the position and orientation are set to standard deviations of 5 m and 3 degrees, respectively.

Fig. 10 shows the fraction of free voxels and the distance travelled by the UAV for the three planning schemes. Under the 30 degrees maximum angle constraint, the total of 12 viewpoints is planned from the first viewpoint located at (0 m, 42m, 21m). As the number of views increases, the visual hull becomes the closest approximation of the tree whereas the rate of carved volume decreases. It is desirable to have a large fraction of the visual hull modeled with small displacement. The resulting modeling error can be measured with respect to the visual hull estimated by all possible viewpoints. The relative errors of random, sequential, and optimal planning methods are 1.83%, 1.79%, and 1.69%, respectively. From the results, it can be seen that our method provides better final reconstruction accuracy and shorter total distance travelled by the UAV. Fig. 11 shows path of viewpoints planned by the three selection methods and Fig. 12 depicts the 3D occupancy grid models for the visual hull made by all viewpoint candidates and the proposed view planning method. While the proposed method allows to capture a proper set of images via probabilistic occupancy estimation, this approach remains the problem of handling uncertainty in a view plan. Reducing pose uncertainty, however, can be achieved by a localization and mapping approach with prior knowledge of known landmarks. Interactive or learning techniques for the image segmentation stage also can be a practical alternative method of dealing with uncertainty in silhouette segmentation.

6 Conclusions

There have been research efforts to select the next best viewpoint of 2D image sensors in unknown objects via automated process. In this paper, we proposed a view planning algorithm for image-based tree modeling using a camera-equipped multi-rotor UAV. The focus is on view planning of the passive sensor for improving reconstruction quality because most of the previous works have dealt with range images from active sensors. The possible next viewpoint is guided by the geometric concept of visual hull obtained by backprojecting the tree's silhouettes at the corresponding viewpoints. The outlined approach provides a useful framework for both modeling of the tree and planning of viewpoints by constructing of the space occupancy grid from multiple different views. The numerical results in virtual environments confirm that the proposed approach has capability to locate efficient next viewpoint in terms of expected new information and total distance travelled by the multi-rotor UAV.

Acknowledgement. This work is supported by the ARC Centre of Excellence programme, funded by the Australian Research Council (ARC) and the New South Wales State Government. Also, part of this work is sponsored by the Korea Aerospace Research Institute (KARI) under the Vision-Based Collision Avoidance Techniques for UAVs Project.

References

1. Göktoğan, A.H., Sukkarieh, S., Bryson, M., Randle, J., Lupton, T., Hung, C.: A Rotary-wing Unmanned Air Vehicle for Aquatic Weed Surveillance and Management. J. Intell. Robot. Syst. 57, 467–484 (2012)
2. Shlyakhter, I., Rozenoer, M., Dorsey, J., Teller, S.: Reconstructing 3D Tree Models from Instrumented Photographs. IEEE Computer Graphics and Applications 21, 53–61 (2001)
3. Reche-Martinez, A., Martin, I., Drettakis, G.: Volumetric "Reconstruction and Interactive Rendering of Trees from Photographs". ACM Trans. Graph. 12, 720–727 (2004)
4. Tan, P., Zeng, G., Wang, J., Kang, S.B., Quan, L.: Image-based tree modeling. ACM Trans. Graph. 26 (2007), doi:10.1145/1276377.1276486
5. Neubert, B., Franken, T., Deussen, O.: Approximate Image-based Tree-modeling Using Particle Flows. ACM Trans. Graph. 26 (2007), doi:10.1145/1275808.1276487
6. Binney, J., Sukhatme, G.S.: 3D Tree Reconstruction from Laser Range Data. In: Proceedings of the 2009 IEEE International Conference on Robotics and Automation, Kobe, Japan, pp. 3183–3188 (2009)
7. Remondino, F., El-Hakim, S.: Image-based 3D Modelling: A Review. The Photogrammetric Record 21, 269–291 (2006)

8. Snavely, N., Seitz, S.M., Szeliski, R.: Photo tourism: exploring photo collections in 3D. ACM Trans. Graph. 25, 835–846 (2006)
9. *Photofly*, Autodesk, `http://labs.autodesk.com/technologies/photofly/`
10. *Photosynth*, Microsoft, `http://photosynth.net/`
11. Laurentini, A.: The Visual Hull Concept for Silhouette-Based Image Understanding. IEEE Trans. Patt. Anal. Machine Intell. 16, 150–162 (1994)
12. Lin, Y., Hyyppa, J., Jaakkola, A.: Mini-UAV-Borne LIDAR for Fine-Scale Mapping. IEEE Geoscience and Remote Sensing Letters 8, 426–430 (2011)
13. Scott, W.R., Roth, G., Rivest, J.F.: View Planning for Automated Three-dimensional Object Reconstruction and Inspection. ACM Comput. Surv. 36, 64–96 (2003)
14. Sujan, V.A., Dubowsky, S.: Efficient Information-based Visual Robotic Mapping in Unstructured Rnvironments. International Journal of Robotics Research 24, 275–293 (2005)
15. Paul, G., Liu, D., Kirchner, N., Dissanayake, G.: An Effective Exploration Approach to Simultaneous Mapping and Surface Material-type Identification of Complex Three-dimensional Environments. J. Field Robotics 26, 915–933 (2009)
16. Schmid, K., Hirschmuller, H., Domel, A., Grixa, I., Suppa, M., Hirzinger, G.: View Planning for Multi-View Stereo 3D Reconstruction Using an Autonomous Multicopter. J. Intell. Robot Syst. 65, 309–323 (2011)
17. Shanmukh, K., Pujari, A.K.: Volume intersection with optimal set of directions. Pattern Recognition Letters 12, 165–170 (1991)
18. Nitya, V.B., Sridevi, N., Pujari, A.K.: Linear octree by volume intersection using perspective silhouettes. Pattern Recognition Letters 13, 781–788 (1992)
19. Bottino, A., Laurentini, A.: What's NEXT? An interactive next best view approach. Pattern Recognition 39, 126–132 (2006)
20. Szeliski, R.: Rapid Octree Construction from Image Sequences. CVGIP: Image Understanding 58, 23–32 (1993)
21. Fitzgibbon, A.W., Cross, G., Zisserman, A.: 3D Structure from Multiple Images of Large-Scale Environments, pp. 23–32. Springer, Berlin (1998)
22. Wurm, K.M., Hornung, A., Bennewitz, M., Stachniss, C., Burgard, W.: OctoMap: A probablistic, flexible, and compact 3D map representation for robotic systems. In: Proc. of the ICRA 2010 Workshop on Best Practice in 3D Perception and Modeling for Mobile Manipulation (2010)
23. Rother, C., Kolmogorov, V., Blake, A.: "GrabCut": interactive foreground extraction using iterated graph cuts. ACM Trans. Graph. 23, 309–314 (2004)
24. Yuri, B., Gareth, F.: Graph Cuts and Efficient N-D Image Segmentation. International Journal of Computer Vision 70, 109–131 (2006)
25. Bishop, C.M.: Pattern Recognition and Machine Learning. Springer (2009)
26. Thrun, S., Burgard, W., Fox, D.: Probabilistic Robotics, pp. 279–398. MIT Press (2005)
27. Yguel, M., Aycard, O., Laugier, C.: Update policy of dense maps: Efficient algorithms and sparse representation. Field and Service Robotics 42, 23–33 (2008)

Planar Evasive Aircrafts Maneuvers Using Reinforcement Learning

Dongjin Lee and Hyochoong Bang

Korea Division of Aerospace Engineering, School of Mechanical, Aerospace & Systems Engineering, Korea Advanced Institute of Science and Technology, Daejeon, Korea
{Djlee,hcbang}@ascl.kaist.ac.kr

Abstract. In this paper, the reinforcement learning technique is proposed to implement evasive strategies for aircrafts during engagement. A simplified point-mass model is used to describe the aircraft and the missile equations of motion. The missile follows the pure proportional navigation guidance (PPNG) law to attack the aircraft. Q-learning algorithm which is a form of reinforcement learning is suggested to learn the evasive maneuvers. The performance of the proposed approach is analyzed with numerical simulations. It is shown that the aircraft evades from a missile properly by reinforcement learning with bang-bang type action profiles.

Keywords: missile evasive maneuvers, reinforcement Leraning, Q-learning, pure proportional navigation.

1 Introduction

Many researchers have investigated about missile evasion problems using differential game analysis or optimal control technique. The formulation of a one-sided optimal control problem can be enabled by fixing the missile navigation guidance. Ben-Asher and Cliff studied evasive tactics by assuming linearized kinematics with constant speeds [1]. And more complete model for the missile evasion problem is investigated by Imado and Miwa in [2]. Ong and Pierson treated the evasion problem as an approximated parameter optimization problem [3]. From the previous work, in order to solve optimal control problem, we need to setup a performance index such as the relative distance at the time of closest approach and maximize it [3]. However, the exact evasion maneuvers are the strategies that just avoid conflictions each other. In other words, it is not necessary to maximize the final miss distance to achieve the missile evasion. Moreover, the optimization method requires much computation time to find the best solution, and therefore, the optimal solution is difficult to implement as a real-time guidance law [10]. To overcome these issues, we suggest a reinforcement learning technique to derive intelligent evasive maneuvers. After the learning converged, the aircraft can easily exploit the evasive strategies based on the learned policy.

In this study, we employ a simplified point-mass model with constant speed to describe the aircraft and the missile dynamics. We consider the planar engagement and assume that the missile follows pure proportional navigation guidance law to

S. Lee et al. (Eds.): Intelligent Autonomous Systems 12, AISC 193, pp. 533–542.
springerlink.com

capture the aircraft with a constant navigation gain. Q-learning [4] is introduced as a reinforcement learning algorithm in order to obtain the optimal policy that maximizes the expectation of rewards. Finally, the numerical simulation is constructed to evaluate the proposed approach using various interception scenarios.

2 Problem Formulation

2.1 Evader Equations of Motion

The evading aircraft equations of motions can be represented by a simplified fourth-order point mass model. The governing equations in a horizontal plane are

$$\begin{aligned} \dot{x}_e &= V_e \cos\psi_e \\ \dot{y}_e &= V_e \sin\psi_e \\ \dot{V}_e &= 0 \\ \dot{\psi}_e &= \frac{a_e}{V_e} \end{aligned} \tag{1}$$

where x_e, y_e, ψ_e, V_e, a_e are the states of the aircraft, which are the horizontal coordinates, heading angle, velocity, lateral acceleration command respectively. In this study, we assume that $V_e = 200m/s$ is constant and the aircraft lateral acceleration command is limited by

$$-n_e g \le a_e \le n_e g \tag{2}$$

where $g = 9.81m/s^2$ is the acceleration of gravity and $n_e = 6$ is the maximum gravity loading. The subscript m and e represent the missile and the evading aircraft respectively.

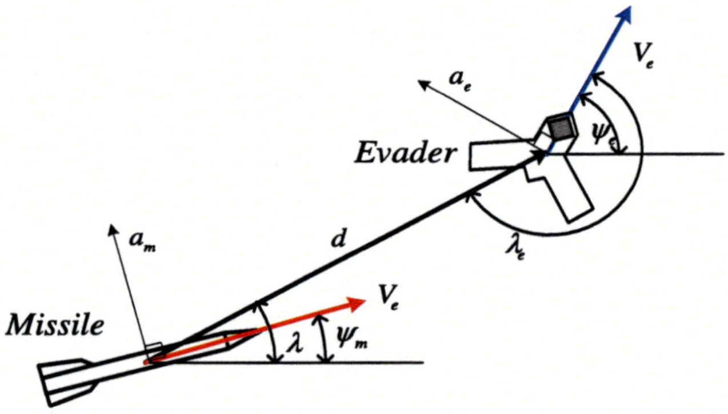

Fig. 1. Interception Geometry

2.2 Missile Equations of Motion

The missile dynamic model used in this study is a simplified fourth-order point-mass model. The equations of motion are

$$\begin{aligned} \dot{x}_m &= V_m \cos\psi_m \\ \dot{y}_m &= V_m \sin\psi_m \\ \dot{V}_m &= 0 \\ \dot{\psi}_m &= \frac{a_m}{V_m} \end{aligned} \tag{3}$$

where x_m, y_m, ψ_m, V_m, a_m are the states of the missile, which are the horizontal coordinates, heading angle, velocity, lateral acceleration command respectively. We also assume that $V_m = 500 m/s$ is constant and the missile lateral acceleration command is limited by

$$-n_m g \le a_m \le n_m g \tag{4}$$

where $n_m = 10$ is the maximum gravity loading. The missile lateral acceleration command is generated by the pure proportional navigation guidance law. Pure proportional navigation (PPN) has been implemented for terminal guidance in many of the currently operational tactical guided missiles. Since PPN guidance law requires no forward velocity variation, it is more appropriate for aerodynamically controlled pursuers than true proportional navigation (TPN) guidance law [6].

$$a_m = N V_m \dot{\lambda} \tag{5}$$

where $N = 3$ is the navigation constant and $\dot{\lambda}$ is the line-of-sight (LOS) rotation rate given by (6).

2.3 Interception Equations of Motion

We can consider the interception equations of motion as follows

$$\begin{aligned} d &= \sqrt{(x_m - x_e)^2 + (y_m - y_e)^2} \\ \dot{d} &= V_e \cos(\lambda - \psi_e) - V_m \cos(\lambda - \psi_m) \\ \lambda &= \arctan\left((x_e - x_m)/(y_e - y_m)\right) \\ \dot{\lambda} &= \frac{V_m \sin(\lambda - \psi_m) + V_e \sin(\lambda - \psi_e)}{d} \end{aligned} \tag{6}$$

where $d, \dot{d}, \lambda, \dot{\lambda}$ are the relative distance, closing speed, LOS angle, LOS rotation rate respectively. We can introduce the interception state vector $\mathbf{X}_I$ using the LOS angle with respect to the aircraft.

$$\mathbf{X}_I = \begin{bmatrix} d & \dot{d} & \lambda_e & \dot{\lambda} \end{bmatrix}^T \tag{7}$$

3 Reinforcement Learning

3.1 Q-Learning

Q-learning is a recent form of reinforcement Learning algorithm that does not need a model of its environment and can be used on-line [5]. Its simplest form, one-step Q-learning, is defined by

$$Q(s,a) \leftarrow Q(s,a) + \alpha\left(r + \gamma \max Q(s_{t+1}, a') - Q(s,a)\right) \tag{8}$$

where r, α, γ are the reward signal, learning rate, discount rate respectively. In this study, we set the learning rate $\alpha = 0.5$ and the discount rate $\gamma = 0.99$. The action value function $Q(s,a)$ denotes the expected return when taking action a at state s under policy τ as follows

$$Q^\tau(s,a) = E\left\{\sum_{k=0}^{\infty} \gamma^k r_{k+t+1} \mid s_t = s,\ a_t = a\right\} \tag{9}$$

and the optimal policy τ^* can be derived by

$$\begin{aligned} Q^*(s,a) &= \max_\tau Q^\tau(s,\ a) \\ &= E\left\{r_{t+1} + \gamma \max_{a'} Q^*(s_{t+1},\ a') \mid s_t = s,\ a_t = a\right\} \end{aligned} \tag{10}$$

$$\tau^* = \arg\max_a Q^*(s,\ a) \tag{11}$$

3.2 State and Action Space

Before applying Q-learning to the continuous missile evasion problem, it is necessary to define the state and action to be learned and divide the state and the action space into a finite number of space. The interception state and the aircraft lateral acceleration are assigned to state $\mathbf{X}_I = s \in S$ and action $a_e = a \in A$ respectively. Action space is divided into two segments $A = \{-6g,\ 6g\}$, which represents the maximum right and left turn. And there are 10^8 states in state space, which are divided into 100 segments on each interception variable. The state space is bounded by

$$\begin{aligned} d &\in [0, 2000]\,(m), \quad \dot{d} \in [-700, 100]\,(m/s) \\ \lambda &\in [0, 2\pi]\,(rad), \quad \dot{\lambda} \in [-1, \; 1]\,(rad/s) \end{aligned} \tag{12}$$

3.3 Reward Function

To accomplish the escape from the missile capture, we need to define a reward function properly. We design the simplest reward function as follows

$$r = \begin{cases} -1 & d < 20m, \; \dot{d} < 0m/s \\ 0 & otherwise \end{cases} \tag{13}$$

where $d < 20m, \; \dot{d} < 0m/s$ is supposed as the condition of failure of evasion. Based on this reward function, the action value function is trained and the action value converges to the expectation of failure.

4 Simulation

4.1 Learning Phase

The aircraft takes actions based on the current action value function and updates the function simultaneously throughout the learning phase. Note that the reward signal is given at terminal states and propagates to the action value function at initial states. Learning is finished if the criterion as (14) is satisfied at all states.

$$\left| r + \gamma \max Q(s_{t+1}, a') - Q(s,a) \right| \le \varepsilon \tag{14}$$

Table 1. Initial conditions

Scenario	Vehicle	Initial Value			
		$y(m)$	$x(m)$	ψ(deg)	$V(m/s)$
scenario #1	aircraft	0	0	90	200
	missile	-992	1125	149	500
scenario #2	aircraft	0	0	90	200
	missile	-992	1125	159	500
scenario #3	aircraft	0	0	90	200
	missile	-992	1125	139	500

where $\varepsilon = 10^{-6}$ is small positive value. During learning phase and the simulation, the system dynamics are integrated using Euler integration. Hereafter, three scenarios are simulated to evaluate the learning policies. The initial conditions of all scenarios are presented in Table 1.

4.2 Scenario #1

In scenario #1, the initial heading angle of the missile is aligned with LOS. The simulation results are shown in figure 2 and figure 3. From the results, we can see that the aircraft achieved the evasion. The miss distance at final time is $r(t_f) = 70.618m$. We can also figure out that the evasive maneuvers are performed around final time.

4.3 Scenario #2

In scenario #2, the initial heading angle of the missile is deviated $+10\deg$ from the LOS line. The results are shown in figure 4 and figure 5. The evasion against the missile is accomplished and the miss distance at final time is $r(t_f) = 373.161m$. The history of interception states is apparently different from the previous scenarios. The trajectory of the aircraft is similar with the one in scenario #1 and the missile lateral acceleration is saturated by the limit.

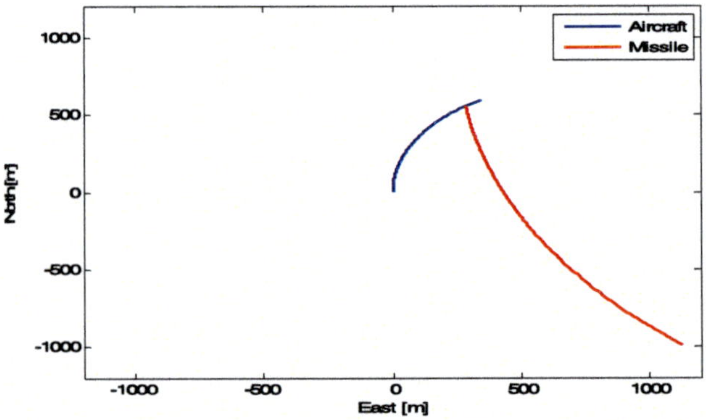

Fig. 2. Planar trajectory

4.4 Scenario #3

In scenario #3, the initial heading angle of the missile is deviated $-10\deg$ from the LOS line. The results are shown in figure 6 and figure 7. The aircraft could not find any evasive strategies in this scenario. This means that the condition is the subpart of the capture set of the missile.

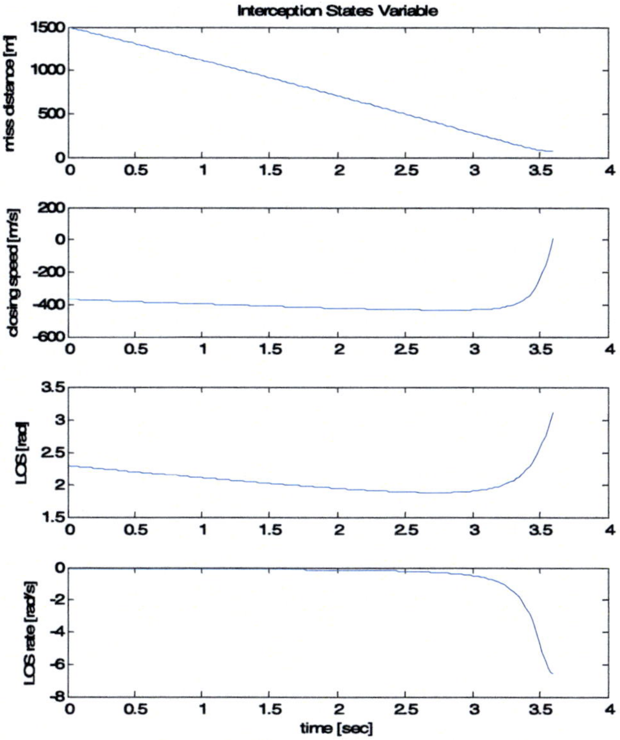

Fig. 3. History of interception states

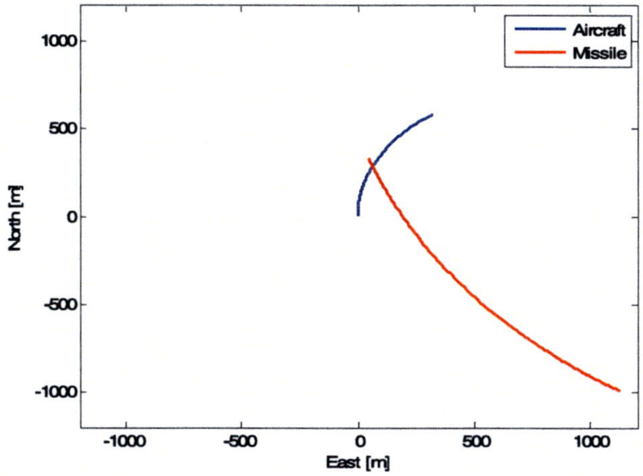

Fig. 4. Planar trajectory

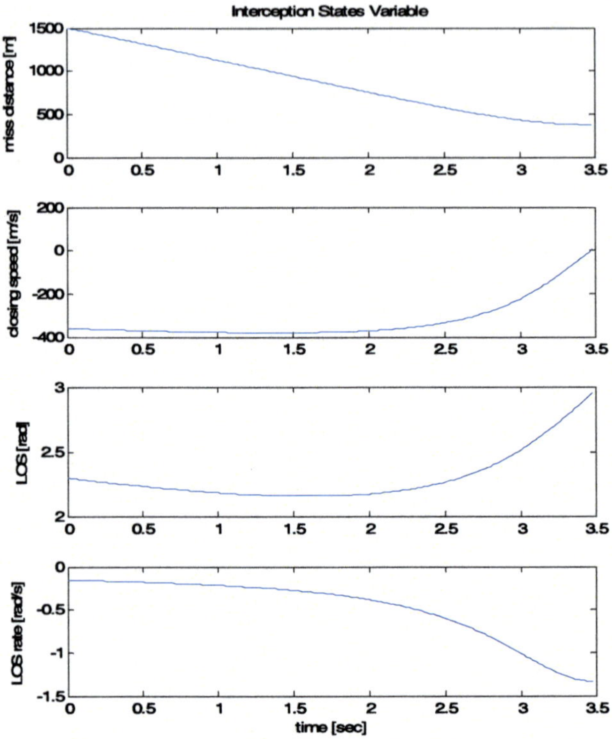

Fig. 5. History of interception states

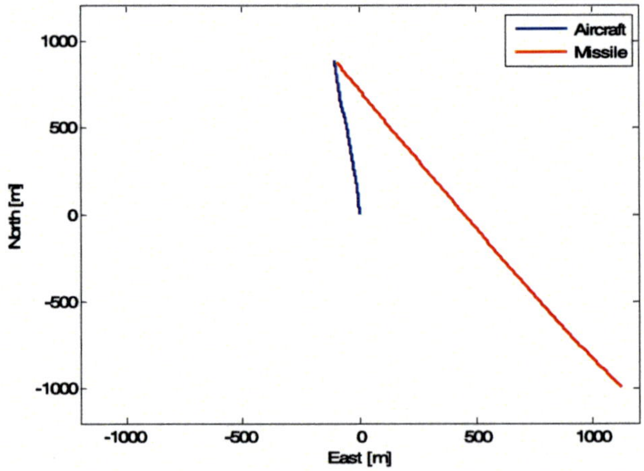

Fig. 6. Planar trajectory

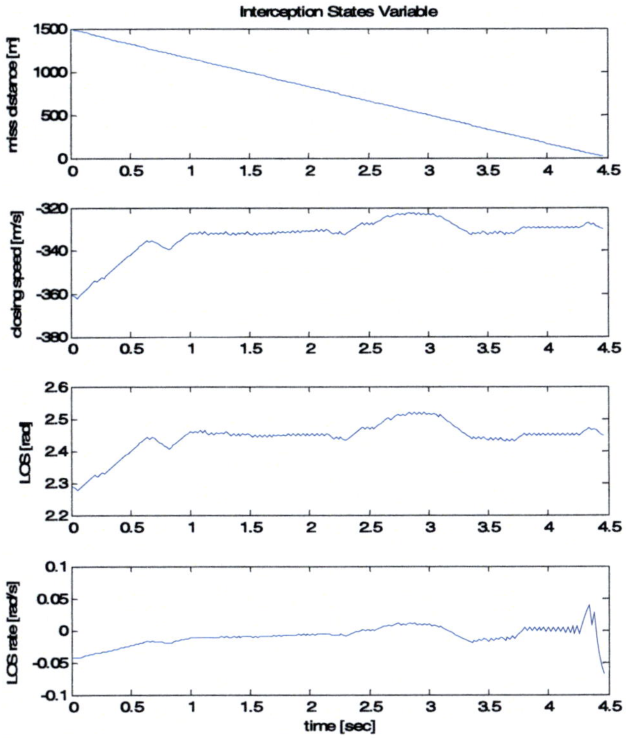

Fig. 7. History of interception states

5 Conclusion

In this paper, we considered missile evading problems and suggested Q-learning as a reinforcement learning algorithm in order to find intelligent evasive maneuvers. After learning is converged, we constructed numerical simulations to evaluate the proposed approach. The simulation results show that the evasive strategies are achieved and can be applied to all learned cases without optimization computing.

References

1. Ben-Asher, J., Cliff, E.M., Kelley, H.J.: Optimal evasion against a proportional guided pursure. Journal of Guidance, Control, and Dynamics 12(4), 598–600 (1989)
2. Imado, F., Miwa, S.: Fighter evasive maneuvers against proportional navigation missile. Journal of Aircraft 23(11), 825–830 (1986)
3. Ong, S.Y., Pierson, B.L.: Optimal planar evasive aircraft maneuvers against proportional navigation missiles. Journal of Guidance, Control, and Dynamics 19(6), 1210–1215 (1996)
4. Watkins, C.: Learning from delayed rewards, PhD. Thesis, Cambridge, England (1989)

5. Manju, M.S.: An analysis of Q-learning algorithms with strategies of reward function. International Journal on Computer Science and Engineering 3(2), 814–820 (2011)
6. Yang, C.D., Yang, C.C.: Optimal pure proportional navigation for maneuvering targets. IEEE Transactions on Aerospace and Electronic Systems 33, 949–957 (1997)
7. Raivio, T.: Capture set computation of an optimal guided missile. Journal of Guidance, Control, and Dynamics 24(6), 1167–1175 (2001)
8. Sutton, R.S., Barto, A.G.: Reinforcement Learning. The MIT Press (1998)
9. Lee, D., Bang, H., Baek, K.: Autorotation of an Unmanned Helicopter by a Reinforcement Learning Algorithm. In: AIAA Guidance, Navigation and Control Conference and Exhibit, Honolulu, Hawaii (2008)
10. Jung, B., Kim, K.S., Kim, Y.: Guidance law for evasive aircraft maneuvers using artificial intelligence. In: Guidance, Navigation, and Control Conference and Exhibit, Austin, Texas (2003)

Design and Implementation of Sensor Modules Enabling Round-the-Clock Underwater Operations

Joongki Park, Juchan Sohn, and Sunghoon Kim

Intelligent Robot Control Research Team,
Department of Robot/Cognitive Convergence System, ETRI
138 Gajeongno, Yuseong-gu, Daejeon, 305-700, Korea
{jkp,jcsohn,saint}@etri.re.kr

Abstract. The legacy technologies of water quality measurement show the weakness that it manually or semi-automatically measures various water quality information at just only small parts of a large area. In this paper, we propose new water quality sensor apparatus which can consistently provide high resolution water quality data as well as attachable and PnP type deployable to underwater vehicles. The water quality sensor module, as a result of going through two water system tests in Han River and Gabchun and benchmarking with the commercial water quality meter, shows not only outstanding performance but also receives an official experimental achievement by the supervisor of the Korea Testing Laboratory.

Keywords: component, sensor system, robotic fish, biomimetic.

1 Introduction

The legacy water quality monitoring systems just measures manually in extremely small parts of a large location and measures through a semi-automated method [1][2][6]. For example, currently in 52 measurement stations in the 4 major rivers, only part items are measured amongst 5 common items (temperature, electric conductivity, pH, dissolved oxygen, total organic carbons) and 11 selective items (biology surveillance, flammable organic compounds, total nitrogen levels, total phosphorus levels, degree of turbidity etc) and once a month water samples are manually measured through 1500 water quality measurement stations in national river/lakes/city/industrial areas. There are difficulties in maintenance and management as the sensors used for the measurement are high priced foreign-made [9].

Of foreign cases, in the case of New York's Hudson river in the US, a REON ((River & Estuary Observatory Network) project was jointly conducted from 2008 to implement a USN water pollution monitoring demonstration service by the Beacon laboratory and IBM, and apart from the wireless communication system, solar battery based self-charging mobile sensor nodes and the embedded compound sensors ((water temperature, EC, pH, DO, TOC)) are currently being developed [3][4][7].

S. Lee et al. (Eds.): Intelligent Autonomous Systems 12, AISC 193, pp. 543–550.
springerlink.com

Fig. 1. Domestic examples of water quality monitoring systems

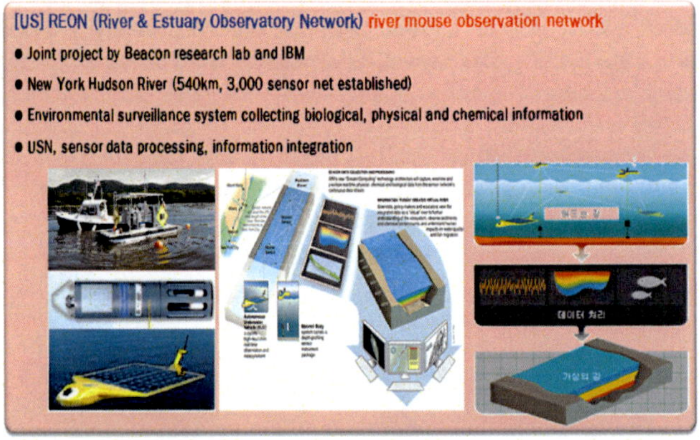

Fig. 2. New York's Hudson River's monitoring system

In this paper, we recognize the problems of legacy technologies as mentioned above, use a biomimetic robot which can freely navigate underwater on not just parts of the river but the entire river area, and design a small size and easy to use water quality sensor module which is requested for continuous and high resolution water quality surveillance.

For example, our sensor modules are installed into a robotic fish as underwater vehicles to monitor water quality and pollution surveillance [5][8][10].

2 Design of Sensor Modules

Our apparatus of water quality measurement is a water quality sensor module in which continuous operation underwater is possible and is composed of the sensor module's fixed shape, infrared light communication module, magnet, sensor module, sensor and sensor protection case.

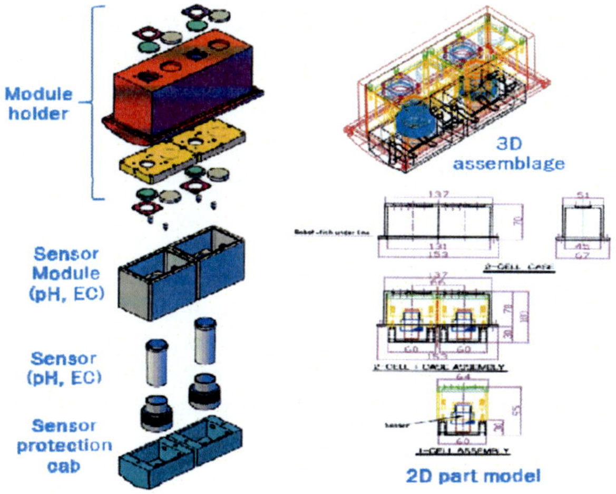

Fig. 3. Assemblage and 2D/3D CAD Model

The infrared light communication module is equipment which communicates sensing information and water quality sensor module and is designed with a wireless infrared communication module, not through a communication cable method connected with a cable. This is done to strengthen the operation ability under water and especially overcome invasions such as electric static of cables, thereby preventing distortion of electric signal treatment of water quality sensors.

The sensor outline is a fixed type outline in order to place sensors, sensor protection case, sensor modules and is installed in the underwater moving object which is loaded with the water quality sensor module equipment The underwater moving object refers to all moving objects that can independently swim underwater like biomimetic robot fish.

The magnet makes adhesion between the sensor module and module holder possible. A necessary and important requirement when constantly operating underwater is the simple maintenance capability. Taking these requirements into consideration, the magnet is used to organically connect the sensor module to the module holder. Screws instead of magnet can be installed because of prevention of interference to position sensor like gyro sensor or digital magnet compass.

The sensor module includes the sensor ' s control board which observes the water quality sensor types rooted in the sensor module and performs the program and circuit drive which fits in each sensor. The sensor ' s analogue signal is extracted and the operation and operated data is transmitted through the sensor fixed outline and infrared light method. Furthermore, it forms the sensor ' s overall information status (temperature, remaining battery life, and sensor is workable or not).

The sensor protection case is to protect the sensor and is composed of a cover form in order to continuously operate underwater. For example in the case of pH sensor, when operating by holding the standard solution in a glass tube, the glass tube can be damaged due to the obstacles underwater or water pressure. In this regards, the sensor protection case mounts by taking these points into consideration and is designed to eliminate if necessary.

3 Implementation of Sensor Modules

At the issue of implementation point of view, we have to set up principles in regarding with module size and power source. The board size is minimized by package on package and system in package strategies respectively. The most important consideration in design of power source is to improve the easy to use in real environment. Execution model of our sensor module is designed by PnP style. As the sensor module should be attached and run by oneself to any kind of unmanned underwater vehicles, battery source is embedded into the sensor module. Currently, Li-Ion battery with 3.7v/970mAh is embedded.

Sensor modules are consisted with 3 on-board modules, which are signal detection and amplification board, signal processing board and communication relay board. The role of SDAB is to detect and amplify the analog signal come from sensor such as pH and EC. SPB as a main sensor board processes signals from SDAB and make the measured value of each sensor. SPB supports dual-mode signal processing, that is, if pH or EC sensors connect to the SPB, then SPB run as pH or EC modes. The measured values are transferred to CRB via IR communication protocol. CRB is basically installed robotic fish and fulfills role of communication repeater between sensor modules and main controller of robotic fish. Key features and specifications are as follows.

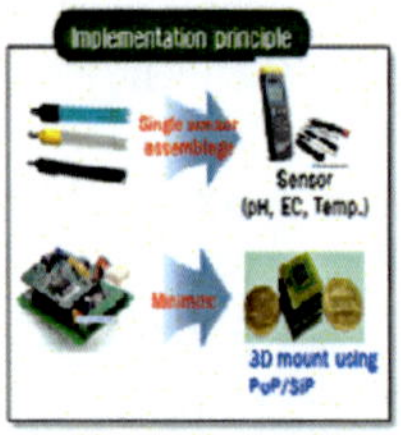

사항	개발 목표치
구조	단일센서 조합
측정항목 (개)	3
신호처리부 크기 (cm³)	8 x 4 x 2
수온 범위/정밀도 (℃/℃)	-5~80 / ±0.2
EC 범위/정밀도 (ms/cm/%)	0~2000 / ±0.5
pH 범위/정밀도 (pH/pH)	0~14 / ±0.1

Fig. 4. Major specification and features of sensor module

4 System Integration

As a moving object which can make the water quality sensor module equipment to freely move underwater, the robotic fish is shown as a real life example. The sensor module equipment is installed in the abdomen of the robotic fish's body as the following.

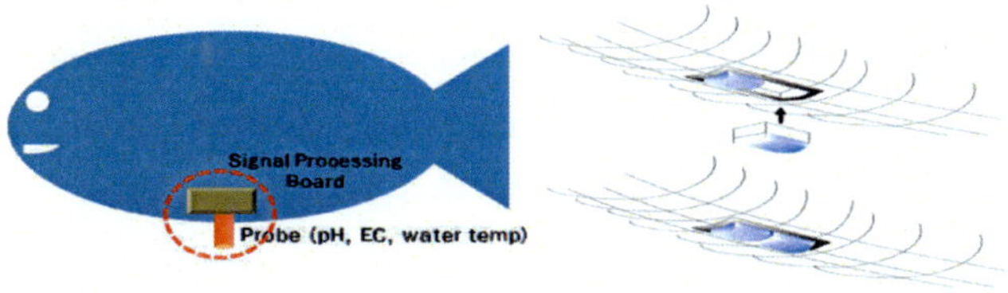

Fig. 5. Conceptual view of installing sensor modules

The reason of installing in the abdomen is because the pH sensor and the dissolved oxygen sensor have transition standard solution, and in order to measure correctly, the standard solution should always be installed below the sensor [11][12]. Furthermore, if the light sensor such as the turbidity sensor or suspended solid sensor observes the water surface upside down or through its side, it is affected by the natural light (sunlight) and causes a defect in the measurement.

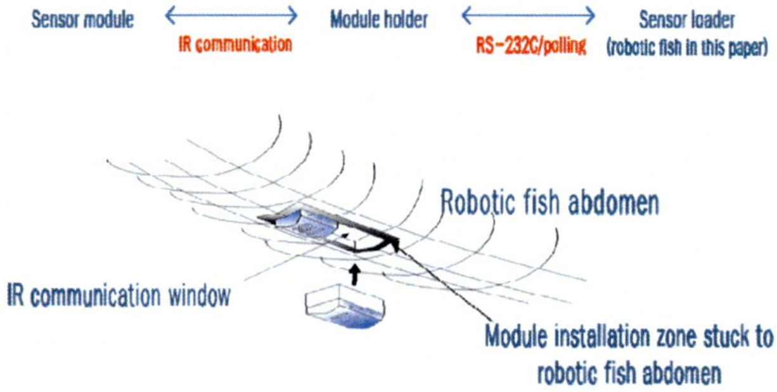

Fig. 6. Integration view of sensor modules

Module installation zone contains at most two different sensor modules respectively among pH, EC, DO, turbidity with marine thermometer. The module installment zone is fixed, and the fixed magnet is adhered to the sensor module and module installation zone. Communication between robotic fish and water quality sensor module is carried out by RS-232 Polling method. The communication protocol is designed as following.

4.1 Communication Protocol

Classification	Specifications
Communication method	RS-232C asynchronous method
Transmission specifications	9600 bps, 8bit, Parity None, Stop bit 1

movement method	Polling method (robot fish-host, measurement tool-slave)
Data transmission method	Half Duplex, block transmission
Mark code	ASCII

4.2 Command Code

Command code (host to sensor module)	**Specifications**	**Sensor module response (sensor module to host)**
0x02DATA0x03XX	request for real time measurement data transmission	data transmission (FORMAT-1)
0x02CCHK0x03XX	search Correction value (Request for transmission of corrected data)	data transmission (FORMAT-2)

5 Real World Experiments and Results

In order to analyze the performance and on-site operational capability between our sensor module and the commercial multiple water quality meters, benchmark tests are conducted in real environment such as Daejeon's Gapchun and Seoul's Han River. Table 1 shows the results as follows.

Table 1. Results of benchmark test in real field

Daejeon Gapchun	pH	Water Temp.	EC
Development Module	8.6 pH	11.7	320.40 μS/cm
Commercial Meter	8.85pH	11.7	347.40 μS/cm
Error Rate	2.8%	0%	7.7%

Han river Yeoui-naru	pH	Water Temp.	EC
Development Module	7.1 ~ 7.3 pH	16.4	189.0 ~ 190.5 μS/cm
Commercial Meter	7.4 ~ 7.6 pH	17.3	193.2 ~ 195.4 μS/cm
Error Rate	4.05% ~ 3.95%	5.20%	2.17% ~ 2.51%

According to the result of benchmark test in real field, it is convinced that both of pH and electric conductivity sensor module satisfies the targeted performance. Especially in the case of Han-river's test site, the measurement value which was tough on the pH sensor's water pressure and water depth changes in velocity 1.5m/s, water depth 1m – 6m was shown to be within 20% of the relative accuracy requested for the formal approval of the environmental measurement equipment by the Agency for Technology and Standards regarding hydrogen ion concentration which uses electrodes.

In the case of Gapchun's test site, the difference in the measurement value of the development module and the commercial meter lies in the differences in measured water depth. The development module is measured in water depth of at least 1m and the commercial meter is measured in water depth around 70cm due to limited cable length.

In the case of the Han River test, the development module is measured by changing the water depth from 0.3m to 6m and the commercial product is measured in water depth of around 70cm.

In order to simulate the robotic fish, the Gapchun test preceded the experiment by loading the sensor modules in the RF based remote control ship and the Han River test loaded the sensor module in a motor-embedded duck boat.

Our water quality sensor module received an authorized test report from the Korea Testing Laboratory and was recognized for the credibility on the measurement result value. The official test certification numbers are 11-1596-114-1 and 2 respectively.

6 Conclusion

Based on this research, compared to the legacy water pollution monitoring system which measured manual work of various river information in extremely small parts of a large location or measures through a semi-automated method, this system can be used as the main sensor equipment which can conduct water quality surveillance by attaching the equipment on moving objects which can freely move continuously underwater to conduct water quality surveillance such as the robotic fish. Accordingly, by persistently managing the river's water quality information as well as getting high resolution data of water quality, we can look forward to making prior actions before the river's environmental changes spread widely.

Future works are to develop a compound sensor module which includes not only temperature, conductivity, and pH, but also DO and Turbidity. Regarding the DO and turbidity, currently we have finished the 3D design and are in the process of implementing prototypes.

Acknowledgements. This research has been supported by the Korea Research Council for Industrial Science and Technology (Project No. B551179-10-02-00). We thank the Daeyoon Scale Cooperation for support, especially real world experiment in Han River in Seoul and Gapchun in Daejeon.

References

1. Sub, A.G.: Revised and resolved problems with the pH roof. Understanding Sensors and Main Technology 22(3), 281–290 (2010)
2. Shin, D., Na, S.Y., Kim, J.Y., Baek, S.-J.: Fish robots for water pollution monitoring using ubiquitous sensor networks with sonar localization. In: ICCIT, pp. 1298–1303 (2007)
3. Yoon, H.S.: Measurement of newly advancing water quality multichannels. C&I, 15–19 (2008)
4. Kolar, H.R., Cronin, J., Hartswick, P., Sanderson, A.C., Bonner, J.S., Hotaling, L., Ambrosio, R.F., Liu, Z., Pawwow, M.L., Reath, M.L.: Complex real-time environmental monitoring of the Hudson River and estuary system. IBM Journal of Research and Development 53(3), 4:1-4:10 (2009)
5. Wen, L., Wang, T., Wu, G., Li, J.: A novel method based on a force-feedback technique for the hydrodynamic investigation of kinematic effects on robotic fish. In: ICRA, pp. 203–208 (2011)
6. Rickerby, D.G., Skouloudis, A.N.: Biosensor networks for monitoring water pollution. In: GHTC, pp. 276–282 (2011)
7. Zhang, S., Zhang, L.: Water pollution monitoring system based on Zigbee wireless sensor network. In: ICECC, pp. 1775–1779 (2011)
8. Waff, W.B., Wolfgram, P.A.: Field testing of a water quality sensor package designed for long-term operation from buoys and other unattended marine platforms. In: OCEAN, pp. 380–384 (1975)
9. Water Quality Measurement Management Team, National Water Quality Automatic Measurement Operation Status, Environment Management Corporation, pp.1-3 (2008)
10. Ye, X., Su, Y., Guo, S.: A Centimeter-scale autonomous robotic fish actuated by IPMC actuator. Robotics and Biomimetics, 262–267 (2007)
11. Zaborowski, M., Jaroszewicz, B., Tomaszewski, D., Prokaryn, P., Malinowska, E., Grygolowicz-Pawlak, E., Grabied, P.: Fabrication of MOS-Compatible Ion-sensitive devices for water pollution monitoring (warmer). In: MIXDES, pp. 477–481 (2007)
12. Sun, Z., Zhao, Y., Li, S.: Research on polarized remote sensing of monitoring of water pollution. In: ICBBE, pp. 1–5 (2010)

A Standard Error Detection Mechanism for Underwater Acoustic Sensor Networks

Imtiaz Ahmed Khan[1], Nam-Yeol Yun[1], Sardorbek Muminov[1], Soo-Hyun Park[1,*], and Chang-Hwa Kim[2]

[1] Ubiquitous System Lab., Graduate School of BIT, Kookmin University, Seoul, Korea
{fish786,anuice,smuminov,shpark21}@kookmin.ac.kr
[2] Underwater Sensor Network System Research Center, Gangneung-Wonju National University, Gangneung, Korea
kch@gwnu.ac.kr

Abstract. Underwater Acoustic Sensor Networks (UWASNs) have some constant factors like energy limitations and node mobility. Energy is a very crucial issue as any kind of recharging is not possible in underwater environment. Energy consumption in each section is important. Saving a small amount of energy helps the life span of the sensor nodes as well as the lifetime of the networks. Especially for static UWASNs, long-term non-time critical application energy efficiency is of major concern. Beside energy efficient protocol design, error control is important in communication as the issue of data integrity is becoming increasingly important. Allowable error rate and undetected error probability in a network is also calculated for noisy channel. Error control is a system design technique that can fundamentally change the trade-offs in a communication system design. An error detection mechanism with low power consumption is also important for saving energy. In this paper, we have tried to find out how we can save our precious energy during communication among sensor nodes and bio mimetic fish-robots. Especially in security and error checking operation in underwater environment based on different error detection and correction mechanisms.

Keywords: UWASNs, Error Check, Energy Efficiency, Underwater Communications, CRC (Cyclic Redundancy Check).

1 Introduction

Ocean is called the world of wealth and our natural resources are limited. Nowadays the oceanic monitoring, UWASNs and its related technology are becoming more and more vital due to the environmental, economic, political and military issues [1]. Every space on the earth is now in range of communication by the help of the satellite communication technology except the underwater part. 70% of the earth consists of water. So, a large amount of area is still out of range. Research works are still going around the world to develop a standard communication system for UWASNs. Communication technology is developing day by day. Moreover, it is also considered

* Corresponding author.

S. Lee et al. (Eds.): Intelligent Autonomous Systems 12, AISC 193, pp. 551–558.
springerlink.com

the futuristic research related to the basic communication technology for ocean sensor networks.

In underwater sensor networks, acoustic wave is the only signal that can travel rapidly in underwater environment than other signals. Radio waves can propagate at long distances through conductive sea water only at extra low frequencies, which require large antennae and high transmission power. Optical, Blue-Green ray is only suitable for the short distance and surface communication. Moreover, transmission of optical signals requires high precision in pointing the narrow laser beams. For signaling, different physical waves (like sound, radio & optical) due to the extraordinary characteristics of sound inside underwater such as propagation speed, multipath, fading and absorption problems. The acoustic signal is only suitable for underwater communication [2].

UWASNs are not secure due to various characteristics of underwater environment. UWASN is a phenomenon where a lot of factors impact. It is totally different than RF based wireless sensor network. Long propagation delay and high bit error rate are the main problems for secure communication among nodes. Underwater environment also has its own peculiar-characteristics which are different based on time, depth and location [3]. In underwater communication, so deployment of sensor network requires pre-localization and environmental modeling information. The acoustic propagation pattern changes due to changing temperature conditions and show the impact of such variability on the network performance of different type of protocols [4].

UWASNs are novel networks which are also vulnerable to various problems due to the harsh environment and outside enemy attacks. The best way to monitor our system is to use a distributed underwater wireless sensor communication systems which consists of tens or hundreds of (battery) operated sensor nodes [5]. Sensor nodes depend on the losses of energy and become out of network due to transmitting weak bacon message. For making protocol design our main target is to design low energy consumption and the high data throughput mechanism.

Our main study is different error detection techniques and develops the sensor nodes lifetime by consuming minimum energy. The mechanism also should achieve high transmission efficiency. Layer wise error detection and correction is a common routine work in every protocol. We want to save a small amount of energy from error detection and security mechanism from MAC (Media Access Control). The strong error checked suite is in physical layer FEC (Forward Error Correction) mechanism, in MAC layer CRC and in network layer checksum mechanism [6].

Nowadays, UWASNs are considering on the one hop and multi hop routing protocols for large range of sensor networks. Moreover, robot-fishes AUV/UUVs are integrating with the underwater sensor networks. These types of networks and operation have unique communication requirements [7]. Collection and gathering data from sensors that distribute to the base station for rapid operation in real time. Also required an advanced security integrated protocol design with low power consumption. Sometimes AUV/UUVs are required to deploy in dense ocean sometimes in the enemy territory for collection of information.

In Static UWASNs for long-term non time-critical application(S-LT-UWASNs) energy is a vital issue [7-8]. To avoid retransmission as it consumes more energy very high accuracy for error detection is not required. Synchronization two robot-fish or UAV/UUVs operation is a great challenge and required minimum error detection mechanism. For security and secure communication it requires checking the packets and insures that there are no malicious activities inside the networks.

In this paper, first we will describe the constant problems for UWASNs, the second chapter will describe the performance of different error detection mechanisms, the third chapter will describe about the CRC degree polynomial selection, the fourth part will discuss the performance analysis and finally, we find out a universal size CRC for Static long term non time critical application(S-LT-UWASNs) and sparse mobile AUV/UUVs robot fish for short-time time critical application in UWASNs.

2 Related Works

UWASN is a peculiar network due activities. Factors like different kind of noise, multipath, signal attention and the properties of the medium is directly put a negative impact in communication. We have shipping noise, rain noises and noise of turbulences.

Our sensor nodes also have some constant problems. One of the major problems is the energy. The sensors work on limited energy. We can't recharge them. Sensors lifetime is their power. When the energy of the cell finishes, the sensor becomes dead and out of network. Another feature is node mobility due to water condition. The mobility rate is (3-6) Km/h. Due to node mobility topology of the network is changed which can also make the protocol design more complex. The medium can also be changed due to some natural clarity. So monitoring underwater environment is essential.

A real time protocol design helps to improve throughput, low energy consumption and high network efficiency in underwater communication with robot-fish. So each part of the network, the physical construction of sensor nodes and the protocols of communication are analyzed with every sector to build a robust underwater sensor network. The total architecture is 2D vs. 3D. Signal comes from different speed in physical layer as bit by bit received by the transducer. After decoding the datagram is sent to the MAC layer, the main error checking is done in this layer. Moreover, if the packet is good enough after error checking, it sends to the network layer. So our main concern is how efficiently we can check the error depends on different error detection mechanism.

The three main error detection mechanisms are parity bit, checksum and CRC. Depends on the underwater factors which mechanism will better for our error detection works and our previous study "A reliable error detection mechanism in underwater sensor network" one byte CRC performs better among Checksum, Parity bit [9]. We had analysis last time about 3 different types of error detection mechanisms depend on some underwater factors like; the high data throughput, network overhead, energy efficiency, error detection rate, burst signal error detection performance, long distance noisy channel performance depends on Link quality Indicator (LQI). From our figure 1, It was found that most of the cases the 1byte CRC-8 mechanism's performance is better than parity bit, check sum and FEC (forward Error Correction) mechanisms. But FEC mechanism is good for long distance and noisy channel communication and suitable for physical layer [6]. Most of the FEC mechanisms use 30~ 40% extra bits that enlarge our Mac pay load size. FEC mechanisms are difficult to apply in our sensor nodes as small memory and computational power is used.

Error detection/ Correction Mechanism	Accuracy Rate	Network Overhead	Transmission Efficiency	Energy Efficiency	Data Throughput	Burst Signal Errors Detection Performance	On Long Distance & noisy channel (LQI)
CRC-8	High	Low	High	High	High	High	Medium
CHECKSUM	Medium	Low	High	High	High	Low	Low
FEC	Medium	High (30~40% extra bit)	Low	Low	Low	Medium	High

Fig. 1. Error detection mechanism

Both Checksum and CRC-8 have same performance due to the network overhead, energy efficiency and data throughput. However the accuracy rate of checksum is lower than CRC-8. The CRC mechanisms have different construction based on the modulo-2 operation and finite algebra theory.

Depending on the FEC and ARQ, these two types of mechanisms are used in lot of RF based networks. But due to the propagation delay, the ARQ mechanism is not used in underwater communications. UWASNs are also different from RF based network depending on packet size and network protocol design. In underwater network packet size is an important issue as it cannot transmit quickly and the receiver also receives it slowly. Node's mobility is a factor and error rate increases when the packet size is increases.

We are searching a way to save our most precious power which is our sensor node's energy. As error checking is a routine work and we have to do this task many times. So, small amount of energy is a big save at total. We try to investigate about mechanism which has a faster error detection capability and the low power consuming rate. Different type of CRCs, CRC-4 to CRC-16 is in our experimental scale. We will try to experiment and analysis depending on some mathematical equation and calculation.

3 Conceptual Idea

RF based networks and underwater networks are totally different. In our underwater environment we have used small size of packets such as 22~50 bytes. Beside this, for error detection we use 1 byte CRC. On the other hand, the RF based network used big size packet up to 2040 byte. For error detection they used 32 to 64 bit CRC. The IEEE and ISO fixed CRC-16, CRC-32 and CRC-128 as standards.

CRC error detection rate will be high near about 99% when CRC generator polynomial size 1 byte less than the codeword size (n-1). We used 1 bit less than the (nR -1) / nR. Where R= CRC degree polynomial. If you use 16 bit polynomial, R-1 / nR, then use the value of generator polynomial n 16- 1 / n 16, Then accuracy rate is 65535/ 65636 = 99.9984 % . But most of the cases it is difficult to maintain for network efficiency. RF based networks also use different ideology for error detection by CRC mechanisms. Compared to RF based network, the data packet size is bigger than the underwater data packet. The ratio of CRC size to payload size (2010:128) = 15:1. Beside this, in underwater network the packer size must be up to 50bits, the ratio of CRC size and the packer size is (50:8) = 6:1, the main trade-offs between the transmission efficiency, poor link quality indicator (LQI) and energy efficiency for underwater environment we can use CRC-4.The undetected error probability and the bit error probability are very low as used in the digital links for CRC (binary redundancy codes) for feedback communication over noisy digital links [10].

Transmission Efficiency: In underwater acoustic sensor network, the high bit error rate and packet size is only 23 bytes. Depends on network efficiency, the ratio of data bits and the ratio of extra information bits can calculate as network efficiency. So, we can use CRC-4 in MAC for transmission efficiency. This is a vital issue when the robot fish, UAV/UUVs are commercially used.

Energy Efficiency: In UWASNs, energy efficiency is another important factor and we have already described it earlier in this document. When we use CRC-4 instead of CRC-8, then we can save 50% of our energy during data transmission. During each transmission we save small faction of energy in each cycle and at the end of the day it will deposit a large amount of energy. This will reduce the payload size of 4 bit. It is difficult and complex to calculate the total power consumption rate. We can calculate the effect that the transducer consumes during data exchange.

Data Throughput: We can save 4 bit which is smaller than 8 bit. These 4 bits also helps us to integrate other information for AUV/UUV and robot-fish. The signal processing, transmitting and receiving time will be decreased. On the other hand, the data exchange time will be completed earlier (saves small fraction of time) but at the end of the day the subtotal of total complete data cycle will be high than before. We can calculate the throughput normally with the equation below.

4 Performance Analysis

For simulation, our packet size is 22 byte, modem speed is 200bps, modem consumes 2watt during signal transmission, during receiving 158 mile watts, sound speed 1500m/s. Simulation area is 100m by 100m.

The performance of different types of error detection mechanism is analyzed. It is found that CRC-4's check bits the lowest payload size among others and the FEC mechanism is the highest payload size. CRC-8 checksum and parity bit all are the same in pay load size.

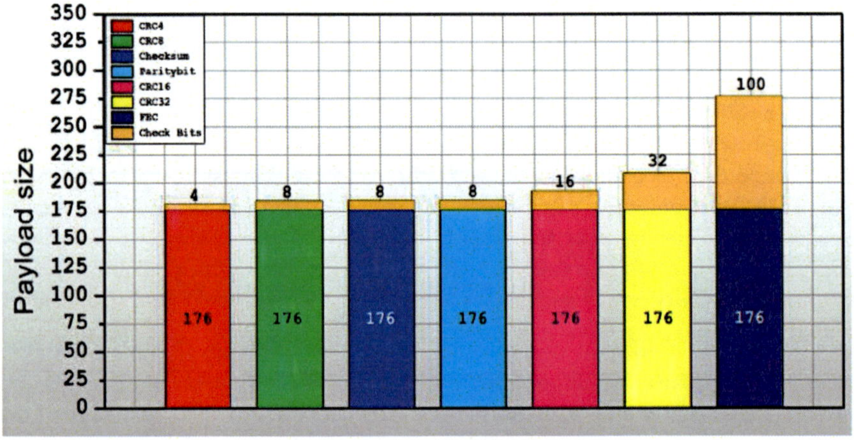

Fig. 2. Payload size with check bits

From Figure3, we can find that duration of data transmits, CRC-4 consume only 9 sec. time for transmitting the packet but CRC-8 consumes 9.2 sec. and FEC mechanism consumes 13.8 sec.

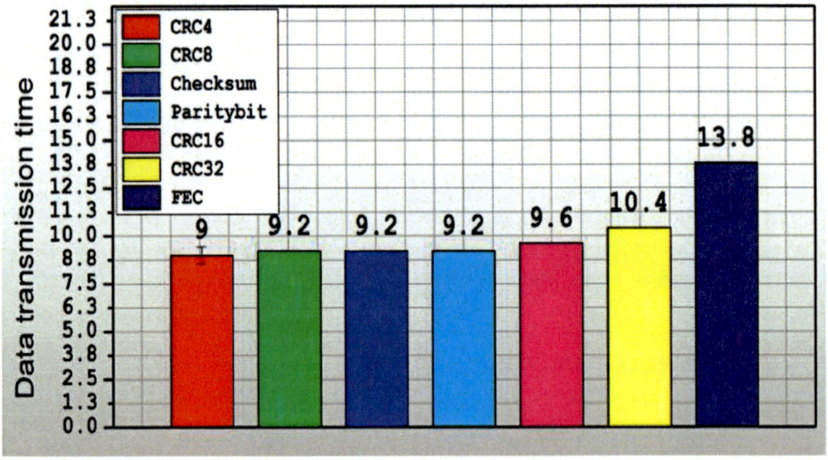

Fig. 3. Data transmitting time

Figure 4 depends on the power consumption of the modem it was found that CRC.

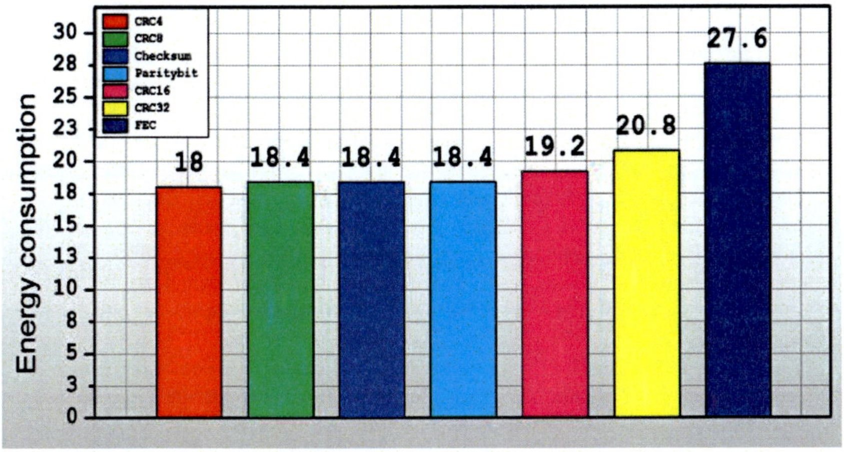

Fig. 4. Energy consumption rates of different error detection mechanism

In figure 4, CRC-4 consumes less energy than CRC-8 and 0.4watts extra power consume by the CRC-8 and FEC consume 27.6watts of energy.

5 Conclusion

It was found that CRC-4 consumes less energy, also required lowest amount of time for data sending, receiving and network efficiency rate is also high. We want to find small gaps in design so that we can save small amount of energy and save it for sensor nodes. RF based networks use large packet size but for underwater sensor network we use small packet size. Depending on transmission efficiency, high data throughput and energy CRC-4 is suitable one for underwater error detection mechanism. Though the accuracy rate is not near the 99.99% but it is better than other error detection mechanism like; checksum or parity bit.

CRC-4 can be used to detect error in RF based networks. CRC-4 may be a better choice for UWASNs as it is well enough for detecting any burst errors and small in size in respect to packet size. It is also suitable for static UWASNs for long term non time-critical application (S-LT-UWASNs). Moreover, small autonomous networks for AUV/UUVs and robot fishes for special purpose operations. CRC-4 is better than CRC-8 and others mechanisms. There is some undetected error probability as underwater environment error rate is very high. Depending on link quality indicator for long distance communication some FEC mechanisms are required for secure communication. In the future we would like to implement it in our robot fishes and observe the real performance in the Han River. It is for the first time in the world, that such experiment is being under taken. It is also our first attempt to use nibble size CRC in communication. We analysis the real performance and develop a standard error detection mechanisms for underwater.

References

1. Akyildiz, I.F., Pompili, D., Melodia, T.: Underwater acoustic sensor networks: research challenges. Ad Hoc Networks (Elsevier) 3(4), 257–279 (2005)
2. Liu, L., Zhou, S., Cui, J.-H.: Prospects and problems of wireless communication for underwater sensor networks. Wireless Communication and Mobile Computing (2008)
3. Etter, P. C.: Underwater acoustic modeling and simulation, 3rd edn.
4. Azad, S., Casari, P., Petrioli, C., Petroccia, R., Zorzi, M.: On the Impact of the Envronment on mAC and Routig in shallow water Scenarios. In: IEEE Santander Conference, OCEANS 2011, Santander (Cantabria), Spain, June 6th- 9th (2011)
5. Cong, Y., Yang, G., Wei, Z., Zhou, W.: Security in underwater Sensor network. In: International Conference on Communication and Mobile Computing (2010)
6. Khan, I.A., Yun, N.-Y., Kim, Y.-P., Park, S.-H.: Layer Wise Error Detection Model for Underwater Sensor Networks. In: Proceedings of the Digital Contents Society Conference, vol. 12(1) (2011)
7. Liu, L., Zhou, S., Chi, J.-H.: Prospects and problems of Wireless Communication for Underwater sensor Networks. Wireless Communications and Mobile Computing Special Issue: Underwater Sensor Networks: Architectures and Protocols 8(8), 977–994 (2008)
8. Cui, J.-H., Wills, J., Syed, A., Li, Y.: Research challenges and applications for underwater sensor network. In: Proceedings of the IEEE Wireless Communication and Networking Conference (2006)
9. Khan, I.A., Yun, N.-Y., Muminov, S., Park, S.-H.: A Reliable Error Detection Mechanism in Underwater Acoustic Sensor Networks. In: Asia Simulation Conference (2011)
10. Koopman, P., Chakravarty, T.: Cyclick redundancy code (CRC) polynomial selection for embedded networks. In: The International Conference on Dependable Systems and Networks, DSN 2004 (2004)
11. Merkey, P., Posner, E.C.: Optimum Cyclic Redundancy Codes for Noiser Channels. TDA progress Report42-76 (October-December 1983)

Analysis on the Robotic Fish Propulsion for Various Caudal Fin Shapes

Dongwon Yun, Jinho Kyung, and Chanhum Park

Dept. of Robotics and Mechatronics, Korea Institute of Machinery & Materials
156 Gajeongbuk-Ro, Yuseong-Gu, Daejeon, 305-343 Korea
dwyun@kimm.re.kr

Abstract. In this paper, propulsion characteristic of various shape caudal fin is analyzed. Using added mass theory, some caudal fins for robotic fish propulsion are analyzed. From this analysis, torque and propulsion characteristic for caudal fins with various shapes can be investigated. This method can be adopted to identify the caudal fin performance more simply and faster than computational method, although with less accuracy.

Keywords: caudal fin, added mass theory, robotic fish.

1 Introduction

Recently, studies on the biomimetic robots have been increased. Biomimetics set its sights on developing better man-made systems by copying the shape and motion of animals and plants for more improved performance, because it is believed that all creatures have evolved for a long time to survive in wild life. For such a reason, studies on mimicking fish have also been increased and conducted widely in the field of biology, robotics, marine engineering and oceanography for underwater mission [1], [2], [3].

In nature, each fish has its own shape and motion underwater. The shape of a caudal fin and swimming mode varies for each fish [4]. This uniqueness is a product of long-time revolution according to its living habits and environments. Propulsion efficiency of about 91% of fish was reported [5]. Accordingly, various researches about robotic fish have been conducted with various shape, motion and functions. Previous studies on robotic fish mainly contain the mechanism, analysis, control, localization and sensor fusion for robotic fish [6], [7]. Each fish can propel itself in different ways. Some use a caudal fin and other use a pectoral fin and so on. Fish motion can be classified according to the body part used for propulsion.

In this paper, a caudal fin of fish was studied. Especially, the propulsion characteristics of various shape caudal fins are analyzed. Actually, there are too many kinds of fins fish uses for propulsion to consider all cases of fins. So, in this paper, three shapes of a caudal fin brought from caudal fins of a carp and a shark are investigated. To do this, added mass theory was used and torque and propulsion characteristic for some caudal fins with various shapes were investigated. This

S. Lee et al. (Eds.): Intelligent Autonomous Systems 12, AISC 193, pp. 559–566.
springerlink.com

method can be adopted to identify the caudal fin performance more simply and faster than computational method, although it shows less accuracy surely.

2 Theory for Fish Propulsion

To analyze the actuation of caudal fin of fish, several methods can be used. Among those, CFD shows the most correct results, however it is not easy to model and simulate the caudal fin by using CFD also it spends much time to calculate the performance. So, lumped approach for analysis of fish swimming has been also studied widely. In this approach, water around the fins is modeled as mass or volume of water, and dynamics of water movement is investigated. In these researches, power and torque was calculated using simplified model [8], [9]. According to this theory, oscillating angle of fin can be expressed as follows

$$\theta = A\sin 2\pi ft \tag{1}$$

where A, f are oscillatory amplitude and frequency respectively. The time differential of formula (1) becomes

$$\dot{\theta} = 2\pi fA\cos 2\pi ft \tag{2}$$

Then the velocity at the mass center of caudal fin can be

$$v_G = r\times\omega = \dot{\theta}r = 2\pi rf\,\cos 2\pi ft \tag{3}$$

From above equations, force F which acts on the fin in the opposite direction of v_G and torque T at the fin hinge can be evaluated from equations,

$$F = C_D\rho V_G{}^2 S_A/2 = C_D\rho(2\pi rf\,\cos 2\pi f)^2 S_A/2 \tag{4}$$

$$T = r\times F = 2C_D\rho S_A(\pi rfA\cos 2\pi ft)^2 r \tag{5}$$

where C_D represents the drag coefficient; ρ is water density; S_A is the largest cross section area of fin.

The mechanical power P for a fish can be calculated as [9]

$$P = mWwU - \frac{mw^2U}{2\cos\theta} \tag{6}$$

where m represents added mass; W is rms of the lateral speed of the trailing edge; w is water velocity at the trailing edge; U is average speed of fish.

Thrust force T_h of a fish is evaluated from formula 6.

$$T_h = \frac{P_t}{U}. \tag{7}$$

In this paper, this added mass approach is used to calculate for the performance of the caudal fin with various shapes.

3 Analysis for Caudal Fin with Various Shapes

Figure 1 shows the view of a caudal fin of a fish. In this figure, y axis lies in the direction of head and y axis lies in the direction of the span.

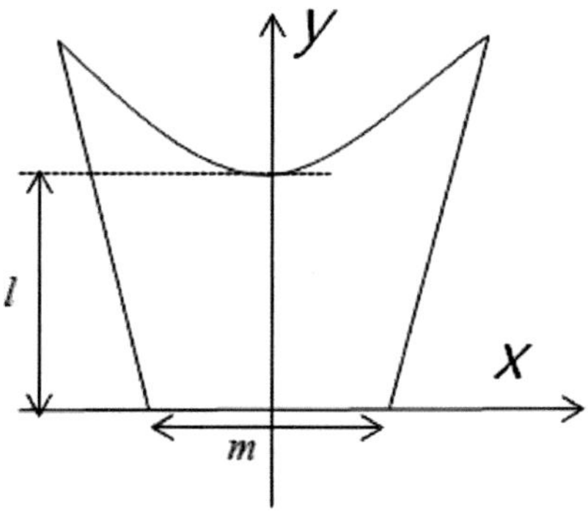

Fig. 1. Caudal fin

In this study, three kind shape of caudal fin are considered. One is symmetric and the others are asymmetric in the shape. To consider this shape differences, the model of figure 1 is assumed as assembly of lines and curves as figure 2. $f_1(x)$ and $f_2(x)$ are straight line and $f_3(x)$ is a quadratic curve. In this figure, straight lines determine the boundary shape of a caudal fin and quadratic curve changes the shape of trailing edge. If the curve f_2 moves in the x-direction, which means the center of curve moves, the leading edge shape can be asymmetric shape like dotted line in figure 2. So, by adjusting the center position of curve, various shape of caudal fins can be obtained.

Let *c* as a x-coordinate of the center of the curve, the quadratic curve can be expressed as

$$y = a(x-c)^2 + b \tag{8}$$

where *a, b* is constants. And straight lines are also expressed

$$y = \pm dx - e\,. \tag{9}$$

In this paper, the length of *m* is assumed as 40mm and *l*, 70mm. And the other parameters are assumed as Table 1. According to *c*, the shape of a caudal fin changes and the shape variations are shown in figure 3. As shown in these figures, we can get the different shapes of fins with different values of *c*, where *c* is chosen to mimic the shape of caudal fins of a carp and a shark.

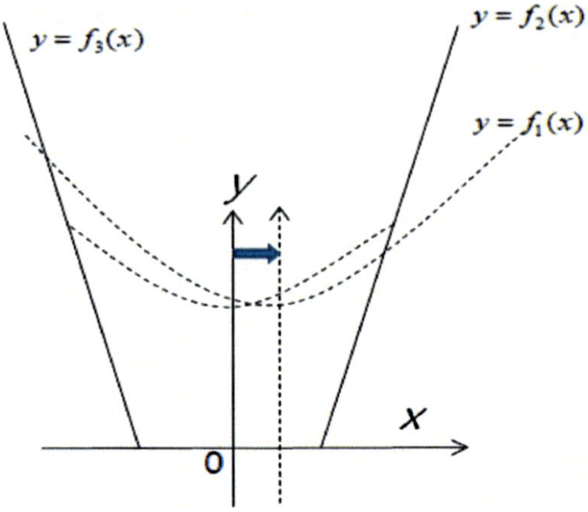

Fig. 2. Modeling of caudal fin

Table 1. Parameters of equations

Parameters	Value
a	12.5
b	0.07
d	4.5
e	0.09

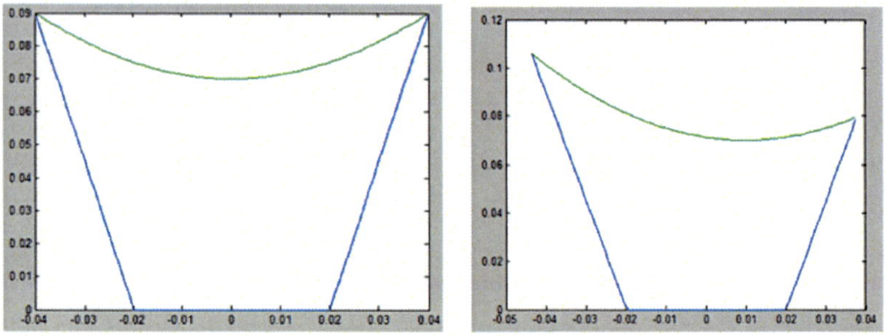

Fig. 3. Cases of various fin shapes, case A:c=0mm(*top left*), case B:c=10mm(*top right*), case C:c=25mm(*lower center*)

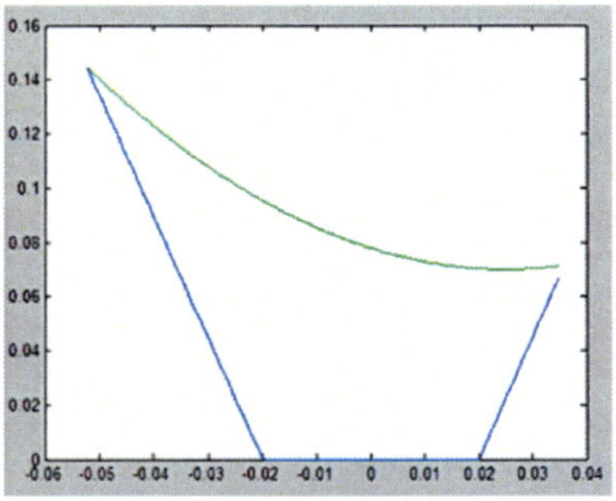

Fig. 3. *(continued)*

To evaluate the torque and thrust using formula (5)~(7), the area of the fin should be calculated. For this, it is assumed that the whole area of the fin is composed of a small area, a_i like figure 4.

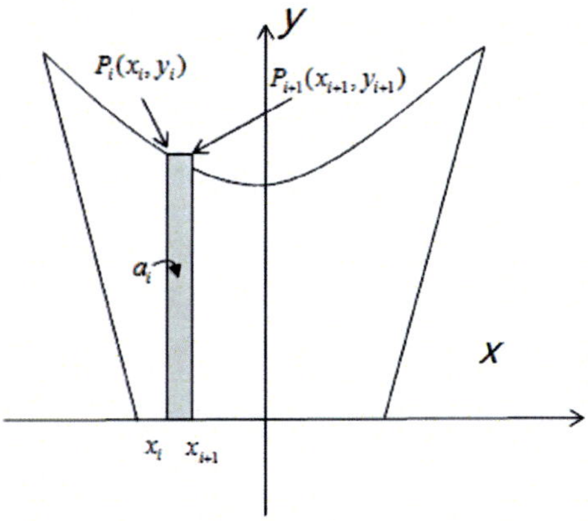

Fig. 4. Calculation for area

From Fig. 4, the area of a caudal fin, A can be

$$A=\sum_{i=1}^{N} a_i=\sum_{i=1}^{N}(x_{i+1}-x_i)f_2(x_i) \quad (-\frac{m}{2}<x_i<\frac{m}{2}),$$
$$\sum_{i=1}^{N}(x_{i+1}-x_i)(f_2(x_i)-f_{1,3}(x_i)) \quad (-\frac{m}{2}>x_i \ or \ x_i>\frac{m}{2}). \tag{10}$$

In this paper, the area of the caudal fin is divided into 88 sub areas. Figure 5 shows the divided area for calculation.

Now, we can get the torque and thrust through formula (1) ~ formula (10) for the fins of figure 3. Calculated area of each fin is shown in Table 2.

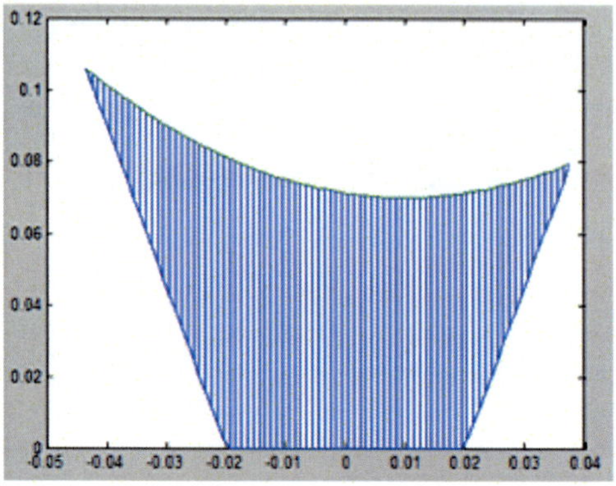

Fig. 5. Example of divided area of case B

Table 2. Calculated area of each fin

Case	Area [m^2]
A	0.0044
B	0.0048
C	0.0044

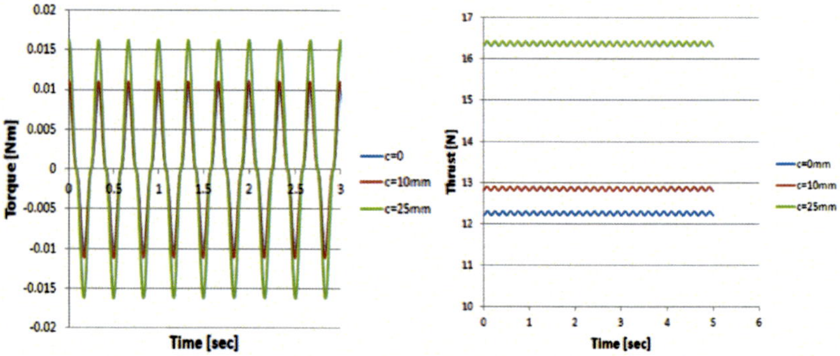

Fig. 6. Analysis results for torque(*left*) and thrust(*right*)

Torque and thrust calculated from formula (1) ~ formula (10) are shown in figure 6, and the averaged thrust and maximum of torque are also summarized in Table 3 and 4. From these results, it can be known that Case C shows the biggest torque and thrust among those. But, Table 2 shows that the area of each fin is not same. So, to compare those results more clearly, normalized torque and thrust are calculated by dividing each value by each area. These normalized values are also shown in formula 7 and the values are summarized in Table 5 and 6.

Table 3. Average of thrust

Case	Torque [N]	Ratio
A	12.26	1
B	12.86	1.05
C	16.36	1.33

Table 4. Peak of torque

Case	Torque [Nm]	Ratio
A	0.010	1
B	0.011	1.082
C	0.016	1.584

Figure 7, Table 5 and 6 show that normalized torque and thrust of case C are the biggest among those and case A shows the smallest torque and thrust. From above analysis, we can calculate for thrust and torque of various shapes by using added mass theory.

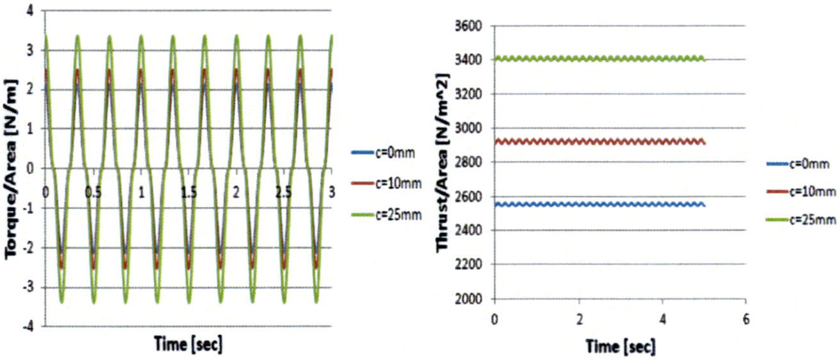

Fig. 7. Normalized analysis results of torque(left) and thrust(right)

Table 5. Normalized average of thrustss

Case	Torque [N/m]	Ratio
A	2554.9	1
B	2922.9	1.14
C	3409.4	1.33

Table 6. Normalized peak of torque

Case	Torque [N/m^2]	Ratio
A	2.129	1
B	2.513	1.18
C	3.374	1.58

4 Conclusion

In this paper, some analyses for a caudal fin of fish with various shapes have been performed. Using added mass theory, torque and thrust of fish propulsion were calculated and compared. From this analysis, we can conclude with certainty that the shape of a caudal fin influences on the fish propulsion performance. Future study includes that the CFD analysis or real water tank experiments for three kind of caudal fin shown in this paper will be performed and compared.

References

1. Lauder, G.V., Maddeny, P.G.A.: Learning from fish: kinematics and experimental hydrodynamics for roboticists. International Journal of Automation and Computing 3, 325–335 (2006)
2. Lauder, G.V., Madden, P.G.A.: Fish locomotion: kinematics and hydrodynamics of flexible foillike fins. Experiments in Fluids 43, 641–643 (2007)
3. Low, K.: Modelling and parametric study of modular undulating fin rays for fish robots. Mechanism and Machine Theory 44, 615–632 (2009)
4. Lindsey, C.: Form, function, and locomotory habits in fish. Fish Physiology 7, 1–100 (1978)
5. Sfakiotakis, M., Lane, D.M., Davies, J.B.C.: Review of Fish Swimming Modes for Aquatic Locomotion. IEEE Journal of Oceanic Engineering 24(2) (April 1999)
6. Yu, J., et al.: Development of a biomimetic robotic fish and its control algorithm. IEEE Transactions on Systems, Man, and Cybernetics, Part B: Cybernetics 34, 1798–1810 (2004)
7. Morgansen, K.A., et al.: Geometric methods for modeling and control of free-swimming finactuated underwater vehicles. IEEE Transactions on Robotics 23, 1184–1199 (2007)
8. Chen, W., Xia, D., Liu, J.: Modular Design and Realization of a Torpedo-Shape Robot Fish. In: Proceedings of 2008 IEEE International Conference on Mechatronics and Automation, pp. 125–130 (2008)
9. Hiara, K., Kawai, S., `http://www.nmri.go.jp/eng/khirata/fish/experiment/upf` (November 26, 2001)
10. Webb and P, Is the high cost of body caudal fin undulatory swimming due to increased friction drag or inertial recoil Journal of experimental biology, vol.162, pp. 157–166 (1992)

Development of a 3-DOF Fish Robot 'ICHTHUS V5'

Gi-Hun Yang, Hyunjin Lee, and Young Sun Ryuh

Biomimetic Robot Research Group, Korea Institute of Industrial Technology
1271-18, Sa-3-dong, Sangrok-gu, Ansan, 426-910, Korea
{yanggh,hjlee22,ysryuh}@kitech.re.kr

Abstract. In recent, there is a rising interest on studying fish-like underwater robots because of real fish's great maneuverability and high energy efficiency. However, researches about the fish-like underwater robots have not been investigated so much and there are still diverse problems in respect of using of the fish robot in the real environment such as in the river for detecting water pollution. For example, the fish robot has a short operating time and cannot move narrow passage such as swimming between aquatic plants. Therefore, this paper mainly describes a development of robotic fish which can be used for water quality monitoring system. The fish robot 'Ichthus V5' has a 3-DOF serial link-mechanism and is developed in KITECH. Furthermore, we propose a dynamic equation of the fish robot to use the underwater environment. We added several sensors to navigate autonomously in the real environment like river. Also, we added two kinds of sensor to detect temperature, electric conductivity, pH (hydrogen ion concentration) of water. Therefore, the developed system can be applied to environmental monitoring system for detect pollution or quality of river.

Keywords: robotic fish, dynamic analysis, propulsion, propulsive.

1 Introduction

In General, underwater mobile robots have been widely used for the purpose of the marine exploration and resource extraction. In order to perform these goals, the propulsion of the high efficiency and the excellent dynamic property of the robots are needed [1]. However, when it comes to the efficiency and a moving ability, there are no positive advantages for the underwater vehicle that has pre-existing types of propeller to move forward. The driving efficiency of the vehicle does not exceed 70 % and this could lead to a short searching distance of the system. Also, the increasing capacity of the battery could cause its ability to carry the apparatuses decreasing. Instead of conventional rotary propellers used in ships or other underwater vehicles, undulatory movement provides the main energy source for robotic fish [14,15]. An observation of real fish shows that this kind of propulsion is less noisy, more effective, and more maneuverable than propeller-based propulsion [13]. Moreover, the radius of the rotational motion is too big and evading speed is too slow. For this reasons, it is not appropriate in moving or exploring in the packed area

S. Lee et al. (Eds.): Intelligent Autonomous Systems 12, AISC 193, pp. 567–576.
springerlink.com

[2]. As a solution for this, there is a study underway and the study is related with the underwater mobile robots which mimic a fish's swimming movement.

First of all, the swimming mechanism of the fish is well known as more than a 20% efficient way rather that the way of propeller proceeding method which is more general one. For instance, MIT research team conducted an experiment about the comparing a performance between the fin which looks like a tuna's tail fin and a propeller in 1994 [9]. And they found that the fins are more efficient method to proceed in the water. The fins of fish could repulse the water more than the square of propeller's blade. And this makes more power to proceed forward. As a result the energy efficiency of the propeller became 70% but the robot tuna's fin recorded 87 % [3]. Several researchers developed various kinds of robot fish. In Japan, Nagoya University developed a micro robotic fish using ICPF actuators [11], while Tokai University produced a robot Blackbass [12] in order to research the propulsion characteristics of pectoral fins. North-western University applied shape memory alloy (SMA) to produce a robot lamprey [10] which aimed to provide mine countermeasures.

Second, fishes usually have more excellent rapid movement rather than other underwater vehicle. So they can turn their way even though there is only 1/10 length of their body. However, submarines which developed by human need almost 10 times of the returning distance, because it needs a time to slow down and turning their body [4].

In our laboratory, we already suggested and experimented about the new mechanism of a side-fin which is designed to enhance the movement of the underwater fish-like robots[14]. In this paper, the development of robotic fish will be introduced from its design stage to the calculation of the Ichthus's dynamic equation which was executed to make more efficient swimming mechanism of the robot 'Ichthus V5'. The developed system has autonomous navigation ability and the water quality monitoring capability. Sensors and related signal processing systems are integrated to the developed system for achieving its abilities. The simulation of the torque calculation and the swimming velocity has been completed. To make those calculations done, the model for the fluid force was derived.

2 Dynamic Equation of Robotic Fish

2.1 Mechanical Link Model

The target model as shown figure 1 of this paper is ‘Ichthus V5’ which has 3 Joints in the body and the tail. Also, it swims and moves its tail to the left and right side as like a real fish. On each joint, the servo motors are connected to the body frame of the robot. Therefore, by using the servo motors, we are able to control frequency and amplitude of the robot. Table 1 and Figure 2 show specifications of “Ichthus Ver. 5” and actual implementation of the robotic fish.

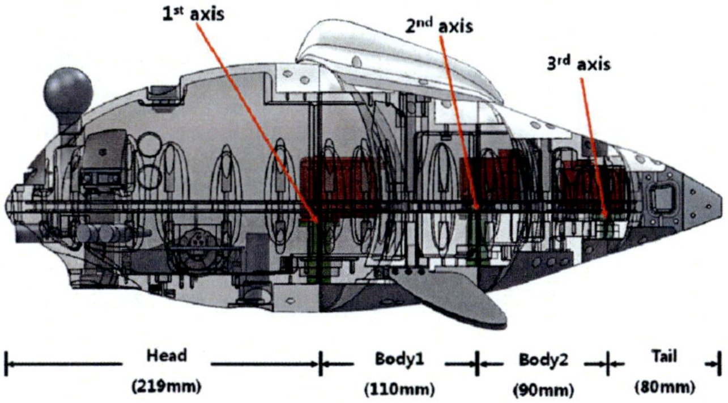

Fig. 1. Mechanism model of 'Ichthus'

Table 1. Specifications of 'Ichthus Ver. 5'

Body size	500mm(L) x 130mm(W) x 170mm(H)
Weight	4.3 kg
DOF	6 (Tail 3, Side Fin 2, mouth and gill 1)
Battery	14.8V Li-Io Battery

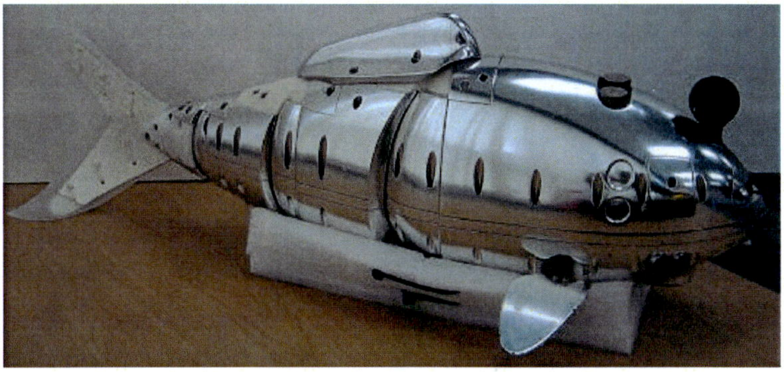

Fig. 2. Actual Implementation of Robotic Fish 'Ichthus V5'

2.2 Sensors

The developed system embedded several kinds of sensors for autonomous navigation and quality detection of water. Sensors to be used for autonomous navigation are IR sensors, ultrasound range sensors, GPS sensor, water pressure sensor, INS(Inertial

Navigation System), USBL. Basically the developed robotic fish uses GPS sensor for localization when the robot is on the surface of water. In case that the robotic fish is located in the underwater environment, GPS data is not accessible. Therefore the robot needs to use water pressure sensor for depth detection and INS for localization. To compensate the error of INS, USBL location data is transmitted through an ultrasound modem. With these features, the robotic fish can move autonomously in the underwater environment as well as the surface of water. Since one of main purpose of the robotic fish is to prevent water pollution, the developed robotic fish has three kinds of sensors measuring temperature, EC(electric conductivity), pH(hydrogen ion concentration) of water. These features can be monitored anytime the operator wants to.

2.3 'Ichthus' Dynamic Model

To make the simulation simple, we take several reasonable assumptions. The Thrust Force F_F and Lateral Force F_C are only applied to the Link 3 that is a tail fin and a yawing motion and a motion along the y-axis direction are fixed as shown in Figure 3 [6].

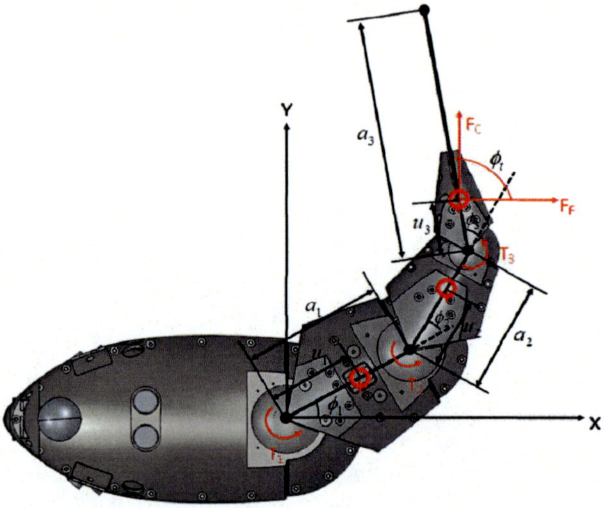

Fig. 3. Analytical Model of 'Ichthus'

With these assumptions, the dynamic equation can be applied to our developed robotic fish, and we can constitute the dynamic model as same as a method of the Serial Link robot which contains a 3-DOF(degrees of freedom) links with applied external forces[14].

Therefore, the torque value which exists on each joint could be obtained by the following equation.

$$T_{\Phi} = \left[I^{*}_{\Phi\Phi}\right]\ddot{\Phi} + \dot{\Phi}^{T}\left[P^{*}_{\Phi\Phi\Phi}\right]\dot{\Phi} - \left[G^{U}_{\Phi}\right]^{T} T_{U}. \qquad (1)$$

In these equation, the $I^{*}_{\Phi\Phi}$ means that the effective inertia, and the $P^{*}_{\Phi\Phi\Phi}$ represents a changing system inertia.

$$\left[I^{*}_{\Phi\Phi}\right] = \begin{bmatrix} I^{*}_{11} & I^{*}_{12} & I^{*}_{13} \\ I^{*}_{21} & I^{*}_{22} & I^{*}_{23} \\ I^{*}_{31} & I^{*}_{32} & I^{*}_{33} \end{bmatrix}$$

$$\begin{aligned}
I^{*}_{11} &= ([\bar{I}_1 + \bar{M}_1 u_1^2] + [\bar{I}_2 + \bar{M}_2 u_2^2] + [\bar{I}_3 + \bar{M}_3 u_3^2] + \bar{M}_2 l_1^2 + \bar{M}_3 [l_1^2 + l_2^2]) \\
&\quad + (2\bar{M}_2 l_1 u_2 \cos\Phi_2 + 2\bar{M}_3 [l_1 l_2 \cos\Phi_2 + l_2 u_3 \cos\Phi_3 + l_1 u_3 \cos(\Phi_2 + \Phi_3)]) \\
I^{*}_{21} &= I^{*}_{12} = ([\bar{I}_2 + \bar{M}_2 u_2^2] + [\bar{I}_3 + \bar{M}_3 u_3^2] + \bar{M}_3 l_2^2) \\
&\quad + (2\bar{M}_2 l_1 u_2 \cos\Phi_2 + \bar{M}_3 [l_1 l_2 \cos\Phi_2 + 2 l_2 u_3 \cos\Phi_3 + l_1 u_3 \cos(\Phi_2 + \Phi_3)]) \\
I^{*}_{31} &= I^{*}_{13} = ([\bar{I}_3 + \bar{M}_3 u_3^2]) + (\bar{M}_3 [l_2 l_3 \cos\Phi_3 + l_1 u_3 \cos(\Phi_2 + \Phi_3)]) \\
I^{*}_{22} &= ([\bar{I}_2 + \bar{M}_2 u_2^2] + [\bar{I}_3 + \bar{M}_3 u_3^2] + \bar{M}_3 u_2^2) + (2\bar{M}_3 l_2 u_3 \cos\Phi_3) \\
I^{*}_{32} &= I^{*}_{23} = ([\bar{I}_3 + \bar{M}_3 u_3^2]) + (\bar{M}_3 l_2 u_3 \cos\Phi_3) \\
I^{*}_{33} &= ([\bar{I}_3 + \bar{M}_3 u_3^2])
\end{aligned}$$

$[P^{*}_{\Phi\Phi\Phi}]$ is a matrix of three-dimensional array and represents as follows.

$$[P^{*}_{n\Phi\Phi}] = \begin{bmatrix} P^{*}_{n11} & P^{*}_{n12} & P^{*}_{n13} \\ P^{*}_{n21} & P^{*}_{n22} & P^{*}_{n23} \\ P^{*}_{n31} & P^{*}_{n32} & P^{*}_{n33} \end{bmatrix} = n^{th} \; plane \; of \; [P^{*}_{\Phi\Phi\Phi}]$$

$$[P^{*}_{1\Phi\Phi}] = \begin{bmatrix} P^{*}_{111} & P^{*}_{112} & P^{*}_{113} \\ P^{*}_{121} & P^{*}_{122} & P^{*}_{123} \\ P^{*}_{131} & P^{*}_{132} & P^{*}_{133} \end{bmatrix} = \begin{bmatrix} 0 & A & B \\ A & A & B \\ B & B & B \end{bmatrix}$$

$$[P^{*}_{2\Phi\Phi}] = \begin{bmatrix} P^{*}_{211} & P^{*}_{212} & P^{*}_{213} \\ P^{*}_{221} & P^{*}_{222} & P^{*}_{223} \\ P^{*}_{231} & P^{*}_{232} & P^{*}_{233} \end{bmatrix} = \begin{bmatrix} -A & 0 & C \\ 0 & 0 & C \\ 0 & C & C \end{bmatrix}$$

$$[P^{*}_{3\Phi\Phi}] = \begin{bmatrix} P^{*}_{311} & P^{*}_{312} & P^{*}_{313} \\ P^{*}_{321} & P^{*}_{322} & P^{*}_{323} \\ P^{*}_{331} & P^{*}_{332} & P^{*}_{333} \end{bmatrix} = \begin{bmatrix} -B & -C & 0 \\ -C & -C & 0 \\ 0 & 0 & 0 \end{bmatrix}$$

$$\begin{aligned}
A &= -\bar{M}_2 l_1 u_2 \sin\Phi_2 - \bar{M}_3 (l_1 l_2 \sin\Phi_2 + l_1 u_3 \sin(\Phi_2 + \Phi_3)) \\
B &= -\bar{M}_3 (l_2 u_3 \sin\Phi_3 + l_1 l_3 \sin(\Phi_2 + \Phi_3)) \\
C &= -\bar{M}_3 l_2 u_3 \sin\Phi_3
\end{aligned}$$

The distance between the center of mass, the inertial moment of each link, and a mass were calculated from the 3D CAD (SolidWorks™) model.

In the equation (1), $[G_{\Phi}^{U}]$ means the Jacobian matrix which represents the correlation between the angular velocity and the velocity of the tail fin. So we can transfer the Thrust Force F_F and Lateral Force F_C which are applied to the fluid into the torque of each joint.

2.4 Model of Fluid Force

In the modeling of fluid force, a thrust force is mainly generated from the tail fin. Therefore we could suppose that the drag force by the fluid could be applied to the body of the robotic fish.

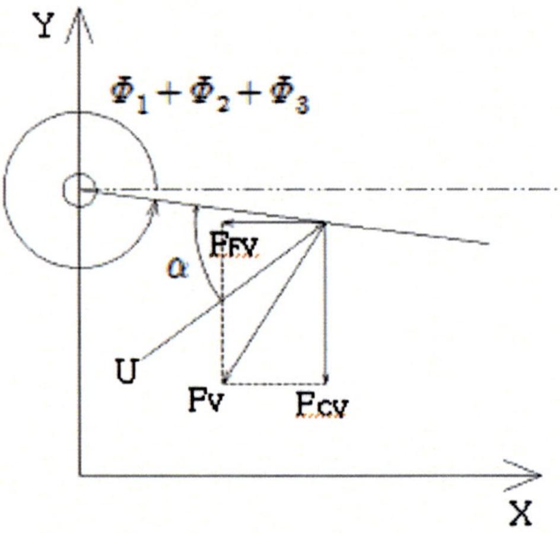

Fig. 4. Model of inertial fluid force

In figure 4, the F_V means an inertia which generated by the movement of a tail fin. The magnitude is proportioned with the angular velocity's value and has opposite direction.

$$F_V = -\pi\rho LC^2(U\sin\alpha + \alpha U\cos\alpha). \tag{2}$$

Where, U means the relative velocity generated from the tail fin, C represents the chord length of tail fin, and the L means span. The density of water (ρ) is 998 kg/m^3. The floating force generated by the U on the occasion of rotation in the tail fin is only applied as the way of vertical direction. Thus, the real force, F_J could be expressed as the equation (3).

$$F_J = 2\pi\rho LCU^2 \sin\alpha\cos\alpha. \tag{3}$$

Therefore, the thrust force element F_F and the lateral force element F_C could be expressed as (4) and (5).

$$F_F = F_{FV} + F_{FJ}. \tag{4}$$

$$F_C = F_{CV} + F_{CJ}. \tag{5}$$

In the mean time, we assumed that the drag force F_D was only applied to the main body.

$$F_D = \frac{1}{2} C_D \rho V^2 S. \tag{6}$$

In the above equation, the C_D represents drag coefficient, the V means comparative trust velocity and the S means an area which are directly projected along with the flow of fluid, the C_D was assumed as 0.5.

Also if we confine the proceeding direction of the robot as the x-axis direction, the relative velocity along the y-axis direction in the middle of tail fin could be derived as follows.

$$\begin{aligned} u = {} & l_1 \cos\Phi_1 \dot{\Phi}_1 + l_2 \cos(\Phi_1 + \Phi_2)(\dot{\Phi}_1 + \dot{\Phi}_2) + \\ & a_3 \cos(\Phi_1 + \Phi_2 + \Phi_3)(\dot{\Phi}_1 + \dot{\Phi}_2 + \dot{\Phi}_3). \end{aligned} \tag{7}$$

Because the U_m which represents the velocity of a moving fluid along x-axis direction and the u are directly crossing, So U could be represented as follows.

$$U^2 = U_m^2 + u^2. \tag{8}$$

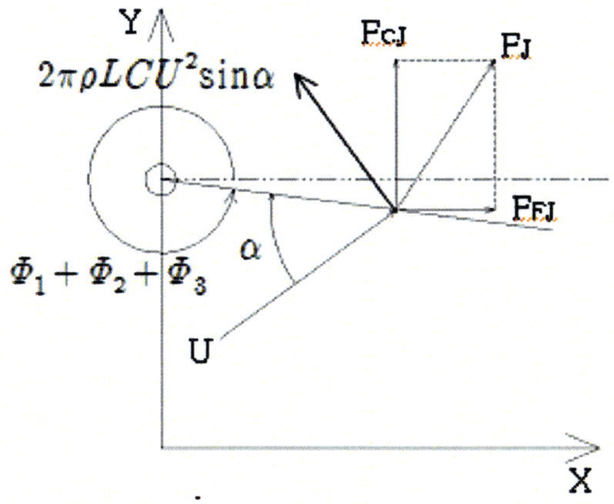

Fig. 5. Model of lift force

2.5 Simulation Design

The simulation was conducted through the Matlab™ and the acceleration of the robot was calculated from the equation (9), and the velocity of robot was derived by the equation of Runge-Kutta.

$$M\ddot{x}_G = F_F - F_D. \tag{9}$$

At this calculation, M means the total mass of the robotic fish, x_G means position of the mass center, F_F means the total thrust force, and F_D means the total drag force. In the simulation, the amplitude and the period are the input value, therefore, the swimming velocity along x-axis direction could be calculated. The figure 6 shows the simulation results.

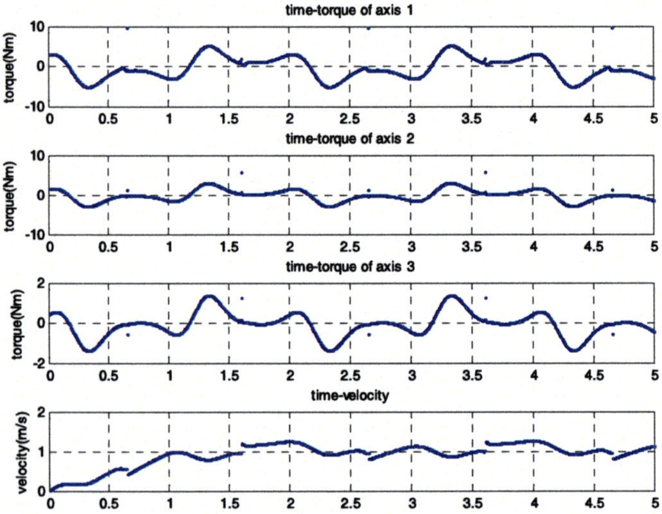

Fig. 5. Results of the simulation. In order joint 1 torque, joint 2 torque, joint 3 torque, velocity(Input : 0.8 Hz, Amplitude 20 deg Um 0.03m/s).

3 Conclusion

In this paper, we introduced our research work on the design and construction of robotic fish 'ICHTHUS V5'. The developed robotic fish has embedded many kinds of sensors to navigate autonomously in the real environment like river. Also, three kinds of sensors are integrated to measure temperature, electric conductivity and pH of water for monitoring the quality of water. We were derived the mathematical model of the fish robot which swims using tail fin of 3-DOF link mechanism. Since we conducted parameter optimization with our previous research and confirmed that the optimized input value compared to the existing value that can produce faster velocity by using smaller torque value, we can apply the input parameters which include the minimum RMS Torque value. These parameters were found to increase energy efficiency. We can improve the energy efficiency and operating time of the robotic fish by using this research result. Also, the developed robotic fish is able to avoid unknown obstacles in the real environment through those experimental results. In the future, the fish robot should be improved its performance. For example, it will be achieved high speed swimming like human. We validate the simulation and will complement by constructing the dynamical experimental environment. And, in many case of occurring in real-world environment, we would study how to optimize energy and prove through an experiment of field test of the robot.

Acknowledgements. This research was supported by Korea Research Council for Industrial Science and Technology, research grant B551179-10-02-00.

References

1. Hirata, K., Takimoto, T., Tamura, K.: Study on turning performance of a fish robot. In: Proc. 1st Int. Symp. Aqua Bio-Mech, pp. 287–287 (2000)
2. Cho, K., Park, H., Kim, S.-W., Yang, H., Park, Y.-P.: Development of robot mimicking propulsion of a fish. In: Korean KSME, pp. 40–45 (2007)
3. Kim, Y.-H.: Robotic fish, the prince of ocean. Dong-A Science (Magazine in Korean) 8, 54–59 (2005)
4. Ryuh, Y.-S.: Development of swimming mechanism and algorithm of fish-like underwater robot. In: KROS, pp. 43–48 (2009)
5. Chung, C., Lee, S.-H., Kim, K.-S., Cha, Y.-S., Ryuh, Y.-S.: Design of side fin mechanism of fish-like underwater robot. In: ICROS, pp. 582–585 (2009)
6. Nakashima, M., Ohgishi, N., Ono, K.: A study on the propulsive mechanism of a double jointed fish robot utilizing self-excitation control. JSME International Journal, Series C 46(3), 982–990 (2003)
7. Chung, G.B., Yi, B.-J., Lim, D.J.: An efficient dynamic modeling methodology for general type of hybrid robotic systems. In: Proc. of the 2004 IEEE International Conference on Robotics & Automation (April 2004)
8. Lee, S.-H.: Data analysis of engineering statistics using minitab. EreTech (2008)
9. Streitlien, K., Triantafyllou, G.S., Triantafyllou, M.S.: Efficient foil propulsion through vortex control. AIAA Journal 2319, 2315–2319 (1996)
10. `http://www.dac.neu.edu/msc/burp.html` (2004)
11. Guo, S., Fukuda, T., Kato, N., Oguro, K.: Development of underwater microrobot using ICPF actuator. In: Proceedings of the 1998 IEEE International Conference on Robotics & Automation, pp. 1829–1834 (1998)
12. Kato, N.: Control performance in the horizontal plane of a fish robot with mechanical pectoral fins. IEEE Journal of Oceanic Engineering 129, 121–129 (2000)
13. Hu, H.: Biologically Inspired Design of Autonomous Robotic Fish at Essex. In: Proceedings of the IEEE SMC UK-RI Chapter Conference 2006 on Advanced Cybernetics Systems, pp. 1–8 (2006)
14. Yang, G., Choi, W., Lee, S., Kim, K., Lee, H., Choi, H., Ryuh, Y.: Control and Design of a 3 DOF Fish Robot ICHTHUS. In: Proceedings of 2011 IEEE International Conference on Robotics and Biomimetics, pp. 2108–2113 (2011)

15. Yamamoto, I., Terada, Y.: Robotic fish and its technology. In: Proceedings of SICE Annual Conference in Fukui, August 4-6, pp. 73–76. Fukui University, Japan (2003); Tapia, E.M., Intille, S.S., Larson, K.: Activity Recognition in the Home Using Simple and Ubiquitous Sensors. In: Ferscha, A., Mattern, F. (eds.) PERVASIVE 2004. LNCS, vol. 3001, pp. 158–175. Springer, Heidelberg (2004)
16. Yamamoto, I.: Research on an oscillating fin propulsion control system. In: Proc. IEEE OCEANS 1993, vol. 3, pp. 259–263 (1993)

Research on Bio-mechanism Robotics by Robotic Fish Fin Technology

Zusong Gu, Ikuo Yamamoto, and Tomokazu Hiratsuka

The University of Kitakyushu
1-1, Hibikino, Wakamatsu-Ku, Kitakyushu-Shi, Fukuoka-ken, Japan
guzusong@gmail.com, yamamoto@env.kitakyu-u.ac.jp,
hiratsuka@brach.co.jp

Abstract. In the surgical operation, the research and development of the surgical robot that has both safety and effectiveness is needed in consideration of the physical mental strain decrease of the surgeon and the patient. Through the observation of the motion of fish, it has been found that fish swim efficiently using flexibility of pliable fins, giving the actuator the name of flexible oscillating fin propulsion system. The author proposed the application to the realm of healing the elastic vibration wing promotion system. The flexible forceps robot for the surgical operation of a vibrating flexible fin promotion system application was developed.

Keywords: Robotic Fish, new underwater vehicle, flexible oscillating fin propulsion system, Forceps robot, Sea bream robot, Model analysis.

1 Introduction

The development of underwater robot has been actively pursued in recent years. A bio-maneuvering type underwater vehicle has been developed through new and different approaches, namely the observation of marine creatures such as fish and the engineering application of their maneuvering mechanism. This robot is widely known as a "robotic fish."

It is well known that marine creatures such as fish swim using small power even at high speeds (dolphin: 60km/h, swordfish: 80km/h) and sweetfish, etc. are superior in their resting state maintenance characteristics. Through the observation of the motion of fish, it has been found that fish swim efficiently using flexibility of pliable fins.

In this study, the authors developed a robotic forceps for surgical operation by applying an flexible oscillating wing propulsion system.

2 Drive Principle of Flexible Oscillating Fin Propulsion System

Figure. 1 shows the drive principle of early flexible oscillating fin propulsion system [1,2,3,4,5]. The flexure like a fish fin was produced by variously changing elastic modules of the flexible oscillating fin. The optimal shape of the fin was also chosen

S. Lee et al. (Eds.): Intelligent Autonomous Systems 12, AISC 193, pp. 577–584.
springerlink.com

through numerical simulation and tank test [6]. Then, the maneuverability was tested with a small craft, on which the oscillating fin propulsion system was installed. Here, it was found that the oscillating fin propulsion system enabled the craft to move back and forth as well as turn right and left using a fin.

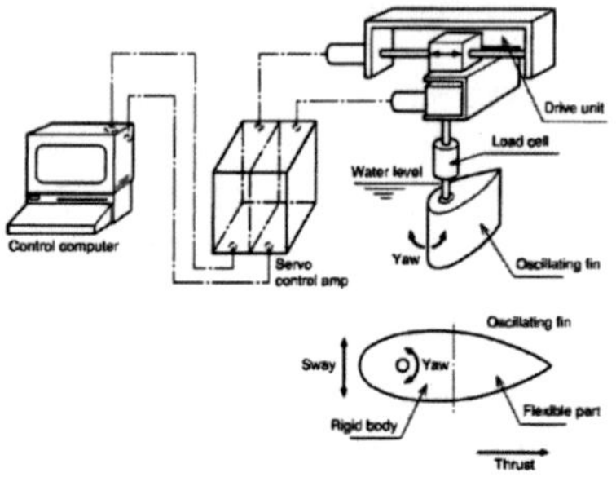

Fig. 1. Drive principle of flexible oscillating fin propulsion system

3 Robotic Fish

As directly using the sensor data is an issue to consider, we need a framework to quantize the data. The flow of the system is like figure 3. The Armband collects the log data based on the user's activity, and a PDA is used for the user to annotate the current activity he or she is doing. These two log data are collected simultaneously and save the same format of the current time. Then, the log integrator will merge these data into entire integrated log data. The preprocessor will use this data and generate a fuzzy integrated data for each sensor log. With this data the fuzzy Bayesian network will train the conditional probability tables, and the inference also uses this fuzzy integrated data.

3.1 Robotic Sea Bream

Based on the basic technical research of the flexible oscillating fin propulsion system, artificial fish have been developed as an applied product of this system. The first developed robotic fish, a sea bream, has gained a reputation in the world for swimming just like the real thing without power and communication cable [7]. Robotic fish has the propulsion mechanism with flexible/Rigid oscillating fin, and can swim in water quietly as. The flow of a surrounding fluid when the fish was swimming was made visible with the robot. The Kalman whirlpool can be done backward. It shows in Figure 2.

The photograph of the development model of Robotic fish is shown in Figure 3. It is composed of Flexible/Rigid Oscillating Fin, actuator, and battery, charge unit, hydraulic pressure sensor, sealing water pump, waste water pump, ballast tank, control unit, and wireless telecommunications unit. [8]

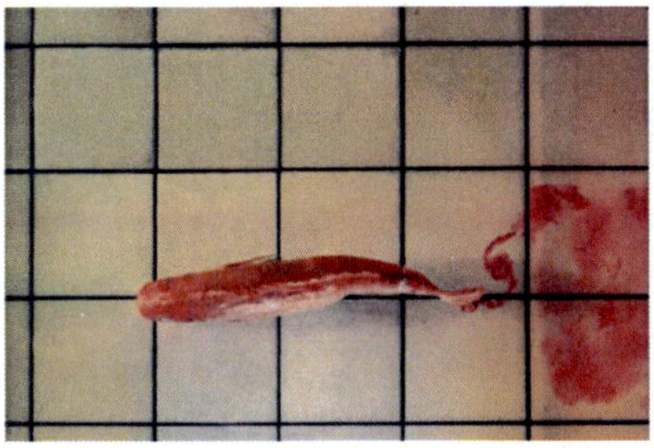

Fig. 2. low test of robotic fish

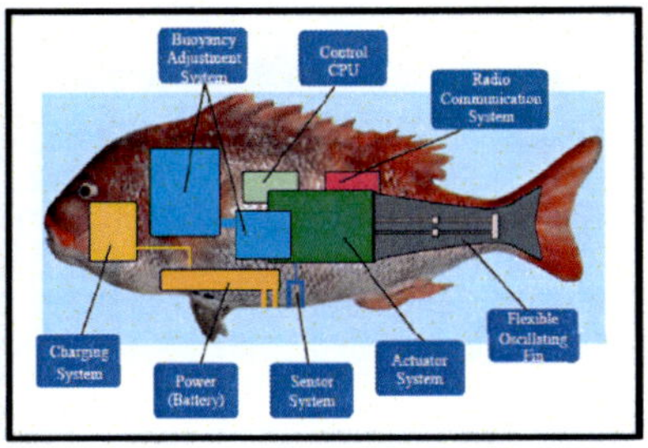

Fig. 3. Internal Structure of Robotic Fish

3.2 Flapping Wing Type Robotic Fish

The manta type robot moves a fin so that a bird flaps the wings, and it is the fish robot which swims flexibly to fly in the water. The elastic vibration wing promotion system demonstrates a big impellent and the lift, and achieves a high performance.

The multiple joints propulsion system has twisting mechanism to realize higher maneuverability of underwater vehicle. Figure 4, Figure 5 shows its photo.

Fig. 4. Structure of flapping wing

Fig. 5. Flapping wing with multiple joints system (twisted mode)

3.3 New Underwater Vehicle Based on Shark Ray

The new underwater vehicle has wing and new tail fin structure for strong propulsion force higher maneuvering characteristic of turning round, etc. This new biomimetic underwater vehicle is developed based on study of shark ray (Rhina ancylostoma). The shark ray has unique structure. Its shape of head and wing like shark, and its shape of tail like ray. They can be stabilizer, propulsion and rudder, in each as shown in Figure 6. Authors are planning to combine these functions in one robotic fish to realize high maneuverability. Figure 7 shows an external view of prototype underwater vehicle with a single plate fin mechanism used on performance evaluation tests. No.1 & 2 motors provide thrust force, and No.3 & 4 motors keep balance of vehicle as main fins. In this performance evaluation test, we used only No.1 & 2 motors.

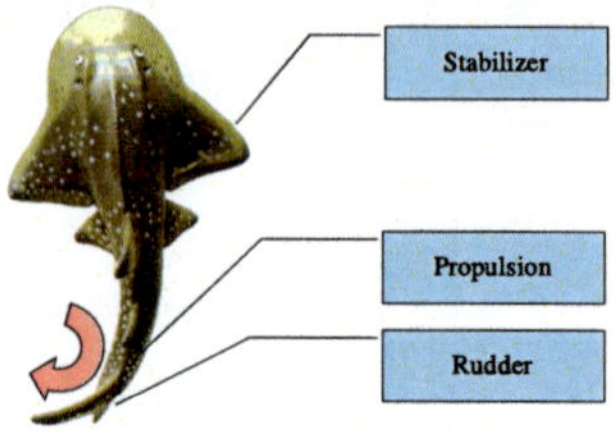

Fig. 6. Unique structure of shark ray

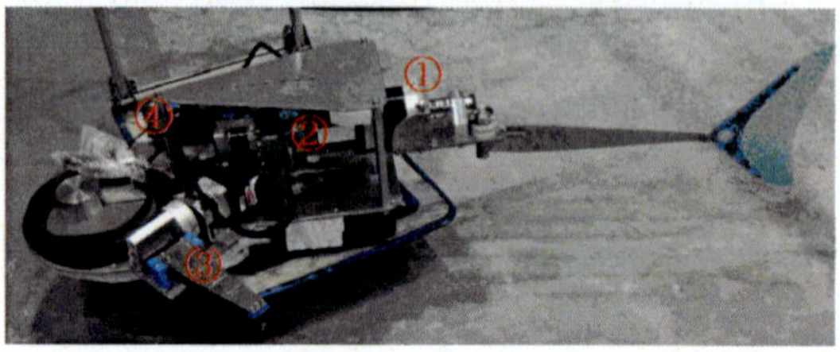

Fig. 7. The external view of prototype underwater vehicle

3.3.1 Experimental Methodology

Figure. 7 shows the experiment system for the propulsion force measurement.. The system is able to measure propulsion force without fixing the underwater vehicle. Figure 8 shows the experiment system for the propulsion force measurement. Specifically, first, a position of the center of gravity of the vehicle is fitted with wire-rope in water tunnel. Second, the wire-rope is divided into two directions, and is retracted backward. Finally, the thrust force is measured by a force gauge.

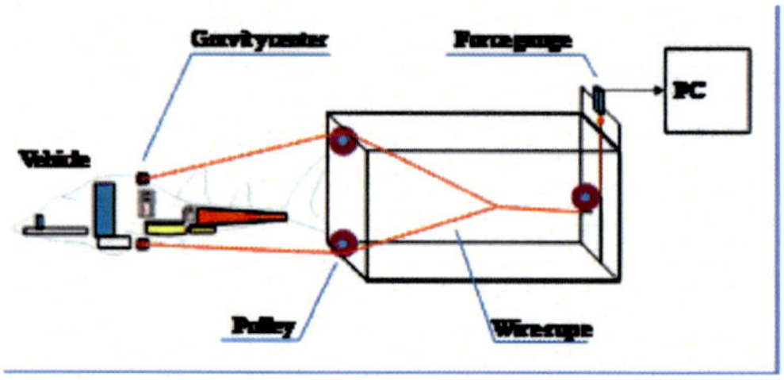

Fig. 8. An experiment system for the propulsion measurement

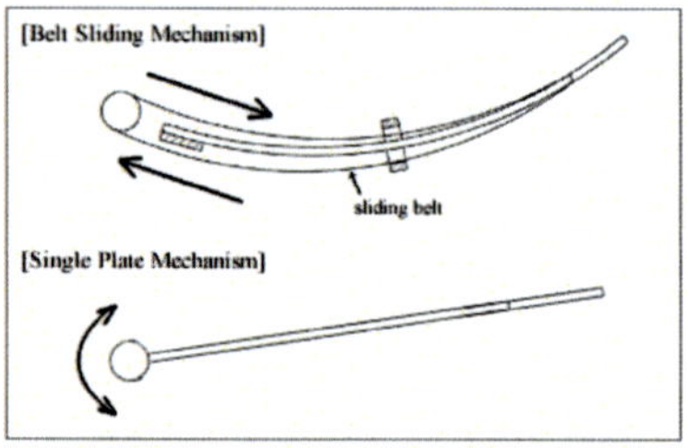

Fig. 9. The comparison of the belt sliding mechanism and the single plate mechanism

3.3.2 Experimental Result

This mechanism is difficult to oscillate at high velocities, and is not able to generate the force required to sail at high velocities. Meanwhile, the single plate mechanism, which mounted on the new prototype underwater vehicle, is oscillated by motor as shown in Figure 9. Therefore the single plate mechanism is able to oscillate at high velocities, and is able to generate the propulsion force larger than the belt sliding mechanism. As a result, the experiment showed that the structure of the prototype underwater vehicle was effective.

4 Development of Flexible Forceps Robot for Surgical Operation

The development of the robotic fish started with the development of an actuator that could realize fishlike fin movement. Through the observation of the motion of fish, it has been found that fish swim efficiently using flexibility of pliable fins, giving the actuator the name of flexible oscillating fin propulsion system.

This chapter is about the development of the forceps robot by the method of the finite element, the forceps robot is modeled and is analyzed. The authors develop the forceps robot that when operating on the intestines tube, the forceps robot is used to hold internal organs such as intestines tube.

4.1 Production of Forceps

New forceps have the function that can apply the elastic vibration wing system and the regulating system, and correspond to patient's blood pressure variation under the operation. The forceps robot system that had this function was made for trial purposes in the present study. Figure. 10 and Figure.11 show a whole system and the grippers.

When the blood pressure rises, increase the holding pressure to prevent the content leaking, when the blood pressure descends, decrease the holding pressure to decrease the load of the organization and internal organs. This regulating system is possible in our research.

4.2 Comparison of Conventional Forces and New Forceps

As shown in Figure 12, the conventional forceps are grasping at anything with two stainless steel plates which have a bulge in the center. It is not out of shape for the grip. It is ease to focus on specific portions of the gripping pressure, increase the gripping internal organ or organization's burden.

As shown in Figure.13, for the new forceps and previous forceps, we measured holding pressure and holding force (pull force), compared with the previous one, the

Fig. 10. Whole system

authors can fully prove that the hold pressure of the new forceps was less 60% than the previous one (average: previous forceps 32.2N, new forceps 12.9N). There is 7mm grip in the center of the new forceps to make load no change and maintain a constant gripping force, even when it can be used in the surgical operation.

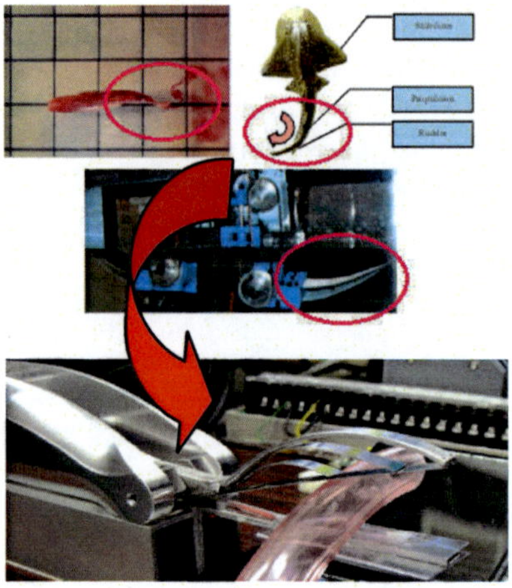

Fig. 11. New forceps

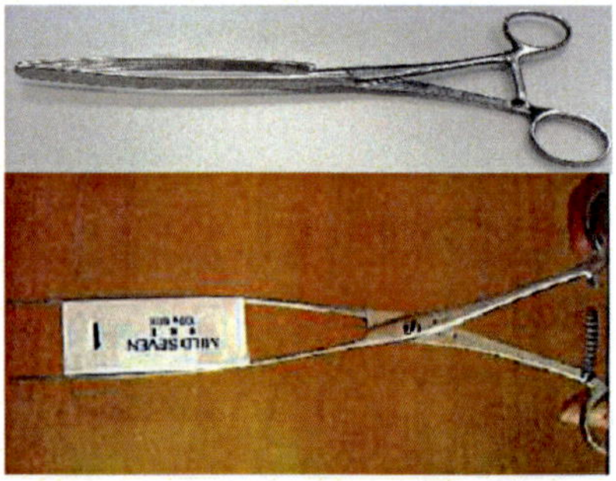

Fig. 12. Conventional forceps

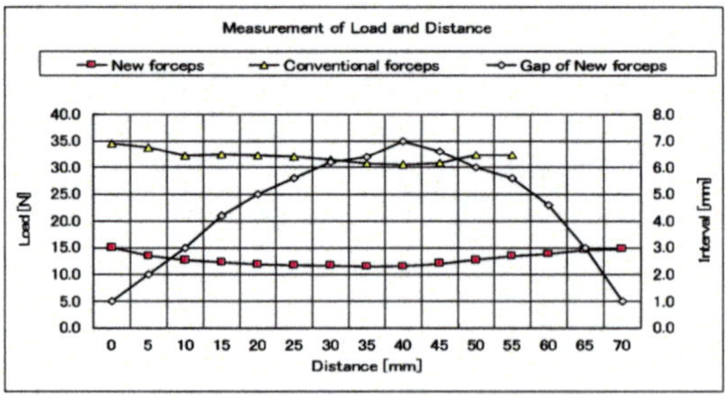

Fig. 13. Comparison of performances

Fig. 14. Grip force measurement

According to Figure 13, we could prove that the holding force was the smallest in the middle of holding part of the new forceps. The experiment is shown as Figure 14. The authors can find that the pig's small intestine that is like the characteristic of human small intestine doing holding experiment with the middle of the new forceps can be utilized.

There was a report that pressure in small intestines was 0.53×10^5Pa(40mmHg) or more (about 544mm H_2O), and, then, the experiment poured water from the place of 550mm in height in small intestines [9]. Pressure F that had acted on the turning angle degree θ and contact side in the gripper when the gripper of new forceps was shut and water which did not leak was measured. The result is shown in Table 1.

Table 1. Exprimental results

Rotation θ[°]	Normal stress F[N]	Contact area S [mm^2]
4.3	4.46	16.1

5 Conclusion

According to the finite element method analysis, like the new forceps as shown in Figure. 15, it is possible to hold for making the power distribution average and wrapping soft internal organs. Power can be concentrated on the hand tip, and it is not necessary to powerfully hold and pull out.

According to the date of the experiment, the authors analyze and get the results shown in Figure 16. On Y axis, the stress is about 4.26N, this date is approach the experiment. When the error margin make the microindentation of actual small intestines not be reproduced well by CAD, and caused by of the error of margin.

Results of the research, the authors can develop the new forceps that can correspond the shape of held internal organs, and fix the internal organs which of the shape can be easily changed with soft and average power.

The fin's flexibility can be utilized actively. It is therefore possible to improve the propulsion performance. Realistic movement becomes possible by making the most of the oscillating fin propulsion system for artificial fish, etc.

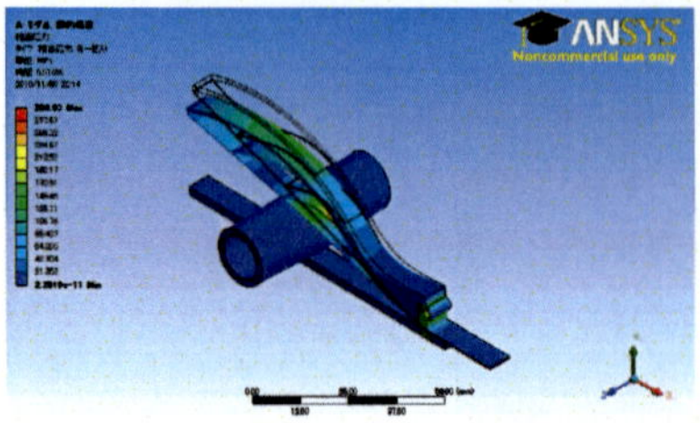

Fig. 15. Equivalent stress distribution

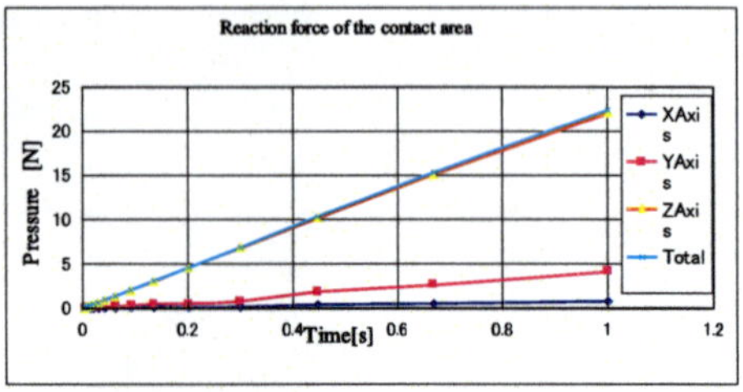

Fig. 16. Measurement of Load and Distance

As for the application to a land robot's flexible hand, a space machine, an airplane, etc., a fish robot's technology is considered.

Acknowledgements. Authors would like to thank the Knowledge Cluster Initiative of MEXT (Ministry of Education, Culture, Sports, Science and Technology) for a grant that made it possible to complete the study.

The present study, the authors wish to express our gratitude to Mr. Chihiro Koga, who was in Ikuo Yamamoto's laboratory of Kitakyushu University, for measuring basic data of the holding characteristic of forceps. The research was funded by Robot Creation Project of Kitakyushu Foundation for the Advancement of Industry, Science and Technology (FAIS).Thanks are extended to the related persons of the project. Dr. Koji Yamaguchi, Kazunori Shibao, Atsushi Nagata, University of Occupational and Environmental Health for medical advices, and Kihara Ironworks for manufacturing.

References

[1] Yamamoto, I.: Marine Control Systems. International Journal of Robust and Nonlinear Control, IFAC 11(13) (2001)
[2] Hess, J., Smith, A.: Calculation of potential flow about arbitrary bodies. Progress in Aeronautical Science 8 (1996)
[3] Kubota, A., et al.: Study on propulsion by partially elastic oscillating foil. Journal of Naval Architects of Japan (156) (1985)
[4] Yamamoto, I., Terada, Y., Kobayashi, E., Togita, T.: Design of a maneuvering control system for a ship. Transactions of the West-Japan Society of Naval Architects (84) (1994)
[5] Yamamoto, I., Terada, Y., Nagamatu, T., Imaizumi, Y.: Development research of oscillating fin propulsion system. In: Proceedings SIC 1994 (1994)
[6] Yamamoto, I.: Propulsion System with Flexible / Rigid Oscillating Fin. IEEE Journal of Oceanic Engineering 20(1) (1995)
[7] Madeen, E.: Doctor Robopet, ANA Wingspan, vol. 416, pp. 24–29 (2004)
[8] Ichikizaki, T., Yamamoto, I.: Development of High Performance Robotic Fish. In: Proceedings of Techno-Ocean 2006 / 19th JASNAOE Ocean Engineering Symposium Kobe, Japan, October 18-20 Paper No.207 (2006)
[9] Oppenheimer, M.J., Mann, F.C.: Intestinal capillary circulation during distention. Surgery 13, 548–554 (2015)

Simox: A Robotics Toolbox for Simulation, Motion and Grasp Planning

N. Vahrenkamp[1,2], M. Kröhnert[1], S. Ulbrich[1], T. Asfour[1], G. Metta[2], R. Dillmann[1], and G. Sandini[2]

[1] Institute for Anthropomatics, Karlsruhe Institute of Technology (KIT), Adenauerring 2, Karlsruhe, Germany
[2] Robotics, Brain and Cognitive Sciences Department, Istituto Italiano di Tecnologia (IIT), Via Morego 30, Genova, Italy

1 Introduction

Software development plays a major role besides hardware setup and mechanical design when it comes to building complex robots such as mobile manipulators or humanoids. Different requirements have to be addressed depending on the application. A low-level controller for example must be implemented for real-time use, whereas a task planning component will interact with the robot on a higher abstraction level. Hence, developing robotics software is subject to several constraints such as performance and robustness.

The selection of libraries or frameworks for software development is influenced by the application and the existing software environment. Therefore, several aspects have to be considered when building and/or choosing software components for developing robotics applications:

- Performance and Robustness: Although increasing CPU speed allows to run complex algorithms in reasonable amounts of time, performance is always an issue in the context of mobile manipulation and planning. Therefore often C++ is the preferred choice in robotics since it combines high level programming language features with efficient compilation techniques. Robustness can be achieved by relying on advanced coding techniques like smart pointers, exception handling and automated test cases.
- Extensibility: A library should offer easy extension capabilities. One possible way to achieve this is offering a plugin mechanism which uses common interfaces to provide uniform extension points to developers.
- Platform-independence and Dependencies: Since software development takes place on different platforms, a framework should at least offer support for the main operating systems Windows™, Unix/Linux, and Mac OS X™. Additionally, a low number of necessary compile-time dependencies ease the setup and compilation process of a library.

Furthermore, the license model might play an important role in selecting software libraries. In the scientific context, open-source software is preferred since it allows both reusing results from others as well as sharing own solutions with

S. Lee et al. (Eds.): Intelligent Autonomous Systems 12, AISC 193, pp. 585–594.
springerlink.com

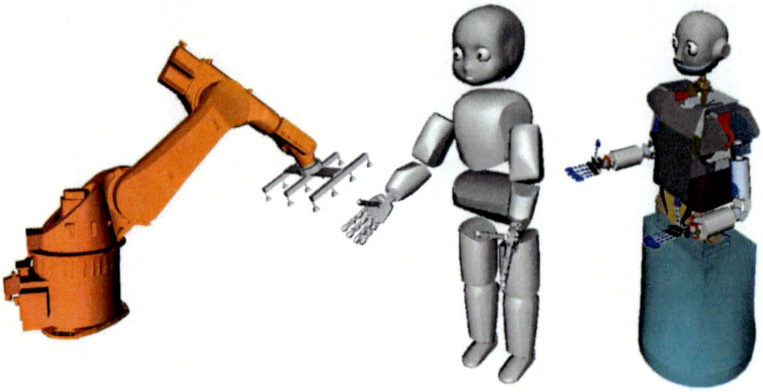

Fig. 1. Robot models realized with *Simox*: Kuka KR60-3, iCub and ARMAR-III

the community. The widely used GNU General Public License (GPL) is one possible open-source license, but it enforces projects that use GPL libraries to be published as GPL as well. Therefore, many libraries apply the GNU Lesser General Public License (LGPL), since it does not require derived products to be released under LGPL terms, thus allowing the usage of the library in commercial products [1].

Several open source frameworks exist in the context of robot simulation, differing in complexity, licensing, and objectives. Physics simulation libraries such as *ODE* [2], *IBDS* [3] or *Bullet* [4] usually offer possibilities to define constrained objects which can be used to setup joints of a robot. Based on these joints the kinematic structure of a robot can be defined. Motion planning libraries (e.g. *OMPL* [5], *MSL* [6], *MPK* [7]) and grasp planning frameworks such as *GraspIt* [8] or *OpenGRASP* [9] either rely on external frameworks for defining robot kinematics or a custom kinematic representation is used. With *OpenRAVE* [10] a comprehensive simulator for motion and manipulation planning is available under an open-source license. Although a lot of features are offered with *OpenRAVE*, a large number of external libraries have to be installed in order to take advantage of the full functionality. Users can write plugins in order to interact with the simulator and therefore the whole system needs to be running. In contrast Simox explicitly offers several lightweight libraries that can be used within a custom project without handing over the control to a simulator. The *Robotics Toolbox* offers a set of useful MATLAB^{TM} scripts covering a wide range of algorithms related to robotics [11].

In contrast to existing frameworks for robot simulation, *Simox* tries to achieve both: Covering a wide range of simulation features while not relying on a large number of dependencies to other libraries. Having lots of dependencies may cause difficulties in installing and using the framework when the execution environment is limited, e.g. when the framework should be running on a robot.

2 Simox

With *Simox* we tried to cover many of the above mentioned requirements. The library is completely implemented in C++ while using modern programming paradigms to achieve both efficiency and robustness. Reliability being a critical aspect, especially in robotic applications, the library comes with numerous unit test cases and relies on smart pointer usage to avoid memory leaks. *Simox* is released under the LGPL license and can therefore be used both in open-source and commercial products. The low number of dependencies required to compile and use the library ensures a convenient setup and a straightforward integration into any type of project.

Simox itself is divided into the three main libraries: *VirtualRobot* provides functionality for describing robot models including kinematic chains, end effectors and visualizations. These models can be read from XML files and afterwards be used for collision detection, inverse and direct kinematics calculation as well as reachability analysis. *Saba* provides sampling-based motion planning algorithms. The library holds several configuration space representations that are used by the planning algorithms. Additionally, path processing approaches and visualization routines are included. *GraspStudio* is a library offering grasp planning capabilities. Methods for generating grasping hypotheses, evaluating grasp qualities, and planning grasps are available. Visualization methods for all generated results are included additionally.

2.1 3D Visualization

Providing meaningful and easy to use visualization methods is crucial for a good framework dealing with geometric data. In *Simox* we provide abstract visualization interfaces which are independent of the underlying 3D libraries. To support a new 3D library it is only necessary to provide the matching models and to implement the visualization interfaces using the API of the specific library. Up to now, support for the two visualization frameworks *OpenSceneGraph* [12] and *Coin3D* [13] is realized with *Simox*.

2.2 Collision Detection

Fast collision detection is important for efficient motion and grasp planning, since sampling-based approaches need to perform numerous calls to the collision detection engine to validate sampled configurations. Hence, an efficient approach is needed, preferably not limiting the shape of the models (e.g. only convex shapes). The PQP library [14] offers efficient collision detection for arbitrary 3D models and an extended version, that is able to handle multi-threaded collision queries, is fully encapsulated by *Simox*. For collision detection we provide the same functionality as for visualization. Therefore, it is possible to use other collision detection engines by implementing the interface classes for collision detection.

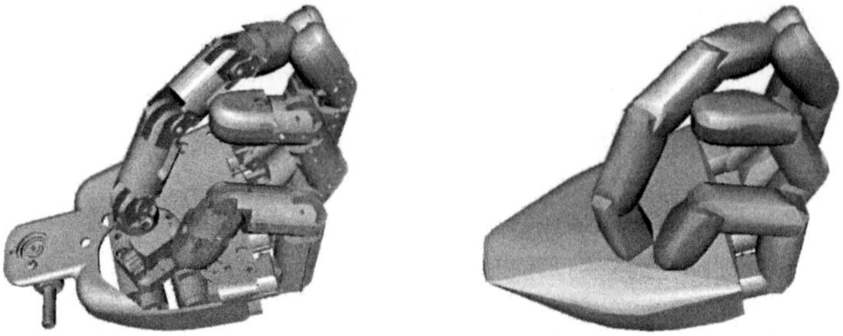

Fig. 2. The visualization model of iCub's hand (left) and its reduced model (right) that is used for efficient collision detection

3 VirtualRobot

The *VirtualRobot* library offers methods to define robots and environments and numerous simulation tools. Furthermore, advanced algorithms such as collision detection, Jacobian calculations or reachability analysis are covered.

Robot Modeling
A robot is defined via its kinematic structure resulting in a tree-like structure of connected *Robot Nodes*, whereas parameters describing the dynamics can be optionally specified. In Fig. 1, three exemplary models are shown.

Visualization and Collision Detection
Based on the kinematic definition, 3D models for visualization and for data processing can be attached to the *Robot Nodes* resulting in two models: one in high resolution for appealing visualizations and a reduced one that can be used for efficient collision detection and distance calculations (see Fig. 2).

Kinematic Chains and Robot Node Sets
Usually when a robot is used, several kinematic chains, as logical sets of joints, are needed. Therefore, so-called *Robot Node Sets* can be defined in *VirtualRobot*, allowing to specify a collection of *Robot Nodes* or a strictly defined kinematic chain.

End Effectors
End effectors play an important role in grasp and manipulation planning in the context of humanoid robots. Hence, a convenient end effector definition is provided by *VirtualRobot*, allowing to easily open and close hands while considering self-collisions and collisions with the environment.

Jacobian Calculations and Differential IK
The Jacobian of a robot manipulator is required for many complex calculations in a robotic application and, hence, has to be calculated very efficiently.

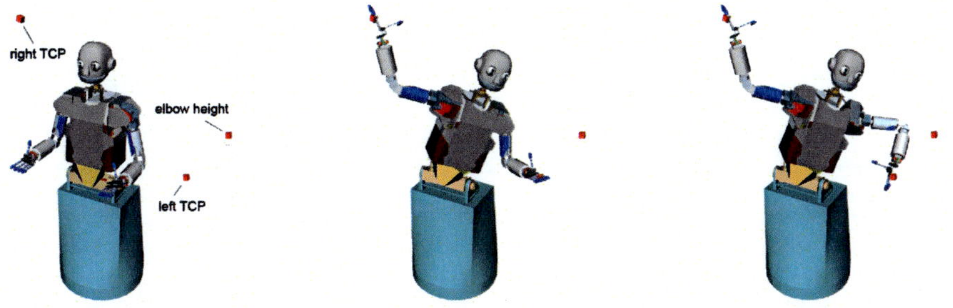

(a) The target locations for the body parts.

(b) Solution without elbow target height.

(c) Solution considering the elbow.

Fig. 3. Results of differential inverse kinematics queries

The most prominent example is the numerical solution of the inverse kinematics (IK) but it also plays a crucial role in calculating the dynamics of a robot. In short, the Jacobian matrix holds the partial derivatives of an end effector's Cartesian position in its columns and can be used to transform joint angle velocities into Cartesian movement. In Simox, special attention has been paid to the construction of the Jacobian and its convenient usage. It is well integrated into its differential inverse kinematics algorithm – a numerical IK solver – but it can also be used easily in other applications. For any node of a node set, the Jacobian as well as its pseudo inverse can be obtained conveniently in its reference frame. In addition, it is possible to exclude either the position or orientation and even individual coordinates from the matrix. This can be useful, for instance, if one is only interested in the vertical acceleration of the end effector while balancing.

In Fig. 3, a use-case for the algorithm is shown. The differential IK is used to calculate the bimanual robot configuration covering 17 degrees of freedom (DoF) of ARMAR-III (see Fig. 3(a)). Two IK requests have been performed, where both TCPs should reach a given position. Firstly, the left elbow may move freely and has no assigned target location (see Fig. 3(b)). For the second IK-query, the desired height for the robot's elbow is considered additionally to the TCP positions (see Fig. 3(c)). For both queries the differential IK approach is able to serve a solution after few iterations.

An application of Jacobian calculations for stability analysis can be seen in Fig. 4(a). Here, the center of mass (CoM) Jacobain is used to move the robot until the surface projection of its CoM lies within the 2D support polygon to ensure static stability.

Reachability Analysis

Having a representation of the reachability of a kinematic chain, e.g. a robot arm, helps to fulfill several tasks in the context of grasp planning, inverse kinematics (IK) solving and mobile manipulation. The reachability is defined as a

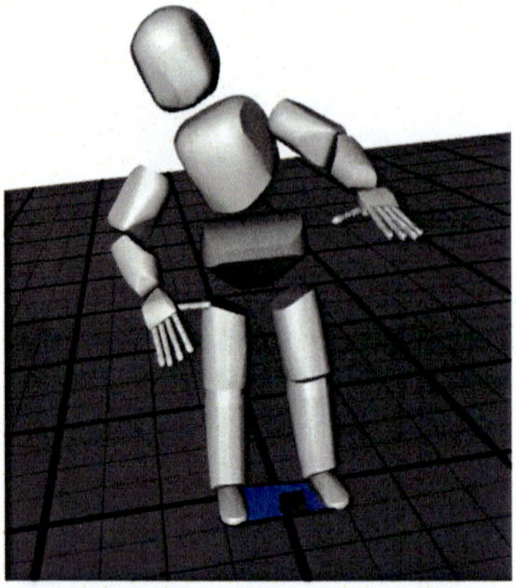

(a) The support polygon together with the projected center of mass.

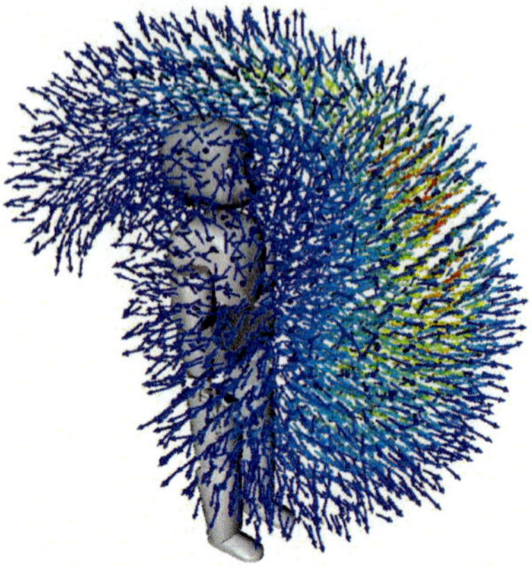

(b) The 6D reachability considering 10 DoF (torso and left arm).

Fig. 4. Stability and reachability computation

3D or 6D volume being reachable by the end effector. In 6D, the orientation is explicitly considered while it is ignored in the 3D case. Several approaches exist to build a representation of the reachability [15,10,16], most of which use a voxelized approximation of the 6D workspace. Reachability structures in *VirtualRobot* can be created for arbitrary kinematic chains considering joint limits and self collisions. A visualization of iCub's reachability for the left end effector is shown in Fig. 4(b). The kinematic chain used for building the reachability data covers three torso and seven arm joints.

4 Motion Planning

Saba, a sampling-based motion planning library, is part of *Simox*.

Generic interfaces are provided for collision detection setups, configuration spaces and planners in order to offer a comprehensive and extensible implementation.

4.1 Setup of Collision Detection

Static and moving environmental objects as well as parts of the robot are handled uniformly. All objects of these types which are considered for collision detection are thus grouped into collision sets. Mutual collision detection is achieved by defining pairs of collision sets.

4.2 Representation of Configuration Spaces

The configuration space C (C-space) is defined based on a collision detection setup and a set of *Robot Nodes*, which in most cases form a kinematic chain of

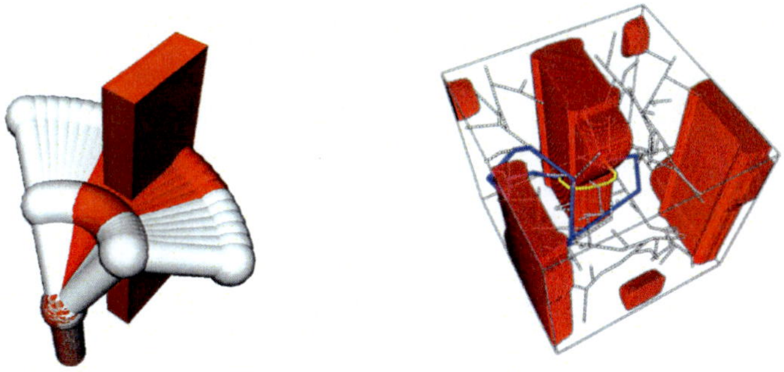

Fig. 5. Extended visualization features allow to analyze planned motions as well as low dimensional configuration spaces

the robot. A configuration $c \in C$ can either be valid or invalid, i.e. resulting in a collision in the workspace. In derived implementations, additional constraints or quality measures can be incorporated. When considering paths from $c_0 \in C$ to $c_1 \in C$, several implementations for collision detection are offered. In most cases, a sampling-based approach, where intermediate samples on the path are checked for collisions will be sufficient (see Fig. 5 left). Nevertheless, different implementations, e.g. relying on continuous collision detection, are explicitly supported, as we showed in [17], where continuous collision detection was realized by Quinlan's *Free-Bubble* approach [18].

4.3 Motion Planners

Motion planners in *Saba* basically rely on the definition of a C-space, which implicitly defines the incorporated joints and the collision setup. Several ready-to-use implementations, related to Rapidly-exploring Random Trees (RRT), are offered as well as advanced planners, such as the *IK-RRT* [19] approach for the planning of reaching and grasping motions (see Fig. 6). Further, several visualization features allow analyzing the results in the workspace and in low-dimensional C-spaces (see Fig. 5).

4.4 Path Processors

A well-known issue in motion planning with sampling-based approaches is that the resulting trajectories are not optimal. In order to create appealing movements, the trajectories have to be smoothed. This has been done by classes implementing the path processor interface of *Saba*. For example, one processor named *ShortcutProcessor* searches collision-free shortcuts in C-space in order to smooth the result. A resulting path in C-space can be seen in Fig. 5 on the right. Here the planned path is shown in blue and the smoothed one in yellow.

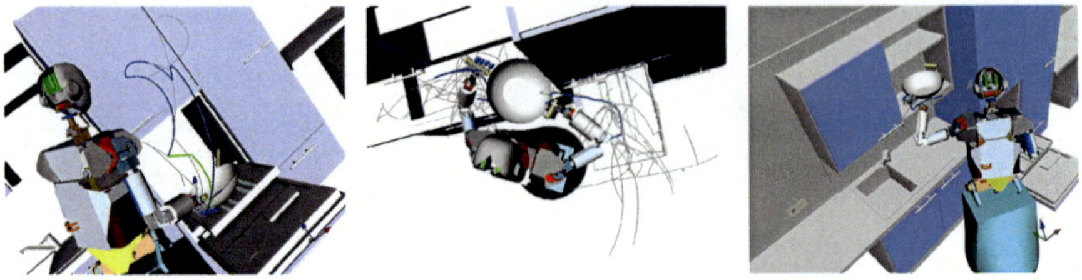

Fig. 6. The bimanual IK-RRT approach is used to plan hand-over motions

5 Grasp Planning

Grasp planning can be performed with the *GraspStudio* library which is part of *Simox*. Based on the robot definitions, any end effector can be decoupled from the model and considered for grasp planning. The Grasp Center Point (GCP) of an end effector defines the favorite grasping position and an approach direction. A generic grasp planner consists of a module for creating approaching motions and a second module for evaluating the grasp quality. Both components are exchangeable by custom implementations of the provided interfaces. Further, custom grasp planners, which do not rely on the presented planning loop, are explicitly supported. When using the generic planning approach, the following steps are executed until the number of requested grasps or a timeout is reached:

- A collision-free grasping hypothesis is generated by an implementation of the *Approach Movement Generator* interface.
- The fingers are closed and all contacts are stored.
- The contact information is passed to the *Grasp Evaluation* module in order to compute the grasp quality and/or force closure information.
- Depending on the quality, the grasp is discarded or added to the set of valid grasps.

5.1 Grasping Hypotheses: Approach Movement Generator

GraspStudio provides an already built-in implementation for generating grasping hypotheses. This *Approach Movement Generator* randomly creates grasping positions based on the object's triangle model:

- The normal information of the object's surface is used to sample potential approach directions, which are aligned with the favorite approach movement of the end effector as given with the GCP definition. Additionally, the remaining DoF (rotation around the approach direction) is randomly sampled.
- The end effector is moved towards the object until the GCP is reached by the object's geometry or a collision is detected.
- In case a collision was detected, the end effector is moved backwards until it is collision-free again.

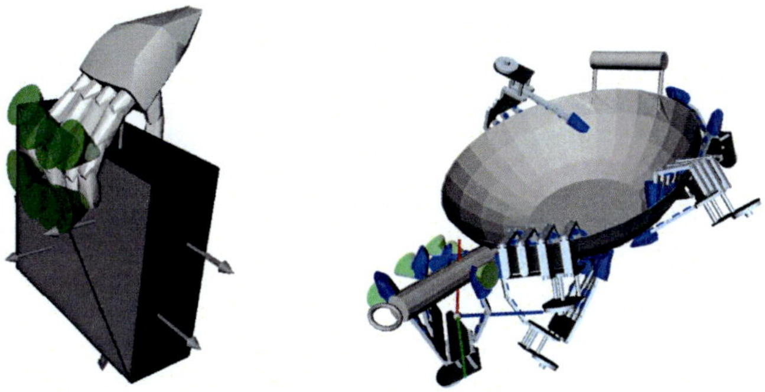

Fig. 7. Several planned grasps for the hand models of iCub and ARMAR-III

This approach is suitable for most applications, although the interface definition for *Approach Movement Generators* allows convenient realizations of custom approaches.

5.2 Grasp Evaluation: Grasp Wrench Space Computation

A common approach in grasp quality measurement is the grasp wrench space computation, where 6D wrenches are constructed from the contact information [20,21,8]. The 6D wrenches represent the contact force and torque and a quality measure is given by analyzing the convex hull of all contact wrenches. Checking whether the origin is inside the convex hull gives an initial indication if the resulting grasp will be a force closure or not, i.e. the object will be fixed in the hand. Further, the minimum distance from the surface of the convex hull to the wrench space origin gives a quality evaluation since it describes the ability of compensating external disturbances. On the left of Fig. 7, a grasping configuration with corresponding friction cones is shown. The right picture shows several grasps that have been planned with *GraspStudio*.

Note that neither the contact model nor the grasp wrench space computation can perfectly represent the reality, but these approaches are meant to approximate the grasping process and to allow efficient planning or efficient first guesses to be refined through learning. Due to the expandable design, *GraspStudio* offers the possibility to also implement more sophisticated planning approaches.

6 Conclusion

With *Simox*, we provide a collection of efficient algorithms to be used in the context of robot simulation, motion and grasp planning. This open source toolbox supports a wide range of applications by offering both state-of-the-art implementations and extendable basic functionalities which can be used to realize more complex algorithms. Future work will address the support for industry standards, such as the COLLADA file format. Further, network transparency and support for shareable objects to be handled within a network of loosely coupled components will be emphasized.

Acknowledgements. The research leading to these results has received funding the German Research Foundation (DFG: Deutsche Forschungsgemeinschaft) under the SFB 588 and from the European Union Seventh Framework Programme under grant agreement 270273 (Xperience).

References

1. Laurent, A.M.S.: Understanding Open Source and Free Software Licensing. O'Reilly Media, Inc. (2004)
2. Open Dynamics Engine, ODE (2011), `http://www.ode.org`
3. Bender, J.: Computer Animation and Virtual Worlds, vol. 18(4-5), p. 225 (2007), `http://dx.doi.org/10.1002/cav.v18:4/5`
4. Bullet Physics Engine (2011), `http://bulletphysics.org`
5. Open Motion Planning Library, OMPL (2010), `http://ompl.kavrakilab.org`
6. Motion Strategy Library (MSL) (2003), `http://msl.cs.uiuc.edu/msl`
7. Motion Planning Kit (MPK) (2006), `http://ai.stanford.edu/~mitul/mpk`
8. Miller, A.T.: Graspit!: a versatile simulator for robotic grasping. Ph.D. thesis, Department of Computer Science, Columbia University (2001)
9. Ulbrich, S., Kappler, D., Asfour, T., Vahrenkamp, N., Bierbaum, A., Przybylski, M., Dillmann, R.: IEEE/RSJ International Conference on Intelligent Robots and Systems (IROS), pp. 1761–1767. IEEE (2011)
10. Diankov, R.: Automated construction of robotic manipulation programs. Ph.D. thesis, Carnegie Mellon University, Robotics Institute (2010)
11. Corke, P.I.: IEEE Robotics Automation Magazine 3(1), 24 (1996)
12. Open Scene Graph (2011), `http://www.openscenegraph.org`
13. Coin3D (2011), `http://www.coin3d.org`
14. Larsen, E., Gottschalk, S., Lin, M.C., Manocha, D.: Fast proximity queries with swept sphere volumes. Tech. rep., Department of Computer Science, University of North Carolina (2000)
15. Zacharias, F., Borst, C., Hirzinger, G.: IEEE/RSJ International Conference on Intelligent Robots and Systems, IROS 2007, pp. 3229–3236 (2007)
16. Vahrenkamp, N., Wieland, S., Azad, P., Gonzalez, D., Asfour, T., Dillmann, R.: 8th IEEE-RAS International Conference on Humanoid Robots, Humanoids 2008, pp. 406–412 (2008)
17. Vahrenkamp, N., Kaiser, P., Asfour, T., Dillmann, R.: International Conference on Robotics and Automation (ICRA 2011), Shanghai, China, pp. 715–722 (2011)
18. Quinlan, S.: Real-time modification of collision-free paths. Ph.D. thesis, Stanford University (1994)
19. Vahrenkamp, N., Berenson, D., Asfour, T., Kuffner, J., Dillmann, R.: Intelligent Robots and Systems, IROS (2009)
20. Pollard, N.: Technical Report AI-TR 1464, MIT, Artificial Intelligence Laboratory (1994)
21. Kirkpatrick, D., Mishra, B., Yap, C.: Proc.of the 20th ACM Symp. on Theory of Computing, pp. 341–351 (1990)

Directional Manipulability to Improve Performance Index of Dual Arm Manipulator for Object Grasping

Hu Chen, Suk-In Lee, Jin-Hyun Do, and Jang-Myung Lee

Dept. of Electrical Engineering, Pusan National University
Jangjeon2-dong, Geumjeong-gu, Pusan, 609-735, Korea
{chenhu,bszsk0,jhdo,jmlee}@pusan.ac.kr

Abstract. Directional manipulability measure is used to improve the performance index of dual arm manipulator, instead of general manipulability measure. In this paper, a performance index of dual arm manipulator has been proposed and manipulability measure is an important part of it. General manipulability measure evaluates for the ability of manipulator to act in either direction. But, for a given task, abilities in either direction are desired and this is called desired manipulability measure. Although the general manipulability measure probably is bigger than desired manipulability measure, only the intersection part of actual and desired manipulability ellipsoids applies to positively influence performance index of dual arm manipulator. Simulation results show the difference and efficiency of these manipulability measures.

Keywords: performance index, directional manipulability, dual arm manipulator, manipulability ellipsoid.

1 Introduction

In service robotics, a dual arm manipulator which seems like human beings constitutes one possible solution to work instead of human beings in the complex or dangerous environment. Some dual arm manipulators had been developed already. Motoman-DA20 can weld and transport objects. RIBA can lift patients and move them to wheel chairs. AIC-AI can cook meals. A robot arm like Motoman-DA20 is introduced in [1]. And a dual arm mobile manipulator like RIBA is presented in [2]. Coordination control [3,4] is a very import issue for a dual arm manipulator. In order to solve this issue, a cost function is defined in this paper. There are two main parts in this function: manipulability measure and energy consumption. Yoshikawa defined the manipulability measure [5-8] as ability of manipulator to act in either direction without taking task. For a given task, the difference of manipulability measure which is based on task orientation and general manipulability measure has an adverse effect on cost function. A new measure called directional manipulability measure [9-12] is used to improve the cost function.

This paper is organized as follows. Section 2 presents the kinematic model of the dual arm manipulator. Section 3 introduces the general manipulability measure and ellipsoid that defined by Yoshikawa and the dual arm manipulability measure and ellipsoid. Section 4 lays out a cost function that can be used for choosing a better

S. Lee et al. (Eds.): Intelligent Autonomous Systems 12, AISC 193, pp. 595–602.
springerlink.com

posture when the manipulator grasps an object. Section 5 proposes a new measure that improves the cost function, called directional manipulability measure. An illustration of manipulability measure and ellipsoid of simulation results will be in Section 6.

2 Kinematic Model

Figure 1 provides an illustration of the dual arm manipulator which is used in the experiment and simulation with Denavit-Hartenberg (D-H) convention. There are 5 DOF in each arm, total 10 DOF, 2 DOF in shoulder and 1 DOF in upper arm, elbow and lower arm. The base coordinate is in the middle of the shoulders and coordinate of each joint satisfied the D-H convention. Because the systems of two arms are almost the same, only the analysis of left arm has been provided as an example. D-H parameters are shown in table 1.

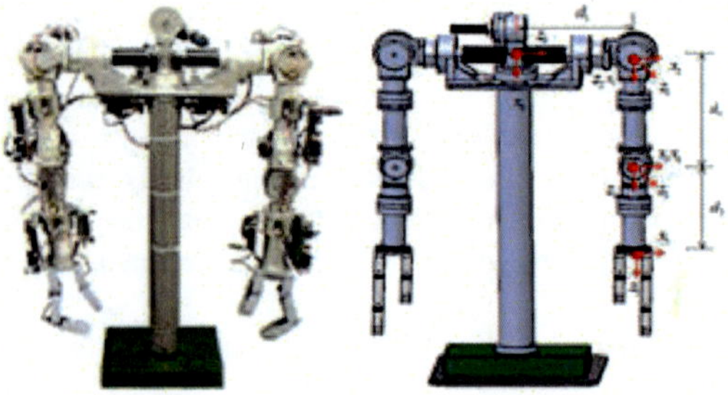

Fig. 1. Dual arm manipulator and simulation

Table 1. DH parameters

i	θ	d	a	α
1	θ_1	d_1	0	90
2	θ_2	0	0	90
3	θ_3	d_2	0	-90
4	θ_4	0	0	90
5	θ_5	d_3	0	0

3 Manipulability Measure and Ellipsoid

3.1 Single arm Manipulability Measure (MM) and Ellipsoid

The relationship between joint space and Cartesian space is expressed by formula (1), where X contains position vector and orientation matrix.

$$X = f(\theta) \tag{1}$$

The relationship between Cartesian velocity and joint velocity is as followed:

$$\dot{X} = J(\theta)\dot{\theta} \tag{2}$$

$J(\theta)$ is the Jacobian matrix.

Manipulability measure evaluates a measure for the ability of manipulator to act in either direction. Yoshikawa defined the manipulability measure in 1984, as followed:

$$MM = \sqrt{\det(J(\theta)J(\theta)^T)} \tag{3}$$

When $J(\theta)$ is square matrix, the manipulability measure can be simplified as followed:

$$MM = \left|\det J(\theta)\right| \tag{4}$$

For a given $J(\theta)$, there are many methods to obtain the manipulability ellipsoid. In this paper, singular value decomposition (SVD) is used, as given by:

$$J(\theta) = USV^T \tag{5}$$

$U \in R^{m \times m}$ is the direction vector of each axis, $V \in R^{n \times n}$ is orthogonal matrix and S is

$$S = \begin{bmatrix} \sigma_1 & 0 & 0 & 0 & \vdots & 0 \\ 0 & \sigma_2 & 0 & 0 & \vdots & 0 \\ 0 & 0 & \cdots & 0 & \vdots & 0 \\ 0 & 0 & 0 & \sigma_m & \vdots & 0 \end{bmatrix} \in R^{m \times n} \tag{6}$$

where $\sigma_1 \geq \sigma_2 \geq \cdots \geq \sigma_m$. σ is axial length of manipulability ellipsoid.

Figure 2 provides an illustration of a manipulability ellipsoid in 2-dimensional space.

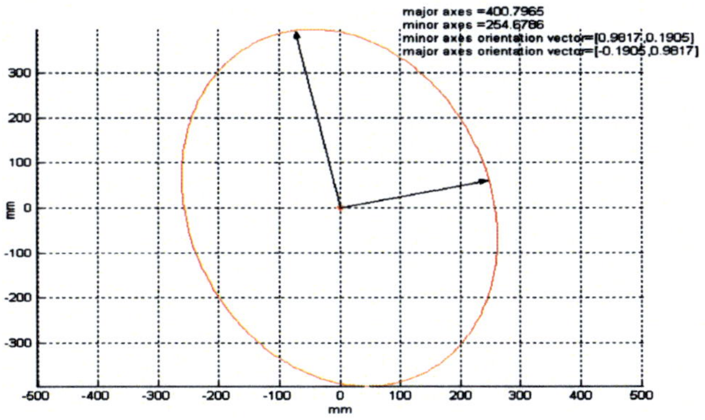

Fig. 2. Manipulability ellipsoid in 2-dimensional space

3.2 Dual Arm Manipulability Measure (DAMM) and Ellipsoid

Dual arm manipulability measure can be defined as the volume of intersection area of manipulability ellipsoid for each arm.

In order to obtain the dual arm manipulability measure, two reference ellipsoids have been defined as followed:

Ellipsoid1: Major axis is from origin to the intersection point of left arm manipulability ellipsoid and major axis of right arm manipulability ellipsoid. Minor axis is same as right arm manipulability ellipsoid.

Ellipsoid2: Major axis is from origin to the intersection point of right arm manipulability ellipsoid and major axis of left arm manipulability ellipsoid. Minor axis is same as left arm manipulability ellipsoid.

Figure 3 provides an illustration of these two reference ellipsoid in 2-dimensional space. Dual arm manipulability measure is the maximum volume between two reference ellipsoids. When the volumes of two reference ellipsoids are same, the dual arm manipulability ellipsoid is same as left arm manipulability ellipsoid.

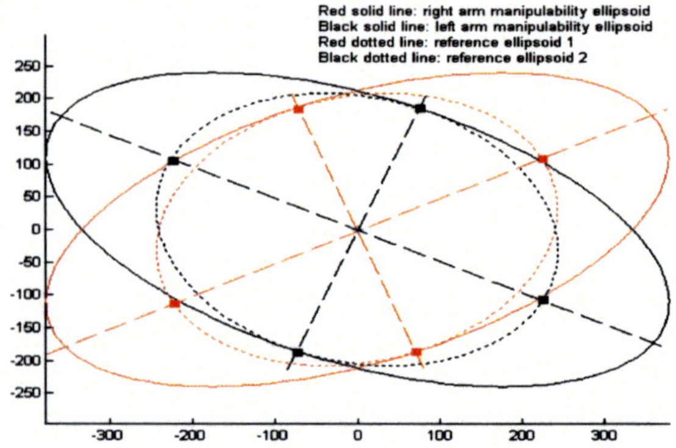

Fig. 3. Manipulability ellipsoid in 2-dimensional space

Then add the third axis, which can be defined as followed:

$$\delta_D = \min(\delta_L, \delta_R) \tag{7}$$

where δ_L is the length of minor axis of the left arm manipulability ellipsoid and δ_R is the length from origin to intersection point of right arm manipulability ellipsoid and z axis.

To facilitate the calculation of manipulability measure, left arm manipulability ellipsoid can set as general ellipsoid. Figure 4 provides an illustration of the ellipsoids in 3-dimensional space.

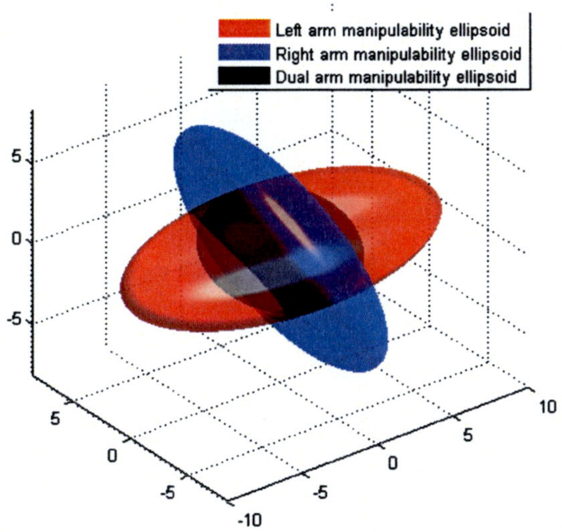

Fig. 4. Manipulability ellipsoids in 3-dimensional space

4 Performance Index Function

When the manipulator is used for grasping an object, there are many postures. Choose a better posture is an important issue. A cost function can be defined to solve this issue. There are two parts in this function: manipulability measure and energy consumption. The cost function can be written in the following form:

$$Cost = \alpha \frac{1}{MM} + \beta P_{power} \tag{8}$$

where α,β are weight of each part.

In order to compare different postures, voltage is constant 24 V. And current can be measured by sensors. So the power can be calculate by formula (9).

$$P = UI \tag{8}$$

5 Directional Manipulability Measure (DMM) and Ellipsoid

Yoshikawa defined the manipulability measure and ellipsoid in either direction. But, in fact, abilities in directions have been desired for a given task. So there are some adverse effects for performance index of manipulator, although the measure of general manipulability is bigger than that of directional manipulability.

For a given task, the manipulability measure and ellipsoid are known, called desired manipulability measure and ellipsoid. Directional manipulability measure can be defined as the volume of intersection area of actual manipulability ellipsoid and desired manipulability ellipsoid.

The method to obtain the directional manipulability measure and ellipsoid is same as dual arm manipulability measure and ellipsoid as figure 4.

6 Simulation Results

When the dual arm manipulator grasps a ball which radius is 10 cm, there are many postures. Figure 5 provides an illustration of three typical postures. Figure 5 provides an illustration of three typical postures.

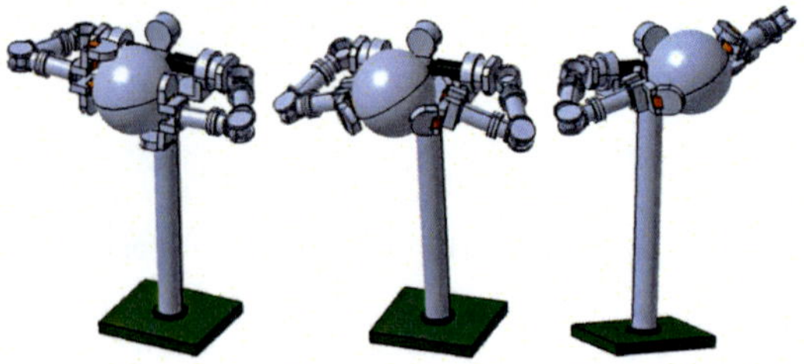

(a) (b) (c)

Fig. 5. Three postures when dual arm manipulator grasps a ball

Figure 6 provides an illustration of desired manipulability ellipsoid. Figure 7 provides an illustration of dual arm manipulability ellipsoids in 3-dimensional space for these cases (left green ellipsoids) and directional dual arm manipulability ellipsoids (right yellow ellipsoids).

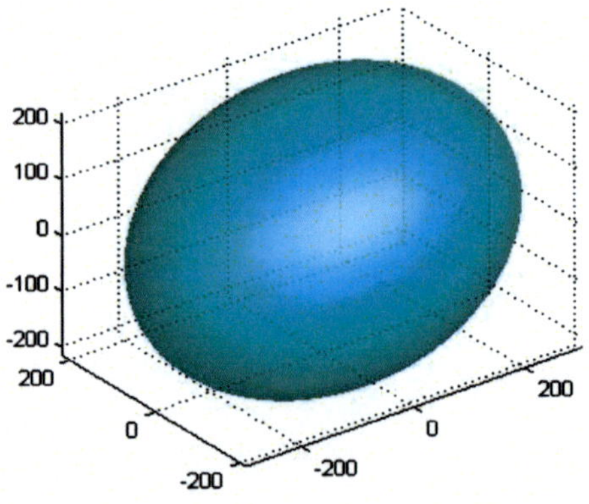

Fig. 6. Desired manipulability ellipsoids

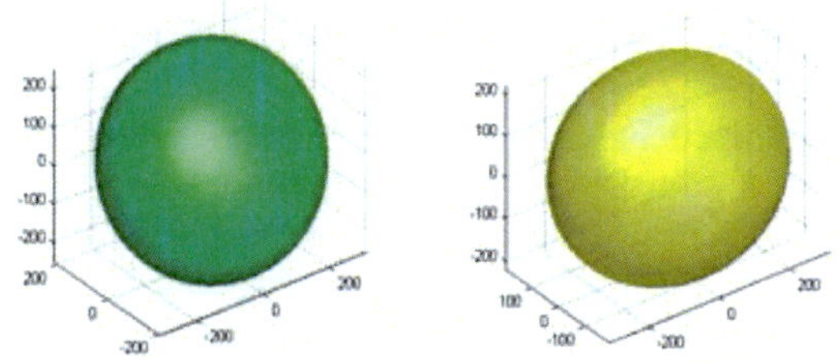

(a)

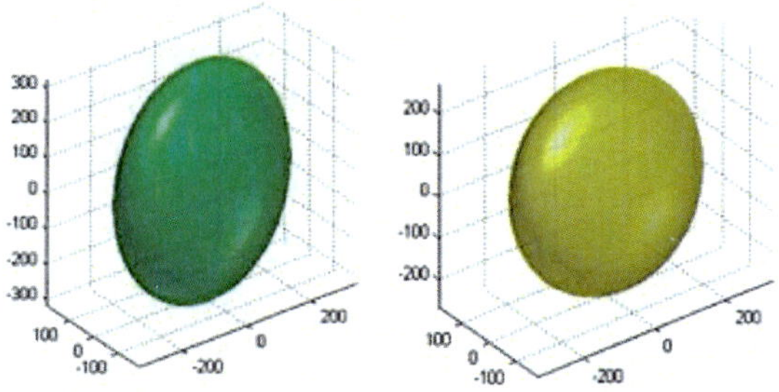

(b)

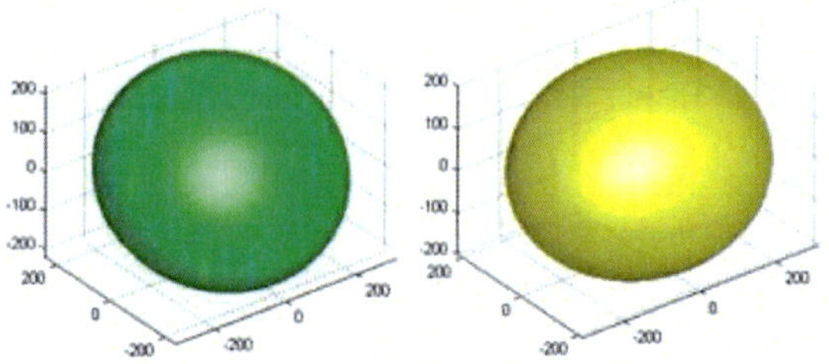

(c)

Fig. 7. Dual arm manipulability ellipsoids and directional dual arm manipulability ellipsoids

Table 2 shows the energy consumption, dual arm manipulability measure and directional manipulability measure for these cases. Values of α,β are constant 1 and d_1,d_3,d_5 are constant 260, 250, 200 cm.

Table 2. Cost of each case

	Power	DAMM	DMM	Cost with DAMM	Cost with DMM
Case(a)	346.5	147.42	126.71	493.92	473.21
Case(b)	342	160.88	126.38	502.88	468.38
Case(c)	349.5	150.32	123.47	499.82	472.97

7 Conclusion

A new measure called directional manipulability measure is proposed. The purpose of this paper is use directional manipulability measure instead of general manipulability measure to improve performance index of dual arm manipulator for object grasping. As shown above, for a given task, directional manipulability measure is more efficiency than general manipulability measure on the cost function.

Acknowledgements. This research was supported by Basic Science Research Program through the National Research Foundation of Korea(NRF) funded by the Ministry of Education, Science and Technology(2011-0026949).

This research was supported by the MKE(The Ministry of Knowledge Economy), Korea, under Specialized Field Navigation/Localization technology research center support program supervised by the NIPA(National IT Industry Promotion Agency).

References

1. Park, C., Park, K., Kim, D.: Design of dual arm robot manipulator for precision assembly of mechanical parts. In: International Conference on Smart Manufacturing Application, pp. 424–427 (April 2008)
2. Takahama, T., Nagatani, K., Tanaka, Y.: Motion planning for dual-arm mobile manipulator – realization of "tidying a room motion". In: International Conference on Robotics and Automation, vol. 5, pp. 4338–4343 (April 2004)
3. Bayle, J., Fourquet, Y., Renaud, M.: Manipulability Analysis for Mobile Manipulators. In: ICRA 2001, Seoul, South Korea, pp. 1251–1256 (May 2001)
4. Nagatani, T., Hirayama, A., Gofuku, Y., Tanaka: Motion planning for Mobile Manipulator with keeping Manipulability. In: IROS 2002, Lausane, Switzerland, pp. 1663–1668 (October 2002)
5. Yoshikawa, T.: Analysis and control of Robot manipulators with redundancy. In: Brady, M., Paul, R. (eds.) Robotics Research: The First International Symposium, pp. 735–747. MIT Press (1984)
6. Yoshikawa, T.: Manipulability of Robotic Mechanisms. International Journal of Robotics Research 4(2), 3–9 (1985)
7. Yoshikawa, T.: Foundation of robotics: Analysis and control. The MIT Press (1990)
8. Lema, V.C.K., Laub, A.J.: The Singular Value Decomposition: Its Computation and Some Application. IEEE Transactions on Automatic Control (1980)
9. Lee, Y.I.: A Study on the Manipulability Measure of Robot Manipulators. Journal of Control, Automation and Systems Engineering 4(1) (February 1998)
10. Zhang, Q., Chen, A.: Kinematics and Task-Oriented Manipulability Measure for the Bionic Cricket Hopping Robot. In: Xiong, C.-H., Liu, H., Huang, Y., Xiong, Y.L. (eds.) ICIRA 2008, Part I. LNCS (LNAI), vol. 5314, pp. 342–351. Springer, Heidelberg (2008)
11. Lee, S.: Dual redundant arm configuration optimization with task-oriented dual arm manipulability. IEEE Transations on Robotics and Automation, 78–97 (February 1989)
12. Yao, J., Ding, X., Zhan, Q.: On task-based directional manipulability measure of redundant robot (2000)

Gait Analysis for a Human with a Robot Walking Helper

Chun-Hsu Ko[1], Kuu-Young Young[2], and Sunil K. Agrawal[3,*]

[1] Dept. of Electrical Engineering, I-Shou University, Kaohsiung, 84001, Taiwan
chko@isu.edu.tw
[2] Dept. of Electrical Engineering, National Chiao Tung University, Hsinchu, 300, Taiwan
kyoung@mail.nctu.edu.tw
[3] Dept. of Mechanical Engineering, University of Delaware, Newark DE, 19716, USA
agrawal@udel.edu

Abstract. With the growth of elderly population in our society, intelligent walking aids will play an important role in providing functional mobility to humans. In this paper, we propose a model to compute gait of humans walking with a robot helper. This model is aimed at designing a control system for the robot walking helper. The human model includes both the single support phase and impacts. Since a human will be walking along with the robot with its help, geometrical constraints and interaction forces are included. To achieve stable walking, zero moment point (ZMP) is utilized in the analysis and friction constraint is included within the reaction force from the ground. Simulations are performed to obtain optimal gait trajectories, the human applied joint torques, and the supporting forces from the robot walking helper.

Keywords: walking helper, human, robot, gait analysis, passive.

1 Introduction

As the elderly population is growing at a rapid pace today, walking assist robots continue to be an important research topic [1-3]. Elder people often have reduced balance due to decline of muscular strength. Walking aids can provide assistive force to the elderly during walking. To safeguard their safety, intelligent walking aids are needed. It is timely to design intelligent walking helpers for the elderly.

Several robot walking helpers have been developed [1-10]. Spenko et al. [7] used a variable damping model to increase walking stability. Hirata et al. [8] proposed an adaptive motion control algorithm for obstacle/step avoidance and gravity compensation. Chuy et al. [9] used passive behavior to enhance the interaction between the user and support system. Agrawal et al. [8,10] proposed passive control algorithms for guiding a user to attain desired position and orientation. These control systems can provide the functions such as guidance [1,8,10], obstacle avoidance [2,7], gravity compensation [7,10], and user intention detection [1,2,4,8-10] in walking. It is also important to have synergistic control with the robot walker.

If the user states, such as the human limb angles, and joint torques are known during walking, the walker can provide more appropriate supporting force to maintain

---

* Corresponding author.

S. Lee et al. (Eds.): Intelligent Autonomous Systems 12, AISC 193, pp. 603–612.
springerlink.com

stability and performance. Kosuge et al. [11,12] used sensors to detect the user body positions within a human model for fall prevention and motion control. Nakano et al. [13] used the model of a biped robot with a wheelchair to design a gait training system. It is important to control the robot according to the user walking states.

In this paper, we propose a dynamic model for a human walking with a robot helper for gait analysis and optimization. The human model extends biped walking models [14-19] that include single support and impact phases. Our human model includes body segments from both the upper and lower body. Since a human walks synergistically with the robot, interaction forces and geometric constraint are included in the model during support phase. Furthermore, impact forces are considered in the model. To achieve stable walking, zero moment point (ZMP) is utilized and friction constraint is considered on reaction force from the ground. Finally, simulations of the human with robot walker, using the model, are performed to obtain optimal gait trajectories, human applied joint torques and helper robot support forces.

2 Model of Human with Robot Walking Helper

A conceptual diagram of a human walking with a robot helper is shown in Fig.1 (*left*). The human model is composed of a torso, an arm and two identical legs. The arm is composed of an upper arm and a forearm. Each leg is composed of two links and a foot [14]. The robot walking helper [9,10] consists of a support frame and two wheels with motors. It provides good support and stability for human walking. For the design of the robot walking helper, the dynamics of human with helper is described below.

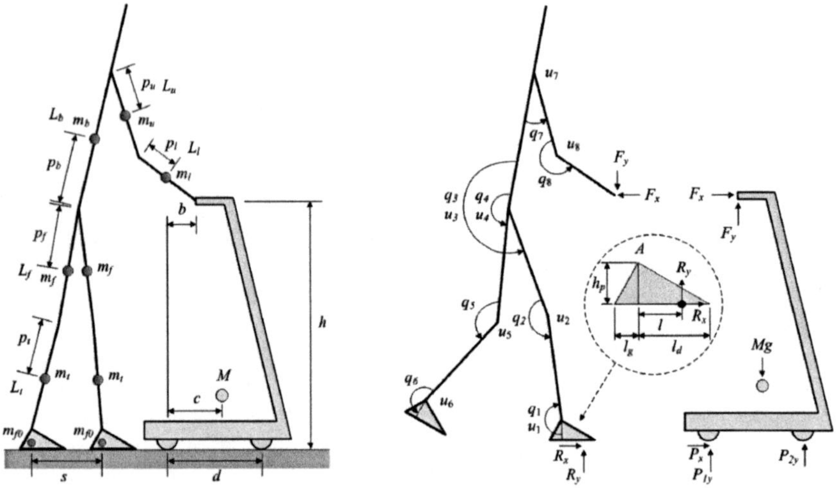

Fig. 1. A human with a robot walking helper (*left*) and the configuration of a human with a robot (*right*)

2.1 Dynamic Model

Fig. 1 (*right*) shows the configuration of human with a robot walking helper in joint coordinates given by $q=[q_1 \quad q_2 \quad \dots \quad q_8]^T$. From Lagrange's equations, the dynamic equations of motion of a human in single support phase can be expressed as

$$D(q)\ddot{q}+C(q,\dot{q})\dot{q}+G(q)=u-E_xF_x-E_yF_y \tag{1}$$

where $u=[u_1 \quad u_2 \quad \dots \quad u_8]^T$ is the human joint torques, D is the inertia matrix, C is the centrifugal and Coriolis matrix, G the gravity matrix, $[F_x\ F_y]$ is the human applied force on helper, and $[E_x\ E_y]$ the corresponding Jacobian matrices. The helper robot can only move in x-direction, hence its equations of motion are

$$\begin{aligned} M\dot{v} &= F_x + P_x \\ F_y + P_{1y} + P_{2y} - Mg &= 0 \\ bF_y + dP_{2y} - Mgc - F_xh &= 0 \end{aligned} \tag{2}$$

where P_x is the robot applied force, P_{1y}, P_{2y} are the reaction forces from ground, and v the robot speed. To make the robot passive and dissipative, the brake force with a constant gain is applied on the wheel and P_x can then be calculated as $P_x=-kv$. The parameter k is a nonnegative number. The human applied force F_x is then calculated as

$$F_x = M\dot{v} + kv \tag{3}$$

The reaction force P_{1y} and P_{2y} can be solved as

$$\begin{aligned} P_{1y} &= \frac{h}{d}P_x + (\frac{b}{d}-1)F_y + \frac{Mg(d-c)-hF_x}{d} \\ P_{2y} &= (Mgc + hF_x - bF_y)/d \end{aligned} \tag{4}$$

2.2 Impact Model

After the swing leg touches the ground, we assume that the stance leg will leave the ground so that the human will still remain in single support but with the other leg. In the event of swing phase touching the ground, the impact forces are generated at all joints. The robot's configuration is assumed to be unchanged before and after impact, denoted by q^-. The velocity is assumed to jump from $\dot{\tilde{q}}^-$ to $\dot{\tilde{q}}^+$. The impact model [19,20] can be written as

$$\tilde{D}(q^-)(\dot{\tilde{q}}^+ - \dot{\tilde{q}}^-) = E^T(q^-)[V_x \quad V_y]^T + H^T(q^-)H_y \tag{5}$$

where V_x, V_y are the horizontal and vertical impact forces of ground, respectively, H_y is the vertical impact force between the hand and walker, $\tilde{D}$ the mass matrix, and

E, *H* are the Jacobian matrices with respect to the swing foot contact point and at the hand. Since the velocities of the foot after the impact and hand along the vertical direction after the impact are zero, we get

$$E(q^-)\dot{\tilde{q}}^+ = 0, H(q^-)\dot{\tilde{q}}^+ = 0 \tag{6}$$

Eqs. (5) and (6) can be written as

$$\begin{bmatrix} \tilde{D}(q^-) & -E^T(q^-) & -H^T(q^-) \\ E^T(q^-) & 0_{2\times 2} & 0_{2\times 1} \\ H^T(q^-) & 0_{1\times 2} & 0_{1\times 1} \end{bmatrix} \begin{bmatrix} \dot{\tilde{q}}^+ \\ V_x \\ V_y \\ H_y \end{bmatrix} = \begin{bmatrix} \tilde{D}(q^-)\dot{\tilde{q}}^- \\ 0_{2\times 1} \\ 0_{1\times 1} \end{bmatrix} \tag{7}$$

With the above set of equations, the velocity $\dot{\tilde{q}}^+$ and the impact forces V_x, V_y, H_y can be calculated.

2.3 Kinematical Constraint

When the human grasps the robot helper, the height of his/her hand remains fixed in the inertial frame and is expressed as $F(q) = 0$, with

$$\begin{aligned} F(q) = hp - L_t \cos(q_1) + L_f \cos(q_{12}) - L_b \cos(q_{123}) \\ + L_u \cos(q_{123} - q_7) - L_l \cos(q_{123} - q_7 - q_8) - h \end{aligned} \tag{8}$$

where $q_{12} = q_1 + q_2, q_{123} = q_1 + q_2 + q_3$. The angle q_8 can be solved as

$$q_8 = \cos^{-1}\left[\begin{pmatrix} hp - L_t \cos(q_1) + L_f \cos(q_{12}) - \\ L_b \cos(q_{123}) + L_u \cos(q_{123} - q_7) - h \end{pmatrix} / L_l \right] + q_{123} - q_7 \tag{9}$$

With Eqs. (9), the degree-of-freedom of the system can be reduced to seven and the configuration can be described with joint angles $q_i, i = 1,2,\cdots,7$.

3 Gait Optimization

To obtain the optimal walking gait, we first perform the joint trajectory planning, and then analyze the ZMP equilibrium and friction constraints. Finally, we can find the optimal trajectory by solving the nonlinear constrained optimization problem.

3.1 Trajectory Planning

The angles $q_i, i = 1,2,\cdots,7$ are expressed as polynomial functions of time:

$$q_i(t) = a_{i0} + a_{i1}t + a_{i2}t^2 + a_{i3}t^3 + a_{i4}t^4, i = 1,2,\cdots,7 \tag{10}$$

The coefficients $a_{ij}, j = 0,1,\cdots,4$ can be solved with the initial conditions $q_i(0), \dot{q}_i(0)$, the angle $q_i(t_f/2)$, and final conditions $q_i(t_f), \dot{q}_i(t_f)$. Once $q_i(t), i = 1,2,\cdots,7$ are obtained, q_8 can be calculated using Eq. (12).

3.2 Zero Moment Point

To achieve stable walking, the human dynamic equilibrium must be satisfied. With the equilibrium in translation, the reaction forces R_x and R_y are calculated as

$$\begin{aligned} R_x &= F_x + m\ddot{x}_g \\ R_y &= F_y + m\ddot{y}_g + mg \end{aligned} \tag{11}$$

where (x_g, y_g) is the position of the center of mass of the human that can be calculated as

$$x_g = x_g(q), y_g = y_g(q) \tag{12}$$

With the equilibrium in rotation, the rate of change of angular momentum σ_A of the human about A is expressed as

$$\dot{\sigma}_A = -mgx_g + lR_y + h_pR_x + F_xy_h - F_yx_h \tag{13}$$

where h_p is the foot height, x_h and y_h the horizontal and vertical distances from the hand to point A, respectively. The angular momentum is linear with respect to the joint velocities and can be calculated as $\sigma_A = N(q)\dot{q}$. With Eqs. (11) - (13), the location of ZMP is obtained as

$$l = \frac{F_yx_h + P}{F_y + Q}, \text{ with } \begin{aligned} P &= \dot{\sigma}_A + mgx_g - h_pm\ddot{x}_g - F_x(y_h + h_p) \\ Q &= m\ddot{y}_g + mg \end{aligned} \tag{14}$$

The ZMP position must lie within the support foot, $-l_g \le l \le l_d$. We get

$$F_y \ge F_{y1}, F_{y1} = -\frac{(P + l_gQ)}{(x_h + l_g)} \tag{15}$$

and

$$\begin{cases} F_y \le F_{y2}, & x_h > l_d \\ F_y \ge F_{y2}, & x_h < l_d \end{cases}, F_{y2} = -\frac{P - l_dQ}{x_h - l_d} \tag{16}$$

3.3 Friction Constraints

When the foot is supported on the ground, the reaction force must satisfy the friction constraint. The impact force V_x and V_y between the foot and ground can be obtained using Eq. (7), hence

$$e \le 0, \text{with } e = |V_x| / \mu - V_y \tag{17}$$

where μ is the friction coefficient. Furthermore, the reaction forces between the support foot and ground obtained with Eq. (11) must also satisfy the friction constraint, we have

$$F_y \ge F_{y3}, F_{y3} = |F_x + m\ddot{x}_g| / \mu - m\ddot{y}_g - mg - F_y \tag{18}$$

On the other hand, the reaction force between the wheel and ground satisfies

$$P_{1y} \ge |P_x| / \mu, P_{2y} \ge 0 \tag{19}$$

Substituting Eqs. (3) and (4) into Eq. (19), we obtain

$$\begin{aligned} F_y &\le F_{y4}, F_{y4} = \frac{Mg(d-c) - hF_x - d|P_{1x}|/\mu}{d-b} \\ F_y &\le F_{y5}, F_{y5} = \frac{Mgc + hF_x}{b} \end{aligned} \tag{20}$$

3.4 Optimal Gait Generation

The gait optimization is performed to obtain the optimal joint trajectories that satisfy the constraints above. By minimizing the human applied torque *u*, the optimization problem is written as

$$\begin{aligned} &\min \frac{1}{t_f} \int_0^{t_f} (wu)^T (wu) dt \\ &\text{subject to } \; e \le 0 \end{aligned} \tag{21}$$

where *w* is the weight. The human applied torque *u* can be obtained from Eq. (1), and calculated as

$$u = AF_y + B, \quad A = E_y, B = D(q)\ddot{q} + C(q,\dot{q})\dot{q} + G(q) + E_x F_x \tag{22}$$

The optimal value of the force is obtained as $F_y = -A^T w^T wB / (A^T w^T wA)$. This F_y must also satisfy Eqs. (15), (16), (18), and (20). If not, we find the minimum from the minimum $(wu)^T (wu)$ values of boundary points $F_{yi}, i = 1,2,\cdots,5$. Once the optimal F_y value is found, the penalty method is then utilized to express the optimization problem (21) as an unconstrained optimization problem, described as

$$\min \int_0^{t_f} (wu)^T (wu) dt + \lambda (e)^+ \tag{23}$$

where λ are the penalty numbers, and $(.)^+$ defined as

$$(e)^+ = \begin{cases} e, \text{if } e > 0 \\ 0, \text{otherwise} \end{cases} \tag{24}$$

Finally, we can solve the unconstrained optimization problem (23) by using *Matlab fmincon* function.

4 Simulation Results

We performed the simulations based on the proposed model, with the human parameters listed in Table 1. The height and weight of the human are 1.7 m and 63.58kg, respectively. The parameters for the robot are M=30 kg, b=0.1 m, c=0.25 m, and d=0.5 m. The average walking speed was set to be 0.5 m/s. Fig. 2 (*left*) shows the human walking gait with a robot walking helper with h=0.86 m and k=0. The results show that the arm was almost straight for reducing the torque. The support forces are shown in Fig. 2 (*right*). The robot walking helper provided pull and push force in horizontal direction, and upward force in vertical direction. Figs. 3 (*left*) and (*right*) show the joint angles trajectories and human applied torques during walking, respectively. The human walked stably and smoothly with the robot walking helper. Since joints 1-3 apply the torques for support leg, they require larger torques than the other joints.

Table 1. Human parameters for simulation

	torso	tibia	femur	feet	upper arm	forearm
mass (kg)	33.3	2.51	7.82	1.15	4.78	2.52
length (m)	0.54	0.35	0.44	0.07	0.33	0.25

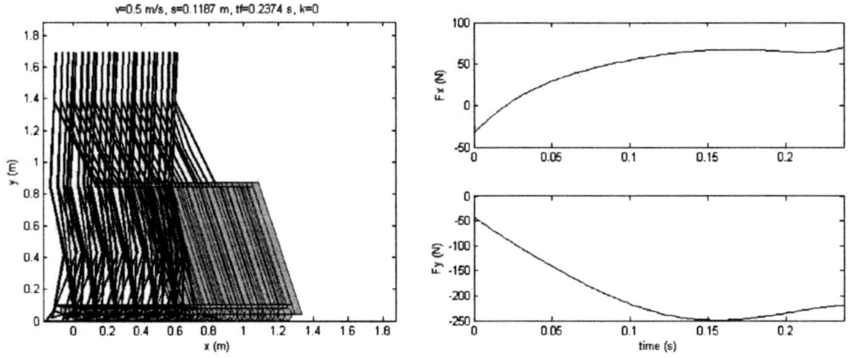

Fig. 2. A human walking with a robot helper with h=0.86 m and k=0 (*left*) and the support forces F_x and F_y (*right*)

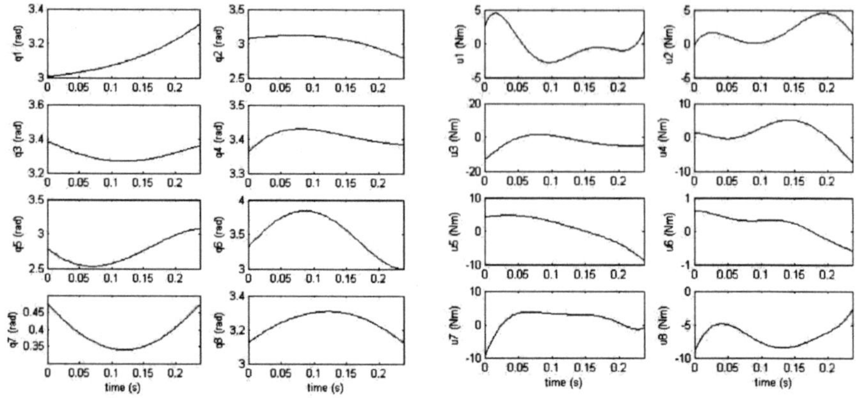

Fig. 3. The joint angle trajectories (*left*) and the human applied torques (*right*)

The effect of the passive force on human applied torque was studied with k set to be 0, 20, and 50, respectively, as shown in Fig. 4. It was found that knee torque u_2 was increased and knee torque u_5 was decreased slightly by increasing k from 0 to 20. And, the torques of $k = 0$ and $k = 50$ were almost the same.

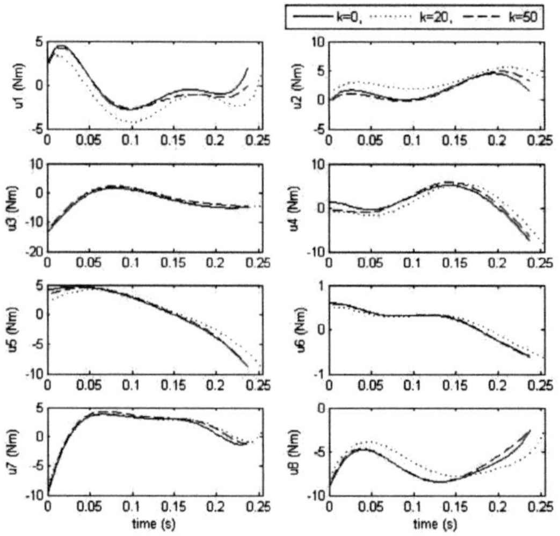

Fig. 4. The human applied torques with k=0, 20, and 50

5 Conclusions and Further Researches

In this paper, we have presented a dynamic model of a human and a robot walking helper. The dynamic model includes single support phase and impact. Geometrical constraint has been established. And, the ZMP equilibrium and friction constraints have also been analyzed within the framework proposed in the paper. Accordingly, optimal trajectories can be obtained by solving the nonlinear constrained optimization problem. The simulation results show that the robot walking helper can provide the support necessary to help the human walk stably. Since the moment between the user hand and the handle may affect the human walking, future work includes the analysis of the moment effect on walking.

Acknowledgements. This work was supported in part by the National Science Council of Taiwan, ROC. under the Grant NSC 100-2221-E-214-016. The third author also gratefully acknowledges partial support from World Class University program.

References

1. Yu, H., Spenko, M., Dubowsky, S.: An adaptive shared control system for an intelligent mobility aid for the elderly. Auton. Robots 15(1), 53–66 (2003)
2. Chuy, O., Hirata, Y., Wang, Z., Kosuge, K.: A control approach based on passive behavior to enhance user interaction. IEEE Trans. Robotics 23(5), 899–908 (2007)
3. Rentschler, A.J., Cooper, R.A., Blaschm, B., Boninger, M.L.: Intelligent walkers for the elderly: Performance and safety testing of VA-PAMAID robotic walker. J. Rehab. Res. Dev. 40(5), 423–432 (2003)
4. Wasson, G., Sheth, P., Alwan, M., Granata, K., Ledoux, A., Huang, C.: User intent in a shared control framework for pedestrian mobility aids. In: Proc. IEEE/RSJ Int. Conf. Intell. Robots Syst., pp. 2962–2967 (2003)
5. Sabatini, A.M., Genovese, V., Pacchierotti, E.: A mobility aid for the support to walking and object transportation of people with motor impairments. In: Proc. IEEE/RSJ Int. Conf. Intell. Robots Syst., pp. 1349–1354 (2002)
6. Spenko, M., Yu, H., Dubowsky, S.: Robotic personal aids for mobility and monitoring for the elderly. IEEE Transactions on Neural Systems and Rehabilitation Engineering 14(3), 344–351 (2006)
7. Hirata, Y., Hara, A., Kosuge, K.: Motion control of passive intelligent walker using servo brakes. IEEE Trans. Robotics 23(5), 981–990 (2007)
8. Ryu, J.C., Pathak, K., Agrawal, S.K.: Control of a passive mobility assist robot. Journal of Medicial Devices 2, 011002 (7 pages) (2008)
9. Yu, S.H., Ko, C.H., Young, K.Y.: On the design of a robot walking helper with human intension and environmental sensing. In: Proc. CACS Int. Auto. Cont. Conf. (2008)
10. Ko, C.H., Agrawal, S.K.: Control and path planning of a walk-assist robot using differential flatness. In: IEEE/RSJ International Conference on Intelligent Robots and Systems, pp. 6016–6021 (2010)
11. Hirata, Y., Komatsuda, S., Kosuge, K.: Fall prevention control of passive intelligent walker based on human model. In: IEEE/RSJ International Conference on Intelligent Robots and Systems, pp. 1222–1228 (2008)
12. Hirata, Y., Komatsuda, S., Iwano, T., Kosuge, K.: Motion control of walking assist robot system based on human model. In: International Conference on Biomedical Engineering, pp. 2232–2236 (2008)

13. Nakano, K., Murakami, T.: An approach to guidance motion by gait-training equipment in semipassive walking. IEEE Trans. Industrial Electronics 55(4), 1707–1714 (2008)
14. Chevallereau, C., Djoudi, D., Grizzle, J.W.: Stable bipedal walking with foot rotation through direct regulation of the zero moment point. IEEE Trans. Robotics 24(2), 390–401 (2008)
15. Agrawal, A., Agrawal, S.K.: An approach to identify joint motions for dynamically stable walking. Journal of Mechanical Design, Transactions of ASME 128, 649–653 (2006)
16. Tlalolini, D., Aoustin, Y., Chevallereau, C.: Design of a walking cyclic gait with single support phases and impacts for the locomotor system of a thirteen-link 3D biped using the parametric optimization. Multibody Syst. Dyn. 23(1), 33–56 (2010)
17. Chevallereau, C., Aoustin, Y.: Optimal reference trajectories for walking and running of a biped robot. Robotica 19(5), 557–569 (2001)
18. Wu, T.-Y., Yeh, T.-J.: Optimal design and implementation of an energy-efficient semi-active biped. Robotica 27(6), 841–852 (2009)
19. Grizzle, J.W., Abba, G., Plestan, F.: Asymptotically stable walking for biped robots: analysis via systems with impulse effects. IEEE Tran. on Automatic Control 46(1), 51–64 (2001)
20. Hurmuzlu, Y., Marghitu, D.B.: Rigid body collisions of planar kinematic chains with multiple contact points. Int. J. Robot. Res. 13(1), 82–92 (1994)

Global Pose Estimation with Adaptive GPS/IMU Fusion

Taek Jun Oh and Myung Jin Chung

Dept. of Electrical Engineering, KAIST
291 Daehak-ro, Yuseong-gu, Daejeon, Republic of Korea
buljaga@rr.kaist.ac.kr, mjchung@ee.kaist.ac.kr

Abstract. The aim of this study is to fuse GPS(Global Positioning System)/IMU(Inertial Measurement System) by using Kalman filter. We acquired the 6D pose data and compared the accuracy of 3D world model by using the data with Kalman filter and 3D world model by without filtering. Using proposed Kalman filter method, we obtain the exact pose data. This indicates that reconstructed 3D world model, using the proposed method is more accurate.

Keywords: Kalman Filter, GPS, IMU, 3D World Model, Sensor System.

1 Introduction

3D world model has received much attention recently because studies of unmanned vehicles has been developed by many institutions. 3D world model is 3D reconstruction of the data which is obtained from sensors such as cameras, laser scanners, etc. Over the past few years, several studies have been made on 3D world model in complex environment [1][2][3][4].

It is important to know an exact current location of a mobile robot to create 3D world model. In order to create 3D world model, it is necessary to know 6D pose data, which are latitude, longitude, altitude, roll, pitch and yaw. GPS provides latitude, longitude and altitude, and IMU provides roll, pitch and yaw. However, under certain circumstances the sensors provide incompletely accurate data. By using inaccurate data, reconstructing 3D world model also cannot be made entirely accurate. The data that GPS and IMU provide are complementary [5], so we can acquire accurate data by fusing the data from GPS and IMU. With the accurate data, we can reconstruct the accurate 3D world model.

In this paper, we proposed Kalman filter based GPS/IMU fusion method to correct the pose data. With the corrected 6D data (latitude, longitude, altitude, roll, pitch and yaw), we reconstruct 3D world model. To verify if the 6D data acquired through Kalman filter is accurate, we compared the accuracy of 3D world model by using the data without filtering and 3D world model by using the data corrected.

The remainder of this paper is organized as follow. The next section describes adaptive GPS/IMU fusion using kalman filter, the section 3 represents experimental results, and the section 4 represents conclusion.

S. Lee et al. (Eds.): Intelligent Autonomous Systems 12, AISC 193, pp. 613–620.
springerlink.com

2 Adaptive GPS/IMU Fusion Using Kalman Filter

GPS provides pose data. $S_k(x_k, y_k)$ is the kth pose data and $S_{k+1}(x_{k+1}, y_{k+1})$ is the k^{th+1} pose data where x is the longitude and y is the latitude. dx is the difference of x value and dy is the difference of y value. θ (yaw) can be represented as $\theta = \tan^{-1}(dy/dx)$.

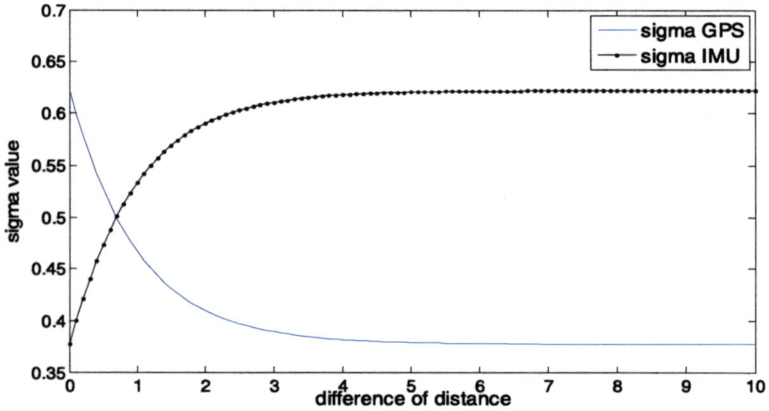

Fig. 1. Distribution σ_k^{GPS} and σ_k^{IMU}

Generally, update rate of GPS is longer than update rate of 2D laser scanner. Therefore the GPS pose data between the update rates, cannot be trusted. Also, if moving vehicle stop, GPS pose data have a value similar to its previous value, it occurs the GPS yaw error. Because of the IMU drift phenomenon, the IMU sensor data accumulates a small error. The IMU sensor data's error enlarges gradually, so IMU yaw cannot be trusted [6]. To solve these problems we propose Kalman filter based GPS/IMU fusion method, for the modeling each step's GPS error can be represented as follows:

$$e_k^{GPS} = \frac{1}{\exp(\sqrt{dx^2 + dy^2})} \tag{1}$$

$$\sigma_k^{GPS} = sigmoid(e_k^{GPS} - 0.5) \times K_G \tag{2}$$

$$\sigma_k^{IMU} = (1 - sigmoid(e_k^{GPS} - 0.5)) \times K_I \tag{3}$$

Where K_G and K_I are the tuning constants and figure 1 shows the distribution of σ_k^{GPS} and σ_k^{IMU} when K_G and K_I are 1 respectively.

System equation and measurement equation can be derived as follows:

$$x_{k+1} = Ax_k + w_k, \quad w_k \sim N(\vec{0}, Q) \tag{4}$$

$$z_{k+1} = Hx_k + v_k, \quad v_k \sim N(\vec{0}, R) \tag{5}$$

where state vector is $x_k = [\omega_k \quad \theta_k]^T$, and measurement vector is $z_k = [\omega_k^{IMU} \quad \theta_k^{GPS}]^T$. ω_k and θ_k are the angular velocity in the direction of yaw, and ω_k^{IMU} and θ_k^{GPS} are the angular velocity in the direction of yaw from IMU, and yaw from GPS. A is the system matrix where $A = \begin{bmatrix} 1 & 0 \\ \Delta t & 1 \end{bmatrix}$ and w_k is the process noise vector with zero mean. H is the measurement matrix where $H = diag(1,1)$ connecting measurement and states, and v_k is a measurement noise vector with zero mean. Initial state vector is $x_0 = [0 \quad 0]^T$, and initial noise covariance matrix is $Q_0 = 0$, and $R_0 = diag(1,1)$., w_k and v_k has a covariance that can be written as $Q_k = 0, \quad R_k = diag(\sigma_k^{IMU}, \sigma_k^{GPS})$.

3 Experiment Result

Figure 2 shows the overall experimental environment of our sensor system in (a) and our environment in (b). Our sensor system, Newxe was developed in our previous

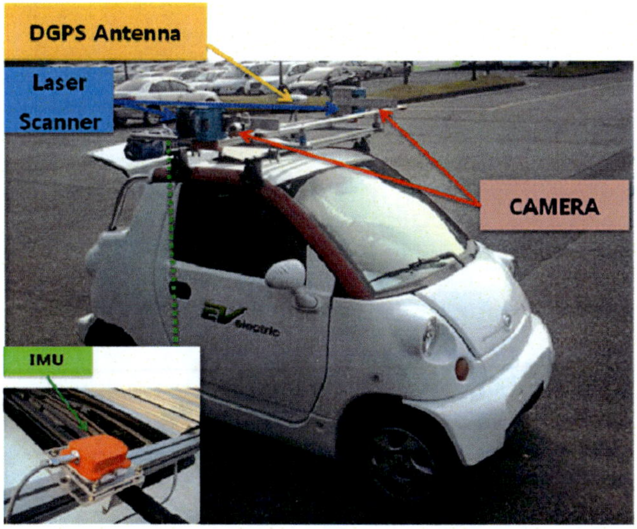

(a)

Fig. 2. (a) Sensor system, Newxe and (b) experimental environment

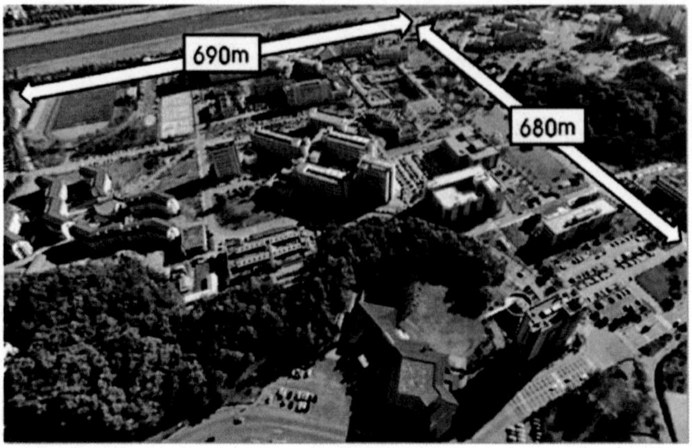

(b)

Fig. 2. *(continued)*

research [7]. Newxe is an electrical vehicle has two 2D laser scanners(SICK LMS291) and one GPS(HUACE DGPS B20 system), and one IMU(XSens MTi). Fig. 2 (b) is the campus of KAIST [8], which is the place we experimented this test. In order to verify the accuracy of the proposed method, we obtained raw data by driving at KAIST using the Newxe. Experimental configuration is shown in table 1.

Table 1. Experimental configuration

	Experimental Configuration
Driving velocity	20km/h(average)
2D Laser scanner	Update rate : 37.5Hz Max range : 80m Scanning angle : 180° Angle resolution : 0.5°
DGPS	Update rate : 1Hz
IMU	Update rate : 50Hz

In our experimental setup, DGPS has 1Hz update rate so in order to apply the Kalman filter, we set $\Delta t = 1$, $K_G = 100$, $K_I = 1$ and error covariance $P = diag(5,1)$.

Figure 3 shows GPS yaw, IMU yaw and GPS/IMU yaw using Kalman filter results. By integrating the angular value, IMU yaw was obtained. Overall, the yaw data showed that both the IMU and the GPS had similar variation, but there was a bias between IMU data and GPS data. Because the vehicle stopped, errors occurred on GPS yaw from step 20 to 30. Because of the errors, the value of GPS yaw became not reliable. Therefore we corrected the yaw error using the Kalman filter.

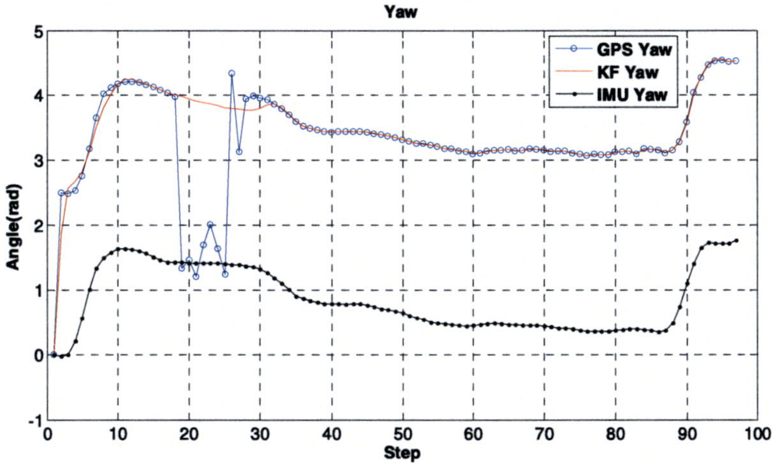

Fig. 3. GPS/IMU fusion data using Kalman filter

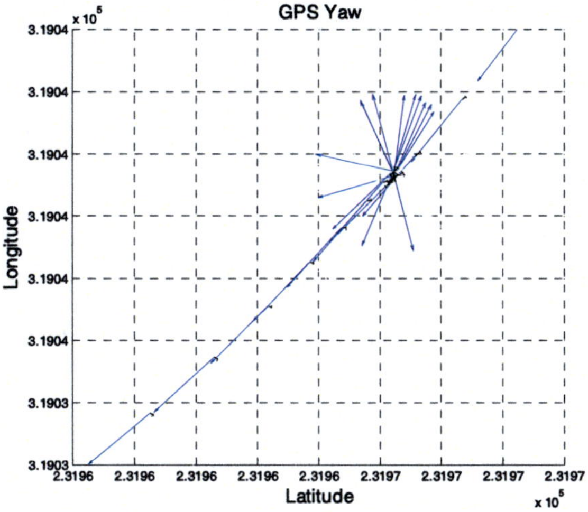

(a)

Fig. 4. Direction of yaw (a) GPS (b) IMU (c) GPS/IMU fusion

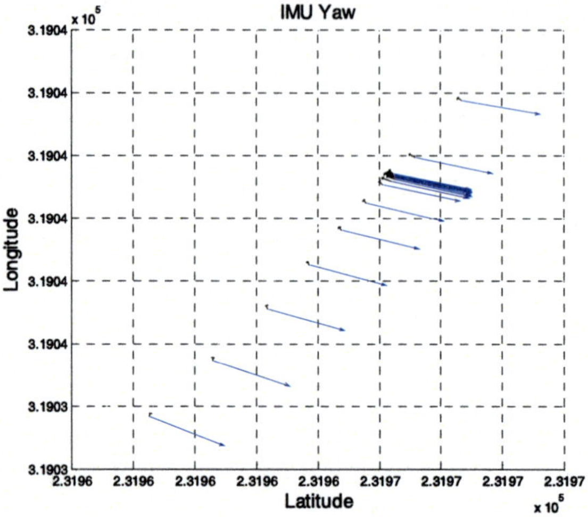

(b)

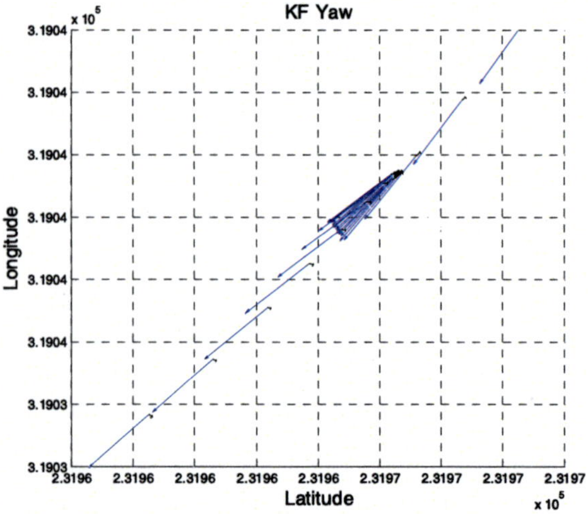

(c)

Fig. 4. *(continued)*

Figure 4 shows the direction of yaw. In (a), the direction of GPS yaw and the direction of the vehicle were the same. However, an error occurred when the vehicle stopped. In (b), the direction of IMU yaw and the direction of vehicle were not the same, due to the IMU error. In (c), that result is pose data using Kalman filter. The direction of yaw data and the direction of vehicle were the same. In this case, errors rarely occurred.

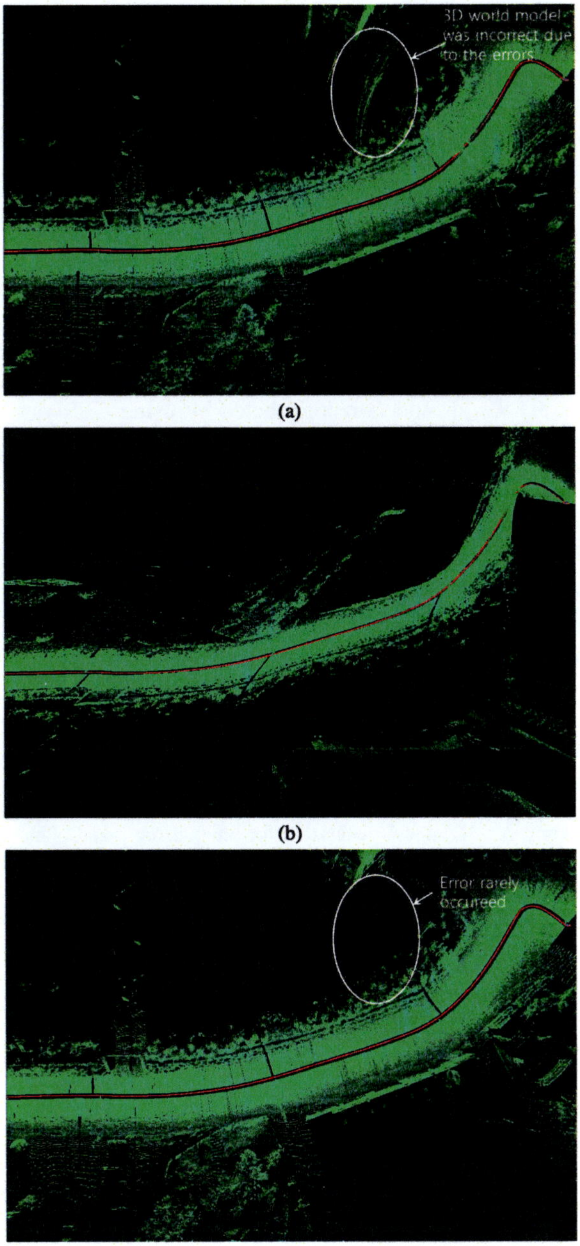

Fig. 5. 3D world model (a) GPS (b) IMU (c) GPS/IMU fusion

Figure 5 shows the 3D world model using 6D pose data and laser scanner data. 3D world model using GPS yaw in (a). Errors occurred when the vehicle stopped, and eventually 3D world model was incorrect due to the errors. Using the IMU yaw for 3D world model is inadequate in (b). Because the direction of yaw data and the direction of vehicle were not the same, 3D world model was incorrect. In (c), the 3D world model using the GPS/IMU yaw. Errors rarely occurred, so 3D world model was more accurate than the 3D world models of in (b) and in (c).

4 Conclusion

In this paper we proposed pose estimation method, which is based on data fusion of GPS/IMU, to obtain the exact pose data to reconstruct accurate 3D world model. We compared three 3D world models reconstructed by using GPS yaw data, IMU yaw data, and Kalman filter yaw data. We confirmed that the 3D world model by using pose from the proposed method was the most accurate. Therefore, to reconstruct 3D world model, using the yaw data from GPS/IMU fusion method is more suitable and reliable for 3D world model than the other two methods.

Acknowledgements. This research was supported by the MKE(The Ministry of Knowledge Economy), Korea, under the Human Resources Development Program for Convergence Robot Specialists support program supervised by the NIPA(National IT Industry Promotion Agency) (NIPA-2011-C7000-1001-0007).

References

1. Fruh, C., Zakhor, A.: 3D model generation for cities using aerial photographs and ground level laser scans. In: Proc. Computer Vision and Pattern Recognition (2001)
2. Georgiev, A., Allen, P.: Localization methods for a mobile robot in urban environments. IEEE Trans. Robotics 20(5), 851–864 (2004)
3. Nuchter, A., Surmann, H., Hertzberg, J.: Planning robot motion for 3d digitization of indoor environment. In: Proc. 11th Int. Conf. on Advanced Robotics (2003)
4. Thrun, S., Burgard, W., Fox, D.: A real-time algorithm for mobile robot mapping with applications to multi-robot and 3D mapping. In: IEEE Int. Conf. on Robotics and Automation (2000)
5. Caron, F., Duflos, E., Pomorski, D., Vanheeghe, P.: GPS/IMU data fusion using multisensor Kalman filtering: introduction of contextual aspects. Information Fusion 7(2), 221–230 (2006)
6. `http://www.xsens.com/en/general/mti`
7. Choe, Y.G., Shim, I.W., Ahn, S.U., Chung, M.J.: Geometrical Featured Voxel Based Urban Structure Recognition and 3D Mapping for Unmanned Ground Vehicle. Journal of Institute of Control, Robotics and Systems 17(5), 436–443 (2011)
8. Kim, S.J., Kang, J.W., Choe, Y.G., Park, S.U., Shim, I.W., Ahn, S.U., Chung, M.J.: Development of Sensor System and 3D World Modeling for Autonimous Vehicle. Journal of Institute of Control, Robotics and Systems 17(6), 531–538 (2011)

Sensor-Based Incremental Boustrophedon Decomposition for Coverage Path Planning of a Mobile Robot

Dugarjav Batsaikhan, Adiyabaatar Janchiv, and Soon-Geul Lee

Dept. of Mechanical Engineering, Kyung Hee University
1732 Deogyeong-daero, Giheung-gu, Gyeonggi-do, Yongin-shi, 446-701, Korea
{dbachka10,sglee}@khu.ac.kr, a_mz_s86@yahoo.com

Abstract. This paper presents a sensor-based coverage algorithm with which a robot covers an unknown rectilinear region while simultaneously constructing cell decomposition. In this algorithm, cell boundaries are indicated by combined lines of critical edges those are sensed partial contours of wall and objects in the workspace. The robot uses the laser scanner to sense critical edges. The proposed algorithm incrementally (one by one) decomposes unknown environment into cells. The constructing order of the cells is very important in this incremental cell decomposition algorithm. In order to decide next target cell from candidate cells, the robot checks redundancy of the planned path and possible position of ending points of the current cell. The key point of the algorithm is memorizing the covered space to define the next target cell from possible cells more than one. Path generation within the defined cell is determined to minimize the number of turns because the number of turns is the main factor to save time for coverage. Therefore, the long boundary of cell should be chosen as a main path of the robot. Verification of this algorithm is done by the simulation under LABVIEW environment.

Keywords: sensor-based, boustrophedon decomposition, incremental cell decomposition, complete coverage path planning.

1 Introduction

The task of covering a bounded region of space is common to many applications as including de-mining, floor cleaning, and other similar tasks. For mobile robots, cell decompositions are often employed in solving coverage problems, where a target position has to be reached without colliding with obstacles during movement. The main idea of the cell decomposition is to decompose a given bounded workspace into a set of non-overlapping regions, since each region is termed *cell*. The union of these regions should cover or approximate a subset of interest from the workspace, the regions not occupied by obstacles.

In early researches of cellular decomposition, one popular technique, which can yield a complete coverage path solution, is the trapezoidal decomposition [1] in which the robot's free space is decomposed into trapezoidal cells. Since each cell is a

S. Lee et al. (Eds.): Intelligent Autonomous Systems 12, AISC 193, pp. 621–628.
springerlinkc.com

trapezoid, coverage in each cell can easily be achieved with simple back and forth motions. Unfortunately, the trapezoidal approach has too much redundancy for the back and forth motions to guarantee completeness. In order to overcome this problem, H. Choset [2] introduced boustrophedon decomposition that was an enhancement of the trapezoidal decomposition. He merged narrow cells into one cell to reduce the excessive motions that designed to minimize the number of excess lengthwise motions.

Another development of the cell decomposition technique presented by Acar [3] was the Morse decomposition algorithm. He indicates the location of cell boundaries with Morse functions. Instead of analyzing the vertices of the given workspace, this algorithm looks for connectivity changes of the slice in the free space to locate the boundaries of the cells. These decomposition methods are formed by sweeping a line over the known workspace, and a new cell boundary is created whenever the sweep line encounters a vertex or critical point. However, a robot can also decompose the cells form an initially unknown workspace by using incoming sensor data to incrementally plan path and execute coverage. This task is so called sensor-based coverage and it directs a robot operating in such environment to explore every point of the workspace.

The method presented by Choset [4] incrementally generates an explicit representation of the workspace as a graph as it performs coverage, and uses the graph to direct coverage and decide when the task is completed. This method creates a set of cells (which collectively cover the environment when coverage is complete) while simultaneously exploring each cell. More aspects on the general idea of incremental cell decompositions can be found in [5, 6, 7]. These algorithms were based on terms of critical point of Morse function. To achieve coverage in an unknown workspace, the robot simultaneously covered the workspace and incrementally constructed the Reeb graph representation that represents the critical points as nodes and cells as edges. Furthermore, another algorithm, which is the hierarchical decomposition for coverage with an extended-range detector, was introduced in [7, 8]. This decomposition combines Morse decompositions and Voronoi diagram. The critical point sensing method was also developed. Notwithstanding, previous researches have several disadvantages such as redundancy for the coverage and creating next target cell where new cells were built automatically whenever sweeping line encounters a vertex or critical point.

In this paper, we propose a sensor-based single robot coverage algorithm, used to guarantee completeness, which covers an unknown workspace while simultaneously constructing a cell decomposition that can increase time efficiency and to minimize redundancy in coverage. This paper is organized as follows: Section 2.1 includes a brief overview of the boustrophedon cellular decomposition. Section 2.2 presents detailed description of the proposed algorithm. Simulation implementation is demonstrated in Section 3 and conclusions are presented in Section 4.

2 Algorithm

2.1 Boustrophedon Decomposition

Boustrophedon decomposition [2] that is an extended version of the trapezoidal decomposition, and it is designed to minimize the number of excess lengthwise motions. With the boustrophedon decomposition approach, the two narrow cells are merged into one cell to reduce the excessive motions. Then robot can cover each cell with back-and-forth boustrophedon motions. The difference between the trapezoidal decomposition and boustrophedon decomposition approach is the MIDDLE events: at the MIDDLE event in the trapezoidal decomposition, the current cell is closed, and the new one cell is opened as Fig. 1(a). However, the boustrophedon decomposition does not open nor close a cell, but rather simply update the current cell. When the IN and the OUT events occur, cells are opened and closed as Fig. 1(b).

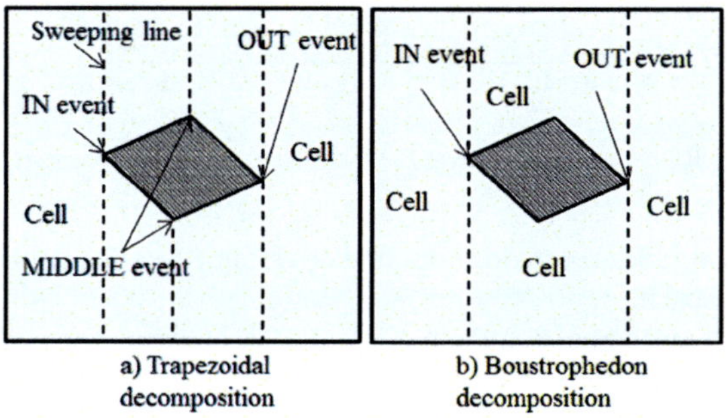

Fig. 1. Trapezoidal and boustrophedon decomposition

2.2 Algorithm Description

The main procedure of the proposed algorithm is totally different from the boustrophedon decomposition. It only adopts the boustrophedon motion which exactly covers a cell with simple back and forth motions. If a robot is equipped with a laser range finder (LRF), the robot can sense contours of wall and/or objects in the workspace which are termed *critical edges* as depicted in Fig. 2. Those are composed of points measured by the LRF and are represented as lines. Then line extraction is used to identify the critical edges those are used to create the new cell boundaries. If the current cell and the next candidate cells are composed, the redundancy of the planned path and the possible location of the ending point of the current cell are examined to determine constructing order of the cells. By managing efficiently constructing order of the cell for the coverage path planning, the proposed algorithm can reduce redundancy in coverage, guarantee completeness and minimize the covering time. The generated inside path of the cell should include minimum number of turns. Therefore, the long side of cell should be chosen as a main path for the robot to cover. Since the number of turns required to cover the cell is the main factor for

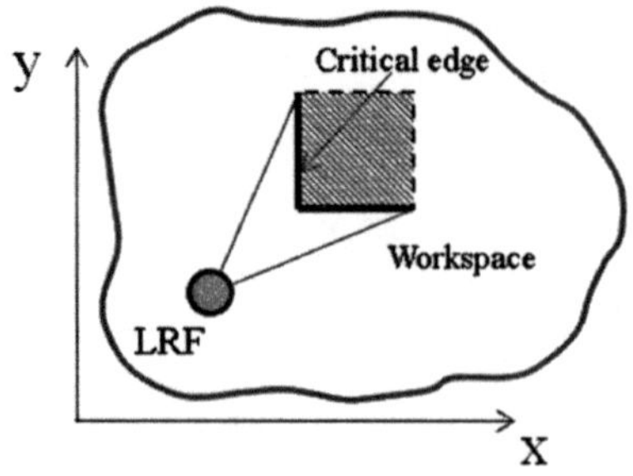

Fig. 2. Principle of the proposed algorithm

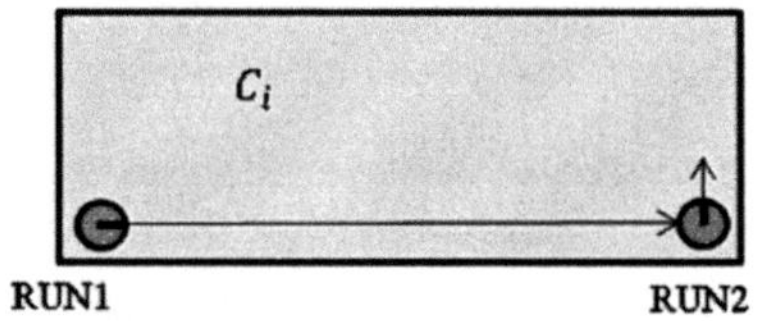

Fig. 3. RUN1 and RUN2 events for the robot motion in current cell C_i

saving time, the algorithm provides optimal covering times for not only single cell but also whole workspace.

The main difference between the boustrophedon and the proposed algorithm is that there are no IN events and no OUT events. Instead of them, RUN1 event is defined as longitudinal one and RUN2 event is perpendicular to RUN1 for robot motion. Another difference is that the proposed algorithm is working on online process. Because we assume to use rectilinear model for the workspace, each cell has rectangular shape. Therefore, the part of boundaries of the cell can be determined by critical edges of the workspace those are obtained by line extraction. Split and merge algorithm is used to obtain critical edges because it is probably the most popular and much faster than other algorithms for the line extraction [9].

After defining critical edges, each intersection points P_i of two adjacent critical edges is obtained. The virtual line L_i corresponding to the critical edge of an obstacle is extracted to determine cell boundary as shown in Fig. 4. By using alternatives to create boundaries of the cells, the proposed algorithm will decomposes workspace into cells. One important factor of proposed algorithm is the detecting range of the laser scanner. If the detecting range is not sufficient to acquire the information of the workspace, the robot should travel to get knowledge about the workspace, following one boundary of the cell that information is not completed. However, we supposed that detecting range of the laser scanner is sufficient long enough to scan the workspace in this paper.

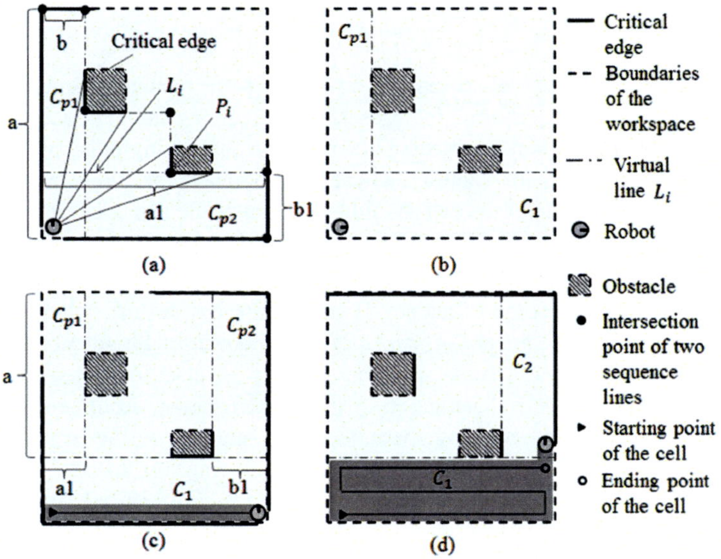

Fig. 4. Sensor-based incremental cell decomposition

The algorithm maintains three lists, D (discovered cells), C (current cell) and C_{pi} (next possible cells). The list D stores all discovered cells that have been visited, and the list C contains the current cell that the robot is covering currently. When the robot finishes covering the current cell, then the covered cell is removed from the list C and added to the list D. The list C_{pi} stores all the next possible cells that recognized with the laser scanner.

The main procedure of the proposed algorithm is shown in Fig. 4. The robot is assumed to start from one corner of the workspace and the laser scanner senses critical edges of the workspace. Then the robot detects that C_{p1} and C_{p2} are the two next possible cells as shown in Fig. 4(a). To determine the first cell to be covered, the robot compares lengths (a, a1) and widths (b, b1) of the next possible cells, C_{p1} and C_{p2}. It is assumed that information of both cells is completely known for the robot to compare the dimension. The longer and wider cell C_1 is chosen as the first cell as shown in Fig. 4(b) and, in this case, the longitudinal direction is the horizontal axis. Now the robot plans coverage path of the current cell C_1 with back and forth motions that causes the ending point. After the robot travels along the longer boundary of the cell until it reaches one corner of the cell or until RUN2 event occurs, the robot can gather information about the next possible cells, C_{p1} and C_{p2} in Fig. 4(c). The robot plans coverage path again and that eventually causes the ending point. Then the robot decides the path, which includes few or no redundant coverage depending on an ending point. The next target cell C_2 is chosen from possible candidates by the ending point of the current cell as shown in Fig. 4(d).

3 Simulation

Fig. 5 shows each step of complete coverage path planning simulation after Fig. 4, while the robot performs incremental cell decomposition. It shows that the next target cell is dynamically selected according to the ending point of the current cell, and the planned path should include few numbers of turns for each cell. Therefore, the longer side of the cell should be chosen as the main path of the robot as C_2 of Fig. 5(a).

The number of turns required to cover the cell is the main factor to save time. For the coverage algorithm, time efficiency is one of the key factors, which should be considered to improve the performance of the algorithm. Simultaneously, when RUN1 and RUN2 events occur, the robot scans around and repeats the same procedures; so incremental construction of the boustrophedon decomposition can be achieved. If there is a no remaining space, then the algorithm will be terminated. This means that the algorithm can guarantee complete coverage so that the robot passes over all free space in the given unknown workspace. Fig. 5(d) shows completed simulation result for the given unknown workspace. You can see sequence of the cells, $C_1 \rightarrow C_2 \rightarrow C_3 \rightarrow C_4 \rightarrow C_5 \rightarrow C_6 \rightarrow C_7$, that have been covered by the robot.

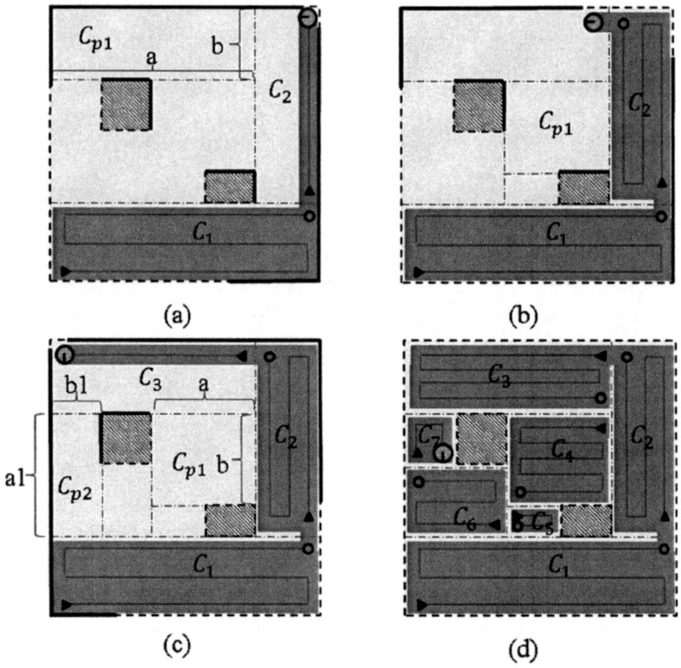

Fig. 5. Sensor-based complete coverage path planning

The proposed path planning algorithm is verified with simulations for the other two different workspace configurations that have two rectangular obstacles. Simulation space is 500cm × 500cm. As shown Fig. 6, the robot starts from the origin of the workspace's reference frame that is denoted as a red solid circle and completely covers the whole workspace with apparently time efficient paths for both simulations. In the figure, the black solid rectangles are representing obstacles, the red dotted blocks denote decomposed cells and the black solid line is the path of the robot.

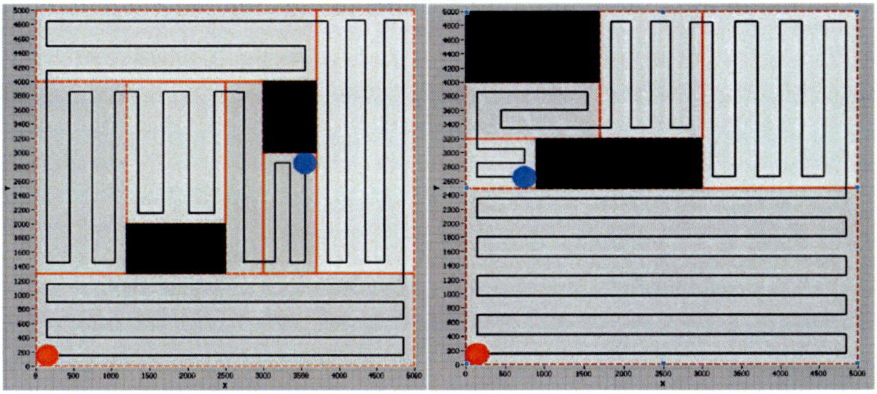

Fig. 6. Simulation for two different workspace configurations

4 Conclusion

This paper presents a sensor-based incremental boustrophedon decomposition algorithm based on the critical edges of the workspace. Most previous researches about cell decomposition are done with assumption that information of the workspace is initially given. Even though few of them develop cell decomposition algorithm for an unknown workspace, their algorithm is based on ultrasonic sensors or contact sensors. In those researches, the robot can directly recognize the critical points from sensors during traveling without any information about the possible cell. That is, if the robot detects critical point, new cells are created unconditionally.

Our proposed algorithm is incrementally (one by one) decomposed unknown workspace into cells based on efficient selection from possible alternatives of the cells. The new cell is created after identifying critical edges and the current cell is closed when the robot finishes covering the current cell. Hence, it was more applicable and efficient for coverage path planning of the mobile robot. That is, this algorithm can reduce revisiting cells that are already covered. In addition, robot planed its path inside the cell with few numbers of turns. This means that time for coverage of the single cell is efficient. Based on the assumption mentioned above, we can conclude that time for coverage of the whole workspace is optimum.

Acknowledgements. This research was partially supported by the Ministry of Knowledge Economy, Korea supervised by the KEIT (Korea Evaluation Institute of Industrial Technology) (No.: 10035544) and the MKE (The Ministry of Knowledge Economy), Korea, under the ITRC (Information Technology Research Center) support program supervised by the NIPA (National IT Industry Promotion Agency)(NIPA-2012-H0301-12-2008).

References

1. Latombe, J.C.: Robot motion planning. Kluwer Academic Publishers, Boston (1991)
2. Choset, H.: Coverage of known space: the boustrophedon cellular decomposition. Autonomous Robots 9, 247–253 (2000)
3. Acar, E., Choset, H., Rizzi, A., Atkar, N., Hull, D.: Morse decompositions for the coverage task. The Int. Journal of Robotics Research 21(4), 331–344 (2002)
4. Choset, H., Acar, E., Riazi, A., Luntz, J.: Exact cellular decompositions in terms of critical points of morse functions. In: The Int. Conf. on Robot Automation, San Francisco (2000)
5. Acar, E., Choset, H.: Robust sensor-based coverage of unstructured environment. In: The Int. Conf. on Intelligent Robots and Systems, Maui, Hawaii, USA (2001)
6. Acar, E., Choset, H.: Sensor-based coverage of unknown environments: incremental construction of Morse decompositions. The Int. Journal of Robotics Research 21, 345–366 (2002)
7. Choset, H., Burdick, J.: Sensor-based motion planning: incremental construction of the hierarchical generalized Voronoi graph. The Int. Journal of Robotics Research 19(2), 126–148 (2000)
8. Acar, E., Choset, H., Ji Yeong, L.: Sensor-based coverage with extended range detectors. IEEE Trans. on Robotics 22(1) (2006)
9. Nguyen, V., Martinelli, A., Tomatis, N., Siegwart, R.: A Comparison of line extraction algorithms using 2D laser range finder for indoor mobile robotics. In: The Int. Conf. on Intelligent Robots and Systems (2005)

Distributed Traffic Signal Control Using PSO Based on Probability Model for Traffic Jam

Cheng-You Cui and Hee-Hyol Lee

Graduate School of Information, Production and Systems, Waseda University, 2-7 Hibikino, Wakamatsu-ku, Kitakyushu, Fukuoka 808-0135
cuichengyou@ruri.waseda.jp, hlee@waseda.jp

Abstract. In this article, a new traffic signal control method is proposed. The Bayesian Network (BN) model and the Cellular Automaton (CA) model are used to build up a probability model for traffic jam. And then using the Particle Swarm Optimization (PSO) based on the probability model, the optimal traffic signal can be obtained. Finally, the effectiveness of the proposed method is shown with a micro-traffic simulator.

Keywords: traffic jam, bayesian network, cellular automaton, particle swarm optimization, probabilistic distribution.

1 Introduction

In recent years, traffic congestion has become a serious problem with the exponential growth in the number of vehicles. In urban areas, almost all of traffic congestion was occurred in intersections. The best way to solve this problem is to extend roads, but it is not easy to be realized in urban areas. In such cases, traffic signal control is considered to be an effective way to solve the problem.

Traffic signal can be divided into two types. One is the off-line (Pre-timed) control, and the other is the on-line (Adaptive) control. In pre-timed traffic signal control, Webster's formula is used to calculate the traffic signal timing off-line using historical traffic data. Pre-timed control cannot handle variations in traffic flow. On the other hand, adaptive control can overcome this limitation by adjusting the traffic signals on-line in response to changing traffic flows. Such as the SCOOT [1], SCATS [2], of centralized systems have been implemented on urban traffic networks successfully. However, centralized systems increase the requirement for extensive data processing and computational requirement for calculate optimal traffic signals. The distributed control systems can solves this limitation.

Some approaches such as [3] calculate optimal traffic time according to observed traffic data with empirical equations. The problem is the saturation flow and speed of traffic flow is variable under the different road and other factors such as weather. Therefore, to determine the parameters of the empirical equations are difficult.

Other approaches [4-5] calculate optimal traffic time according to observed traffic data with traffic flow model. They use the traffic model to predict future arrivals of the

S. Lee et al. (Eds.): Intelligent Autonomous Systems 12, AISC 193, pp. 629–639.
springerlink.com

different traffic phase timings at the intersection, and to obtain the best traffic signal by minimize the traffic delay time. If the road length between two intersections is very long or there are some minor roads, the traffic flow prediction is difficult.

Recently, the various intelligence techniques such as Fuzzy concept, Neural Networks, and Reinforcements Learning agent [6-11] have been used to implement distributed control systems. However, the fuzzy concept needs to obtain the appropriate rule base and the NN needs training data, but it is very difficult to obtain. The Reinforcements Learning agent needs long learning time.

In this paper, a new real-time traffic signal control method is proposed. In the system, any rule base and training data are not required. The Bayesian Network (BN) model is used to predict the probabilistic distribution of the standing vehicles of the roads with the observed traffic flows. The Cellular Automaton (CA) traffic model is used to update the priori probability of the traffic flows of the different traffic signals. And then, Particle Swarm Optimization (PSO) is used to calculate the optimal traffic signals according to minimize the probability of traffic jam.

2 Structure of Proposed Traffic Signal Control System

The structure of the proposed system is shown in Fig.1. The procedure of calculation is as following:

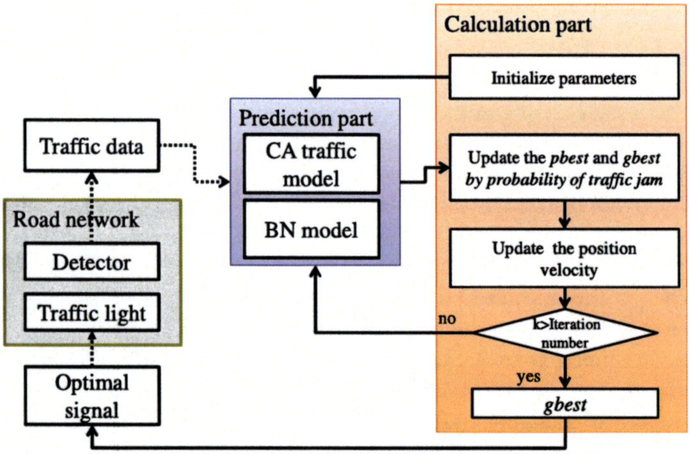

Fig. 1. Structure of the proposed traffic control system (CAPSOBN)

Step1: to initialize the particle position and velocity within the adjustable range. Each particle represents a traffic signal.

Step2: using the CA traffic model to estimate the traffic data at different traffic signal (particle), and to update priori probabilistic distribution of the inflow and outflow.

Step3: using the BN model to predict probabilistic distribution of the standing vehicles to calculate traffic jam probability.
Step4: according to the velocity and fitness (traffic jam probability) to update particle.
Step5: if reached maximum iteration number, the searching will be stopped, and the best particle as the optimal traffic signal.

2.1 Prediction of Probabilistic Distribution of Standing Vehicles Using Bayesian Network Model

The Bayesian network is a directed and acyclic graphical model, each node represents variables of the given problem. The relationship between each variable is evaluated quantitatively using the conditional probability.

Here, we consider a crossroads as shown in Fig.2. The random variables of the inflows and the outflows of the crossroad and the standing vehicles between the two intersections are represented as nodes. The Bayesian network model of the standing vehicles is shown in Fig.2.

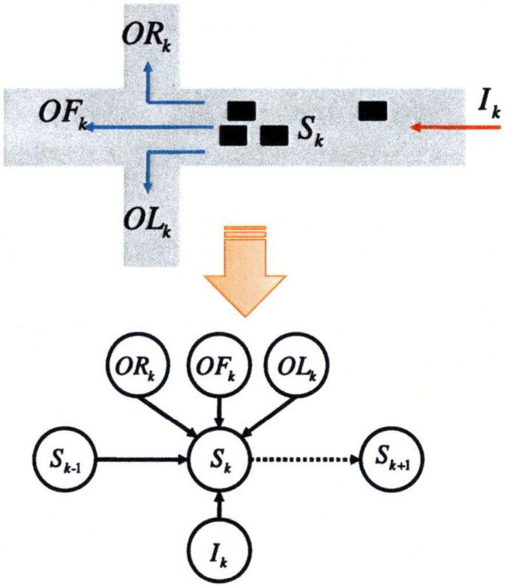

Fig. 2. Crossroad and BN model

The number of the standing vehicles of k-th cycle can be calculated as equation (1).

$$S_k = S_{k-1} + I_k - (OR_k + OF_k + OL_k) \tag{1}$$

The probabilistic distribution of the standing vehicles is obtained by summing over all joint probability distribution of the other variables. With the chain rule, the joint probabilistic distribution is represented as the product of conditional probability as following:

$$P(S_k, S_{k-1}, I_k, OR_k, OF_k, OL_k) = P(S_k \mid S_{k-1}, I_k, OR_k, OF_k, OL_k) \\ \times P(S_{k-1} \mid I_k, OR_k, OF_k, OL_k) \times P(I_k \mid OR_k, OF_k, OL_k) \\ \times P(OR_k \mid OF_k, OL_k) \times P(OF_k \mid OL_k) \times P(OL_k) \tag{2}$$

And according to the d-separation, equation (2) can be represented as

$$P(S_k) = \sum_{S_{k-1}} \sum_{I_k} \sum_{OR_k} \sum_{OF_k} \sum_{OL_k} P(S_{k-1}) \times P(I_k) \times P(OR_k) \times P(OF_k) \times P(OL_k) \tag{3}$$

According to above equation, the probabilistic distribution of the standing vehicle is calculated.

2.2 Estimation of Traffic Flows Using Cellular Automaton Traffic Model and Updating Rule of Priori Probability

Generally, the priori probability of the traffic flows is variable under the different traffic signal. Hence, we use the CA traffic model to estimate the traffic flow of the different traffic signals. And then, according to the estimated traffic flows the priori probabilistic distributions will be updated. The updating process is as following:

- Using the observed traffic data of before the k-th cycle to calculate the probabilistic distribution of inflow $\sum_u P(I_{k-1} = u)$ and outflow $\sum_u P(O_{k-1} = u)$.
- Using the CA traffic model to estimate the traffic inflow I_k^e and outflow O_k^e of the different traffic signal of the k-th cycle.
- To calculate the priori probabilistic distributions of inflow $\sum_u \hat{P}(I_k = u)$ and outflow $\sum_u \hat{P}(OR_k = u)$ $\sum_u \hat{P}(OF_k = u)$ $\sum_u \hat{P}(OL_k = u)$ by the equation (4)-(11).

$$\hat{P}(I_k^i = I_{k-1}^i - E(I_{k-1}^i) + I_k^e) = P(I_{k-1}^i) \tag{4}$$

$$E(I_{k-1}^i) = \sum_i I_{k-1}^i \times P(I_{k-1}^i) \tag{5}$$

$$\hat{P}(OR_k^i = OR_{k-1}^i - E(OR_{k-1}^i) + OR_k^e) = P(OR_{k-1}^i) \tag{6}$$

$$E(OR_{k-1}^i) = \sum_i OR_{k-1}^i \times P(OR_{k-1}^i) \tag{7}$$

$$\hat{P}(OF_k^i = OF_{k-1}^i - E(OF_{k-1}^i) + OF_k^e) = P(OF_{k-1}^i) \tag{8}$$

$$E(OF_{k-1}^i) = \sum_i OF_{k-1}^i \times P(OF_{k-1}^i) \tag{9}$$

$$\hat{P}(OL_k^i = OL_{k-1}^i - E(OL_{k-1}^i) + OL_k^e) = P(OL_{k-1}^i) \tag{10}$$

$$E(OL_{k-1}^{i}) = \sum_{i} OL_{k-1}^{i} \times P(OL_{k-1}^{i}) \quad (11)$$

$$(i \in u)$$

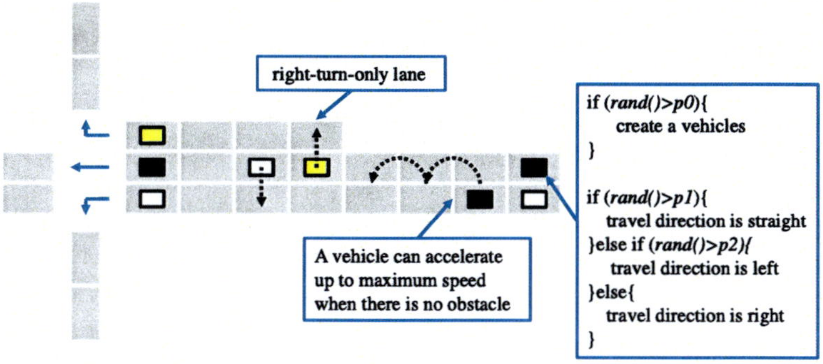

Fig. 3. Rules of the vehicles movements of the micro simulator

In order estimate traffic flow, the CA traffic model is established based on the Na-Sch model [12]. The rules of vehicles movements on the road network are given below, and as an example it is shown in Fig.3. The parameters of the CA model *p0*, *p1* and *p2* are determined by the observed data.

Input to cell: According to the generation of a random number with a inflow rate, a vehicle will be generated and the direction of travel through the next intersection will be determined. The inflow rate of the each road can be obtained by the observed traffic data.

Movements: A vehicle can accelerate up to maximum speed when there is no obstacle. Considering the mean speed on the urban area, the maximum speed is setup to 2cell/1step (1cell=7.5 m; 1step=1sec). According to the density of the road the speed can be changed randomly. On the multi-lanes, vehicles can move in parallel lanes. If the direction of travel is right turn, the vehicle moves to the right-turn-only lane.

Intersection: When the traffic signal is green, the vehicle will be allowed to cross the intersection according to its direction of travel, and the direction of travel on the next intersection will be reset.

2.3 Optimization of Traffic Signal Using PSO

The objective of the proposed system is to search the optimal traffic signals of the adjustable range to minimize probability of the traffic jam. In order to search optimal traffic signals, the Particle Swarm Optimization is utilized. Due to speed of convergence of PSO is very fast, real-time traffic signal control can be achieved.

PSO was introduced by Kennedy and Eberhart [13], it uses a simple mechanism that mimics swarm behavior in birds flocking and fish schooling to guide the particles to search for globally optimal point. The initial particle is created with random solutions. Each particle flies in D-dimensional problem space by a velocity, which is adjusted at each step. According to its own past best position and the position of the best of the swarm, the velocity is updated. The quality of a particle position depends on a fitness function. Fig.4 shows the procedure of PSO algorithm.

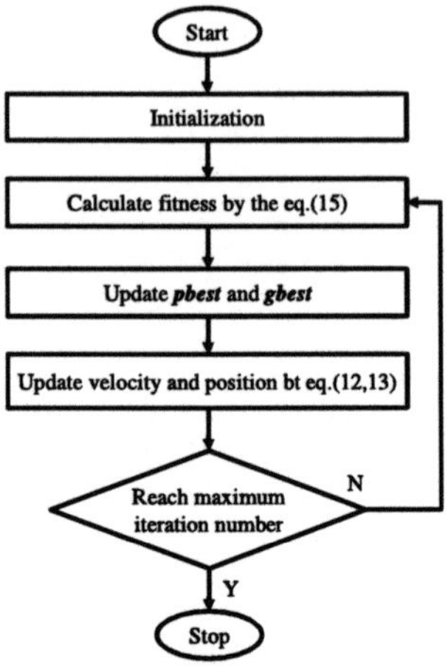

Fig. 4. Procedure of PSO algorithm

In the proposed method, each particle represents a cycle of the traffic signal, and each dimension is determined by the number of the different phases of the traffic signal. Particle's position and velocity are updated according to equation (12, 13). The parameter w_l is changed by equation (14), l is the iteration number, $rand()$ is the random value from 0 to 1.

$$v_{l+1}^d = w_l \times v_l^d + c_0 \times rand() \times (pbest^d - x_l^d) + c_1 \times rand() \times (gbest^d - x_l^d) \tag{12}$$

$$x_{l+1}^d = x_l^d + v_{l+1}^d \tag{13}$$

$$w_l = w_{\max} - \frac{w_{\max} - w_{\min}}{l_{\max}} \times l \tag{14}$$

And, the other parameters of the equations are setup as following;

$$c_0 = c_1 = 2 \quad w_{\max} = 0.9 \, ; w_{\min} = 0.9 \, .$$

When the position of the particle is changed, the priori probabilities of the flows will be updated. And then, using the equation (3) the predicted probabilistic distributions of the standing vehicles of the road can be obtained. After that, using the equation (15) the fitness of the particle is calculated.

$$F(x) = w_{ew} \times \sum_{S=S_{\max}}^{S=+\infty} P_{ew}(S_k) + w_{ns} \times \sum_{S=S_{\max}}^{S=+\infty} P_{ns}(S_k) \tag{15}$$

In order to reduce the calculation time, only one road of east and west roads or north and south road is considered to predict traffic jam probability. Because of the signals are the same on the east and west road or north and south road. For example, if the traffic of the east road is heavy than west road, $P_{ew}(S_k)$ is represents the predicted probabilistic distribution of the east road. If the number of standing vehicles is bigger than $S_{\max}$, the road is considered to have a traffic jam. And $S_{\max}$ is determined by the road lengths. The set values w are determined by the importance degree of the road.

3 Simulation

To prove the effectiveness of the proposed method, an experiment was carried out with a micro-traffic simulator. The micro-traffic simulator is developed based on the CA traffic model. We assume a road network and detector systems. The road network of the experiment includes 16 intersections, and all traffic data of the intersections can be observed by the detector systems. The simulator frame is shown in Fig.5. The conditions of the traffic flows and road networks are shown in table 1. The parameters of the fitness function are set up as following: $w_{ew} = 1$; $w_{ns} = 0.5$; $S_{\max} = 30$[cell]

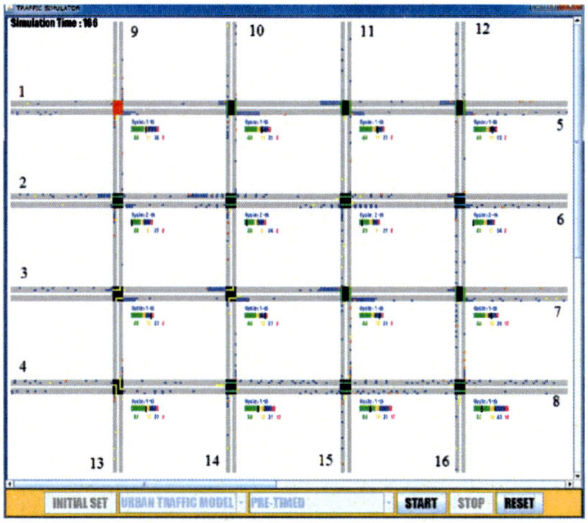

Fig. 5. Frame of the micro simulator

Each pre-timed signal of the intersections was calculated according to the Webster's formula. As an example, pre-timed signals of the red intersection of the figure 5 are shown table 2.

Table 1. Traffic inflows of the road network

Direction	Road length	Lane	direction of ravel straight	right	left	No.	Inflows (pcu/h)
						1	1300
West to East	80[cell]	3	90%	4%	6%	2	1100
						3	1400
						4	1500
						5	1400
East to West	80[cell]	3	90%	4%	6%	6	1300
						7	1000
						8	900
						9	600
South to North	70[cell]	2	60%	10%	30%	10	400
						11	500
						12	500
						13	400
North to South	70[cell]	2	60%	20%	20%	14	500
						15	500
						16	600

* pcu : passenger car unit ; 1 cell=7.5m

Table 2. Pre-timed signals of the intersection 1

	EW green	EW yellow	NS green	EW yellow
phases	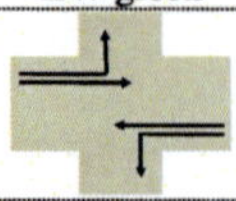	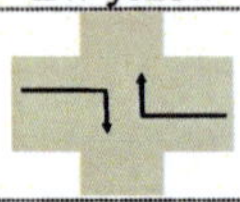	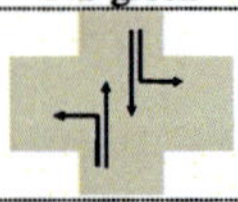	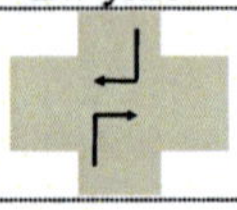
Time (sec)	40	10	15	5

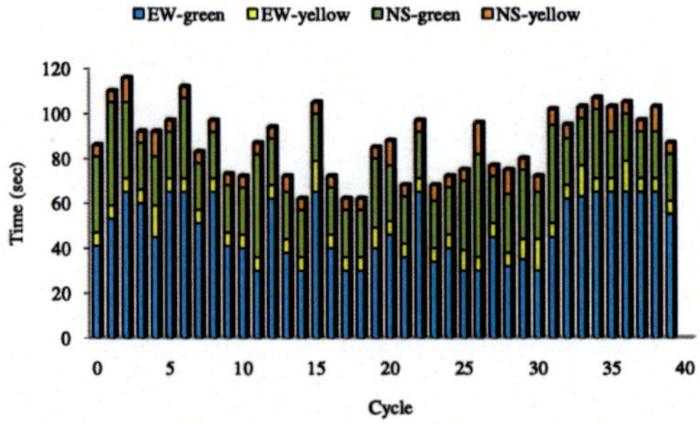

Fig. 6. Traffic signals of the red intersection by proposed method

And, the traffic signals of the intersection 1 suggested by the proposed method are shown in Fig.6. Compared with pre-time signals, the signals by the proposed method are updated according to the changing traffic flows. In addition, the predicted probabilistic distributions of the intersection at the first cycle are shown in Fig.7 and Fig.8. Fig.7 shows the probabilistic distributions of the standing vehicles on the pre-timed signals, and Fig.8 shows the probabilistic distributions of the standing vehicles of on the suggested signals by the proposed method. The probability of the traffic jam on the suggested signals is smaller than on the pre-timed signals. In the same way, the traffic signals of the other intersections will be updated according to the optimization for traffic-jam probability.

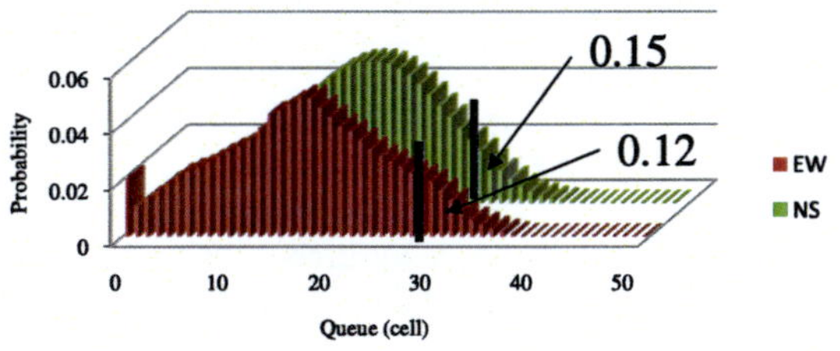

Fig. 7. Predicted probabilistic distributions of the road on pre-timed signal

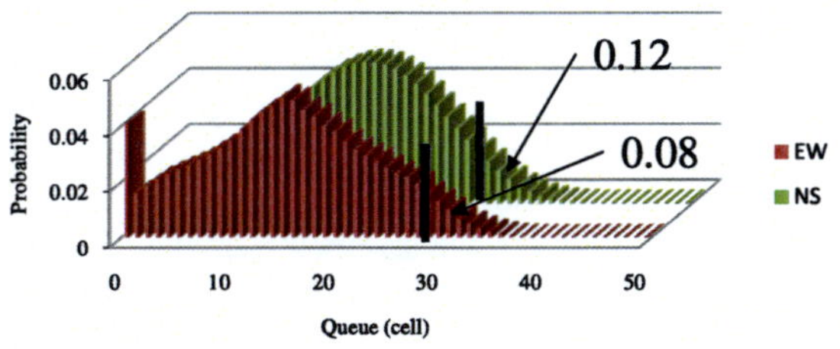

Fig. 8. Predicted prob. dist.of the road the signal on proposed method

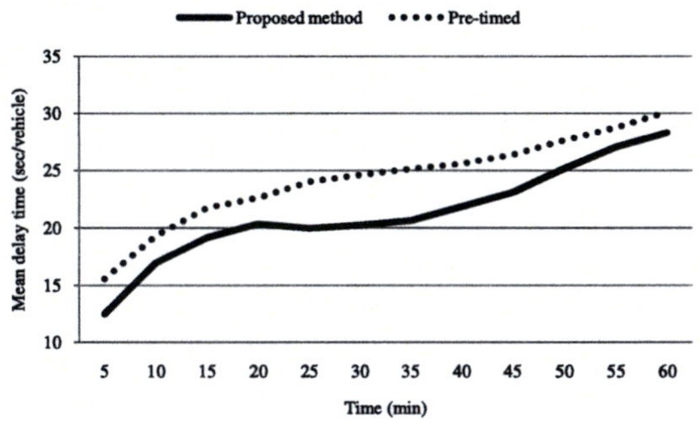

Fig. 9. Mean delay time

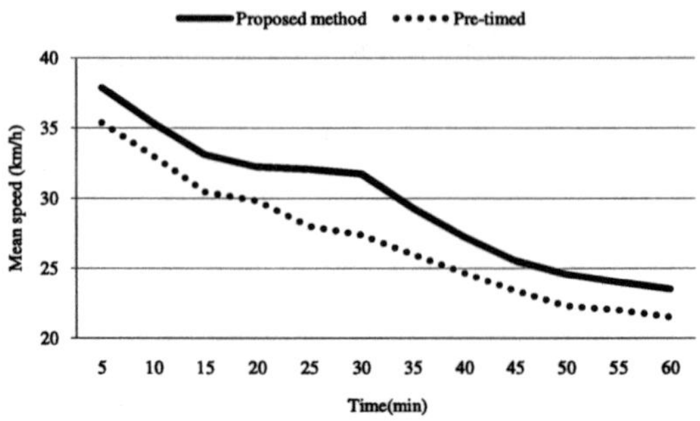

Fig. 10. Mean speed

By the proposed method, the mean delay time is decreased about 15% compared with pre-timed control, and the result is shown in Fig.9. And, the comparison of mean speed of proposed method and pre-timed control is shown Fig.10. By the proposed method, the mean speed is increased.

4 Conclusion

We have proposed a new traffic signals system, it based on the optimization model of the traffic-jam probability, and using the PSO method to calculate the optimal signals. Finally, through an experiment with the micro simulator the effectiveness of the proposed method is shown. However, we assumed a simple road network to carry out an experiment not according to the real traffic data. So, as the future work, we will use the real traffic data to test the proposed system, and to discuss the results sequentially to improve the proposed method.

References

1. Peck, C., Gorton, P.T.W., Liren, D.: The Application of SCOOT in developing countries. In: Third Int. Conf. on Road Traffic Control (1990)
2. Sims, A.G., Dobinson, K.W.: The Sydney Coordinated Adaptive Traffic (SCAT) system philosophy and benefits. IEEE Trans. Veh. Technol. 29, 130–137 (1980)
3. Kouvelas, A., Aboudolas, K., Papageorgiou, M., Kosmatopoulos, E.B.: A Hybrid Strategy for Real-Time Traffic Signal Control of Urban Road Networks. IEEE Trans. Intelligent Transportation System 12(3), 884–894 (2011)
4. Lee, J., Abdulhai, B., Shalaby, A., Chung, E.-H.: RealTime Optimization for Adaptive Traffic Signal Control Using Genetic Algorithms. Journal of Intelligent Transportation System 9(3), 111–122 (2005)
5. Dong, C., Huang, S., Liu, X.: Urban Werea Traffic Signal Timing Optimization based on Sa-PSO. In: 2010 International Conference on Artificial Intelligence and Computational Intelligence (2010)
6. Murat, Y.S., Ergun, G.: A fuzzy logic multi-phased signal control model for isolated junctions. Transportation Research Part C 13(1), 19–36 (2005)

7. Gokulan, B.P., Srinivasan, D.: A fuzzy logic multi-phased signal control model for isolated junctions. IEEE Trans. intelligent Transportation System 11(3), 714–727 (2010)
8. Chao, K.-H., Lee, R.-H., Wang, M.-H.: An intelligent traffic light control based on extension neural network. In: Lovrek, I., Howlett, R.J., Jain, L.C. (eds.) KES 2008, Part I. LNCS (LNAI), vol. 5177, pp. 17–24. Springer, Heidelberg (2008)
9. Dipti, S., Choy, M.C., Cheu, R.L.: Neural Network for Real-Time Traffic Signal Control. IEEE Trans. Intelligent Transportation System 7(3), 261–272 (2006)
10. Dai, Y., Hu, J., Zhao, D., Zhu, F.: Neural Network Based Online Traffic Signal Controller Design with Reinforcement Training. In: 2011 14th IEEE Confer. on Intelligent Transportation System (2011)
11. Balaji, P.G., German, X., Srinivasan, D.: Urban traffic signal control using reinforcement learning agents. Intelligent Transport Systems, IET 4(3) (2010)
12. Nagel, K., Schreckenberg, M.: A cellular automaton model for freeway traffic. J. Phys. I France 2(12), 2221–2229 (1992)
13. Kennedy, J., Eberhart, R.: Particle swarm optimization. In: Proceedings of the IEEE International Conference on Neural Networks (1995)

Design of a Sensing Limited Autonomous Robotic System

Lim Han Yang Benjamen[1] and Marcelo H. Ang Jr.[2]

[1] Anglo Chinese Junior College
Singapore, 139745
[2] Department of Mechanical Engineering,
National University of Singapore, Singapore 117576
olive_tree132@hotmail.com, mpeangh@nus.edu.sg

Abstract. This paper describes the development and implementation of a behavioral-based solution for a sensing-limited robotic system in area coverage problem using the LEGO Mindstorms NXT robotics kit. The main aim is to investigate how area coverage algorithms can be implemented on a robot with limited sensing and processing capabilities to cover a given area efficiently without localization or map building as well as to compare these algorithms with each other. Interestingly, there has been limited research done in this aspect, particularly in the efficiencies of current heuristic based algorithms in commercially available robots. In this project, three behaviors: Random Walk, Spiral, and Weave are proposed. A robot was constructed to mimic a sensing limited robot and it was used to carry out the proposed behaviors to determine the most efficient behavior. During experiment, it was found that aspects such as the parameter being measured and placement of obstacles in the environment affected the perception and performance by the robot. Results showed that the zigzag motion of Weave was the most efficient movement of the three, performing consistently well under varied environments and measurements.

Keywords: sensing-limited robotics, behavior-based robotics, Lego Mindstorms NXT, area coverage, heuristics.

1 Introduction

Autonomy has always been at the core of robotics. One such problem is area coverage, which has found many important applications such as cleaning [1] or searching an area. In area coverage, there are a variety of solutions available. A popular method is to use Simultaneous Localization and Mapping (SLAM) which allows the robot to create and update a map of an unknown environment as well as to use landmarks to locate its position within the map. From these, the robot can then plan its path according to its objectives. SLAM is highly accurate at determining an unknown area and mapping it out, so that it becomes increasingly efficient in subsequent iterations. However, a few aspects such as the size of the robots, as well as the cost of their sensors, usually laser range finders, limit their appeal outside of the

S. Lee et al. (Eds.): Intelligent Autonomous Systems 12, AISC 193, pp. 641–648.
springerlink.com

academic arena. The practicality of such an approach in a dynamic and cluttered environment is questionable.

Smaller robots might be the key to this problem, as they usually require fewer materials to construct, reducing cost, and are more flexible at navigating smaller spaces typical in homes. However, robots such as the MICAbot [2] usually do not possess the sensing or processing capabilities to implement complex area-coverage algorithms, and this will prove more challenging as the size of the robot shrinks.

A related approach to the problem of area coverage is the use of a multi-agent robotic system, as they have algorithms such as the use of potential fields which spread out robots uniformly across a given environment. They are highly efficient at covering a given area as multiple robots can explore and cover a greater area in the same time as compared to a single robot. However, the problem arises when considering how cost and complexity of communication, coordination, and navigation might become prohibitive with a large number of agents working explicitly with each other.

One way to overcome this problem is to use a behavioral approach to area coverage which might be comparatively more cost efficient and simpler to implement [3,4]. This can be implemented in sensing-limited robots as these heuristics are usually neither memory nor sensing intensive in operation. Interestingly, comparisons of the efficiencies of such methods in the context of area cleaning do not appear to have a significant amount of scientific literature relating to them, despite the proliferation of commercially available floor cleaning robots which utilize such methods for area coverage in the recent years.

2 Research Methodology

A simulation was felt to be insufficient for this study as it does not account for real world considerations such as the noise of the sensors and actuators which would require redundancies to be built into the behavior. Furthermore, since this study is application-based, it would be more appropriate to develop a physical system that would be able to closely model the desired end-product. For this project, LEGO Mindstorms NXT robotics kit was chosen as it allowed for fast prototyping and easy changes to the robot's structure. The robotics kit provides both flexibility and speed in construction as it contains basic requirements of mechanical, actuators, sensors, and software tools needed to construct a robotics system without the need for machining these parts together. The robot was designed to be compact and stable, as shown in the design in Fig. 1.

Two third-party sensors were implemented to extend the capabilities of the robot beyond that provided by its native sensors. The HiTechnic compass sensor was placed 22cm away from the robot body to prevent magnetic interference generated by the processing unit and the motors. It was used for determining the orientation of the robot at any given time and to improve the odometry of the robot. The SumoEyes sensor, which has a tested range of 16cm to 24cm with a field of vision of 60 degrees, depending on the intensity of light reflected, was attached to the front of the robot to allow basic obstacle-avoidance routines to be programmed.

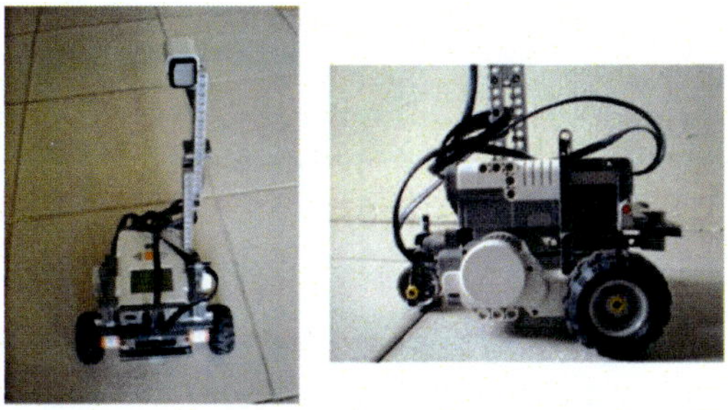

Fig. 1. Top and side views of the robot respectively

Three behaviors were proposed and tested respectively (Fig. 2): Weave, a behavior that would cause the robot to move in a zigzag motion, sweeping across an area to cover it, and would rotate its axis of motion by 90 degrees whenever it meets a corner; Random Walk, in which the robot moves forward in a given direction until it meets a wall, whereby it would turn by a randomly generated value between 80 – 135 degrees [5]; and Spiral, in which the robot spirals out from its original position, the pattern of the NXT brick modeling the Archimedean spiral, until it meets a wall, whereby it will follow the wall for a random distance before moving perpendicular to the wall and beginning to spiral again.

Fig. 2. Illustration of the proposed behaviors

Some considerations have to be made when the robot reaches the boundary of the area for Spiral. As seen in Fig. 3 (left), areas where the robot has not covered are shaded. Instead of choosing a random walk sequence which would only allow a robot to visit points of the boundaries, a wall-following sequence was chosen as the robot would then be able to traverse deeper into the uncovered regions as seen in Fig 3 (right). Similarly, there must be a time when the robot has to return to the center, as it might have just covered a small part of the room with its spiral. Hence moves perpendicular to the wall it is following after a random distance between 0.80m – 1.20m, which was experimentally determined to be the average distance between each spiral, and was implemented to prevent the robot from continuously covering the same area. In Spiral, no special provision was made for obstacles as the wall-following behavior allowed it to detect obstacles and circle them, ensuring that the areas around obstacles are also covered.

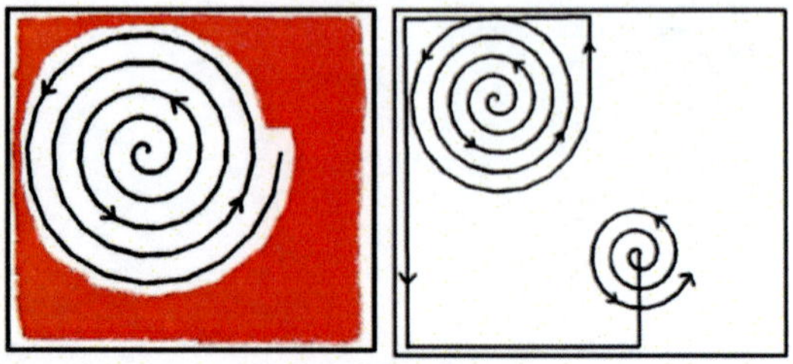

Fig. 3. Illustration of Spiral programming considerations

Weave was chosen as it allows the systematic area coverage [6]. However, instead of the robot returning the way it came by, it was decided that rotating the axis of the robot's zigzag motion after it senses a corner can increase efficiency of coverage, especially in areas with obstacles as seen in Fig. 4. Moving in an opposite orientation would allow the robot to cover areas that it did not do so before, which might have been due to internal odometry errors or the environment itself.

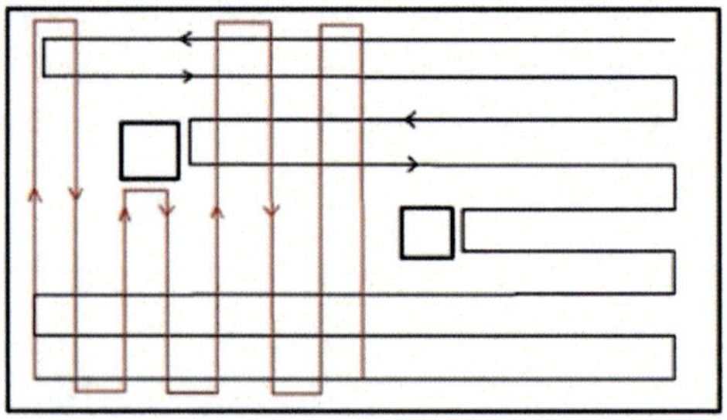

Fig. 4. Illustration of rotation of axis of movement for Weave

3 Results

Four fields of testing were applied to measure the effectiveness of the behaviors in area coverage: 1.20m x 1.20m (field 1), 2.40m x 2.40m (field 2), 1.20m x 1.20m with obstacles (field 3), and 2.40m x 2.40m with obstacles (field 4). Each obstacle was 0.30m x 0.30m in area. The degree of clutter was 4.69% for field 3 and 6.25% for field 4 (Figure 5). Thirty sets of data were taken for each combination of behavior, measurement and field. The robot was started from the center of each field, and to ensure that the conditions remain constant, the robot was charged and started at the same point after each run.

Two measurements were taken for each dataset to measure efficiency: the time taken for the robot to completely cover the given area (measurement 1) and the time taken for the robot to return to its origin three times (measurement 2). The rationale for the second measurement is that in the real world, area coverage usually takes place within a localized area of a given space, and hence, a suitable measurement would be to measure how often a robot returns to a given area. In applications such as cleaning where the probability of success is less than one, multiple passes might be needed to ensure that an area is cleaned properly. If a robot covers an area quickly, but its algorithm causes it to take a long time in coming back to the same area, it can hardly be called efficient in completing its task of cleaning the area thoroughly. Figs 6 to 7 and Table 1 show the results.

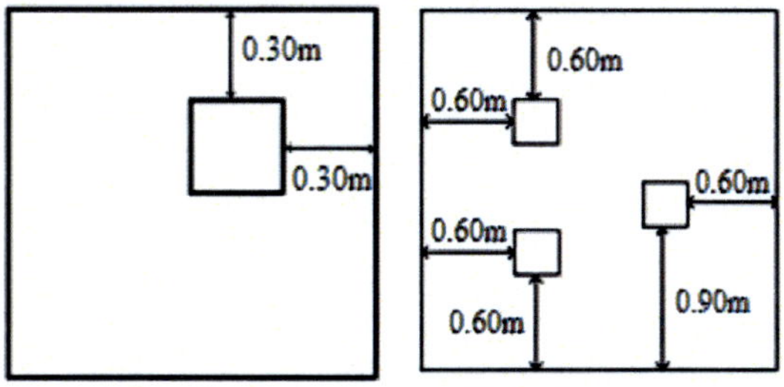

Fig. 5. Specifications for fields 3 and 4

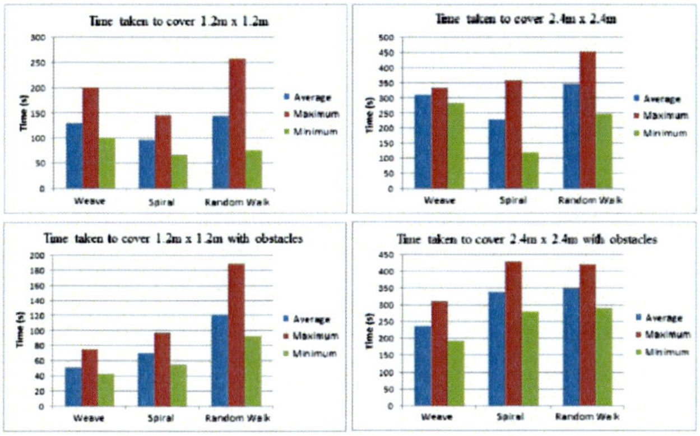

Fig. 6. Results of behavior testing (measurement 1)

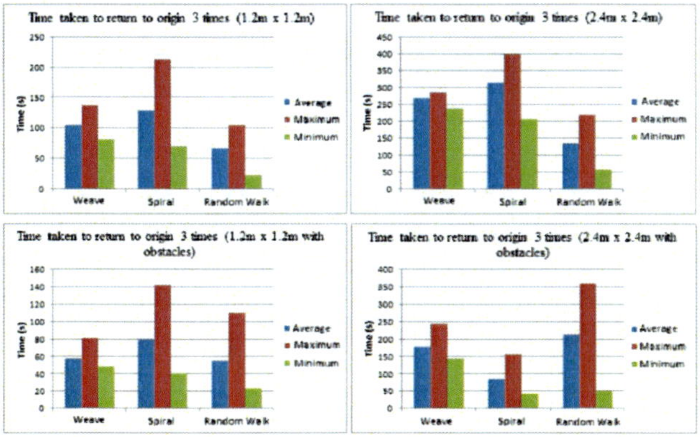

Fig. 7. Results of behavior testing (measurement 2)

During experiment 2, it was noted that there was still some residual interference in the surrounding environment with the robot's compass sensor, resulting in some curved movements of the robot where it was supposed to travel in a straight line. The compass was recalibrated to no effect. Also, the field of vision appeared to be inadequate as the robot occasionally bumped into the walls that it was supposed to detect.

Table 1. Variation of timings

Conditions	Behaviors		
	Weave (s)	Spiral (s)	Random Walk(s)
Field 1, Measurement 1	4.07	3.50	8.32
Field 1, Measurement 2	2.89	6.60	3.73
Field 2, Measurement 1	2.55	9.56	9.59
Field 2, Measurement 2	2.19	9.62	9.36
Field 3, Measurement 1	4.86	6.13	7.13
Field 3, Measurement 2	4.79	5.01	17.0
Field 4 Measurement 1	1.33	1.73	5.14
Field 4, Measurement 2	1.49	4.37	3.52
Average Variation, Measurement 1	3.20	5.23	7.55
Average Variation, Measurement 2	2.84	6.40	8.40

While Random Walk was able to cover the middle portions of each field relatively well, it did not cover the sides of the field as efficiently. This resulted in it taking a much longer time than it should have as seen by the higher average time of Random Walk in Fig 6. Conversely, Spiral covered the sides of the field with a high efficiency due to the inclusion of a wall-following algorithm into its behavior, and this is reflected in the significantly lower timings it had for field 1 shown in Fig 6.

The time by which Spiral was to follow the wall played a significant role in determining the timings for area coverage: if it was too short, the robot would begin Spiral in the same area where it had covered earlier, if it was too long, the robot might cover the area near the wall too frequently, and this will result in redundancy of area covered. The random distance the robot would have to travel before it exited the wall-following algorithm was tested before being adjusted to a moderate value within the range of 0.80m – 1.20m.

4 Discussion

A trend appears if we compare the performance of the behaviors using the two measurements, i.e., the longer the robot takes to cover an area, the shorter the time taken to return to the same area. This might be related to the varying degrees of redundancy in area coverage of each behavior, which is measured by the length of time the robot takes to cover the area. Thus, the time taken for the robot to efficiently cover an area depends very much on the measurement being measured. For example, Spiral might be more efficient in covering Field 1 under measurement 1, but not so under measurement 2.

However, some exceptions do occur. The trend for measurement 2 changed when obstacles were introduced into the environment as they prevented Random Walk from moving directly back towards the center after sensing the boundaries, greatly increasing its timings. Spiral experienced a significant drop in the timing for measurement 2, field 4 due to its wall-following algorithm: It quickly sensed the obstacles, and circled them. When the robot left the walls, it headed straight towards the region in the center, thus the timing was reduced significantly as its program effectively directed it back towards the center. Interestingly, this does not hold true for field 3 as the placement of the obstacle resulted in the robot running Spiral circling the obstacle more than once, and hence resulting in its longer timing. Therefore, we can also conclude that the placement of the obstacles in the environment does affect the efficiency of the behavior adopted by the robot.

It should be noted that Weave takes less time in covering an area when obstacles are added into the environment. This might be due to a greater economy of movement resulting from the reduction in the duplication of movement. Weave usually takes two repetitions to cover the same area to account for the areas missed. With obstacles, it is able to complete these repetitions faster and it takes the same amount of repetitions to cover the same area fully, thus decreasing the time taken.

If both measurements are considered, Weave is the behavior that optimizes between the need to return to the same area, and to cover the given area within a reasonable time limit. It has performed relatively well as compared to the other two behaviors for each measurement and field. It also has another added advantage of having the lowest variation of timings as seen in Table 1. Hence, we can conclude that Weave is the most reliable and efficient behavior amongst the three that can be implemented for area coverage given limited resources of sensing and processing.

5 Conclusion and Further Applications

This paper has provided an analysis and comparisons of various behavioral methods for area coverage. The conclusion is that Weave can be best implemented in commercial robot applications such as floor cleaning where the precision of area coverage is not so stringent as compared to the other behaviors which are not as efficient.

A limitation of this experiment is that the small range of environments tested does not accurately reflect the complexities of environments where there are narrow or low spaces which might hinder robot movements. Thus, the robot might possibly take much more time than that measured to completely clean the area.

It was also discovered that the measurements used to determine efficiency also have a significant effect of changing the perception of effectiveness of a certain method used, and thus the development and implementation of more performance metrics in area coverage tasks would be useful in determining the overall efficiency of an algorithm in the future.

Further improvements to the values used to measure efficiency can be made, for example, measuring the longest time taken for the robot to return to any given area, and other parameters can be implemented in measuring efficiency [7]. Other behaviors such as a random walk combined with wall-following could be explored. Factors such as the placement of obstacles in the environment and the random value of the distance travelled in Spiral could be explored further to see their effects on the performance of the algorithms.

References

1. Siciliano, B., Khatib, O. (eds.): Springer Handbook of Robotics, pp. 1278–1291. Springer, Berlin (2008)
2. McMickell, M.B., Goodwine, B., Montestruque, L.A.: MICAbot: A Robotic Platform for Large-Scale Distributed Robotics. In: Proceedings of the 2003 IEEE International Conference on Robotics and Automation, Taipei, Taiwan (2003)
3. Choset, H.: Coverage for Robotics – A survey of recent results. Annals of Mathematics and Artificial Intelligence 31, 113–115 (2001)
4. Balch, T.: The case for randomized search. In: Workshop on Sensors and Motion, the 2000 IEEE International Conference on Robotics and Automation, San Francisco, CA (2000)
5. Ribes, M.T.: Optimization of Floor Cleaning Coverage Performance of a Random Path-Planning Mobile Robot, `http://www.robotica.udl.cat/tfc/pdf/marceltresanchez2.pdf`
6. Butler, Z.J., Rizzi, A.A., Hollis, R.L.: Contact sensor-based Coverage of Rectilinear Environments. In: Proceedings of the 1999 IEEE International Symposium on Intelligent Control/Intelligent Systems and Semiotics, Cambridge, MA, USA (1999)
7. Wong, S.C., Middleton, L., MacDonald, B.A.: Performance metrics for robot coverage tasks. In: Proceedings of the 2002 Australasian Conference on Robotics and Automation (ACRA), Auckland, New Zealand (2002)

On Sub-modularization and Morphological Heterogeneity in Modular Robotics

A.H. Lyder[1], K. Stoy[3], R.F.M. Garciá[2], J.C. Larsen[3], and P. Hermansen[3]

[1] Danish Technological Institute
ahl@teknologisk.dk
[2] University of Tarapacá, Chile
rmendozag@uta.cl
[3] University of Southern Denmark
{kaspers,jcla,pahe}@mmmi.sdu.dk

Abstract. Modular robots are a kind of robots built from mechatronic modules, which can be assembled in many different ways allowing the modular robot to assume a wide range of morphologies and functions. An important question in modular robotics is to which degree modules should be heterogeneous. In this paper we introduce two contributing factors to heterogeneity namely morphological heterogeneity and sub-functional modularization. Respectively, the ideas are to create modules with significantly different morphologies and to spread sub-functionality across modules. Based on these principles we design and implement the Thor robot and evaluate it by participating in the ICRA Planetary Robotic Contingency Challenge. The Thor robot demonstrates that sub-functional modularity and morphological heterogeneity may increase the versatility of modular robots while reducing the complexity of individual modules, which in the longer term may lead to more affordable modular robots.

1 Introduction

Modular robots are robots built from modules that each encapsulates some of the complexity of the robot. This is in contrast to the conventional way of building robots as one complex, monolithic structure. A modular design potentially has several advantages over a non-modular approach in terms of versatility, robustness, and reduced cost [11].

Modular robots can be *reconfigurable* or *self-reconfigurable*. Self-reconfigurable robots can change their own shape by rearranging modules, but reconfigurable robots need human assistance (or assistance from another robot) to change their shape. Most works in modular robotics have focussed on self-reconfigurable robots e.g. M-TRAN [6], ATRON [7], and SuperBot [9] and others [10]. For self-re-configurable robots, it is convenient to use a homogeneous modular design because it allows modules to replace each other during a self-reconfiguration process. However, it has been pointed out that in the longer term self-reconfigurable robots need to be heterogeneous because it is not efficient in terms of cost and space to equip all modules with all the sensors a robot needs [2]. For reconfigurable robots, the motivation for moving towards heterogeneous systems is even higher because the removal of the self-reconfiguration ability removes

S. Lee et al. (Eds.): Intelligent Autonomous Systems 12, AISC 193, pp. 649–661.
springerlink.com

one of the motivations for a homogeneous design. However, an important, unanswered question and the focus of this paper is how to increase heterogeneity and if it is beneficial for the resulting modular robot.

Fig. 1. A rendering of a mobile manipulator built from Thor modules. The robot has four independently controlled wheels and a manipulator with five degrees of freedom; two twisting joints, two rotational joints, and one end-effector. The robot consists of 28 modules of 6 kinds: 9 motor, 4 wheel, 8 angle-90, 2 node, 4 rotation-165, and 1 gripper.

In order to answer this question we have designed and implemented a sub-functionally modularized and morphologically heterogeneous, reconfigurable robot named after Thor, the god of thunder in Norse mythology. In contrast to both existing self-reconfigurable robots and its ancestor Odin [5,3] and other heterogeneous reconfigurable robots such as CKBot [12,8], Molecubes [13] and Symbrion [4], the Thor modules are both 1) morphologically heterogeneous (currently Thor consists of eighth modules with significantly different morphologies) and 2) sub-functionally modularized in the sense that each module is simple and only provides a sub-functionality. As such Thor represents an extreme in terms of heterogeneity and modularization compared to existing modular robots.

We participated with the Thor modular robot in the ICRA Planetary Robotics Contingency Challenge (see Section 4). The main feature of this challenge is that tasks are not known in advance and thus the robots cannot be designed beforehand. This gears the competition towards modular robots and for us represents a way to evaluate the versatility of Thor. In one competition the Thor robot was assembled into a mobile manipulator and in another into a climbing robot, in both cases with a minimal use of modules and extra materials. Thor was mildly successful, but was hampered by basic technical problems (details in Section 4).

While Thor and the experience from the challenge are too limited a basis to draw a general conclusion, they indicate that morphological heterogeneity and sub-functional modularization provide the benefits of increased versatility, reduced mechatronics overhead, and reduced mechatronics complexity of the individual modules. If we, in the future, can generalize this result, it may lead to both more practically useful and more affordable modular robots.

2 Design Principles

2.1 Morphological Heterogeneity

The main motivation for designing a system with morphological heterogeneity was based on our experience in building modular robots.

Our first modular robot was the homogeneous ATRON self-reconfigurable robot [7]. ATRON successfully demonstrated three-dimensional self-reconfiguration, which also was its main purpose, but it had two problems. It was highly complex, which made it costly and time-consuming to develop, but more problematic the ATRON modules weigh 0.85 kg and have a diameter of 11 cm making robots built from ATRON modules too bulky to be practically useful.

Our second modular robot, ODIN [5,3], was an attempt to address these problems. In ODIN we removed the ability to self-reconfigure since it was a significant contribution to the size and weight of ATRON. However, theoretically more importantly we made it heterogeneous. It was our experience that these two changes combined made ODIN modules simple and light. However, ODIN remained practically of limited use because modules were morphological homogeneous, which made it difficult to build a robot morphologically suitable for a specific task.

Based on these experiences we decided to increase the morphological heterogeneity of modules to allow us to build robots with a wide range of morphologies. In hindsight, it is trivial to see that given the same number of connectors per modules you can build a wider range of morphologies with n heterogeneous modules than with n homogeneous modules, because in the homogeneous case permutations do not represent new morphologies, but they do in the heterogeneous case. A theoretical understanding of the relationship between morphological diversity, heterogeneity, and number of connectors represents important future work.

2.2 Sub-functional Modularization

The fundamental idea behind sub-functional modularization is that one functionality can be spread across several simpler modules rather than one module. For instance, instead of building a functional gripper module that includes a gripper, actuator, and batteries, we implement sub-functional modules, that is, battery, motor, and a wide range of grippers, joints, and wheels. The advantages of sub-functional modularization are substantial: the individual sub-modules are simple and easy to design, which allows for fast development of new types of sub-modules. Costly sub-modules such as motor sub-modules only need to be attached where needed and hence allow us to better exploit expensive sub-modules. This also reduces the number of modules and hence reduces the overall weight of the robot, which often cancels out the mechanical overhead of having more connectors. Sub-functional modularization was also employed in the ODIN robot and other reconfigurable robots, but in Thor this is taken further by separating out the motor, which is possible due to its connector's novel ability to transfer torque between modules through a shaft passing through the connector surface.

3 Implementation

Due to the sub-functional, morphological heterogeneous design, the Thor modules provide very different sub-functionalities. The implementation of some modules may require little space, such as sensors, and some modules may require more space, such as actuators. For that reason we define a unit of measure called *Thor Cubic Unit* (TCU). The Thor connector is defined to be 4x4 TCU. This means that a node holding six connectors in a cubic arrangement must at least be 4x4x4 TCU. Let us say that another module holds only two connectors, one at each end. The length of such a module may vary. For example, two 4x4x2 TCU modules in series are the same size as one 4x4x4 TCU module. The TCUs define the space requirements of a module within a simple cubic lattice. A TCU unit corresponds to 18 mm.

3.1 Connector

Probably the most crucial element of all modular robots is their connector since it is a relatively complex piece of mechatronics and, in addition, is the only mechatronic component that has to be integrated into all modules. The Thor connector is a genderless magnetic 8-redundant connector with an optional rotational drive shaft. It consists of a few simple components, a base ring, a printed circuit board (PCB) with 6 spring contacts, a bearing and 16 magnets. Fig. 2 shows the connector attached to a motor module.

The base ring has 8 symmetrically placed pins and matching holes. Each pin and hole have a small 5x2mm disc magnet with opposite polarity. This allows two connectors to magnetically connect and align at 8 different angles. The connector also has 8 screw holes, in case the magnets are not strong enough.

Inside the base ring sits a PCB with 6 spring contacts. Each spring contact is assigned to establish dedicated electrical connection between two connected modules, as follows: 2 Power lines, Vcc and Ground, 2 lines for a RS-485 differential bus, 1 line for detecting the angle of connection allowing the robot to detect its topology, and 1 line is still unassigned. The hole in the center of the PCB can either be blinded or it can fit a bearing. The bearing can hold a 5-mm hexagonal socket, which in combination with a 5-mm hexagonal shaft can transfer torque between two modules.

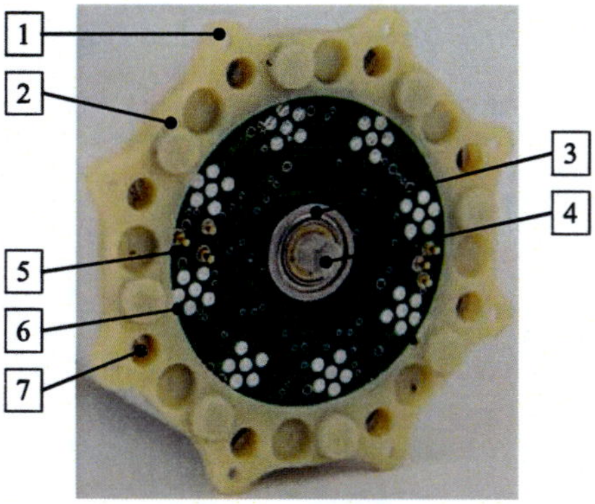

Fig. 2. A genderless magnetic 8 redundant connector with optional rotational drive shaft – [1]: Screw holes; [2]: Pin and hole with magnets; [3]: Bearing; [4]: 5mm hex drive/output shaft; [5]:Spring contacts; [6]: Receptor pads; [7]: Mounting screws.

Each module may have several connectors, and from the backside of the PCB all of the connectors of a module can be interconnected with power and communication. The backside of the PCB also allows us to attach a microcontroller board facilitating an ATMEL AT91SAM7S256 microcontroller with an ARM-7 core, and RS-485 communication. The microcontroller board can be combined with additional electronics for specific modules, such as a motor controller and power electronics for a motor module. Though the connector is prepared for topology detection, the electronics is not fully developed and does not enable this yet.

The Thor connector has a diameter of 70 mm and is centered on a face which is 4x4 TCU.

3.2 Motor Module

The motor module in Fig. 3a is able to drive other modules, such as the rotation-165, wheel and gripper modules (described below). Having separate motor modules simplifies developing new actuated modules, since they do not need to have their own motor. The motor module has a 5:1 planetary gearing driven by a Maxon EC45-flat brushless DC-motor. The brushless DC-motor has a stall torque of 255 mNm, which should give us at least a torque of 1 Nm on the output shaft. It is all contained, including electronics, in a 3D-printed (ABS) cylinder with a connector at each end. The cubic dimensions of the motor module are 4x4x4 TCU.

3.3 Node Module

The cubic node module in Fig. 3c is a passive element. The node consists of two identical shells milled in aluminum. It holds six connectors, one on each face. The cubic dimensions of the node module are 4x4x4 TCU.

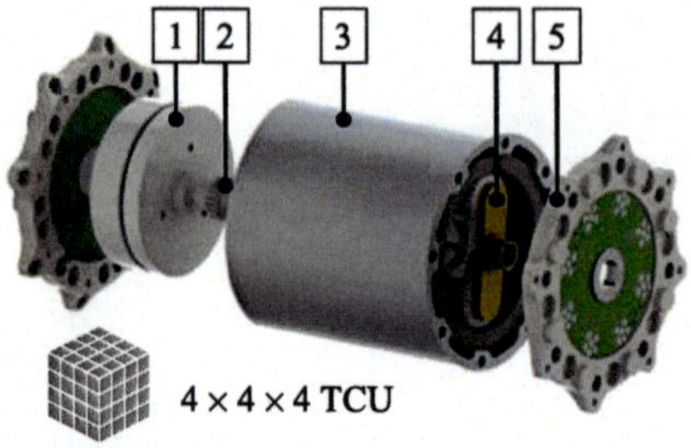

(a) Thor Motor – 1: Maxon EC45-flat brushless DC-motor; 2: Pinion gear (20 teeth); 3: Shell; 4: Planet-gears (30 teeth) on mount with output shaft; 5: Internal gear (80 teeth).

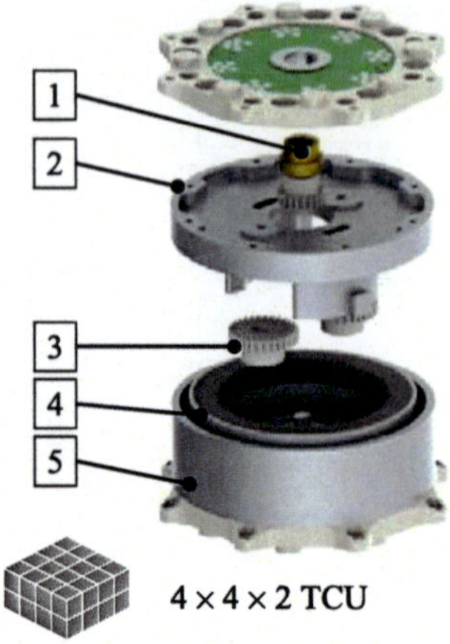

(b) Thor Rotate 165 – 1: Drive shaft with pinion gear (20 teeth). 2: Fixed shell; 3: Planet gears (30 teeth); 4: Internal gear; 5: Rotation shell.

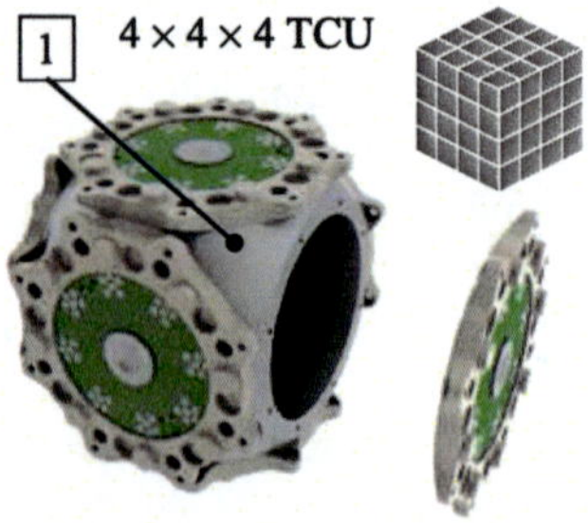

(c) Thor Node – 1: Aluminum shell.

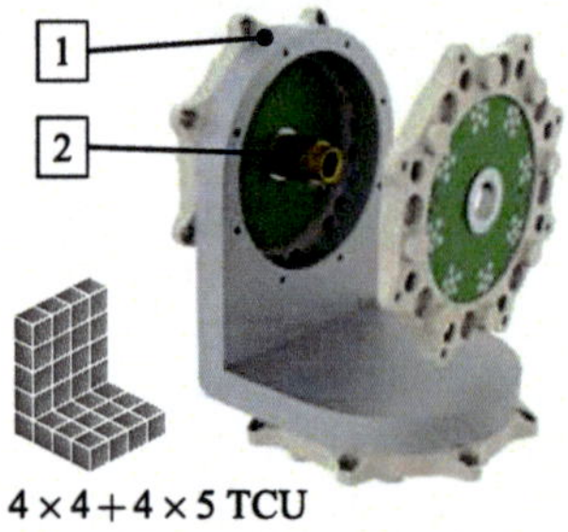

(d) Thor Angle – 1: Shell; 2: Drive/output shaft.

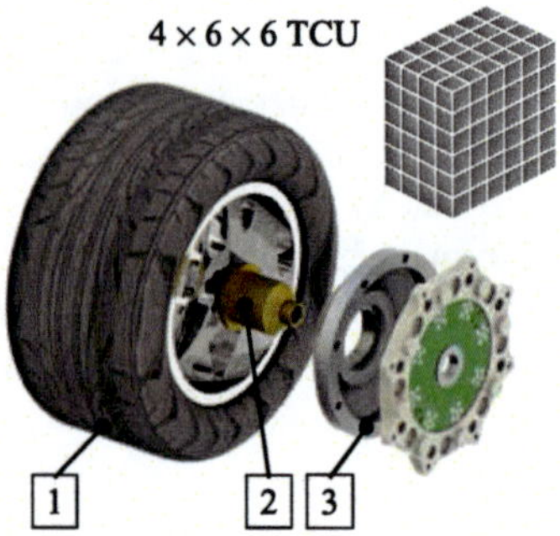

(e) Thor Wheel – 1: Wheel; 2: Drive shaft; 3: Shell.

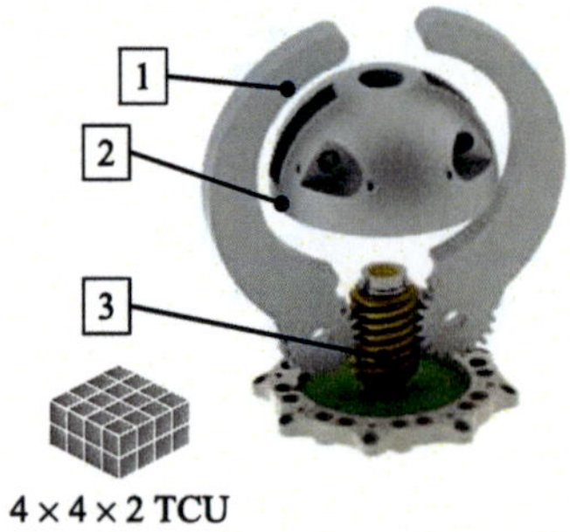

(f) Thor Gripper – 1: Gripper arms; 2: Shell; 3: Double-threaded worm gear.

Fig. 3. Thor Modules

3.4 Rotation-165 Module

The rotation-165 module in Fig. 3b can be driven by a motor module to rotate a third module ±165 degrees. Torque is provided on the drive shaft which holds a pinion gear. The fixed shell holds two planet gears, which drive the internal gear on the rotation shell. This results in a gear-ratio of 4:1. The rotation-165 module is limited in rotation to prevent breaking the wires connecting the two connectors. The shells are produced on a 3D printer. The cubic dimensions of the node module are 4x4x2 TCU.

3.5 Angle-90 Module

The Angle-90 module in Fig. 3d has an L-shape and can connect up to three modules at a 90-degree angle. Two of the connectors are directly opposite each other on the side face and a shaft allows an axel to pass through them. The third connector is placed on the bottom face at a 90-degree angle. Two angle modules can be used in combination with a rotation module to create a bending joint. The dimensions of bottom face are 4x4 TCU, and of the side face 4x5 TCU. The angle module is produced on a 3D printer.

3.6 Wheel Module

The wheel Module in Fig. 3e is rotated by a motor module. The cubic dimensions of the wheel are 4x6x6 TCU, where the connector is centered on the 6x6 TCU face concentric with the wheel. The wheel is bought off-the-shelves from an RC-car shop and mounted on a drive shaft. The drive shaft is contained within a shell holding the connector. The shell is produced on a 3D printer.

3.7 Gripper Module

The gripper module shown in Fig. 3f can be driven by a motor module to grasp objects or other modules. The drive shaft holds a double-threaded worm gear, enabling the identical gripper arms to move symmetrically. The cubic dimensions of the base of the gripper modules are 4x4x2 TCU. Since the gripper module is an end-effector, the dimensions of the gripper arms do not need to be restricted. All of the components for the gripper is 3D printed.

3.8 Other Modules

The battery module provides battery power to the robot. By connecting a 12V power supply the robot can run tethered, or the battery can be charged directly.

The wireless module allows for PC-to-robot or robot-to-robot communication using wireless ZigBee.

3.9 Kinematics and Control

In the current iteration Thor robots are remote-controlled. The wireless module receives commands from a host computer and broadcasts to all modules in the system using the internal RS-485 bus. However, in the long term it is the intension to make it autonomous. An important aspect of this is to be able to discover the kinematics of the robot during assembly — a topic we are addressing in related work [1].

4 The ICRA Planetary Robotic Contingency Challenge

The goal of participating in the ICRA Planetary Robotics Contingency Challenge was to demonstrate that through sub-functional modularization and morphological heterogeneity versatile and practically useful modular robots can be developed.

The challenge takes place in a simulated planetary environment, with a planetary surface and a human habitat. The planetary surface is a 6 m by 6 m sandbox with 5 cm deep gravel. The gravel is approximately 1/4 size pieces. Humans can only exist in the habitat. Thus, the participating teams are situated in the habitat and can only view the planetary surface from small windows. Only the robots can enter the planetary surface through the designated airlock. The robots must be placed in the airlock and move out to the planetary surface by itself, but can be remotely controlled which all have been so far (while Thor has the potential to work autonomously). For a robot to return to the habitat, it must also do so through the airlock. The airlock is 1.5 m long, 1 m wide, and 1m tall. Its doors are 1m by 1m and placed at each end of the long dimension. Each team is only allowed to bring a container with outside dimensions adding to 150 cm or less, and weighing a maximum of 25 kg. The limitation in equipment simulates the real-world size and weight restrictions on a space mission. Each day of the competition an unexpected problem occurs. The problem can be anything one can imagine happening on a planetary habitat. The problem is announced to all teams simultaneously, and must be solved within 4-6 hours. Below we will describe the two tasks of the 2010 edition of the challenge and the corresponding Thor solutions.

4.1 Task: Restore Communication

*"**Scenario:** You are on Mars. A Martian storm has tipped over and damaged your base's antenna. The mast has split into parts and the antenna has fallen to the planetary surface. Within 5 hours a space ship will arrive with important supplies. You must restore communication to the space ship for it to be able to land. Communication can be restored by picking up and lifting the antenna as high as possible."*

Using four wheel modules, the plan was to create a wheeled base able to drive on the planetary surface. The wheeled base was created using two motors actuating the front left and right wheels. The back wheels are passive. The wheels are mounted on a passive articulated chassis consisting of two node modules and two angle-90 modules in combination with a rotation-165 module. The passive back wheels and the passive articulated chassis allow the actuated front wheels to steer. We made two attempts picking up the antenna using the wheeled base, with two different manipulator configurations on top. Fig. 4 shows the two configurations.

The first configuration was based on a manipulator with three degrees of freedom. The manipulator can bend down using two angle-90 modules in combination with a rotation-165 and a motor module. The gripper is mounted at a 90-degree angle on the

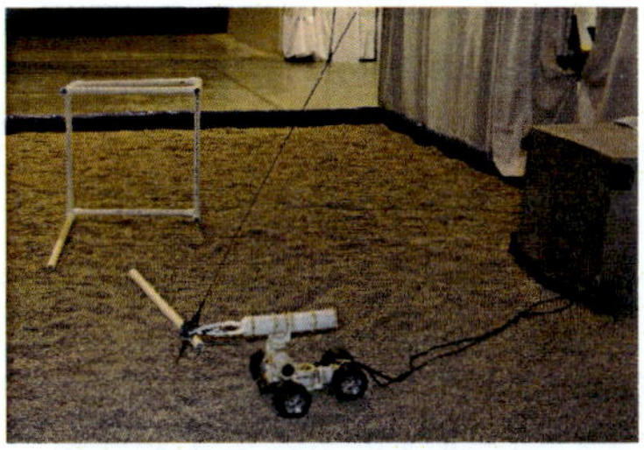

(a) Mobile robot with a gripper raising an antenna.

(b) Mobile robot with modified wheels and a scoop.

(c) Mobile robot with a magnetic end-effector picking up a solar panel.

Fig. 4. Thor Robots used in the Contingency Challenge

bending joint using an angle-90 module. The gripper can rotate using a rotation-165 module and a motor module. The gripper module is actuated by a motor module. The gripper is modified with a few LEGO components for grasping the antenna. In the first attempt, this robot succeeded in driving to the antenna, grasping and lifting the tip of the antenna 1.72 m above the ground.

In an overly ambitious second attempt, we tried to use the damaged mast. In styrofoam we created a scoop mounted on a bending joint on top of the wheeled base. The bending joint consists of two angle-90 modules in combination with a rotation-165 module and a motor module. The plan was to scoop up the antenna and place the bottom in the hollow end of one of the mast pieces. Once the antenna is placed, the mast and the antenna could potentially be pushed up, sliding on the scoop, assuming that the opposite end of the mast would stay fixed on the ground. This did not succeed, in hindsight not surprising.

Despite the successful first attempt, the wheeled base did not drive very well in the gravel. The actuated wheels tended to dig down in the gravel, until the robot ended up on its belly unable to move. We tried to increase the diameter of the driving wheels on site, but only with limited success.

4.2 Task: Solar Panel Recovery

*"**Scenario:** You are still on Mars. A second storm has blown off the solar panels from the structures on which they were mounted. In 5 hours the batteries will run out. Power has to be restored before the oxygen level gets too low and kills everyone. The solar panels must be retrieved from the planetary surface and mounted on the structures. There are two different structures on which the solar panels have to be mounted. The first is a square metal post about 2.5 m high. The other is a plastic structure covered in a plastic netting, also about 2.5 m high. It has a vertical side and a sloped side. On the vertical side, the netting is tightly fastened to the structure. On the sloped side, the netting is hanging loose, like a hammock. The normal airlock is still functioning. Two additional sliding airlocks are able to bring the robot close enough to the structures, so that they can touch them. Another team is working on fixing old solar panels, which can be mounted after 3.5 hours."*

The plan was to retrieve the solar panel using the same wheeled base as in the first task. We placed a wooden rod on a bending joint, created from two angle-90 modules in combination with a rotation-165 and a motor. At the tip of the wooden rod we placed two strong magnets to pick up a solar panel. Due to poor mobility and broken axles in the wheel modules caused by the 3D-printed plastic shafts not being strong enough, we did not succeed in retrieving the solar panels and had to wait for the repaired ones being available.

For placing solar panels, the plan was to create a configuration capable of vertically climbing the plastic netting. The configuration consists of a gripper module actuated by a motor module. Using a few LEGO components, we created linear motion from opening and closing the gripper. At the tip of the linear mechanism a hook made from a large paperclip is mounted. A rotation-165 module is placed on an angle-90 module at the bottom of the configuration. The rotation-165 module is actuated using a motor module, with a rotation axis perpendicular to the netting. On the rotation-165 module we placed a wooden rod horizontally with hooks on each end. This enabled us to steer the climbing robot.

The robot was able to climb vertically by pushing the top hook up, and when it was pulled down again the hook got stuck in the netting and pulled the entire robot up. When the top hook was being pushed up the two bottom hooks prevented the robot from falling down again. By rotating the wooden rod, on which the two bottom hooks were mounted, the top hook could be pointed and steered in the right direction. To prevent the robot from tipping outward and keeping it close to the vertical netting, a wooden rod was placed vertically at the bottom of the robot. Fig. 5 shows the vertical climber on the plastic netting.

The vertical climber was deployed on the plastic netting structure by placing it vertically on the sliding airlock. As the airlock was pulled back, the hooks got the robot stuck to the plastic netting. A solar panel was mounted on the body of the robot, so the robot had to climb slightly higher than the mounting marks. The robot climbed roughly 1 m above ground before falling down.

Fig. 5. Vertical climbing Thor robot with controllable direction

5 Discussion

The most interesting observation from the competition is that the solutions created with the Thor robot only required very few accessories: we tried to increase the diameter of the wheels on the wheeled base using a bit of styrofoam and duct tape. The gripper was modified with a few LEGO components for grasping the antenna, and creating linear motion. The linear motion of the vertical climber could have been optimized by having linear actuators. In addition, a few, short wooden rods and a few paperclips were used. As the problems are unexpected it is impossible to prepare the robot for any situation. Thus, it is practically impossible to create a robot that can perform any unexpected task without utilizing any additional accessories. However, the fact that we only required few accessories for the Thor modular robot to solve the tasks of the challenge, demonstrates that the initial set of modules for the Thor modular robot are versatile.

In order to create a versatile modular robot, we need a variety of functionalities, such as actuated joints, end-effectors, wheels, legs, sensors, and batteries. The functionality needs to vary depending on the task and environment, neither of which is necessarily known beforehand. If a homogeneous modular robot on a cm-scale were to incorporate all of this functionality, each module would become complex and large. The Thor modular robot demonstrates that by using the principle of morphological heterogeneity we can reduce the complexity of the individual modules while improving versatility.

Since Thor is sub-functionally modularized, only a fraction of the modules will typically be used for a single task, leaving a number of unused modules. The actuated modules are usually the most mechanically complex and expensive modules in a modular robot, partially due to high-grade motors. Therefore, if each actuated module has an integrated motor, the overhead in cost and complexity from the unused modules could be fairly high. By developing a connector capable of transferring mechanical rotation, we have removed the integrated motor from the actuated modules. Instead we have developed a simple motor module, which is able to drive a variety of actuated modules. This has made the actuated modules simpler and easier to develop and produce, and it allows us to have a larger variety of actuated modules. However, the number of actuated modules, which can be used in a single configuration of the Thor modular robot, are limited to the number of motor modules available. The total size of an actuated module plus a motor module will also be larger than an integrated module. Dividing the functionality of the Thor modules into simpler modules with less functionality requires at least two or more modules to create an actuated joint or end-effector. This will increase the size of the robot, and also require a higher number of connectors. However, by making the connector simple with cheap components, and easy to produce, the overhead in cost and assembly is minimized. Furthermore, the mechatronics overhead is also reduced by only having to include the sub-modules required for a specific robot.

6 Conclusion

In this paper we introduced the concepts of morphological heterogeneity and sub-functional modularization in the context of modular robots. Morphological heterogeneity implies that not only are modules heterogeneous, they also have significantly different morphologies. Sub-functional modularization is the idea that a module does not necessarily have to equal a functional unit, instead sub-functionalities can be spread across several modules. These principles are implemented in the Thor reconfigurable robot along with a novel connector able to transfer torque between modules. While the experimental basis is limited, we conclude that compared to earlier robots these principles led to a more versatile modular robot, simpler modules, and reduced mechanical overhead since sub-modules could be combined as needed. In conclusion, the principles of morphological heterogeneity and sub-functional modularization may in the longer term lead to simple, affordable and versatile modular robots.

Acknowledgements. The research leading to these results has received funding from the European Community's Seventh Framework Programme FP7/2007-2013 - Future Emerging Technologies, Embodied Intelligence, under grant agrement no. 231688.

References

1. Bordignon, M., Schultz, U.P., Stoy, K.: Model-based kinematics generation for modular mechatronic toolkits. In: Proceedings of the Ninth International Conference on Generative Programming and Component Engineering, pp. 157–166 (2010)
2. Fitch, R.C.: Heterogenous Self-Reconfiguring Robotics. PhD thesis, Dartmouth College, Hanover, New Hampshire, USA (2004)
3. Garcia, R.F.M., Lyder, A., Christensen, D.J., Stoy, K.: Reusable electronics and adaptable communication as implemented in the ODIN modular robot. In: Proceedings, IEEE International Conference on Robotics and Automation (ICRA 2009), Kobe, Japan, pp. 1152–1158 (2009)

4. Kernbach, S., Meister, E., Scholz, O., Humza, R., Liedke, J., Ricotti, L., Jemai, J., Havlik, J., Liu, W.: Evolutionary robotics: The next-generation-platform for on-line and on-board artificial evolution. In: Proc. of the IEEE Congress on Evolutionary Computation (IEEE CEC 2009), Trondheim, Norway, pp. 1079–1086 (2009)
5. Lyder, A., Garcia, R.F.M., Stoy, K.: Mechanical design of ODIN, an extendable heterogeneous deformable modular robots. In: Proceedings, IEEE/RSJ International Conference on Intelligent Robots and Systems, Nice, France, pp. 883–888 (2008)
6. Murata, S., Yoshida, E., Kamimura, A., Kurokawa, H., Tomita, K., Kokaji, S.: M-TRAN: self-reconfigurable modular robotic system. IEEE/ASME Transactions on Mechatronics 7(4), 432–441 (2002)
7. Østergaard, E.H., Kassow, K., Beck, R., Lund, H.H.: Design of the ATRON lattice-based self-reconfigurable robot. Autonomous Robots 21(2), 165–183 (2006)
8. Park, M.G.: Configuration Recognition, Communication Fault Tolerance and Self-reassembly for the CKBot. PhD thesis, University of Pennsylvania (2009)
9. Salemi, B., Moll, M., Shen, W.-M.: SUPERBOT: A deployable, multi-functional, and modular self-reconfigurable robotic system. In: Proceedings, IEEE/RSJ Intl. Conf. on Intelligent Robots and Systems, Beijing, China, pp. 3636–3641 (2006)
10. Stoy, K., Christensen, D.J., Brandt, D.: Self-Reconfigurable Robots: An Introduction. MIT Press (2010)
11. Yim, M., Shen, W.-M., Salemi, B., Rus, D., Moll, M., Lipson, H., Klavins, E., Chirikjian, G.S.: Modular self-reconfigurable robot systems. IEEE Robotics & Automation Magazine, 43–52 (March 2007)
12. Yim, M., Shirmohammadi, B., Sastra, J., Park, M., Dugan, M., Taylor, C.J.: Towards robotic self-reassembly after explosion. In: Proceedings, IEEE/RSJ International Conference on Intelligent Robots and Systems, San Diego, CA, USA, pp. 2767–2772 (2007)
13. Zykov, V., Chan, A., Lipson, H.: Molecubes: An open-source modular robotics kit. In: Proceedings of the 2007 IEEE/RSJ Int. Conference on Robots and Systems, Self-Reconfigurable Robotics Workshop, San Diego, CA, USA (2007)

Balancing Control of Unicycle Robot

In-Woo Han, Jae-Won An, and Jang-Myung Lee

Department of Electrical Engineering, Pusan National University
Jangjeon2-dong, Geumjeong-gu, Pusan 609-735, Republic of Korea
{Supergksdn,jaewon1696,jmlee}@pusan.ac.kr

Abstract. This paper proposes controlling balance of unicycle robot by using PID control. The robot consists of the Pitch and Roll. And the robot measures the Pitch and Roll angle which are obtained from IMU sensor. In this paper, Assuming that the Pitch and Roll is decoupled, design the controller separately. Because controller's performance is confirmed from simulation, the dynamic equations of the robot are derived for simulation. Finally, designed controller is applied to have a test of the unicycle robot.

Keywords: PIDcontrol, Unicycle robot, dynamics, balancing.

1 Introduction

This paper discusses the balancing control of unicycle robot. The balancing control of unicycle robot is composed of Pitch control for two wheeled inverted pendulum robot and Roll control for maintain attitude of reaction wheel pendulum. The angular data of body is measured by an IMU sensor. PID controller is widely used in industries for it is simple and implement easily. And the controller has properties which include keeping control performance effectively in various environments. For designing the PID controller, the most important step is to get a proper P(Proportional) gain, I(Integral) gain and D (Derivative) gain. In this paper, those are obtained from repetitive experiments.

Research on the single-wheel (unicycle) robot has been ongoing since the 1980's in the U.S. and Japan. A. Schoonwinkel of Stanford University proposed a linear dynamic model of the robot and presented its optimal motion control in his Ph. D. thesis in 1987 Prof. Yamafujii of Tokyo University also proposed a dynamic model of the robot as an upper turntable and a lower rotating wheel. He also implemented PI motion control for it in 1997. A dynamic model of the robot has proven too complex to be implemented in real time [1]. Therefore, without system modeling, we aim to keep balancing of unicycle robot by using the PID controller which can control a system according to the specific controller gain. In this paper, Roll angle, Pitch angle and the speed of rotation of Disc are fed to generate the controlled variable.

Fig 1 is the unicycle robot. In Figure 1, we have made a decision form of the Disc to generate greater rotational inertia. And we have designed the system for its central axis to be center of gravity.

S. Lee et al. (Eds.): Intelligent Autonomous Systems 12, AISC 193, pp. 663–670.
springerlink.com

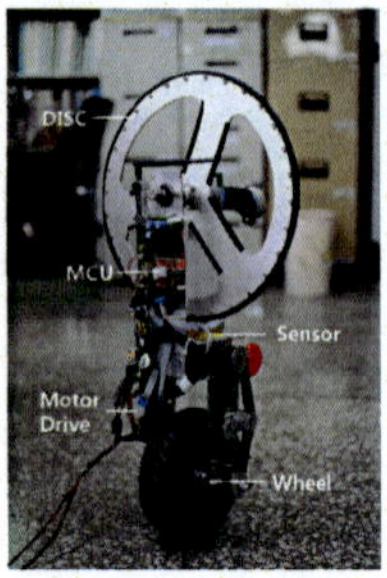

Fig. 1. The overview of the system flow

2 Consist of Robot

2.1 Controller

In Roll and Pitch control, we use the DC motor to rotate the disc and the wheel which are installed at the top part and the bottom part respectively. The DC motor is driven by motor-drive module which consists of H-bridge circuit. The motor-drive's input is PWM (Pulse Width Modulation). The motor's torque and speed of rotation are changed by input PWM's width.

At interval of 10mS, the MCU transmits control signal to motor-drive to keep balance according to the designed controller. The control signal of the MCU is PWM and the motor-drive generates power for DC motor to rotate according to the input signal.

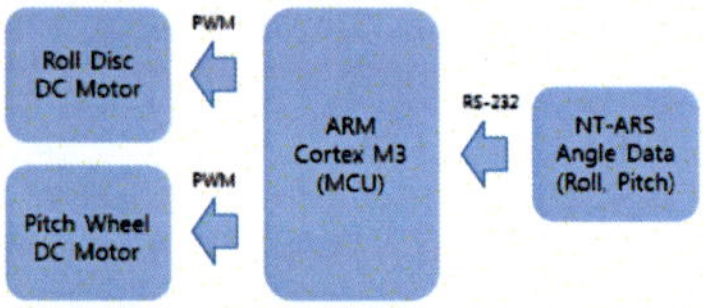

Fig. 2. Describes the overall system structure of unicycle robot

3 Dynamics of Unicycle Robot

In this section, the dynamic equations are derived to confirm a performance and controllability of designed controller by simulating the controlled system. Because the Pitch and Roll are controlled separately, we assume that the Pitch and Roll are decoupled. And then we derive the dynamic equations. The dynamic equations are derived by using Lagrange equation.

3.1 Roll Dynamics

Roll dynamics, we assumed that robot's body and disk as a reaction wheel pendulum.[2,4]

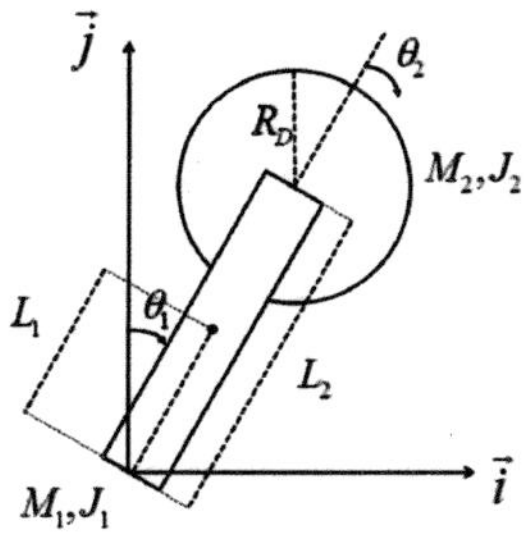

Fig. 3. Roll model

Set the axis in Fig 3. L_2 denotes the length from the origin to the center of disk. L_1 denotes the length from the origin to the center of the mass which includes Disc and body. R_D denotes the radius of the Disc, θ_1 denotes the angle of the Roll axis, θ_2 denotes the angle of the disc M_1 denotes the sum of the body mass and wheel mass.

Lagrange equations use Kinetic energy and potential energy to derive the dynamic equation of a system. Therefore, Kinetic energy and potential energy of the unicycle robot need to obtain. First, set a position vector to get a kinetic energy the $\vec{r}_1$ is the position vector of the body. And the $\vec{r}_2$ is the position vector of the Disc.

Position vector is

$$\vec{r}_1 = L_1 \sin\theta_1 \vec{i} + L_1 \cos\theta_1 \vec{j} \tag{1}$$

$$\vec{r}_2 = L_2 \sin\theta_1 \vec{i} + L_2 \cos\theta_1 \vec{j} \tag{2}$$

T, the kinetic energy of Roll is

$$T = \frac{1}{2}M_1(\vec{v}_1 \cdot \vec{v}_1) + \frac{1}{2}M_2(\vec{v}_2 \cdot \vec{v}_2) + \frac{1}{2}J_1\dot{\theta}_1^2 + \frac{1}{2}J_2(\dot{\theta}_1 + \dot{\theta}_2)^2 \tag{3}$$

where $\vec{v}_i = \dfrac{d\vec{r}_i}{dt}$, J_1 denotes the rotational inertia of the body, J_2 denotes the rotational inertia of the Disc.

V, The potential energy of Roll is

$$V = M_1 g L_1 \cos\theta_1 + M_2 g L_2 \cos\theta_1 \tag{4}$$

Now, we can get the Lagrangian which is defined as $L = T - V$. L , The Lagrangian is

$$\begin{aligned} L &= T - V \\ &= \frac{1}{2}M_1(\vec{v}_1 \cdot \vec{v}_1) + \frac{1}{2}M_2(\vec{v}_2 \cdot \vec{v}_2) + \frac{1}{2}J_1\dot{\theta}_1^2 \\ &+ \frac{1}{2}J_2(\dot{\theta}_1 + \dot{\theta}_2)^2 - M_1 g L_1 \cos\theta_1 - M_2 g L_2 \cos\theta_1 \end{aligned} \tag{5}$$

Finally, Lagrange Equation is

$$\frac{d}{dt}\frac{\partial L}{\partial \dot{\mathbf{q}}} - \frac{\partial L}{\partial \mathbf{q}} = \tau_q \quad where, \mathbf{q} = [\theta_1 \quad \theta_2]^T \tag{6}$$

Substitute the Lagrangian(L) into the Lagrange equation to obtain a dynamic equations. The dynamic equations of Roll are

$$(J_1 + L_1^2 M_1 + L_2^2 M_2)\ddot{\theta}_1 - g(L_1 M_1 + L_2 M_2)\sin\theta_1 = -\tau \tag{7}$$

$$J_2(\ddot{\theta}_1 + \ddot{\theta}_2) = \tau \tag{8}$$

Where τ denotes the torque which is generated by rotation of Disc.

3.2 Pitch Dynamics

Pitch dynamics, we assumed that robot's body and wheel as an inverted pendulum.[3,4]

Pitch dynamics are also derived by using the Lagrange equation.

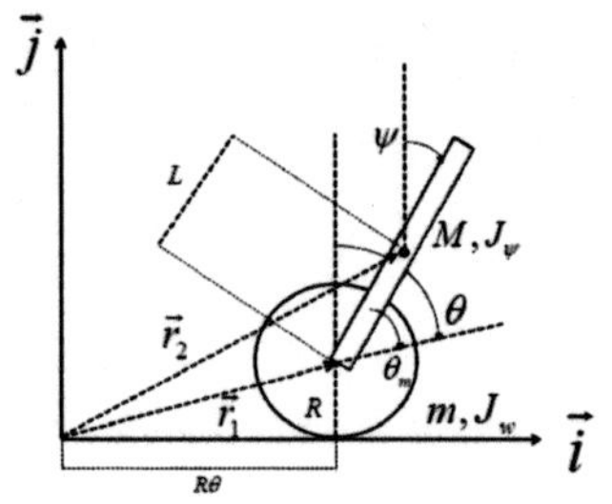

Fig. 4. Pitch Model

Set the axis in Fig 4. Position vector is

$$\vec{r}_1 = R\theta\vec{i} + R\vec{j} \tag{9}$$

$$\vec{r}_2 = (R\theta + L\sin\psi)\vec{i} + (R + L\cos\psi)\vec{j} \tag{10}$$

where R is radius of wheel, θ is the rotating angle when the system move from origin to specific point in normal coordinate axis, ψ is the angle which means how much slanting. Differentiate the obtained position vector, then the velocity vectors are obtained.

Velocity vector is

$$\vec{v}_1 = R\dot{\theta}\vec{i} \tag{11}$$

$$\vec{v}_2 = (L\cos\psi\dot{\psi} + R\dot{\theta}) + (-L\sin\psi\dot{\psi})\vec{j} \tag{12}$$

Kinetic energy of Pitch is

$$T_1 = \frac{1}{2}m(\vec{v}_1 \cdot \vec{v}_1) + \frac{1}{2}M(\vec{v}_2 \cdot \vec{v}_2) \tag{13}$$

$$T_2 = \frac{1}{2}J_w\dot{\theta}^2 + \frac{1}{2}J_\psi\dot{\psi}^2 + \frac{1}{2}n^2J_m(\dot{\theta} - \dot{\psi})^2 \tag{14}$$

where T_1 is translation kinetic energy and T_2 is rotational kinetic energy. In equation (13), m is the mass of wheel, M is the mass of body, J_w is the rotational inertia of wheels, J_m is the rotational inertia of motors and J_ψ is the rotational inertia of body. Thus total kinetic energy can be obtained by adding the translation and rotational kinetic energy. Therefore, in Lagrange equation T is sum of T_1 and T_2.

Potential energy of the Pitch is

$$V = mgR + Mg(R + L\cos\psi) \tag{15}$$

where g denotes acceleration of gravity.

The system dynamics can be obtained by substituting formerly obtained kinetic and potential energy to Lagrange equation

Found dynamic equations of Pitch are

$$\begin{aligned}(J_w + J_m n^2 + (m+M)R^2)\ddot{\theta} + (LMR\cos\psi \\ - J_m n^2)\ddot{\psi} + LMR\sin\psi\dot{\psi}^2 = \tau_p\end{aligned} \tag{16}$$

$$
\begin{aligned}
(LMR\cos\psi - J_m n^2)\ddot{\theta} + (J_\psi + L^2 M \\
+ J_m n^2)\ddot{\psi} - gLM\sin\psi = -\tau_p
\end{aligned}
\tag{17}
$$

where τ_p represents the torque generated by wheel motor. And it acts as torque of the opposite direction and the same size for inverted pendulum.

4 Dynamics of Unicycle Robot

Ziegler-Nichols method, Modified Ziegler-Nichols method and Chien Hrones Reswick method are mostly and widely used to obtain the PID gains in PID control. In this paper, we only obtain the PID gains through repetitive experiments. Assuming that the Pitch control and the Roll control are decoupled, design the controllers separately.

4.1 Pitch Control

In Pitch control, we only use state of a pitch angle to control attitude. Pitch control is the same as the balancing control of the two wheeled inverted pendulum. As mentioned earlier, the PID gains are obtained from repetitive experiments.

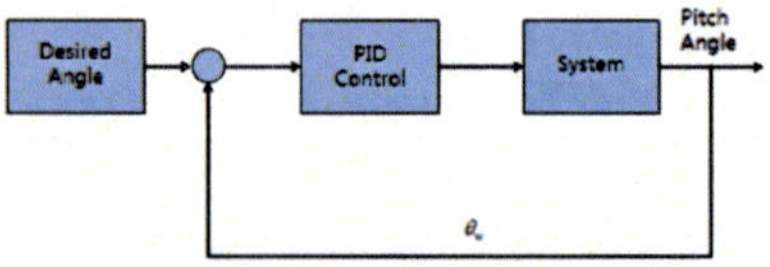

Fig. 5. Pitch controller

Fig 5 indicates the Pitch controller. Where θ_w denotes Pitch angle. The input of the controller is an error between current pitch angle and desired angle (0 degree).

4.2 Roll Control

The concept of balancing control of the Roll is similar to attitude control of spaceship, which applies the Law of Conservation of Angular Mometum.

In Roll control, PID controller's inputs consist of a Roll angle and a speed of rotation of the disc. If we only use the Roll angle to control attitude then the designed controller doesn't reflect state of disc. Consequently the overall system becomes unstable. Therefore, we feedback state of Disc which is not considered to the Pitch control. Another fed state of the Disc is speed of rotation of Disc.

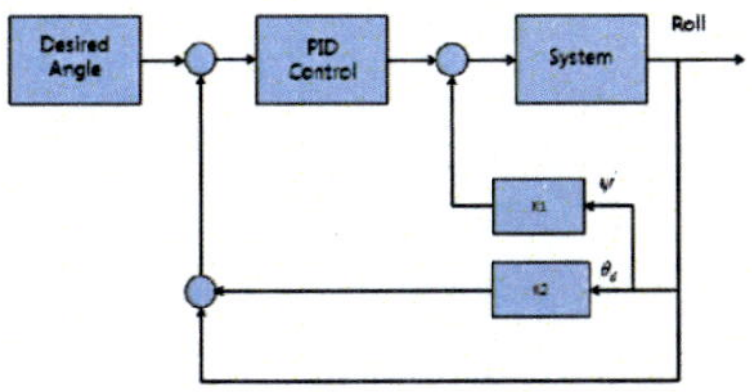

Fig. 6. Roll controller

Fig 6 indicates the Roll controller. Where ψ' is the speed of rotation of the Disc, θ_d is the measured Roll angle.[5]

5 Experiment

In Fig 7, Roll angle A denotes measured roll angle from proposed PID controller which feedbacks state of the Disc. And Roll angle B denotes measured roll angle from simple PID controller which only feedbacks state of Roll angle.

In Fig 8, this result demonstrates that Pitch control which only feedback state of angle is enough to control attitude of the robot.

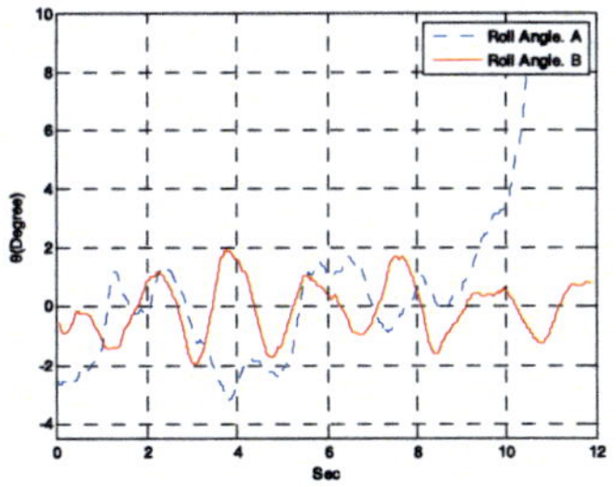

Fig. 7. Roll angle data

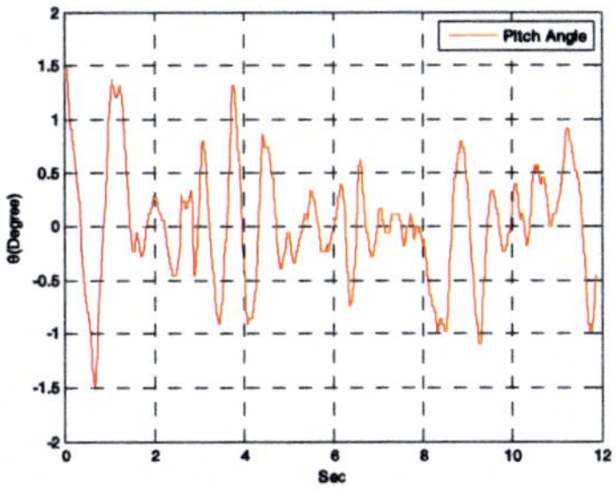

Fig. 8. Pitch angle data

Table 1. Parameters of the unicycle robot

Symbol	Value	Symbol	Value
M_2	$1.225[kg]$	L_1	$0.285[m]$
m	$1.300[kg]$	L	$0.330[m]$
M_1	$4.889[kg]$	J_2	$0.0402\left[kg\,m^2\right]$
M	$4.964[kg]$	J_1	$0.5295\left[kg\,m^2\right]$
R	$0.110[m]$	J_ψ	$0.7208\left[kg\,m^2\right]$
R_D	$0.200[m]$	J_w	$0.0079\left[kg\,m^2\right]$
L_2	$0.570[m]$	J_m	$0.0001\left[kg\,m^2\right]$

Acknowledgements. This word was supported by the Brain Korea 21 project.

This research was supported by the MKE(The Ministry of Knowledge Economy), Korea, under Specialized Field Navigation/Localization technology research center support program supervised by the NIPA(National IT Industry Promotion Agency).

References

1. Lim, H., Hwang, J.-M., Ahn, B.-H., Lee, J.-M.: Robust yaw Motion Control of Unicycle Robot. Journal of Institute of Control, Robotics and Systems 15(11), 1130–1136 (2009)
2. Block, D., Åström, K., Spong, M.: The Reaction Wheel Pendulum. In: Synthesis Lectures on Control and Mechatronics. Morgan & Claypool Publishers, Princeton (2007)
3. Technical report of LEGO Mindstorm, `http://www.mathworks.com/matlabcentral/fileex-change/19147`
4. Lee, J.-O., Han, I.-W., Lee, J.-M.: Fuzzy Sliding Mode Control of Unicycle Robot. In: International Conference on Ubiquitous Robots and Ambient Intelligence, Songdo connentia, Incheon, Korea, November 23-26 (2011)
5. Lee, S.-I., Han, I.-W., Lee, J.-O., Lee, J.-M.: Balancing Control of Bicycle Robot. In: SPENALO International Symposium, Bexco, Busan, Korea, September 30 (2011)

Autonomy for Mobility on Demand

Z.J. Chong[1], B. Qin[1], T. Bandyopadhyay[2], T. Wongpiromsarn[2], B. Rebsamen[2], P. Dai[2], E.S. Rankin[3], and M.H. Ang Jr.[1]

[1] National University of Singapore
{chongzj,baoxing.qin,mpeangh}@nus.edu.sg
[2] Singapore-MIT Alliance for Research and Technology, Singapore
{tirtha,nok,brice,peilong}@smart.mit.edu
[3] DSO National Laboratories, Singapore
erankin@dso.org.sg

Abstract. We describe the development of our autonomous personal vehicle that attempts to provide mobility on demand service to address the first- and last-mile problem. We discuss the challenges faced for such a system in a campus environment and discuss our approach towards mitigating them. The autonomous vehicle has operated over 30km of autonomous operation in a campus environment interacting with pedestrian and human driven vehicles.

1 Introduction

As the use of private vehicles starts approaching its limits to effectively meet the demand for personal mobility in densely populated cities, mobility-on-demand systems emerge as a more economical and sustainable alternative [3]. These systems rely on the deployment of a fleet of vehicles at different stations that are distributed throughout the city. The customers simply have to walk to a station near their origin, pick up a vehicle, drive it to the station near to their destination and drop it off. Electric ultra-small vehicles or bicycles may be utilized for systems that primarily aim at serving short trips. Such systems can supplement and stimulate the use of public transport by providing a convenient mean for the first- and last-mile transportation (e.g., from home to a transit station and back); thus, improving public transportation accessibility. The feasibility of mobility-on-demand systems that employ traditional bicycles has been demonstrated in many cities [1].

One of the main challenges in managing mobility-on-demand systems is in keeping a balanced distribution of the vehicles among different stations to ensure minimal waiting time for the customers at sustainable cost. This problem is critical especially for the cities where some origins and destinations are more popular than others, leading to an unbalanced distribution of the vehicles throughout the city. Hence, most of the existing vehicle sharing systems only offer round-trip service, forcing the customers to return the vehicle only at their origin. In [2], an optimal, real-time rebalancing policy that determines a proper distribution of the vehicles in the anticipation of future demand is proposed. However, a means

S. Lee et al. (Eds.): Intelligent Autonomous Systems 12, AISC 193, pp. 671–682.
springerlink.com

of transporting the vehicles for the re-balancing trips remains an open problem. In this paper, we propose the use of autonomy to implement the proposed policy and allow efficient operation of mobility-on-demand systems and enable a one-way vehicle sharing option.

Autonomy can play an important role, not only for the re-balancing trips but also for transportation from a pick-up point to a delivery point. This allows the customers to be picked up at their actual origin or dropped off at their actual destination, instead of requiring the customers to walk to or from a station. This problem is closely related to Dynamic one-to-one Pick-up and Delivery problems [4,5]. We show how this can be accomplished in a fully automatic manner without any human assistance.

Our system aims at providing transportation over a relatively short distance. In particular, the vehicles mainly operate in crowded urban environments that are typically equipped with sensors on the infrastructure including cellular networks, traffic cameras, loop detectors and ERP (Electronic Road Pricing) gantries. The detailed road network and many features of the environment in which the vehicles operate can also be obtained a priori. As opposed to existing autonomous vehicles such as those in the 2007 DARPA Urban Challenge and Google driverless car [6], we take a minimalistic approach and exploit the prior knowledge of the environment features and the availability of the existing infrastructure to ensure that the system is economically feasible.

The rest of the paper is organized as follows. Our mobility-on-demand system is described in Section 2. Section 3 describes our autonomous vehicle, including both the hardware and software components. The operation of the system is demonstrated in Section 4. Finally, Section 5 concludes the paper and discusses future work.

2 Mobility-on-Demand System

The components of our mobility-on-demand system is shown in Figure 1. First, the customers may request or cancel services, specify their pick-up and drop-off locations and view useful service information through a personal electronic device such as a smart phone or through our web interface. Sample snapshots of our Android mobile phone application and web interface are shown in Figure 2.

The service requests and cancellation are then added to the SQL database on our server. A scheduler, which is the main component of the server, interacts directly with the database. It determines the order in which the requests will be serviced and assigned each request to a vehicle. Once the assignment is made, the scheduler populates the database with useful service information such as service status, expected waiting time and the vehicle that is assign to each request. This service information will be transfered to the customers through our web API. The server also communicates with each vehicle to obtain its current status and provide the information about its next task (e.g. pick-up and drop-off locations of the next customer the vehicle is supposed to serve).

A vehicle completes each task as follows. First, it has to drive autonomously to a specified pick-up location, come to a complete stop and wait until the customer

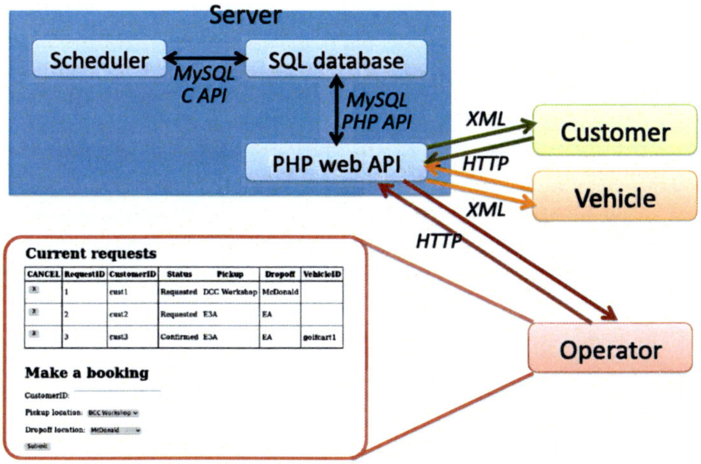

Fig. 1. Mobility-on-demand system

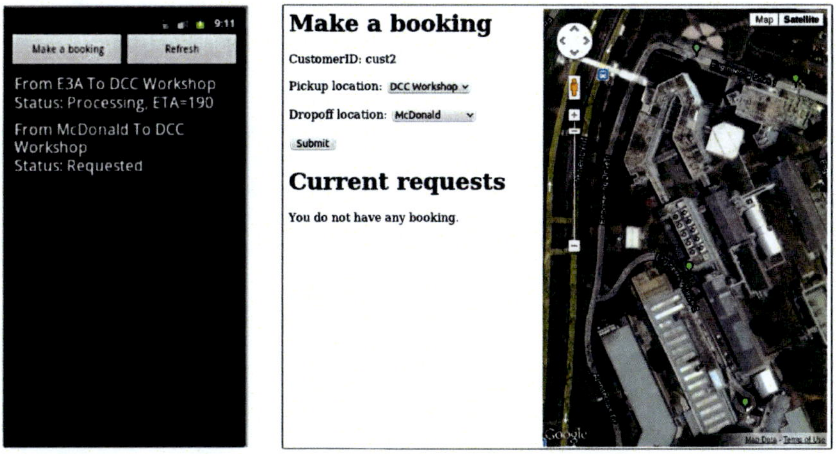

(a) Android mobile phone application

(b) Web interface

Fig. 2. Snapshots of service information available

successfully boards the vehicle. It then goes to the drop-off location. The task is completed when the vehicle reaches the drop-off location, comes to a complete stop and after the customer alights. The vehicle may not start the next task until the current task is completed. More detail on the autonomous operation of our vehicles is provided in the next section. Lastly, a human operator may monitor, add, cancel and modify service requests and access the status of each vehicle through our secure web interface (Figure 2b).

3 Autonomous Personal Transporter

3.1 Hardware Architecture

Our platform is based on a Yamaha G-Max 48 Volt Golf Car G22E. It has a seating capacity of 2 persons with maximum forward speed of 24 km/h. Current setup of the platform is shown in Figure 3 which features the placement of sensors and other hardware. The multifunction frame structure provides flexible sensor configuration which allows changes to the sensors configuration to be done quickly and easily. All hardware are powered by the onboard 6 x 8V US 8VGC deep cycle batteries. Some of the devices which requires AC, i.e. computers and motors get the power supplied from an inverter which draws power directly from the onboard batteries.

Actuators

The golf car has been modified to be able to drive by wire for computer control. An AC motor is connected to the steering column by bevel gear to enable automatic steering. It is designed such that the bevel gear can be disengaged to allow switching back to manual driving. Another AC motor is fitted near the brake pedal to actuate the brake directly. Finally, direct electronic interface into the throttle signal is made to achieve complete control of the vehicle's speed and direction.

The low-level controls, which comprise the controls of steering, throttle and brake, are handled by a realtime system to provide necessary signals as required by different actuators. The 2 AC motors have been configured to receive pulse signals with position controls similar to a stepper motor where the amount of

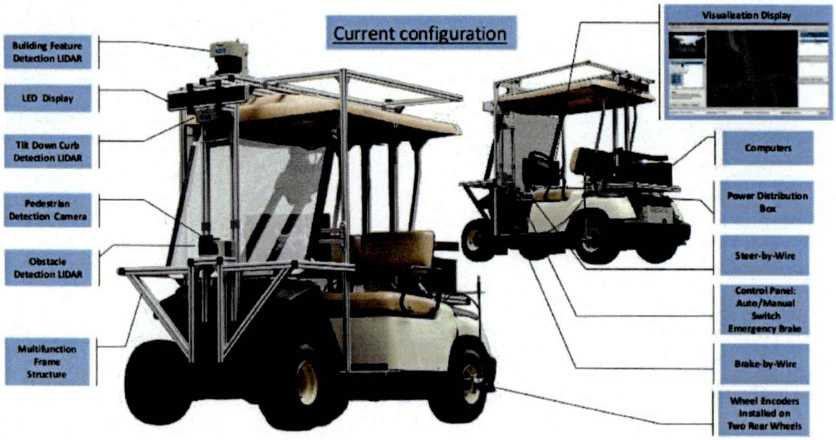

Fig. 3. Autonomous vehicle hardware architecture

rotation is proportional to the number of pulses. On the other hand, a PWM signal of 3.3 V is used for the throttle to regulate the speed of the vehicle.

Sensors

Both rear wheels of the golf car are mounted with encoders that provide an estimate of the distance traveled. An Inertial Measurement Unit (IMU) MicroStrain 3DM-GX3-25 is mounted at the center of the rear axle to provide attitude and heading of the vehicle. The encoders and IMU are combined to provide odometry information for the vehicle in 6 DOF.

For external sensing, a variety of LIDARs are used. There are 2 SICK LMS 291 mounted in front of the golf car. The SICK LMS 291 provides a single plane range measurement of 180 degree field of view. Both of the LIDARs are connected through USB-COMi-M, which enables high speed connection to the LIDARs, providing measurement rate at 75Hz with 0.25 degree of resolution. The top LIDAR is mounted horizontally to provide measurements of stable building features to allow accurate localization within a known environment. The second LIDAR is mounted looking downward and is used to detect the curb lines along the road for navigation purposes.

Additionally, a 4-layer LIDAR, SICK LD-MRS400001 mounted at the waist level provides additional information about the environment. The data returns at the rate of up to 50 Hz with the total operating angle of 110 degree. Just on top of the LIDAR, a USB camera Logitech HD Pro Webcam C910 is placed. The camera is calibrated with the 4-layer LIDAR to provide pedestrian detections. The combination of LIDAR and camera is with the objective of extracting each sensor's different capabilities to achieve a robust detection system. This way, the excellent tracking performance of the LIDARs and the ability of vision to disambiguate different objects can be fully utilized.

Computing

There are 2 regular desktop PCs fitted with Intel i7 quad-core CPUs and interface card. All computers run Ubuntu 10.04 with Robot Operating System (ROS) installed. One of the computers is installed with RealTime Application Interface (RTAI), a real-time extensions for Linux Kernel to provide the low level control to the steering, brake and throttle of the golf car. Modular software architecture has been developed for ease in incorporating additional functionality without modifying the core system, as detailed in the next section.

3.2 Software Architecture

Figure 4 shows a high level view of the software components currently setup in our vehicle. In the following we briefly describe the navigation, localization and perception module.

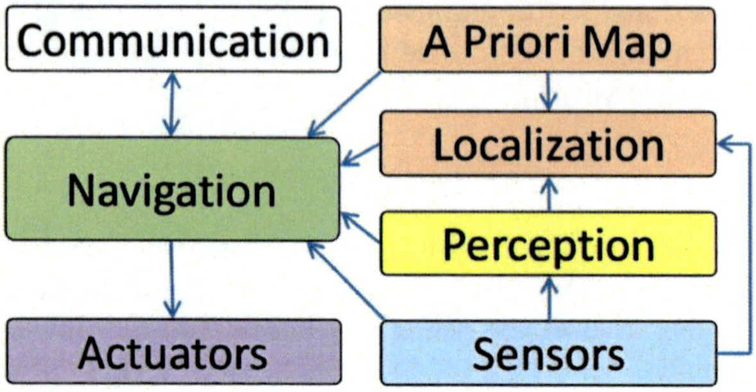

Fig. 4. Overview of the software modules implemented on the autonomous vehicle

Navigation module

Since the vehicle navigates on a known road network, all routes from any origin to any destination are generated a-priori as a set of waypoints. The choices of routes are made online depending on the request from the mobility on demand scheduler. The obstacles detected from the sensors are incorporated as a rolling cost map centered on the vehicle. The cost is propagated radially outward with an exponential function. At the low level, speed and steering control are separated. For the speed control, the vehicle considers the following input before planning for next action: the average cost function that is present within a defined area in front of itself and the curvature of the path [7]. The waypoint follower is implemented using a pure pursuit control [9].

Localization

Localization is very important for autonomous navigation. Most of the popular approaches for localization in autonomous navigation outdoors depend heavily on GPS based localization. In fact the DARPA challenge was based on GPS based waypoints as input. However, GPS is not very reliable in urban areas due to satellite blockage and multi-path propagation effect caused by tall buildings. An alternative approach to using GPS is to generate a high fidelity map of the area to be navigated using a high resolution range scanner. This approach is used by Google driverless car where the car, mounted with high fidelity Velodyne 3-D range sensor collect data of the road networks from various runs. Subsequently, when the vehicle travels it matches scans from its range sensors to the collected data and infers its location. However, collecting such a-priori information requires significant investment in terms of cost and manpower.

In line with our vision of lowering the cost of the autonomous vehicle, we use a single 2-D range sensor to detect roadside curb features. We use Adaptive Monte-Carlo Localization scheme to match the detected curb features to a road network map known a-priori. Figure 5 shows the basic components of this algorithm. This

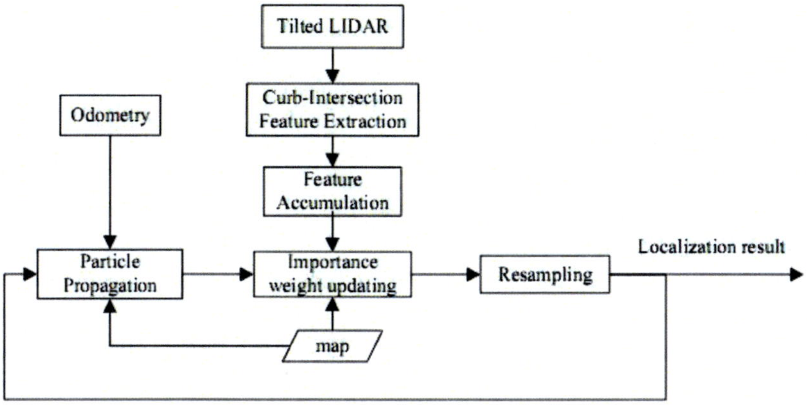

Fig. 5. Localization algorithm flowchart

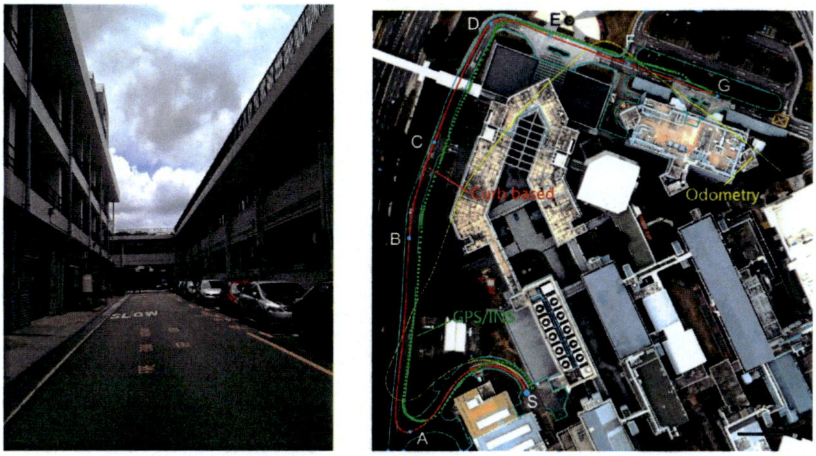

(a) A scene from the road segment traversed by the vehicle, typical road segments in Engineering Campus

(b) Comparing different localization approaches. Green shows GPS, Yellow shows IMU odometry and Red shows the estimate of the curb based localization

Fig. 6. Localization experiments

algorithm is tested through experiments in the campus environment as seen in Figure 6(a). A snapshot of the result is shown in Figure 6(b), where our vehicle drives from starting point S to goal point G. We can see that our curb based localization (shown in red) outperforms odometry (shown in yellow) and GPS based localization (shown in green).

The errors in location estimate are plotted in Table 1. It can be seen that position error of our algorithm is usually small, less than 0.6 meter; and the

Table 1. Localization error at several marked points

Marked Points	A	B	C	D	E	F	G
Position Error (m)	0.20	0.55	0.06	0.20	0.32	0.06	0.08
Orientation Error (deg)	< 3						

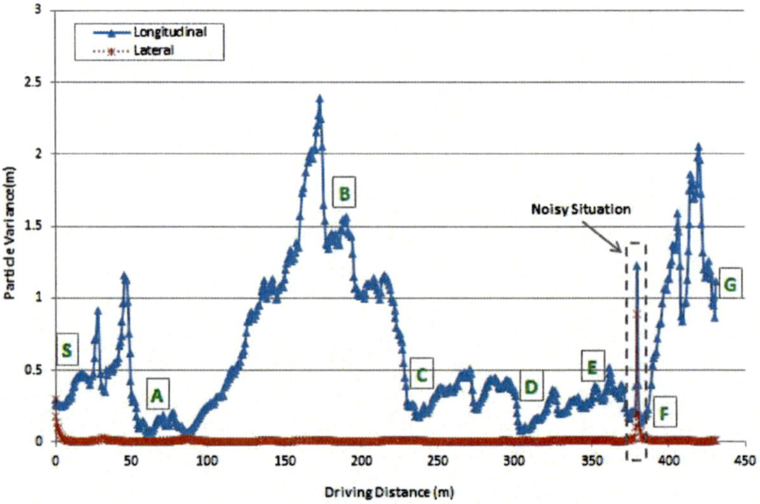

Fig. 7. Position estimation variance

Fig. 8. Vehicle navigating the dynamic environment

orientation estimation is quite accurate, less than 3 degrees to the ground truth. From Table 1, one can also observe that position errors at some critical points of intersections and turnings (like A, C, D, F) are much smaller than that of the straight road (like B). Fig. 7 shows "estimation variance" vs. "driving distance" in road longitudinal and lateral direction. It can be concluded that, while curb features on straight roads help to estimate the lateral position, the intersection and tightly curved curb features contribute very much to the longitudinal positioning. In our operations, we augmented the curb map with patches of laser map in areas where curb information was not available, e.g. pick up and drop off lobbies. Details are mentioned in [8].

Perception

For autonomous driving, having a good perception of moving objects on the road is extremely important. Figure 8 shows a typical scenario the vehicle has to navigate in. The problem of detecting pedestrians, moving vehicles and static obstacles in cluttered, dynamic and time-varying lighting conditions is extremely complex. In addition, often the on-board sensing is occluded due to the presence of big trucks, buses or other environmental features. While vision systems can detect features more reliably, often ascertaining the distance of the features becomes difficult. On the other hand, while the laser range finders are quite accurate in detecting the distance to the obstacle, they are not well suited to disambiguate similar shaped obstacles like moving pedestrians or a static lamp-post.

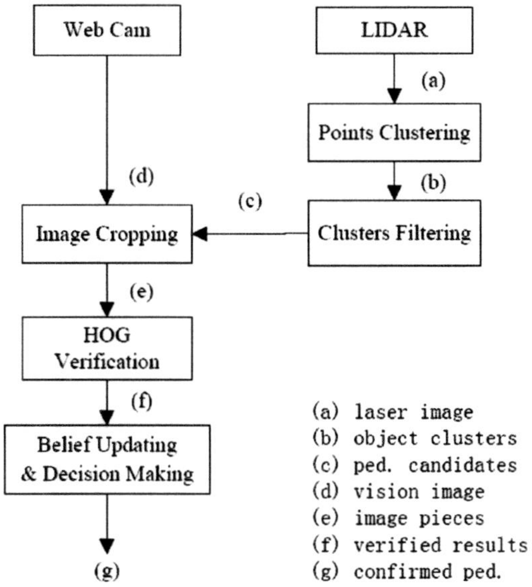

Fig. 9. Pedestrian detection algorithm flowchart

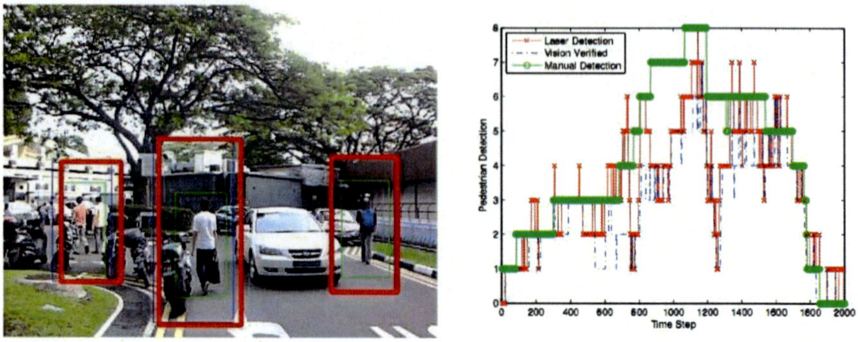

(a) Vehicle interacting with incoming pedestrians

(b) Detection of the pedestrians

Fig. 10. Pedestrian detection in a dynamic, cluttered environment

We use a the combination of a single laser range finder and a simple web camera calibrated properly to detect pedestrians on the road. Figure 9 shows the basic components of the pedestrian detection algorithm. The laser clusters the sensor information based on the proximity, and the corresponding sub images are sent to a HoG SVM classifier to detect a person. A resulting snapshot of the vehicle while in operation is shown in Figure 10(a), where pedestrians are boxed. Fig.10(b) shows the number of objects tracked by LIDAR, pedestrians verified by webcam, and the ground truth number of pedestrians. In the test, most pedestrians got detected, whether as an individual, or as a group, making safe autonomous driving of our vehicle. Frequency of this detection system is up to 37Hz, limited by scan frequency from LIDAR. Range of effective detection is about 15 meters, limited by resolution of webcam. Details of the algorithm and implementation can be found at [7].

4 Demonstration

The autonomous vehicle covered over 30km autonomously in the engineering section of National University of Singapore (NUS) campus, the route as shown in Figure 11. The selected route has representative segments of a typical road network, while being on campus the vehicle has to be more conservative in dealing with incoming student pedestrians and other vehicles. There are 4 pickup and drop off stations present in this section of the campus. The customer requests a pickup and drop-off location from either the mobile phone or the web interface shown in Figure 2. It is able to detect pedestrians and other vehicles and safely stop when the pedestrians or the vehicles are within a safety threshold along the vehicle's immediate path, preventing any collision. The videos of the operation are uploaded at (http://bit.ly/xiJDmZ).

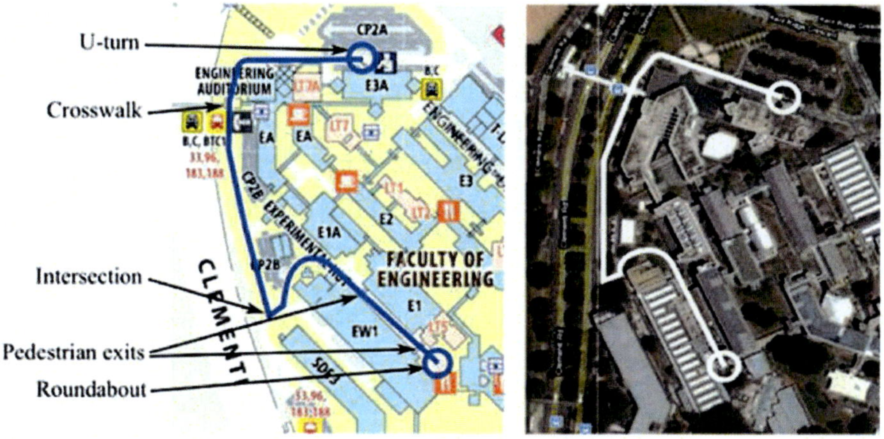

Fig. 11. Route of the autonomous vehicle

5 Conclusion and Future Work

In this paper we present an autonomous vehicle implementing mobility on demand in a campus environment. Successful operation of the system has been demonstrated where the customers request mobility service and vehicle autonomously picks them up from their desired origin and drops them off at their requested destination.

We are currently incorporating mechanism for the vehicle to interact more meaningfully with other vehicles and pedestrians on the road, to predict their intentions and generate decisions accordingly. We are also including additional personal transport platforms to utilize multiple autonomous vehicles in our mobility on demand setup. We are also looking into incorporating infrastructure sensors to augment the vehicle's perception.

Acknowledgement. The research described in this project was funded in whole or in part by the Singapore National Research Foundation (NRF) through the Singapore-MIT Alliance for Research and Technology (SMART) Center for Future Urban Mobility (FM). The research was partnered with Defence Science Organisation (DSO). We acknowledge the valuable guidance provided by Prof. Emilio Frazzoli, Prof. Daniela Rus and Prof. David Hsu in developing the mobility on demand platform.

References

1. Shaheen, S.A., Guzman, S., Zhang, H.: Bikesharing in Europe, the Americas, and Asia. Transportation Research Record: Journal of the Transportation Research Board, 159–167 (2010)

2. Pavone, M., Smith, S.L., Frazzoli, E., Rus, D.: Load Balancing for Mobility-on-Demand Systems. In: Robotics: Science and Systems (2011)
3. Mitchell, W.J., Borroni-Bird, C.E., Burns, L.D.: Reinventing the Automobile: Personal Urban Mobility for the 21st Century. The MIT Press, Cambridge (2010)
4. Berbeglia, G., Cordeau, J.-F., Laporte, G.: Dynamic pickup and delivery problems. European Journal of Operational Research 202(1), 8–15 (2010)
5. Pavone, M., Treleaven, K., Frazzoli, E.: Fundamental Performance Limits and Efficient Polices for Transportation-On-Demand Systems. In: IEEE Conf. on Decision and Control (2010)
6. Markoff, J.: Google Cars Drive Themselves, in Traffic. The New York Times (October 9, 2010)
7. Chong, Z.J., Qin, B., Bandyopadhyay, T., Wongpiromsarn, T., Rankin, E.S., Ang Jr., M.H., Frazzoli, E., Rus, D., Hsu, D., Low, K.H.: Autonomous Personal Vehicle for the First- and Last-Mile Transportation Services. In: IEEE International Conference on Robotics, Automation and Mechatronics, RAM (2011)
8. Qin, B., Chong, Z.J., Bandyopadhyay, T., Ang Jr., M.H., Frazzoli, E., Rus, D.: Curb-Intersection Feature Based Monte Carlo Localization on Urban Roads. In: IEEE International Conference on Robotics and Automation (2012)
9. Kuwata, Y., Teo, J., Fiore, G., Karaman, S., Frazzoli, E., How, J.P.: Real-Time Motion Planning With Applications to Autonomous Urban Driving, `http://dspace.mit.edu/openaccess-disseminate/1721.1/52527`

Interleaving Planning and Control of Mobiles Robots in Urban Environments Using Road-Map

Guillaume Lozenguez[1,2], Lounis Adouane[2], Aurélie Beynier[3], Abdel-Illah Mouaddib[1], and Philippe Martinet[2,4]

[1] GREYC, Campus Côte de Nacre, Bd Marchal Juin, BP 5186, 14032 Caen Cedex France
firstname.lastname@info.unicaen.fr
[2] Institut Pascal - 24 Avenue des Landais - 63177 Aubiere Cedex France
firstname.lastname@univ-bpclermont.fr
[3] LIP6, University Pierre & Marie Curie, Boîte courrier 169, 4 place Jussieu, 75005 Paris France
firstname.lastname@lip6.fr
[4] IRCCYN, 1 rue de la Noë, 44312 Nantes France
firstname.lastname@irccyn.ec-nantes.fr

Abstract. This paper presents a robot solution that allows to automatically reach a set of goals attributed to a robot. The challenge is to design autonomous robots assigned to perform missions without a predefined plan. We address the stochastic salesman problem where the goal is to visit a set of points of interest. A stochastic Road-Map is defined as a topological representation of an unstructured environment with uncertainty on the path achievement. The Road-Map allows us to split deliberation and reactive control. The proposed decision making uses a computation of Markov Decision Processes (MDPs) in order to plan all the reactive tasks to perform while there are goals not yet reached. Finally, from a brief explanation on how the approach could be extend to multi-robot missions, experiments in real conditions permit to evaluate the proposed architecture for multi-robot stochastic salesmen missions.

1 Introduction

Designing fully autonomous robots to achieve complex goals requires the fusion of severals capabilities from perception to control. Such design induces the need of engineers from different domains to work together regarding the type of used robots, their equipments and the targeted environment. For ground robots, the notion of mobility requires mapping the environment in order to localize the robot regarding its goals' positions. Two kinds of map models are provided in the literature, they permit to connect perception, decision and control: The occupancy grid maps [1] which are discretized metric representations of the robot's environment in a global basis; and the topological maps [2] which are graphs where nodes match particular locations and arcs represent path.

The robot's behavior has to converge efficiently towards goals while reacting safely to unexpected events. The most common deliberative approaches for robot are based

S. Lee et al. (Eds.): Intelligent Autonomous Systems 12, AISC 193, pp. 683–691.
springerlink.com

on grid representation [3] [4]. The discretization is suitable to compute the policies. So, the occupancy grid map is used as a unified representation of the environment for the perception (localization and mapping), the deliberation and a step by step control. Grid representation strongly depends on the robot configuration. This kind of map is hard to produce off line with different data sources from the robot sensors. Furthermore, they are not efficient for large outdoor environments with difference in altitude and limited range sensors.

Reactive control approaches allow the robot to make safe and smooth movements using a simple and on-line local representation. Reactive control requires supervision to optimize the overall task achievement. Hierarchical architectures [5] [6] [7] allow to separate decision making from control with several levels of abstraction.

The paper focuses on mobile ground robots and presents a topological map (*Road-Map*) in a hierarchical architecture to allows the deliberation to plan high level actions. Considering the reactive control capabilities, we suppose that it is possible to decrease the deliberation input data size in order to focus on abstract and efficient maps for control supervision. The *Road-Map* which represents the environment has to integrate perception and control constraints and has to be suitable for decision making. For somes applications, it is interesting to be able to produce the *Road-Map* from aerial data.

The proposed approach is illustrated by the need of a fleet of cooperative robots to visit a set of point of interest given an initial approximate map (Fig. 1). Such situation can be met for search and rescue applications where an UAV (Unmanned Aerial Vehicle) can evaluate the situation and send an abstract map to the robot. Once the points of interest are allocated among the robots, each robot computes its own plan to visit its attributed points of interest. The paper focuses on a single robot in the fleet and the control supervision from an initial plan. The plan is produced from the abstract *Road-Map* based on the UAV information. We show in this paper that the proposed approach works for urban environments.

The paper is organized as follows: Section 2 presents the robot navigation mission and the proposed solution. Experimental results are discussed in Section 3. Finally, the paper ends by future research directions.

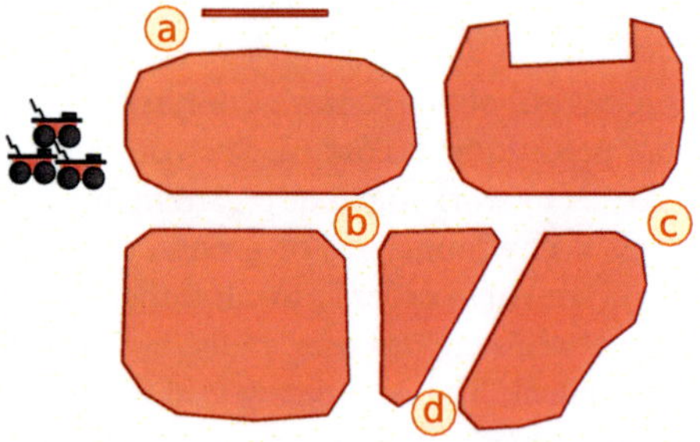

Fig. 1. Exploration area with 4 points of interest

2 Robot Navigation and Decisional Strategy

This section present the proposed architecture for the autonomous mobile robots control. It is organized in two levels: the reactive and the cognitive. The reactive level includes perceptions and local controls of the vehicle. The cognitive level includes a model of the environment, the decision making algorithm to plan movements and a supervision part to connect the built plan to local controls.

2.1 Reactive Level for Navigation

The approach is evaluated with a simple perception module without global sensors as GPS or external global camera. Each robot is equipped to estimate its pose and to detect the local free navigation space. The localization is done by odometry in a global Cartesian basis and an error grows with movements. The local free navigation space is defined in the robot basis by the list of limit points. From the limit points, semantic perception states (*PS*) are defined as: • *"left to obstacle"* or *"right to obstacle"* if one obstacle is detected in the left or the right of the robot • *"corridor"* if obstacles are detected with a unique way between them • *"intersection"* if there are more than one possible ways • *"blocked"* if obstacles prevent any movement (like a dead-end corridor with a robot without a rear perception). • *"undefined"* if no obstacle is detected or the obstacle configuration is unstable

The perception is designed with three refiners: global localization (odometer), local free navigation space detection and perception state estimation. In fact, we consider that the recognition of particular places or landmarks in the sensors input stream allows a moving robot to, punctually, localize itself [2][8][9] and re-initialize its odometry.

In parallel, we define several controllers to: reach a local target, navigate on the voronoi or circle an obstacle. The reaching controller is the default task of our robots. We design it as Multi-controller [10] to allow the robot to smoothly switch between the attraction to a target and obstacle avoidance. Our voronoi controller reaches the middle of a path in corridor perception state. We also define a left or right circle controller to allow the robot to navigate along an obstacle by keeping a safe distance. This controller is available in all perceptive states where at least one obstacle is detected.

2.2 Road-Map Definition

A *Road-Map* $< W, P >$ (Fig 2) is a topological map where: the nodes represent particular way-points; the edges represent the paths connectivity between nodes. The knowledge is defined regarding the capacities of our perception and control modules. We define a way-point $w \in W$ regarding the expected robot pose given by the odometer localization (x_w, y_w, θ) and a path $p \in P$ is characterized by the attached active controller regarding the expected perceptive state $s_p \in PS$.

Several attributes complete the path p knowledge from w_p to w'_p in order to permit efficient control and decision making. An attribute c_p gives the associated movement cost; it depends on the distance and the quality of the path. In fact, the movement cost between two positions varies according to the type of ground, the slope or the obstruction of the path. According to the possible error during a movement, a normalized function d_p (deviation) gives the probability to reach other way-points. The d_p function is initialized proportionally to the distances from w'_p to all other connected way-points to w_p. The *Road-Map* $< W, P >$ is defined as:

$$W = \{ \quad (x_w, y_w, \theta_w) \in \mathbb{R}^3 \quad \}, \ P = \{ \quad (w_p, w'_p, s_p, c_p, d_p) \\ w_p, \ w'_p \in W, \ s_p \in PS, \ c_p \in \mathbb{R}, \ d_p : T \rightarrow [0,1] \quad \}$$

Actually, the designed *Road-Map* could be defined as a mix of a voronoi graph and a visibility graph when the voronoi could not be locally detected. Structuring knowledge in a stochastic collision-free connectivity permits to plan a safe and efficient path between two positions [11]. Given a global topology as the *Road-Map*, the deliberation module has to cooperatively plan a policy to reach several goals.

2.3 Policies Computation

The aim of decision making is to associate a reactive task to perform (action) to each robot state in order to allow the robot to reach its mission by minimizing the needed resources. Due to the uncertainties related to the *Road-Map* and the reactive control, we chose to use MDPs solved by each robot. An MDP is defined as a tuple $< S, A, t, r >$ with S and A are respectively, the state and the action sets that define the system and its control possibilities. t is the transition function defined as $t : S \times A \times S \rightarrow [0,1]$ that gives the probability $t(s,a,s')$ to reach the state s' from s by doing action $a \in A$. The reward function r is defined as $r : S \times A \rightarrow \mathbb{R}$, $r(s,a)$ gives the reward obtained by executing a from s.

Regarding the treated problem, an MDP state s needs to include the last recognized position $w_s \in W$ of the robot and the set of already visited points of interest I_s. An action a is added for each path $p_a \in P$ for reaching a way-point w'_{p_a}. Furthermore, the action of visiting a point of interest w_s is modeled by a loop path p_s.

$$S = \{ \ (w_s, I_s) \mid w_s \in W, \quad I_s \subset I \ \} \\ A = \{ \ (p_a) \mid p_a \in P \ \} \cup \{ \ (p_s) = (w_s, \ w_s) \mid w_s \in I \ \}$$

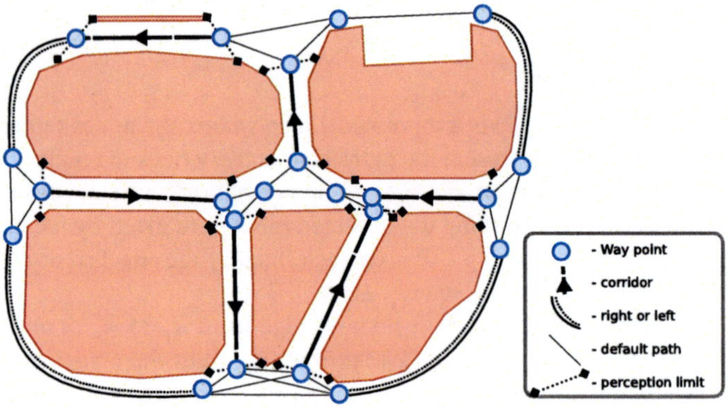

Fig. 2. Corresponding *Road-Map* to the environment described in Fig 1

As the states and actions are directly computed from the Road-Map, the transition and reward functions are build from the attribute attached to each corresponding path p_a of an action a. When executing the available action $a = p_a$ from the way point w_s, the transition function t returns the probabilities to reach neighbor positions according to the deviation function d_{p_a}. Deterministic transition are added for action corresponding to visit a new points of interest (p_s).

The reward function returns a negative value regarding the movement cost (linked to the path), and a positive constant gain g if a new point of interest is reached. Optimally solving the MDP consists in searching an optimal policy π^* that maximizes the expected gain. A policy $\pi : S \rightarrow A$ is a function which associate an action to each state. The value iteration algorithm [12] permits to compute the optimal policy π^* and the associated value function V^{π^*}. Previous study [13] allows the robots to compute, on-line, sub-optimal policies involving a large number of point to visit. It is done by splitting the *Road-Map*, into regions and applying a hierarchical solving.

2.4 Supervision and Localization

The individual policy gives the succession of paths to reach in the *Road-Map* in order to visit the attributed points of interest. Each path involves the next targeted way-point, the expected perception state along the path and the controller to activate. The control supervision monitors the odometrer position and the perceptive state to verify which way-point is reached by the reactive control. While a way-point is crossed, the reactive task is upgraded regarding the policy. This allows the robot to adapt its actions in function of the deviation due to the reactive control. A way-point is considered crossed when the robot is in a given safe distance around this way-point.

In parallel, the robot has to re-initialize the odometer. In fact, while reaching a way-point, the distance to it decreases. In a close radius around the reached way-point, the instant where the distance between the robot and the way-point start growing, the odometer is automatically re-initialized by considering the robot is just-on it. The re-localization is particularly used with voronoi or circle control where the way-point is not directly reached but the trajectory of the robot is defined by its perception. For example (Fig 3), if the robot reaches a target position before the end of a corridor, regardless the odometry, the robot has to cross the position by the voronoi controler.

In practice, the *Road-Map* has to be predefined with way-points involving coherent positions and orientations. That means the way-points are in the voronoi of corridors and at a safe distance of circled obstacles. Furthermore, the way-points have to be defined in local stable orientation area to correct the robot orientation with the best possible accuracy while the robot is not exactly on the targeted way-point.

We propose a navigation process as simple as possible in order to illustrate the capability of the architecture to connect the reactive level with the cognitive level. By using more sophisticated refiner, it is possible to develop more accurate landmark localizations or to work in more complex environments (dynamic, with map error).

3 Experiments

The proposed solution is tested by computer simulation and on real mobile ground robots in an experimental outdoor urban area (Fig 4). For this experiments, the *Road-Map* is defined by an human operator. The free navigable space is delimited by side-walks and 2 pioneer robots which have to visit a set of 12 points of interest and return to their initial positions. The robots are equipped with a laser at 45 degrees to detect sidewalks that are around one meter in front of the robot and establish the perception states. The area size is 32×26 meters and the robot are parametrized by 0.8 and 4.0 meters for the safe and close distances (*cf.* video[1]).

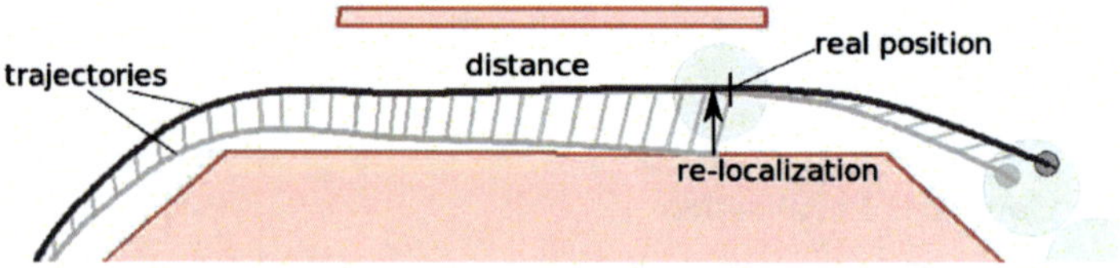

Fig. 3. Robot re-localization after a voronoi control

Fig. 4. One pioneer robot, the experimental area and the aerial view

The multi-robot coordination is done by dividing the set of points of interest in order to allocate part of mission to each robot of the team. Then each robot could perform its own sub-mission alone. Previous study [13] allows a group of robots to compute the allocation in a centralized way where a leader robot search to maximize the sum of individual expected gains.

The communication between the robots and the operator PC is done by Wifi. The operator initializes the robot positions and orientations, builds the map and selects the points of interest (Fig 2) before sending data to all the robots. A way-point in the *road-map* is defined regarding the radius of robot's perception and an aerial snapshot where

[1] www.greyc.fr/node/1629

the sidewalk is easily detectable. It is important to respect the robot's perception and control modules to draw the expected way-point position closer as possible to real robot position.

Experimental results (Fig 5 and 6) show that the *road-map* allows the robots to maintain their localization despite the few numbers of information and their inaccuracy. In this paper, we are more interested in the connexion between cognitive and reactive levels. We observe in simulation (Fig 5) and in real condition (*cf.* the attached video) that the control supervision using the deliberative policy allows the robot to smoothly move between two distant positions involving several way-points.

We notice that the third robot with the 3 points of interest in the south of the map (Fig 5) starts returning to its initial position by circling the local obstacle from the right. At this moment, the policy gave a task of reach a way-point behind the robot. By turning around, the reactive control circles the obstacle and reaches a bad way-point.

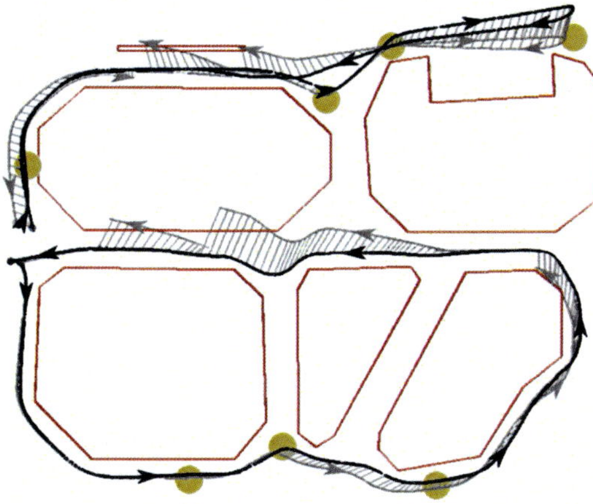

Fig. 5. Real (dark) and supposed (gray) trajectories of the 2 robots (Computer simulated result)

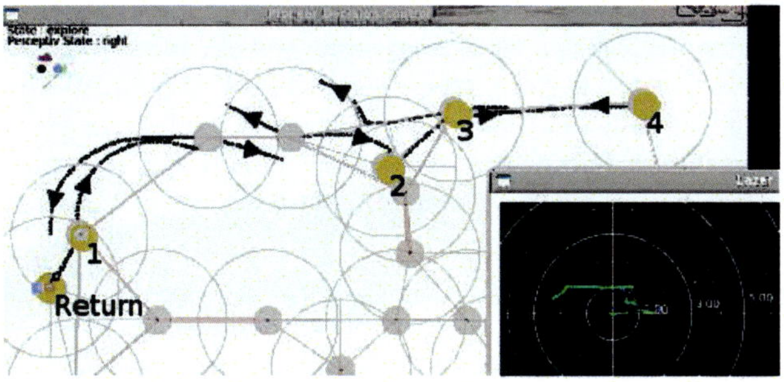

Fig. 6. One robot belief state at the end of a real experiment (*Road-Map*, attributed points of interest, supposed trajectory and laser snapshot)

We observe (as expected) that the supervision adapts correctly the planned path to the initial position according to the policy.

In fact, in this experiments, the re-localisation process has a blind trust in the operator *Road-Map*. The *Road-Map* parameters (way-points positions and orientations, safe and close distances) needed adjustments to assume the robot re-localisation at any position. This is notably true in the real experiments where the robot has to go and come back from the point of interest 2 to the corner north-east only with the odometer (Fig 6). This round trip measures 2×18 meters and the robot reaches the target (point of interest 4) with an error around 3 meters and come back with an error around 3.8 meters. This is translated by a gap of the curve trajectory resulting of the avoided sidewalk near the point of interest 2 (Fig 6). At the end, the robot corrects its gap to finish with an error of 0.8 from its start position.

The odometer re-initialization is correct but not perfect (Fig 5 and 6). The process shares an error in the orientation of the obstacle limit (in relation of the Degrees of freedom of the used control) and the error grows while the robot moves in the same direction (Fig 3).

In conclusion, without considering an uncertainty about the *Road-Map*, its definition by the operator has a great importance in the success of the robot movements. Regarding our reactive level, a way-point has to be positioned in retreat of intersection areas to assume stability on the orientation and anticipate a switch between controllers. The error on the map coupled to the odometric error has to be bounded to assume that the robot is able to reach the farthest next way-point where the re-localization involve a different orientation.

4 Conclusion

A control architecture based on a *Road-Map* to connect deliberation and reactive control is presented in this paper. Splitting robot control into 2 levels provide flexible solution in regard to the needed robot functionality. The *Road-Map* allow us to merge decision making and reactive control modules developed separately. The use of this solution is illustrated by computing and executing collaborative policies in order to visit a set of points of interest with a fleet of robots. The proposed distributed solution allows the fleet to build a valid task achievement. Experiments in real conditions validate the approach in an urban environment using a *Road-Map* based on UAV information.

In future work, we want to study the consequences of map actualizations on the decision making and introduce uncertainty on the map knowledge. The perceptive module will also be increased to detect robots and humans, the goal is to permit the robots to maintain their localization and their policies in dynamic environment.

References

1. Elfe, A.: Sonar-based real-world mapping and navigation. Robotics and Automation 3, 249–265 (1987)
2. Kuipers, B., Byun, Y.-T.: A robot exploration and mapping strategy based on a semantic hierarchy of spatial representations. Journal of Robotics and Autonomous Systems 8, 47–63 (1991)

3. Goldman, C.V., Zilberstein, S.: Decentralized control of cooperative systems: Categorization and complexity analysis. Journal of Artificial Intelligence Research 22 (2004)
4. Burgard, W., Moors, M., Stachniss, C., Schneider, F.: Coordinated multi-robot exploration. IEEE Transactions on Robotics 21 (2005)
5. Alami, R., Chatila, R., Fleury, S., Ghallab, M., Ingrand, F.: An architecture for autonomy. Journal of Robotics Research 17(4), 315–337 (1998)
6. Volpe, R., Nesnas, I., Estlin, T., Mutz, D., Petras, R., Das, H.: The claraty architecture for robotic autonomy. In: Aerospace Conference, vol. 1, pp. 121–132 (2001)
7. Mourioux, G., Novales, C., Poisson, G.: Control robot by a generic control architecture. In: IEEE International Conference on Intelligent Robots and Systems, pp. 3050–3055 (2007)
8. Thrun, S.: Learning metric-topological maps for indoor mobile robot navigation. Artificial Intelligence 99(1), 21–71 (1998)
9. Ulrich, I., Nourbakhsh, I.: Appearance-based place recognition for topological localization. In: International Conference on Robotics and Automation, vol. 2, pp. 1023–1029 (2000)
10. Adouane, L.: Hybrid and safe control architecture for mobile robot navigation. In: Autonomous Robot Systems and Competitions (2009)
11. Alterovitz, R., Siméon, T., Goldberg, K.: The stochastic motion roadmap: A sampling framework for planning with markov motion uncertainty. In: Robotics: Science and Systems (2007)
12. Puterman, M.L.: Markov Decision Processes: Discrete Stochastic Dynamic Programming. John Wiley & Sons, Inc. (1994)
13. Lozenguez, G., Adouane, L., Beynier, A., Martinet, P., Mouaddib, A.-I.: Map partitioning to approximate an exploration strategy in mobile robotics. In: Advances on Practical Applications of Agents and Multiagent Systems, vol. 88, pp. 63–72 (2011)

Aggressive Manuevering of Unmanned Helicopters: Learning from Human Based on Neural Networks

Dalei Song, Chong Wu, Juntong Qi, and Jianda Han

State Key Laboratory of Robotics
Shenyang Institute of Automation, Chinese Academy of Sciences
Shenyang, China
daleisong@sia.cn

Abstract. "Teaching by Showing" control of a small helicopter's aggressive maneuvering often needs inner aided controllers based on helicopter's dynamics, which is very complex to identify. In this paper, a neural network based control is proposed, based on the identification of the relationship between the pilot's control and flight states, and it is a model-free control method. Flight test is done in simulation environment based on real flight data. The results show the effectiveness of the neural network based controller for aggressive flight control.

Keywords: neural network, unmanned helicopter, learning, flight test.

1 Introduction

Unique capabilities, such as hover, low-speed flight, vertical takeoff and landing, of autonomous helicopters provide indispensable function in many applications. On the other hand, autonomous helicopter control is a very challenging problem since the complexity in its architectures and dynamics. The control of autonomous helicopters provides an important test-bed for control algorithms. Many unmanned helicopters' control algorithms involve complex dynamics that are hard to model, the PID algorithm is simple but the performance is limited. Although stabilization and autonomous flight has been achieved, the flight performance of current control algrithms is still modest compare to the expert human pilot; control systems still do not cover the full flight envelope of unmanned helicopter.

"Teaching by Showing" control for a small helicopter's attitude control was firstly demonstrated in simulation by Montogomery [1], a fuzzy logic controller is augmented with a neural network, and the fuzzy controller's number and type of membership functions were decided upon by a competent human pilot based on their perceptions of how they flew a helicopter. In [2], a feed-forward artificial neural network (ANN) with 10 hidden units was trained to mimic the actions of a human pilot. Only the aileron servo was driven by the ANN in the flight test, the other servos were controlled by the pilot, and only a hover of 15 seconds was accomplished. A reinforcement learning policy for an autonomous helicopter was demonstrated in [3]

S. Lee et al. (Eds.): Intelligent Autonomous Systems 12, AISC 193, pp. 693–703.
springerlink.com

by Bagnell. A pair of very simple feed-forward networks with 5 weighted connections for each of the pitch and roll control channels were used. With this controller, the R-50 helicopter was able to track moving position set points in the presence of strong wind gusts. In 2004, Ng. published the results for a simple neural network controller for a Yamaha R-50 helicopter using another reinforcement learning technique [4]. In 1994, Calise published a scheme for adaptively correcting the inversion errors for an attitude controller based on an inverted model of the helicopter attitude dynamics [6]. Using the same control structure as Corban et al, Johnson and Kannan have obtained excellent tracking results on the GTMAX unmanned helicopter [7]. Dynamic inversion and neural-network-based adaptation was used to increase performance of the attitude control system and the method of pseudocontrol hedging (PCH) was used to protect the adaptation process from actuator limits and dynamics.

The works mentioned above mainly concentrated on the hovering control and trajectary flight control, in contrast, aerobatic flight has seldomly been investigated. Gavrilets [9] investigated a limited set of autonomous aerobatic flights include a stall-turn, a split-S, and an axial roll. In [5], autonomous control of sustained inverted hover was successfully demonstrated using the reinforcement learning technique. Abbeel [8] described a generative probabilistic model that enabled them to extract an expert helicopter pilot's intended trajectory from multiple suboptimal demonstrations; the experimental results include the first autonomous execution of a wide range of maneuvers, including in-place flips, in-place rolls, loops and hurricanes, and even auto-rotation landings, chaos and tic-tocs, which only exceptional human pilots can perform. However, these control algorithms all need aided attitude controllers based on helicopters' dynamics for inner-loop stabilization, which is very complex to identify.

In this paper, a neural network based control of unmanned helicopter in circular maneuver is investigated. A double hide layer structure is proposed for unmanned helicopters' control by learning from the pilots, and the experiment process is described in detail in this paper. Compared with aforementioned algorithms, it is a neural network based model-free controller, and is able to finish circular maneuvering with a high speed.

2 Experimental Setup

The goal of the experimental setup was to identify the aggressive maneuvers flown by pilots on small actual and simulated R/C helicopters. In the experiment, the pilot drives the actual helicopter to track circle trajectories under high speed by remote controller (RC). In the procedure, all the flight states are recorded and the strategy of the pilot is identified to build a neural networks based controller. After that, the neural networks based controller is tested in the simulation environment for safety concerns. The following will introduce the details of the R/C helicopter and the simulation environment.

First, the frame of unmanned helicopter and corresponding variables are introduced. A helicopter in flight is free to simultaneously rotate and translate in six degrees of freedom. Figure 1(*left*) shows the helicopter variables in a body-fixed frame with origin at the vehicle's center of gravity. In the following, some notations are used for controller identification. u, v, and w are longitudinal (along X-axis), lateral (along Y-axis) and heave (along Z-axis) velocities, respectively; p, q, and r are roll, pitch and yaw angular rates in the helicopter's reference frame; φ and θ are the angles of roll and pitch, respectively; a and b are the lateral and longitudinal flapping angles of main rotor, c and d are the lateral and longitudinal flapping angles of stabilizer bar; δ_{lat} is the lateral control input, δ_{lon} is the longitudinal control input, δ_{ped} is the yawing control input, and δ_{col} is the heave control input.

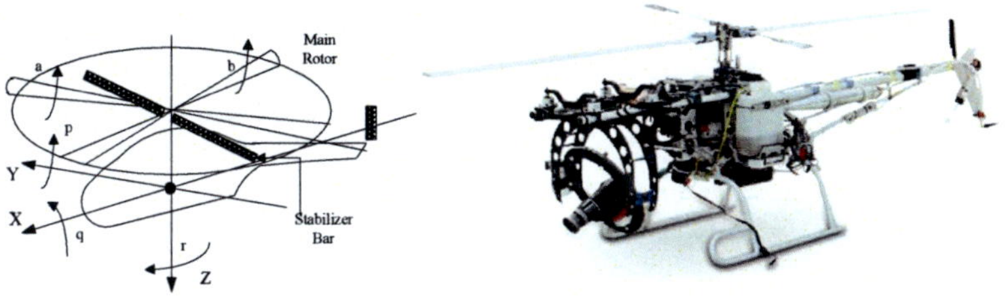

Fig. 1. Helicopter with its body-fixed reference frame (*left*) and The SERVOHELI-40 small-size helicopter (*right*)

2.1 Flight Platform

SERVOHELI-40 helicopter, which was built by Shenyang Institute of Automation Science as shown in figure 1(*right*), was chosen as the platform for testing the autonomous aerobatics algorithms. The helicopter weights close to 20kg empty, carries a 15kg avionics payload and 4kg of fuel, which lasts for about 45 minutes. A hinge-less main rotor is equipped with a Bell-Hiller stabilizer bar [11], which provides lagged rate feedback and augments the servo torque with aerodynamic moment to change the cyclic pitch of the blades. The helicopter is powered by a KOMATSU aero-engine running on a mixture of methanol and oil. It is equipped with a 3-axis gyro, a 3-axis accelerometer, a compass and a GPS. The sensory data can be sampled and stored into an SD card through an onboard DSP. Table 1 lists the physical characteristics of the SERVOHELI-40 small-size helicopter.

Table 1. Physical parameters of SERVOHELI-40 helicopter

Length	2.12m
Height	0.73m
Main rotor diameter	2.15m
Stabilizer bar diameter	0.75m
Rotor Rev.	1450rpm
Net weight	20kg
Engine	2-stroke, air cooled
Flight duration	45 min

2.2 Hardware-in-the-Loop Simulation

The simulated environment was a high-level-of-detail model of the SERVOHELI-40 helicopter. The model includes full helicopter dynamics, such as rigid-body, aerodynamic, rotor and servo dynamics. An essential tool in the development of a real-time control system is a hardware-in-the-loop simulation (HILSim). It allows the engineer to test the actual software running on the target hardware in conditions as close to a real flight as is achievable in the lab.

A prototype Hardware-in-the-loop (HIL) simulation and test platform for RUAVs has been developed in Shenyang Institute of Automation, Chinese Academy of Science. It is composed of 3D flight and scenery simulator, flight control system (FCS), ground control station (GCS), and a RC pilot controller. 3D flight and scenery simulator runs a RUAV aerodynamic model, which is identified from real RUAV flight data, and generates simulated navigational sensors information includes GPS data, 3-axis flight velocity data, 3-axis RUAV acceleration data, 3-axis angular data, and 2-axis wind disturbances at 50Hz. It also displays 3D flight progress in Flightgear[@], an open source flight simulator [12]. The flight control system (FCS) is the core avionics of the system, which communicates the simulated navigational information from 3D flight and scenery simulator and executes the autonomous control algorithm iteratively in real time.

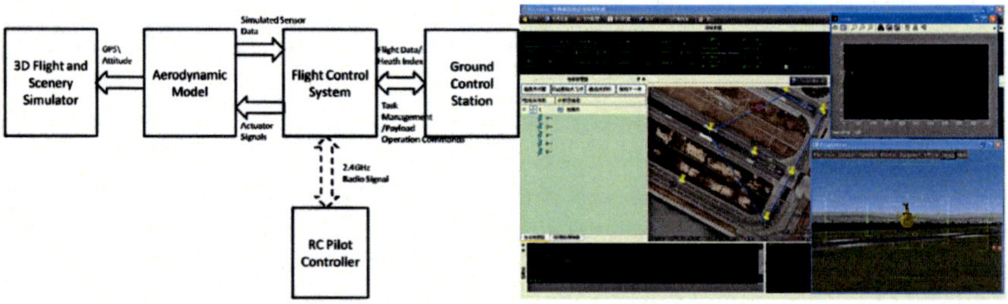

Fig. 2. The structure of HIL simulation environment (*left*) and The windows of the HIL simuation environment (*right*)

2.3 Verification of the Simulation Environment

A series of control experiments have proved that RC pilot has a considerable better performance than traditional PID control in unmanned helicopter autonomous control of circular flights [8]. Heading and aileron are weighted relatively significant than velocity and height in order to produce a quality circular flight with high speed [7]. In this paper, the performance of PID controller and RC pilot are compared when executing circle tracking on SERVOHELI-40 platform, and for verification of the effectiveness of the HIL simulation, the output of the HIL simulation environment are recoded under the same control input of the RC pilot when doing the real circle tracking.

In figure 3(*left*), the red dash-dot curve is set to be a reference circular flight trajectory with 20 meters in radius and 5 m/s in velocity.

Firstly, a traditional PID controlled autonomous flight was conducted and shown with blue dash curve. In this flight, the programmed traditional PID algorithm controlled all four control loops discussed. The heading and aileron control progressively drifted and there is about 10 meters off in the trajectory control.

Then, a semi manual flight was conducted. In this flight, a set of markers were placed along the circular path on the ground and a PID algorithm was implemented in the controller to facilitate the pilot to control the longitudinal flight velocity and altitude, so that the pilot could focus on the manual control of heading and aileron channels. Compared to the traditional PID flight trajectory, the blue dot curve in figure 3(*right*), pilot manual flight, follows the reference circle only with 2 to 3 meters offset. That is because that pilot had to spot the markers to perform a circular flight. Obviously, manual control demonstrated significant improvements in heading and aileron control of a circular flight.

A third experiment has been done in the hardware-in-the-loop (HIL) simulation platform by using the previous pilot manual control outputs to drive the aerodynamic model. The green curve, in figure 3(*left*), shows that the simulated flight trajectory reconstructed the manual flight only with trivial bias in heading and aileron. Figure 3(*right*) shows the control inputs of RC pilot, figure 4(*left*) and figure 4(*right*) show the comparison of speed and attitude under the RC control inputs. Clearly seen in figure 4(*left*) and figure 4(*right*), HIL simulation has a good tracking for flight data, and roll control must be implemented effectively to maintain proper centripetal acceleration for realizing a good circle tracking. These imply that the aerodynamic model for simulation has high fidelity performance to represent the real flight experiments.

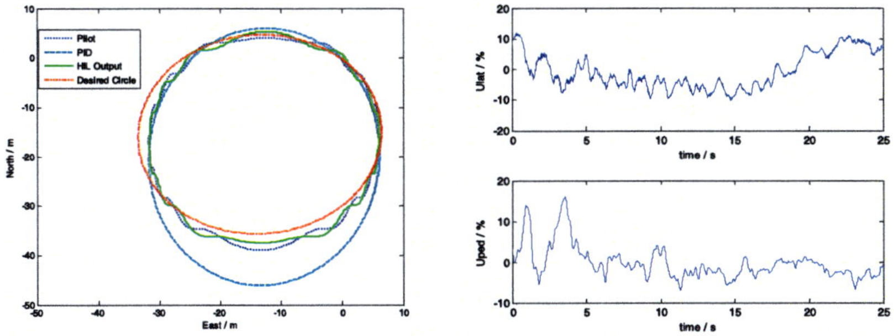

Fig. 3. The recorded position in circle tracking(*left*) and The lateral and yaw control inputs in circle tracking by RC pilot(*right*)

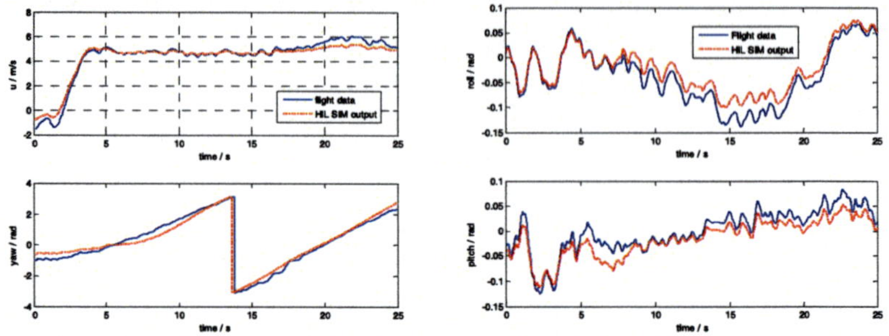

Fig. 4. The longitudinal speed and yaw in circle tracking (*left*) and the roll and pitch in circle tracking (*right*) by pilot and HIL simulation environment

3 Maneuver Identification Based on Neural Networks

When the pilot's performing the circle tracking maneuvering, the flight controller is the pilot's brain and the feedbacks are flight states, measured by pilot's vision. This control structure is shown as figure 5(*left*). In this process, the human's strategy of aggressive maneuvering by remote control is a non-linear projection, and it has more advantages than traditional controller when doing circle tracking for helicopters as discussed above. Our aim is to identify the expression of the brain controller using neural network, which is shown in figure 5(*right*).

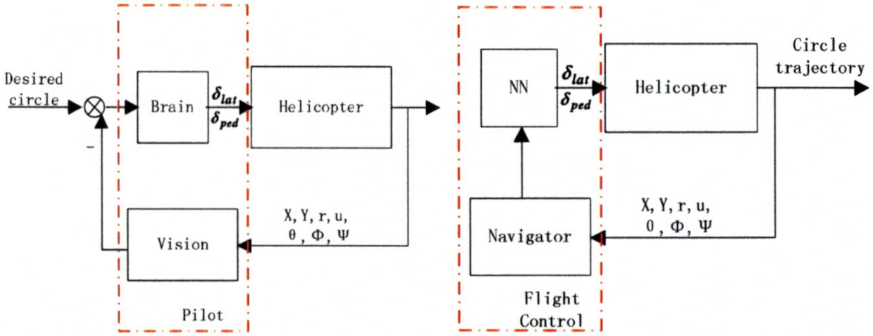

Fig. 5. The controller structure in circle tracking maneuvering by pilot (*left*) and by neural network controller (*right*)

The neural networks can realize any smooth non-linear function in theory [2]. Hence, based on the recorded flight states and control inputs, we use the neural networks to identify the mathematic model for pilot's control strategy.

A three-layer network with two hide layers is used as the model's structure, which is shown in figure 6. The first layer is the input layer, the second layer is hide layer, and the third layer is the output layer. Here N_1 is the number of nodes in input layer,

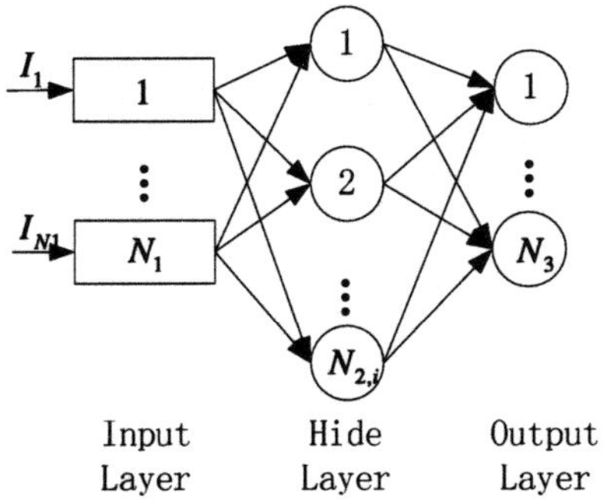

Fig. 6. The structure of neural network for identification

$N_{2,i}$ is the number of nodes in the i-th layer in hide layer, and N_3 is the number of nodes in output layer.

By communicating with the pilot, we learned that when he performed the circle tracking in section II.C, the pilot gave out the heading control input δ_{ped} based on position X and Y, longitudinal velocity u and yaw angle speed r; based on longitudinal velocity u and attitude φ, θ, and ψ, he gave out the lateral control input δ_{lat}. The longitudinal control input δ_{lon} and heave control input δ_{col} are generated by PID controller for reducing the complexity of operation for pilots as above. Thus, we select X, Y, u, r, φ, θ, and ψ as the input for the neural network in figure 6, and δ_{lat} and δ_{ped} are the output for the network. This means N_1=7, N_3=2, and $N_{2,1}$=30 with $N_{2,2}$=30 is selected in the identification process in the next part.

For node i in this hide layer, it is selected as the following normal structure, shown in figure 7, where

$$y_i = f_i(z_i) = f(\sum_{j=1}^{N} w_{i,j}x_{i,j} - \bar{\theta}_j) \tag{1}$$

$$f_i(z_i) = \frac{1-e^{-z_i}}{1+e^{-z_i}} \tag{2}$$

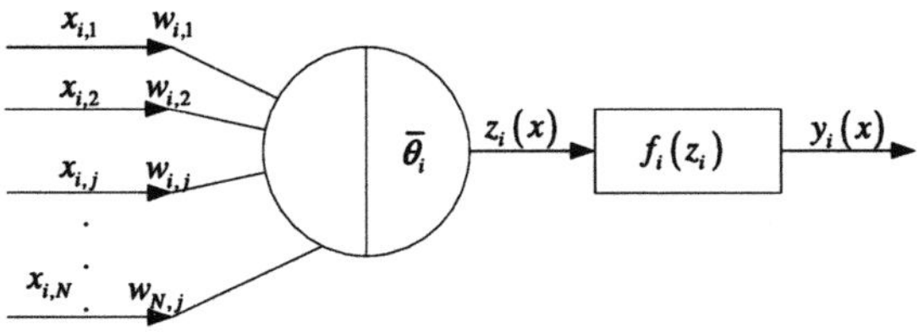

Fig. 7. The node structure of neural network for identification

To identify the weight parameters in this neural network, error back propagation neural network algorithm [10] is adopted. In the training mode, the quadratic sum of the internal error from all neurons in the output layer of Layer L forward feedback network is

$$J_{Tp} = \sum_{i=1}^{M} J_{pLi} \tag{3}$$

Assuming the output layer (Layer L) has M nodes. And the internal state error of the node n in this output layer is

$$\delta_{pLn} = d_{pLn} - z_{pLn} \tag{4}$$

As described, the relationship between d_{pLn} and neuron expected output d'_{pLn} is

$$d_{pLn} = f^{-1}(d'_{pLn}) \tag{5}$$

Because of w_{Ln} is biased from the ideal value, which makes z_{pLn} and J_{Tp} deviated from the ideal values. Thus δ_{pLn} also can be denoted as

$$\begin{aligned} \delta_{pLn} &= -\frac{\partial J_{Tp}}{\partial z_{pLn}} \\ &= -\frac{\partial J_{Tp}}{\partial y_{pLn}} \frac{\partial y_{pLn}}{\partial z_{pLn}} \\ &= -\sum_{r} \left\{ \frac{\partial J_{Tp}}{\partial z_{p(L+1)r}} \frac{\partial z_{p(L+1)r}}{\partial y_{pLn}} \right\} \frac{\partial y_{pLn}}{\partial z_{pLn}} \end{aligned} \tag{6}$$

In this equation, r is denoted as the number of notes in layer L+1. From equation (4), (5), (6), we can derive recursion equation

$$\delta_{pL\mathrm{n}} = f'(z_{p\ln}) \sum_{r} \delta_{p(L+1)r} w_{(L+1)rn} \tag{7}$$

where

$$w_{(L+1)rn} = \frac{\partial z_{p(L+1)r}}{\partial y_{pLn}} \tag{8}$$

$$f'(z_{p\ln}) = \frac{\partial y_{pLn}}{\partial z_{pLn}} \tag{9}$$

To calculate the recursion equation, firstly, using (7) to get $\delta_{pL\mathrm{n}}$, then reversely propagates to (L-1) hidden layer. Progressively, when error reaches a certain node, this error is utilized to correct the node weight.

Assuming the weight is $w_{Ln}(k)$ after k recursive calculations, then according to gradient method

$$
\begin{aligned}
w_{Ln}(k+1) &= w_{Ln}(k) + \rho(-\frac{\partial J_{Tp}}{\partial w_{Ln}}) \\
&= w_{Ln}(k) + \rho\delta_{p\ln}x_{pl}
\end{aligned}
\tag{10}
$$

here ρ is the recursion step increment.

The calculation steps of the back-propagation algorithm, are listed as follow

1. Randomly set the network weights $w_{Ln}(0)$;
2. Apply the known input x_p to the network, and calculate z_p;
3. Calculate J_{Tp};
4. According to equation (10), adjust the weights $w_{Ln}(k)$. Hidden layer δ_{pLn} is calculated by equation (7) for hidden layer, and is calculated by equation (4) in output layer;
5. Back to step 2, repeat the recursion.

4 Aggressive Maneuver Demonstration

Using the approach described in the previous paragraphs, the SERVOHELI-40 helicopter is flown to a forward speed of 10m/s to then perform a circle tracking maneuver with radius 20m for 9 times, shown in figure 8. The control inputs and states are all recorded. As described in 2.3, in this process, the longitudinal speed and height tracking are also implemented by PID controller. The pilot only gave the lateral and heading control input based on the marks, placed on the ground, and attitudes from his vision.

Using the neural network identification method, described in III, we finished identifying the pilot's controller in circle tracking maneuvering. Here, two hide layers were defined, and the node number in each hide layer is set as 30 and 10 respectively. Inputs to the network are X, Y, u, r, φ, θ, and Ψ, and the outputs from the network are δ_{lat} and δ_{ped}. The maximum epoch set for training is 100, and the goal of mearn-square-error is set as 0.1 since the control output is in the range of -1000 to +1000 that 0.1 is sufficient accurate for control purpose. In order to capture the nonlinear property between the input and output, the first hide layer's neuron transfer function is set as 'Tansig' with the input unified between -1 and 1; on the other hand in order not to limit the output, the second hide layer's transfer function is set as (2). Error back-propagation algorithm is used to update the weight parameters in each training epoch, as described in part III. To compare the identified neural network's accuracy, different samples are introduced as neural network training samples. We split the 9 times circle tracking maneuver data into 9 samples and each sample represents a completing circle tracking for one time.

Figure 4 (*left*) shows the performance of the neural network in trainings. Training for one time means that using one sample of circle tracking to adjust the weight parameters, as we can see from the training result, as the samples increase, the neural network can gradually adjust its weight parameters and obtain a better mean-square-error (MSE) between the neural network's outputs and the real outputs. Figure 9 (*right*) shows the neural network circle tracking performances in the HIL simulation environment after trainings. The tracking trajectory is closer to the desired circle as more training samples are introduced to the neural network's training procedure, and this matches the results in figure 9(*left*), and the reliability for high speed circle tracking is better than human pilots in figure 8.

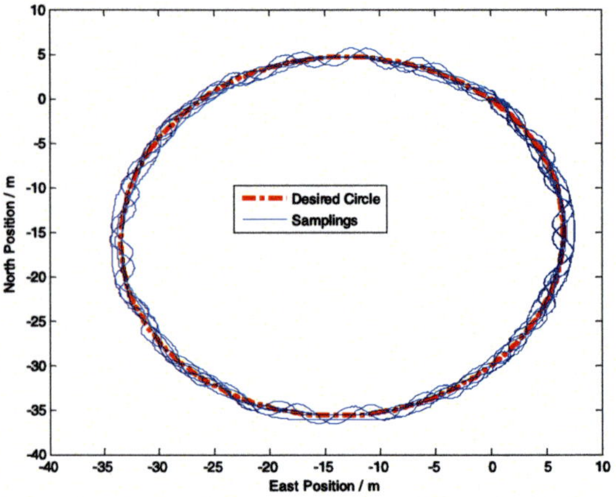

Fig. 8. The training sample for neural network

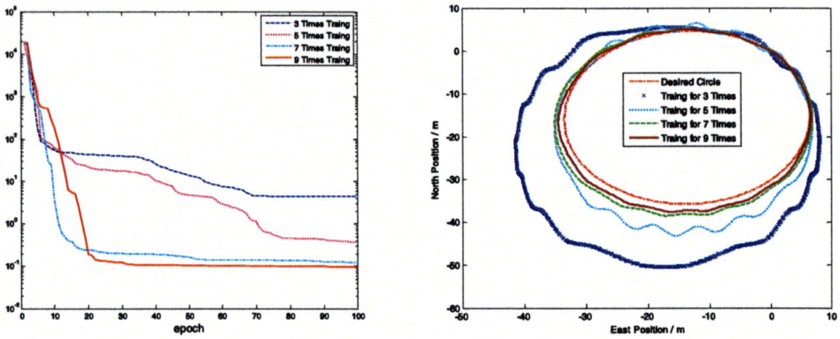

Fig. 9. MSE of Trainings with Different Samples (*left*) and tracking performances after trainings(*right*)

5 Conclusions

Based on neural network, the pilot's controller in brain is identified in this paper, and the experiment is described in detail. The neural network gives out control inputs to the helicopter directly without aided inner-controller, and realizes circle maneuvering in HIL simulation based on real flight data. It is a model-free control strategy for aggressive flight control. The results demonstrate the effectiveness of the training method for helicopter control.

Acknowledgements. This paper is sponsored by national nature science foundation 61005086 and the national science and technology support 2011BAD20B07.

References

[1] Montgomery, J.F., Bekey, G.A.: Learning helicopter control through teaching by showing. In: Proceedings of the 37th IEEE Conference on Decision and Control, vol. 4 (1998)
[2] Wyeth, G., Wyeth, G., Roberts, J.: Autonomous helicopter hover using an artificial neural network. In: Proceedings 2001 ICRA. IEEE International Conference on Robotics and Automation, vol. 2, pp. 1635–1640 (2001)
[3] Bagnell, J.A., Schneider, J.G.: Autonomous helicopter control using reinforcement learning policy search methods. In: Proceedings of the 2001 IEEE International Conference on Robotics and Automation, Seoul, Korea (May 2001)
[4] Ng, A.Y., Kim, H.J., Jordan, M.I., Sastry, S.: Autonomous helicopter flight via reinforcement learning. In: Advances in Neural Information Processing Systems, vol. 16 (2004)
[5] Ng, A.Y., Coates, A., Diel, M., Ganapathi, V., Schulte, J., Tse, B., Berger, E., Liang, E.: Autonomous inverted helicopter flight via reinforcement learning. In: International Symposium on Experimental Robotics (2004)
[6] Calise, A.J., Kim, B.S., Leitner, J., Prasad, J.V.R.: Helicopter adaptive flight control using neural networks. In: Proceedings of the 33rd IEEE Conference on Decision and Control, vol. 4 (1994)
[7] Johnson, E.N., Kannan, S.K.: Adaptive trajectory control for autonomous helicopters. Journal of Guidance, Control and Dynamics 28(3) (2005)
[8] Abbeel, P., Coates, A., Ng, A.Y.: Autonomous helicopter aerobatics through apprenticeship learning. The International Journal of Robotics Research (2010)
[9] Gavrilets, V.: Autonomous aerobatic maneuvering of miniature helicopters. Massachusetts Institute of Technology, Dept. of Aeronautics and Astronautics (2003)
[10] Rumelhart, D.E., Hintont, G.E., Williams, R.J.: Learning representations by back-propagating errors. Nature 323(6088), 533–536 (1986)
[11] Qi, J., Song, D., Dai, L., Han, J., Wang, Y.: The New Evolution for SIA Rotorcraft UAV Project. Journal of Robotics 1(9) (2010)
[12] Olson, C.L.: FlightGear Flight Simulator (2010), `http://www.flightgear.org/Downloads/aircraft`

Robust Object Recognition in Unstructured Environments

Ester Martínez-Martín[1] and Angel P. del Pobil[2]

[1] RobIn Lab, Universitat Jaume-I, Castellón, Spain
emartine@icc.uji.es
[2] RobIn Lab, Universitat Jaume-I, Castellón, Spain, and Dept. Interaction Science, SungKyunKwan University, Seoul, Korea
pobil@icc.uji.es

Abstract. One of the main goals of Robotics research is to help human beings in their daily tasks. However, physical interaction in everyday human scenarios requires robot systems endowed with rich sensory-motor skills and multisensory feedback that enables them to exhibit levels of adaptability high enough to achieve their goals. In this context, vision plays a main role since it provides rich information about the state of the environment. Although it is an active research area, there are still some challenging issues to be solved. Among them, the research presented in this paper addresses the object recognition for manipulation tasks in unstructured environments from a visual input. We have designed a mechanism that provides a *background model* covering the whole system's peripersonal space such that the objects of interest can be always identified without any information about the system. Different objects and scene conditions have been considered to assess approach's performance by showing robust object recognition in all the cases.

1 Introduction

One of the main goals of Robotics research is to help human beings in their daily tasks. However, physical interaction in everyday human scenarios requires robot systems endowed with rich sensory-motor skills and multisensory feedback that enables them to exhibit levels of adaptability high enough to achieve their goals.

Focusing on perception, vision plays a main role since it is undoubtedly the most important sense for the information it can provide. So, knowledge about physical objects and scenes can be obtained from a visual input. As a result, a wide range of robotic tasks use it to reach their goals such as navigation, for detecting and properly avoiding obstacles [4, 8, 10, 23, 27]; manipulation, for identifying objects to interact with [16, 30]; cooperative behaviour, for identifying the activities of other team members [5, 9]; or, human-robot interaction, for properly recognizing the human being with who interacts [18, 29].

However, there are still some challenging issues to be solved. Among them, the research presented in this paper addresses the object recognition for manipulation tasks in unstructured environments from a visual input. This can be a difficult task because of noise in images, nature of objects, partial occlusions, complex object shapes, scene illumination changes and real-time processing requirements.

S. Lee et al. (Eds.): Intelligent Autonomous Systems 12, AISC 193, pp. 705–714.
springerlink.com

Keeping in mind an autonomous system, no constraints about the targets should be established and, consequently, model-based vision approaches are not considered, despite the fact they are the most general way to recognize objects and shapes [1,2,19,21, 25]. Mainly, in model-based vision, the features extracted from the objects in a scene are matched against features of previously stored object models. Note that the chosen features must be invariant with respect to various transformations (e.g. changes in illumination, scale and/or view direction) and be also robustly extracted, which cannot be guaranteed in real scenarios. Moreover, for $3D$ recognition, a three-dimensional formulation of these features is required. Although these methods work well under certain conditions, they commonly have drawbacks with the need of manual intervention for creating representations for new considered objects, the fine-tuning of these representations and the lack of representations in a general context.

As an alternative, the concept of *Object Action Complexes* (OACs) can be used. In this case, it is assumed that objects and actions are inseparably intertwined. So, OACs are obtained from instantiated actions in robot control-space and they can be used as an interface between the very different representation languages of robot control and Artificial Intelligence planning (e.g. [13, 24]). In addition, robot actions can be linked to the visual and haptic perception of objects to interact, such as in [12, 17]. The same idea underlies in [14, 15], where a process to segment interest objects and to extract their shape becomes realized through active visual exploration. Despite the exploration system is completely autonomous, the system still requires a significant amount of *prior knowledge* about the world (in terms of a sophisticated visual feature extraction process in an early cognitive vision system), *knowledge* about its body schema and *knowledge* about geometric relationships such as rigid body motion. That is, it is necessary to know the system's visuomotor map in order to be successful. For that reason, this approach has been also discarded.

In addition, a significant issue arises when unstructured environments are used. The reason lies on their high-dimensional state space, the inherent uncertainty and the lack of specific patterns. Thus, a complete knowledge about an unstructured environment cannot be achieved. As a consequence, robots must continuously acquire sensory data to support decision making and adapt to changes resulting from their actions. Some effort has been carried out on this topic. For instance, Yinka-Banjo et al. [32] proposed a collision avoidance strategy based on a Behavioural Bayesian Network approach for autonomous robots in an unstructured environment with static obstacles.

Therefore, a new approach is needed to achieve our goal. Keeping in mind that no information about the system is available, the required knowledge to object recognition can be only obtained from the visual input. As a solution and under the hypothesis that a *background image* can be obtained in some way, the interest objects could be detected as those regions of change in the subsequent captured images by a *modified background* subtraction technique. In other words, we have designed a mechanism that provides a *background model* covering the whole system's peripersonal space such that the objects of interest can be always identified. When several potential targets are visible in the scene, two different ways can be used to determine the interest object to interact with: (1) the user can choose it by means of a *Graphical User Interface* (GUI), or (2) a *Task manager* module randomly selects one. At that moment, the system builds a model of the chosen object that will be used to recognize it in the following frames.

In that way, the robot's ability in interacting with the environment is improved. An advantage of the proposed approach is that the robotic system does not require previous calibration. Thus, the main contributions of this paper is an approach that successfully handles challenging issues such as:

- Segmentation process in real environments
- Segmentation process when the fixation point is continuously changing over time
- Estimation of target's positions without requiring precise calibration information (in terms of the relative orientation of the cameras)
- Robust object recognition under different conditions

This paper is structured as follows: a general overview of the proposed approach is described in Section 2; some experimental results are presented in Section 3 and discussed in Section 4.

2 Object Recognition

Keeping in mind that no information about the system is available, the required knowledge to object recognition can be only obtained from the visual input. Therefore, the first stage of this process consists of detecting the presence/absence of interest objects in a scene. For that, under the statement that interest objects are those which are not part of the background, a *background subtraction technique* could be used. That is, under the hypothesis that a *background image* can be obtained in some way, the interest objects could be detected as those regions of change in the subsequent captured images by a *modified background subtraction technique.*

Actually, in the existing approaches (e.g. [3,11,20,28,31]), basically, the scene background is modelled by means of a frame or a statistical model, that can be updated or not over time. So, a pixel is considered as part of an interest object when its value is considerably different from that in the reference model. It is worth noting that this kind of methods has been designed for static cameras and is aimed at motion detection. On the contrary, in our case, the perceived scene can continuously be changing since the fixation point can be modified over time to foveate on the target's centroid before manipulating it.

As a result, we have designed a mechanism that provides a *background model* covering the whole system's peripersonal space such that the objects of interest can be always identified. Nevertheless, instead of taking a single image of the background, a series of *saccadic* movements provides the data necessary to build the *peripersonal background model*. Thus, as these images are superimposed at any point, correspondences between images are established by using the scale -and rotation- invariant SURF detector [6]. Note that an image *stitching* process cannot be used since it requires nearly exact overlaps between images and identical exposures to produce seamless results, and information about the performed camera movements and calibration are not provided.

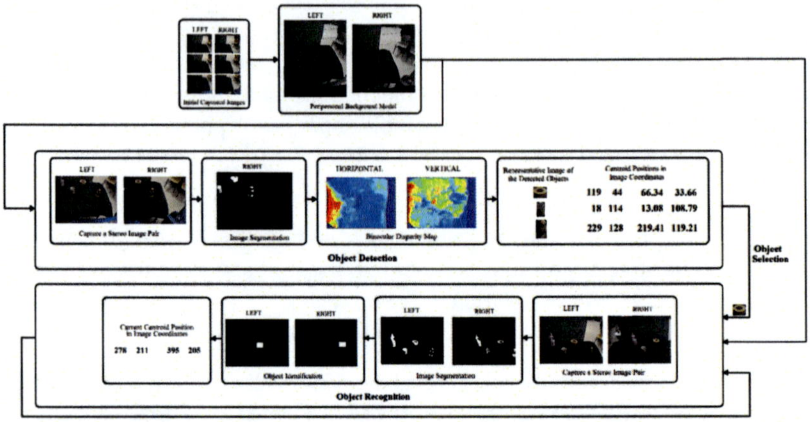

Fig. 1. Flowchart of the *Object Recognition* approach performance through its different stages

Once feature descriptors have been found, the next step is to find the perspective transformation between the different image planes. This perspective transformation is given by:

$$\begin{bmatrix} x'_i \\ y'_i \\ 1 \end{bmatrix} \sim H \begin{bmatrix} x_i \\ y_i \\ 1 \end{bmatrix} \tag{1}$$

where $\begin{bmatrix} x'_i \ y'_i \ 1 \end{bmatrix}^T$ represents the image coordinates in one image; $\begin{bmatrix} x_i \ y_i \ 1 \end{bmatrix}^T$ is the image coordinates in the *peripersonal background model*; and H is the homography matrix. The back-projection error is minimised by means of a simple least-squares scheme and further refined with the Levenberg-Marquardt method. Thus, the system is initially provided with a *peripersonal background model* composed by a single *macro*-image spanning the whole peripersonal space for each visual sensor as those depicted in Fig. 1. Note that no calibration information is required.

At this point, a new issue arises: how to properly choose the threshold value to properly detect the objects of interest in the scene. Actually, the threshold is a key parameter in the segmentation process because it can affect quite critically the performance of the successive steps. For that reason, it has to be carefully set. Despite the wide variety of possibilities to automatically determine its value, the existing methods only work well when the images to be thresholded satisfy their assumptions about the distribution of the gray-level values over the image. So, situations such as shape deformations of the interest object, relationship of the foreground object size with respect to the background, or overlapping of background and target gray-level distributions, make them fail. As a solution, we propose an *adaptive dynamic thresholding method* capable of adapting to non-uniform-distributed resolution, inadequate illumination gradient in the scene, shadows, and gradual as well as sudden changes in illumination. The main idea is to divide each *peripersonal background model* in regions such that the image features are kept, especially in terms of resolution distribution. Then, a threshold is obtained for each region based on *region* histogram properties [26].

After completely initializing the system, the objects of interest will be located at any point in the system's peripersonal space to be detected. Then, the robotic vision system is randomly moved. Thus, the potential targets are detected as follows:

1. Estimation of the perspective transformation between the stereo image pair and their corresponding *peripersonal background model*
2. Image segmentation through background subtraction
3. Image segmentation refinement by applying two consecutive morphological operations ($3x3$ erosion to erase isolated points and/or lines, followed by a $3x3$ expand filter for foreground region recovery)
4. Representation of potential targets composed of an object *id* and a representative *image* of the detected target (for subsequent recognition)

Finally, since no calibration information is provided, a reliable matching between the two image points corresponding to potential targets is required. As only a representative *image* of the detected objects is available, it is not possible to extract features distinctive enough to be detected and matched by using an automatic detector like SURF. Consequently, the phase-based binocular disparity estimation approach presented in [22], has been used. In this way, the image position of the detected potential targets in both stereo images, is known.

At the last stage, the designed method should recognize the chosen object in the subsequent images. For that, after image segmentation, the target's representation is used. Basically, the stored representative *image* of the chosen target is compared with all the potential targets detected in the current frame through image convolutions. Note that, at each time, the background information is being updated with the information of pixels classified as background.

3 Experimental Results

3.1 Robotic Set-Up

Tombatossals is a humanoid torso (see Fig. 2) endowed with a pan-tilt-vergence stereo head and two multi-joint arms. The head mounts two cameras with a resolution of 1024x768 pixels that can acquire colour images at 30 Hz. The baseline between cameras is 270 mm and the motor positions are provided by high resolution optical encoders. In this work, we have employed three degrees of freedom (d.o.f.) for the head (tilt, left pan and right pan).

3.2 Experiments Setup and Results

Experiments were conducted to show how the proposed system is able to detect and recognize objects under different environmental conditions. Thus, two different kind of experiments were carried out. Firstly, the system's workspace consisted of a semi-structured environment, in the sense that some conditions could be controlled. That is,

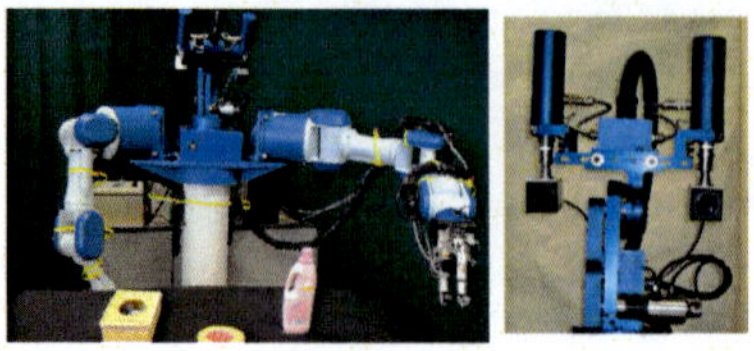

Fig. 2. Experimental Setup: external view of the used humanoid robot (left) and a detail view of pan/tilt/vergence head (right)

the system was located in front of a table where the objects were placed. Note that, although the remaining workspace was not controlled due to laboratory members' activity, the detection procedure was considerably reduced since only the objects to be detected were placed on the table. However, this kind of experiment allows to analyze the method's effectiveness under certain uncontrolled factors such as, for instance, blinking light sources, textured surfaces (e.g. that of the table), light reflexes, etc. On the contrary, the robustness of the presented method was tested in an unstructured environment. For that, a different scenario was considered. In this case, the robotic system was observing a desk where everyday objects such as, for instance, books, pens, clocks and calendars, are sharing the space with the objects to be recognized. In that way, the method's performance was evaluated at a real scenario by allowing to know its applicability to autonomous systems.

In addition, the proposed object/recognition approach is assessed when different fixation points were considered. To this end, the task to be performed in both experiments was fixating the vergent stereo cameras at the visual stimulus, that is, the visual object chosen as target.

So, the vision system captures several images to build a *peripersonal background model* for each visual sensor. Then, the first stage of the system starts, that is, the detection of all the potential targets in the scene. A stereo image pair is taken and a segmentation process is applied to one image. That is, the homography matrix between the captured image and the *peripersonal background model* built for the corresponding visual sensor, is obtained. Then, the *modified* background subtraction technique provides the first background/foreground distinction. After that, the phase-based binocular disparity estimation is performed to establish image correspondences between the potential targets and to discard those that are not visible in both images. Note that the disparity estimation error is less than 1.0 pixel. To end this stage, a representation of each detected object as well as its centroid position in both images (in terms of image coordinates) is provided to the *Task Manager* with the aim of selecting the target to be foveated. As abovementioned, the target choice is randomly performed although different criteria could be established to automatically carry it out or, it could be the user who would make the choice.

Therefore, the proposed object recognition approach has been tested with different objects and different scene conditions in order to study its robustness. Some examples of the object recognition performance are depicted in Figs. 3 and 5. Fig. 3 refers to the experimental results obtained when a semi-structured environment was considered. In this case, three different objects have been used as targets: a toy car, a bottle and a

Fig. 3. Examples of the object recognition performance in a semi-structured environment

money box. As it can be observed, despite the presence of different factors that can make the system fail (e.g. flickering light sources, variable light reflexes, (partial) occlusions, targets with different features (in the sense of colour, shape, size, etc.), etc.), all the objects were successfully recognized even when they have changed their orientation or location in the scene.

These qualitative results are supported by a quantitative evaluation. For that, the *recall* and *precision* measurements have been used [7]. Note that recall (also known as true positive rate (TPR) or sensitivity) is computed as the ratio of the number of foreground pixels correctly identified to the number of foreground pixels in the ground truth; whereas precision or positive predictive value (PPV), is obtained as the ratio of the number of foreground pixels properly identified to the number of foreground pixels detected. Therefore, a good performance is obtained when both measurements are close to 1. The quantitative analysis is depicted in Fig. 4, where the red line refers to the obtained results, while the blue line divides the recall-precision space such that curves above it mean a good segmentation and curves below indicates bad results. Therefore, it can be concluded that the system was successful in object detection/recognition.

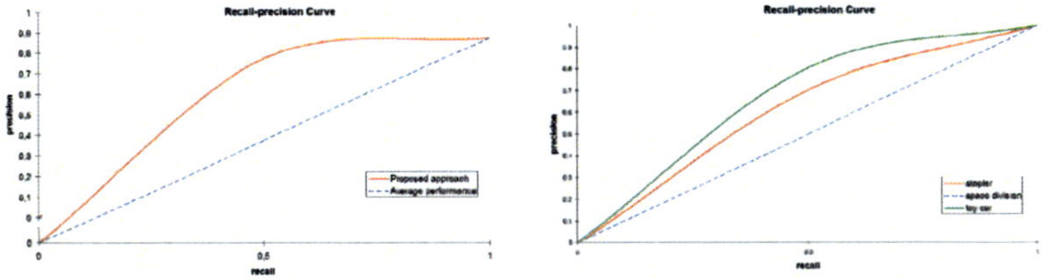

Fig. 4. Recall-precision curve for the evaluation of the proposed approach in semi-structured environments (left) and unstructured environments (right)

On the other hand, a real desk represents the scene where objects to be detected and recognized were located. That unstructured environment is composed of objects of different nature and features such as textured books, pens, clock, etc. In this context, three different objects were used: a toy car, a stapler and a revolution object. These objects were located at different positions and/or orientations within the described scenario, resulting (partially) occluded in some cases. Despite the nature of the environment, the objects were properly detected in all situations as shown in Fig. 5.

Fig. 5. Examples of the object recognition performance in an unstructured environment

4 Conclusions and Future Work

The current trend in robotics is toward the capability of a robotic system to autonomously operate in real scenarios. Among its multiple applications, the manipulation task provides a general framework since it requires object detection, segmentation, recognition, servoing, alignment and grasping. In this context, we proposed an approach which provides a robotic system with the ability of detecting and recognizing objects present in its peripersonal space, as well as estimating their location.

The proposed approach has been evaluated by considering different parameters which can make it fail such as object geometry, surface characteristics, illumination, changes in the visual fixation point, presence/absence, etc. That large number of parameters allows us to analyze the robustness of the method. As shown in experimental results, the performance was successful. As a result, it provides a robotic system with the ability of operating in everyday, dynamic environments.

Acknowledgments. This research was partly supported by WCU (World Class University) program through the National Research Foundation of Korea funded by the Ministry of Education, Science and Technology (Grant No. R31-2008-000-10062-0), by Ministerio de Ciencia e Innovación (DPI2011-27846), by Generalitat Valenciana (PROMETEO/2009/052) and by Fundació Caixa Castelló-Bancaixa (P1-1B2011-54).

References

1. Humanoid mobile manipulation using controller refinement. In: RSS Workshop: Manipulation for Human Environments, Philadelphia, PA (2006)
2. Operating humanoid robots in human environments. In: RSS Workshop: Manipulation for Human Environments, Philadelphia, PA (2006)
3. Abbott, R.G., Williams, L.R.: Multiple target tracking with lazy background subtraction and connected components analysis. Machine Vision and Applications 20, 93–101 (2009)
4. Antich, J., Ortiz, A., Oliver, G.: A control strategy for fast obstacle avoidance in troublesome scenarios: application in underwater cable tracking. In: 7th IFAC Conf. on Manoeuvring and Control of Marine Craft (MCMC), Lisbon, Portugal (2006)
5. Asada, M., Uchibe, E., Hosoda, K.: Cooperative behavior acquisition for mobile robots in dynamically changing real worlds via vision-based reinforcement learning and development. Artificial Intelligence 110, 275–292 (1999)

6. Bay, H., Ess, A., Tuytelaars, T., Gool, L.V.: Surf: Speeded up robust features. CVIU 110(3), 346–359 (2008)
7. Benezeth, Y., Jodoin, P.M., Emile, B., Laurent, H., Rosenberger, C.: Review and evaluation of commonly-implemented background subtraction algorithms. In: 19th ICPR, Tampa, Florida, pp. 1–4 (December 2008)
8. Bonin-Font, F., Ortiz, A., Oliver, G.: Visual navigation for mobile robots: A survey. Journal of Intelligent and Robotic Systems 53, 263–296 (2008)
9. Chang, O.: Evolving cooperative neural agents for controlling vision guided mobile robots. In: 9th Int. Conf. on Cybernetic Intelligent Systems (CIS), Reading, UK, pp. 1–6 (2010)
10. Choset, H., Lynch, K., Hutchinson, S., Kantor, G., Burgard, W., Kavraki, L., Thrun, S.: Principles of Robot Motion. Theory, Algorithms, and Implementations. In: Intelligent Robotics and Autonomous Agents. MIT Press (2005)
11. Elgammal, A., Harwood, D., Davis, L.: Background and foreground modeling using nonparametric kernel density estimation for visual surveillance. IEEE Proceedings 90, 1151–1163 (2002)
12. Fitzpatrick, P., Metta, G.: Grounding vision through experimental manipulation. Philosophical Transactions of the Royal Society A: Mathematical, Physical and Engineering Sciences 361(1811), 2165–2185 (2003)
13. Geib, C., Mourão, K.M., Petrick, R., Pugeault, N., Steedman, M., Krueger, N., Wörgötter, F.: Object action complexes as an interface for planning and robot control. In: Workshop: Towards Cognitive Humanoid Robots at IEEE RAS Int. Conf. on Humanoid Robots, Genoa, Italy (2006)
14. Kraft, D., Detry, R., Pugeault, N., Başeski, E., Piater, J., Krüger, N.: Learning Objects and Grasp Affordances through Autonomous Exploration. In: Fritz, M., Schiele, B., Piater, J.H. (eds.) ICVS 2009. LNCS, vol. 5815, pp. 235–244. Springer, Heidelberg (2009)
15. Kraft, D., Pugeault, N., Baseski, E., Popović, M., Kragić, D., Kalkan, S., Wörgötter, F., Krüger, N.: Birth of the object: Detection of objectness and extraction of object shape through object action complexes. Int. Journal of Humanoid Robotics 5(2), 247–265 (2008)
16. Kragic, D., Christensen, H.: Survey on visual servoing for manipulation. Tech. rep., Computational Vision and Active Perception Laboratory (2002)
17. Krüger, N., Ackermann, M., Sommer, G.: Accumulation of object representations utilising interaction of robot action and perception. Knowledge-Based Systems 15(1-2), 111–118 (2002)
18. Lee, D.-H., Kim, J.-H.: A framework for an interactive robot-based tutoring system and its application to ball-passing training. In: ROBIO, Tianjin, China, pp. 573–578 (2010)
19. Lee, S., Lee, S., Lee, J., Moon, D., Kim, E., Seo, J.: Robust recognition and pose estimation of 3d objects based on evidence fusion in a sequence of images. In: ICRA, Rome, Italy, pp. 3773–3779 (2007)
20. Mahadevan, V., Vasconcelos, N.: Background subtraction in highly dynamic scenes. In: CVPR, Anchorage, Alaska, pp. 1–6 (2008)
21. Mamic, G., Bennamoun, M.: Representation and recognition of 3d free-form objects. Digital Signal Processing 12(1), 4776 (2002)
22. Martínez-Martín, E., del Pobil, A., Chessa, M., Solari, F., Sabatini, S.: An integrated virtual environment for visual-based reaching. In: The 5th ACM ICUIMC, Seoul, Korea (2011)
23. Morita, H., Hild, M., Miura, J., Shirai, Y.: Panoramic view-based navigation in outdoor environments based on support vector learning. In: IROS, Beijing, China, pp. 2302–2307 (2006)
24. Petrick, R., Kraft, D., Mourão, K., Pugeault, N., Krüger, N., Steedman, M.: Representation and integration: Combining robot control, high-level planning, and action learning. In: Sixth CogRob at European Conf. on Artificial Intelligence, Patras, Greece, pp. 32–41 (2008)

25. Rothganger, F., Lazebnik, S., Schmid, C., Ponce, J.: 3d object modeling and recognition using affine-invariant patches and multi-view spatial constraints. In: CVPR, Madison, Wisconsin, USA, vol. 2, pp. 272–280 (2003)
26. Shapiro, L.G., Stockman, G.C.: Computer vision. Prentice-Hall (2001)
27. Shen, J., Hu, H.: Visual navigation of a museum guide robot. In: The Sixth WCICA, Dalian, China, vol. 2, pp. 9169–9173 (2006)
28. Stauffer, C., Grimson, W.: Learning patterns of activity using real-time tracking. PAMI 27(5), 747–757 (2000)
29. Turaga, P., Chellappa, R., Subrahmanian, V., Udrea, O.: Machine recognition of human activities: A survey. IEEE Trans. on Circuits and Systems for Video Technology 18, 1473–1488 (2008)
30. Whitcomb, L., Yoerger, D., Singh, H., Mindell, D.: Towards precision robotic maneuvering. survey, and manipulation in unstructured undersea environments. In: Eight Int. Symp. on Robotics Research, pp. 45–54. Springer, London (1998)
31. Wren, C., Azarbeyejani, A., Darrell, T., Pentland, A.: Pfinder: Real-time tracking of the human body. PAMI 19(7), 780–785 (1997)
32. Yinca-Banjo, C., Osunmakinde, I., Bagula, A.: Collision avoidance in unstructured environments for autonomous robots: a behavioural modelling approach. In: ICCRC, New Delhi, India, pp. 297–303 (2011)

Mapping of Incremental Dynamic Environment Using Rao-Blackwellized Particle Filter

Alper Öner

Control and Automation Engineering Department / Mechatronics Education and Research Center, Istanbul Technical University, Istanbul, Turkey
alperoner@yahoo.com

Abstract. This research is a preliminary research for real time autonomous robot in unknown incremental dynamical environment. A general method for mapping incremental dynamic environment using Multiple Target tracking (MTT) was proposed in this research. Rao-Blackwellized Particle Filter (RBPF) was used for the multiple moving obstacles tracking problem. Firstly data association problem was solved via Multiple Hypothesis Tracking (MHT) data association by a new method. The new MHT method can use extra information except using only position of targets. Particle Filter is used in the method. Each particle is assumed as an obstacle map. Then tracking problem for each obstacle in the particle is solved by Extended Kalman Filter (EKF). Finally the particle which has highest weight is assumed as the dynamic map. Additionally a new resampling method was proposed in this research. The algorithm can cope with new obstacles and false detection according to the pure particle filter. Obstacles are assumed as human in this research hence their velocities are determined randomly up to human walking speed. Furthermore the robot moves approximately at human walking speed. A graphical user interface program was constituted in MATLAB so different states are surveyed.

Keywords: Rao-Blackwellized Particle Filter, Multiple Hypothesis Tracking, Data Association, Dynamic Obstacle Detection in Unknown Environment, Mapping.

1 Introduction

Detection of dynamic obstacles is always a problem for autonomous robotic systems. Moreover there are a lot of studies in the literature on multiple target tracking. When a robot moves in an unknown cluttered environment, it encounters so many dynamic obstacles. While these obstacles can be seen by sensors, the following time step they can disappear. This circumstance is a problem to make a decision for dynamic mapping [1, 2].

Usually the robot determines obstacles as features by this way it can understand its localization and features are accepted as a map of unknown environment [2]. This is straightforward for static obstacles but in dynamic environment determining features is challenging so localization and mapping are difficult. Hence detection and tracking of dynamic obstacles is important for robotic systems [1].

S. Lee et al. (Eds.): Intelligent Autonomous Systems 12, AISC 193, pp. 715–724.

In tracking literature there are some methods for data association such as multiple hypotheses tracking (MHT) [3, 4] and joint probabilistic data association (JPDA) [5, 6]. MHT associates each measurement to one of the existing tracked obstacle or form a new obstacle from the measurement. MHT has not to be unique; several hypotheses are continuously constituted and maintained.

JPDA approximates the posterior distribution of the tracked obstacle as separate Gaussian distributions for each obstacle. Sequential Monte Carlo (SMC) as multiple dynamic obstacles tracing method generally belongs to the class of unique-neighbor data association methods. They represent the data association and state posterior as a discrete set of hypotheses. SMC methods can be accepted as generalizations of MHT [7].

Multiple hypotheses tracking (MHT) is a method for data association is proposed by Reid [3]. In the conventional MHT method only position of the obstacles are used for data association. In this research additional information such as color and size are used for increasing the accuracy of the method. Furthermore addition and deletion of obstacles is advantage of MHT according to the other methods.

The aim of this research is to determine dynamic motion model of moving obstacles in the research. Odometer motion model and velocity motion model are proposed in the literature [1, 2]. Velocity motion model of each obstacle is learned by the algorithm. The more truly motion model prediction is, more successful posterior coordinate prediction for moving obstacles occurs hence understanding the motion model of obstacles is crucial. Velocity of the moving obstacle must be calculated quickly and accurately in unknown cluttered environment. This significant necessity can be assumed as a learning problem.

Particle filter is a more effective filter according to EKF in unknown environment because of the fact that it uses randomness property when new particle is assigned [8]. Hence multiple moving obstacle detection is done by a new RBPF. Data association methods increase performance of MHT [5, 7]. Hence data association is done by particle filter. Each particle is a hypothesis for the map of all obstacles. Each hypothesis or particle includes different motion models for each obstacle which are in the particle. Finally best particle, which means the dynamic map, includes true motion model and position information is selected as the map by the algorithm.

Resampling is one of the most important steps for a successful method. If proposed particle value, which means value of the proposed map, is lower than a threshold resample step deletes weak particles and generates new ones. However resampling frequently causes loss of past measurement in the process. Some researchers did not use resampling [9]. But if resampling does not implemented to the particle filter success of the algorithm is decreasing in time. The proposed RBPF has better dynamic obstacle detection and tracking according to pure particle filter through resample unsuccessful predictions are deleted and new ones are constituted.

The environment of work has been explained in section two. MHT data association and RBPF have been stated in section three. Results were represented in section four. Finally conclusion and future works were stated in section five.

2 Environment

Before testing the algorithm some arrangement must be done for RBPF on the program. Fig. 1 shows graphical user interface of the program. The arrangements are explained in this section.

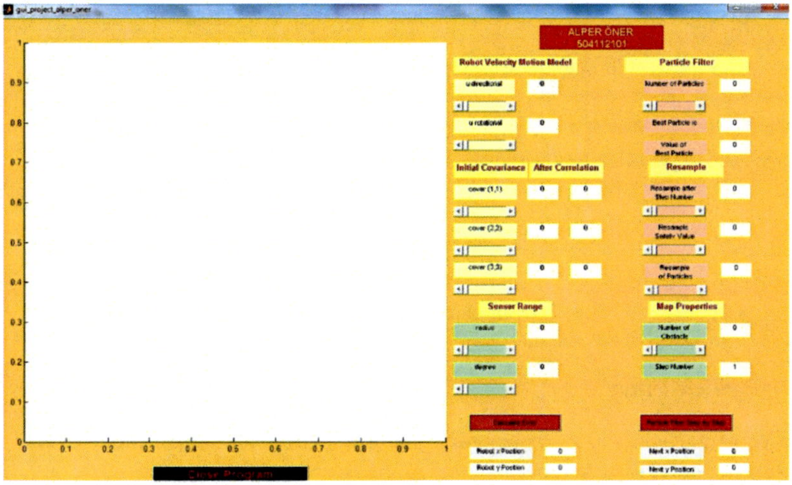

Fig. 1. Graphical user interface

2.1 Robot Motion Model

The robot uses dead reckoning velocity motion model for its movement [2].

$$u = \begin{pmatrix} v_t \\ w_t \end{pmatrix} \tag{1}$$

Robot states are x position, y position and θ rotational direction of the robot. u directional is directional movement of the robot. u rotational is rotational movement of the robot. Equation 2 and figure 2 show robot states for dead reckoning motion model.

$$\begin{pmatrix} x' \\ y' \\ z' \end{pmatrix} = \begin{pmatrix} x \\ y \\ z \end{pmatrix} + \begin{pmatrix} -\frac{v}{w}\sin\theta + \frac{v}{w}\sin(\theta + w\Delta t) \\ \frac{v}{w}\cos\theta + \frac{v}{w}\cos(\theta + w\Delta t) \\ w\Delta t \end{pmatrix} \tag{2}$$

Initial covariance is covariance matrix of the robot states. After correlation is covariance matrix after correlation. When the robot sees a feature it correlates its positions and covariance matrix.

The robot moves with these parameters. Then it calculates its position using Extended Kalman Filter (EKF) and it is assumed that its position is true position of the robot.

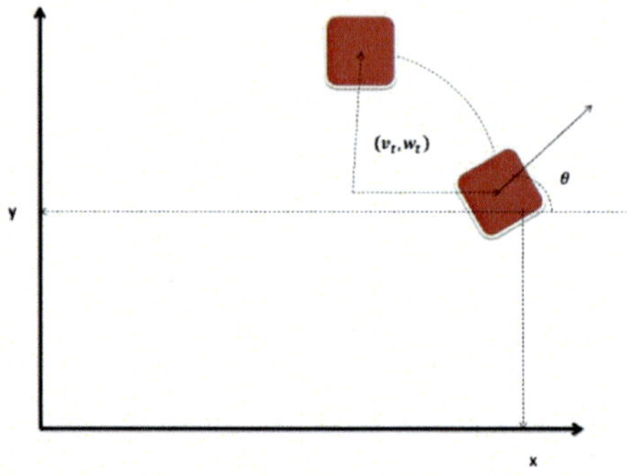

Fig. 2. Dead reckoning velocity motion model

2.2 Particle Filter

Number of particle shows that particle number of the data association particle filter which is potential map number.

Best particle is the particle which has the highest data association value between the obstacles that are seen by the sensor of the robot and obstacles that are tracked in each particle.

Resample is deleting particles which have low value and adding new ones.

Resample can be done either determining step number which is mod of resample or assign a value when particle value is lower than the value resample occurs.

Resample of particles determines how many particles are deleted after resample. After determining the ratio some particles are deleted according to it.

2.3 Sensor Range and Map Properties

Sensor behaves as a laser scanner. Range and scanning degree are added as input of the sensor. Furthermore reflectivity value of some of laser sensor can give an opinion about color of the obstacle.

Number of obstacles is moving obstacle number. Size, color and velocity of obstacles are determined randomly.

Step number shows that how many step moved the robot.

Obstacles are assumed as human hence their speed is determined randomly up to human walking speed. Moreover the robot moves at human walking speed. A speedy walking speed is assumed as 2 m/s (7.0 km/h) [10].

In this research one step is assumed 0.5 second so robot directional movement is 1 m at one step, robot rotational movement is 11 degree, variance of states is 0.1, sensor range is 6 m, sensor degree is 144 degree and finally number of obstacle is 5.

3 Proposed Algorithm

In order to detect moving obstacle three problems must be solved these are: detection, data association of obstacles in cluttered environment, modeling moving obstacle [1].

3.1 Multiple Hypotheses Tracking Data Association

Overall form of multiple hypotheses tracking algorithm is shown in fig. 3. MHT involves heuristic methods for hypotheses reduction. The proposed algorithm applies a gating technique, in the light of that measurements are only associated with targets, whose anticipated prior position falls within a certain neighborhood around the measurement. Besides target size and color are evaluated at gating procedure.

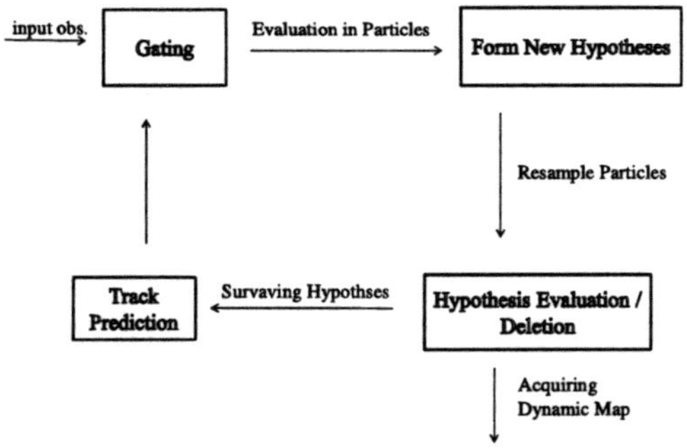

Fig. 3. The overview of the system flow

Data association problem is modeled as a Bayesian estimation problem and posterior distribution is calculated with MHT method. This value is

$$c_k^{p,s,c} \sim p\left(c_k^{p,s,c} \middle| y_{1:k}, c_{1:k-1}^{p,s,c}\right) \tag{3}$$

$c_k^{p,s,c}$ is probabilistic value of data association according to position, size or color of the target. c_k^i is product of each one. α is normalization constant.

$$c_k^i \sim \alpha c_k^p c_k^s c_k^c \tag{4}$$

Each particle is

$$\begin{array}{c} particle\ 1\colon \left\{D_{1,k}^1, D_{2,k}^1, \ldots, D_{T,k}^1, \omega_k^1\right\} \\ \cdot \\ \cdot \\ \cdot \\ particle\ N\colon \left\{D_{1,k}^N, D_{2,k}^N, \ldots, D_{T,k}^N, \omega_k^N\right\} \end{array} \tag{5}$$

N is number of particle and k is number of obstacles in the particle, ω is weight of the particle and D is data information of the obstacle. D includes x position, y position, rotational direction of the obstacle, color of the obstacle, directional movement of the obstacle, rotational movement of the obstacle, covariance matrix of states of the obstacle and size of the obstacle.

Each particle represents a dynamic map. MHT associates each measurement with obstacles in each particle. New measurements are added to the particle. Finally weight of particle is calculated according to the measurement. Best particle assumed as dynamic map.

New weight of the particle is

$$\omega_k^i \propto \omega_{k-1}^i \times \frac{p\left(y_k \middle| c_k^i, y_{1:k-1}, c_{1:k-1}^i\right) p\left(c_k^i \middle| c_{1:k-1}^i\right)}{p\left(c_k^i \middle| y_{1:k}, c_{1:k-1}^i\right)} \quad (6)$$

3.2 Rao-Blackwellized Particle Filter

1. Sensor results are obtained from the sensor.
2. Particles are constructed according to number of particles. Each particle has one obstacle prediction initially.
3. Each obstacle in the particle moves according to EKF prediction step.
4. Obstacles match with sensor results according to the data association algorithm. This algorithm shows position, size and color of the obstacle.
5. Each obstacle in the particle is correlated according to EKF correlation step.
6. Add new obstacles to the particle.
7. Combine same obstacles in the particle. So complexity of algorithm is reduced.
8. Step 3 to 7 is done for each particle.
9. Then particles are sorted according to their weight and resample weak particles. If particle value is lower than particle safety value and above 0.2, particles are resampled using best particle which means that best particle is parent of new particles and if particle value is lower than 0.2, new particles are resampled. Resampling percentage of particles quantity is determined by the resampling ratio.

4 Results

Figures show x and y position of objects. Filled color objects shows real position of obstacles and empty color objects shows best hypotheses of the algorithm. Robot moves four steps. This corresponds to two seconds in real time.

Results figures show that number of obstacles in the best particle. The obstacle has zero error means that the obstacle fails at data association process from the measurement. Even though it can not associate any measurement it can be tracked from past measurement.

4.1 Pure Particle Filter with 1000 Particles

Figure 4, 5 show results for 1000 particles for pure particle filter. This type of particle filter starts to search five obstacles. Each particle has five objects and no resample occurs. Best hypotheses value is 0.5 at first step and 0.02 at fourth step. It is clearly understood that pure particle filter is unsuccessful for multiple moving obstacles.

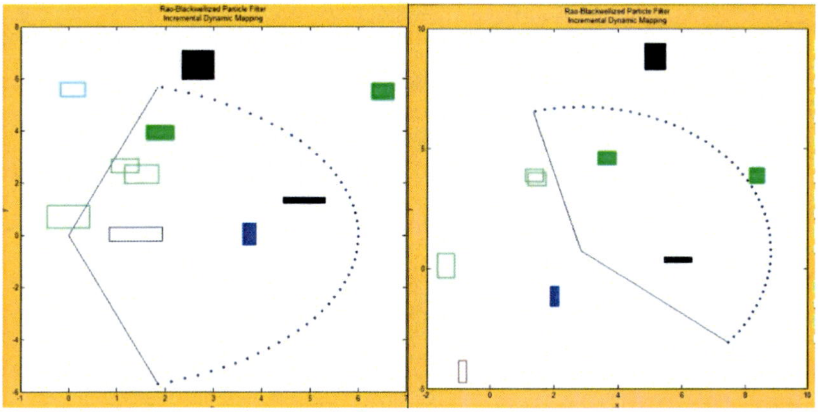

Fig. 4. Pure Particle Filter with 1000 particles step 1 and step 4

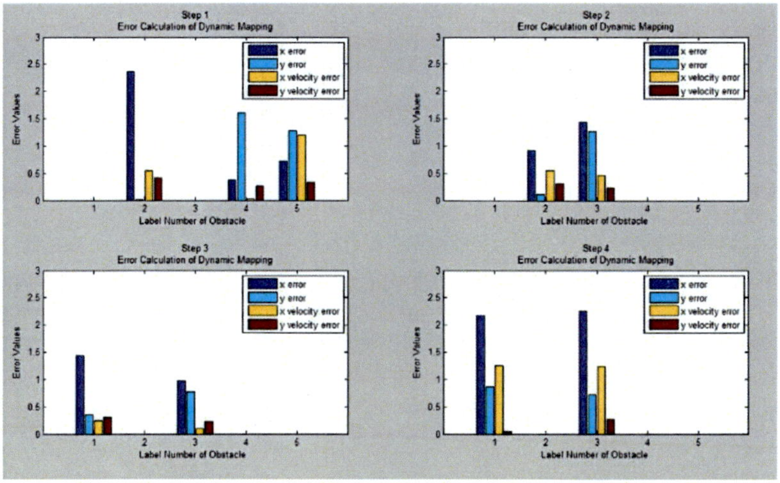

Fig. 5. Pure Particle Filter with 1000 particles result

4.2 RBPF with 100 Particles

Figure 6, 7 show results for 100 particles. % 80 of particles is regenerated. Best hypotheses value is 0.85 at first step and 0.28 at fourth step. Resample safety value is 0.8 so after second and third step resample occurs using best hypotheses. At last step one of obstacle goes away from sensor measurement range but it can be seen at fig 6b that the algorithm still tracks the obstacle. According to fig. 7 after resample at step 2 resampling of particles using best particle as parent decreases error. Model prediction is more successful than previous steps.

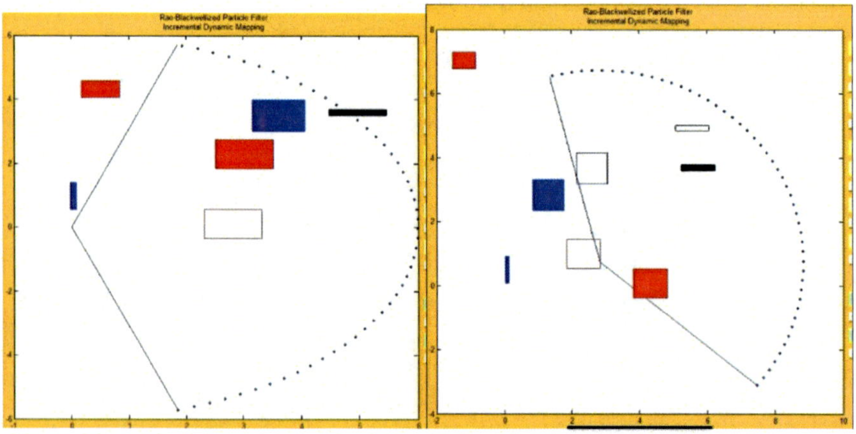

Fig. 6. RBPF with 100 particles step 1 and step 4

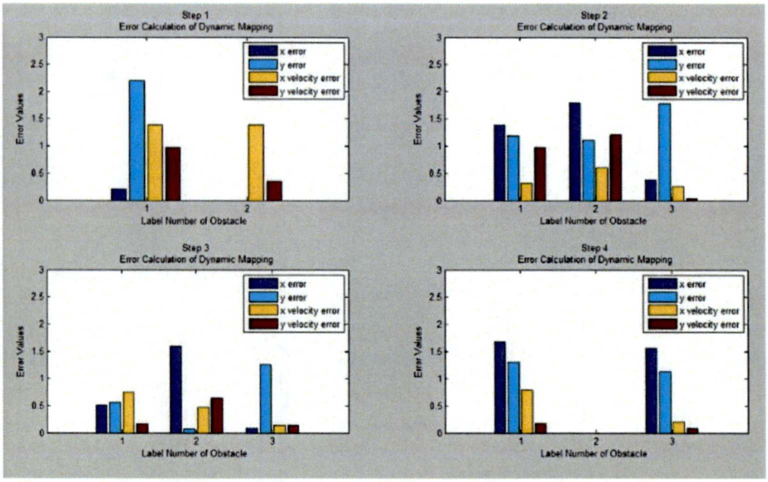

Fig. 7. RBPF with 100 particles result

Figure 8, 9 show results for 100 particles for different conditions. Best hypotheses value is 0.92 at first step and 0.94 at fourth step. The value increases due to random resampling from best particle of previous step. At this condition two obstacles get in range of sensor measurement after first step. The algorithm resample % 80 of particles using best particle at second and third step. Finally a better results even first step is achieved.

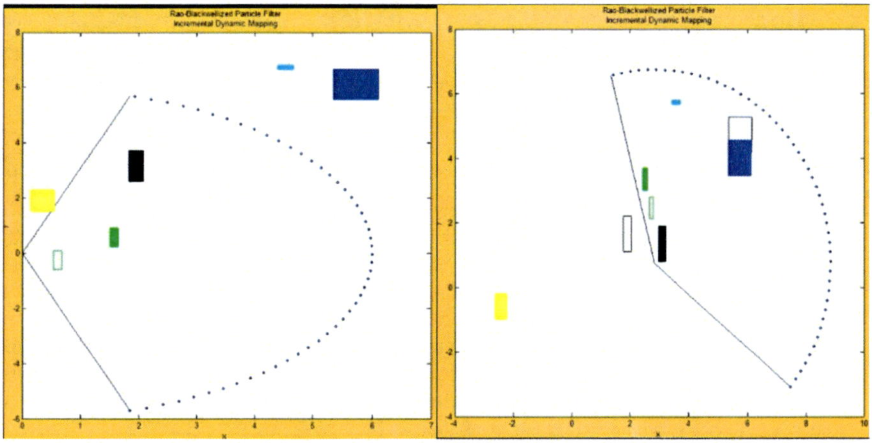

Fig. 8. Another RBPF with 100 particles step 1 and step 4

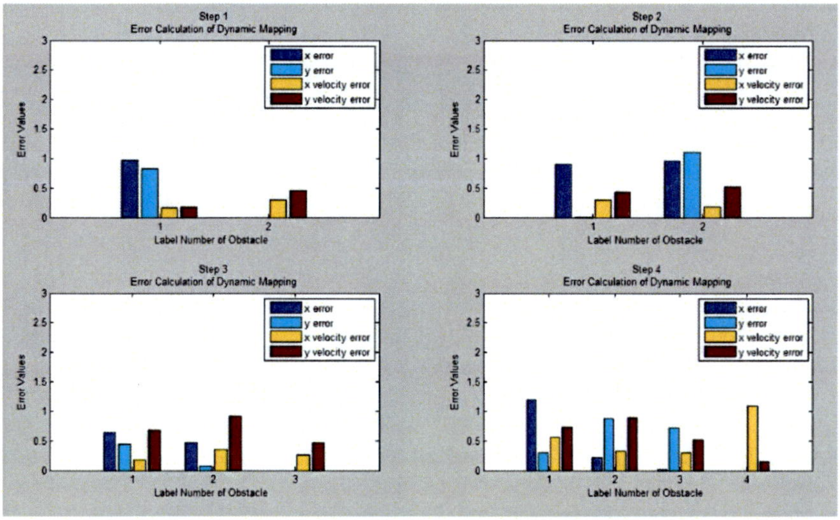

Fig. 9. Another RBPF with 100 particles result

5 Conclusion

This research is a preliminary work for real time autonomous robot in unknown incremental dynamical environment. Hence computational burden must be decreased. The algorithm that is used in this research success better results with 100 particles than 1000 particle. Data association of each particle is achieved. Even though an obstacle goes away from sensor measurement range, tracking is done. Furthermore when new obstacle gets in range of the sensor measurement, the algorithm adapted quickly which is a necessity for autonomous vehicles.

Resample is necessary for renewing the particles. If resample ratio and frequency are high some past measurement data is lost. Hence a hybrid method between generating new particle and using best particle as parent is created. So while hypotheses value is decreasing, a better particle can be generated using randomness.

Acknowledgements. This research has been supported in part by T.R. Ministry of Development.

References

1. Wang, C.: Simultaneous localization, mapping and moving object tracking. PhD Thesis, Robotics Institute, Carnegie Mellon University, CMU-RI-TR-04-23 (April 2004); Tapia, Intille, S. Larson, K.: Activity recognition in the home using simple and ubiquitous sensors. Lecture Notes in Computer Science, Int. Conf. on Pervasive Computing, vol. 3001, pp. 158–175
2. Thrun, S., Burgard, W., Fox, D.: Probabilistic Robotics. The MIT Press, Cambridge (2005)
3. Reid, D.B.: An algorithm for tracking multiple targets. IEEE Trans. on Automatic Control 24(6) (December 1979)
4. Cox, I.J., Hingorani, S.L.: An efficient implementation of Reid's multiple hypothesis tracking algorithm and its evaluation for the purpose of visual tracking. IEEE Trans. on Pattern Analysis and Machine intelligence 18(2) (February 1996)
5. Schulz, D., Burgard, W., Fox, D., Cremers, A.B.: Tracking Multiple Moving Targets with a Mobile Robot using Particle Filters and Statistical Data Association. In: Proc. of the 2001 IEEE International Conference on Robotics & Automation (ICRA 2001), Seoul, Korea (2001)
6. Fortmann, T.E., Bar-Shalom, Y., Scheffe, M.: Sonar tracking of multiple targets using joint probabilistic data association. IEEE Journal of Oceanic Engineering OE-8, 173–184 (1983)
7. Sarkka, S., Vehtari, A., Lampinen, J.: Rao-blackwellizedmontecarlo data association for multiple target tracking. In: Svensson, P., Schubert, J. (eds.) Proceedings of the Seventh International Conference on Information Fusion, Mountain View, CA, vol. 1, pp. 583–590 (2004)
8. Gordon, N., Clapp, T.: A Tutorial on Particle Filters for Online Nonlinear/Non-Gaussian Bayesian Tracking, M. SanjeevArulampalam, Simon Maskell. IEEE Transactions on Signal Processing 50(2) (February 2002)
9. Miller, I., Campbell, M.: Rao-Blackwellized Particle Filtering for Mapping Dynamic Environments. In: 2007 IEEE International Conference on Robotics and Automation, Roma, Italy, April 10-14 (2007)
10. Browning, R.C., Baker, E.A., Herron, J.A., Kram, R.: Effects of obesity and sex on the energetic cost and preferred speed of walking. Journal of Applied Physiology 100(2), 390–398 (2006)

Texture-Based Crowd Detection and Localisation

Stefano Ghidoni[1], Grzegorz Cielniak[2], and Emanuele Menegatti[1]

[1] Intelligent Autonomous Systems Laboratory (IAS-Lab)
Department of Information Engineering
The University of Padova
via Gradenigo, 6/B – 35131 Padova – Italy
{ghidoni,emg}@dei.unipd.it
[2] School of Computer Science
University of Lincoln
Brayford Pool, Lincoln, LN6 7TS, UK
gcielniak@lincoln.ac.uk

Abstract. This paper presents a crowd detection system based on texture analysis. The state-of-the-art techniques based on co-occurrence matrix have been revisited and a novel set of features proposed. These features provide a richer description of the co-occurrence matrix, and can be exploited to obtain stronger classification results, especially when smaller portions of the image are considered. This is extremely useful for crowd localisation: acquired images are divided into smaller regions in order to perform a classification on each one. A thorough evaluation of the proposed system on a real world data set is also presented: this validates the improvements in reliability of the crowd detection and localisation.

Keywords: Crowd detection, intelligent video surveillance.

1 Introduction

Crowd monitoring is a very important topic in the field of video surveillance. Crowds represent a potentially dangerous environment, due to the difficulty of controlling a high number of people. Crowds often set up during events like football matches and concerts, that take place inside large venues, under the control of a security staff. Such a large public is constantly monitored by security officers that look at the whole crowd from a distant point of view, and try to detect unusual or potentially dangerous situations. They coordinate a number of stewards that operate in the field in order to prevent dangerous situations and keep under control crowd density, that is crucial for safety considerations. Crowds can create and grow also unexpectedly, under several exceptional conditions; this second case is extremely dangerous, because of the lack of management of such large groups of people.

Over the years, technology has been used to help security officers: nowadays, several active cameras are installed in the venues and the control room of a football stadium is usually full of monitors and camera controls. This virtually enables security staff to analyse every part of the venue, inside and outside, so that the whole crowd can be observed. However, the huge amount of data provided by such surveillance systems would require several operators to be completely analysed, leading to very high costs.

S. Lee et al. (Eds.): Intelligent Autonomous Systems 12, AISC 193, pp. 725–736.
springerlink.com

On the other hand, if just one or two people are devoted to the task of analysing video streams provided by all cameras, they will likely miss several situations of interest.

To cope with this issue, a system capable of analysing multiple video streams and automatically detecting situations that require human intervention would be extremely effective. Such a system should be able to detect the crowd inside a venue, analyse its behaviour, and detect situations that could be dangerous: in that case, a security officer could be warned, in order to adopt proper countermeasures. In other words, such an automatic system is asked to select suspicious situations, and let security staff focus on them. The first step in developing a system capable of understanding dangerous situations inside large venues is detection and localisation of the crowded areas, together with crowd density estimation.

In this paper, we present a crowd monitoring system that is able to detect and localise crowded areas in the image. The core of our approach is based on a machine learning classifier which can categorise texture features extracted from image regions into a set of classes corresponding to different crowd density. We extend the feature set proposed by [1] to improve the classification performance in small image areas, demonstrate the ability of the method to localise the crowd and present a thorough evaluation of the system on a challenging data set acquired during a real crowded event. We first present state of the art in the area of crowd monitoring, then describe different components of our system, followed by experiments and finally briefly conclude and discuss possible future work.

2 Related Work

The first thorough study on crowd analysis dates back to the mid '90s: [2] discusses the importance of such topic in order to enhance public safety, and proposes a way for describing crowd at a high level that is inspired by gas dynamics laws. At a lower level, the paper deals with several computer vision techniques for detecting crowds.

Over the years, several approaches have been developed to detect crowd. For instance, some methods are based on motion analysis: in [3] a technique based on the observation of the motion of particles is described; such particles are initially evenly placed over the image domain, and then moved according to optical flow. In [4] a technique based on motion heat maps is presented, together with a set of indicators for measuring motion entropy and, finally, classifying motion as normal or abnormal. Finally, in [5] a method is presented, that is capable of detecting the precise contour of a crowd.

2.1 Texture Analysis

The main approach is based on the evaluation of GLDM (Gray Level Dependency Matrix): this is a method, dating back to the '70s, that aims at measuring texture content, and was originally developed for satellite image processing [6]. This first paper presenting GLDM discussed also a set of features for characterising it, and consequently measure image texture. The proposed feature set is quite large, but only few of the features were exploited when the co-occurrence matrix was employed for detecting crowds: for example in [1] four features called contrast, homogeneity, energy, and entropy were used as input to a neural network, developed for classifying crowd density. In [7] the same indicators were exploited, again for measuring crowd density.

Beside optical flow-based and GLDM-based methods, some works tackle the problem in other ways, like edge density [8,9], HOG descriptor [10], SIFT feature density [5]. In [11], a number of texture classification techniques is considered.

2.2 Crowd and Groups

A point that needs to be highlighted is the difference between a crowd and a group. Usually, people detectors (or pedestrian detectors) focus on detecting every person, that is supposed to be fully visible. However, people often walk in groups, that are composed of a number of people that are fully visible, and just side by side, or cause little occlusion [12,13,14]. A crowd is, on the other hand, a group in which people are so close that the occlusion level is sensibly high, and a few people (or even none) are fully visible [11]. This definition is coherent with computer vision techniques that are used in these two scenarios: when dealing with groups, a common people detector can be used to analyse the scene, while when dealing with crowds, such techniques fail and specific algorithms need to be created in order to understand the scene. Of course, intermediate scenarios also exist, like in [10].

3 Crowd Detection System

The developed system aims at detecting crowd, defined, as previously described, as a large group of people characterised by a high level of occlusion, that makes it impossible to see every single human body. The system has been developed starting from the GLDM-based texture analysis technique already proposed in the literature; however, deeper studies on the GLDM revealed that such matrix contains a great amount of information, that is only partially extracted by features commonly used in other systems. Therefore, we proposed a novel method for crowd analysis.

Our system is based on the GLDM approach since this offers several advantages over other methods. First, it does not require the crowd to generate sensible motion cues, as it is needed by systems based on motion observation; moreover, it runs faster than methods based on feature density: in this case, performance depends on the computational complexity for extracting each feature; in general, however, this time is not negligible and performance tends to drop when high concentrations of features are found in the images, that is, just when crowd is present in the scene. This makes features-based approaches not best-suited for real-time applications. On the other hand, GLDM-based approaches suffer from the complexity of extracting salient features once such histogram is available: this one of the main topic addressed in this paper.

3.1 GLDM Indicators

The GLDM can be seen as a two-dimensional histogram, that is created by setting up a grid of 256×256 locations (in the common case of 8-bit image depth), one for each greyscale value. The image for which the GLDM should be evaluated is then analysed by considering pixel couples, that are usually two pixels that are adjacent in a row, column or along a diagonal. For each couple, the two grey-scale values are used as the coordinates of the GLDM, and the corresponding bin value is increased of one unity. A GLDM depends on two parameters: d – the distance between the pair of pixels and the orientation (horizontal, vertical or diagonal). Once the whole image has been scanned and each pixel couple considered, the GLDM represents how pixels change: if only smooth variations can be found in the image, the GLDM will be concentrated towards the diagonal, while abrupt changes will lead to peaks that have a certain distance from the diagonal.

Since abrupt pixel changes increase the bins that are far from the diagonal, several indicators exploited so far are ways of performing a weighted sum of the bins. The weights depends on the distance to the diagonal; this is the case, for example, of contrast and homogeneity. Other indicators measure how even the distribution among the bins is, as it is the case of energy and entropy. These four features were exploited in [1].

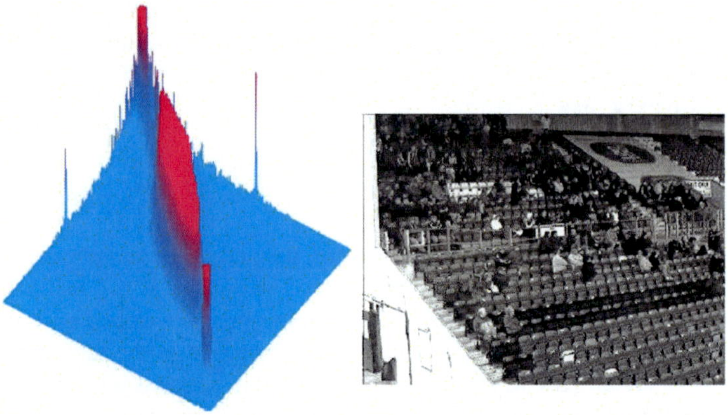

Fig. 1. An example of GLDM and the frame it is evaluated from

In Figure 1 a three-dimensional plot of a GLDM is shown: it presents several peaks of different size and shape: it is clear that such histogram is only partially measured when discussed indicators are evaluated. For example, a high contrast indicates that a sensible amount of the histogram is localised at some distance to the diagonal, but it is then impossible to know if this component is concentrated in some small area, or it develops over a wide region. For example, consider Figure 2 where two frames with substantially different texture contents are represented, together with the corresponding GLDMs that appear different as well. However, by considering the four mentioned indicators, it is possible to observe that just one of them (i.e. energy) undergoes significant change in value when comparing the two histograms.

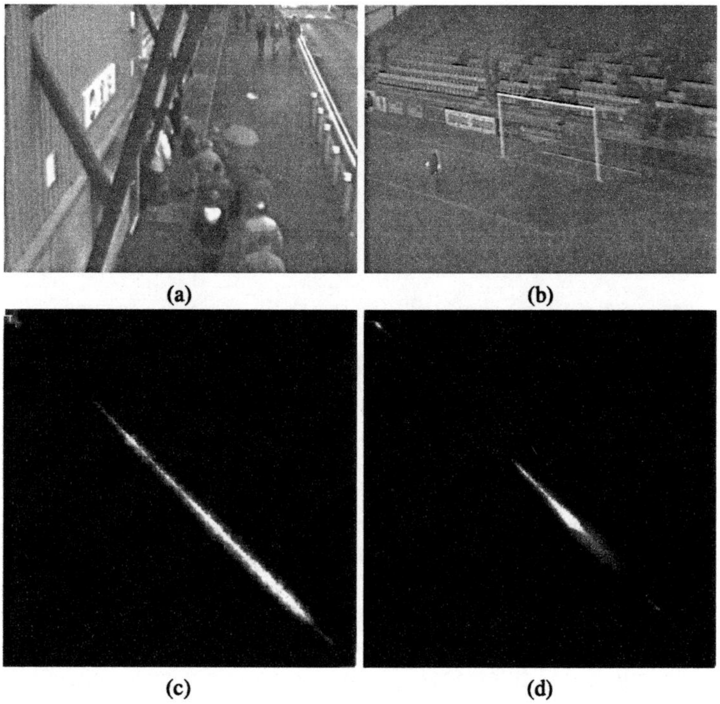

Fig. 2. Example of two different frames (a) and (b) characterised by different GLDMs (c) and (d). However, values of indicators commonly used in the literature for the two images are similar. GLDMs are drawn in two dimensions: the height of each bin is coded by the grey-level of the corresponding pixel.

3.2 GLDM Shape Analysis

To obtain a better characterisation of GLDMs, a set of new features has been developed, whose aim is to describe in more detail the shape of the histogram. From a detailed shape description it is then possible to obtain much more data than that provided by commonly used indicators.

To analyse the shape of a GLDM, a technique based on its contour lines at several heights has been developed: such lines are evaluated, and some parameters are exploited to measure the properties of their shapes; finally, parameters referring to a single line are compared to those evaluated for contour lines at other heights, and from the observed pattern, information is gathered about the shape of the GLDM.

3.2.1 Single Contour Line Description

A single contour line usually appears as a set of curves, one of which is clearly larger than the others, and is therefore called the principal component. This is the only part to be analysed, as smaller components are usually similar for every image. Each contour line is evaluated by intersecting the GLDM with a horizontal plane at a given height: by choosing several planes at different heights it is possible to provide a detailed description of the two-dimensional histogram.

Once the principal component is found, the least squares fitting algorithm is used to find the best fitting ellipse; the axes lengths are then evaluated. The ellipsis was chosen because it is the best suited shape for representing the typical contour line provided by GLDMs.

In all cases of interest, the ellipse has its major axis placed along the diagonal of the GLDM; its length tends to be maximum when the level curve is at a low height, unless the original image is not extremely uniform.

3.2.2 Multilayer Shape Analysis

The GLDM is analysed at several heights; ten height values were enough to precisely characterise the shape in all cases we observed. Values used for the actual heights depend on the size of the original image, as larger image produce higher histograms, due to the higher number of pixels; however, this problem can be solved by evaluating the normalised GLDM.

Once the result of the ellipse fitting is available for every contour line, some indicators are derived from the comparison of all of them. Such indicators are:

- **Evenness** depends on the ellipse minor axis at different heights. If w_i represents the minor axis at the i-th layer, N the total number of layers, and $\overline{\Delta w}$ is the average of the minor axis difference between ellipses belonging to consecutive layers, evenness is then defined to be:

$$\mathrm{EV} = \sqrt{\frac{1}{N} \frac{\sum_{i=1}^{N-1} \left[(w_i - w_{i+1}) - \overline{\Delta w} \right]^2}{i}} . \quad (1)$$

  Evenness is a measure of how evenly the ellipse minor axis shorten its length as long as the height at which the contour line is found is increased; the division by i is necessary to cope with the fact that the difference between subsequent height values does not increase linearly.

- **Minor axis spread** measures how steep the GLDM is, and is evaluated as the difference between the maximum and minimum lengths of the minor axis over all contour lines:

$$\mathrm{MAS} = w_{max} - w_{min} . \quad (2)$$

- **Minimum width** of the minor axis over all contour lines, which provides information about the smoothness of the upper part of the GLDM.
- **Total volume** included under the analysed level curves measures how spread the GLDM is outside the main component. If h_i and w_i are the major and minor axis of the ellipse found at the i-th height h_i, it is evaluated as:

$$\mathrm{TV} = w_1 h_1 l_1 + \sum_{i=2}^{N} w_i h_i (l_i - l_{i-1}) . \quad (3)$$

- **Width-height ratio** measures the mean of the width-height ratio of all ellipses, describing how spread over dark and light regions the texture is. It is evaluated as:

$$\mathrm{WHR} = \frac{1}{N}\sum_{i=1}^{N}\frac{w_i}{h_i}\,. \tag{4}$$

- **Volume of the peak** is the volume of the portion of the GLDM that is above the highest level curve, and represents another descriptor of the GLDM shape in its peak. If L is the highest level at which the GLDM is analysed, and there are M bins of height $H_i \geq L$, the peak volume is then evaluated as:

$$\mathrm{PV} = \sum_{i=1}^{M}(H_i - L)\,. \tag{5}$$

  This is a way of taking into consideration the part of the histogram that is neglected by the analysis of the contour lines.
- **Maximum-minimum area ratio** is the ratio between the area of the largest ellipse, and that of the smallest one, and describes the gradient of the GLDM; this can appear similar to the minor axis spread, but the major axis of each ellipse is also involved in this case. Since ellipses found at lower levels are larger than the ones found at higher levels, this ratio is evaluated as:

$$\mathrm{MMAR} = \frac{\mathrm{Area}_1}{\mathrm{Area}_N}\,. \tag{6}$$

- **Minimum area** is the area of the smallest ellipse, another descriptor of the GLDM peak.
- **Blank GLDM locations** is the number of empty bins in the GLDM. This is not directly related to the shape of the GLDM; however, this indicator is useful because it describes the histogram at the lowest level, as if a level curve at height 1 would be considered. The number of blank locations has been chosen instead of such level curve, because this second solution would have neglected a high number of blank locations that are inside a region composed of bins that are low, but anyway greater than zero. This situation is often the case in the region where the GLDM is very low.

These features represent a novel way for analysing GLDMs, that takes into consideration a high number of geometrical characteristics. It is then possible to use them as input for a classifier, together with previously used features, in order to obtain more accurate results.

3.3 Classification

Features described in the previous section form input data for a machine learning classifier. We use AdaBoost [15] to categorise these features into classes corresponding to different density of the crowd. AdaBoost combines so called "weak classifiers" into one final strong classifier that performs better than any of the weak classifiers alone (see [15] for details of the training algorithm). We use the Real AdaBoost variant of the algorithm [16] that provides a lower error rate by allowing weak classifiers to vote by their individual degree of certainty instead of making simple binary decision. We use binary decision trees [17] as weak classifiers. A single decision tree is constructed recursively based on the weighted training examples and in result only the relevant input features are selected. Therefore such a classifier can be considered as a feature selector.

The standard AdaBoost algorithm performs the binary classification only; to perform multi-class classification, we adopt the AdaBoost.MH algorithm [16] which creates a separate strong classifier for each class. This method requires the training examples to be presented individually for each class with amended labels indicating their affiliation to this specific class.

3.4 Crowd Localisation

A standard GLDM captures the global properties of the entire image, and is therefore unable to localise the detected crowd. In order to obtain localisation information, we applied a pyramidal image division into a set of regions: first, the entire image is divided into four equal parts, then each part is divided into four regions. After splitting, a separate GLDM is calculated and a set of the proposed texture features is extracted from each region. Each splitting step increases the number of regions but at the same time reduces the region area. In result, the localisation ability of the algorithm improves, while its discriminative power (i.e. ability to categorise) decreases. This trade-off should be considered when deciding for the optimal number of the splitting steps. In our work, we analyse results obtained for 0, 1 and 2 splitting steps which corresponds to division into $1, 4$ and 16 regions.

The pyramidal division is a key factor of the developed system, since it enables a GLDM-based method to provide localisation information. It has been observed that when the GLDM is evaluated over a portion of the image, the new descriptors based on multilayer shape analysis provide better classification results with respect to the ones used in the literature: the new features are therefore required in order to let GLDM provide strong result with localisation information.

4 Experiments

4.1 Experimental Data

The presented system has been tested on a set of challenging sequences recorded during the real football matches at the local stadium arena. The acquisition system consisted of the four best placed PTZ cameras installed at the venue: three cameras directed at the spectator areas and one placed outside the stadium. Data collection started one hour before the match and ended half an hour after the game was over.

The collected video sequences include people getting into the venue, watching the game and leaving the stadium. Available scenes include several scenarios: the public framed at several distances watching the game, the game field with players, people queuing up for coffee and crowds flowing into and out from the venue. The crowd itself does not always present the same texture density due to the different zoom levels and camera orientations used during the acquisition. As it can be seen, the system is required to correctly face large variety of different situations, with several crowd densities, scales and viewpoints; performed tests are therefore able to measure the generality of the proposed method.

The video resolution was set to 640×480 pixels. From all available data, we have selected 22 different video sequences resulting in 901 images in total that were used for our experiments.

The ground truth data were collected by manually annotating the number of persons in each image region corresponding to image division into 16 parts. These examples were later categorised into a set of classes representing different crowd density according to the criteria presented in Table 1. Ground truth for other image divisions (i.e. into 4 and 1 region) was calculated in a similar way.

Table 1. Experimental data: crowd categories, the corresponding number of persons per image region and a number of examples per class

Category	No crowd	Low	Med	High
Persons	0-1	2-4	5-9	10 and more
Examples	9954	2072	1489	901

4.2 Classification Results

4.2.1 Training and Testing

To train AdaBoost we have used 300 training examples for each class. The number of weak classifiers was set to 50 and the depth of the decision tree set to 3. Each experiment was cross-validated 20 times. The performance of the classifier was verified on 400 testing examples for each class and compared to the ground truth information.

4.2.2 Classification Performance

We have run several tests to investigate the main characteristics of our system. Table 2 presents the classification results obtained when using the original and proposed feature sets alone and using both sets together. In addition, to check the ability of the system to discriminate between different crowd categories we have run the tests for different number of classes including 4 original crowd categories, 3 categories where the Mid and

Table 2. Classification results for different feature sets and different number of classes. The feature sets include: MAR - original feature set, MD - the proposed feature set, MAR&MD - combination of all features.

Classes	MAR	MD	MAR&MD
4	$84.0 \pm 1.0\%$	$84.3 \pm 1.2\%$	$89.4 \pm 0.8\%$
3	$83.2 \pm 1.3\%$	$83.6 \pm 0.9\%$	$88.4 \pm 1.0\%$
2	$85.6 \pm 2.0\%$	$85.2 \pm 1.5\%$	$89.8 \pm 1.4\%$

Table 3. Classification results for different number of image regions

Regions	MAR	MD	MAR&MD
16	84.0±1.0%	84.3±1.2%	89.4±0.8%
4	95.9±0.9%	96.0±0.8%	97.3±0.9%
1	98.7±0.6%	98.7±0.6%	98.3±1.1%

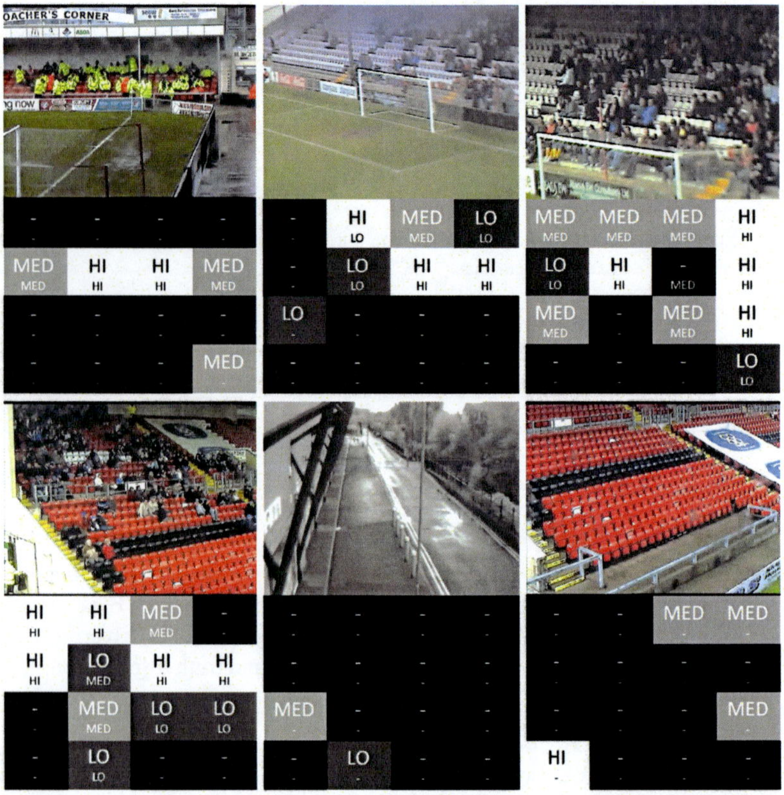

Fig. 3. Six frames presenting different situations of crowd-no crowd. These six frames are correctly classified when they are considered as one region or divided into four regions. However, when divided into 16 regions, the system correctly classifies most of the blocks (as shown by the black blocks in the lower part of the images), correctly locating the crowd. Only few blocks are problematic and misclassified (the grey blocks in the lower part of the images – brighter the block, larger the difference between the classifier output and the ground truth). Each block also contains two words: the classification output of the system (upper row) and the ground truth (lower row).

High class were combined into one and 2 categories with Low, Mid and High classes combined into one.

While performance for these two features sets is comparable (~84%) combination of all features results in a significant improvement (~89%). We can also see that this trend is similar for different number of classes and that merging different crowd classes did not significantly affect the results.

To test the trade-off between the system's ability to localise the crowd and accurately estimate its density we run the classification tests for different number of splitting steps (i.e. image regions): default division into 16 regions, division into 4 regions and for the entire image. The results presented in Table 3 indicate that the classification performance drops significantly (by $\sim$8%) when increasing the number of regions from 4 to 16. We can also see that differences in performance for different feature sets are less pronounced for lower number of regions and almost the same when analysing the entire image.

The main sources of misclassification were caused by rich texture content of non-crowded regions, significant image blur and the rough division into image regions. Some of the problematic situations are presented in Fig. 3, where the analysed images are coupled with the classification output, shown as a image that is black when the classification is correct, and that becomes lighter for increasing error entity. The results of the classification provided by our system and the ground truth are also reported, in the first and second row of each block, respectively.

We have also investigated the importance of each input feature, by analysing the weights of individual weak classifiers. In our tests, the features were ranked in the following order, starting from the most important one: homogeneity, entropy, total volume, contrast, blank locations, energy, minimum width, minor axis spread, volume of the peak, max-min area ratio, width-height ration and evenness. Two of the most important features turned out to be from the original set, while the most relevant features from the proposed set include total volume and blank locations. The evenness and width-height ratio were the last in the rank; however these features were also contributing to the final outcome of the classifier.

5 Conclusions and Future Work

We have presented a texture-based crowd detection system that is able to categorise image regions into classes corresponding to different crowd density and to roughly localise the crowd in the image. The presented results demonstrate that the additional proposed features improve the classification performance in smaller image areas. We have also demonstrated a trade-off between the localisation and classification ability of the system and investigated the importance of individual features.

Possible extensions to the system include incorporation of motion cues (e.g. based on motion flow analysis) for detection of moving crowds, the use of additional features (e.g. Gabor filters), combining information from multiple sensors, tracking individuals and groups of people in crowds, etc. It should also be possible to apply regression models to obtain continuous density estimates instead of discrete density categories.

References

1. Marana, A.N., Velastin, S.A., Costa, L.F., Lotufo, R.A.: Estimation of crowd density using image processing. In: IEE Colloquium on Image Processing for Security Applications (Digest No.: 1997/074), pp. 11/1–11/8 (March 1997)

2. Davies, A.C., Hong Yin, J., Velastin, S.A.: Crowd monitoring using image processing. Electronics Communication Engineering Journal 7(1), 37–47 (1995)
3. Mehran, R., Oyama, A., Shah, M.: Abnormal crowd behavior detection using social force model. In: IEEE Conference on Computer Vision and Pattern Recognition, CVPR 2009, pp. 935–942 (June 2009)
4. Ihaddadene, N., Djeraba, C.: Real-time crowd motion analysis. In: 19th International Conference on Pattern Recognition, ICPR 2008, pp. 1–4 (December 2008)
5. Arandjelović, O.: Crowd detection from still images. In: Proc. British Machine Vision Conference (BMVC) (September 2008)
6. Haralick, R.M.: Statistical and structural approaches to texture. Proceedings of the IEEE 67(5), 786–804 (1979)
7. Rahmalan, H., Nixon, M.S., Carter, J.N.: On crowd density estimation for surveillance. In: The Institution of Engineering and Technology Conference on Crime and Security, pp. 540–545 (June 2006)
8. Kong, D., Gray, D., Tao, H.: Counting pedestrians in crowds using viewpoint invariant training. In: Proc. British Machine Vision Conference (BMVC) (September 2005)
9. Kong, D., Gray, D., Tao, H.: A viewpoint invariant approach for crowd counting. In: 18th International Conference on Pattern Recognition, ICPR 2006, vol. 3, pp. 1187–1190 (September 2006)
10. Gárate, C., Bilinsky, P., Bremond, F.: Crowd event recognition using hog tracker. In: 2009 Twelfth IEEE International Workshop on Performance Evaluation of Tracking and Surveillance (PETS-Winter), pp. 1–6 (December 2009)
11. Marana, A.N., Costa, L.F., Lotufo, R.A., Velastin, S.A.: On the efficacy of texture analysis for crowd monitoring. In: Proceedings of SIBGRAPI 1998, International Symposium on Computer Graphics, Image Processing, and Vision, pp. 354–361 (October 1998)
12. McKenna, S.J., Jabri, S., Duric, Z., Rosenfeld, A., Wechsler, H.: Tracking groups of people. Computer Vision and Image Understanding 80(1), 42–56 (2000)
13. Gennari, G., Hager, G.D.: Probabilistic data association methods in visual tracking of groups. In: Proceedings of the 2004 IEEE Computer Society Conference on Computer Vision and Pattern Recognition, CVPR 2004, vol. 2, pp. II–876–II–881 (June 2004)
14. Lau, B., Arras, K.O., Burgard, W.: Tracking groups of people with a multi-model hypothesis tracker. In: IEEE International Conference on Robotics and Automation, ICRA 2009, pp. 3180–3185 (May 2009)
15. Freund, Y., Schapire, R.E.: A Decision-Theoretic Generalization of On-line Learning and an Application to Boosting. In: Vitányi, P.M.B. (ed.) EuroCOLT 1995. LNCS, vol. 904, pp. 23–37. Springer, Heidelberg (1995)
16. Friedman, J.H., Hastie, T., Tibshirani, R.: Additive logistic regression: a statistical view of boosting. Technical report, Dept. of Statistics, Stanford University (1998)
17. Breiman, L., Friedman, J., Olshen, R., Stone, C.: Classification and Regression Trees, 1st edn. Wadsworth and Brooks, Monterey (1984)

Polar Histogram Based Sampling Method for Autonomous Vehicle Motion Planning

Dmitriy Ogay, Jee-Hwan Ryu, and Eun-Gyung Kim

Korea University of Technology and Education, Cheonan, South Korea
{cactus,jhryu,egkim}@koreatech.ac.kr

Abstract. In this paper we present a sampling based motion planning algorithm for an autonomous vehicle, which allowed our vehicle to navigate smoothly at high speed with limited computation resources. A new sampling method, limiting candidate states, is introduced to reduce computation burden, associated with sampling based motion planning algorithms. The proposed method is experimentally evaluated driving an autonomous vehicle at speeds up to 60 km/h. It showed how advantages of both sampling based and space discretization based planning algorithms can be combined in one method, providing short planning time in higher dimensional configuration space and good performance in narrow and cluttered environment.

1 Introduction

Recently there have been many research activities in the area of autonomous vehicles. Being integrated complex systems, autonomous vehicles consist of many functional subsystems, and their development brings challenges in different fields of research. One of the problems is safe and reliable planning of vehicles motion. Development of our planning algorithm was motivated by Hyundai-Kia Autonomous Vehicle Competition.

Even though much work has been done in the field of obstacle avoidance and path planning, simple obstacle avoidance algorithms such as potential fields [2] and vector field histogram (VFH) could not be applied for navigation of autonomous vehicles because of their local nature, local minima problem, and inability to take dynamic constraints of autonomous vehicle into account. There were also many path-planning algorithms that work in a discrete configuration space[7], but, because of continuous nature of car motion, and non-holonomic constraints, generated paths are not satisfactory for use [1].

There were some hybrid (discrete-continuous) approaches, some of which were successfully implemented for autonomous vehicles[1]. But this method needs considerable high resolution for certain environments such as narrow passage. The specificity of the competition mission was passing in narrow passages. Obstacles were not complicated enough to require backward motion, and the car speed in narrow lanes was important. If we have used discrete approaches, then we would have to have considerably high resolution of space discretization, which in turn would have led to high computation burden and high latencies in the planner. Another problem with such algorithms is that in higher-dimensional hyperspaces their complexity grows exponentially with added dimensions.

S. Lee et al. (Eds.): Intelligent Autonomous Systems 12, AISC 193, pp. 737–744.
springerlink.com

The other type of path-planning algorithms is continuous space planning algorithms, among which Rapidly Exploring Random Trees (RRT)[6] based ones are well known, and studied a lot. One of the weaknesses of RRT algorithm is its performance in a cluttered environment and narrow passages.

We have selected a sampling based approach, and have developed a method to limit sampling area for candidate states, to achieve a reasonable computation time. This method is based on our a priori knowledge, that in most cases a vehicle would have to move in the direction of a given reference path. Our algorithm is a search tree, gradually growing in the direction of a reference path. Sampling is made in such a way that states belonging to the propagating front can be clustered in every step. We use the centers of these clusters to represent the higher probability state, where vehicle would have a better visibility of further environment. And then use a polar histogram to configure a probability distribution function for the sampling at the next step.

Sampling biasing for RRT like algorithms is not a new approach and the use of visibility information is neither. However, to the best of our knowledge, there is no such approach that combines sampling and space discretization approaches using clustering of states, belonging to the propagating front. The main advantage is that we can maintain good performance of sampling based methods in higher dimensional hyperspaces, at the same time providing fast navigation in narrow passages and cluttered environment, as with space discretization based methods.

2 Motion Planning Overview and Method Description

2.1 Motion Planning Overview

Most of up-to-date motion planning algorithms can be divided into two groups by the type of expansion. The first group is where nodes to be expanded are first selected from the search tree, and then expanded. Expansion is done by integrating some control input $u(t)$ for a period of time from the selected state in a tree. Priority queue is used in the most cases. The second group is where a new candidate state is first sampled in a configuration space, and then based on that information, an existing node is selected, then $u(t)$ is found, and the new state is added to the search tree. In this case nodes to be expanded are chosen based on the nearest neighbor search, like in RRT.

These two approaches have different behavior in road-like environment. If we represent a road as a space bounded by walls from the both sides, then the first type of algorithms would expand faster in forward direction because space is limited. But sampling based algorithms would have many collision check failures, which would lead to slow expansion.

But the first type approach, from the other side, has the following disadvantage: due to the large range of road curvature change, discretization of the control input $u(t)$ should be high enough to fit the shape of the road curve.

2.2 Target Environment and Conditions

For the competition, we participated in, the main task was to follow a reference path, and avoid obstacles if they exist. Reference path was provided. Obstacles were not

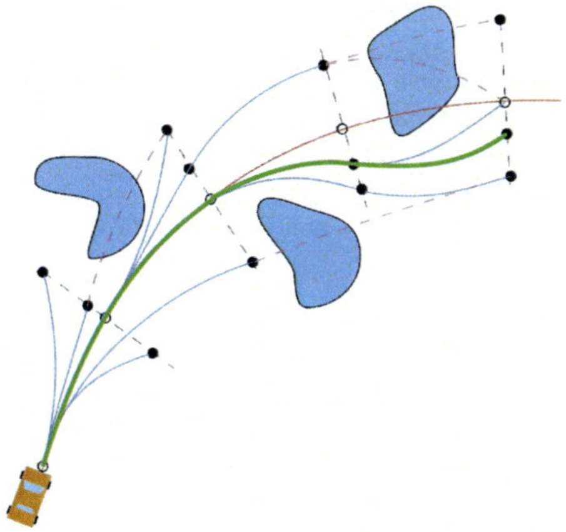

Fig. 1. Search tree growth is performed in steps, target points are sampled at normals of the reference path at reference way points

complicated enough to require backward motion planning or intricate manoeuvres, and there is no such obstacle that requires driving in the direction opposite to the reference path.

These conditions allowed us to make simplifications to our algorithm. Our algorithm works a growing tree, with a propagating front. At every cycle we process a propagating front. Taking into account above-mentioned conditions we assume that normals to the reference path would be good candidates for a propagated front, as it is seen on Fig. 1. The completeness of the planner will be discussed later.

In general our method is not limited to such of representation of a wave front. But for the simplicity of implementation we let it be so by now.

2.3 Planning Algorithm

By the expansion approach our planning method is close to a well-known RRT algorithm. Initial and goal states in the configuration space $\mathscr{C}$ given as an input. A search tree is grown from the initial state.

Unlike in RRT algorithm where candidate states are sampled from a configuration space in every cycle, we make a tree expansion in steps to provide more deterministic behavior, so it is better suited for real-time planning. At every step we propagate the expansion front of our tree by some distance. As we mentioned before, for the reasons of simplicity we have selected normals to the reference path to be candidates of a tree front. So at every cycle we make sampling from the normal taken at the next reference point.

After a new candidate is sampled, the nearest neighbor state from the tree is chosen, a trajectory linking two states is calculated, and a collision check is performed. If collision check passes successfully then the new state is connected to the tree. At every cycle, for every propagation front we perform a number of sampling, enough to provide smooth motion. And also we require enough samples to give a good representation of an actual propagated front.

It is necessary to distinguish a candidate propagated front, which is determined heuristically, and an actual propagated front, which depends on an actual environment and is represented by a set of states.

3 Sampling Method

3.1 Issue with Sampling

In previous section a simple sampling strategy, sufficient for motion planning of a vehicle, was described. However such a simple strategy may sometimes show a poor performance because. Tree expansion is computationally expensive; it includes two operations, which is nearest neighbor search, and collision check.

Problem with the nearest neighbor search is that for planning with differential constraints, Euclidean distance in configuration space is not enough, and algorithms which provide $\mathcal{O}(n \log n)$ look up doesnt work. So in real environment we have to deal with $\mathcal{O}(n^2)$ lookups. Collision check is a costly operation in general.

We have investigated the candidate samples, which cause most of the collision failures in an original algorithm, and detected a road pattern where algorithm performed poorly. This was a path bounded by walls from both side, like on Fig.3. The problem is that nave sampling from the normals may pick candidate states from the area behind the wall. And the narrower the road, the more probably such a failure state would be sampled.

3.2 Polar Histogram Based Sampling

To solve the problem of inefficient sampling our task was to find a probability distribution function that would give better failure rate. As we mentioned before, we distinguish

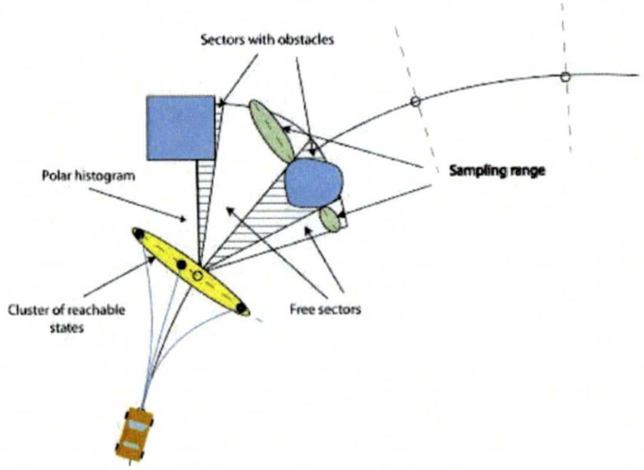

Fig. 2. Clustering of actually propagated front states, and a polar histogram, built from the center of a cluster

candidate propagation front and actual propagation front. The latter one is represented by a set of states, connected to the tree. At this step we find clusters of those states. This clustering is performed to determine discrete obstacles on the road, i.e. obstacles are discontinuities in an actually propagated front. By making clustering we find continuous segments of a front, and then choose the center of a cluster to be a good point for visibility lookup. We can justify this approach because if we were located between two obstacles, then the widest view angle would be from the point in the middle between two obstacles.

After the viewpoint is determined we build a polar histogram to determine a probability distribution function for sampling. So the output of a histogram is an approximation of $P\{successful\ motion|current\ position = view\ point\}$. One of the good approximation is an intersection of histogram sectors with candidate propagation front, as shown in Fig. 2. But this approach is too limiting. So the edges of intersection segments can be expanded by some margin.

Such an approximation of probability of successful motion, is justified by the fact that at a high speed vehicle is more probable to move forward in the direction of road, and if such a motion is not feasible, then the speed of a vehicle can be reduced, and thus some more time is reserved to investigate less probably chosen directions.

3.3 Generalization and Completeness

It may be noticed that in such an implementation, the following assumption was made. Since road width is not large, propagated front states clusters are also not wide in comparison to the distance to the next candidate front. To generalize the method for wider applications we can define a maximum size of a cluster, and divide large clusters into small subclusters, and perform histogram based lookup from the centers of the latter.

The second issue that can be noted is that points are not sampled in between the reference points. Such an algorithm may work if required maneuvers are larger than distance between propagated fronts, but this problem is solved incrementally in a following way. If propagation fails, then a reference point is chosen in the middle between the current reference point and the next failed reference point, and sampling is made on the normal to it. Correspondingly, maximum width of a cluster should also be reduced by two. Following these steps iteratively we can cover the whole space with desired density.

In our implementation candidate propagation fronts are represented by normal to the reference path, but in general they can be constructed using heuristic function, or be constructed interactively online, from the clustering information.

Our method works because two-dimensional workspace can be a good approximation of a projection of higher dimensional configuration space. This gives us opportunity to perform preliminary approximate lookups using fast algorithms like vector field histogram in two-dimensional space. The innovation of our method is that we combine sampling based approach with wave front propagation approach using clustering in a reduced space.

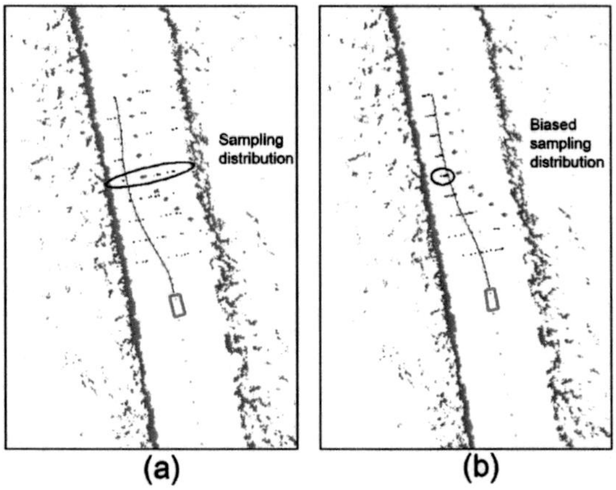

Fig. 3. Comparison of sampling distribution (black dots) of the algorithm with uniform sampling (a), and polar histogram based sampling (b). Passing artificially made narrow path.

4 Experiment

We implemented the proposed algorithm and tested it on our Hyndai SantaFe SUV based autonomous vehicle, driving in real environment. We have experimentally proved the efficiency of our algorithm. Seen on Fig.5 is a comparison of a nave sampling and sampling based on our method. As it is seen sampled candidate states (black dots), are not located beyond the reachability of a vehicle. We provide a comparison graph of algorithm execution time over an experimental track, seen on Fig. 3.

4.1 Comparison to Existing Approaches

To estimate efficiency of our algorithm to the existing ones, we made a comparative analysis. In comparison to original simple RRT with uniform distribution in configuration space our algorithm performs better because we use some a priori known information about environment, and limited the sampling range. We also used the fact that vehicle travels faster in longitudinal direction than in lateral, especially at high speed. The strongest factor in favor of our algorithm is that it performs better in narrow passages.

In comparison to space discretization methods our algorithm performs better, because sampling defines the discretization of the space and by increasing sampling density in lateral direction of car motion, and decreasing in longitudinal direction we avoid large amount of discretized cells. Another factor in favor of our algorithm is that obstacles are present in robot workspace, and we make histogram lookups in a workspace, while sampling is performed in higher dimensional hyperspace, where sampling based algorithms perform better.

There is also another class of path following algorithms that beat our algorithm in terms of computation time, for example spatiotemporal state lattices[8] approach, but

Fig. 4. Hyundai SantaFe SUV based autonomous vehicle

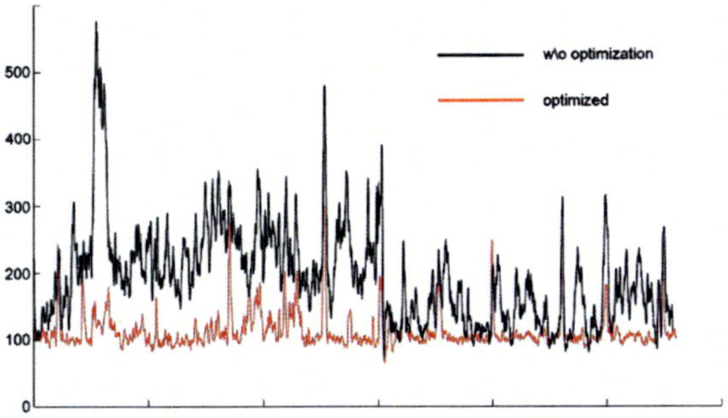

Fig. 5. Algorithm execution time(ms). Depicted by black is algorithm with nave sampling, by red is the proposed method.

such algorithms have a pre-processing stage, and the reference path should be known a priori. In case of our algorithm, the reference path can be changed dynamically for example from more abstract higher level planner.

5 Conclusion and Further Work

In this paper we have presented a new approach for sampling based motion planning, which allows us to drive our autonomous vehicle at high speed with low computation power requirements.

The method provides a sampling probability distribution function for more efficient search tree expansion by utilizing some assumptions and approximations about car motion. We experimentally prove the efficiency of proposed method by comparing a planner with nave sampling and the polar histogram based sampling. Comparison study over existing approaches is provided as well.

Our planning algorithm was tested and proved well performing in real driving environment. The peculiarity of real environment is that it is cluttered and consists mostly of narrow passages, from the other side planning with differential constraints requires higher order configuration space.

The proposed method is a synergy of RRT-like sampling based approach, which performs well in high dimensional hyperspaces and has a very flexible space discretization, and Dijkstra-like approaches based on wave front propagation, which perform well in narrow passages. The innovation of our method is that we combine sampling based approach with wave front propagation approach using clustering in a reduced space.

Presented motion planner was used by our team during Hyundai-Kia Autonomous Vehicle Competition 2010. It let our vehicle successfully pass all planning related missions.

Acknowledgement. The Authors are grateful of the following support: Ministry of Knowledge and Economy of Korea grant as part of the Development of a Tele-Service Engineand Tele-Robot Systems with Multi-Lateral Feedback project.

References

1. Dolgov, D., Thrun, S., Montemerlo, M., Diebel, J.: Practical Search Techniques in Path Planning for Autonomous Driving. In: Proceedings of the First International Symposium on Search Techniques in Artificial Intelligence and Robotics (STAIR 2008 (June 2008)
2. Duda, R., Hart, P., Stork, D.: Pattern classification, 2nd edn. Wiley, New York (2001) ISBN 0-471-05669-3
3. Frazzoli, E., Dahleh, M., Feron, E.: Real-time motion planning for agile autonomous vehicles. AIAA Journal of Guidance and Control 25(1), 116–129 (2002)
4. Konolige, K.: A gradient method for realtime robot control. In: Proceedings of the IEEE International Conference on Intelligent Robots and Systems, IROS (2000)
5. Kuwata, Y., Fiore, G., Teo, J., Frazzoli, E., How, J.: Motion planning for urban driving using RRT. In: Proc. IROS, pp. 1681–1686 (2008)
6. LaValle, S., Kuffner, J.: Randomized kinodynamic planning. International Journal of Robotics Research 20(5), 378–400 (2001)
7. LaValle, S.: Planning Algorithms. Cambridge University Press, Cambridge (2006), `http://planning.cs.uiuc.edu/`
8. Ziegler, J., Stiller, C.: Spatiotemporal state lattices for fast trajectory planning in dynamic on-road driving scenarios. In: Proc. IROS, pp. 1879–1884 (2009)

A Multiresolution Approach for Real-Time Motion Planning under Differential Constraints

Dmitriy Ogay, Jee-Hwan Ryu, and Eun-Gyung Kim

Korea University of Technology and Education, Cheonan, South Korea
{cactus,jhryu,egkim}@koreatech.ac.kr

Abstract. In this paper we present a complex approach to resolve a contradiction in planning between space discretization based algorithms and sampling RRT-like algorithms in a way, appropriate for a real time motion planning with limited computation resources. The former type planning algorithms have advantage that they can systematically explore discretized space, and are less vulnerable to the problem of navigation in narrow passages, but from the other side, to achieve satisfactory results in terms of path smoothness and traversability, higher resolution and additional dimensions in configuration space might be required, which makes use of such algorithms problematic for real time motion planning, because of curse of dimensionality. RRT-like sampling based algorithms have advantage, that they can explore fast in high dimensional configuration space, but they may spent indefinitely long time if they encounter narrow passages. For the safe motion it is desired that motion planner gives response to the plan execution part in a limited time. The proposed solution implements this feature in a more efficient way than existing solutions. The proposed solution was tested in a real autonomous driving, including parking with backward motion.

1 Introduction

Recently there has been much work done in both real time motion planning and planning for robots with differential constraints [16]. Especially much work devoted to navigation of autonomous vehicles [6,3,5]. Though much success was achieved in navigation, and safe motion planning with kinodynamic constrains, there is still one problem that was not completely explored safety of planning in terms of computation time with limited computation power.

Methods, proposed recently for motion planning of autonomous vehicles and wheeled mobile robots, may be divided in at least three categories: propagated wavefront approach [8], graph search in discretized space [3], and sampling based approaches [6,10,16]. The use of first type methods is limited by simplicity of cost functions that can be used. There are also methods based on artificial potential fields [16], but their use in planning under differential constraints is limited. Discretized space approaches were extensively investigated in the past. These methods are mainly based on discretizing configuration space by a high-dimensional grid into hyper-parallelograms.

The limitation of grid-based discretization is that motion with differential constraints is continuous in its nature and pure discrete cells result in kinematically intraversable paths. There are several approaches that address this problem, in particular Field-D*

S. Lee et al. (Eds.): Intelligent Autonomous Systems 12, AISC 193, pp. 745–754.
springerlink.com

[4] and Hybrid-A* [3] algorithms. But still those algorithms might need path post-processing to generate smooth trajectories. In addition, when dynamic behavior should be considered in path planning, it might require additional dimensions, so classical grid based algorithms fall victims of curse of dimensionality. The other group of path planning approaches are sampling based planners in continuous space, they have some advantages for planning in high-dimensional configuration space. Among them RRT [10] should be mentioned for its ability to explore rapidly in unexplored parts of the space. But one of the biggest issues of RRT is poor performance in presence of narrow paths, as algorithm can take indefinitely long time to find the way through, resulting in unpredicted path planner response. The problem of narrow paths also affects grid based path planners, as it requires higher resolution, and thus longer computation time.

As real-time applications require response in a limited time, several approaches are generally used. For the case of grid-discretization based algorithms there exist so-called anytime algorithms [12], which give fast response, with feasible path, sacrificing optimality in the beginning and refining at following steps. For the case of RRT-like algorithms various approaches may be generalized to running an algorithm for a limited time, and then taking corresponding action, if solution is found or not [6].

Our approach is based on parameterization of reachability graph expansion strategy on speed of a robot,with higher priority of expansion given to nodes corresponding to the states with higher velocities. But we sacrifice resolution and completeness if computation resources are not enough to meet deadline at a corresponding higher speed. This lets planner give early command to the execution part to reduce the speed of a robot, thus giving more time for planning at increased resolution, and providing completeness at lower speeds, when fast decision is not that critical and regions of ICS (inevitable collision states) [14] become smaller.

2 Related Work and Analysis

2.1 Multiresolution Approach and Advantages of RRT

Multiresolution approach is not new in motion planning; there exist many methods for grid-based space discretization algorithms. The main task in developing such algorithms is to control level of discretization, so that resolution could be increased, but search graph would also be explored systematically.

RRT gained its popularity because of its ability to rapidly explore large configuration spaces. In fact it has a built-in multiresolution feature. It can expand in large steps in a free space, making coarse plans, and then refining the search by increasing density of sampling. Size of Voronoi cells can be considered as a measure for resolution. Hence, with more new states added to the search graph, density grows and average resolution grows with time, providing probabilistic completeness of a method. But in planning resolution completeness is usually preferred over probabilistic completeness, because planner can give response in a finite time, that solution does not exist, while for probabilistically complete planner this time goes to infinity.

The weak point of multiresolution approach implementation in RRT is that resolution is randomized. Though on average resolution is increasing with time, on a particular stage of planning resolution is random in different states of a search graph. The problem of narrow paths could be considered as a problem of not sufficient resolution, and this is common for both sampling based approaches and space discretization based approaches. RRT-like motion planner can find a way through a narrow path, when density of sampling near the path approaches the width of the path. But density is proportional to the volume of Voronoi regions in sampled space, and this volume decreases exponentially with dimensionality of space, so again we have curse of dimensionality.

There exist some approaches, which address problem of narrow paths in RRT, in particular OBRRT [15] and PCA-RRT [2]. They main idea is to bias sampling distribution in such a way, that interesting regions (narrow paths) are sampled with higher probability, thus increasing rate at which density and thus resolution grows in those regions.

2.2 An Approach of Lindemann and LaValle

In [13] Lindemann and LaValle introduce their modified version of Barraquand and Latombes planner [1]. The original planner builds a reachability graph in configuration space by expanding visited states. As parameters it takes Δt - time discretization, H depth of a search tree, and R space discretization (the space is decomposed into a grid having $2R$ cells per axis). The nodes of search tree are expanded, integrating input controls $u(x) \in U$ for the period of time Δt, where U is a finite set of control inputs. They prove, that with Δt small enough, planner becomes complete. Lindemann and LaValle introduce a multiresolution feature of a planner. Their proposition is to gradually decrease time discretization in such a way that would allow systematic search graph exploration with increasing resolution. They introduce rank of a node, with time period Δt decreased by two with the increase of a rank. Thus planner convergence may be achieved by gradually increasing time discretization.

2.3 Real-Time Extensions for Sampling Based Planners

Real-time systems require response in a limited time. There are several proposed real-time frameworks, which use sampling based motion planners as their underlying planner [7]. The main idea can be generalized to running an underlying planner for a predefined amount of time, and then making a decision about following steps in plan execution part. As it was mentioned earlier, in RRT-like sampling based approaches resolution is randomized and no preferences are given to the states, where fast response of a planner is critical, this weakness led us to invent our approach, where resolution could be controlled, prioritizing more critical areas of configuration space for expansion.

3 Method Description

3.1 Parameterization on Speed

It was mentioned in the previous paragraph that there should be something what measures, how critical and important fast feedback from motion planner is. For a dynamic

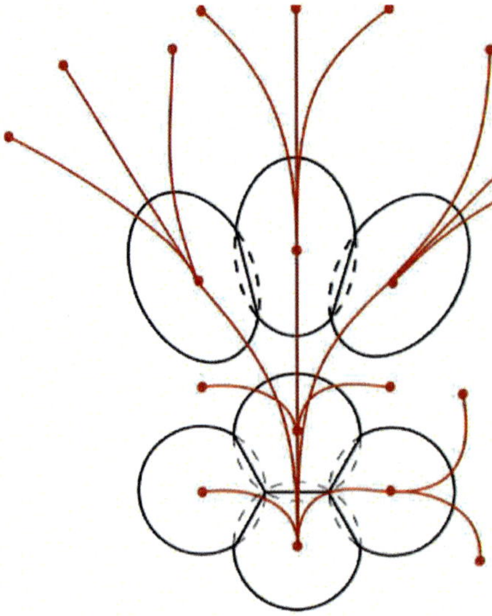

Fig. 1. Nodes are expanded at various resolutions. Circles and ellipses show boundaries of cells. If during expansion nodes gets into a cell it means that this state is occupied. Shape of the cells can be parameterized, for example longitudinal displacement at high speed is not as critical as lateral displacement.

system, we could say that it is speed $|\bar{v}|$. Speed is important because of the following reasons: when perception range of a robot is limited, it will take less time for robot to reach the limits of its perception range, because of the delays at perception, planning, and execution stages, second, regions of ICS are larger at higher speeds, third, momentum is larger at higher velocities while inertia is constant, so it requires larger control forces to be applied. Robot may actually run into an ICS, if previous plan is being executed, and delay from replanning is high, then execution part may bring robot into a state, which could not have been detected as an ICS at previous step because of perception range limit. Thats why we choose robots speed as a parameter for our motion planner.

Then second question is, what can be parameterized by speed. The answer is planning step, and resolution. What can be sacrificed is accuracy and completeness. It is true for biological intelligence, when for example we drive a car. We can drive at a very high speed without worrying, if surface is wide and smooth. We dont concentrate on details. Then what happens, when we notice a narrow passage or we want not to miss a turn we brake. This gives us more time to make a decision and generate a sequence of control with higher accuracy.

So the first feature of our motion planner is a parameterization of a planning step by speed.

$$x_{(k+1)} = \int_0^{(t_{|\bar{(v)}|})} f(x_k, u(t), \bar{v}_k)dt \tag{1}$$

Algorithm 1. Main Body of Algorithm

```
for v = v_max → v_min do
    while !solutionFound & !limitingCondition do
        x ← chooseNode();
        for all u ∈ U_v do
            x_new ← ∫ u(t)dt;
            if trajectoryIsCollisionFree() then
                T ← T ∪ x_new;
                Q ← Q ∪ x_new;
            end if
        end for
        if x is completely expanded then
            remove x from Q;
        end if
    end while
    if solutionFound then
        stop and return solution;
    else
        continue;
    end if
end for
report failure
```

Where $t_{|\bar{(v)}|}$ is a time of integration, which is parameterized by speed. In many cases this time may be even the same for different speeds, it will generate proportionally larger steps for higher speeds.

3.2 Space Discretization

As it is easy to notice for a proposed motion planner, number of states in a search tree grows exponentially with the number of expansion steps. Motion planners from previous works used space discretization, to address this problem, as nodes may not be added to a search tree if a cell in a configuration space has already been visited and is occupied.

In the original method Barraquand and Latombe use grid-based discretization [1]. And in the modified version Lindemann and LaValle used a more general approach, which is to use Voronoi regions formed by points, placed according to some distribution P. This approach gives more flexibility in managing volume of cells while preserving full coverage of a configuration space. In our planner we have used a modified version of the latter approach. Instead of using predefined distribution to be the centers of Voronoi cells, we use states, present in search tree as centers of cells. We also should define some measure of distance in $\mathbb{R}^N$, which is not necessarily be Euclidean, to be used as a radius of a cell. For practical use we can have a weighted Euclidean distance, or some other. For example for the case of autonomous vehicle, which has at least three variables in its state that are Cartesian coordinates and heading angle, the measure can be a weighted Euclidean distance rotated to the heading angle. Then the cell, bounded by an equidistant surface, would be a hyper ellipsoid oriented along vehicles direction. So our cells will be limited by some radius δ.

Such a discretization does not give us full coverage of a configuration space, because cells can be disjoint. But the advantage is that planning error would be bounded by. This means that at some certain speeds, if we design motion-planning steps too long, and choose relatively small, we will get an incomplete planner. But still in many cases such a planner would work.

3.3 Completeness

In practical implementations motion planning problem cannot always be solved exactly in a limited amount of time. We could say that planner found a solution when distance between one of nodes of reachability graph and the target state is less than some small value . Contribution of Lindemann and LaValles approach is that they manage to gradually increase the resolution of a planner, while guaranteeing exploration of reachability graph.

In our motion planner however we use speed as a parameter for resolution, rather than time. The problem is that if we split time discretization t by two, and expand search tree in two steps with the same set of controls, then the states of coarse resolution graph nodes will be a subset of nodes in higher resolution graph. It means that if we increase time discretization by two we will still get all the states, present in original graph. But in case if we increase resolution parameterizing by speed, we lose such a property of a motion planner. This is also true for Lindemann and LaValles planner if motion primitives are used instead of constant controls. Anyway if we can design motion primitives in such a way that with increase of discretization (time or space) we still can guarantee that generated states will be a superset of states in coarse search tree, even with some small error, then the motion planner would be complete, as we can choose infinitely small step of discretization.

However with speed based approach it is strictly impossible to make a higher resolution graph contain lower resolution graph because of the fact, that velocities are part of the state space in motion planner. So at these step one may think that the proposed motion planner is inferior to the one being extended. It is not true, because the main reason to introduce speed based resolution parameterization were limits on computing resources. So we want to introduce another type of obstacles in the configuration space, as States, Unreachable under Computation Resource Limits (SUCRL). One may think of them as of virtual obstacles, like ICS, which appear at certain speeds in presence of limits on computation power.

As a result we cannot use the same expansion strategy as was used in original planner. At first the highest speed, and thus the lowest resolution is taken, then search graph is expanded unless some limiting condition is reached. Then the speed is lowered and resolution is refined, and the space is searched again. At these step we should also investigate all the nodes at lower resolution, to be able to return to higher velocities. But this is not a problem in terms of algorithm complexity, as number of nodes in search graph is on exponential order with resolution.

Fig. 2. Motion planning at turn, planner reports that goal is unreachable at high resolution, drops the speed and refines a search

3.4 Control Input Design

The other advantage in using speed as a parameter for a motion planner is of more practical reason. As we experienced while developing an autonomous vehicle in our university, we could notice, that dynamics of a vehicle depends on speed. And breaking and accelerating can directly control the speed. So motion primitives are much easier to be designed or recorded, based on speed. As it was mentioned in [13] motion primitives are much more useful in real implementation of such a path planner. They are also useful because collision check over pre-computed parameterized trajectories could be done in parallel.

4 Experiments

To evaluate performance of the developed planner we decided to test it for motion planning of an autonomous vehicle, for which we have developed an RRT based path following algorithm, before. As it was expected the new path planner has shown much better performance in narrow paths. It should be also noted that in practical implementation, in comparison to RRT the proposed path planner scales much better. This is because of the reason, that RRT based path planners require sophisticated distance measure for the nearest neighbor search in presence of differential constraints, for example weighted combination of arc curvature and local path length, which makes it impossible to use fast nearest neighbor search algorithms, like kd-trees, as it was also noted in [16].

Fig. 3. Motion planning at a corner, if planned at a high speed, planner fails and reports absence of road

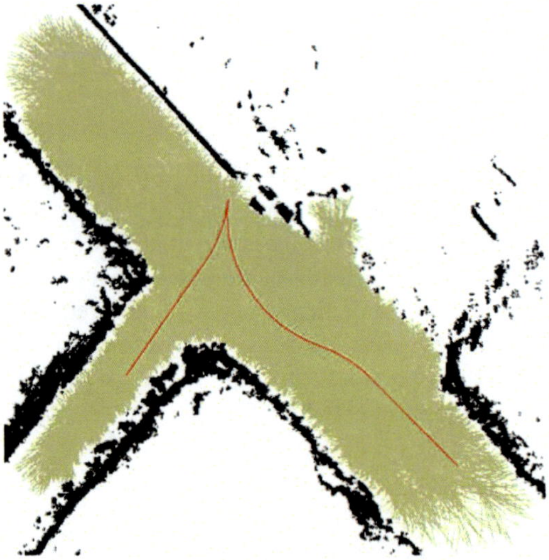

Fig. 4. Example of backward motion planning. Vehicle starts motion from the right bottom corner on the map.

It should also be noted that to achieve good performance in real applications, algorithm requires tuning, thorough modeling of motion primitives or control inputs and choice of cell size.

5 Conclusion

In this paper we have analyzed existing motion planning approaches for non-holonomic robots. We have outlined an aspect of motion planning safety, which was not fully addressed in previous works, and proposed a solution that tries to resolve the contradiction between the two main types of motion planners, used for navigation of car-like robots. The contribution of our work is that we propose a speed based parameterization for a multiresolution motion planner developed by Lindemann and LaValle. Advantage of our approach is that at high speeds planner provides fast feedback, while at lower speeds it provides completeness. We changed the expansion strategy to adapt it to the speed based parameterization. We have applied our approach for navigation of an autonomous vehicle, and give a comparison with previously developed RRT-based motion planner, based on a practical experience. The efficiency of algorithm was tested, running on an autonomous vehicle, with various scenarios, including autonomous parking.

Acknowledgements. Authors are grateful of the following support: Ministry of Knowledge and Economy of Korea grant as part of the Development of self-initiated and hands-on e-training service technology for the car mechanics project.

References

1. Barraquand, J., Latombe, J.-C.: Nonholonomicmultibody mobilerobots: Controllability and motion planning in the presence of obstacles. Algorithmica 10, 121–155 (1993)
2. Dalibard, S., Laumond, J.-P.: Control of probabilistic diffusion in motion planning. Presented at the International Workshop on the Algorithmic Foundations of Robotics (WAFR) (2008)
3. Dolgov, D., Thrun, S., Montemerlo, M., Diebel, J.: Practical Search Techniques in Path Planning for Autonomous Driving. In: Proceedings of the First International Symposium on Search Techniques in Artificial Intelligence and Robotics, STAIR 2008 (June 2008)
4. Ferguson, D., Stentz, A.: Field d*: An interpolation based path planner and replanner. In: Proceedings of the Int. Symp. on Robotics Research, ISRR (2005)
5. Ferguson, D., Howard, T., Likhachev, M.: Motion Planning in Urban Environments. Journal of Field Robotics (JFR) 25(11-12), 939–960 (2008)
6. Frazzoli, E., Dahleh, M.A., Feron, E.: Real-time motion planning for agile autonomous vehicles. AIAA Journal of Guidance and Control 25(1), 116–129 (2002)
7. Hauser, K.: Adaptive Time Stepping in Real-Time Motion Planning. In: WAFR (2010), `www.indiana.edu/~motion/`
8. Konolige, K.: A gradient method for realtime robot control. In: Proceedings of the IEEE International Conference on Intelligent Robots and Systems, IROS (2000)
9. Kuwata, Y., Fiore, G.A., Teo, J., Frazzoli, E., How, J.P.: Motion planning for urban driving using RRT. In: Proc. IROS, pp. 1681–1686 (2008)
10. LaValle, S.M., Kuffner, J.J.: Randomized kinodynamicplanning. International Journal of Robotics Research 20(5), 378–400 (2001)
11. LaValle, S.M.: Planning Algorithms. Cambridge University Press, Cambridge (2006), `http://planning.cs.uiuc.edu/`
12. Likhachev, M., Ferguson, D., Gordon, G., Stentz, A., Thrun, S.: Anytimedynamic a*: An anytime, replanning algorithm. In: Proceedings of the International Conference on Automated Planning and Scheduling, ICAPS (2005)

13. Lindemann, S.R., LaValle, S.M.: Multiresolution Approach for Motion Planning under Differential Constraints. In: Proc. ICRA, pp. 139–144 (2006)
14. Petti, S., Fraichard, T.: Safe motion planning in environments. In: IEEE International Conference on Intelligent Robots and Systems (IROS), pp. 3726–3731 (2005)
15. Rodriguez, S., Tang, X., Lien, J.-M., Amato, N.M.: An Obstacle-Based Rapidly-Exploring Random Tree. In: Proc. IEEE Int. Conf. Robot. Autom. (ICRA), Orlando, FL, pp. 895–900 (May 2006)
16. Ziegler, J., Stiller, C.: Spatiotemporal state lattices for fast trajectory planning in dynamic on-road driving scenarios. In: Proc. IROS, pp. 1879–1884 (2009)

A Study of Path Planning Algorithm Based on the Survival Probability

Min-Ho Kim, Chi-Beom Noh, Jung-Hun Heo, and Min-Cheol Lee

Dept. of Mechanical Engineering, Pusan National University
Jangjeon 2-dong, Geumjeong-gu, Busan, 609-735, Korea
xho1995@gmail.com, {nochichi,gasmaster}@naver.com,
mclee@pusan.ac.kr

Abstract. After the development of fusion technology of mechanics, electronics and IT, there are a lot of researches about the autonomous vehicle. For this kind of vehicle to move automatically, it needs the reference input that the vehicle can follow, and this reference input comes from a path planning algorithm. A* is one of the well-known global path planning algorithm that finds the optimal path on the given map using the heuristic cost function. However, A* algorithm gives the vehicle some reference points on the path not the continuous reference path. And A* may not guarantee the path will not collide the corner of the obstacle. To overcome this problem, in this paper, the hybrid path planning algorithm is suggested using B-spline equation and new heuristic cost function with survival probability. At last, we developed the simulation program and the results are shown.

Keywords: autonomous vehicle, A* algorithm, B-spline curve equation, path planning algorithm.

1 Introduction

After the development of fusion technology of mechanics, electronics and IT, there are a lot of researches about the autonomous vehicle. For example, today Google has tested an autonomous vehicle on the real road environment. For this kind of vehicle to move automatically, it needs the reference input that the vehicle can follow, and this reference input comes from a path planning algorithm. Therefore, researches on the path planning algorithm have increased.

Among these, A* is one of the well-known path planning algorithm developed by Peter Hart, Nils Nilsson, Bertram Raphael in 1968. [1] It is the global path planning algorithm that finds the optimal path on the given map using the heuristic cost function. Usually the map of A* algorithm consists of several grid cells, and then each cell has a cost value to be compared and evaluated to find the optimal path.

However, due to the grid map, A* algorithm gives the vehicle some reference points on the path not the continuous reference path. And A* may not guarantee the path will not collide the corner of the obstacle. To overcome this problem, in this paper, the hybrid path planning algorithm is suggested that combines A* algorithm, B-spline equation and survival probability.

S. Lee et al. (Eds.): Intelligent Autonomous Systems 12, AISC 193, pp. 755–763.
springerlink.com

In the hybrid path planning algorithm, the B-spline interpolation curve equation is used to make the continuous path. Previously we developed the continuous path generation algorithm using B-spline interpolation equation. [2] B-spline interpolation curve is generated with way points which are the points lying on the curve. Using A* algorithm, the center points of selected grid cells become the way points.

And the probability theory is used to evaluate the hazardousness of some region. The cell nearby the obstacle or having some unknown hazard may have low survival probability. In this paper, we suggest a new heuristic cost function to evaluate hazardousness. This heuristic cost function consists of hazardous probabilities on each cell.

In chapter 2, we explain the theory of A* algorithm and the B-spline equation to plan the path. Next, New heuristic cost function using survival probability is suggested. And, to verify our algorithm, we developed the simulation program and the results are shown.

2 Path Planning Algorithm

2.1 The Cost Function of the A* Algorithm

Before introducing the hybrid path planning algorithm, first we look around the normal A* algorithm. The A* algorithm uses the cost function like as; [3]

$$F(n) = g(n) + h(n) \tag{1}$$

In this equation, g(n) is the cost value from the start position to current position on the map. And h(n) is the heuristic cost value from the current position to the goal

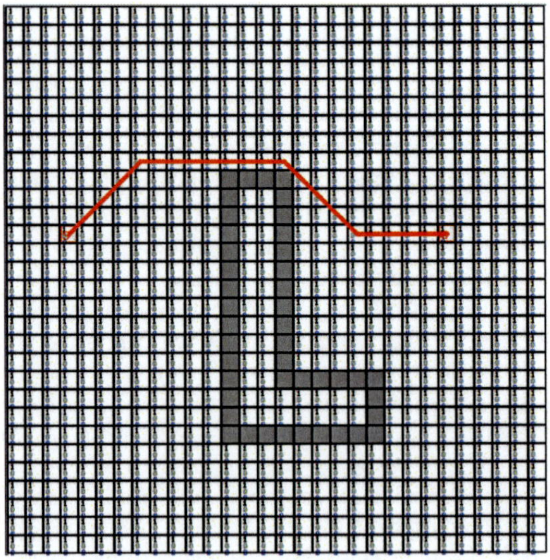

Fig. 1. The Path generated with A* Algorithm contacts with obstacle at the corner

position. F(n) is the total cost value to be compared and evaluated for optimal path planning.

Figure 1 shows one of the result path of A* algorithm. The path consists of several discontinuous lines, and at the corner of obstacle the path may contact with the obstacle. So if we want to use A* algorithm into the real vehicle, collision problem should be resolved. Later, we will show the new heuristic cost function to solve this problem.

2.2 The B-Spline Curve Equation

To build a continuous path from some points, we researched about the B-spline curve equation. Because, the B-spline curve can be generated by a blending function which connects and smoothes the vertices of a polygon. [4, 5] Cox and de Boor suggested a blending function $\mathbf{N}(u)$ shown as in equation (1). So if we know the some points, then we can build up the continuous curve path with those points.

$$N_{i,k}(u)=\frac{(u-t_i)N_{i,k-1}(u)}{t_{i+k-1}-t_i}+\frac{(t_{i+k}-u)N_{i+1,k-1}(u)}{t_{i+k}-t_{i+1}}$$
$$N_{i,1}(u)=\begin{cases}1 & t_i\le u\le t_{i+1}\\ 0 & others\end{cases} \tag{2}$$

Where, t_i is called as knot value, which is a parameter that makes blending function nonzero. Using the above blending function, B-spline curve equation $\mathbf{P}(u)$ can be defined as;

$$\mathbf{P}(u)=\sum_{i=0}^{n}\mathbf{P}_i N_{i,k}(u) \qquad (t_{k-1}\le u\le t_{n+1}) \tag{3}$$

Where, each $\mathbf{P}_i$ means pre-defined control points. At last, the following relation can be obtained.

$$\mathbf{P}(u)=\sum_{i=l-k+r+1}^{l}\mathbf{P}_i^r N_{i,k-r}(u)$$
$$\mathbf{P}_i^r=\frac{u-t_i}{t_{i+k-r}-t_i}\mathbf{P}_i^{r-1}+\left(1-\frac{u-t_i}{t_{i+k-r}-t_i}\right)\mathbf{P}_{i-1}^{r-1} \tag{4}$$

The point $\mathbf{P}_i^r$ is called as de Boor. r is the order of the equation. Usually r is '3' because a mechanical system like an autonomous vehicle has 2nd order equation, so it needs a continuous of 2nd derivatives. Plugging (k-1) to r, the following relation can be obtained.

$$\mathbf{P}(u) = \mathbf{P}_l^{k-1} \tag{5}$$

Following the above sequence, a curve $\mathbf{P}(u)$ could be obtained from the given points $\mathbf{P}_i$. Fig. 2 shows the curve path obtained from the B-spline curve equation.

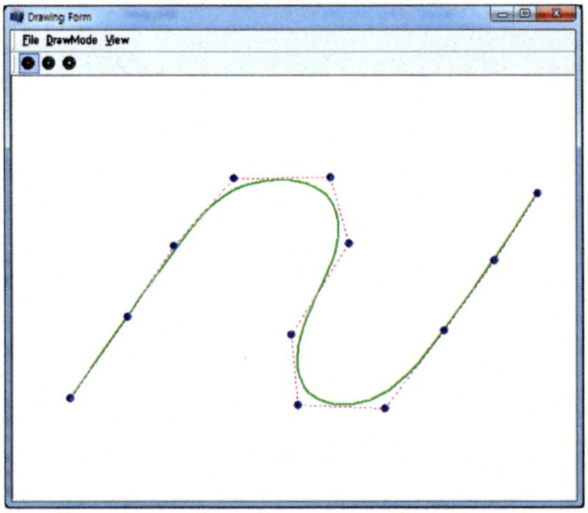

Fig. 2. The Path generated with B-spline curve equation. The dark blue points are the control points to make the green curve path.

2.3 The Interpolation Curve Equation

But that continuous path isn't sufficient to apply a vehicle, because the vehicle might know the way points, but can't know the control points. So we should make the method generating the unknown control points with the known way points. The way points go through the path, so we can call them "the interpolation points". [2]

To express a B-spline curve equation passing the interpolation points $\mathbf{Q}_j$, the adjusting points $\mathbf{P}_i$ should be defined. The interpolation points $\mathbf{Q}_j$ and the control points $\mathbf{P}_i$ must satisfy the following condition. Because, actually all interpolation points are the solution of B-spline equation.

$$\mathbf{Q}_j = \sum_{i=0}^{n} \mathbf{P}_i N_{i,k}(u_j) \qquad (j = 0, 1, \cdots, n) \tag{6}$$

To find adjusting control points with interpolation points, users should specify parameter values and knot vector. To define parameter values, chord length method was usually used, and averaging method was for knot vectors. With a chord length method, parameter values can be defined from the following equation;

$$u_j = u_{j-1} + \frac{|\mathbf{Q}_j - \mathbf{Q}_{j-1}|}{\sum_{i=1}^{n} |\mathbf{Q}_j - \mathbf{Q}_{j-1}|} \quad (j = 1, \cdots, n-1)$$
$$u_0 = 0, \quad u_n = 1 \tag{7}$$

An averaging method to define knot vectors is written as;

$$t_{k-1+j} = \frac{\sum_{i=j}^{j+k-2} u_i}{k-1} \quad (j = 1, \cdots, n-k+1) \tag{8}$$
$$t_0 = \cdots = t_{k-1} = 0, \quad t_{n+1} = \cdots = t_{n+k} = 1$$

Using equations (8) and (9), equation (7) can be written as a polynomial as;

$$\begin{bmatrix} N_{0,k}(u_0) & \cdots & N_{n,k}(u_0) \\ N_{0,k}(u_1) & \cdots & N_{n,k}(u_1) \\ \vdots & \ddots & \vdots \\ N_{0,k}(u_n) & \cdots & N_{n,k}(u_n) \end{bmatrix} \begin{bmatrix} \mathbf{P}_0 \\ \mathbf{P}_1 \\ \vdots \\ \mathbf{P}_n \end{bmatrix} = \begin{bmatrix} \mathbf{Q}_0 \\ \mathbf{Q}_1 \\ \vdots \\ \mathbf{Q}_n \end{bmatrix} \tag{9}$$

Then we can obtain the control points with the interpolation points using linear algebra. Fig. 3 shows an output obtained from the above algorithm, in which dark points and light points mean interpolation points and control points, respectively.

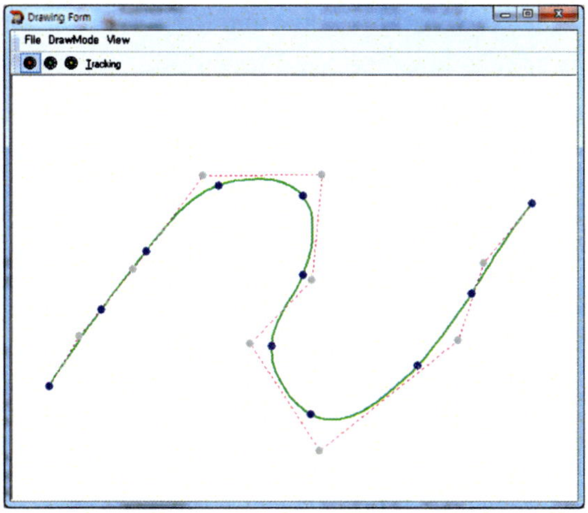

Fig. 3. The Path generated with B-spline interpolation curve equation. The dark blue points are the interpolation points and the light gray points are the control points.

3 The New Cost Function Using Survival Probability

Nearby obstacles and some unknown hazardous environment, the vehicle may encounter with some dangerous situation, such as collision. To avoid the hazardousness, the vehicle has to find not only the fast path, but also the safe path.

Finding safe path needs the new cost value on the each cell of the grid map. To evaluate this value, we suggest a new heuristic cost function using survival probability.

The hazard might hardly be detected. So survival probability consists of Gaussian distribution like this;

$$P(z) = \int_{-z}^{+z} \frac{1}{\sqrt{2\pi}} e^{-\frac{z^2}{2}} \tag{10}$$

If $P(z)$ is bigger than other cells, then that cell position is more dangerous. Each cell on the grid map has this probability value. And with this value we can define the cost function as a summation of each survival probabilities like this;

$$F_{new}(n) = \sum_{k=0}^{n} P(z) \tag{11}$$

4 Simulation and Results

4.1 B-Spline Path Tracking Simulation and Results

For the tracking the path, a PID controller was used. To define the parameters in the PID controller, the Ziegler-Nichols method was used. The transfer function of the Ziegler-Nichols method can be written as;

$$C(s) = K_p(1 + \frac{1}{T_i s} + T_d s) \tag{12}$$

Where, K_p , T_i , and T_d mean proportional constant, integration time, and differential time interval, respectively.

The result is shown in "Fig. 5". The maximum error occurred at 0.04 second with the magnitude of 0.027m. It takes 0.115 second to reduce the error within 1% of the maximum value.

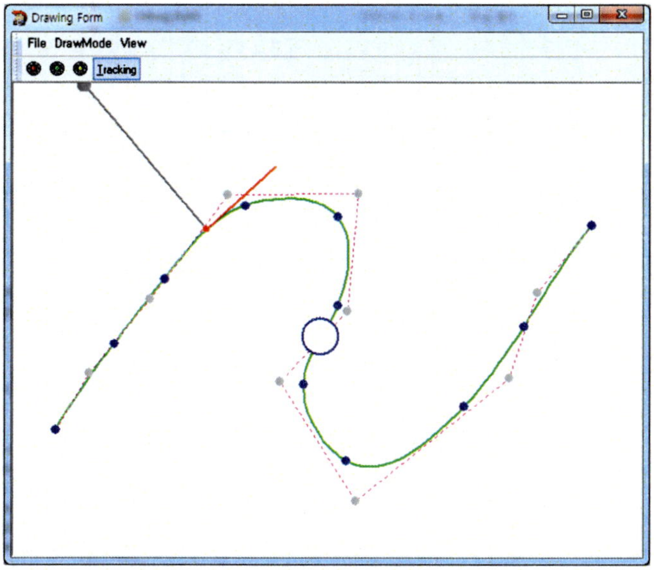

Fig. 4. The Simulation Program shows the vehicle model track the continuous path. In this Figure, the dark blue points are the interpolation points and the light gray points are the control points to make the green curve path.

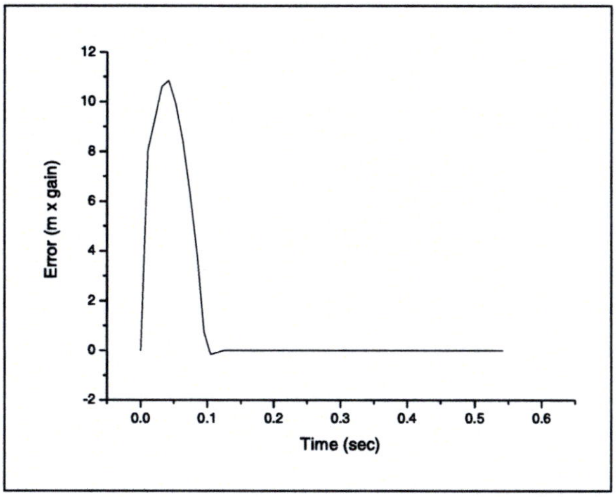

Fig. 5. This is the one of the results of path tracking simulation

4.2 Survival Probability Simulation and Results

The result of improved A* algorithm is shown in "Fig. 6". The path is different with in "Fig. 1" with survival probability. Using new cost function nearby the corner this algorithm finds more safe path than the normal A* algorithm.

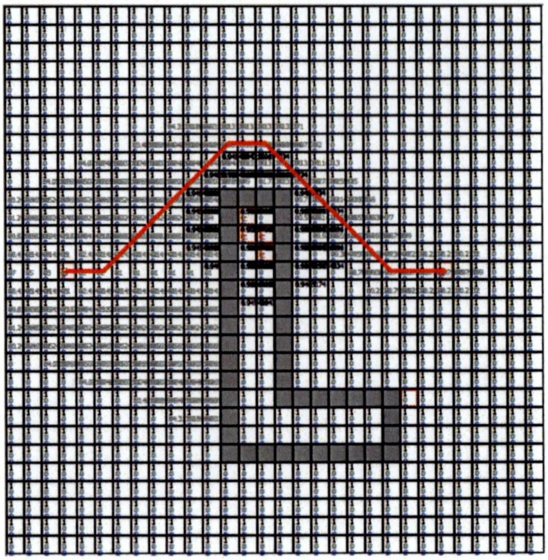

Fig. 6. The Path generated with B-spline curve equation. The dark blue points are the control points to make the green curve path.

5 Conclusion

In this paper, we suggested the improved A* algorithm based on the survival probability for the autonomous vehicle in the unknown hazardous environments. For this we developed the survival probability map and the new cost function to find the optimal path on the probability map. And the hybrid path planning algorithm is developed with the B-spline curve equation. With this algorithm, the continuous path can be generated with the way points which are from the improved A* algorithm.

At last, the simulation programs and the results were shown to verify our research.

Acknowledgements. This research was supported by the MKE(The Ministry of Knowledge Economy), Korea, under the Specialized Field Navigation/Localization Technology Research Center support program supervised by the NIPA (National IT Industry Promotion Agency) (NIPA-2010-(C7000- 1001-0004)).

References

1. Hard, P.E., Nilsson, N.J., Raphael, B.: A Formal Basis for the Heuristic Determination of Minimum Cost Paths. IEEE Transactions of Systems Science and Cybernetics ssc-4(2) (July 1968)
2. Lee, K.: Principles of CAD/CAM/CAE systems. Addison Wesley (1999)
3. Lester, P. (July 2005), `http://www.policyalmanac.org/games/aStarTutorial.htm`

4. Cox, M.G.: The Numerical Evaluation of B-splines. J. Inst. Maths Applics. 15, 95–108 (1972)
5. de Boor, C.: On Calculating with B-spline. Journal of Approx. Theory 6, 52–60 (1972)
6. Press, W.H., Teukolsky, S.A., Vetterling, W.T., Flannery, B.P.: Numerical Recipes in C++, 2nd edn. Cambridge University Press (2002)
7. Kim, M.-H., Lee, M.-C.: A Path Generation Method for Path Tracking Algorithms that use the Augmented Reality. In: International Conference on Control, Automation and Systems, pp. 1487–1490 (2010)

Gaze Control-Based Navigation Architecture for Humanoid Robots in a Dynamic Environment

Jeong-Ki Yoo and Jong-Hwan Kim

Department of Electrical Engineering, KAIST,
335 Gwahangno, Yuseong-gu, Daejeon 305-701, Republic of Korea
{jkyoo,johkim}@rit.kaist.ac.kr

Abstract. Due to the limited information from the environment using a local vision sensor, gaze control research is very important for humanoid robots. In addition, multiple objectives for navigation have interactive relationships among them. From this point of view, this paper proposes a gaze control-based navigation architecture using fuzzy integral and fuzzy measure for humanoid robots. Four criteria are employed along with their partial evaluation functions in order to determine the final gaze direction. By employing fuzzy integral approach for the global evaluation for candidate gaze directions, effective gaze control considering the interactive phenomena among criteria is accomplished and verified through a simulation using a developed simulator for HanSaRam-IX (HSR-IX).

Keywords: gaze control, Choquet fuzzy integral, preference-based selection algorithm, univector field method.

1 Introduction

Researches for humanoid robots have been started from walking pattern generation researches [1, 2]. Based on the improvement of such researches, the extended research issues such as navigation, human-robot interaction, active vision-based 3D object localization, etc. have been performed [3–5]. In particular, gaze control issues were emerged to be important for the robust navigation due to the limited capacity of visual perception of a humanoid robot [6]. In this manner, information-centered approach for gaze control was proposed [7, 8]. Even though these various issues for navigation are closely dependent each other, the integration aspect has not been fully considered so far. Based on the importance of gathering surrounding information, this paper proposes a gaze control-based navigation architecture for humanoid robots dealing with navigation in a dynamic environment. Four criteria are employed to consider the local map confidence area, self-localization uncertainty, obstacle-centered surrounding situation, and the capacity to gaze at the waypoint direction along with their corresponding partial evaluation functions and representative gaze directions. As mentioned above, user's preference and interactive relationships among criteria are considered by employing fuzzy measure and fuzzy integral [9]. The effectiveness of the proposed architecture is verified through a computer simulation using the developed simulator for the small-sized humanoid robot, HSR-IX. This paper is organized as follows. Section 2 presents the proposed gaze control-based navigation architecture, four criteria for gaze control

S. Lee et al. (Eds.): Intelligent Autonomous Systems 12, AISC 193, pp. 765–774.
springerlink.com

and their corresponding partial evaluation functions. Simulation environment and results are presented in Section 3, and concluding remarks follow in Section 4.

2 Gaze Control-Based Navigation Architecture

This section describes the gaze control-based navigation architecture for humanoid robots employing fuzzy measure and fuzzy integral for a preference-based gaze selection algorithm incorporated with the navigation architecture.

2.1 Proposed Gaze Control Architecture

In order to determine the most appropriate gaze direction considering user's preference and the interactive phenomena among criteria, various types of information such as distance and velocity of surrounding obstacles, the localization errors of surrounding obstacles and the robot itself, etc. have to be considered simultaneously. For those purpose, the gaze control-based navigation architecture, as shown in Fig. 1, is proposed for humanoid robots with four criteria considering above information and their corresponding partial evaluation functions, $E_\alpha, \alpha = lm, sl, wp, ob$. Considering a set of candidate gaze directions, X_{gd}, defined by the capacity of local vision system, the partial evaluation values over criteria are assigned to each visibility-checked candidate gaze direction in $\bar{X}_{gd}$ if the corresponding representative gaze directions are visible through the visibility check process. The user-defined preference is represented by the fuzzy measures [10]. The global evaluation of $\bar{\boldsymbol{\theta}}_i^{pt}$ in $\bar{X}_{gd}$ is calculated using the Choquet fuzzy integral of partial evaluation values corresponding to visible representative gaze directions for $\bar{\boldsymbol{\theta}}_i^{pt}$ with respect to their fuzzy measures. Then, the candidate gaze direction having the maximum global evaluation value is selected as the final gaze direction $\boldsymbol{\theta}^*$.

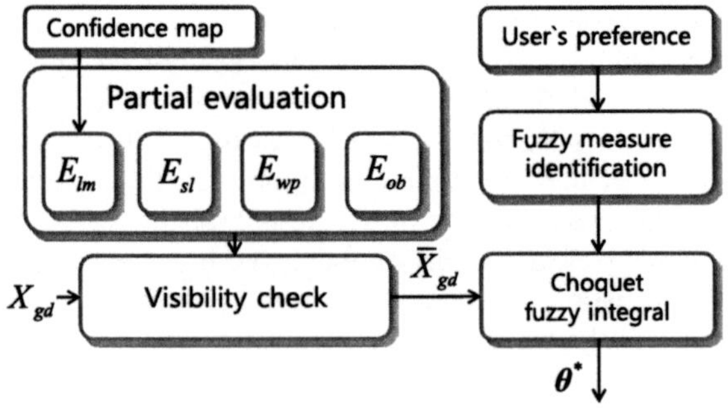

Fig. 1. Gaze control-based navigation architecture. X_{gd} is a candidate gaze direction set, and $\boldsymbol{\theta}^*$ is the final gaze direction.

2.2 Criteria for Gaze Control

The following describes the four criteria and their corresponding partial evaluation functions along with their representative gaze directions in detail.

2.2.1 Local Map Confidence Area-Based Criterion (C_{lm})

This criterion drives gaze selection tendency to un-explored or out-of-date explored directions by managing once explored gaze directions using the confidence values assigned to each grid of the local map [11–13]. Thus, the corresponding partial evaluation function estimates the necessity to gaze at unconfident area according to their confidence-based evaluations. The representative gaze direction for this criteria is generated to gaze at the mean direction of unconfident sample points.

Using the sampled points of robot-centered occupancy grid map, the partial evaluation for C_{lm} is defined as the ratio of the number of unconfident sample points N_{up} to the number of sample points N_{sp} excluding the current gaze area as follows:

$$E_{lm} = n(\{\boldsymbol{p}_i^s | K(\boldsymbol{p}_i^s) < \lambda_{cf}, \boldsymbol{p}_i^s \in R_{s-g}\}) / n(\{\boldsymbol{p}_i^s | \boldsymbol{p}_i^s \in R_{s-g}\}) \tag{1}$$

where λ_{cf} is a user-defined confidence threshold for counting unconfident sample points, $\boldsymbol{\theta}_c^{pt}$ is the current gaze direction, R_{s-g} is the sampling area excluding the current gaze area R_g^c, $\boldsymbol{p}_i^s$ denotes the ith sample point in R_{s-g}, the function $K(\cdot)$ returns the confidence value of the corresponding sample point, and the function $n(A)$ counts the number of elements in a set A.

2.2.2 Waypoint-Based Criterion (C_{wp})

The gaze control should be performed to keep the tendency of maintaining its gaze direction toward the destination direction. Thus, the representative gaze direction for this criterion is defined as the direction of waypoint. In this manner, the partial evaluation for this criterion is defined to evaluate the capacity to gaze at the waypoint direction. Using the obstacle-free distance d_{of} which is the maximum obstacle-free distance from the robot, the partial evaluation function for C_{wp} is defined as follows:

$$E_{wp}(d_{of}) = \begin{cases} d_{of}/d_{of}^{max} & \text{if } d_{of} \leq d_{of}^{max} \\ 1 & \text{otherwise} \end{cases} \tag{2}$$

2.2.3 Self-localization-Based Criterion (C_{sl})

For the localization of the robot and the detected obstacles, this paper uses the Unscented Kalman filter-based SLAM, and provides the error covariance matrix Q_r of the robot position [14]. The magnitude of Q_r, σ_r is used to represent the localization uncertainty of the robot. To reduce the localization error, the robot has to gaze at the surrounding static obstacles as often as possible to update their positions. In this manner, the representative gaze directions for this criterion are defined as the directions of surrounding static obstacles. The partial evaluation function is defined as follows:

$$E_{sl}(\sigma_r) = \begin{cases} \sigma_r/\tau_{sl} & \text{if } \sigma_r \leq \tau_{sl} \\ 1 & \text{otherwise} \end{cases} \tag{3}$$

where τ_{sl} is a user-defined threshold for the maximum magnitude of σ_r.

2.2.4 Obstacle-Based Criterion (C_{ob})

The movement of surrounding obstacles has to be considered with the top priority for the dynamic obstacle avoidance problem [6]. Since the objective of this criterion focuses on the obstacle avoidance in a dynamic environment, this criterion uses representative gaze directions gazing at surrounding obstacles including dynamic ones. To evaluate the necessity to gaze at the surrounding obstacles, the partial evaluation function for this criterion is defined as

$$E_{ob}(\sigma_{o_i}) = \begin{cases} \alpha_{o_i}/T_o & \text{if } \alpha_{o_i} \leq T_o \\ 1 & \text{otherwise} \end{cases} \tag{4}$$

with

$$\alpha_{o_i} = \sigma_{o_i}\omega_{ob}(\omega_{ro}{}^R\hat{V}_{o_i} + \omega_{vo})\|\mathbf{v}_{o_i}\|_2)/2\|{}^R\mathbf{p}_{o_i}\|_2,$$

where α_{o_i} denotes the arranged terms related to the velocity and position of the ith obstacle, $\mathbf{v}_{o_i}$ means the absolute velocity of the ith obstacle, ω_{ob} is a weight adjusting the applicable range of this partial evaluation function, and ω_{ro} and ω_{vo} are user-defined weights managing relative ratio between two velocity-related terms, ξ_{o_i} and $\|\mathbf{v}_{o_i}\|_2$.

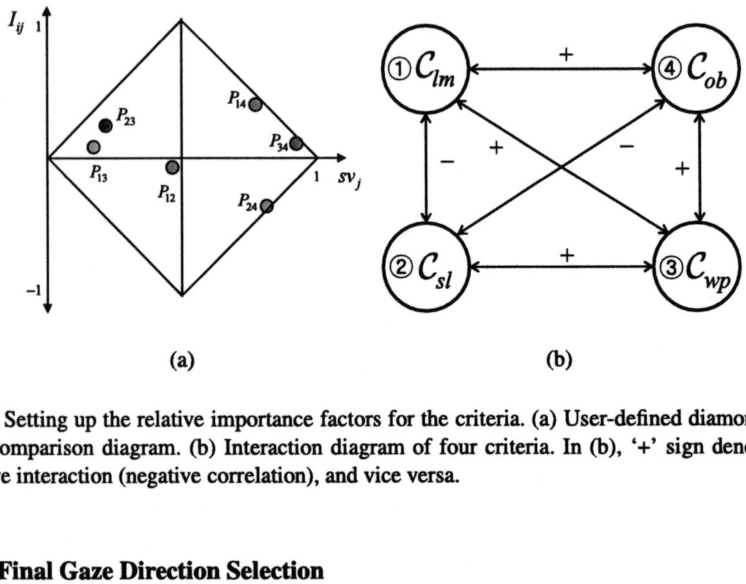

Fig. 2. Setting up the relative importance factors for the criteria. (a) User-defined diamond pairwise comparison diagram. (b) Interaction diagram of four criteria. In (b), '+' sign denotes the positive interaction (negative correlation), and vice versa.

2.3 Final Gaze Direction Selection

The global evaluation is performed for each of the candidate gaze directions $\bar{\boldsymbol{\theta}}_i^{pt}$ in $\bar{X}_{gd}$ using the determined fuzzy measures and partial evaluation values. User's preference is considered through fuzzy measures which are calculated by Grabish's graphical interpretation and identification method [9,10,15]. Fig. 2 shows user's preference for the four criteria. For the ith criterion, $\bar{C}_i, i = 1,\ldots,4$, in the criteria set $\mathbf{C}_g = \{C_{lm}, C_{sl}, C_{wp}, C_{ob}\}$, P_{ij}, $i, j = 1,\ldots,4$, $i \neq j$, means user-defined preference points comparing the ith criterion with the jth one. Table 1 is the calculated values of c_{ij} and ξ_{ij}, which are derived

Table 1. Weights of criteria pairs and their calculated interaction degree values

i vs. j	Horizontal			Vertical	
	sv_i	sv_j	c_{ij}	I_{ij}	ξ_{ij}
1 vs. 2	0.5490	0.4510	1.2174	-0.0902	0.5458
1 vs. 3	0.8358	0.1642	5.0909	0.0328	0.4707
1 vs. 4	0.2732	0.7268	0.3758	0.4917	0.0955
2 vs. 3	0.8070	0.1930	4.1818	0.2573	0.3230
2 vs. 4	0.2359	0.1930	0.3087	-0.4246	0.7436
3 vs. 4	0.0688	0.9313	0.0738	0.1238	0.1295

Table 2. Fuzzy measures

A	$g(A)$	A	$g(A)$
{}	0.0000	{4}	0.2750
{1}	0.0453	{1,4}	0.7494
{2}	0.1980	{2,4}	0.4589
{1,2}	0.2409	{1,2,4}	0.9089
{3}	0.0087	{3,4}	0.3644
{1,3}	0.0544	{1,3,4}	0.8454
{2,3}	0.2063	{2,3,4}	0.5455
{1,2,3}	0.2496	{1,2,3,4}	1.0000

from the values of sv_i and I_{ij} as described in [10]. Using this result, the fuzzy measure values are finally identified as in Table 2.

If a representative gaze direction is not located within the visible gaze area of $\bar{\boldsymbol{\theta}}_i^{pt}$, its corresponding partial evaluation value are excluded from the ones which are to be used for the global evaluation for $\bar{\boldsymbol{\theta}}_i^{pt}$. Thus, the relevant partial evaluation values for criteria are used to obtain $\bar{\boldsymbol{\theta}}_i^{pt}$ if their corresponding representative gaze directions are visible by $\bar{\boldsymbol{\theta}}_i^{pt}$ using the following visibility check function $V_j(\cdot)$:

$$V_j(\bar{\boldsymbol{\theta}}_i^{pt}, \hat{\boldsymbol{\theta}}_\alpha^{pt_j}) = \begin{cases} E_\alpha & \text{if } \bar{\boldsymbol{\theta}}_i^{pt} \in G(\hat{\boldsymbol{\theta}}_\alpha^{pt_j}) \\ 0 & \text{otherwise} \end{cases} \tag{5}$$

where the function $G(\cdot)$ generates a gaze area for the input gaze direction, and $\hat{\boldsymbol{\theta}}_\alpha^{pt_j}$ denotes the *j*th representative gaze direction for F_α. If there are multiple representative gaze directions for one criterion as the cases of C_{sl} and C_{ob}, the corresponding partial evaluation values for one criterion over $\bar{\boldsymbol{\theta}}_i^{pt}$ are added up and bounded by 1.0 using the following the normalizing function $U_\alpha(\cdot)$:

$$U_\alpha(\bar{\boldsymbol{\theta}}_i^{pt}) = \min\left(\sum_{j=1}^{N_\alpha} V_j(\bar{\boldsymbol{\theta}}_i^{pt}, \hat{\boldsymbol{\theta}}_\alpha^{pt_j}), 1.0\right), \tag{6}$$

where N_α is the number of representative gaze directions for C_α. Then, the global evaluation for $\bar{\boldsymbol{\theta}}_i^{pt}$ considering four criteria is computed by the Choquet fuzzy integral as follows:

$$E_{global}(\bar{\boldsymbol{\theta}}_i^{pt}) = \sum_{i=1}^{4} \{U_i^*(\bar{\boldsymbol{\theta}}_i^{pt}) - U_{i-1}^*(\bar{\boldsymbol{\theta}}_i^{pt})\} g(E_i^C), \tag{7}$$

where $U_i^*(\cdot)$, $i = 1, \ldots, 4$, are reordered partial evaluation values for criteria in ascending order, and $E_i^C = \{C_i^*, \ldots, C_4^*\}$ is a corresponding reordered criteria set excluding lower $i-1$ criteria. This calculation is performed for all of the candidate gaze directions in $\bar{X}_{gd}$. If there are multiple directions with the same maximum global evaluation value in E_{global}, the mean gaze direction corresponding to those maximum ones is selected as the final one $\boldsymbol{\theta}^*$.

2.4 Coordination with the Navigation Framework

The proposed gaze control-based navigation architecture uses modified univector field-based in order to avoid collisions against static and moving obstacles [16–18]. By using dynamically generated virtual obstacle and velocity modification schemes, the proposed gaze control-based navigation architecture copes with dynamic obstacle avoidance problem. Since the results of UKF-SLAM are used for the obstacle avoidance and gaze control, the whole architecture can efficiently deal with gaze control along with navigation simultaneously.

3 Simulation

This section describes the developed simulation environment and results. By using the developed simulator, the proposed architecture was verified through the simulation in a dynamic environment.

3.1 Simulation Environment

For the simulation, OpenInventor-based navigation simulator was used for the simulation as shown in Fig. 3 [2]. The simulator uses the same frame design files and walking pattern generation algorithm with ones used for the small-sized humanoid robot, HSR-IX.

3.2 Simulation Results

In this simulation, eight waypoints were traversed in order. 10 obstacles including three dynamic ones were set up. The static ones among them were located randomly, on the other hand, moving ones were assigned not to traverse the paths of the others. To show the capability of the collision avoidance scheme against moving obstacles, one of the three moving obstacles was assigned to move toward the robot at the beginning of the simulation. Fig. 5 shows the simulation result graph, where the red rectangle denotes the current Center-of-Mass (CoM) of the robot, rectangles following the footstep denote footstep history, double circles denote detected obstacles, the trapezoid in front of the robot depicts the current gaze area, and dots around the robot denote unconfident

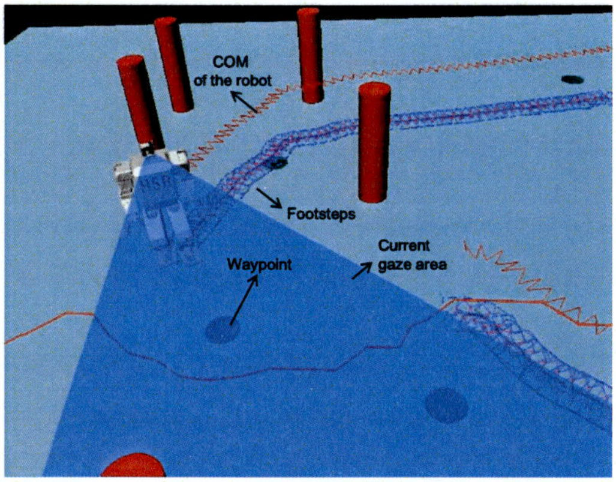

Fig. 3. HSR-IX simulator

sample points used for calculating E_{lm} in Section 2.2.1. Fig. 4 shows the true and estimated paths which are calculated by the process of SLAM. Fig. 5 shows the process of the simulation. After avoiding the first approaching moving obstacle from 19.5 sec. to 44.5 sec., The robot distributed its gaze direction toward surrounding obstacles in order to decrease the localization uncertainty of the obstacles and itself during the traverse process between waypoints. At 259.0 sec., the robot successfully detected and avoided the collision against the approaching moving obstacle by the virtue of the proposed architecture.

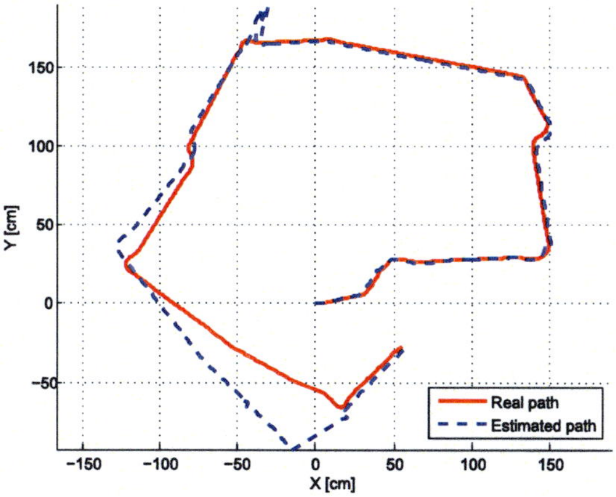

Fig. 4. Real and estimated paths of the robot

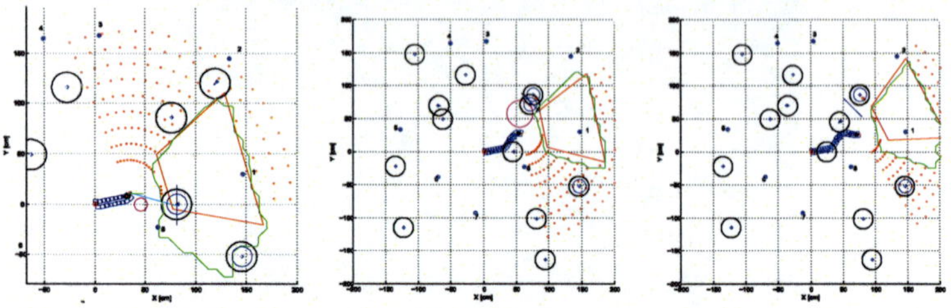

(a) 19.0 sec. (b) 35.5 sec. (c) 44.5 sec,

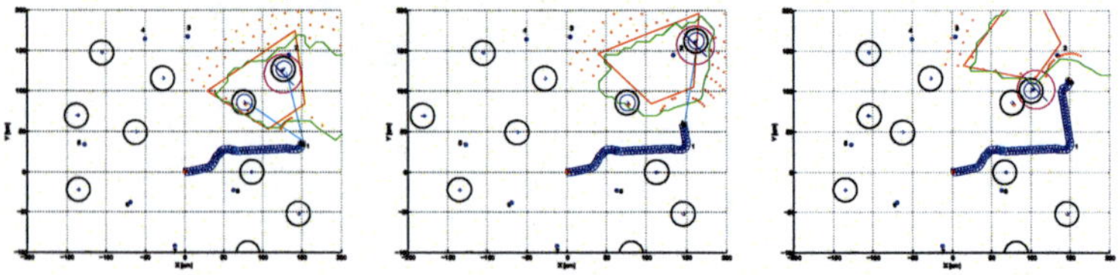

(d) 84.5 sec. (e) 96.5 sec. (f) 125.5 sec.

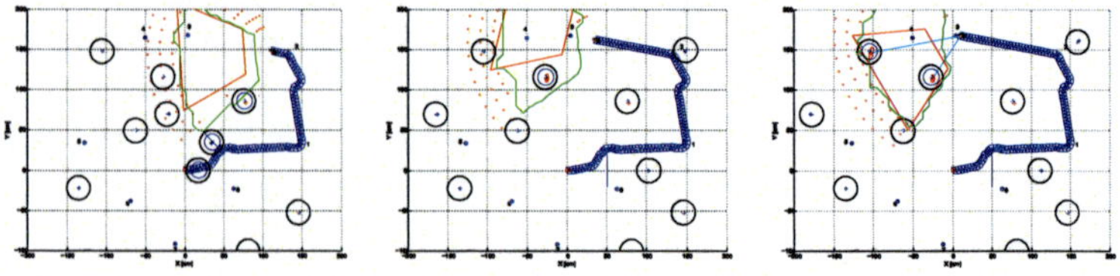

(g) 153.0 sec. (h) 192.0 sec. (i) 205.5 sec.

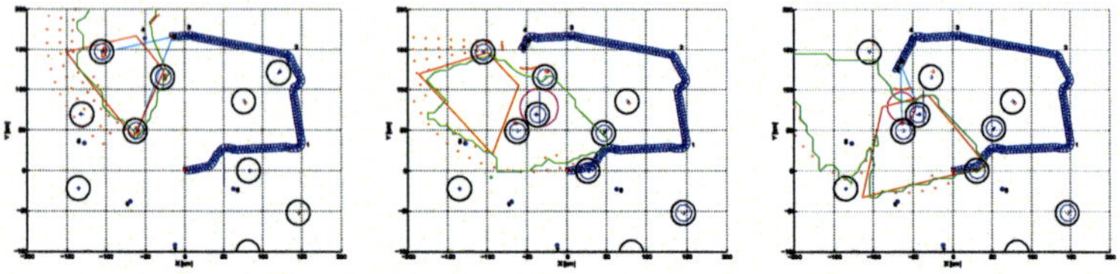

(j) 218.5 sec. (k) 224.0 sec. (l) 259.0 sec.

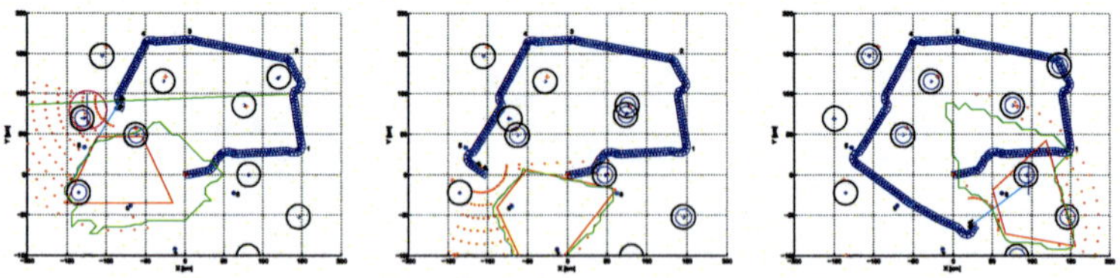

(m) 282.5 sec. (n) 334.5 sec. (o) 413.5 sec.

Fig. 5. Sequential simulation results by the proposed architecture for 413.5 ms

4 Conclusions

This paper proposed the gaze control-based navigation architecture for humanoid robots using the fuzzy integral approach. For the effective distribution of gaze direction and dynamic obstacle avoidance, four criteria were employed along with their corresponding partial evaluation functions and representative gaze directions. User's preference was considered by adopting fuzzy measures of criteria, and the global evaluations of all the candidate gaze directions were calculated through the Choquet fuzzy integral. The effectiveness of the proposed architecture was demonstrated by the simulation. In the simulation, humanoid robot could effectively dealt with obstacle avoidance and gaze control in a dynamic environment by using the proposed architecture.

Acknowledgment. This research was supported by Basic Science Research Program through the National Research Foundation of Korea (NRF) funded by the Ministry of Education, Science and Technology (2012-0000150).

References

1. Akachi, K., Kaneko, K., Kanehira, N., Ota, S., Miyamori, G., Hirata, M., Kajita, S., Kanehiro, F.: Development of humanoid robot HRP-3P. In: Proc. IEEE-RAS Int. Conf. Humanoid Robots, Tsukuba, Japan, pp. 50–55 (December 2005)
2. Yoo, J.-K., Lee, B.-J., Kim, J.-H.: Recent progress and development of the humanoid robot HanSaRam. Robotics and Autonomous Systems 57(10), 973–981 (2009)
3. Gutmann, J.-S., Fukuchi, M., Fujita, M.: 3D perception and environment map generation for humanoid robot navigation. Int. J. Robot. Res. 27(10), 1117–1134 (2008)
4. Asfour, T., Azad, P., Vahrenkamp, N., Regenstein, K., Bierbaum, A., Welke, K., Schöder, J., Dillmann, R.: Toward humanoid manipulation in human-centred environments. Robot. Auton. Syst. 56(1), 54–65 (2008)
5. Andreopoulos, A., Hasler, S., Wersing, H., Janssen, H., Tsotsos, J. K., Körner, E.: Active 3D object localization using a humanoid robot. IEEE Trans. Robot. 27(1) (February 2011)
6. Henderson, J.M.: Human gaze control during real-world scene perception. TRENDS in Cognitive Sciences 7(11), 498–504 (2003)
7. Seara, J.F., Schmidt, G.: Intelligent gaze control for vision-guided humanoid walking: methodological aspects. Robot. Auton. Syst. 48(4), 231–248 (2004)
8. Lidoris, G., Kuhnlenz, K., Wollherr, D., Buss, M.: Information-based gaze direction planning algorithm for SLAM. In: 2006 6th IEEE-RAS Int. Conf. Humanoid Robots, Genova, Italy, pp. 302–307 (December 2006)
9. Takahagi, E.: A fuzzy measure identification method by diamond pairwise comparisons and ϕ_s transformation. Fuzzy Optim. and Making 7(3), 219–232 (2008)
10. Grabisch, M.: A graphical interpretation of the Choquet integral. IEEE Trans. on Fuzzy Syst. 8(5), 627–631 (2000)
11. Posner, M.I., et al.: Inhibition of return: neural basis and function. Cogn. Neuropsychol. 2, 211–228 (1985)
12. Leek, E.C., Reppa, I., Tipper, S.P.: Inhibition of return for objects and locations in static displays. Percept. Psychophys. 65(3), 388–395 (2003)
13. Thrun, S.: Learning occupany grid maps with forward sensor models. Auton. Robots 15(2), 111–127 (2003)

14. Bailey, T.: Mobile robot localization and mapping in extensive outdoor environments. Doct. Thesis, Univ. Sydney, Australia (2003)
15. Sugeno, M.: Theory of fuzzy integrals and its applications. Doct. Thesis, Tokyo Institute of Technology (1974)
16. Kim, Y.-J., Kim, J.-H., Kim, D.-S.: Evolutionary programming-based univector field navigation method for fast mobile robots. IEEE Trans. Syst., Man, Cybern. B 31(3), 450–458 (2001)
17. Lee, B.-J., Stonier, D., Kim, Y.-D., Yoo, J.-K., Kim, J.-H.: Modifiable walking pattern of a humanoid robot by using allowable ZMP variation. IEEE Trans. Robotics 24(4), 917–925 (2008)
18. Yoo, J.-K., Kim, J.-H.: Navigation framework for humanoid robots integrating gaze control and modified-univector field method to avoid dynamic obstacles. In: Proc. of IEEE/RSJ Int. Conf. on Intelligent Robots and Systems, Taipei, Taiwan, pp. 1683–1689 (October 2010)

A Multimodal Distributed Intelligent Environment for a Safer Home

Salvatore M. Anzalone[1], Stefano Ghidoni[2], Emanuele Menegatti[2], and Enrico Pagello[2]

[1] Intelligent Robotics Laboratory,
Department of Systems Innovation,
Graduate School of Engineering Science,
Osaka University
[2] Intelligent Autonomous Systems Laboratory (IAS-Lab),
Department of Information Engineering,
Faculty of Engineering,
University of Padua
{ghidoni,emg,epv}@dei.unipd.it

Abstract. This paper presents an ambient intelligence system aimed at increasing the security of elderly people in their living environment. The proposed system processes both video and audio signals to detect dangerous events and trigger automatic warnings. The system is implemented through a software middleware for multimedia streaming and processing in which the information is processed in several, distributed software nodes. This paper focuses on the fall detection video processing node and on the sound classification and localization audio processing node. The system has been tested both in a laboratory and in a real-world scenario, demonstrating high performance levels.

1 Introduction

Distributing intelligence among agents and objects that share the same environment is the main idea of Ambient Intelligence [1,2,3]. These kinds of systems are usually based on spreading different kinds of sensors and actuators over the environment in order to capture relevant information and to act accordingly inside the ambient.

In recent years particular attention has given to applications of intelligent ambient aimed at helping elderly people living alone, maintaining their independence, while assuring a high level of assistance when needed.

Here we present a system able to detect video and/or audio events occurring in an assisted house which might be related to dangerous situation for people living inside. In the presented system several dangerous events are recognized through a distributed multimodal approach, salient elements are presented to a human operator that will doublecheck the events retrieved and will accordingly plan an emergency response.

2 System Overview

The intelligent ambient proposed in this paper is composed by several audio and video sensors spread in the environment. Data streams from these sensors are collected and

S. Lee et al. (Eds.): Intelligent Autonomous Systems 12, AISC 193, pp. 775–785.
springerlink.com

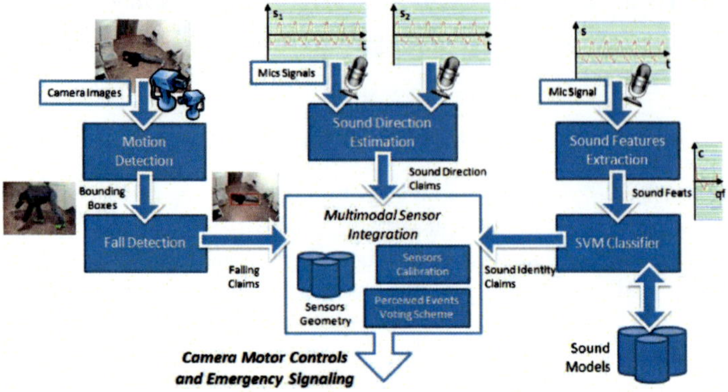

Fig. 1. Scheme of the proposed system. Cameras and microphones placed in the environment are used as input to the fall detector and sound classification modules.

processed in order to present to a remote operator or a remote control room only the relevant (or dangerous) events detected by the system. However, like in traditional CCTV surveillance systems, in case of an emergency the remote operator can have access to the raw streams of audio and video data, in order to validate or not the "dangerous event" reported by the autonomous system. Cameras of different types and microphones collaborate to acquire as much information as possible from the environment. The use of a multimodal and distributed approach implies the management of several computational units capable of cooperating and exchanging information. To achieve this modular approach a distributed framework called Network-Integrated Multimedia Middleware developed by Motama (NMM) has been used [4]. Through this framework it has been possible to develop (see Figure 1) and implement the whole system in a modular way, distributing the algorithms and their computational weights among several smaller software modules and hardware processing units. The system has been organized in modules that can exchange video and audio streams as well as custom messages. Thanks to the network support, modules can be distributed among several computers: this feature becomes crucial in order to guarantee scalability while managing a high number of sensors. Each computer has in fact limitations in terms of number of sensors that can be connected; moreover, when using USB, Firewire or analog cameras, cable length have a limited maximum length, and it becomes therefore difficult to manage a system that is distributed over large areas.

The system is capable of recognizing important events using images and sounds: sensor information are elaborated inside several NMM modules to find and track people and to detect sounds, in order to recognize dangerous events, such as falls of persons, but also screams, bangs or broken glasses. People are detected using cameras through frame differencing and background subtraction and are then tracked. Moreover, from visual information it is possible to detect falls of persons: such kind of event has been achieved by taking into account changes of tracked bounding boxes aspect ratio. According to the multimodal approach, also sound information are used to recognize events in the environment. Sound identification based on short time frequency analysis has been achieved through the classification of some known classes of dangerous sounds. In order to obtain a more exhaustive information, auditive information coming from a couple of microphones have also been used to detect the direction of the perceived sound source.

Events detected by the system can involve one or more of its sensing systems: for instance, a scream will be detected by the sound subsystem only, but a person can also scream while falling down to the ground. On the other hand, a scream and a fall can occur at the same time, but happening to different people at different positions. To understand whether two events are related or not, a data fusion algorithm is employed to associate events detected by each module. In particular, spatial information, available for both modalities used, is analyzed to understand if it is possible to associate the gathered information. This association mechanism is employed to provide a more detailed description of the events.

Three types of video sensors have been tested for surveillance purposes: common perspective cameras, Pan-Tilt-Zoom (PTZ), and omnidirectional cameras. The omnidome sensor [5], composed of an omnidirectional camera and a PTZ unit has also been employed.

3 Intelligent Perception

Audio and video sensors are used to obtain a high-level understanding of the environment, that is exploited for surveillance applications. This is what can be called an intelligent perception system, that is, a system capable of understanding what is happening in the environment and to properly react to specific stimuli.

3.1 Fall Detection

One event that deserves special care with elderly people is the fall: this is a dangerous situations, because after a fall elderly could be unable to move, even for pushing a button on a device. For coping with this kind of problems, smarter devices exist, capable of automatically recognizing falls: they are usually specifically designed for a particular event, and do not require any input from the assisted person. Devices for fall detection commonly available are small boxes equipped with inertial sensors that are able to detect a brief acceleration like in case of a fall [6]; the same application has also been implemented in common mobile phones: this solution exploits an object that people commonly owns and carry in their pocket, without the need for a specific device [7]. All these discussed systems, however, include a device that people need to wear or have in the pocket in order to be assisted, and this might be distressing especially for elderly people.

A different way for addressing the same problem is to use cameras and image processing for detecting falls. Computer vision offers the clear advantage of freeing assisted people from wearing devices: in other words, these kinds of systems assure a high reliability without requiring any kind of people cooperation. Moreover, the same acquisition system could be used for detecting multiple events, each of which would require a specific processing, but not a specific device; since it is a computer and not a human operator that analyzes all images, privacy issues related at installing a camera are solved by destroying images after their processing.

The topic of fall detection using computer vision has already been addressed: the main approach is based on motion analysis, since a fall is a movement that generates a specific motion pattern, as pointed out in [8]. Fall detectors can be based on either motion analysis of the whole body, or focusing on some parts only, as it is the case of the head in [9], or, finally, exploiting features like gradients [10]. In all cases, the aim of the system is to distinguish between a fall and a normal motion, like sitting on a chair or on a sofa, or going to bed. For this reason, both motion direction and speed should be analyzed.

The developed fall detection module is aimed at detecting the particular motion cues of a falling person. The module exploits a motion detection system based on a mixed technique relying on both background subtraction and frame differencing. Such combination is used to compensate the drawbacks they both have; in particular, frame differencing turns out to be particularly efficient when coping with illumination changes, as discussed in [5].

The fall detection algorithm can be executed indifferently on images provided by both perspective cameras or omnidirectional ones, after the unwarping process. The algorithm considers all blobs found by the motion detector, and performs a further processing based on three factors: aspect ratio, gradient and movements during and after a possible fall. Aspect ratio provides information about the shape of the moving object: such parameter is tracked over time in order to detect possible changes in orientation.

The second parameter considered by the fall detector is the gradient. For performing gradient analysis, both horizontal and vertical Sobel edge detectors are exploited for evaluating corresponding edge images I_x and I_y. Two features, called horizontal (G_x) and vertical (G_y) gradients are then evaluated by summing all pixel values inside the considered motion blob in the two images:

$$G_{\mathrm{dir}} = \sum_{i=0}^{W} \sum_{j=0}^{H} I_{\mathrm{dir}}(i,j)\,, \tag{1}$$

where dir indicates the direction (either x or y), while W and H are the width and height of the bounding box containing the analyzed motion blob. This could seem similar to aspect ratio, but it represents a different approach, because gradient analysis considers the content of the bounding box, neglecting the ratio between width and height: an object can present a strong horizontal gradient even though it is vertically placed, and vice-versa. Horizontal and vertical gradients provide information because people that are walking or standing have a vertical component that is sensibly stronger with respect to the horizontal one, while fallen people have their main component in the horizontal direction. The fall detector then compares the two gradients with their evolution over time, for understanding if the moving blob corresponds to a person that is falling.

3.2 Sound Identification

Sound information is also important to reveal dangerous events. Audio is useful especially when other sensors, such as video, fail to correctly detect the events. For example, when objects are occluded or in the dark, audio sensors can be more appropriate in detecting a "noisy" event. Conceptually, there are many events which can be better detected using audio rather than other sensors, like screams or gunshots, broken glasses, barking dogs, etc. Finally, audio sensors are a cost accessible resource.

Traditional approaches to sound identification are based on the extraction and classification of the signal features obtained through a spectral analysis of short time slices of the audio wave. Mel-Frequency Cepstral Coefficients (MFCCs) are the most common features representation for audio recognition [11]. Other well known sound features used to classify audio sources are the Linear-Log Frequency Cepstral Coefficients (LLFCC), or the Perceptual Linear Prediction (PLP) coefficients, or Noise Reduction Auditive Features (NRAF) [12]. A learning system is used to classify features according to the source. Different kind of standard classification methods have been used to classify these features, such as hierarchical sets of cascaded Gaussian Mixture Models (GMM), or Support Vector Machine (SVM) [13].

Focusing on the problem of sound identification inside an intelligent environment, some assumptions can be done: first of all a closed set of "dangerous" sounds will be identified, such as screams, shots, bangs or breaking glasses. Moreover, the system will work in real-time. According to these constraints, sound classification will be performed through a short time analysis of the perceived sound, whose spectrum will be processed to extract relevant features that can be used as "signature" of the sound class. Mel-Frequency Cepstrals Coefficients (MFCC) are used as fast, real-time and reliable sound features.

The sound recognition system is articulated in several steps: in the first one, sound features, the MFCC, are extracted; then the sound identity claim is calculated using a classification system based on these features, previously trained.

More in detail, during the features extraction step short frames of 30 milliseconds of the audio signal are captured from the microphone, collected and filtered through a raised cosine Hamming window to prevent aliasing. Each window is overlapped with the following one by 10 milliseconds. The frequency spectrum is then calculated, applying the FFT algorithm. At last, MFCC are calculated: the module in decibel of the frequency spectrum is multiplied by a Mel filter bank and then, finally, the Discrete Cosine Transform is applied. The Mel filter bank is a set of band-pass filters localized according to the Mel scale, an auditory scale capable to model the response of the human ear. The Discrete Cosine Transform is applied to the results of the Mel filter bank, to obtain the Mel-Cepstral vectors, that are capable of representing the short-term power spectrum of a sound, that can in turn be used as a feature of the current sound.

Features are classified in a second stage: during an offline training step, cepstral vectors are collected and classified to build a model for each sound; in the identification phase the cepstral vectors are compared to these models in order to find an identity claim. The classification process of cepstral vectors is carried out by the Support Vector Machine technology. SVM is a two-class classifier: to handle different classes the problem has been reformulated according to a strategy "One vs. All", in which for each target identity all remaining non-target information are modeled as a "world" class, and then the target model is trained. This operation is performed for each identity, in order to have a complete set of sound models.

In order to prevent fluctuations and to obtain more stable performances, a final refinement stage has been applied: through a statistical analysis of the last 50 identity claims, the most frequent identity is claimed as the identity of the current sound perceived.

3.3 Sound Localization

In order to have a complete knowledge of the environment, detecting the identity of a perceived sound is not enough: it is important to know its direction to allow the system react in a correct way. Standard methodologies use two or more microphones spread over the environment in order to estimate the time delay of arrival of the sound wave. According to this delay and to the geometrical position of the microphones, it is possible to accurately calculate not only the direction but also the three-dimensional position of the sound source. Commonly used techniques to calculate the time delay are based on the Generalized Cross-Correlation (GCC) function between couples of signals. In particular, in order to take advantage of redundant information provided by multiple sensors, the Multichannel Cross-Correlation Coefficient (MCCC) method can be a good choice to calculate the sound source position [14].

Sound source localization has been achieved through the use of couples of microphones placed in the environment. Each couple will be able to perceive a single sound source. Knowing the distance between the two microphones, the audio source is located through the estimation of the time delay between the waveforms perceived by each microphone (TDOA):

$$\theta = \arcsin\left(\frac{\tau c}{d}\right), \tag{2}$$

where θ is the incoming direction of the sound, d is the inter-aural distance, c is the sound speed and t is the TDOA. The TDOA is evaluated using the Generalized Cross-Correlation. The GCC in the frequency domain is defined as:

$$r_{x_1x_2}^{\text{GCC}}[p] = \sum_{f=0}^{N-1} \Psi[f] S_{x_1x_2}[f] e^{\frac{j2\pi pf}{L}}, \tag{3}$$

where N is the number of samples, $\Psi(f)$ is the frequency domain weighting function, and the cross-spectrum of the two input signals is defined as:

$$S_{X_1X_2}(f) = E\{X_1(f)X_2^*(f)\}, \tag{4}$$

were X1 and X2 are the Discrete Fourier Transform of the input signals and the * denotes the complex conjugate. GCC shows a peak in correspondence of the time delay, while minimizes the influences of uncorrelated noise. In order to mainly use the phase information during the computation of the GCC, the PHAT weighting function (Phase Transform Filter) has been used to normalize the amplitude of the spectral density of the two signals:

$$\Psi_{\text{PHAT}}[f] = \frac{1}{|S_{x_1x_2}[f]|}. \tag{5}$$

The GCC is capable to show consistent information when the sound signal changes over time without any pseudo-periodic feature: otherwise, in this last case, the performances are drastically reduced.

3.4 Sensors Fusion

A system composed of several sensors distributed over the environment enables the monitoring of complex scenarios that would be impossible to keep under control by using a single sensor, such as several rooms in a house, large areas, parking lots. Moreover, the use of different modalities allows the system to capture a wide set of events that the use of a unique modality could be not able to retrieve. But, in order to use in an efficient way these modalities, the system needs to put in relation the events detected at the same time, so that once an event is recognized in one data flow, it is possible to evaluate where it will appear in the others: this can be performed by collecting the geometrical relationships between sensors, obtaining a multi-sensor geometrical binding.

Geometrical information enable the system to easily put in relation information coming from different modalities. It is possible to use a logical disjunction function (Logical OR) to use these information in an efficient way: if some sensors detect events that are near in the time and near in the space then these events may be related, otherwise if only a single modality detects an event, it can drive the second one to focus on it. An instance of simultaneous multimodal detection is the localization of a scream that is confirmed by the detection at the same time of a falling person localized at the same position. On the other side, if only a single modality is able to detect an event, the geometrical descriptors enable to retrieve more information about it: the localization by the audio processing system of a broken glass or of a barking dog can trigger and guide a PTZ camera towards the sound source.

Such kind of geometrical information is particularly useful while dealing with PTZ cameras: they are a natural choice for acquiring highly detailed shots of the environment, but it is impossible to get an overall understanding of the environment from their images. Cooperation can overcome this problem by letting PTZ cameras be controlled by other modules: in this case the geometric information of all the sensors present in the environment is used to point the PTZ cameras on the direction of the perceived event, focusing on it while maintaining the attention to the rest of the environment, through the other sensors.

With this information fusion approach, all the information captured by the system will be used to trigger a warning message that will be sent to an human operator in a remote control room: he will be able to override the system and, using all the provided sensors, to double check the harmful conditions.

4 Experimental Results

Tests of the system have been performed in a laboratory and in a real apartment. Tests conduced in the laboratory Figure 2(a) used a PTZ camera, a Omnidome(R) camera, two perspective cameras and two microphones. More extensive tests have been performed also in a real house for elderly people, jointly with industrial partners in a complex infrastructure, in the context of the project Safe Home. Industrial partners provided several domotic nodes to control air conditioning systems, lights and windows. A sketch of the system is shown in figure 2(b): two perspective Axis M1103 IP cameras have been placed in the kitchen and in the bedroom, respectively. In the Safe Home environment the system has been tested during 10 days of uninterrupted running.

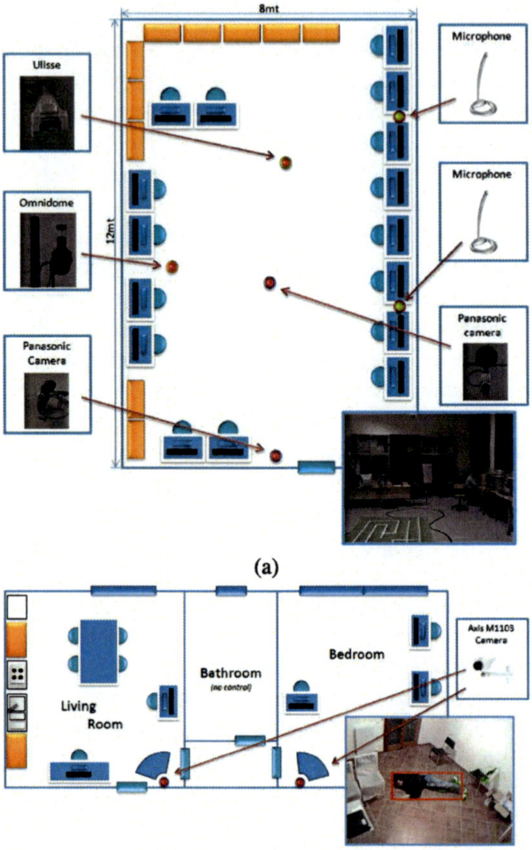

(a)

(b)

Fig. 2. The two testing environments used for experiments: a laboratory equipped with a number of cameras (a) and a real apartment (b)

When a human enters the intelligent environment, the omnidirectional and prospective cameras perceive him and the tracking starts. Each detected person is associated to a numerical identifier and to a bounding box: cameras monitor the variation on the dimension of the bounding boxes in order to detect falls. If one camera reveals a fall of a person, a warning message is issued to a human operator in a remote control room, which in this way can simultaneously monitor several dozens of smart environments. At the same time, if it is connected to the intelligent environment, the PTZ cameras will receive a command to point at the direction of the falling person. The behaviour of the system in response to auditive stimuli is similar. The sound detection system tries to isolate sounds from silence; then, if a sound has been revealed, it is identified to recognize dangerous events, such a scream, a shoot or a broken glass, beside natural human voice.

4.1 Evaluation

Some experiments focused on situations with just one person in the scene, while other were performed with several people. Event-based performance has been measured: test sequences have been divided into events, that are a set of consecutive frames in which one or more people are either walking or falling. In 21 events a person falls, while in 35 no falls happen.

For measuring performance, true positives (TP) and negatives (TN) have been measured, together with false positives (FP) and negatives (FN). All these statistics are event-based: videos exploited for performance measurements have been divided into sub-sequences, each one representing a single event to be recognized. During the experiments, 19 true positives and 28 true negatives were measured, while errors involved 7 false positives and 2 false negatives. This leads to an overall accuracy of 85%, and a precision of 90% (see Table 1). Errors were mainly due to instabilities in the patterns found by the motion detector, and can therefore be corrected by smoothing such motion information.

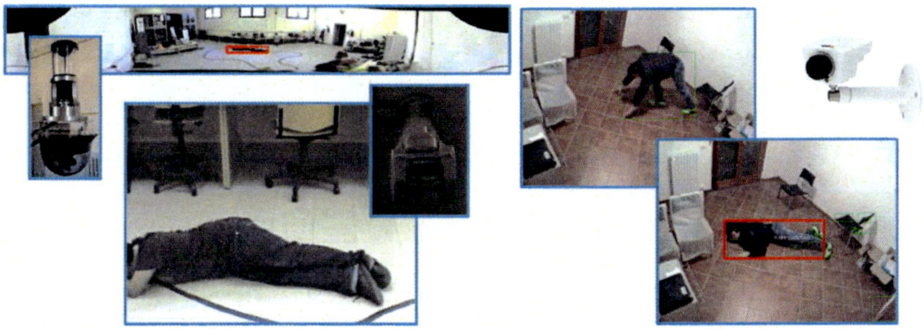

(a) (b)

Fig. 3. Two experiments: in the lab, the omnidirectional camera detects a fall, then the Ulisse PTZ camera focus on it (a); in the real apartment, the prespective camera tracks a person then detects a fall (b)

The sound detection system has been trained using a total of 60 seconds pre-recorded sound data, distributed between screams, bangs, broken glasses, and dog barking samples, together with a special "background" class of human voices in unharmful situations. Experiments for evaluating performance of the sound recognition system reported 83% of true positives and 94% true negatives with 16% of false negatives and 5% of false positives, showing a correct perception of sound events in the 85% of the cases. The system gathered an overall accuracy of 88% and a precision of 83% on its measurements (see Table 1).

However, thanks to the multimodal approach, the complete system has been able to outperform the performances of the single modalities, giving a correct evaluation of true positives and negatives of 98%: when the system is not able to recognize an event through one channel, a second one will recovery the situation, improving the total performances of the whole system. At last, the tracking bounding boxes and the reliable

Table 1. The confusion matrices of (a) the people falling detection system and of (b) the sound identification system

(a) Fall Detection

Actual classes	Predicted classes: Falling	Predicted classes: Not Falling
Falling	0.90	0.09
Not Falling	0.2	0.8

(b) Sound Identification

Actual classes	Predicted classes: Scream	Gunshot	Glass	Bark
Scream	0.85	0.11	0.04	0.00
Gunshot	0.08	0.85	0.07	0.00
Glass	0.02	0.02	0.96	0.00
Bark	0.06	0.27	0.00	0.67

sound localization estimation assured in the 83% of the cases a correct framing of the PTZ cameras to the event sources.

5 Conclusion and Future Work

In this paper, an intelligent perception system has been presented, capable of performing surveillance tasks based on multimodal (audio and video) sensors, that make it possible to perceive different kinds of features on the environment. The system was successfully used during the project Safe Home as a part of a domotic infrastructure provided by industrial partners, in a real environment, to provide assistance to elderly people, by detecting people falls and classifying sounds connected with dangerous situations. The use of the visual and auditive modalities assured an improvement of the performances respect to single channels. The system has been also capable of coordinating active sensors (PTZ cameras) in order to react to perceived stimuli, for instance by letting the active cameras framing a region where a sound triggering an alarm has been perceived.

Future efforts will aim at improving already available modules: regarding the fall detector, a people detector algorithm could be used together with the current motion detector, to strengthen the recognition – but the motion detector will anyway be needed, since motion information is crucial to detect fall events. Moreover, sound processing will be improved by adding a people identification module, capable of verifying the identity of the people in the environment, detecting potential intrusion by unauthorized people.

Acknowledgments. Authors would like to express their gratitute to Riccardo Levorato and Alberto Salamone. This project was partially founded by Regione Veneto Cod. progetto: 2105/201/5/2215/2009 approved with DGR n. 2215, 21/07/2009.

References

1. Aarts, E., de Ruyter, B.: New research perspectives on ambient intelligence. Journal of Ambient Intelligence and Smart Environments 1(1), 5–14 (2009)
2. Kofod-Petersen, A., Aamodt, A.: Case-based reasoning for situation-aware ambient intelligence: A hospital ward evaluation study. In: Case-Based Reasoning Research and Development, pp. 450–464 (2009)
3. Nakashima, H., Aghajan, H., Augusto, J.C.: Handbook of Ambient Intelligence and Smart Environments. Springer (2010)
4. Lohse, M., Repplinger, M., Slusallek, P.: Network-Integrated Multimedia Middleware. Services, and Applications, Department of Computer Science, Saarland University, Germany, Diss. (2005)
5. Ghidoni, S., Pretto, A., Menegatti, E.: Cooperative tracking of moving objects and face detection with a dual camera sensor. In: 2010 IEEE International Conference on Robotics and Automation (ICRA), pp. 2568–2573 (May 2010)
6. Lord, C.J., Colvin, D.P.: Falls in the elderly: Detection and assessment. In: Proceedings of the Annual International Conference of the IEEE. Engineering in Medicine and Biology Society, vol. 13, pp. 1938–1939 (October 1991)
7. Zhang, T., Wang, J., Liu, P., Hou, J.: Fall detection by embedding an accelerometer in cellphone and using kfd algorithm. International Journal of Computer Science and Network Security 6(10), 277–284 (2006)
8. Noury, N., Fleury, A., Rumeau, P., Bourke, A.K., Laighin, G.O., Rialle, V., Lundy, J.E.: Fall detection – principles and methods. In: 29th Annual International Conference of the IEEE Engineering in Medicine and Biology Society, EMBS 2007, pp. 1663–1666 (August 2007)
9. Nait-Charif, H., McKenna, S.J.: Activity summarisation and fall detection in a supportive home environment. In: Proceedings of the 17th International Conference on Pattern Recognition, ICPR 2004, vol. 4, pp. 323–326 (August 2004)
10. Khan, M.J., Habib, H.A.: Video analytic for fall detection from shape features and motion gradients. Lecture Notes in Engineering and Computer Science, vol. 2179, pp. 1311–1316 (2009)
11. Bimbot, F., Bonastre, J.F., Fredouille, C., Gravier, G., Magrin-Chagnolleau, I., Meignier, S., Merlin, T., Ortega-García, J., Petrovska-Delacrétaz, D., Reynolds, D.A.: A tutorial on textindependent speaker verification. EURASIP Journal on Applied Signal Processing, 430–451 (2004)
12. Ravindran, S., Anderson, D.V., Slaney, M.: Improving the noise-robustness of melfrequency cepstral coefficients for speech processing. Reconstruction 12, 14 (2006)
13. Steinwart, I., Christmann, A.: Support vector machines. Springer (2008)
14. Salvati, D., Rodà, A., Canazza, S., Foresti, G.L.: A real-time system for multiple acoustic sources localization based on isp comparison. In: Pomberger, H., Zotter, F., Sontacchi, A. (eds.) Proc. of the 13th Int. Conference on Digital Audio Effects - DAFx 2010, Graz, Austria, pp. 201–208 (September 2010)

Comparison between Photo Interrupter and Giant Magnetoresistive Sensor for Auto Focusing System in the Digital Camera

Sakura Sikander, Han Nam Lee, and Hee-Je Kim

Dept. of Electrical Engineering, Pusan National University
Busan 609-735, Republic of Korea
sakura.sikander@gmail.com, heeje@pusan.ac.kr

Abstract. Photo interrupter (PI) sensors are used to focus the lens in the optical devices such as digital camera, close circuit camera etc. It is cheap and easy to control. Despite of the advantages it has some limitations such as it is difficult to assemble, needs small tolerance and lacks high speed and accuracy. Magnetoresistive (MR) especially giant magnetoresistive (GMR) sensor provides the solution of PI sensor's weak points. Furthermore, using GMR sensor shows better performances than PI sensor for the focusing system.

Keywords: photo interrupter, giant magnetoresistive sensor, auto focus, digital camera.

1 Introduction

Most of the recent mobile devices like cell phones, laptops, digital cameras are designed for better speed. It doesn't matter how big or what purpose it is used for, it should be faster, precise, reliable and cheap. To keep pace with the market and customer need the production method also should be simple and cost effective. All these devices are based on sensors which is one of the key components that need continuous modification.

In optical applications like digital camera, close circuit camera the key component which needs continuous modification is the auto focus (AF) system to make the camera faster. So the sensor involved in AF is very important. AF means proper positioning of the lens so that the focus is sharp. Thus the key task of the sensor is to detect the position of the lens and control the lens movement. Normally for focusing photo interrupter (PI) sensor is used.

PI sensor has three tiny wings inside in it which spins between the emitter and detector inside the device. The detectors count how many times the wings interrupt the light and detect the position of the lens for AF. The PI sensor is easy to design and also cheaper than other sensors. But one of the major limitations is making precise molds of the three wings which have a significant impact on the precession of the position detection. A little tolerance in such a small scale wing can incorporate error in the sensor. Thus while assembling the sensor precession is interrupted due to this

S. Lee et al. (Eds.): Intelligent Autonomous Systems 12, AISC 193, pp. 787–792.
springerlink.com

reason. Another issue is when surface mounting the sensor in the reflow machine the precise positioning of the sensor isn't possible because the solder is liquid for certain period of time. So improper positioning of the sensor causes improper wing alignment which leads to precession problem. So it is very important to choose such a sensor which is independent of mounting, position and movement.

In this paper to find a solution of these drawbacks we have compared several sensors including PI sensor and concluded finding a sensor with superior performance.

2 Selection of the Sensor

In this paper we will concentrate on the AF mechanism of the digital camera which has moving barrels controlled with motors to focus the lens. So controlling the barrel position with the sensor properly is our concern. To overcome the earlier mentioned draw backs in the PI sensor we will focus on a non contact sensor. So we are interested in the sensor that uses magnetic signals. In that case hall sensor and magnetoresistive sensor is a good choice.

As shown in figure 1 we can see that hall sensor detects vertical magnetic component where else the magnetoresistive (MR) sensor detects the horizontal component. So if the target object is not in vertical position then the sensor will not detect properly and will create a weaker signal than normal position which makes it a poor sensing device compared with the MR sensor. So assembly of the hall sensor will cause loss of accuracy and effectiveness of the sensor because the sensor barrel is round in shape where the magnetic strip will be winded around which means the magnetic signal may not be vertical properly and also improper assembly will affect the signal.

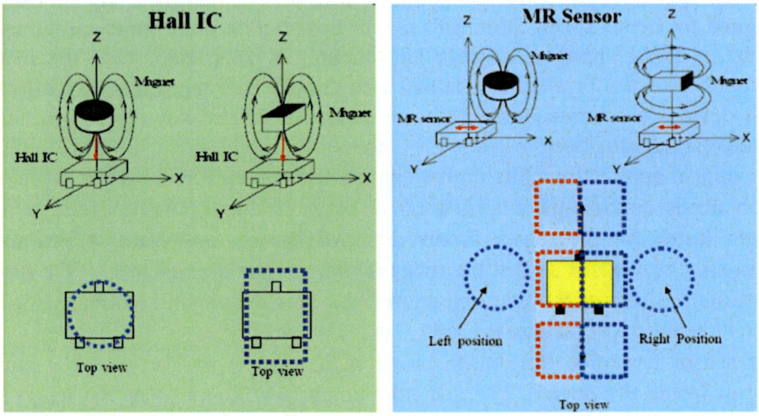

Fig. 1. Magnetic field orientation of a Hall sensor (*left*) and a MR sensor (*right*)

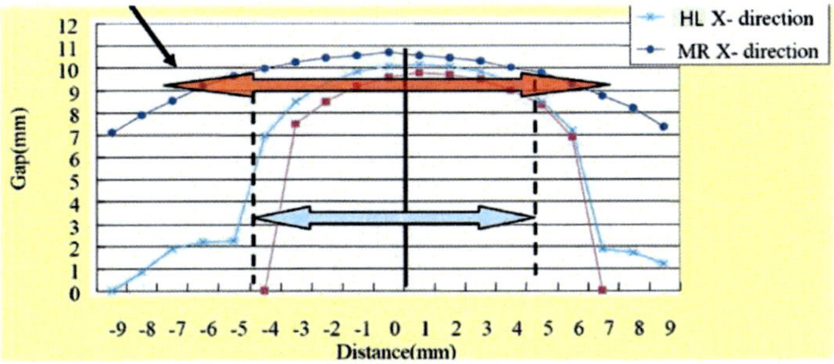

Fig. 2. Range of the sensors

Figure 2 shows the range of the two sensors. In terms of range also the MR sensor has a significant advantage which means MR sensor has larger tolerance for assembling. Compared to sensitivity and signal level giant magnetoresistive (GMR) sensor has the same principle but has higher sensitivity than the MR sensor [1]. GMR sensor has more effect for thin multi-layer structures [2]. From formula (1) we can find the MR ratio where R_{max} and R_{min} are the maximum and minimum magneto resistance respectively. GMR produces MR ratio of 20-50%, which means that 20-50% of the obtained signal can be collected from the sensor [3] which is very satisfactory.

$$MR_{Ratio} = \frac{R_{max} - R_{min}}{R_{min}} \tag{1}$$

3 Results

A Magnetic sheet is installed inside the lens barrel for digital camera and using the barrels spinning the GMR sensor detects the magnetic field from the magnetic sheet as shown in figure 3.

For experiment GMR sensor from "Hitachi Metal" was used [4], which generates a pair of signals. The signals are same except the phases are different as shown in figure 4. The bridge circuit's opposite polarity makes the difference between A and B's phase. If we multiply A and B better resolution can be obtained. Here the pitch distance was set 20um for our experiment and was multiplied with phase difference 0.25 which improved the resolution to 5um. In case of Hitachi Metal GMR the resolution is up to 1.56um. For a PI sensor the resolution is 1mm at 120 rpm. So the resolution can be improved up to 10 to 700 times, compared with the PI sensor.

A circuit was designed to measure the signal for the experiment using formula (2) where $V_{A\text{-phase}}$ and $V_{B\text{-Phase}}$ are phase voltage of A and B signals respectively.

$$\theta = \tan^{-1}\left(V_{B-Phase} / V_{A-Phase} \right) \tag{2}$$

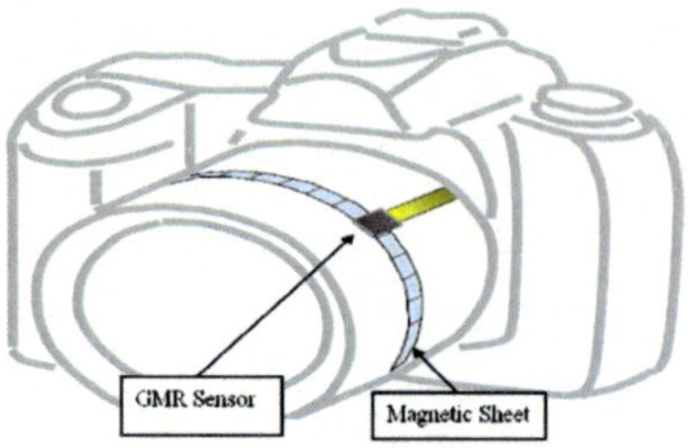

Fig. 3. Position of the GMR sensor and magnetic sheet

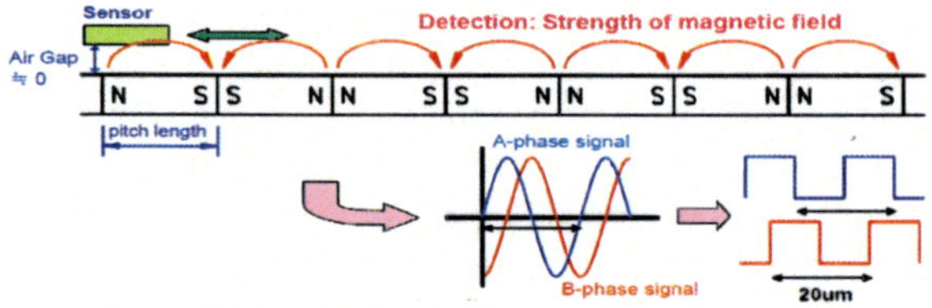

Fig. 4. GMR signal with magnetic sheet

The output of the signal is shown in figure 5 measured using the circuit we have built. The camera lens barrel rotated at a speed of 120 rpm with 20um pitch and found the signal. At 12 KHz the amplitude dropped 25% to 45mV, which is still larger than the PI sensor signal at this speed.

One of the most important factors of a sensor is its repeatability, durability and accuracy. Using the test set in figure 6 we can find the results shown in figure 7 from which we can see that the repeatability and accuracy is very nice.

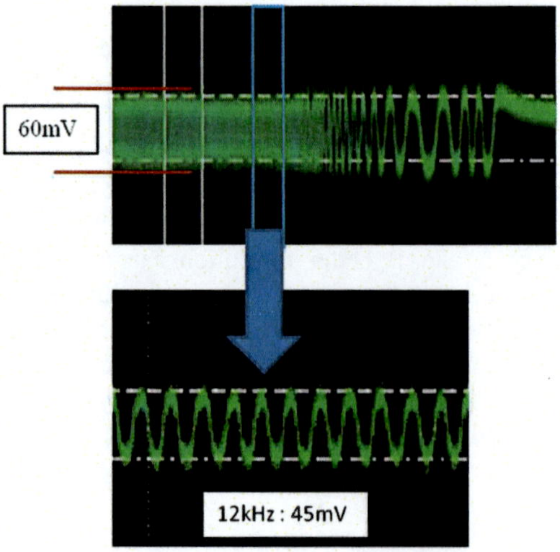

Fig. 5. GMR sensor output frequency

In figure 8 we have collected the signal from GMR sensor after certain period of operation. The GMR sensor was tested for up to 1000,000 cycles. After 1000,000 cycles still the output signal was reasonable where else for PI sensor the cycle is around 200,000. So it is evident that reliability and durability is superior to the PI sensor. The GMR sensors lifetime was calculated using the Arrhenius equation and found 39,000 hours.

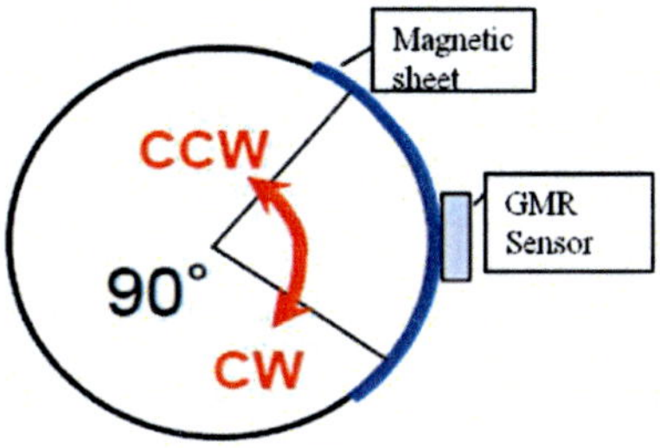

Fig. 6. GMR sensor test setup

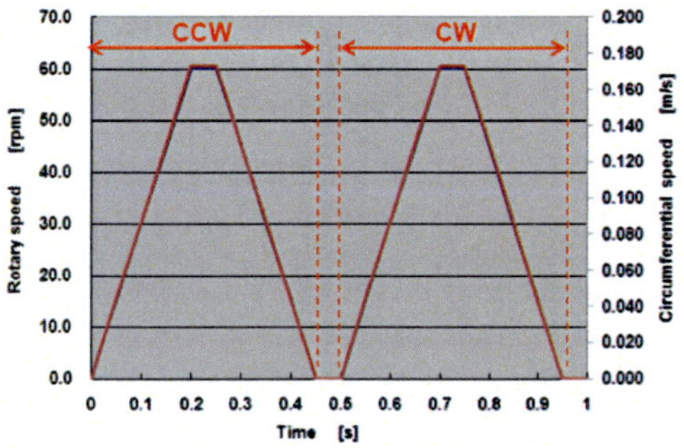

Fig. 7. Rotary speed and cycle time of the GMR sensor

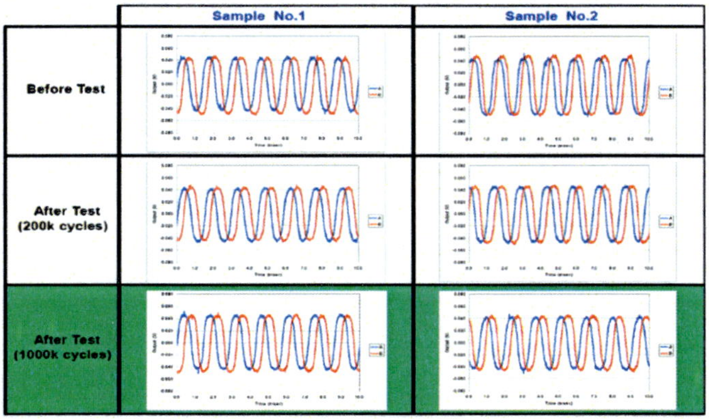

Fig. 8. Result of the durability and reliability test of the GMR sensor

4 Conclusion

For AF system the sensor has a key role. All the test results show that the GMR sensor can play far more superior role than the PI sensor for AF system.

The GMR sensor has larger range which will make it less sensitive to the assembly error and give higher accuracy. Also small defects in the assembly will not cause any error in the sensing.GMR sensor has a higher reliability than the PI sensor and the life time is also significantly higher than the PI sensor. Though the PI sensor is cheaper than the GMR sensor but due to higher life time and reliability the total cost will decrease. Compared to the PI sensor the GMR sensor is a better solution for digital cameras AF mechanism.

Acknowledgements. This research was supported by the MKE(The Ministry of Knowledge Economy), Korea, under the Human Resources Development Program for robotics support program supervised by the NIPA(National IT Industry Promotion Agency) (NIPA-2011-C7000-1001-0009).

References

1. Smith, C.H., Schneider, R.W.: Low-Field Magnetic Sensing with GMR Sensors. Sensors EXPO (1999)
2. Baibich, M.N., Broto, J.M., Fert, A., Nguyen Van Dau, F., Petroff, F.: Giant Magnetoresistance of (001)Fe/(001)Cr Magnetic Superlattices. Phys. Rev. Lett. 61, 2472–2475 (1988)
3. Jander, A., Smith, C., Schneider, R.: Magnetoresistive Sensors for Nondestructive Evaluation. In: SPIE International Symposium, vol. 5770, pp. 1–13 (2005)
4. Data Specification from Hitachi Metals (2011)

Natural Terrain Detection and SLAM Using LIDAR for UGV

Kuk Cho[1], SeungHo Baeg[2], and Sangdeok Park[2]

[1] Dept. of Intelligence Robot Engineering, University of Science and Technology
176 Gajung-dong, 217 Gajungro Yuseong-gu Deajeon, 305-350, Korea
googi33@ust.ac.kr
[2] Dept. of Applied Robot Technology, KITECH
1271-18, Sa-3 dong, Sangrok-gu, Ansan, 426-791, Korea
{shbaeg,sdpark}@kitech.re.kr

Abstract. This paper describes a natural terrain detection algorithm and a SLAM algorithm using a LIDAR sensor for an unmanned ground vehicle. We describe how features are detected from natural terrain, and then we localize the vehicle's position and compose a map with the detected features. The LIDAR equipped on the experimental vehicle to scan natural terrain. The scan data is included many kinds of intrinsic disturbance on uneven terrain: a banded tree, a branch of a tree, uniform size of bush, undefined or unexpected objects. We apply a RANSAC (RANdom SAmple Consensus) algorithm to discriminate ground point cloud data and object point cloud data, and then separate bush point cloud data and tree point cloud data by two combination algorithms; GMM (Gaussian Mixture Model) and EM (Expectation Maximization). Both GMM and EM algorithms are for extracting features and classifying groups, respectively. We propose the double FCM (Fuzzy C-mean clustering) algorithm to robustly estimate the number of trees and its position. The Extended Kalman Filter approach to simultaneous localization and mapping (EKF-SLAM) is composed of extracted tree features. The *mahalanobis* distance is applied to remain consistency for feature correspondence which is for data association. Finally, we show the results which is experienced in a tree-filled mountain.

Keywords: object detection, SLAM, natural terrain, extended Kalman filter, data association.

1 Introduction

Unmanned systems have been studying in various fields. UGVs (Unmanned ground Vehicles), UAVs (Unmanned Aerial Vehicles) and UUVs (Unmanned Underwater Vehicles) are respectively operated on ground, in air and under water. The field of UGVs had drastically been starting to study to achieve missions since the Grand Challenge event in which UGVs follow a designed route in the desert. In the case of urban area research, the vehicles encountered various sizes of objects and used different speeds during the route. This environment is also rapidly started to study

S. Lee et al. (Eds.): Intelligent Autonomous Systems 12, AISC 193, pp. 793–805.
springerlink.com

since the Urban Challenge event. Next, one of other research field is unstructured environment recognition where is about natural terrain.

To achieve even general missions in tree-filled and bush-filled mountain, an UGV should recognize its position and object position: localization of the vehicle and mapping with objects. An UGV driving in natural terrain suffers from environment recognition. It should be based on the feature object detection. The feature can be tree, bush, canopy, shrub, and so on.

In a dense forest and a deep mountain, GPS can't receive the signal from satellites to estimate precise position because of a lot of blockage. Even data from a high-precision INS-GPS is contaminated by uncertain measurement. An image sensor is especially sensitive to light. In the forest, the light is frequently changed by leaves of trees. An image sensor doesn't guarantee performance of environment recognition both day and night. But both LRF (Laser Range Finder) and LIDAR (Light Detection And Ranging) sensors are not sensitive to light and robust to circumstance. Thus LRF and LIDAR devices are primary pieces of equipment to acquire and analyze environment information. Environment resources for detection and tracking are either artificial landmarks or natural landmarks. There is no artificial landmark and no unique landmark in natural terrain. There are only many trees for landmarks which are different sizes and not uniformly round. Also, bushes are not uniform size and different height. Even the sensor provides each piece of scan data of obstacles.

There are many researches using a camera and a LIDAR. But it is only applied to recognize traversability on a trail. There is two methods for measurement devices. In 2D LRF researches, they assume trees are round and have only one pillar. In 3D LIDAR researches, they extract the feature using the nearest neighbor point. Then, they make a Gaussian Mixture Model (GMM) using the Expectation-Maximization (EM) algorithm to classify the object. To recognize a point cloud data group, they use Principle Component Analysis (PCA) and Support Vector Machine (SVM) [1, 2]. 3D cluttered range data processing is restricted with an uneven environment and processing time.

In this paper, we describe both natural terrain detection for feature extraction and SLAM for vehicle localization and object mapping using a LIDAR sensor for an unmanned ground vehicle. In section 2, we propose a method which extracts the significant objects from natural terrain on point cloud data. Section 3 describes the

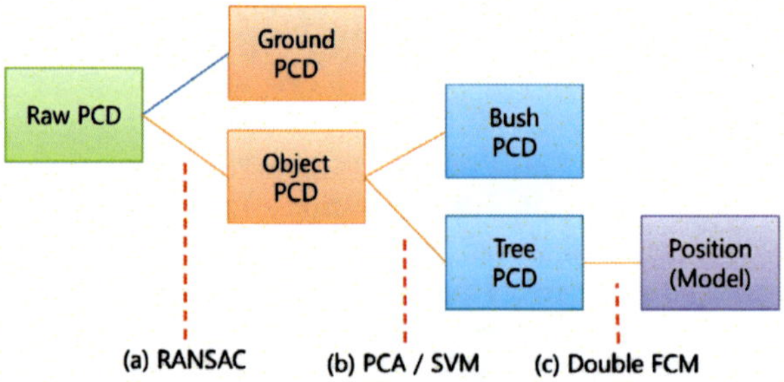

Fig. 1. The general point cloud processing illustrates how features detected from the environment scan data (raw point cloud data) are converted into a map state as landmarks

extended Kalman filter approach to a simultaneous localization and mapping algorithm. Section 4, we will show two implementation results about both an object detection methods and an EKF-SLAM method.

2 Object Detection

An object detection method is a process of a point cloud classification as in Fig. 1 [3]. It is a feature extraction procedure from raw point cloud data which is included both significant information and insignificant information. We acquire the point cloud data by scanning a LIDAR sensor in natural terrain. In this chapter, we classify feature landmarks from natural terrain 3D scanning data.

2.1 Terrain and Object Discrimination

The point cloud scanning data consist of a terrain part and an object part. On the ground, terrain comprises pebble, soil, sand, etc. Over the ground, objects are comprises tree, bush, rock, etc. The point cloud data are composed mostly of terrain points. To reduce data processing and computing time, we discriminate between terrain point and object point with a RANSAC algorithm as in Fig 1-(a) [4].

Let $\mathbf{X} = \{\mathbf{x}_i \mid \{x_1, y_1, z_1\}, \cdots, \{x_N, y_N, z_N\}\}$ be point cloud raw data. $\hat{\mathbf{x}}_m = \{\{x_i, y_i, z_i\} \; : \; i = 1, \cdots M\}$ is the uniformly selected random sample in point cloud data. We assume that a scanning ground shape is defined as a plane model;

$$\hat{\eta} = f(\hat{\mathbf{x}}_m) = \{ax_m + by_m + cz_m + d\}, \quad \{a,b,c,d\} \in \mathbb{R} \tag{1}$$

where the evaluation function (ε) is the distance between the candidate of the ground model and point cloud data. The distance is less then ω_{in}, inliner. If the number of ground inliner point is the enough candidate points, the parameters are stored. After iteration process reaches a certain time, the process is finished as in algorithm 1.

Algorithm 1

Let $\mathbf{X} = \{\mathbf{x}_i \mid \{x_1, y_1, z_1\}, \cdots \{x_N, y_N, z_N\}\}$ be N point cloud raw data

Repeat

1. Select m-points

$$\text{point } \hat{\mathbf{x}}_m, m \sim u(0,N) \begin{cases} 1/N & \textit{for } x \in [0,N] \\ 0 & \textit{otherwise} \end{cases}$$

2. Estimate a model parameter $\hat{\eta} = f(\hat{\mathbf{x}}_m)$ with selected m-points.
3. Evaluate the candidate ground model with the evaluation function (ε) within inliner point threshold ($< \omega_{in}$)

$$\varepsilon_i = \frac{ax_i + by_i + cz_i + d}{\sqrt{a^2 + b^2 + c^2}} < \omega_{in}$$

A. If the number of ground inliner points is more than the remaining parameters' inliner points, keep the parameter.

Until the maximum number of iterations k is reached

2.2 Vegetation and Object Points Discrimination

The object point cloud data consist of the bush point cloud data and the tree point cloud data as in Fig 1-(b). After extracting the feature from the object point cloud data, we applied a Gaussian Mixture Model (GMM) using Expectation Maximization (EM) [4] and a Bayesian classifier to discriminate the tree object and the bush object. First, we made a covariance matrix with the nearest neighbor points for each point, and then extracted the feature as an eigenvalue of the matrix. We made a group for training and a classifier by a GMM with EM. The covariance matrix for N 3D-point set $\mathbf{X}_i = \{x_i, y_i, z_i\}$ with $\bar{\mathbf{X}} = \frac{1}{N}\sum \mathbf{X}_i$ is defined as:

$$\frac{1}{N}\sum (\mathbf{X}_i - \bar{\mathbf{X}})(\mathbf{X}_i - \bar{\mathbf{X}})^T \tag{2}$$

The covariance matrix is decomposed into principal components. We define eigenvalues; e_o, e_1, e_2 are the eigenvectors corresponding to the eigenvalues $\lambda_0, \lambda_1, \lambda_2$ where $\lambda_0 \geq \lambda_1 \geq \lambda_2$, respectively. The bush point cloud data are $\lambda_0 \simeq \lambda_1 \simeq \lambda_2$. In the case of tree point cloud data, the eigenvalue will be $\lambda_0, \lambda_1 \gg \lambda_2$.

Second, we use a parametric model for a classifier by a Bayes classifier. The classification model is with a hand labeled data set. The class parametric Gaussian model expressed a mean and covariance (μ^k, Σ^k) of the k-th class. The k-th class has the probability density function $p(x \mid C_k)$. C is a class. Since we assume that each class has same portion of the entire point cloud data.

$$y(x) = \begin{cases} 1 & \text{if } p(\mathbf{x} \mid C_1) > p(\mathbf{x} \mid C_2) \\ 0 & \text{otherwise} \end{cases} \tag{3}$$

Each class has different mean (μ_k) and covariance (Σ_k). Using Bayes classifier, we can compose the decision rule as

$$y(\mathbf{x}) = \arg\max_k \left\{ (\mathbf{x} - \mu_k)(\Sigma_k)^{-1}(\mathbf{x} - \mu_k)^T \right\} \tag{4}$$

2.3 Object Clustering

The fuzzy c-means (FCM) algorithm is one of the most famous methods in a fuzzy clustering part [6] [7]. Let $\mathbf{X} = \{\mathbf{x}_1, \cdots, \mathbf{x}_n\}$ is a set of given point cloud data, where each point cloud data $\mathbf{x}_k (k=1,\cdots,n)$ is a vector in $\mathbb{R}^P$. U_{cn} be a set of real $c \times n$ matrices and c be an integer, $2 \le c < n$. The fuzzy c-partition space for $\mathbf{X}$ is the set

$$M_{fcm} = \{U \in U_{cn} : u_{ik} \in [0,1], \ \sum_{i=1}^{c} u_{ik} = 1, \ 0 < \sum_{k=1}^{n} u_{ik} < n\} \tag{5}$$

where u_{ik} is the membership value of x_k in cluster i $(i=1,\cdots,c)$. The aim of the FCM algorithm is to find an optimal fuzzy c-partition and corresponding prototypes minimizing the object function:

$$J_m(U,V;X) = \sum_{k=1}^{n} \sum_{i=1}^{c} (u_{ik})^m \left\| x_k - v_i \right\|^2 \tag{6}$$

where $V = \{v_1, v_2, \cdots, v_c\}$ is a matrix of unknown cluster centers, $v_i \in \mathbb{R}^P$. Choose a value c, m and ε, a small positive constant, then, generate randomly a fuzzy c-partition U^0 and set iteration number t=0. A two-step iterative process works as follows. Given the membership values $u_{ik}^{(t)}$, the cluster centers $v_i^{(t)} (i=1,\cdots,c)$ are calculated by

$$v_i^{(t)} = \frac{\sum_{k=1}^{n} \left(u_{ik}^{(t)}\right)^m \mathbf{x}_k}{\sum_{k=1}^{n} \left(u_{ik}^{(t)}\right)^m} \tag{7}$$

Given the new cluster centers $v_i^{(t)}$, update membership value $u_{ik}^{(t)}$

$$u_{ik}^{(t+1)} = \left[\sum_{j=1}^{c} \left(\frac{\left\| x_k - v_i^{(t)} \right\|^2}{\left\| x_k - v_i^{(t)} \right\|^2} \right)^{\frac{2}{m-1}} \right]^{-1} \tag{8}$$

The process stops when $\left| U^{(t+1)} - U^{(t)} \right| \le \varepsilon$, or a predefined number of iterations is reached

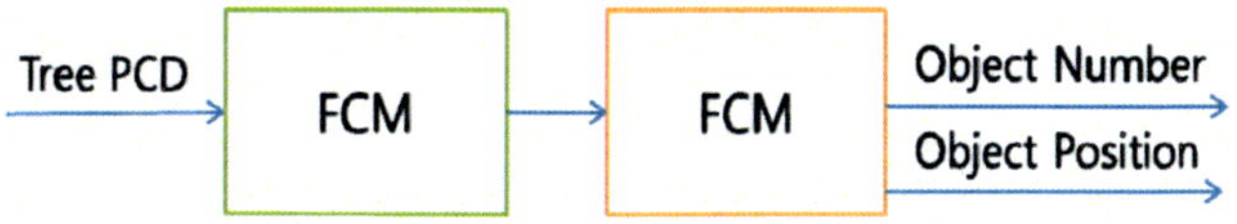

Fig. 2. The double fuzzy c-mean clustering diagram

To make an enhanced clustering, we apply double fuzzy-c mean clustering as in Fig 2. There is many false detections with one clustering processing. The first FCM clustering focuses on reducing the noise and then the second FCM clustering makes extract clusters. First clustering works for under-fitting to object parameter and second clustering works to fit an appropriate parameter to extract the object number and position.

3 SLAM

In this chapter, we use the Extended Kalman Filter approach to simultaneous localization and mapping [7]. To find correspondence feature, the *mahalanobis* distance method is applied which is called NIS (normalized innovation squared).

3.1 System Model

The vehicle state matrix ($\hat{\mathbf{x}}_v$) is represented by it's position and direction based on a Cartesian coordinate frame and its covariance matrix ($\mathbf{P}_v$) represents cross-correlation between a position and a direction.

$$\hat{\mathbf{x}}_v = \begin{bmatrix} \hat{x}_v & \hat{y}_v & \hat{\phi}_v \end{bmatrix}^T, \ \mathbf{P}_v = \begin{bmatrix} \sigma^2_{x_v x_v} & \sigma^2_{x_v y_v} & \sigma^2_{x_v \phi_v} \\ \sigma^2_{x_v y_v} & \sigma^2_{y_v y_v} & \sigma^2_{y_v \phi_v} \\ \sigma^2_{x_v \phi_v} & \sigma^2_{y_v \phi_v} & \sigma^2_{\phi_v \phi_v} \end{bmatrix} \tag{9}$$

The estimate map matrix ($\hat{\mathbf{x}}_m$) consists of distinguishable and static landmark which is the tree position. Its covariance matrix ($\mathbf{P}_m$) represents cross-correlation between the features as :

$$\hat{\mathbf{x}}_m = \begin{bmatrix} \hat{x}_1 & \hat{y}_1 & \cdots & \hat{x}_n & \hat{y}_n \end{bmatrix}^T,$$

$$\mathbf{P}_m = \begin{bmatrix} \sigma^2_{x_1 x_1} & \sigma^2_{x_1 y_1} & \cdots & \sigma^2_{x_1 x_n} & \sigma^2_{x_1 y_n} \\ \sigma^2_{x_1 y_1} & \sigma^2_{y_1 y_1} & \cdots & \sigma^2_{y_1 x_n} & \sigma^2_{y_1 y_n} \\ \vdots & \vdots & \ddots & \vdots & \vdots \\ \sigma^2_{x_1 x_n} & \sigma^2_{y_1 x_n} & \cdots & \sigma^2_{x_n x_n} & \sigma^2_{x_n y_n} \\ \sigma^2_{x_1 y_n} & \sigma^2_{y_1 y_n} & \cdots & \sigma^2_{x_n y_n} & \sigma^2_{y_n y_n} \end{bmatrix} \tag{10}$$

We can make an augmented state matrix with the vehicle state matrix and the feature map state matrix. The vehicle and map have a correlation as $\mathbf{P}_{vm}$.

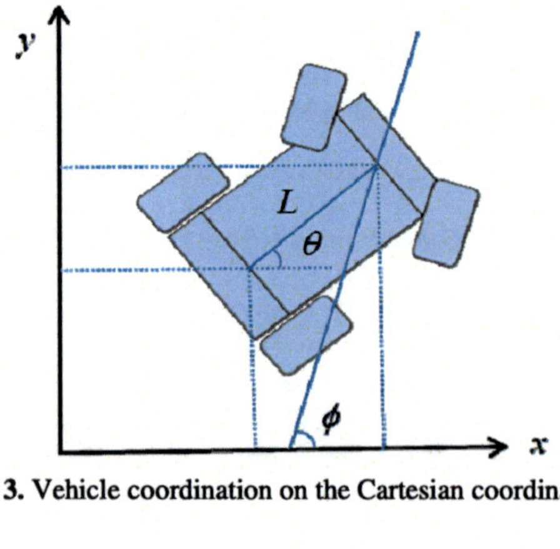

Fig. 3. Vehicle coordination on the Cartesian coordination

3.2 Prediction

The process model specifies that the vehicle moves relatives to its previous pose according to a dead reckoning motion estimate, and the map feature remain stationary. We make a model as a bicycle type and choose a real-wheel driving type car model.

$$\hat{\mathbf{x}}_k^- = f(\hat{\mathbf{x}}_{k-1}, \mathbf{u}_{k-1}) = \hat{\mathbf{x}}_{k-1} + \begin{bmatrix} \cos(\theta) \\ \sin(\theta) \\ \dfrac{\tan(\phi)}{L} \end{bmatrix} v \cdot \Delta t,\ \mathbf{u}_{k-1} = \begin{bmatrix} \phi \\ v \end{bmatrix} \tag{11}$$

where x and y are x-axis translation (forward) and y-axis translation (side). θ and ϕ are the body angle and steering angle, respectively. L is the body length. v is the velocity of the vehicle.

The effect of this model on the state estimate is a change in the $\hat{\mathbf{x}}_v$ portion of the state vector, and in the $\mathbf{P}_v$ and $\mathbf{P}_{vm}$ terms of the state covariance matrix while the $\hat{\mathbf{x}}_m$ and $\mathbf{P}_m$ portions remain constant. An estimate of the vehicle position with covariance is commonly obtained using odometry and a vehicle model.

$$\mathbf{P}_k^- = \mathbf{A}_k \mathbf{P}_{k-1}^- \mathbf{A}_k^T + \mathbf{W}_k \mathbf{Q}_{k-1} \mathbf{W}_k^T \tag{12}$$

$$\mathbf{A}_k = \left.\frac{\partial f}{\partial \mathbf{x}}\right|_{\mathbf{x}=\hat{\mathbf{x}}_{k-1}, \mathbf{u}=\mathbf{u}_{k-1}} = \begin{bmatrix} 1 & 0 & -\sin(\theta)\cdot v \cdot \Delta t \\ 0 & 1 & \cos(\theta)\cdot v \cdot \Delta t \\ 0 & 0 & 1 \end{bmatrix}$$

$$\mathbf{W}_k = \left.\frac{\partial f}{\partial \mathbf{u}}\right|_{\mathbf{x}=\hat{\mathbf{x}}_{k-1}, \mathbf{u}=\mathbf{u}_{k-1}} = \begin{bmatrix} 0 & \cos(\theta)\cdot v \cdot \Delta t \\ 0 & \sin(\theta)\cdot v \cdot \Delta t \\ \dfrac{v}{L}\cdot \sec^2(\phi)\cdot \Delta t & \dfrac{\tan(\phi)}{L}\cdot \Delta t \end{bmatrix}$$

3.3 Update

If a feature already stored in the map, it is observed by a range-bearing sensor, a 3D LIDAR, with the augmented state matrix and the measurement as below

$$\mathbf{z} = \begin{bmatrix} r \\ \theta \end{bmatrix},\ \mathbf{R} = \begin{bmatrix} \sigma_r^2 & \sigma_{r\theta}^2 \\ \sigma_{r\theta}^2 & \sigma_\theta^2 \end{bmatrix} \tag{13}$$

Here is observation using LIDAR (measurement) and observation covariance, and then the sensed information is related to the map by the following equation.

$$\hat{\mathbf{z}}_i = \mathbf{h}_i(\hat{\mathbf{x}}_a) = \begin{bmatrix} \sqrt{(\hat{x}_i - \hat{x}_v)^2 + (\hat{y}_i - \hat{y}_v)^2} \\ \arctan\left(\dfrac{\hat{y}_i - \hat{y}_v}{\hat{x}_i - \hat{x}_v}\right) - \hat{\phi}_v \end{bmatrix} \tag{14}$$

where $\hat{x}_i$ and $\hat{y}_i$ are the extracted feature position in the state vector. And here is an extra kalman process.

$$\begin{aligned} \upsilon_i &= \mathbf{z} - \mathbf{h}_i(\hat{\mathbf{x}}_a^-) \\ \mathbf{W}_i &= \mathbf{P}_a^- \nabla \mathbf{h}_{\mathbf{x}_a}^T (\nabla \mathbf{h}_{\mathbf{x}_a} \mathbf{P}_a^- \nabla \mathbf{h}_{\mathbf{x}_a}^T + \mathbf{R})^{-1} \end{aligned} \tag{15}$$

3.4 Data Association

Since, within the EKF framework, both the target location estimates and target observations are assumed to have Gaussian uncertainty, it is possible that any given observation might correspond to any target. However, to reject unlikely associations, only targets within a reasonable neighborhood of an observation should be considered. Association validation is usually performed in observation space, and a validation gate defines the maximum permissible discrepancy between a measurement $\mathbf{z}_i$ and a predicted observation $\hat{\mathbf{z}}_j = \mathbf{h}(\hat{\mathbf{z}}_j)$ for target $\hat{\mathbf{z}}_j$.

The most common statistical validation gate is based on the *normalized innovation squared* (NIS) also known as the Mahlanobis distance. Given an observation innovation $v_{ij} = \mathbf{z}_i - \mathbf{h}(\hat{\mathbf{x}}_j)$ with covariance, the NIS is defined as

$$M_{ij} = v_{ij}^T \mathbf{S}_{ij}^{-1} v_{ij} < \gamma \tag{16}$$

4 Experiment

In this section, we will describe the implemented system and environment with experimental results. The experimental vehicle is equipped with a laser scanner sensor; SICK LD-MRS. It measure 4-ch scan laser. All devices for navigation the vehicle is installed and process distributed computing; a camera, a LIDAR, a GPS and an INS. Each system communicates though the Ether-Cat communication as real-time. Figure 4 show the experimental vehicle and natural terrain environment. The process is executed with one frame point cloud scan data.

In detection part on the process, the clustering part needs the distinguished feature. The estimating cluster is an important part to determine system performance for simultaneous localization and mapping for an unmanned ground vehicle. We compare three clustering results; general fuzzy c-mean clustering, k-mean clustering, double fuzzy c-mean clustering. The clustering error is from scan disturbance; banded trees, not uniform size tree, not uniform size and height canopy.

Table 1. Object Detection Results Compared to Clustering Methods

Method	True	False	Percent
Single FCM Clustering	145	65	69%
K-Mean Clustering	155	55	73%
Double FCM Clustering	194	16	92%

Figure 5 shows the raw scan point cloud data of one frame which is scanned by 4-ch LIDAR on the vehicle. The figure is x-z axis. The blue point is ground. The red point is tree point, and green point is vehicle's position. The color describes the results of the detection from extraction object feature and classification of tree point cloud data as landmark features. Figure 6 shows the extracted tree features and corresponding trees on the image in the local top view. In Fig 6 (a), the red arrows describe the detected trees with scanning point cloud data. In Fig 6 (b), the red circles represent the extracted tree features. The red line represents the field of view of scan sensor. Figure 7 shows the tree feature extraction result which is based on the map generation. Figure 8 represents the covariance of the vehicle pose; position and direction. It is going increase before loop closing. When the vehicle pose is rapidly down, it is happened the close loop as Fig. 8. It is not much resource to decrease the covariance. Fig. 9 Point cloud feature data for all frames, green color points are the moving path of the vehicle, red color points are feature objects. It is include a lot of disturbance because of the canopy in front of the vehicle, the banded tree and the separated branches to confuse the feature clustering number.

Fig. 4. Experimental vehicle and natural terrain environment

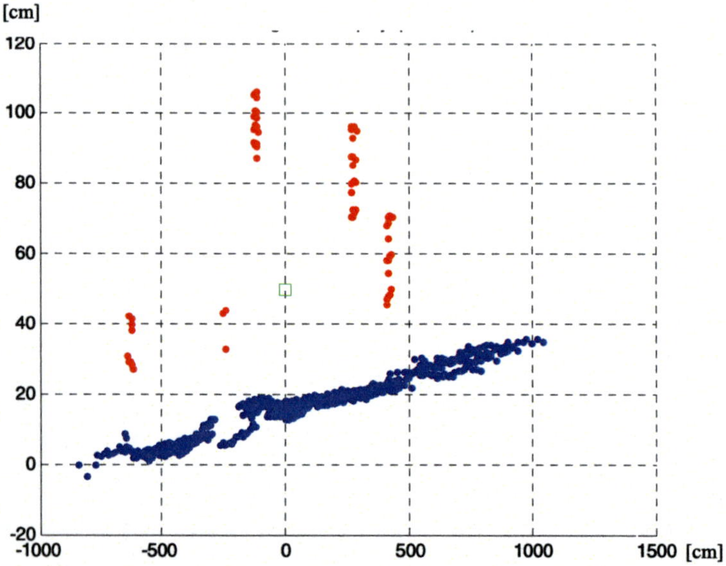

Fig. 5. The local side view (x-z axis) of raw point cloud data of one frame which is scanned by 4ch LIDAR on the vehicle. The blue points are ground scan points. The red points are tree objects. A green point are the vehicle's position.

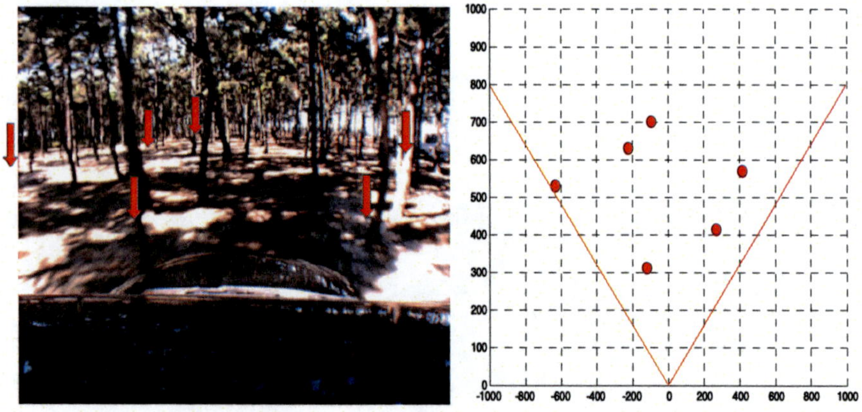

(a) (b)

Fig. 6. The extracted tree features and corresponding trees on the image in the local top view. Left (a): In the synchronized camera image, the red arrows describe the detected trees with scanning point cloud data. Right (b): The red circles represent the extracted tree features. The red line represents the field of view of scan sensor.[unit:cm]

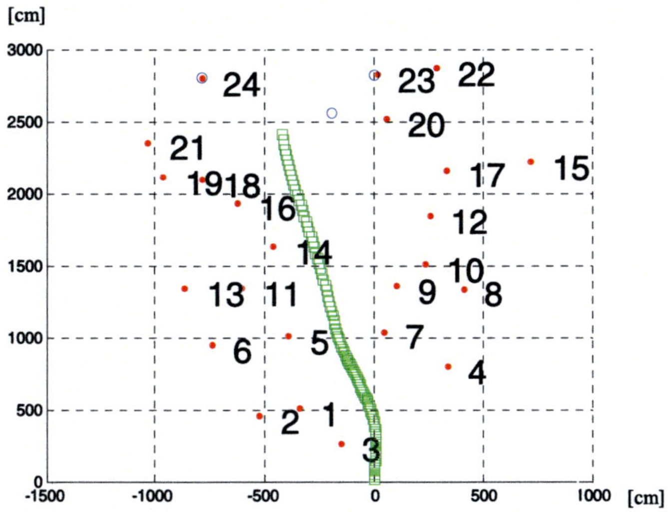

Fig. 7. Feature Extraction Result, red marks are the tree feature on the Cartesian coordinate, a set of text of next the red mark are its identification number. The blue mark is current scanning object. The green mark is the trajectory of the vehicle. [unit:cm]

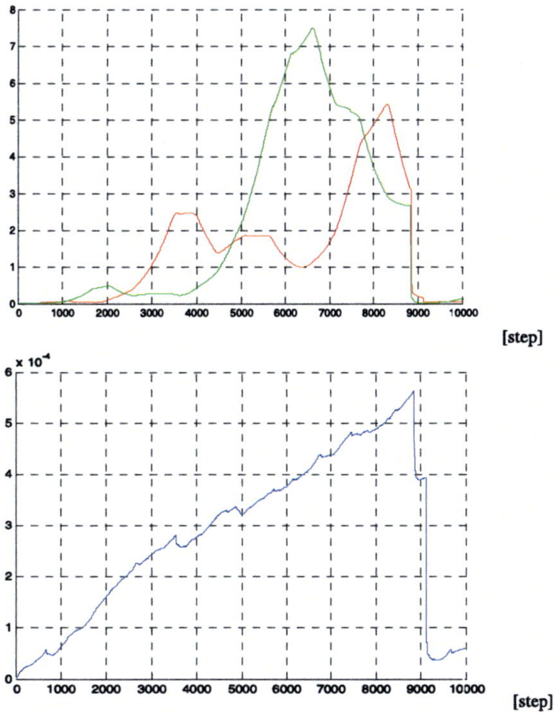

Fig. 8. The covariance of vehicle position (upper), the green line is x-axis and, the red line is y-axis, and direction (lower). When the loop closing, the covariance is rapidly down.

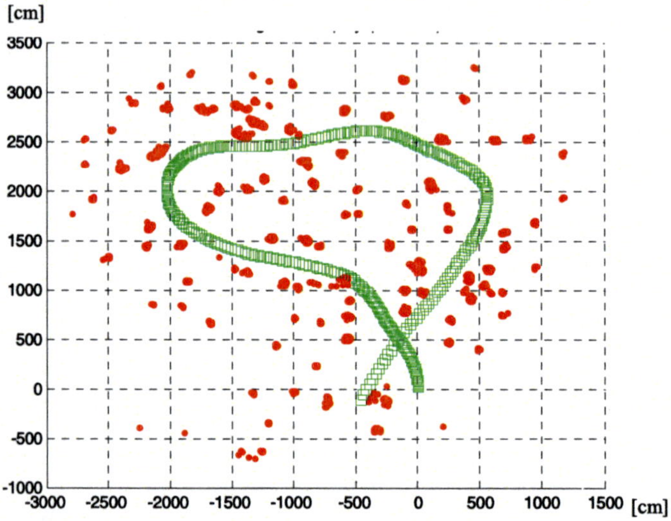

Fig. 9. Point cloud feature data for all frames, green color points are the moving path of the vehicle, red color points are feature objects

5 Conclusion

In this paper, we describe an EKF-SLAM method with feature extraction on natural terrain using LIDAR. It is 4 channel device and we process each one scan frame. In the case of detection, the propose method is better than conventional single fuzzy c-mean clustering and k-mean clustering. In the case of SLAM, it shows the close loop. In the future, we will increase discriminate classes; tree, bush, canopy, etc. The terrain will be influence on the vehicle's localization with disturbance and slip. A critical problem of an EKF-SLAM framework is miss correspondence with feature landmarks. To reduce a miss correspondence, we should increase the map consistency with map joining skill.

Acknowledgements. This work was supported by Dual Use Projects of MKE (Ministry of Knowledge and Economy) contract titled "Development of Quadruped Legged Robot Platform Technology".

References

1. Vandapel, N., Huber, D.F., Kapuria, A., Hebert, M.: Natural Terrain Classification using 3-D Ladar Data. In: Proceeding of IEEE Int. Conf. on Robotics and Automation, ICRA 2004 (2004)
2. Lalonde, J.F., Unnikrishnan, R., Vandapel, N., Hebert, M.: scale selection for classification of point-sampled 3D surfaces, 3-D Digital Imaging and Modeling. In: Fifth International Conference on 3DIM 2005, June 13-16 (2005)

3. Cho, K., Baeg, S.H., Park, S.D.: Multiple object detection and tracking on the uneven terrain using multiple lidar for UGV. In: Proceeding SPIE Defense Sensor and Security, March 26-30 (2012)
4. Fischler, M.A., Bolles, R.C.: Random Sample Consensus: "A paradigm for model fitting with applications to image analysis and automated cartography". Communications of the ACM 24(6), 381–395 (1981)
5. Hebert, M., Vandapel, N.: Terrain classification techniques from ladar data for autonomous navigation. In: Collaborative Technology Alliances Conference (May 2003)
6. Nascimeto, S., Mirkin, B., Moura-Pires, F.: A Fuzzy Clustering Model of Data and Fuzzy C-means. In: S. Proc. 19th IEEE Int. Conf. Fuzzy Syst., vol. 1, p. 302 (2000)
7. Pal, N.R., Pal, K., Keller, J.M., Bezdek, J.C.: A possibilistic fuzzy c-means clustering algorithm. IEEE Trans. Fuzzy Syst. 13(4), 517–530 (2005)
8. Bailey, T.: Mobile Robot Localisation and Mapping in Extensive Outdoor Environments. Dissertation paper (2002)

Positioning Accuracy Improvement of Laser Navigation Using Unscented Kalman Filter

Jungmin Kim, Kyunghoon Jung, Jaeyong Kim, Hajun Song, and Sungshin Kim*

Dept. of Electrical Engineering, Pusan National University
Geumjeong, Busan 609-735, Korea
{kjm16,hooraring,arioner,hjsong,sskim}@pusan.ac.kr

Abstract. This paper presents positioning improvement of a laser navigation system (LNS) using unscented Kalman filter (UKF) for an automatic guided vehicle (AGV). The existing AGVs mainly used a magnetic system or an inductive system as a guidance system. However, those have high cost and difficult maintenance according to change of environment, and can drive only the designated path which sensors are installed on. The laser guidance system is developed to solve those problems, but it has also problems which is slow response time and low accuracy. Therefore, we propose a sensor fusion method for the AGV. The sensors used in this paper are encoders, a gyro and the LNS, and they are fused by UKF. To analyze the performance of the proposed system, we designed a fork-type AGV for ourselves, and performed the experiment that was repeated 5 times under the same working conditions. In experimental results, we verified that the proposed method could improve positioning accuracy of the LNS effectively. In addition, it was appropriate to apply a real AGV system for autonomous driving.

Keywords: unscented Kalman filter, laser navigation, automatic guided vehicle.

1 Introduction

In recent years, the flexible manufacturing system (FMS) has become an important issue in production management. In particular, the automatic guided vehicle (AGV) which is an industrial robot is a new growth industry in several countries and is being researched actively. The AGV can work autonomously, including loading and unloading, storage, and transport [1]–[3]. To develop the AGV, positioning, driving control, and path-planning techniques are necessary. Among them, the positioning and driving control as guidance system are key techniques for autonomous driving techniques [4]–[6].

The AGVs usually uses two kinds of the guidance system: inductive guidance system and magnetic guidance system [7]–[10]. The sensor of the inductive guidance system is placed on the bottom of the AGV, and the wire is placed approximately 1

* Corresponding author.

S. Lee et al. (Eds.): Intelligent Autonomous Systems 12, AISC 193, pp. 807–816.
springerlink.com

inch below the ground. The sensor detects the radio frequency being transmitted from the wire. It has the advantage that it can turn off the wires power in an urgent situation and immediately stop the AGV system. The magnetic guidance system is similar to the inductive guidance system, but it uses transponder of the magnets that are embedded in the floor of the workplace. However, those systems have high cost and difficult maintenance according to changes of the environment, and can drive only the designated path which sensors are placed or embedded in. To overcome those weaknesses, the laser navigation system as a wireless guidance system has been developed. However, the laser navigation system has a slow response time and low accuracy. In addition, it calculates an incorrect position of the AGV because it does not recognize reflectors properly while the AGV is moving at high speed or making around sharp turn. The laser navigation can cause serious problems because the AGV is heavy equipment. Therefore, a robust and accurate laser guidance system is needed. We propose the sensor fusion method for the AGV. The used sensors are encoder, gyro and laser navigation, and fused by unscented Kalman filter (UKF).

2 Measurement System

2.1 Automatic Guided Vehicle

To develop the fork-type AGV, we remodeled the manual forklift production (CRX-10) made by Clark Material Handling into the AGV. Figure 1 shows the size of the fork-type AGV used in this paper.

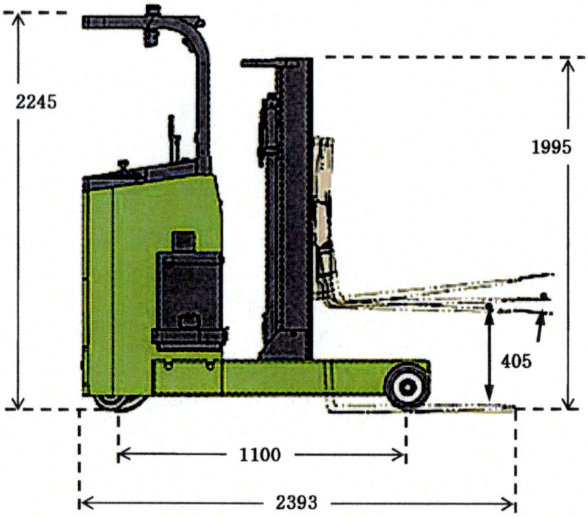

Fig. 1. The size of the fork-type AGV

The fork-type AGV has an axle drive unit with a half-electric power steering (EPS). But the half-EPS is used for manual driving usually. To solve this, we changed it into a full-EPS.

2.2 System Configuration

The system consists of positioning part and driving control part, and the positioning part consist of two encoders, a gyro and laser navigation, and the driving control part has a data acquisition (DAQ) system. In addition, the fork-type AGV is based on the industrial PC for a flexible system and rapid development. Figure 2 shows the major system configuration of the fork-type AGV.

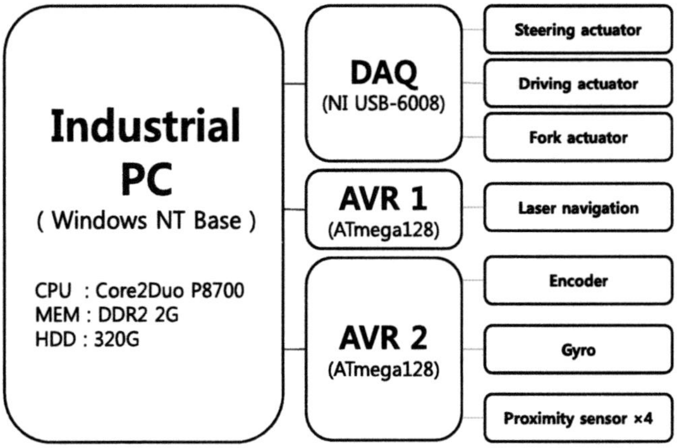

Fig. 2. The major system configuration of the fork-type AGV

The positioning sensors are measured by ATmega128 microcontroller unit (MCU). A NI-USB6008 DAQ (data acquisition) was used for driving control. The major specifications of these units are shown in table 1.

Table 1. The Major Specifications of the Units

Item (model)	Specification	
Laser navigation (NAV 200)	Input voltage	: 24V
	Resolution[a]	: Approx. 15mm
	Typ. Range	: 1.2m to 28.5m
Encoder (drive hall sensor)	Input voltage	: 12V
	Resolution	: 64 pulse
Gyro (myGyro75SPI)	Input voltage	: 5V
	Sensitivity	: ±75°

2.3 Kinematics

The axle drive system of the fork-type AGV consists of a driving and a steering actuat -or situated left of the body centerline. Figure 3 shows the kinematics model of the fork-type AGV with absolute coordinate O-X-Y and relative coordinate o-x-y.

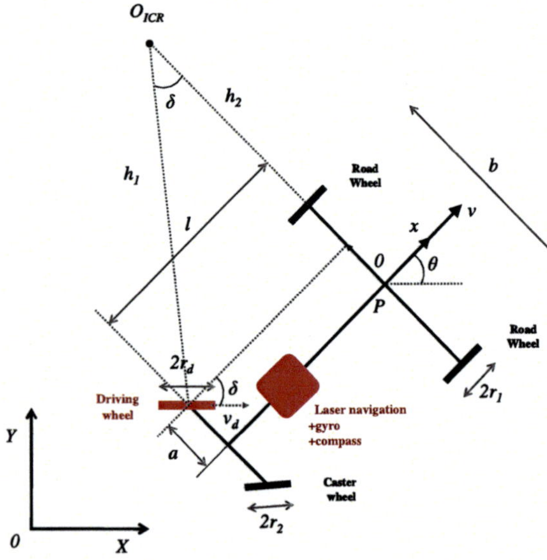

Fig. 3. Kinematics model of the used fork-type AGV

In figure 3, the fork-type AGV is controlled by steering wheel and driving wheel. h_1 represents the distance between the instantaneous center of rotation (*OICR*) to the center of the driving wheel, and is calculated

$$h_1 = \frac{l}{\sin\delta} \tag{1}$$

where a and l are the length of the AGV without a fork and the offset of the center of the driving wheel from the centerline of the body, respectively. The rotational angle on center position between road wheels becomes

$$\dot{\delta} = \frac{v_d}{h_1} = \frac{v}{h_1\cos\delta + a} \tag{2}$$

where v_d is the linear velocity of the driving wheel characterized by the following relationship:

$$v_d = r_d w_d \tag{3}$$

where r_d and w_d are the radius of the driving wheel and the angular velocity of the driving wheel, respectively. The linear velocity of the AGV is calculated by

$$v = r_d w_d \left(\cos\delta + \frac{a \sin\delta}{l} \right) \tag{4}$$

when the linear velocity and the angle of the driving wheel are calculated, the position and orientation of the AGV are given as follows:

$$\begin{aligned} \dot{X} &= v\cos\theta \\ \dot{Y} &= v\sin\theta \\ \dot{\theta} &= \frac{r_d w_d \sin\delta}{l} \end{aligned} \tag{5}$$

3 Positioning Accuracy Improvement Using UKF

The basic premise behind the unscented Kalman filter is it is easier to approximate a Gaussian distribution than it is to approximate an arbitrary nonlinear function. Instead of linearizing using Jacobian matrices, the UKF uses a deterministic sampling approach to capture the mean and covariance estimates with a minimal set of sample points [11]. More details can be found in [12]-[17].

A general nonlinear system described by the difference equation and the measurement model with inputs and additive noises:

$$\begin{aligned} x_{k+1} &= f(x_k) + w_k \\ z_k &= h(x_k) + v_k \end{aligned} \tag{6}$$

where x_k and z_k are the n-dimensional state of the system and the measurement at time step k, w_k and v_k are system noise and measurement noise, respectively. The function f can be used to compute the predicted state from the previous estimate and similarly the function h can be used to compute the predicted measurement from the predicted state.

The system model x_k and the measurement model z_k in the proposed system are defined by

$$\begin{aligned} x_k &= \begin{bmatrix} x_k^{pos} \\ y_k^{pos} \\ \theta_k^{pos} \end{bmatrix} = \begin{bmatrix} x_{k-1}^{pos} + v\cos(\theta_k^{pos}) \\ y_{k-1}^{pos} + v\sin(\theta_k^{pos}) \\ \theta_{k-1}^{pos} + \dot{\theta} \end{bmatrix} + \begin{bmatrix} w_x \\ w_y \\ w_\theta \end{bmatrix} \\ z_k &= \begin{bmatrix} 1 & 0 & 0 \\ 0 & 1 & 0 \\ 0 & 0 & 1 \end{bmatrix} \begin{bmatrix} x_k^{pos} \\ y_k^{pos} \\ \theta_k^{pos} \end{bmatrix} + \begin{bmatrix} v_x \\ v_y \\ v_\theta \end{bmatrix} \end{aligned} \tag{7}$$

The input of the system model is the relative position of the AGV. The relative position is calculated by linear velocity and the angular velocity measured by data from encoders and a gyro as the local positioning sensors. The input of the measurement model is the absolute position measured by the laser navigation installed on the AGV. The process of the UKF consists of a prediction step and a correction step. The prediction step calculates a sigma points χ_i and weights Wi using system model x, error covariance P and constant κ at time step k-1 as follows:

$$(\chi_i, W_i) \leftarrow (x_{k-1}, P_{k-1}, \kappa) \tag{8}$$

The UKF uses a deterministic sampling technique known as the unscented transform to pick a minimal set of sample points. Therefore, the priori state $\hat{x}^-$ and priori error covariance P^- at time step k are estimated by the unscented transform of which inputs are the sigma points χ_i and the weights W_i.

$$(\hat{x}_k^-, P_k^-) = \mathrm{UT}(f(\chi_i), W_i, Q) \tag{9}$$

where Q is the process noise covariance matrix. In the correction step, the columns of sigma points χ_i should be transformed by the measurement function. As the last part of the prediction step, the measurement system is estimated by

$$(\hat{z}_k, P_k) = \mathrm{UT}(h(\chi_i), W_i, R) \tag{10}$$

where R is the measurement noise covariance matrix. With the transformed state vector $\hat{z}_k$, a posteriori state is estimated by

$$\hat{x}_k = \hat{x}_k^- + k_k(z_k - \hat{z}_k) \tag{11}$$

where K_k is Kalman gain. In the UKF algorithm, K_k is computed by

$$k_k = P_{xz} P_z^{-1} \tag{12}$$

where

$$P_{xz} = \sum_{i=1}^{2n+1} W_i \{f(\chi_i) - \hat{x}_k^-\}\{h(\chi_i) - \hat{z}_k\}^T \tag{13}$$

Finally, a posteriori estimate of the error covariance given by

$$P_{xz} = \sum_{i=1}^{2n+1} W_i \{f(\chi_i) - \hat{x}_k^-\}\{h(\chi_i) - \hat{z}_k\}^T \tag{14}$$

4 Experiments

4.1 Experimental Environment

The experiment space used in the paper is shown in the top-right (shade) of the figure 4, and the size is 850 cm×1000 cm. The absolute coordinate is on the top-right. 15 reflectors (dotted line circles) for laser navigation were installed.

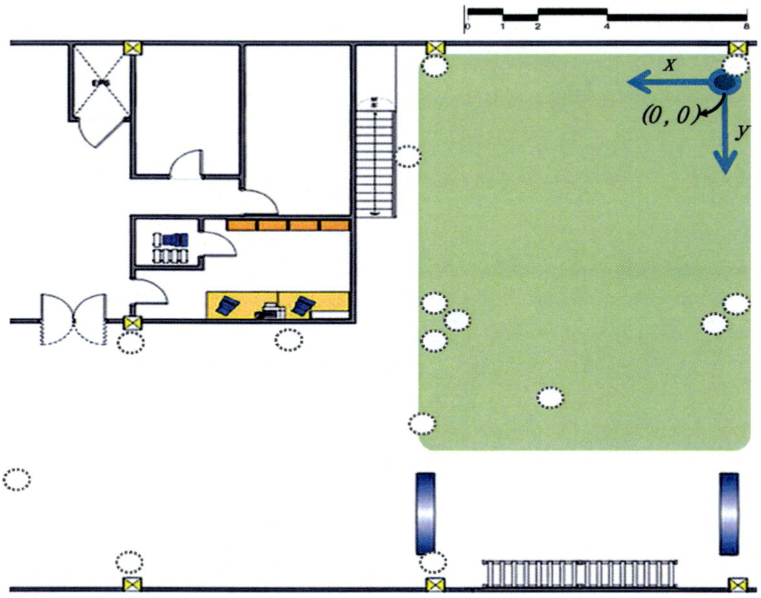

Fig. 4. Experimental environment

4.2 Results and Discussion

We fixed the input signal of the driving actuator of the fork-type AGV and analyzed the positioning accuracy according the input signal of the steering actuator for the experiment of the proposed method. Figure 5 shows the experimental results when the driving input of the fork-type AGV was 3.6 V (approximately 50 cm/s) and steering input was 2.0 V(approximately 20°).

In figure 5(a), the start position is (0, 0), and the black line represents the result which does not contain error because of the simulation using only the Kinematics of the fork-type AGV. The red dots represent the positioning result when only the laser navigation was used. In result, we can see the damage and loss of data of the laser navigation by disturbance. When only the local sensors were used (figure 5(b)), the error was cumulative depending on the time. However, the proposed method could compensate for the defects of the laser navigation and the local sensors as shown in figure 5(c). Table 2 presents the experimental results according to the steering input when the driving input was fixed at 3.6 V (approximately 50 cm/s).

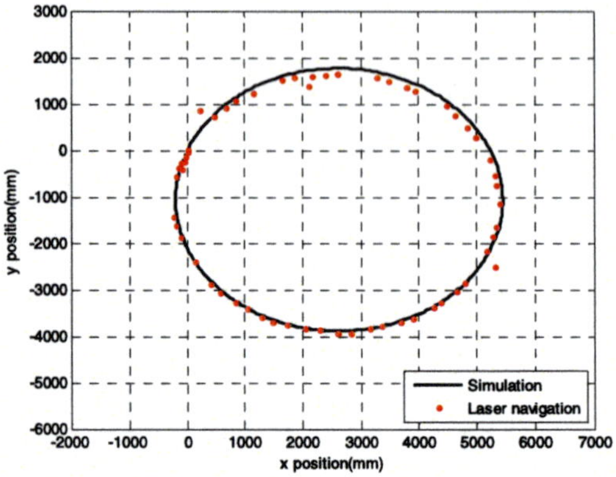

(a) using only the laser navigation

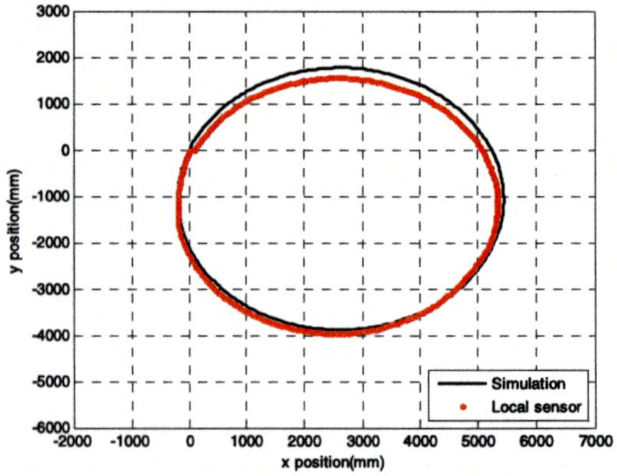

(b) using only the local sensors

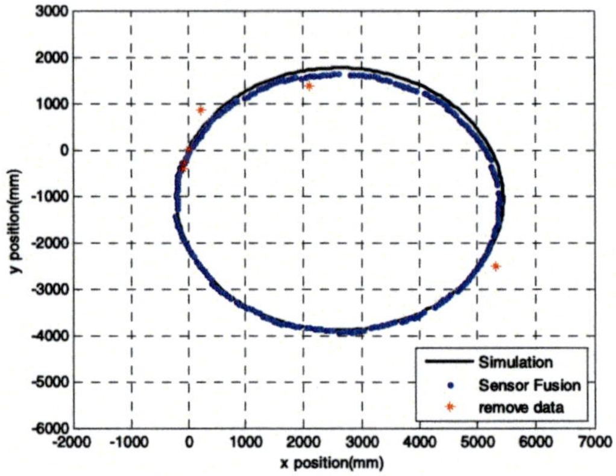

(c) using the proposed method

Fig. 5. Experimental environment

Table 2. Experimental Results

Steering input (degree)	**5 times average positioning error [mm]**		
	Laser navigation	Local sensors	Sensor fusion
1.0 V (60°)	57.76	138.54	49.53
1.2 V (50°)	59.94	171.26	45.71
1.5 V (40°)	40.54	166.07	42.73
1.8 V (30°)	40.54	141.39	50.21
2.0 V (20°)	65.78	97.80	59.02
Total average error	52.91	143.01	49.44

In table 2, the results of the laser navigation and the sensor fusion look similar, but it is because we used the minimum error between the results of the simulation and the actual driving of the fork-type AGV. it is impossible to keep the synchronization between the results of the simulation and the actual driving of the fork-type AGV. In addition, the number of the data from the laser navigation is much less than the others because the response time of the laser navigation is slower than the others.

We verified that the proposed method has better performance than using only the laser navigation, but we wonder if the proposed method works robustly in the real AGV system. Hereupon, we applied the proposed method to the real AGV system. The result can be verified through video [18].

5 Conclusion

This paper presented the performance improvement of the laser navigation using the UKF for the real AGV. The existing laser navigation has had slow response time and low accuracy. In addition, it calculates an incorrect position while the AGV is moving at high speed or making around sharp turn. Therefore, we proposed a positioning approach method using a sensor fusion based on the UKF. To analyze the performance of the proposed method, we designed the AGV by ourselves. In experimental results, we verified that the proposed method could improve positioning accuracy of the laser navigation effectively. In addition, it was appropriate to apply a real AGV system for the autonomous driving. In this paper, we did not describe our experiments on sensor fusion using an extended Kalman filter, but the results were very similar to the proposed method.

Acknowledgements. This research was supported by the MKE(The Ministry of Knowledge Economy), Korea, under the Specialized Field Navigation/Localization Technology Research Center support program supervised by the NIPA (National IT Industry Promotion Agency).

References

1. Gutsche, R., Laloni, C., Wahl, F.M.: A flexible transport system for industrial environments using global sensor and navigation concepts. Robotics and Autonomous Systems 14(2-3), 85–98 (1995)
2. Hung, N., Im, J.S., Jeong, S.K., Kim, H.K., Kim, S.B.: Design of a sliding mode controller for an automatic guided vehicle and its implementation. International Journal of Control, Automation, and Systems 8(1), 81–90 (2010)
3. Vis, I.F.A.: Survey of research in the design and control of automated guided vehicle systems. Eur. J. Oper. Res. 170(3), 677–709 (2006)
4. Xidias, E.K., Azariadis, P.N.: Mission design for a group of autonomous guided vehicles. Robotics and Autonomous Systems 59(1), 34–43 (2011)
5. Jia, Z., Balasuriya, A., Challa, S.: Sensor fusion-based visual target tracking for autonomous vehicles with the out-of-sequence measurements solution. Robotics and Autonomous Systems 56(2), 157–176 (2008)
6. Yang, Z.F., Tsai, W.H.: Viewing corridors as right parallelepipeds for vision-based vehicle localization. IEEE Trans. on Industrial Electronics 46(3), 653–661 (1999)
7. Setlur, P., Wagner, J.R., Dawson, D.M., Braganza, D.: A trajectory tracking steer-by-wire control system for ground vehicles. IEEE Trans. on Vehicular Technology 55(1), 76–85 (2006)
8. Schulze, L., Zhao, L.D.: Worldwide Development and Application of Automatic Guided Vehicle Systems. Service Operation and Informatics 2, 164–176 (2007)
9. Wakaumi, H., Nakamura, K., Matsumura, T.: Development of an automated wheelchair guided by a magnetic ferrite marker lane. Journal of Rehabilitation Research and Development 29(1), 27–34 (1992)
10. Chan, C.Y.: A system review of magnetic sensing system for ground vehicle control and guidance. California PATH Research Report, UCB–ITS–PRR–2002–20 (2002)
11. Julier, S.J., Durrant-Whyte, H.F.: Navigation and parameter estimation of high speed road vehicles. In: Robotics and Automation Conference, pp. 101–105 (1995)
12. Julier, S.J., Uhlmann, J.K., Durrant-Whyte, H.F.: A new approach for filtering nonlinear systems. In: Proceedings of the American Control Conference, pp. 1628–1632 (1995)
13. Julier, S.J., Uhlmann, J.K.: A new extension of the Kalman filter to nonlinear systems. In: The Proceedings of AeroSense: The 11th International Symposium on Aerospace/Defense Sensing, Simulation and Controls, Multi Sensor Fusion, Tracking and Resource Management II. SPIE (1997)
14. Ko, S.I., Choi, J.S., Kim, B.H.: Indoor mobile localization system and stabilization of localization performance using pre-filtering. International Journal of Control, Automation, and Systems 6(2), 204–213 (2008)
15. Kim, K.J., Park, C.G.: Non-Symmetric Unscented Transformation with Application to In-Flight Alignment. International Journal of Control, Automation, and Systems 8(4), 776–781 (2010)
16. Choi, K.S., Lee, S.G.: An Enhanced CSLAM for Multi-robot based on Unscented Kalman Filter. International Journal of Control, Automation, and Systems 10(1), 102–108 (2012)
17. Laneurit, J., Chapuis, R., Chausse, F.: Accurate vehicle positioning on a numerical map. International Journal of Control, Automation, and Systems 3(1), 15–31 (2005)
18. (keyword: AGV, LIT), `http://www.youtube.com/watch?v=inlMp83rTNo`

Improvement of Position Accuracy of Magnetic Guide Sensor Using Kalman Filter

Eunkook Jung[1], Jungmin Kim[1], Hyunhak Cho[2], Junha Lee[2], and Sungshin Kim[1,*]

Dept. of Electrical Engineering[1], Pusan National University
Interdisciplinary Cooperative Course: Robot[2], Pusan National University
Geumjeong, Busan 609-735, Korea
{silverkook,sjm16,darkruby1004,hattie,sskim}@pusan.ac.kr

Abstract. This paper is represented to research of method to improve the performance of magnetic guide sensor using Kalman filter. The magnetic guide sensor is calculating the center position of the AGV (Automatic Guided Vehicle) by to measure the magnetic information of a magnetic substance that is magnetic tape, magnet spot. The existing magnetic guide sensor is the device that calculates the center position of the AGV using the magnetic force of the measured data that is one pole to measure each axis at the histogram algorithm. But, the existing method is unfit for requiring the high accuracy such as industrial setting because of interference between sensors and the effect by disturbance. Therefore, in this paper proposed method that increases the position accuracy of the magnetic guide sensor using Kalman filter. To verify the proposed method, we use the AGV to install magnetic guide sensor. And it compare the positioning accuracy of the propose method and the commercialized magnetic guide sensor. As a result, the proposed method was found 24.78% to improve the positioning accuracy of the proposed method than that of the commercialized magnetic guide sensor.

Keywords: magnetic guide sensor, Kalman filter, AGV.

1 Introduction

Recently, automation industry is increasing in popularity and the methods for improving the performance of the sensors are being studied actively. The AGV controller related to the sensors are classified as the wire guide system and the wireless guide system [1].

The wired sensor is placed on the bottom of the robot and is placed facing the ground. The sensor detects the radio frequency being transmitted from the wire and traces it. The wireless guide system calculate the position of AGV by matching method using the data to measure the receiver or transmitter to attach the walls, pillars, ceilings and the environments. The wireless guide system has advantage that is installation and change [2]-[3]. Because of it can the perfect autonomous driving

* Corresponding author.

S. Lee et al. (Eds.): Intelligent Autonomous Systems 12, AISC 193, pp. 817–824.
springerlink.com

and the virtual driving to trace the virtual path by to do not affect to the surrounding environment.

However, the system had problems that increase the position error by disturbance and the feature of the installed transmitter or receiver to the AGV. Thus, the industrial settings to emphasize safety usually use wire guide system. But, the wire guide system had that is high cost of installation, difficult to maintain and difficult to modify the path due to the buried device to the floor or the attached sensors at the surrounding environment [4]-[5]. But, the magnetic guide sensor unlike the general magnetic sensor can update wire-guide system weaknesses by the installed magnetic substance that having information of driving direction, velocity, angle and etc. However, the sensor requires the high localization accuracy. Because of cannot know the next information of progress when the AGV drop out of the trajectory [6].

The general magnetic guide sensor calculates the position of the AGV which is using histogram algorithm to the sum of the measured magnetic field in each the axis. But, the existing method to use histogram algorithm had a problem that is low the position accuracy of the sensor by interference of sensors and disturbance. To solve the problem, we propose method to improve the position accuracy of the magnetic guide sensor using Kamlan filter that based on probabilistic [7]-[8].

2 Measurement System

2.1 Test AGV

For experiment, we used magnetic guide sensor which is arranged to the magnetic sensor in 8 x 5 and we used the AGV that is designed by ours. The structure of AGV used in this paper is shown in Fig. 1.

Fig. 1. Used AGV

The used AGV is two-wheel type that had two-wheel and use BLDC-motor of 200W. For various experiments, The DSP(TM320C28346) configures the controller, and the used sensors are encoders and a gyro. The entire system configuration of AGV is shown in Fig. 2.

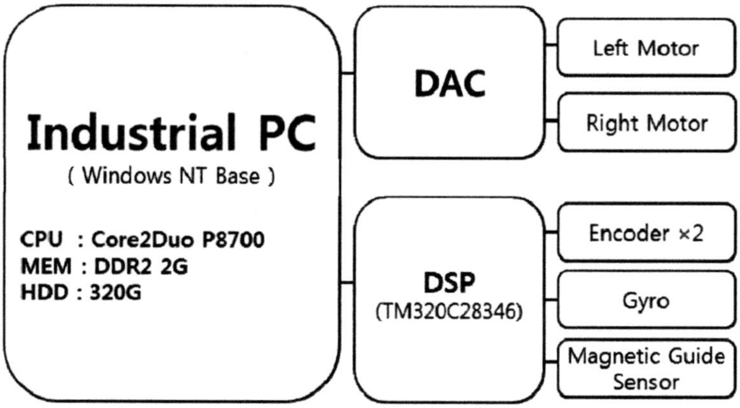

Fig. 2. System configuration of AGV

2.2 Kinematics

Fig. 3 is shown kinematics of the AGV that had the two-wheel.

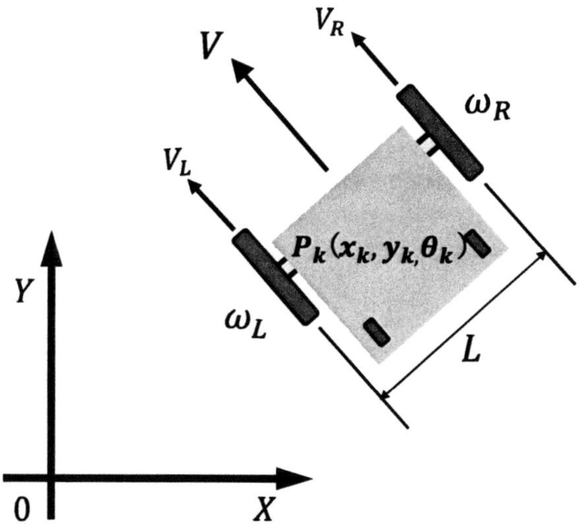

Fig. 3. Kinematics of AGV

The angular velocities w_L and w_R are calculated by data which measured in encoders for the each time as Fig. 3 the length of two-wheel is L, and the radius of wheel is R. the linear velocities V_L and V_R are calculated

$$\begin{cases} v_L = w_L \times R \\ v_R = w_R \times R \end{cases} \tag{1}$$

The linear velocity of the AGV v and the angular velocity of the AGV w are calculated as (1) at this time, the angular velocity of rotation is calculated by the direction of counter clockwise.

$$\begin{cases} v = \dfrac{v_R + v_L}{2} \\ w = \dfrac{v_R - v_L}{L} \end{cases} \tag{2}$$

If to know the location information ($x_{k,}$ $y_{k,}$ θ_k), the position ($x_{k+1,}$ $y_{k+1,}$ θ_{k+1}) of the time k+1 is calculated by (3)

$$\begin{bmatrix} x_{k+1} \\ y_{k+1} \\ \theta_{k+1} \end{bmatrix} = \begin{bmatrix} x_k + v \times \cos(\theta_k + w) \\ y_k + v \times \sin(\theta_k + w) \\ \theta_k + w \end{bmatrix} \tag{3}$$

3 Kalman Filter for Position Measurement

3.1 Kinematics Vehicle Position Measurement of Magnet Using Histogrm

For calculating the position of the measured magnetic field to the magnetic guide sensor, measure the change value of the magnetic field based on the measured data at 0 Gauss. To calculate the accurate position of the magnetic field is as follow steps. Step 1 measure the occurred error by the surrounding environment that is the error of the sensors or the interference of the other device. And the actual magnetic force is calculated by the measured data that is the removed data for the error. Step 2 calculates the center position of the magnetic field using histogram algorithm at the sum of the sensor each axis based on the result of Step 1. But, we calculate the maximum point using the high rank 4 in the value of each axis because of the actual center point do not calculate. The center point of the magnetic field is calculated using the maximum point. The calculated procedure is that the center position of magnet shown in Fig. 4.

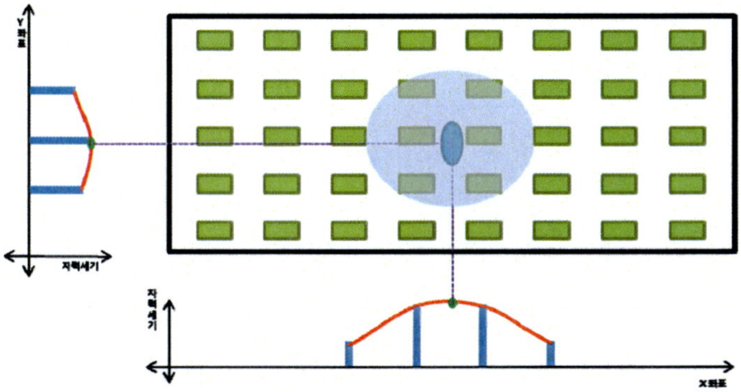

Fig. 4. Position Measurement of magnet using histogram

3.2 Position Measurement of Magnet Using Kalman Filter

Kalman is a recursive filter that follows the state of a linear dynamic system included the error. And that is to predict and to update the measured data included the probabilistic error. In this paper, we use the measurement system as the velocity and position of the AGV for improvement of the positioning accuracy of the sensor. Because of the values are the movable these for one time of the AGV based on the measured data to the magnetic guide sensor. Thus, the position of the AGV is predicted to the velocity and position of the AGV. The observation model use the center position of the calculated result from 3.1. The errors w, v of each system use the calculated values through the error analysis of each system. To update the center point of that using Kalman filter is as Fig. 5.

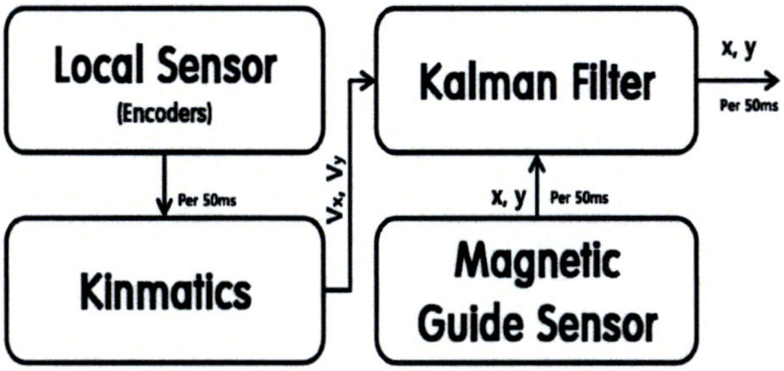

Fig. 5. Position Measurement of magnet using Kalman Filter

4 Kalman Filter for Position Measurement

4.1 Experimental Environment

In this paper, we performance evaluation of proposed magnetic guidance sensor using the AGV mounted developed magnetic guidance sensor. Experiment is done is space consist of 20 m 15mm and used mainly experiment space is 8.4 m x 11 m. The experiment method is AGV driving along by the installed magnet. Also, we are compared with driving trajectory of the proposed method and the existing method. The experiment environment is shown in Fig.6.

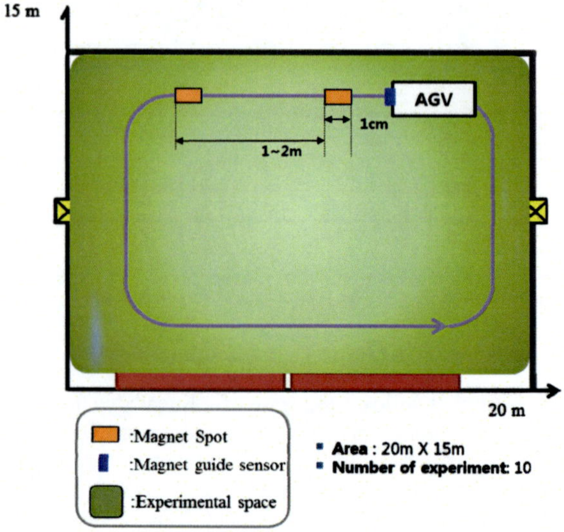

Fig. 6. Experimental Environment

4.2 Experimental Result

The result of the max and the error variance at results of the entire is as Table. 1. And it is shown in Fig. 7. That is driving the rotation trajectory of one of the whole experiment.

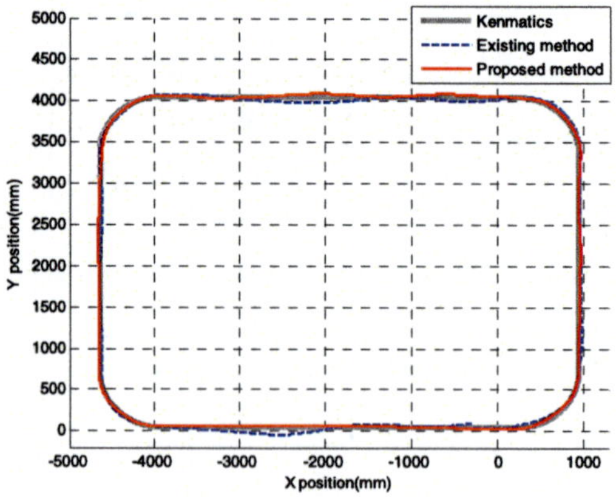

Fig. 7. Results of localization due to driving

Table 1. Experiment Result

#	Existing Method		Proposed Method	
	MAX. err.	Var.	MAX. err.	Var.
1	35.212	60.245	20.123	49.407
2	37.934	69.987	26.375	51.097
3	39.745	57.122	32.109	60.678
4	29.245	50.234	29.915	58.705
5	35.444	60.356	32.543	51.340
6	36.865	65.978	23.257	57.097
7	34.678	53.332	28.069	50.838
8	33.925	68.078	28.903	55.845
9	31.176	59.316	20.715	54.436
10	37.025	70.770	22.207	53.258
AVG.	35.1249	61.5418	26.4216	54.2701

In the experiments, the proposed method and conventional methods was confirmed with the max. position error measured indicates respectively 39.745mm and 32.5432mm. As a result, the proposed method was found 24.78% to improve the positioning accuracy of the proposed method than that of the commercialized magnetic guide sensor.

5 Conclusion

This paper presented the position method of magnet using the Kalman Filter to minimize the errors of magnetic guide sensor. The magnetic guide sensors are usually used for operation and control of AGV. But, the existing method has the problems of interference and disturbance. To reduce the error of sensor, we are used for the Kalman filter. The proposed method is the position of the magnets that were used for Kalman filter where the Input is the measured position of histogram and velocity. For experiment, we tested the accuracy of the localization while AGV moves 10 times, and verified that the proposed method is efficient.

Acknowledgements. This research was supported by the MKE (The Ministry of Knowledge Economy), Korea, under the Specialized Field Navigation/Localization Technology Research Center support program supervised by the NIPA (National IT Industry Promotion Agency).

References

1. Evers, J.J.M., Koppers, S.A.J.: Automated guided vehicle traffic control at a container terminal. Transportation Research Part A: Policy and Practice 30, 21–34 (1996)
2. Setlur, P., Wagner, J.R., Dawson, D.M., Braganza, D.: A trajectory tracking steer-by-wire control system for ground vehicles. IEEE Trans. on Vehicular Technology 55(1), 76–85 (2006)
3. Schulze, L., Zhao, L.D.: Worldwide Development and Application of Automatic Guided Vehicle Systems. Service Operation and Informatics 2, 164–176 (2007)
4. Caruso, M.J., Bratland, T., Smith, C.H., Schneider, R.: A New Perspective on Magnetic Field Sensing. Sensor Magazine 15(12), 34–46 (1998)
5. Ryoo, Y.J., Kim, E.S., Lim, Y.C.: Intelligent Positioning System for Magnetic Sensor Based Autonomous Vehicle. In: SCIS & ISIS (2004)
6. Tsakonas, E.E., Sidiropoulos, N.D., Swami, A.: Optimal Particle Filters for Tracking a Time-Varying Harmonic or Chirp Signal. IEEE Transactions on Signal Processing 56, 4598–4610 (2008)
7. Carrasco, R., Cipriano, A.: Fuzzy logic based nonlinear kalman filter applied to mobile robots modelling. In: IEEE International Conference on Fuzzy Systems, vol. 3, pp. 25–29 (2004)
8. Chatterjee, A., Matsuno, F.: A neuro-fuzzy assisted extended kalman filter based approach for simultaneous localization and mapping (slam) problems. IEEE Transactions on Fuzzy Systems 15, 984–997 (2007)
9. Julier, S.J., Uhlmann, J.K., Durrant-Whyte, H.F.: A new approach for filtering nonlinear systems. In: Proceedings of the American Control Conference, pp. 1628–1632 (1995)

Indoor Flight Testing and Controller Design of Bioinspired Ornithopter

Jun-Seong Lee and Jae-Hung Han

Dept. of Aerospace Engineering, KAIST
291 Daehak-ro, Yuseong-gu, Daejeon 305-701, Korea
{jsl1009,jaehunghan}@kaist.ac.kr

Abstract. Indoor flight testing of a bioinspired ornithopter is conducted in this study and the dominant flight state variables such as body pitch angle, forward flight speed, altitude, wings and tail motions of the freely flying ornithopter are simultaneously measured by using a three-dimensional visual tracking system. A control-oriented system model of the ornithopter in trimmed level flight is established based on the recorded inputs and outputs dataset and the system matrices are fitted in a least-squares sense. To reduce the amplitude of the ornithopter body oscillations, the identified linear time-invariant system model is formulated to a disturbance-rejection problem and an optimal controller minimizing the quadratic performance index is designed. The continuous wing motion defined as the known disturbance deteriorates the pitch balance with respect to the center of gravity; however, the designed feedforward and feedback controller periodically activates the ornithopter tail and successfully reduces the magnitudes of the body oscillations.

Keywords: flapping-wing flight, ornithopter, feedback and feedforward control, visual tracking system, indoor flight testing.

1 Introduction

A bioinspired ornithopter is an engineering realization of nature's flyers with insect-like flexible flapping-wings and a bird-like tail. *Nano hummingbird* (Aerovironment, Inc.) and *SmartBird* (Festo, AG & Co. KG) are recently developed to mimic the flight of insects and birds, respectively; however, the ornithopter herein is much close to *Cybird* (Neuros, Co. Ltd) or Kinkade's *Slowhawk* [1-5].

For a trimmed level flight of the ornithopter, the continuous flapping-wing motion generates the positive net thrust and the ornithopter body is accelerated to gain the forward flight speed. The pitching moment induced by aerodynamic loads on tail makes the body have the positive pitch angle, and the flexible wings are aeroelastically deformed and twisted in spanwise direction to have the positive mean angle of attack. Thus, the ornithopter has an aerodynamic lift to compensate its own weight resulting from the lifting surfaces with the positive angle of attack [6-8]. In the cycle-averaging mean sense, the forces and moments acting on the ornithopter are statically balanced in both translational and rotational directions, and the pre-described wings and tail motions

S. Lee et al. (Eds.): Intelligent Autonomous Systems 12, AISC 193, pp. 825–834.
springerlink.com

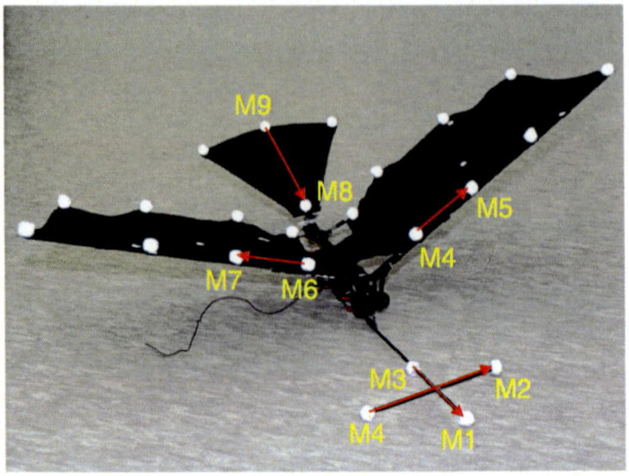

Fig. 1. Tested ornithopter platform and the retro-reflective marker placements for visual tracking system

with aeroelastic responses make the ornithopter have a trimmed level flight without active control of the tail. The previous numerical and experimental studies showed that the forward flight ornithopter has distinct trim conditions corresponding to the fixed wingbeat frequencies and tail trim angles [8-9]. In other words, the cycle-averaged mean values of flight state variables are mostly constant during the trim flight without any active feedback control. The aeroelasticity of wings and tail is known to passively reject the external disturbances and sustain the original flight trajectories in longitudinal direction. Limit-cycle oscillation (LCO) is one of the unique trim characteristics of flapping-wing flight [8, 10]. Note that the body movements of hovering flapping-wing flyers are quite different from those of the forward flight ornithopters; the disturbances on body are inherently damped out due to the symmetric and continuous flapping-wing motions in hover [11]. In case of the ornithopters, the existence of such passive damping has not been experimentally found yet.

In this study, indoor flight testing of the ornithopter was performed and the flight data were recorded by using a three-dimensional visual tracking system. Based on the acquired inputs/outputs dataset, a control-oriented system model in linear time-invariant form was established. The oscillatory behavior of the ornithopter body somehow enhances the flight stability; however, the oscillations should be minimized to collect the best visual information and prevent the malfunction of other sensitive on-board sensors such as inertial measurement unit, GPS, microprocessor. A linear quadratic controller was designed to reduce the amplitude of the ornithopter body oscillations; the combined feedforward and feedback controller periodically activated the tail so that the body pitching was successfully suppressed.

2 Indoor Flight Testing of Bioinspired Ornithopter

KARPE stands for KAIST Arena with Real-time Positioning Environment, which is similar to the RAVEN of MIT ACL [12] and the Micro Air Vehicle Integration and

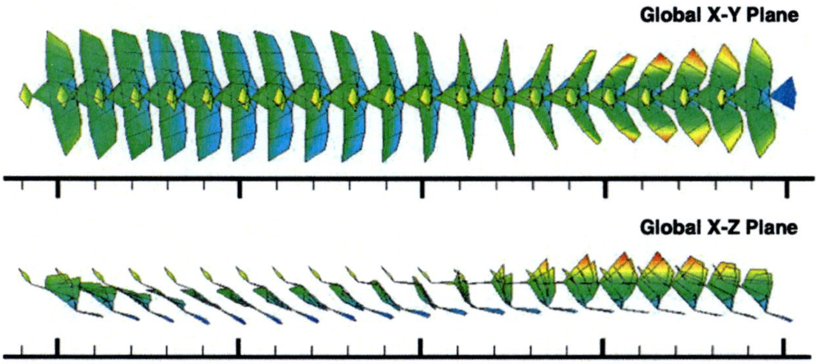

Fig. 2. Dorsal and left-side view of freely flying ornithopter at wingbeat frequency 10Hz and forward flight speed 4.2m/s (frame rate of strobe cameras: 200 fps)

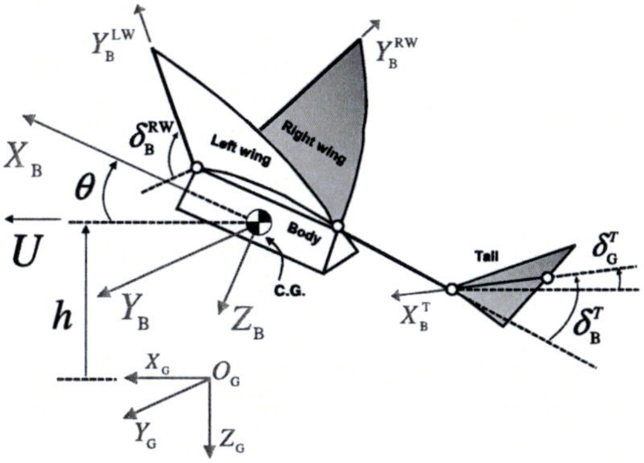

Fig. 3. Definition of flight state variable and coordinate system for free flight testing of the ornithopter flight testing of the ornithopter

Application Research Institute (μAVIARI) at Wright Patterson Air Force Base [13]. KARPE consists of a three-dimensional visual tracking system with 12 strobe IR cameras (Eagle digital, Motion Analysis Corp.). The cameras are installed in an indoor room capturing a volume of 12 m long, 9 m wide, and 5 m height. The system provides sub-mm spatial resolution with a 12 mm diameter retro-reflective marker (sphere-shaped) and the uncertainty of orientation angles such as body pitch angle, tail elevation angle, wings and tail motion are also examined as 0.5 to 1.0 % of the measured values. Traces of the attached marker sets can be recorded and transmitted almost real-time to the NI's PXIE controller or dSPACE's DSP controller for the feedback control applications [7].

Fig. 1 shows the ornithopter platform used for the indoor free flight testing; 20 markers were installed to the wings, tail and body. The overall flight behavior including the detailed aeroelastic deformations on wings and tail is illustrated in Fig. 2. Note that the contour color represents the spatial displacement in global Z direction. The visual tracking system is quite promising to obtain the flight state variables without weight penalty and changes of flight dynamics. To construct a control-oriented system model of the freely flying ornithopter, 9 markers (M1~M9) were chosen and the dominant flight state variables such as the forward flight speed U, the altitude h, the body pitch angle θ, the flapping-wing motion δ_B^{RW} (= u_{wing}), and the tail elevation angle δ_B^T (= u_{tail}) were calculated from the vectors defined in Figs. 1 and 3. To achieve the straight and level flight in KARPE, the ornithopter was hand-launched with the fixed wingbeat frequency of 10 Hz and the tail elevation angle of 33°. After the several times of free flight trials, the cycle-averaged mean of the dominant flight state variables were found as: the mean forward flight speed of 4.2m/s, the mean body pitch angle of 47.3°, and the mean altitude of 1.97m. The ornithopter flight dynamics can be separated into longitudinal and lateral planes; it can also be categorized as the cycle-averaged mean flight dynamics and the zero-mean, time-varying flight dynamics, as described in Ref. [14]. Thus, the measured dataset were offset by the cycle-averaged means and the zero-mean, time-varying flight state variables are prepared for identifying the control system model as in Fig. 4. More details of flight dynamic characteristics of the freely flying ornithopter can be found in Ref. [7].

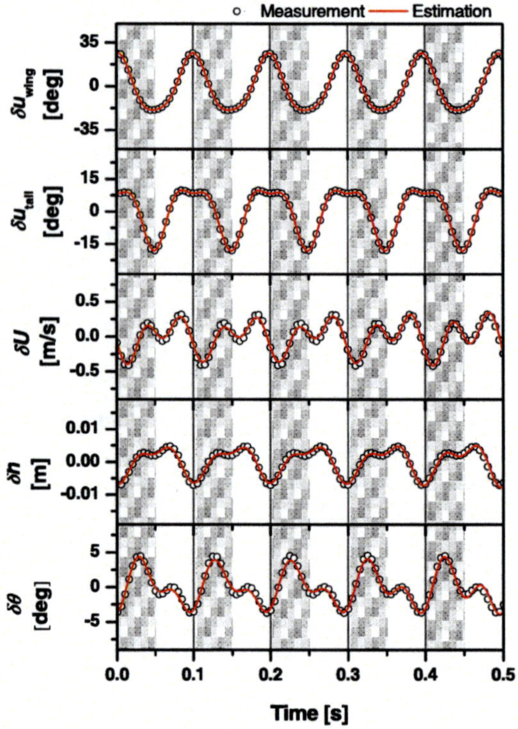

Fig. 4. Time history of zero-mean time-varying dominant flight state variables of freely flying ornithopter (Symbol: measurement data, Solid line: Estimation results)

3 System Identification and Controller Design of Bioinspired Ornithopter

The cycle-averaged mean flight dynamics of the forward flight ornithopter is determined by the constant values of wingbeat frequency and the tail elevation angle. As explained earlier, the mean flight dynamics of the ornithopter shows a stable LCO, believed to be partially induced by the aeroelasticity of wings and tail. This paper focuses more on the zero-mean, time-varying flight dynamics than the cycle-averaged one because the suppression of body oscillation is the control objective. In the previous numerical study on the suppression of the time-varying dynamics [14], the prediction error method was introduced to model the forward flight ornithopter. The modeling error and sensor noise were considered as the Gaussian noise and the linear-quadratic Gaussian regulator was applied to construct the feedback controller. Herein, the system matrices were fitted in least-squares sense and due to the advantage of the visual tracking system, all of the interested state variables were measured with a negligible magnitude of noise and a reasonable sampling rate.

Let the identified system model have the form:

$$\delta\mathbf{x}_{3\times1}[k+1]=\hat{\mathbf{\Phi}}_{3\times3}\delta\mathbf{x}_{3\times1}[k]+\hat{\mathbf{\Gamma}}_{3\times2}\delta\mathbf{u}_{2\times1}[k] \tag{1}$$

where

$$\delta\mathbf{x}_{3\times1}[k]=\left[\delta U[k] \quad \delta h[k] \quad \delta\theta[k]\right]^T$$
$$\delta\mathbf{u}_{2\times1}[k]=\left[\delta u_{\text{wing}}[k] \quad \delta u_{\text{tail}}[k]\right]^T$$

$$\begin{aligned}\hat{\mathbf{\Phi}}_{3\times3}&=\begin{bmatrix}+9.96\times10^{-1} & 3.27\times10^{0} & 2.55\times10^{-3}\\ -7.07\times10^{-4} & 9.85\times10^{-1} & 2.17\times10^{-6}\\ -5.29\times10^{-1} & 1.93\times10^{+1} & 9.91\times10^{-1}\end{bmatrix}\\ \hat{\mathbf{\Gamma}}_{3\times2}&=\left[\hat{\gamma}_{3\times1}^{\text{wing}} \quad \hat{\gamma}_{3\times1}^{\text{tail}}\right]\\ &=\begin{bmatrix}+2.89\times10^{-4} & -6.17\times10^{-6} & +4.61\times10^{-3}\\ +4.99\times10^{-4} & -6.84\times10^{-7} & -8.46\times10^{-5}\end{bmatrix}^T\end{aligned} \tag{2}$$

The responses of the identified system are plotted in Fig. 4 (red solid line) with the measured data (symbol). The time histories of two dataset were sufficiently similar to each other. The eigenvalues of $\hat{\mathbf{\Phi}}_{3\times3}$ indicated that the identified system was stable and the controllability and observability were also satisfied. For a given system model, various control methods can be applied to achieve the desired performance of the control system [15-18]. In this study, the disturbance-rejection problem was formulated by using the identified system model in Eq. (1) while the flapping-wing motion was considered as the known disturbance and the tail was the only control input of the control system. Even the wing motion deteriorates the body dynamics, flapping-wing motion should be continued during the trimmed level flight. Due to the

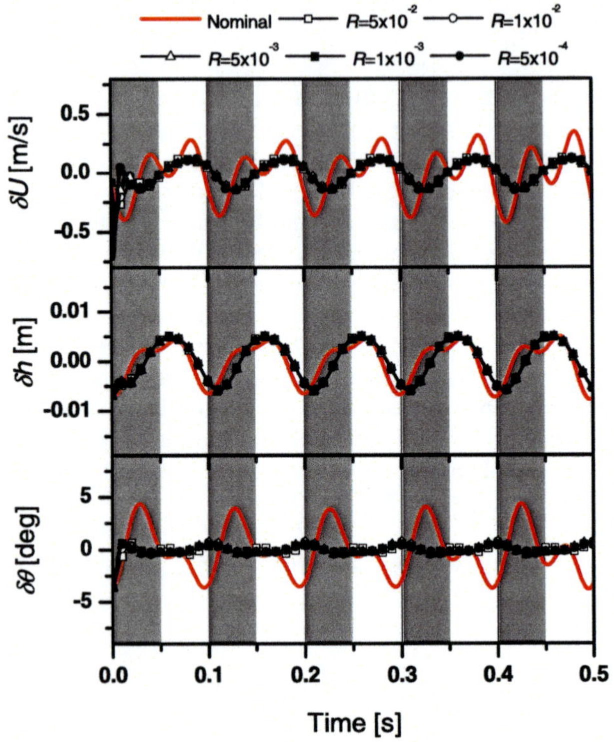

Fig. 4. Time histories of the suppressed zero-mean time-varying dominant flight state variables (Simulation results, Symbol: control activated cases, Solid line: nominal system, control deactivated case)

limited degrees of freedom, the modification of wing kinematics is very difficult for the ornithopter. Thus, active control of tail motion is quite natural rather than the change of wing motion during the trim flight.

Let w[k] is the known disturbance of the system and the input matrix $\hat{\Gamma}_{3\times 2}$ in Eq. (2) is divided as in Eq. (3).

$$\delta\mathbf{x}_{3\times1}[k+1]=\hat{\Phi}_{3\times3}\delta\mathbf{x}_{3\times1}[k]+\hat{\gamma}_{3\times1}^{\text{tail}}\delta u_{\text{tail}}^{*}[k]+\underbrace{\hat{\gamma}_{3\times1}^{\text{wing}}\delta u_{\text{wing}}[k]}_{w[k]} \tag{3}$$

To reduce the magnitude of the oscillations, the performance index to be minimized is defined in Eq. (4).

$$J=\int_0^{\infty}\left(\delta\mathbf{x}_{3\times1}^{T}Q_{3\times3}\delta\mathbf{x}_{3\times1}+\delta u_{\text{tail}}^{T}R_{1\times1}\delta u_{\text{tail}}\right)d\tau \tag{4}$$

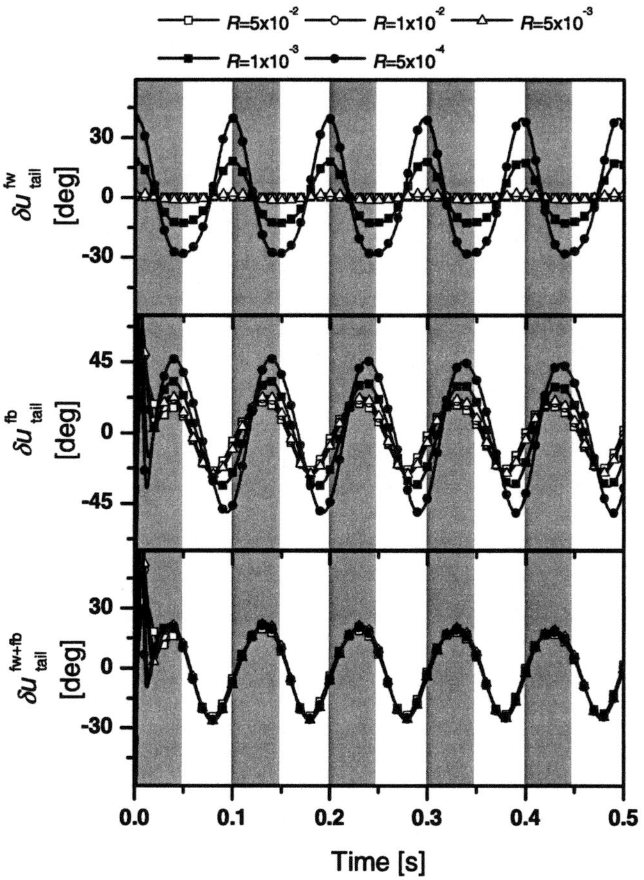

Fig. 5. Time histories of the feedforward, feedback and combined feedforward and feedback control inputs of tail (Simulation results)

Applying the linear quadratic control framework [19], the optimal control solution of the disturbance-rejection problem consists of feedforward control, $\delta u_{\text{tail}}^{\text{fw}}$ and feedback control, $\delta u_{\text{tail}}^{\text{fb}}$. $\delta u_{\text{tail}}^{\text{fw}}$ is the control effort on the known disturbance rejection and $\delta u_{\text{tail}}^{\text{fb}}$ corresponds to the state regulation via full-state feedback (Eqs. (5) and (6)).

$$\delta u_{tail}^{*} = \delta u_{\text{tail}}^{\text{fw}} + \delta u_{\text{tail}}^{\text{fb}} \tag{5}$$

where

$$\begin{aligned} \delta u_{\text{tail}}^{\text{fb}} &= -K_{\text{LQR}}\,\delta \mathbf{x}_{3\times1} = -\{R_{1\times1}^{-1}(\hat{\gamma}_{3\times1}^{\text{tail}})^{T}\,\overline{P}_{3\times3}\}\,\delta \mathbf{x}_{3\times1} \\ \delta u_{\text{tail}}^{\text{fw}} &= -R_{1\times1}^{-1}(\hat{\gamma}_{3\times1}^{\text{tail}})^{T}\,\mathrm{Re}[-\{\hat{\Phi}_{3\times3} - \hat{\gamma}_{3\times1}^{\text{tail}}K_{\text{LQR}} + j\omega\mathbf{I}_{3\times3}\}^{-1}]\,\overline{P}_{3\times3}\,w[k] \end{aligned} \tag{6}$$

For the steady-state case, the matrix Riccati equation in Eq. (7) has time-invariant solution, $\overline{P}_{3\times3}$ and is easily calculated for obtaining the optimal control gain, K_{LQR}.

$$\hat{\Phi}^T_{3\times3}\overline{P}_{3\times3}+\overline{P}_{3\times3}\hat{\Phi}_{3\times3}+Q_{3\times3}-\overline{P}_{3\times3}\hat{\gamma}^{tail}_{3\times1}R^{-1}_{1\times1}(\hat{\gamma}^{tail}_{3\times1})^T\overline{P}_{3\times3}=0 \tag{7}$$

The weighting matrices, $Q_{3\times3}$ and $R_{1\times1}$ were tuned to achieve the reduction of states $\delta\mathbf{x}_{3\times1}$ and the values were finally found as Eq. (8). The range of tail motion was adjusted by the five different weighting values of $R_{1\times1}(=R_j, j=1\sim5)$.

$$\begin{aligned} Q_{3\times3} &= \text{diag}([10 \quad 10 \quad 20]) \\ R_{1\times1} &= 5\times10^{-2}, 1\times10^{-2}, 5\times10^{-3}, 1\times10^{-3}, 5\times10^{-4} \end{aligned} \tag{8}$$

As shown in Fig. 5, the oscillatory behaviors of the states were successfully minimized and 86% of the suppression of body pitch angle, $\delta\theta$ was found at $R_{1\times1}=5\times10^{-4}$. Table 1 summarizes the control results in the root-mean-squared (RMS) sense and the time history of feedforward, feedback and combined of both control inputs were shown in Fig 6. The changes of the weighting value, $R_{1\times1}$ can be used to adjust the portion of the control efforts between feedforward and feedback controls; however, the combined control efforts, δu^{fw+fb}_{tail}, were almost the same regardless of the variation in $R_{1\times1}$ as shown in Fig 6. Note that the tail motion in Figs. 4 and 6 has different phase angles with respect to the wing motion and the simple sweeping of the tail motion might be helpful to reduce the magnitude of body oscillations of the ornithopter [15].

Table 1. Summary of suppression of flight state variables

	Nominal	R_1	R_2	R_3	R_4	R_5
δU [m/s]	0.20			0.10		
δh [m]			3.8			
$\delta\theta$ [deg]	2.31	0.45	0.49	0.48	0.39	0.33

4 Conclusion

An indoor flight testing of a bioinspired ornithopter was conducted at KARPE. The dominant flight state variables of the freely flying ornithopter were measured and the control-oriented system model of the ornithopter in trimmed level flight was established in the least-squares sense. By applying the linear-quadratic control framework, the oscillation of the body pitch angle due to the continuous wing motion was successfully minimized by introducing the active control of the tail. The resultant control input of tail was quite different from the one of the nominal system; it had a dominant single frequency, periodic shape and a certain phase angle with respect to the wing motion. In future work, the designed controller is going to be implemented in the same environment, KARPE and the automation of the indoor ornithopter flight will be experimentally made.

Acknowledgements. This work was supported by Mid-career Researcher Program (2011-0015569) through NRF grant funded by the MEST. The first author would like to thank the Brain Korea 21 Project of 2012.

References

1. Hylton, T., Martin, C., Tun, R., Castelli, V.: The DARPA Nano Air Vehicle Program. In: Proc. of 50th AIAA Aerospace Science Meeting, Nashville, TN, January 06-12, AIAA 2012-0583 (2012)
2. Keennon, M., Klingebiel, K., Won, H., Audriukov, A.: Development of the Nano Hummingbird: A Tailless flapping wing micro air vehicle. In: Proc. of 50th AIAA Aerospace Science Meeting, Nashville, TN, January 06-12, AIAA 2012-0588 (2012)
3. Festo, A.G., Co, K.G.: Brochure of SmartBird (April 2011), `http://www.festo.com/`
4. Han, J.-H., Lee, J.-S., Kim, D.-K.: Bio-inspired flapping UAV design: A university perspective. In: Proc. SPIE, vol. 7295, pp. 7295I1–7295I12 (2009)
5. Park, J.H., Yoon, K.-J.: Designing a biomimetic ornithopter capable of sustained and controlled flight. J. Bionic Eng. 7(1), 39–47 (2008)
6. Pfeiffer, A.T., Lee, J.-S., Han, J.-H., Baier, H.: Ornithopter flight simulation based on flexible multi-body dynamics. J. Bionic Eng. 7(1), 102–111 (2010)
7. Lee, J.-S., Han, J.-H.: Experimental study on flight dynamics of bioinspired ornithopter: Free flight testing and wind tunnel testing. Smart Materials and Structures (accepted for publication, April 2012)
8. Lee, J.-S., Kim, J.-K., Kim, D.-K., Han, J.-H.: Longitudinal flight dynamics of bioinspired ornithopter considering fluid-structure interaction. J. Guid. Control. Dynam. 34(3), 667–677 (2011)
9. Grauer, J., Hubbard Jr., J.: Inertial measurements from flight data of a flapping-wing ornithopter. J. Guid. Control. Dynam. 32(1), 326–331 (2009)
10. Kim, J.-K., Lee, J.-S., Han, J.-H.: Passive longitudinal stability of ornithopter flight. J. Guid. Control. Dynam. 35(2), 669–673 (2012)
11. Cheng, B., Deng, X.: Translational and rotational damping of flapping flight and its dynamics and stability at hovering. IEEE T. Robot. 27(5), 849–864 (2011)
12. How, J., Bethke, B., Frank, A., Dale, D., Vian, J.: Real-time indoor autonomous vehicle test environment. IEEE Contr. Syst. Mag. 28(2), 51–64 (2008)
13. Maniar, G., Randall, R., Shkarayev, S.: Kinematics of Free-flight Ornithopters. In: Proc. of 50th AIAA Aerospace Science Meeting, Nashville, TN, January 06-12, AIAA 2012-0666 (2012)
14. Kim, J.-K., Lee, J.-S., Han, J.-H.: Limit-cycle oscillation suppression of ornithopter longitudinal flight dynamics. In: Proc. of AIAA GNC Conference, Portland, OR, August 08-11, AIAA 2011-6404 (2011)
15. Lee, J.-S., Kim, J.-K., Han, J.-H., Ellington, C.P.: Periodic tail motion linked to wing motion affects the longitudinal stability of ornithopter flight. J. Bionic. Eng. 9(1), 18–28 (2012)

16. Grauer, J., Hubbard Jr., J.: Modeling of ornithopter flight dynamics for state estimation and control. In: Proc. of America Control Conf., Baltimore, MD, June 30-July 02 (2010)
17. Dietl, J., Garcia, E.: Ornithopter control with periodic infinite horizon controllers. J. Guid. Control. Dynam. 34(5), 1412–1421 (2011)
18. Krashanitsa, R., Silin, D., Shakarayev, S., Abate, G.: Flight dynamics of a flapping-wing air vehicle. Int. J. MAV. 1(1), 35–49 (2009)
19. Dorato, P., Abdallah, C.T., Cerone, V.: Linear-Quadratic Control: An Introduction, pp. 43–62. Prentice-Hall, Inc. (1995)

Effect of Passive Body Deformation of Hawkmoth on Flight Stability

Ryusuke Noda, Masateru Maeda, and Hao Liu*

Graduate School of Engineering, Chiba University
1-33 Yayoi-cho, Inage-ku, Chiba 263-8522, Japan
hliu@faculty.chiba-u.jp

Abstract. In this study, the effect of passive body deformation on flight stability during insect flapping flight is investigated numerically. We developed a flexible body dynamic solver for a three-dimensional flexible beam model and coupled it with an *in-house* fluid dynamics solver. With this integrated model, hawkmoth free flights are simulated and analyzed systematically with six cases, in which the joint stiffness between thorax and abdomen varied from extremely rigid to very flexible. Our results indicate that the passive body deformation works likely altering the aerodynamic torque, the body attitude and the flight trajectory. We further found that the most stable flight can be achieved by a moderate joint stiffness, in which the body attitude remains approximately around the initial angle of 40 degree. This points to the importance that the flexible body and its passive deformation during flapping-wing flight are capable to enhance stable flight and flight control.

Keywords: insect flight, flexible body dynamics, stability.

1 Introduction

The effect of wing deformation of insects on their flight ability has attracted attentions from researchers in either biology [1-3] or engineering [4-7]. The work on the body deformation, however, is rare [8, 15], and with our knowledge, the work on the passive body deformation in terms of flight stability or performance is not published. One of the difficulties in treating flexibility in flying insects is that fluid dynamics and passive structural deformation couple each other: fluid force makes the shape of the body changed; deformed body generates altered fluid force. We developed a flexible body dynamics solver where insect body is modeled as a three dimensional beam. By coupling the solver with an existing fluid dynamics solver, numerical simulation of a hovering hawkmoth with flexible body is achieved. We tested several values of elasticity (from rigid to very flexible) for the joint between thorax and abdomen and saw how it affects aerodynamic force/torque generation and especially history of the flier's attitude. The result showed that in the most flexible case the body angle increased 45 degrees in five wingbeat cycles, while in the stiffer cases the trends were similar to that in the rigid case.

---

* Corresponding author.

S. Lee et al. (Eds.): Intelligent Autonomous Systems 12, AISC 193, pp. 835–842.
springerlink.com

2 Method

2.1 Insect Model

Nakata [9] showed that the larger the insect, the smaller the wing dimensionless stiffness. It is not so unnatural to assume this trend also applies for the body. Therefore we used a hawkmoth as a model insect in the current study, where large body deformation compared to the smaller insects is expected, thus suitable for assess its effect. The morphological and wing kinematics models (Fig. 1) are the same as those in Aono *et al.* [10].

2.2 Computational Fluid Dynamics

For the flow field computation around insect body and wings, we used a computational fluid dynamics solver developed by Liu [11]. In the present study the global (background) grid is replaced with the Cartesian structured grid and the body grid block is now a local grid. The major parameters and the grid used in the fluid dynamics computation are shown in Table 1 and Fig. 1, respectively. The number of grid used (i×j×k) are: background grid, 61×61×57; body grid, 37×37×9; each wing grid, 37×33×11.

Table 1. Parameters for fluid dynamics simulation

Reynolds number	Re	5400
Reduced frequency	k	0.35
Mean chord length	c_m	18.6 mm
Reference velocity	U_{ref}	4.34 m/s
Wingbeat frequency	f	26.0 Hz
Initial thorax angle (body angle if rigid body)	β_{ini}	40.0 deg

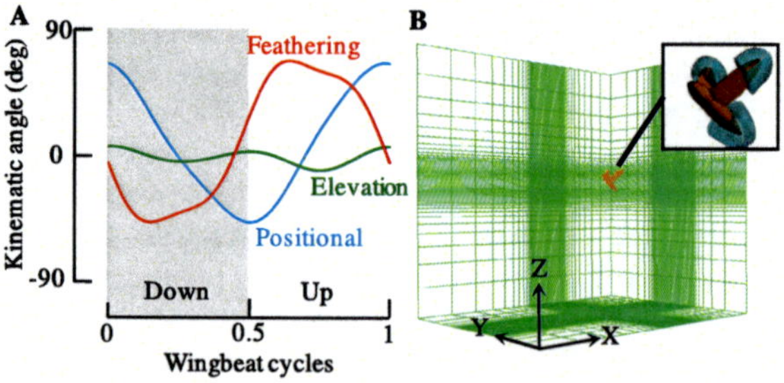

Fig. 1. Hawkmoth model. (A) Wing kinematics. (B) Computational grid for fluid dynamics simulation with the global coordinates.

2.3 Flexible Body Dynamics

In order to reproduce the body deformation, we developed a flexible body dynamics solver, where three dimensional flexible beam with 41 elements takes the load (Figs. 2A and 2B). The deformation of the beam is solved with the finite element method (FEM). Note that in the present computation, we only considered the passive body deformation and not the active body deformation which is generated by its muscle. The aerodynamic force acting on the body itself is too small to provoke the deformation. Rather, the aerodynamic forces on the wings are introduced to the two wing base nodes, becoming the main source for the body deformation. The governing equations of FBD solver is written as

$$\mathbf{M}\ddot{\mathbf{q}}+\mathbf{C}\dot{\mathbf{q}}+\mathbf{K}_\mathbf{L}\mathbf{q}+\mathbf{f}_{\mathbf{int}}(\mathbf{q})-\mathbf{f}_{\mathbf{ext}}=\mathbf{0} \tag{1}$$

where $\mathbf{M}$ is the lumped mass matrix; $\mathbf{C}$ is the damping matrix; $\mathbf{K}_\mathrm{L}$ is the linear stiffness matrix; $\mathbf{f}_\mathbf{int}$ is the nonlinear restoration force; $\mathbf{f}_\mathbf{ext}$ is the external force; $\mathbf{q}$ is the positions and orientations in the global frame. Note that in this problem, each element changes the position not only by deformations but by whole-body motion. Therefore the restoration force is expressed as a nonlinear function of the displacement. For the integration, we use Newmark-β scheme for nonlinear problems. The tangent stiffness matrix $\mathbf{K}_\mathrm{N}$ is required in the formulation, which is expressed as the partial derivative of $\mathbf{f}_\mathbf{int}$ as

$$\mathbf{K}_\mathbf{N}=\partial\mathbf{f}_\mathbf{int}/\partial\mathbf{q}. \tag{2}$$

The $\mathbf{K}_\mathrm{N}$ can be divided into two parts as

$$\mathbf{K}_\mathbf{N}=\mathbf{K}_\mathbf{G}+\mathbf{K}_\mathbf{O} \tag{3}$$

where $\mathbf{K}_\mathrm{G}$ is the geometric stiffness matrix and $\mathbf{K}_\mathrm{O}$ is the elastic deformation matrix. [12] The flexible body dynamics solver developed was validated with both fixed-end and free-end flexible beam benchmark problems [13].

The fluid dynamics solver and the flexible body dynamics solver need to exchange information to achieve the computation. This process is called coupling and we used the following manner [14]. (1) Predict the displacement of each node at time t_{n+1} from the information (displacement, velocity and acceleration) at time t_n; (2) Update the position of the fluid grid based on the increments of each node; (3) Transfer fluid force to the structure; and (4) advance the time in the FBD solver.

The major parameters used in the flexible body dynamics computation are shown in Table 2 and 3. We have chosen the location for the center of mass of the body as follows. The mass of each block (head, thorax, abdomen, and wings) was assigned according to [15]. Next, the density distribution to the nodes in each block was determined so that the rigid case (described below) can maintain relatively stable hovering. Therefore caution has to be taken that the density distribution may not reflect that of the real hawkmoth. In order to see the effect of the different body flexibility on the flight dynamics, several different Young's moduli at the three joint elements between thorax and abdomen (E_{j0} and E_{j1}, where $E_{j1} := 2E_{j0}$. see Fig. 2D) were chosen. The Young's modulus $E_{j0} = 10^{11}$ pascals was selected to represent a rigid body model. In addition, $E_{j0} = 10^7$, 10^5, 5×10^4, 4×10^4, and 3×10^4 pascals were tested. The values of E_{j0} were determined so that the resultant deformation angles (Fig. 2C) fall within an experimental value of approximately five degrees [8]. In addition, we assumed a uniform Young's modulus distribution across each cross section.

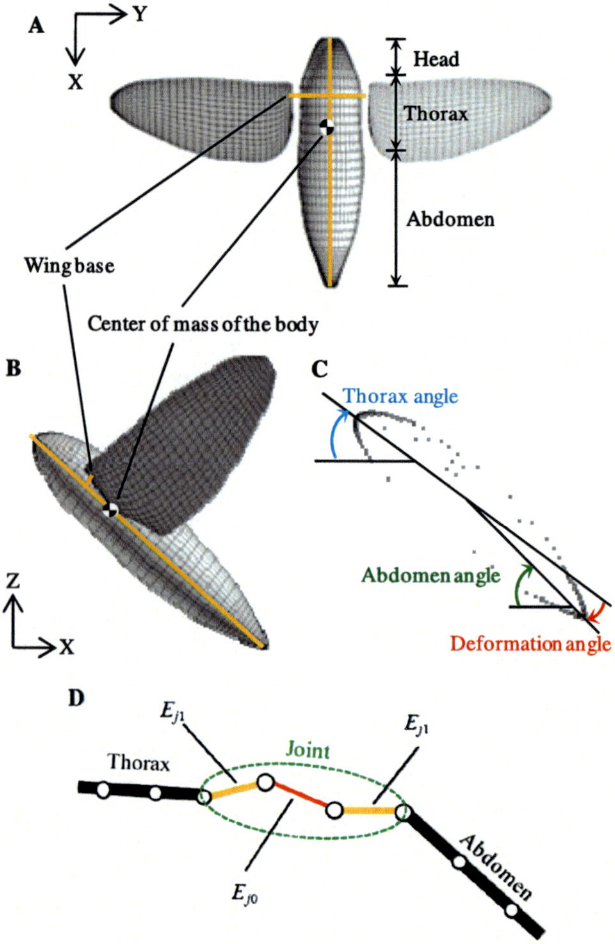

Fig. 2. Flexible body model. (A) and (B) A flexible beam (orange) is placed in the middle of the insect grid which is used fort the fluid dynamics simulation (gray). Note the grid is also used for calculating the moment of inertia. (C) Definitions of thorax, abdomen, and deformation angles. Note thorax and abdomen angles are measured from the horizon. (D) Schematic of the structural elements around the thorax-abdomen joint. Young's modulus of the middle joint element is termed E_{j0}; each side joint element is E_{j1}, defined as $E_{j1} := 2E_{j0}$. All the other elements are given the same (high) Young's modulus.

Table 2. Mass of the hawkmoth model [15]

Body mass		
Head	m_h	107 mg
Throax	m_t	596 mg
abdomen	m_a	818 mg
Wing mass (one wing)	m_w	35 mg

Table 3. Number of nodes and Young's modulus for each structural block

	Number of elements	Young's modulus (Pa)
Head	8	10^{11}
Thorax	7	10^{11}
Joint between thorax and abdomen	3	*E_{j0}, E_{j1}
Abdomen	21	10^{11}
Left wing base	1	10^{11}
Right wing base	1	10^{11}

* See the description in the main text and Fig. 2.

3 Results

Our simulation (summarized in Fig. 3) shows that passive body deformation alters aerodynamic torque, body attitude and flight trajectory. There was almost no difference between the most rigid two cases ($E_{jo} = 1\times10^{11}$ Pa and $E_{jo} = 1\times10^{7}$ Pa). The next rigid case ($E_{jo} = 1\times10^{5}$ Pa) performed the most stable flight, where body attitude remained around initial angle of 40 degrees (Fig. 3A) while horizontal deviation was minimal (Fig. 3E). The most flexible case ($E_{jo} = 3\times10^{4}$ Pa) exhibited the most unstable flight, where body attitude reached 80 degrees nose-up while the body moved backward. The intermediate flexibility cases ($E_{jo} = 4\times10^{4}$ Pa and $E_{jo} = 5\times10^{4}$ Pa) showed moderately stable trends of body attitudes and flight trajectories, similar to those of more rigid cases. The body deformation angle (Fig. 3B) is calculated by subtracting abdomen angle (dotted line in Fig. 3A) from thorax angle (solid line in Fig. 3A).

4 Discussion

Although the link is not sufficiently clear, from our simulation, especially from the result of the most flexible case ($E_{jo} = 3\times10^{4}$ Pa, red lines in Fig. 3) it is assumed that the body deformation (Fig. 3B) leads to the change in the aerodynamic torque (Figs. 3C and 3D), which determines the overall body attitude (Fig. 3A) and flight trajectory (Fig. 3E). One possible story is that the reduction in the moment of inertia in the

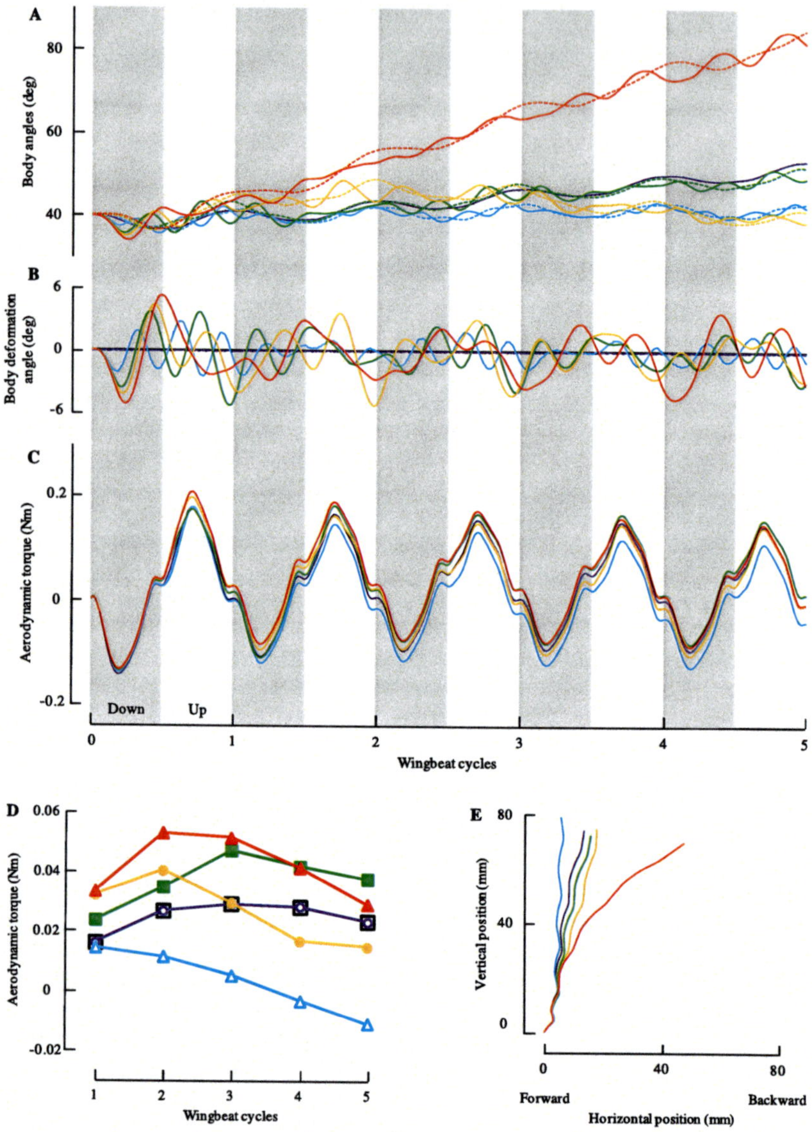

Fig. 3. Time series of simulated flight. (A) Thorax angle (solid lines) and abdomen angle (dotted lines). (B) Body deformation angle. (C) Y component (pitching) of aerodynamic torque around body center of mass. (D) Y component (pitching) of wingbeat cycle-averaged aerodynamic torque around body center of mass. (E) Location of body center of mass in the XZ-plane, starting from origin at time $t = 0$. Key to colors: red, Young's modulus of the joint between throax and abdomen $E = 3\times10^4$ Pa; orange, $E = 4\times10^4$ Pa; green, $E = 5\times10^4$ Pa; blue, $E = 1\times10^5$ Pa; purple, $E = 1\times10^7$ Pa; black, $E = 1\times10^{11}$ Pa.

y-(pitching) axis due to the body deformation led to the increase in the pitching angular velocity, thus increased wing velocity and aerodynamic force, and finally aerodynamic torque. Another possibility includes the reaction of the abdomen rotation revolved the thorax, thus wing base, in the opposite direction of the abdomen motion, which resulted in the increase in the wing velocity. Since there are several factors affecting the aerodynamic force generation, some manipulation may be necessary to separate out which has the largest impact. For example, preparative computation for achieving the "trimmed hovering" by choosing the appropriate wing kinematics by the aid of optimization [16] or via flight control (Gao Na, unpublished work) would be useful.

In addition, we have also tested the effect of the location of the body center of mass (not shown), where it was revealed that the relative location of the body center of mass against wing bases has a large impact on the body attitude and flight trajectory. For the insect flight research, this means the precise measurement of the body center of mass is inevitable. For a flapping-wing micro air vehicle (fMAV) without tail (like insects rather than birds), this means that the designer needs to carefully choose the location of the body center of mass.

Acknowledgements. R. N. thanks Toshiyuki Nakata for checking and commenting on the flexible body dynamics program.

References

1. Combes, S.A., Daniel, T.L.: Into thin air: contributions of aerodynamic and inertial-elastic forces to wing bending in the hawkmoth Manduca sexta. J. Exp. Biol. 206, 2999–3006 (2003)
2. Zhao, L., Deng, X., Sane, S.P.: Modulation of leading edge vorticity and aerodynamic forces in flexible flapping wings. Bioinsp. Biomim. 6, 036007 (2011)
3. Nakata, T., Liu, H.: Aerodynamic performance of a hovering hawkmoth with flexible wings: a computational approach. Proc. R. Soc. B 279, 722–731 (2011)
4. Shyy, W., Aono, H., Chimakurthi, S.K., Trizila, P., Kang, C.-K., Cesnik, C.E.S., Liu, H.: Recent progress in flapping wing aerodynamics and aeroelasticity. Prog. Aerospace Sci. 46, 284–327 (2010)
5. Nakata, T., Liu, H., Tanaka, Y., Nishihashi, N., Wang, X., Sato, A.: Aerodynamics of a bio-inspired flexible flapping-wing micro air vehicle. Bioinspir. Biomim. 6, 045002 (2011)
6. Mahardika, N., Viet, N.Q., Park, H.C.: Effect of outer wing separation on lift and thrust generation in a flapping wing. Bioinsp. Biomim. 6, 036006 (2011)
7. Tanaka, H., Shimoyama, I.: Forward flight of swallowtail butterfly with simple flapping motion. Bioinsp. Biomim. 5, 026003 (2010)
8. Hinterwirth, A.J., Daniel, T.L.: Antennae in the hawkmoth Manduca sexta (Lepidoptera, Sphingidae) mediate abdominal flexion in response to mechanical stimuli. J. Comp. Physiol. A 196, 947–956 (2010)
9. Nakata, T.: Simulation-based study on aerodynamic performance of flexible flapping wings. Chiba University PhD thesis (2012)
10. Aono, H., Shyy, W., Liu, H.: Near wake vortex dynamics of a hovering hawkmoth. Acta. Mech. Sin. 25, 23–36 (2009)

11. Liu, H.: Integrated modeling of insect flight: From morphology, kinematics to aerodynamics. J. Comput. Phys. 228, 439–459 (2009)
12. Chan, S.L., Chui, P.P.T.: Non-Linear Static Analysis of steel Frames with Semi-Rigid Connections. Elsevier Science Ltd, London (2000)
13. Simo, J.C., Vu-Quoc, L.: On the dynamics in space of rods undergoing large motions - A geometrically exact approach. Comput. Methods Appl. Mech. Eng. 66, 125–161 (1988)
14. Piperno, S., Farhat, C.: Partitioned procedures for the transient solution of coupled aeroelastic problems - part ii: energy transfer analysis and three-dimensional applications. Comput. Methods Appl. Mech. Eng. 190, 3147–3170 (2001)
15. Hedrick, T.L., Daniel, T.L.: Flight control in the hawkmoth *Manduca sexta*: the inverse problem of hovering. J. Exp. Biol. 209, 3110–3114 (2006)
16. Wu, J.H., Zhang, Y.L., Sun, M.: Hovering of model insects: simulation by coupling equations of motion with Navier-Stokes equations. J. Exp. Biol. 212, 3313–3329 (2009)

ACS-PRM: Adaptive Cross Sampling Based Probabilistic Roadmap for Multi-robot Motion Planning

Zhi Yan, Nicolas Jouandeau, and Arab Ali Cherif

Advanced Computing Laboratory of Saint-Denis (LIASD), Paris 8 University, 93526
Saint-Denis, France
{yz,n,aa}@ai.univ-paris8.fr

Abstract. In this paper we present a novel approach to multi-robot motion planning by using a probabilistic roadmap (PRM) based on adaptive cross sampling (ACS). The proposed approach, we call ACS-PRM, consists of three steps, which are C-space sampling, roadmap building and motion planning. Firstly, a sufficient number of points should be generated in C-space on an occupancy grid map by using an adaptive cross sampling method. Secondly, a roadmap should be built while the potential targets and milestones are extracted by post-processing the result of sampling. Finally, the motion of robots should be planned by querying the constructed roadmap. In contrast to previous approaches, our ACS-PRM approach is designed to plan separate kinematic paths for multiple robots to minimize the problem of congestion and collision in an effective way so as to improve the planning efficiency. Our approach has been implemented and evaluated in simulation. The experimental results demonstrate the total planning time can be significantly reduced by our ACS-PRM approach compared with previous approaches.

1 Introduction

Motion planning is a fundamental problem in robotics. It could be explained as producing a continuous motion for an agent, that connects a start configuration and a goal configuration, and avoid collision with any static obstacles or other agents in an environment. Agent and obstacle geometry are generally described in a 2D or 3D workspace, and the motion could be represented as a path in configuration space. Motion planning algorithms are widely applied in many fields, such as bioinformatics, robotic surgery, industrial automation, planetary exploration, and intelligent transportation system.

The multi-robot system (MRS) is proposed to deal with some problems that are difficult or impossible to be solved by a single robot, or to improve the system implementation efficiency in some missions completed by multi-robot rather than a single robot [1], [2]. The biggest challenge for the MRS is coordination. Without coordination, it will not only lower the system efficiency, but also lead to the failure of the entire system in extreme cases. In this paper, we consider the issue of coordinated motion planning for a homogeneous team of autonomous mobile robots in structured environments. Most of the proposed approaches for multi-robot motion planning usually have the problems of resource conflict such as congestion and collision [5], [9]. We arrange these cases to the waiting situation problem [10]. To handle this practical problem, this paper presents

S. Lee et al. (Eds.): Intelligent Autonomous Systems 12, AISC 193, pp. 843–851.
springerlink.com

a novel approach for multi-robot motion planning by using a probabilistic roadmap planner (PRM) which is based on manner of adaptive cross sampling (ACS).

The rest of the paper is organized as follows: Section 2 describes an overview of some related works; Section 3 describes our ACS-PRM approach; Section 4 presents the experimental results obtained with our approach; and the paper is concluded in Section 5 at last.

2 Related Work

In our previous work [10], we considered the issue of using separate topological graphs to coordinated multi-robot motion planning for exploration mission. This work aims at solving the waiting situations in the process of the robot motion planning. In particular, if all the robots take the same topological graph derived from grid map, then they might follow the same exploration path partly or wholly, and this contributes to the problem of waiting situation. We proposed an approach based on sampling environment map iteratively to support the coordinated multi-robot exploration. This research is related to the approach proposed in this paper, even if the methodology is substantially different. Compared to our previous approach, we obtain a significantly reduced mapping time by this approach.

Our recent research focuses on the issue of multi-robot goods transportation [11]. The objective is to complete the transportation mission with high efficiency and low cost. We proposed a heuristic method based on the empirical model, which aims at planning the transportation task for each individual robot by estimating the production rate of goods based on multi-robot coordination, so as to improve the system performance. In the module of robot motion planning, we used the wavefront propagation algorithm to global path planning and the nearness diagram algorithm to goal seeking and local obstacle avoidance. Nevertheless, in the experiment, one important reason which influences the system performance is still the waiting situation. Furthermore, if the speed of the robot is too fast, then the goods would be damaged or lost in transit because of the collisions with obstacles or other robots. We limited the speed of the robot to handle this problem, whereas this strategy limited the efficiency of the whole transportation system as well. Therefore, in this paper we discuss the essence of the problem and propose a novel approach to address such problems.

3 ACS-PRM: Adaptive Cross Sampling Based Probabilistic Roadmap

Sampling-based approaches have been proposed to improve the computational efficiency for robot motion planning. The main idea is to avoid the explicit construction of the obstacle region in the C-space (C_{obs}). Unlike the incremental heuristic search such as A* or D* and the topological map methods such as Voronoi diagram or straight skeleton, the sampling-based approaches work well for complex environments and high-dimensional configuration spaces, and they are generally easier to implement. The probabilistic roadmap planner is one of the typical sampling-based approaches. The original PRM technique is introduced by Kavraki *et al.* [4], which has been shown to perform well in a variety of situations. On the basis of this method, different extensions have been proposed [8], [6]. The approach described in this paper is also an extension of PRM, which is aimed at performing multi-robot motion planning efficiently.

3.1 C-space Sampling

The ACS-PRM approach presented in this paper is a multi-query approach. The first step is C-space sampling, in which a sufficient number of points should be generated to represent the free space of the environment. The main idea of this step is to let a random point p retracts to a position $P(q)$ with the distance d to the obstacle C_{obs} along horizontal and vertical directions (i.e., cross direction).

For autonomous nonholonomic mobile robots, in two dimensions, there are three representational degrees of freedom (DOFs) which are one rotational DOF and two translational DOFs (along or across), but only two controllable DOFs which only move by a forward motion and a steering angle, the configuration space C is the special Euclidean group $SE(2) = \mathbb{R}^2 \times SO(2)$ where $SO(2)$ is the special orthogonal group of 2D rotations. To avoid the collision caused by the point retracts too close to the obstacle, we set the distance d as the sum of the positive number w and the radius r of the minimum circle to cover the robot with centering at the rotation center of the robot:

$$d = r + w, \ (w > 0) \tag{1}$$

The set of w is to deal with the negative influence of sensor error and it should be adjusted in practical applications.

C_{obs} represents the set of the obstacle, $\forall q \in C_{obs}$ define a direction r_q, then determine a symmetry point $S(q)$ which is an intersection of the open-ray with end q direction r_q and another C_{obs}:

$$S(q) = \{q + t\overrightarrow{r_q} | t > 0\} \cap C_{obs} \tag{2}$$

where, if $\{q + t\overrightarrow{r_q} | t > 0\} \cap C_{obs} = \{q\}$, then define $S(q) = \infty$. Let $dist(x,y)$ represent the distance between point x and point y, then the retraction function can be described as:

$$P(q) = \begin{cases} q + d\overrightarrow{r_q} & \text{if } dist(q, S(q)) \geq 2d \\ \frac{q+S(q)}{2} & \text{otherwise} \end{cases} \tag{3}$$

where $P(q)$ is the position for the point p to retract. In this way, the random points are adapted around to the obstacle (see Figure 1(b)), then:

$$\text{ACS-PRM} = \{P(q)\} | q \in C_{obs}\} \tag{4}$$

The implementation of this step is summarized in Algorithm 1, where the time complexity is $O(n)$ and the space complexity is $O(1)$. This step corresponds the learning phase of classic implementation of PRM.

Algorithm 1. Adaptive cross sampling

Require: N, the sufficient number of points to generate.
Ensure: N points in C_{free} by adaptive cross sampling.
repeat
  Generate a uniformly random point p in C-space.
  if p is free **then**
    for horizontal and vertical directions **do**
      Find $q \in C_{obs}$ the nearest distance from p.
      if $dist(p,q) \geq d$ **then**
        p retracts to $q + d\overrightarrow{r_q}$.
      else
        Find $\{S(q)\} = \overrightarrow{qp} \cap C_{obs}$.
        p retracts to $\frac{q+S(q)}{2}$.
      end if
    end for
  end if
until N points have been generated.

3.2 Roadmap Building

The second step is roadmap building, in which the potential targets and milestones should be extracted and connected to the roadmap. In the previous step C-space sampling, if there are sufficient points generated, then the points will gather into segments. The main idea of this step is post-processing the graph resulted from the previous step while identifying three types of point as follows:

- In the previous step, if $dist(p,q) < d$, then p will retract to $\frac{q+S(q)}{2}$ and be labeled as the medial axis. Therefore, we find those medial axis segments with length l a small fixed value (in our implementation, we took the thickness of obstacle), and the midpoints of segments are marked as *potential target* for task allocation. Figure 1(c) shows the extracted potential targets which are precisely doorways of the structured environment.
- For those segments without containing the potential target, we extract both of the endpoints and mark them as *milestone* (see Figure 1(d)).
- The points of intersection between two segments are also extracted and marked as *milestone* (see Figure 1(d)). These milestones have not been used in our experiments, but they will be required for the exploration problem.

This step also corresponds the learning phase of classic implementation of PRM. Figure 1 illustrates the process of generating a roadmap for an example occupancy grid map by using our approach with 200,000 random samples.

3.3 Motion Planning

The third step is motion planning, in which each individual robot's kinematic path should be planned by querying the constructed roadmap. The main idea of this step includes the following three points:

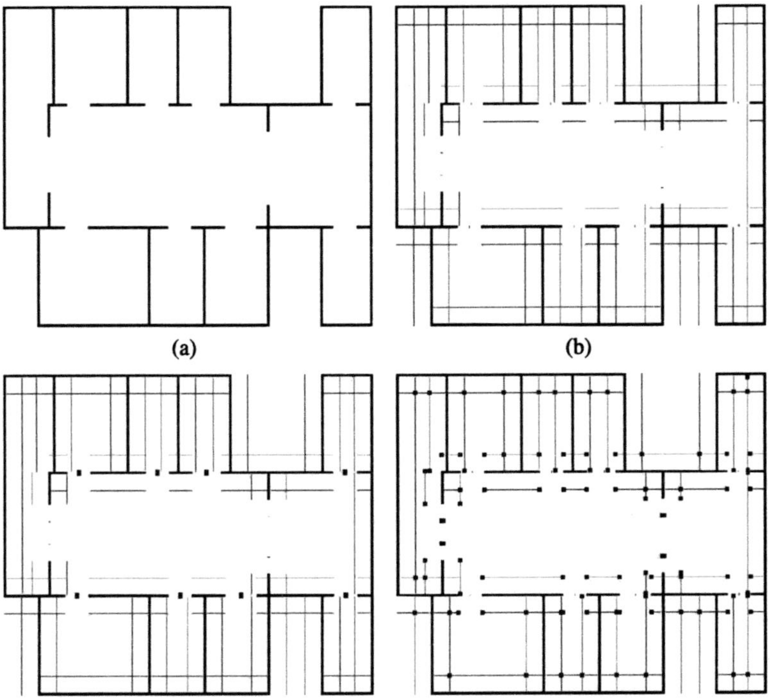
(a) (b) (c) (d)

Fig. 1. Generation of the roadmap based on ACS-PRM. (a) The original gridmap, (b) adaptive cross sampling in C-space, (c) the extracted potential targets (doorways), and (d) the extracted milestones.

- The potential targets $\{t_i\}$ are considered as the goal nodes for path planning and the objects for task allocation as well. Then, the individual $\{r_i\}$ robots are assigned to different potential target:

$$\{r_i\} \mapsto \{t_i\} \tag{5}$$

- To maximize the difference between the paths, we assign the potential target which is the closest from the robot but further from the previously assigned target to the current individual robot:

$$t = further(closest(\{t_i\}, r), t_{i-1}) \tag{6}$$

- Similar to the classic PRM, we use the fast local planning method (i.e., the straight line planner) for the global path planning, except that we choose the path with the minimum number of milestones for the robots invariably.

This step corresponds the query phase of classic implementation of PRM.

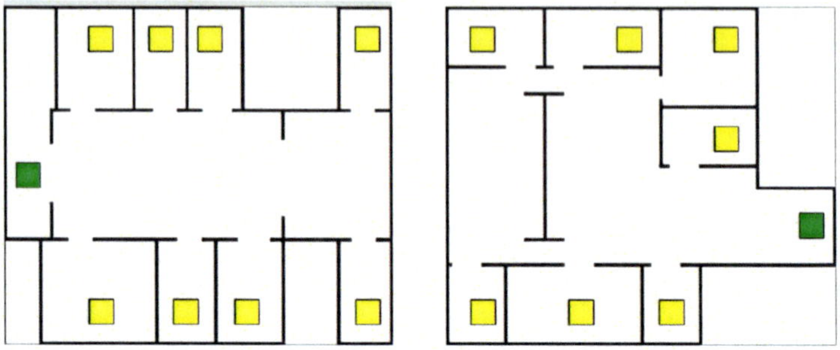

(a) map A (b) map B

Fig. 2. Two environment maps used in our simulation

4 Experiments

To evaluate our ACS-PRM approach, we conducted a series of simulation experiments with the 2D multi-robot simulator Stage [3]. The experiment is to transport a certain amount of goods from one origination to divers destinations by a fleet of mobile robots. The simulated robot is the Pioneer 2-DX robot equipped with a laser range finder providing 361 samples with 180 degrees field of view and a maximum range of 8 meters. Each robot can localize itself based on an abstract localization device which models the implementation of GPS or SLAM. To transport goods, the robots are equipped with a gripper that enable them to sense, pick up and put down the goods, and the carrying capacity is limited to one unit per robot.

We used a different number of robots to conduct several experiments in various environments. Two maps (Figure 2) were used in our simulation which are both structured environments. For each map, the green area signifies the original position of goods, and the yellow areas represent the destinations which are always placed in the rooms. The transportation team size is varied from 2 to 8 robots. On each team size, 10 experimental runs are performed for a transportation mission of 50 goods. The mission objective is to transport the goods to every room equally. The ratio between real-world time and simulation time is about 1:1. We also compared our approach to the commonly used Voronoi-based approach [7] in which a topological map is built on top of the grid map by using the Voronoi diagram, and the *critical points* are extracted like milestones for mobile robot motion planning. All experiments reported in this paper were carried out on a system with an Intel Core 2 Duo E8400 3.00GHz processor, an Intel Q43 Express chipset and two DDR2 800MHz 1024MB dual channel memory.

In the experiments, we assumed that there exists a central server which is able to communicate with all mobile robots and assign the transportation tasks to each individual robot. The results of our experiments are given in Figure 3. We measured the transportation time gained by our approach and compared to the Voronoi-based approach. In each plot, the abscissa denotes the team size of the mobile robots, the ordinate denotes the percentage of the transportation time in the total transportation time, and the error bar indicates the confidence interval of each corresponding gain of robot team

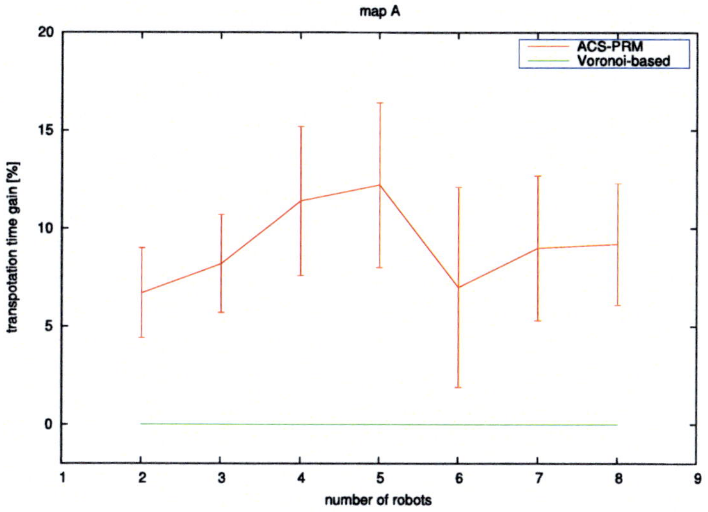

(a) map A

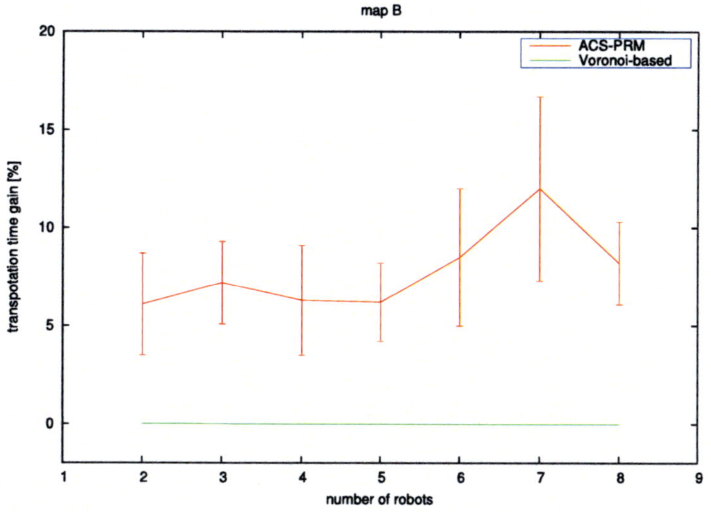

(b) map B

Fig. 3. Transportation time gained by using our ACS-PRM approach compared with the Voronoi-based approach

size with the 0.95 confidence level. Figure 3 shows that, a transportation time saving of 6.7% to 12.2% in map A and 6.1% to 12.0% in map B is obtainable under our ACS-PRM approach compared to the Voronoi-based approach. These results proved that our technique could significantly improve the system planning efficiency.

Table 1. Statistics of the number of occurrences of the waypoint mutex

(a) map A

#robots	2	3	4	5	6	7	8
ACS-PRM	1.6	3.1	6.4	7.5	10.0	11.1	16.2
Voronoi-based	15.3	18.7	26.0	26.8	19.9	23.7	27.2

(b) map B

#robots	2	3	4	5	6	7	8
ACS-PRM	3.8	4.3	7.1	14.9	12.8	10.3	16.7
Voronoi-based	17.1	19.2	19.0	26.5	27.4	27.0	29.9

Moreover, we mentioned earlier that our ACS-PRM approach is more effective than our previous approach because the ACS-PRM spends much less time for the learning phase. The experiments show that, with the new approach, the mapping times are respectively 0.321 seconds and 0.329 seconds for map A and map B with 200,000 random samples, which are averages of the 10 runs. We also counted the average number of occurrences of waypoint mutex in each map as shown in Table 1. This table shows that the problem of waiting situation is significantly reduced by using our ACS-PRM approach, because our approach is able to plan separate paths for robots, especially in the corridor. Unlike the Voronoi-based approach, there is only one path for all robots.

5 Conclusion

In this paper, we presented a novel approach for coordinated motion planning of multiple robots by using the probabilistic roadmap planner based on a manner of adaptive cross sampling, which we called ACS-PRM. The basic thought of the proposed approach is to build separate kinematic paths for multiple robots to minimize the problem of waiting situation such as collision and congestion caused by waypoint mutex in an effective way, thus to improve the efficiency of automated planning and scheduling. In consideration of the context of the issue of multi-robot goods transportation, the experiments were conducted to transport a certain amount of goods by a fleet of mobile robots in structured environments. The experimental results demonstrate that, by using our ACS-PRM approach, the total time needed to complete the transportation mission has been significantly reduced compared to the Voronoi-based approach.

References

1. Cao, Y.U., Fukunaga, A.S., Kahng, A.B.: Cooperative mobile robotics: Antecedents and directions. Autonomous Robots 4(1), 7–27 (1997)
2. Dudek, G., Jenkin, M.R.M., Milios, E., Wilkes, D.: A taxonomy for multi-agent robotics. Autonomous Robots 3(4), 375–397 (1996)
3. Gerkey, B.P., Vaughan, R.T., Howard, A.: The player/stage project: Tools for multi-robot and distributed sensor systems. In: Proceedings of the 11th International Conference on Advanced Robotics (ICAR 2003), Coimbra, Portugal, pp. 317–323 (2003)

4. Kavraki, L.E., Svestka, P., Latombe, J.-C., Overmars, M.H.: Probabilistic roadmaps for path planning in high-dimensional configuration spaces. IEEE Transactions on Robotics and Automation 12(4), 566–580 (1996)
5. Solanas, A., Garcia, M.A.: Coordinated multi-robot exploration through unsupervised clustering of unknown space. In: Proceedings of IROS 2004, Sendai, Japan, pp. 852–858 (2004)
6. Song, G., Amato, N.M.: Randomized motion planning for car-like robots with C-PRM. In: Proceedings of IROS 2001, Maui, HI, USA, pp. 37–42 (2001)
7. Thrun, S.: Learning metric-topological maps for indoor mobile robot navigation. Artificial Intelligence 99(1), 21–71 (1998)
8. Wilmarth, S.A., Amatoy, N.M., Stiller, P.F.: MAPRM: A probabilistic roadmap planner with sampling on the medial axis of the free space. In: Proceedings of the 1999 IEEE International Conference on Robotics and Automation (ICRA 1999), Detroit, MI, USA, pp. 1024–1031 (1999)
9. Wurm, K.M., Stachniss, C., Burgard, W.: Coordinated multi-robot exploration using a segmentation of the environment. In: Proceedings of IROS 2008, Nice, France, pp. 1160–1165 (2008)
10. Yan, Z., Jouandeau, N., Ali Cherif, A.: Sampling-based multi-robot exploration. In: Proceedings of the Joint 41st International Symposium on Robotics and 6th German Conference on Robotics (ISR/ROBOTIK 2010), Munich, Germany, pp. 44–49 (2010)
11. Yan, Z., Jouandeau, N., Ali Cherif, A.: Multi-robot heuristic goods transportation. In: Proceedings of the 6th IEEE International Conference on Intelligent Systems (IS 2012), Sofia, Bulgaria (2012)

On-Line Learning of the Visuomotor Transformations on a Humanoid Robot

Marco Antonelli[1], Eris Chinellato[2], and Angel P. Del Pobil[3]

[1] Robotic Intelligence Laboratory, Jaume I University, 12071 Castellón de la Plana, Spain
antonell@uji.es
[2] Department of Electrical and Electronic Engineering, Imperial College London, UK
e.chinellato@imperial.ac.uk
[3] Robotic Intelligence Laboratory, Jaume I University, Spain and Department of Interaction Science, Sungkyunkwan University, Seoul, South Korea
pobil@uji.es

Abstract. In infant primates, the combination of looking and reaching to the same target is used to establish an implicit sensorimotor representation of the peripersonal space. This representation is created incrementally by linking together correlated signals. Also, such a map is not learned all at once, but following an order established by the temporal dependences between different modalities, which is imposed by the choice of the vision as master signal. Indeed, visual feedback is used both to correct gazing movements and to improve eye-arm coordination. Inspired by these observations we have developed a framework for building and maintaining an implicit sensorimotor map of the environment. In this work we present how this framework can be extended to allow a humanoid robot to update on-line the sensorimotor transformations among visual, oculomotor and arm-motor cues.

1 Introduction

Manipulation tasks, such as reaching or grasping, require knowledge about the position of the target in the peripersonal space. In human beings, this information is obtained by different sensorimotor signals that have to be linked together in order to create an integrated representation of the environment [4]. The brain encodes the sensorimotor transformations among these signals in population of neurons of the parietal cortex (PPC) that exhibit gain fields. The computational model we propose for modeling the kind of transformations performed by the PPC is based on radial basis function networks (RBFNs) [3]. RBFNs were chosen because they are able to naturally reproduce the gain-field effects often observed in parietal neurons [13] and are particularly suitable for maintaining sensorimotor associations [12]. Our conceptual framework [3], and its first implementation on a humanoid robot [2], show how the learning task can be decomposed into incremental phases, where some fundamental competences are learned before more advanced ones [10]. This task decomposition can be observed also in infant development, where a child learns to direct his gaze to an object before learning how to reach to it [1].

In the literature, several works have tackled the problem of hand-eye coordination with the use of Self-Organizing Maps (SOM) [5,7,9], RBFNs [8,15] or other types of

S. Lee et al. (Eds.): Intelligent Autonomous Systems 12, AISC 193, pp. 853–861.
springerlink.com

neural networks [11,14]. However, some major differences can be pointed out between the previous works and our approach. Marjanovic *et al.* [8] did not use stereo vision, while other approaches did not consider eye movements [7,9,15]. In general, sensorimotor associations are created directly between the visuo/oculomotor space and the arm-motor space without considering the advantage (and the biological plausibility) of using the visual to oculomotor transformation [5,7,9,15]. Where this transformation was taken into account, either it was not learned on-line [15,14] or it was not learned at all [11].

In our approach, the sensorimotor framework was built using three RBFNs: one for converting the visual position of a stimulus into an oculomotor position, and the other two for hand-eye movement coordination. This work extends the previous framework [2,3] to allow the robot for incrementally learning to gaze and reach target objects. The extended framework is implemented on an agent based architecture and tested on *Tombatossals* humanoid torso (see Fig. 1).

We describe next the complete sensorimotor framework (Section 2) and how it can be updated during the exploration of the peripersonal space (Section 3). Experiments (Section 4) show that the robot is able to learn the transformations between the different frames of reference and create, in such a way, an implicit mapping of the environment.

2 Encoding of the Sensorimotor Transformations

In our framework, the representation of the peripersonal space is maintained both by limb sensorimotor signals on the one hand and visual and oculomotor signals on the other hand. In this way, different representations of the same target are maintained contextually, and are used to act on the target if required [3]. The relations between these representations, i.e. sensorimotor transformations, are accessed and modified by a conjunction of gazing and reaching movements to the same targets (see Fig. 2).

These movements constitute the basic behaviors employed to incrementally build a sensorimotor representation of the peripersonal space. The learning sequence in our system is inspired by infant development [1]. As a first step, the system learns to associate retinal information and gaze direction (i.e. proprioceptive eye position), through successive foveations to salient points of the visual space. The subject looks around and focuses the eyes on certain stimuli, thus learning the association between retinal information and eye position. Then, gaze direction is associated to arm position, as exemplified in Fig. 1. This can be done by moving the arm randomly and following it with the gaze, so that each motor configuration of the arm joints is associated to a corresponding configuration of the eye motor control, and vice versa. This process allows learning a bidirectional link between different sensorimotor systems. Thus, the robot can look where its hand is but also reach the point in space he is looking at. The learning of the bidirectional oculomotor/arm-motor association between hand and eye positions (O↔Atransformation) requires that the robot is able to fixate its hand. Thus the visual/oculomotor association (V→Otransformation) is learned before.

The location of a visual target is expressed with its position in the cyclopean visual field (c_x, c_y), accompanied by information on binocular disparity (d):

$$c_x = \frac{x_l + x_r}{2} \quad c_y = \frac{y_l + y_r}{2} \quad d = x_l - x_r \tag{1}$$

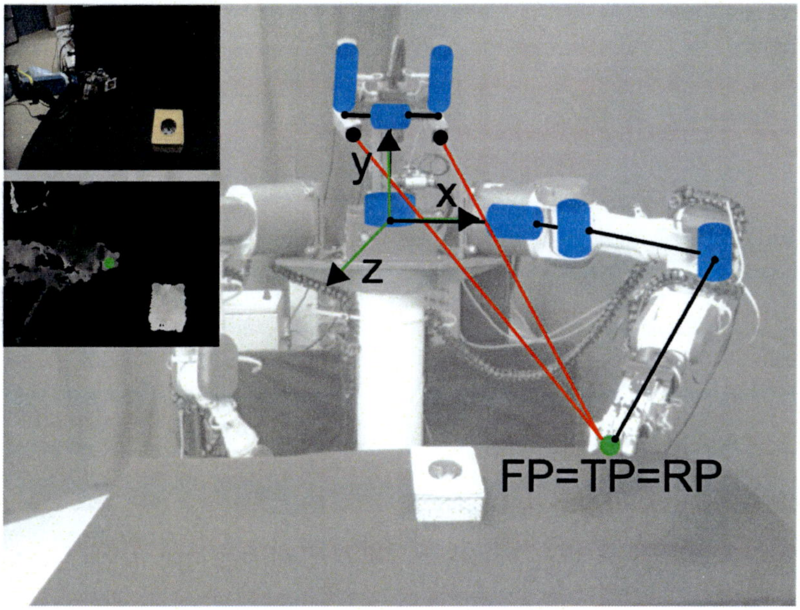

Fig. 1. Association of the oculomotor and arm-motor signals. When the robot reaches (RP) and gazes (FP) towards the same target point (TP), it can update its sensorimotor representation

where x_l, y_l and x_r, y_r provide the position of the target in the left and in the right images. The output of the V→O transformation is the correspondent potential vergence/version movement required to foveate the target (left side of the schema in Fig. 2). This transformation has been implemented with a RBF network, described below. The movement computed by the network, together with the current eye position, provide the head centered representation of the target.

The right hand of Fig. 2 schema shows the oculomotor/arm-motor mapping (O→A), required to code arm movements. This mapping is modified by proprioceptive feedback, and allows reaching towards visual targets. The inverse mapping (A→O) is also stored, so that the robot is able to look at its hand without the need of new visual processing. We use two populations of radial basis function neurons for mapping the vergence/version space (representing oculomotor neurons) into the arm joint space (representing arm-motor neurons) and vice versa. In this work, we have employed three joints for the arm, two for the shoulder and one for the elbow, in order to avoid the redundancy problem. The use of a higher number of joints does not create theoretical problems for the A→O transformation, but increases the search space. On the other hand, with more joints, the O→A transformation should be implemented by more sophisticated mechanisms, such as the inversion of A→O [15] or an associative map [6].

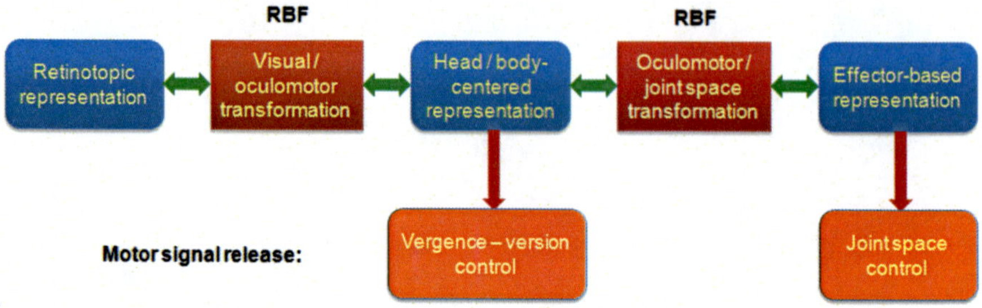

Fig. 2. Computational framework of the sensorimotor model, which allows the robot to perform reaching and gazing movements on demand

3 On-Line Learning

In this work, the radial basis function framework was trained using the delta rule learning algorithm. The application of the algorithm to a real robot raises the problem of how to calculate or estimate the error of the map. Usually, this error is provided by external agents as human users or other systems that control the environment. This type of learning is called supervised. However, given that our model is inspired by primate behavior, we want the robot itself to calculate or estimate the output error. We call this learning method *self-supervised*, because the error term is provided by the robot itself.

3.1 Visual to Oculomotor Transformation

For the visual to oculomotor transformation, the application of the delta rule is not straightforward in the context of self-supervised learning. V→O transformation uses the retinotopic position of the stimulus to update the oculomotor position, in order to keep the stimulus at the center of the retina. A stimulus position different from zero indicates that the gaze position needs to be adjusted. However, the error is observed on the retina, i.e., in the input space, while the application of the delta rule requires the error in the output space, i.e. in terms of vergence and version angles. This error can not be calculated exactly because its computation requires the knowledge of the transformation V→O itself, that is what the robot is trying to learn. Therefore, the output error has to be known a-priori or it has to be estimated from the input.

3.1.1 Learning with Ground Truth

A possible solution to the described problem is to provide the robot with an ad-hoc behavior for the training phase. When the robot is fixating a stimulus, it can perform a random ocular movement towards a different location. After the movement, the position of the visual stimulus can be used to estimate the vergence/version components of the movement that would allow to return to the starting point. If there were no errors, the sum of the two movements would make the robot fixate again the original stimulus. The actual residual error is used to train the network according to the standard delta rule. This algorithm has the advantage of training the network with the exact movement (ground truth) that it is expected to perform. Also, this solution allows learning the visual to oculomotor mapping without any previous knowledge. On the other hand, this adaption process can be applied only when the visual stimulus remains in the field of view after the eye movement.

3.1.2 Learning with Linear Approximation

Another way to estimate the error is to exploit the fact that the transformation can be approximated locally by a linear function. The executed oculomotor movement corresponds to the saccade covering the visual distance between the position of the stimulus before and after the movement. Ideally, after the movement the stimulus position should be zero, but practically it is small but not null. This algorithm works in a goal directed phase, i.e., it trains the network while it is trying to gaze a target stimulus which is out of the fovea. The linear approximation is exploited to adapt the network by using the performed movement. It is important to note that even if the adaptation procedure is based on a linear assumption, the learned transformation is not linear. Moreover, the smaller the error of the performed movement is, the closer the functioning of the algorithm is to the *ground truth method*.

3.2 Oculomotor and Arm-Motor Transformations

As in the V→O case, on-line training of the robot is achieved by means of the *delta rule*. In order to create the eye-arm association it is necessary that the robot fixates the point that it has reached with the hand. This can be done using vision as master signal, under the assumption that the vision system is able to recognize the robot hand. We addressed this problem by placing a marker on the robot's hand. Once the robot is able to fixate the hand, the application of the delta rule becomes straightforward, as the input and the target are given by oculomotor and arm-motor positions, while the output of the networks can be obtained by applying the direct and inverse networks to the input.

The transformations are learned during a free exploration of the peripersonal space. This phase consists of random arm movements and subsequent saccades towards the final hand position, which allows learning the transformation from joint space to oculomotor space and vice versa. To generate the training points, a correct gazing behavior is required. The hand is a visual target that can be used to train the V→O transformation, which thus needs to be learnt before those involved in the eye-hand coordination.

4 Experimental Results

Tombatossals is a humanoid torso endowed with a pan-tilt-vergence stereo head and two multi-joint arms (Fig. 1). During the training of the RBFNs the left hand was equipped with a marker that can be recognized by means of the ARToolkit library. This marker can be used by the robot to identify its own hand, and it represents also a visual stimulus that can be moved very precisely in the three dimensional space. We employed three degrees of freedom (d.o.f.) for both the head (tilt, left pan and right pan) and the left arm (first, second and forth joint).

The parameters of the three RBFNs were chosen using a heuristic search on a simulated model of the robot. The centers of the RBFN that computes the V→O transformation were distributed according to a retinotopic-like criterion, following a logarithmic distribution. A good performance was achieved by distributing the neuron centers on a $7x7x7$ neural lattice in the cyclopean (x, y) and disparity space. Neural activation functions were Gaussians with a spread proportional to the distance between adjacent centers. The centers of the O↔A networks were distributed uniformly on a $7x7x7$ lattice in the input space, i.e. the vergence / version (x, y) space for the direct transformation, and the arm joint space for the inverse one. Again, the activation functions were Gaussians, but the spread was the same for all the neurons.

The weights of the networks were not learned from scratch, but were bootstrapped using those provided by the simulated model of the robot. However, due to the approximation of the latter, the weights had to be adapted to fit the parameters of the robot's body. At each interaction with the environment, the robot autonomously estimated the local error of the sensorimotor transformations and corrected the weights using the *delta rule*. The learning rate of the delta rule was set empirically to 0.01 for both V→O and O↔A transformations.

4.1 Visual to Oculomotor Transformation

The V→O learning process (Section 3.1) can be summarized as follows. The vision system locates saliency points in the left and right images. Based on an attentive model, the most salient stimulus is chosen as saccadic target. Then, the robot executes a sequence of saccades until the target is foveated. After each saccade the new position of the stimulus is used to train the V→O network. Once the target is foveated, the head is moved randomly and another stimulus is chosen as target. In the current setup, markers were used instead of real objects in order to minimize visual errors. The visual target was considered foveated when its distance from the centers of both images was smaller than 10 pixels. The foveating behavior was performed 500 times. The *linear approximation method* was used to estimate the error of the network and its performance was compared with the *ground truth method*.

At the end of the experiment, the performance of the system was evaluated. The head of the robot was placed in an initial position fixating the visual marker. It was then moved on all points of a lattice on the vergence/version space, and the visual position of the marker was recorded for each point. The maximum amplitude of the movement was $\pm 0.3\,rad$ for both vergence and version (x, y) angles. The pairs *visual position - performed movement* were used as test set for the system.

The first line of Table 1 reports the errors of the V→O network with the weights obtained during the simulations, and successively adapted with the *delta rule*, using both the *ground truth* and the *linear approximation* methods. The error of the simulated model is relatively small, but the online adaptation mechanism allows improving the performance of the system by a factor of 2. The linear approximation method provides results close to those obtained with the ground truth, and allows the system to adjust the weights during the normal exploration of the environment. Results obtained with a Sum of Squared Error (SSE) batch learning method are also displayed, and will be commented upon below.

Table 1. Transformation networks performance, mean errors $\mu \pm \sigma$ [degrees]

Transformation	Simulation	Delta Rule	Linear approx.	SSE
V→O	0.91±0.57	0.40±0.26	0.42±0.34	0.02±0.01
A→O	3.98±0.89	0.36±0.26		0.33±0.25
O→A	13.20±6.20	3.89±3.05		0.83±0.84

An more revealing analysis of the error can be performed by plotting the mean of the error as a function of the distance of the visual stimulus from the fovea (Fig. 3). The weights found in the simulation (black line) produce an error that depends on the distance. This is due to the approximation of the model that did not take into account the focal distortion of the lens. The delta rule allows correcting for this error either using the ground truth method (red line) or the linear approximation one (blue line). On the periphery, the correction effect is reduced, due to the smaller number of training points for that region. However, the network is potentially able to approximate correctly the right transformation, as shown by the result obtained training the network by minimizing the sum of squared errors (SSE, green line, see also Table 1). It is not surprising that the minimization of the SSE outperforms the on-line method, as it uses the global information of the training set, while the delta rule uses only the information of each new observed point. On the other hand, the delta rule does not require to store the whole dataset to learn the sensorimotor transformation, which is rather implausible from a biological point of view.

4.2 Oculomotor and Arm-Motor Transformations

The networks that compute the direct and inverse transformations from oculomotor to arm-motor signals require a training set composed of pairs of eye and arm positions. The training set was generated by moving the arm randomly.

Once the arm movement is completed, the robot executes a sequence of saccades to foveate the marker placed on the hand. However, this procedure generates training points that are inherently noisy, due to small imprecisions in identifying the center of the marker in the two images and due to the size of the "fovea", which was set to 10 pixels. The on-line training was iterated for 791 trials. The misalignment between the oculo- and arm-motor position was estimated to be on average 2.08 pixels, as viewed by the robot's cameras. For each transformation, Table 1 reports the mean and standard deviation of the error of the RBFNs bootstrapped from the weights learned during the simulations, and successively adapted on-line. Again, the table shows also the results obtained with the weights learned using the minimization of SSE to show the potential performances of the networks.

The results show that the difference between the simulated model and the robot is remarkable. The weights obtained during the simulations are not good enough for the robot to execute the reaching task. Therefore, the on-line adaptation is necessary. By training the network with 791 points, the system was able to improve its performance by a factor of 10 for A→O and of 3 for O→A. The off-line training performed by minimizing the SSE method shows that the performance of the O→A can improve further

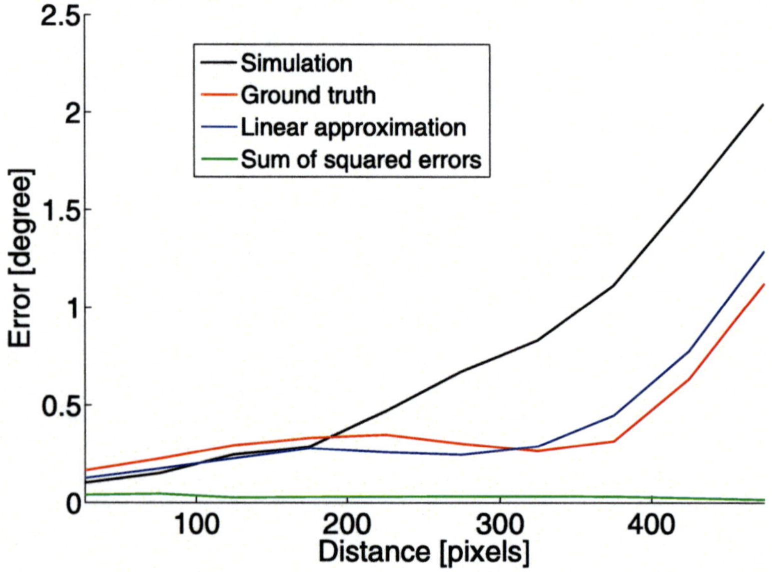

Fig. 3. V→O: Mean error as a function of the distance from the fovea

with more training points, while the A→O seems to have reached its best performance. In general, as observed during the simulations, O→A transformation is harder to learn than A→O. This is probably due to the uniform distribution of the neurons that seems to be not so efficient for the O→A transformation as it is for the A→O one.

5 Conclusions

In this work we have described how the transformations necessary to maintain a sensorimotor representation of the environment can be updated by the robot itself during the exploration of the peripersonal space. By means of gazing and reaching movements, the robot firstly learns to associate visual cues with oculomotor movements (V→O) and then produces arm movements useful to learn the association between the fixation point and the reaching point (O↔A). During such exploration, the weights of the RBFNs that implement the V→O and O↔A transformations are refined using the delta rule. This learning process is the normal behavior of the agent, constituting the most fundamental component of its basic capability of interacting with the world, and contextually updating its representation of it.

The proposed framework works on the assumption that problems such as object recognition and disparity extraction have already been solved. We are currently working on improving the visual system by introducing models, inspired by the primate visual cortex, which will allow the system to deal with real objects instead of visual markers.

Acknowledgements. This work was supported by WCU program through the National Research Foundation of Korea funded by the Ministry of Education, Science and Technology (Grant No. R31-2008-000-10062-0), by Spanish Ministerio de Ciencia e Innovación (FPI Grant BES-2009-027151 and DPI2011-27846), by Generalitat Valenciana (PROMETEO/2009/052) and by Fundació Caixa Castelló-Bancaixa (P1-1B2011-54).

References

1. Adolph, K.E., Joh, A.S.: Motor development: How infants get into the act. In: Slater, A., Lewis, M. (eds.) Introduction to Infant Development, pp. 63–80. Oxford University Press (2007)
2. Antonelli, M., Chinellato, E., delPobil, A.P.: Implicit mapping of the peripersonal space of a humanoid robot. In: IEEE Symposium on Computational Intelligence, Cognitive Algorithms, Mind, and Brain, pp. 1–8 (2011)
3. Chinellato, E., Antonelli, M., Grzyb, B., delPobil, A.: Implicit sensorimotor mapping of the peripersonal space by gazing and reaching. IEEE Transactions on Autonomous Mental Development 3, 45–53 (2011)
4. Chinellato, E., Grzyb, B.J., Marzocchi, N., Bosco, A., Fattori, P., del Pobil, A.P.: The dorsomedial visual stream: From neural activation to sensorimotor interaction. Neurocomputing 74(8), 1203–1212 (2011), doi:10.1016/j.neucom.2010.07.029
5. Fuke, S., Ogino, M., Asada, M.: Acquisition of the head-centered peri-personal spatial representation found in vip neuron. IEEE Transactions on Autonomous Mental Development 1(2), 131–140 (2009)
6. Hoffmann, H., Schenck, W., Möller, R.: Learning visuomotor transformations for gaze-control and grasping. Biological Cybernetics 93(2), 119–130 (2005)
7. Jones, M., Vernon, D.: Using neural networks to learn hand-eye co-ordination. Neural Computing and Applications 2(1), 2–12 (1994)
8. Marjanovic, M., Scassellati, B., Williamson, M.: Self-taught visually guided pointing for a humanoid robot. In: From Animals to Animats 4: Proc. Fourth Intl Conf. Simulation of Adaptive Behavior, pp. 35–44 (1996)
9. Martinetz, T.M., Ritter, H.J., Schulten, K.J.: Three-dimensional neural net for learning visuomotor coordination of a robot arm. IEEE T. Neural Networ. 1(1), 131–136 (1990), doi:10.1109/72.80212
10. McBride, S., Law, J., Lee, M.: Integration of active vision and reaching from a developmental robotics perspective. IEEE Transactions on Autonomous Mental Development 2(4), 355–366 (2010)
11. Nori, F., Natale, L., Sandini, G., Metta, G.: Autonomous learning of 3d reaching in a humanoid robot. In: IEEE/RSJ IROS, International Conference on Intelligent Robots and Systems, pp. 1142–1147 (2007)
12. Pouget, A., Sejnowski, T.J.: Spatial transformations in the parietal cortex using basis functions. Journal of Cognitive Neuroscience 9(2), 222–237 (1997)
13. Salinas, E., Thier, P.: Gain modulation: A major meeting report computational principle of the central nervous system. Neuron 27, 15–21 (2000)
14. Schenck, W., Hoffmann, H., Möller, R.: Learning internal models for eye-hand coordination in reaching and grasping. In: Proceedings of the European Cognitive Science Conference, Osnabrück, Germany, p. 289 (2003)
15. Sun, G., Scassellati, B.: A fast and efficient model for learning to reach. International Journal of Humanoid Robotics 2(4), 391–414 (2005)

Adaptive Face Recognition for Low-Cost, Embedded Human-Robot Interaction

Yan Zhang, Kenneth Hornfeck, and Kiju Lee

Case Western Reserve University, Cleveland, OH, USA
yzhang@propertypanorama.com, ken.hornfeck@gmail.com,
kiju.lee@case.edu

Abstract. This paper presents an accelerated AdaBoost face detection algorithm and an incremental PCA-based face recognition algorithm for human robot interactive applications. The accelerated AdaBoost algorithm utilizes an image resizing technique and a skin color filter for detecting face regions. To track a detected face precisely and efficiently while also recognizing the face, a hybrid face tracking approach is applied based on an adaptive skin color mode and an estimated potential face area. In addition, a novel adaptive face recognition method is implemented by automatically upgrading the set of sample faces of a *known* person and collecting new samples of an *unknown* person with incrementally enhanced recognition performance. These algorithms are well suited for embedded systems, such as socially interactive robots, because of their cost and time efficiency and little pre-training required for reliable performance.

1 Introduction

Human-robot interaction (HRI) research focuses on the relationship between humans and robots, which is a multidisciplinary field involving human-computer interaction, artificial intelligence, robotics, and social sciences. HRI plays an essential role and is a prerequisite for allowing a robot to be a part of human life by assisting with daily tasks and solving practical problems. HRI can be realized and enhanced in different ways. An auditory-based system gives a robot the capability of understanding human's words and thus makes it easy to communicate with. A friendly user interface such as an appealing appearance, an expressive face or animated behavior may also encourage HRI [3]. Vision-based interation also plays an important role in HRI, which is utilized by most interactive robotic applications to perform diverse functions [21,23].

Face detection, tracking, and recognition can be viewed as fundamental capabilities for initializing vision-based interactions. For face detection, the AdaBoost (adaptive boosting) classifier has been regarded as a consistently successful approach [19], as well as various skin color detection techniques [12,13]. However, with increasing image size, the detecting time increases dramatically. To solve this problem, Huang and Li implemented a skin tone filter ahead of the general AdaBoost detection for each sub-window which largely reduced the number of windows scanned by the AdaBoost classifier [10]. For a detected face, the primary purpose of face tracking is to provide the precise position of a target face from given video sequences. Color-based tracking is a widely used technique for real-time face tracking [5,11] . A radial basis function

S. Lee et al. (Eds.): Intelligent Autonomous Systems 12, AISC 193, pp. 863–872.
springerlink.com

Fig. 1. An original 320 × 240 image (left) and selected skin color pixels after applying the RSC-Filter with a step size, $\delta = 8$ (right)

network has been also implemented to predict potential coordinates and width of the face, followed by *k*-means clustering [24]. As another approach, Bradski proposed a fast and efficient object tracking algorithm called CAMSHIFT (Continuously Adaptive Mean Shift), a non-parametric technique based on color distribution [2]. In addition, a modified CAMSHIFT tracking mechanism has been applied to an HSV color space [18]. Self-organized integration of five different cues was also proposed to predict the face area [16]. In terms of face recognition, Principal Component Analysis (PCA) has been proven to be one of the most successful techniques in face recognition [17]. However, PCA requires a controlled environment in order to obtain a high recognition rate, but in the application of HRI, the surrounding environment continually varies. Several techniques have been proposed to overcome this limitation in the PCA algorithm [9,6,7,1,22].

Despite the improved performance of currently available face tracking and recognition techniques, most of these algorithms still require relatively heavy computing resources to process smoothing, morphology, and color correlation especially when the image size is large or real-time processing is required. This paper describes two vision-based algorithms for embedded HRI systems: i) a low-cost face detection and real-time tracking algorithm based on an accelerated AdaBoost Classifier and ii) a face recognition algorithm based on a novel adaptive approach for incrementally improved recognition performance. Based on previous studies and the assumption that only one face is to be tracked and then recognized at a time, we apply a novel AdaBoost face detection technique combined with a resizing skin color filter, which aims to quickly reduce the searching scope. Unlike the algorithm in [10], our skin color filter is applied only once to determine the potential region in which a face may be found. In addition, a novel histogram-based approach integrated with a skin color filter is also proposed. Our face recognition algorithm resembles Huang's method that refines the classifier to improve the recognition rate [9]. Two primary differences in our approach are: first, at most one of the clusters is updated at once; second, PCA recalculation is applied only when a new person whose face images are not in the existing database is detected.

2 Real-Time Face Detection

Real-time face detection is realized by tracking a detected face continuously over time. Face detection using the original AdaBoost classifier [19] may take too long for a low-

Table 1. Average detection times in two different computing environments (BeagleBoard/PC) for four different-sized images with and without the RSC-Filter

Image Size (Face Region)	**Adaboost Classifier**	**Adaboost Classifier with RSC-Filter**
320×240 (50×50)	669ms /10ms	114ms / 2ms
640×480 (100×100)	815ms / 19ms	210ms / 5ms
800×600 (120×120)	975ms / 23ms	302ms / 8ms
1280×1024 (200×200)	1312ms / 35ms	419ms / 10ms

cost, emebedded system with limited hardware capability. To reduce processing time and improve performance, several existing techniques are combined and improved.

2.1 Accelerated Face Detection with RSC-Filter

The resizing color skin filter (RSC-Filter) reduces the searching scope within a given image by determining a region that potentially contains a face. Let $I_{x,y}(r,g,b)$ be the input image, where x and y indicate the coordinate of the pixel within the image. (r,g,b) denote the pixel's coordinate in RGB color space. Since we are only interested in the approximate location of the face, an appropriate step size, δ, may be selected to accelerate the detection procedure. A normalized RGB color space is selected to classify skin color and non-skin color regions, while effectively eliminating distortion from lights or shadows by defining

$$E_{skin} = \{(x,y) | f(I_{x,y}) < \Delta, x \bmod h = 0, y \bmod h = 0\}$$

where Δ is a threshold value for distinguishing skin pixels from non-skin pixels. E_{skin} is a set of pixels that potentially contain a skin color. The matching criterion is defined as the cosine of the included angle between the target color vector and the skin color vector, such that

$$f(I_{x,y}) = 1 - \frac{C \cdot C_0}{|C||C_0|} \tag{1}$$

where $C = (\frac{r}{r+g+b}, \frac{g}{r+g+b}, \frac{b}{r+g+b})$ and C_0 denotes the normalized skin color vector which is initialized by an average human skin color. In (1), a smaller value of $f(I_{x,y})$ indicates that the target pixel color is closer to C_0, and therefore the searching scope is reduced for the areas containing small values of $f(I_{x,y})$. Fig. 1 shows an example face image and the selected skin color pixels after applying the RSC-Filter when $\delta = 8$. Table 1 compares the average detection times for four different-sized images with and without the RSC-Filter prior to applying the AdaBoost classifier. It also compares the detection time in two different computing environments, a BeagleBoard with 720 MHz processor and 520 MB of RAM, as an example of an embedded application, and a PC with an Intel Core 2 Duo^{TM} Processor and 4GB of RAM.

2.2 Adaptive RSC-Filter

Skin color varies depending on individual's race, as well as among people of the same race. Differences in illumination, shadows, and cameras may also have a considerable

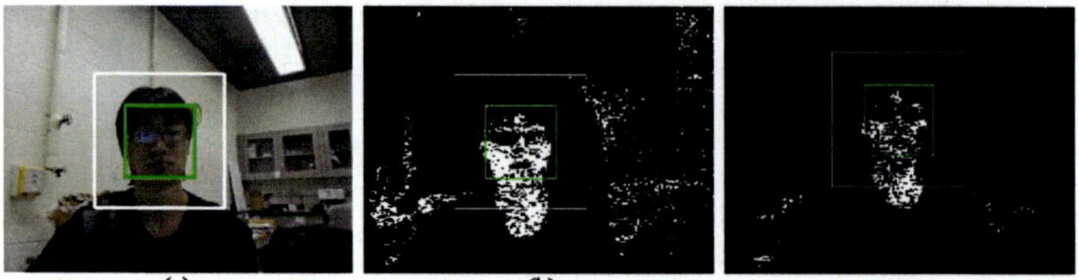

Fig. 2. Face detection using an adaptive RSC-Filter: (a) initial face detection; (b) skin pixels initially filtered from (a); and (c) skin pixels filtered by the updated skin color set

effect. A simple approach to solve this problem is to initialize the system with a set of skin colors including a broad range of variances instead of a single color, and then determine all pixels within the range for each frame. However, this approach may cause the algorithm to incorrectly classify non-skin pixels as skin pixels. The key for obtaining a personalized and adaptive skin color is to identify appropriate skin color pixels that can be used for training. In [15], this issue was addressed by an adaptive skin color histogram. It assumes that the skin color at the current image frame can be described by the skin color histogram from the previous frame while the histogram is updated iteratively.

In our approach, the skin color is extracted when the face is detected and also during the following tracking process. The newly created skin color updates the initial set of skin colors. Furthermore, the skin color set continually updates its elements as the face is being tracked in order to adapt to small changes in brightness or other external factors. The result can be observed in Fig. 2. For a detected face image in Fig. 2(a), Fig. 2(b) shows the skin pixels filtered by the initial set of skin colors and Fig. 2(c) shows the result filtered by the updated skin color set after several iterations. As the algorithm proceeds, most of the erroneously detected skin pixels are removed from the background.

2.3 Face Region Estimation Using H_{SCP} for a Non-frontal Face

The AdaBoost classifier with the adaptive RSC-Filter can effectively detect a face in each frame. However, in cases where the person's face is tilted or turned to a side, this altorithm often fails to detect or will lose track of the face. Assuming that the detected faces in two consecutive frames are located relatively close to each other, the *face region* is defined by a rectangular area surrounding the entire face (e.g., the detected face window established by the AdaBoost classifier with the adaptive RSC-Filter), and the *potential region* is defined by an expanded area containing the face region at the center and also involving potential face regions in the next frame. As shown in Fig. 2(a), the potential region (white square) is configured as an area expanded around the detected face region (green square).

A histogram of the skin color pixel distribution (H_{SCP}) is used to detect a non-frontal face by tracking the high distribution area within the potential region. H_{SCP} projected

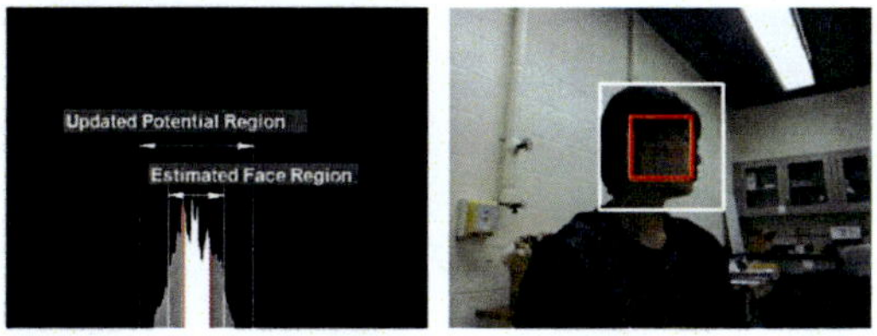

Fig. 3. Face detection using an adaptive skin color and H_{SCP}: (left) estimated face region and updated potential region and (right) the final detection result

onto each axis is divided into three parts, and we call the middle part the *1/2-face region*. The relative percentage of skin color pixels located in each of the three parts (i.e. $\alpha\%$, $1-(\alpha+\beta)\%$, $\beta\%$) is then calculated. We assume that all three parts share a sufficient amount of skin pixels in order to reduce the effect of noise. When the AdaBoost classifier fails to detect the face region, the potential and face regions are estimated by the information learned from H_{SCP} and the 1/2-face region. This approach is demonstrated in Fig. 3. For a filtered image, H_{SCP} is analyzed within the potential region along the x-axis, and the boundaries of the 1/2-face region are estimated. Fig. 3 (left) shows an estimated face region and an updated potential region and Fig.3 (right) shows the final tracking result. The figures only show H_{SCP} projected onto the x-axis, but the same technique is applied for the y-axis.

3 Face Recognition

Face recognition is realized in two steps: 1) creating an initial database by collecting sample face images for each individual, and 2) updating the database by replacing an old sample with a newly collected one. If the detected face does not match any of the samples in the database, the system automatically creates a new set of samples and labels them as a new person. All face sample images for each person are resized and converted to 64×64 matrices and then reorganized into $4096(=64\times64)$-dimensional vectors. By applying PCA, each of these vectors can be expressed as the summation of the average vector and a linear combination of a series of principal eigenvectors (projected vectors).

3.1 Recognition Process

The system is initialized by N individuals, where M face image samples for each are collected. Only the first t principal components are analyzed for matching. The projected vectors from the trained samples are given by $s_i^j = (x_1, .., x_t)^T$ where s_i^j is the jth trained vector of the ith individual. We also define S as the set of s_i^j, such that $S = \{s_i^j|(1 \le i \le N, 1 \le j \le M\}$. Let v be the corresponding projected vector for the face image of a person V. First, the algorithm determines whether v belongs to a known person or not by investigating the *similarity* between v and the elements of S. The similarity between a vector x and a set of vectors, Y, denoted by $r(x,Y)$, is defined by

$$r(x,Y) = \min\{-\frac{d(x,y)}{K}|y \in Y\} \tag{2}$$

where

$$d(x,y) = -\frac{1}{\sqrt{\sum_{i=1}^{t} x_i^2}\sqrt{\sum_{i=1}^{t} y_i^2}}\sum_{i=1}^{t} z_i x_i y_i; \qquad z_i = \sqrt{1/\lambda_i}.$$

K is a constant scalar that scales a simplified Mahalanobis distance to $[-1,0]$ so that $r(x,Y)$ ranges from 0 to 1. The simplified Mahalanobis distance is adopted because it is proven to work better on PCA than the Euclidian distance or other norm distances, according to [20]. λ_i's in the above equation are the corresponding eigenvalues. A large value of $r(x,Y)$ implies a *high confidence level* for identifying a person within the existing database.

Now, let S_p be a subset of S that includes a person P's trained vectors, such that $S_p = \{s_p^1, s_p^2, \cdots, s_p^M\}$. If $r(v,S_p) = r(v,S)$, then P is determined to be the "closest person" to V. For f representing the status of recognition initialized as 0 and iteratively updated by the similarity r, $f \geq 1$ for a known person, $-1 < f < 1$ if further consideration is needed for recognition, and $f \leq -1$ for an unknown person. Table 2 contains pseudocode that demonstrates the face recognition process. Thresholds τ_1 and τ_2 are used to determine how close the currently detected face is to the existing samples where τ_1 is close to 1 and $0 < \tau_2 < \tau_1 < 1$. By using this algorithm, a person can be recognized either by a single "extremely familiar" input face, or by a series of "less familiar" faces.

Table 2. Pseudocode for the face recognition algorithm

```
If τ1 ≤ r(v,S), the person recognized,
    f = 1
Else if τ2 ≤ r(v,S) < τ1
    If P = P', the nearest persons to the current (P) and previous (P') frames are the same,
        f = f + 0.1
    Else
        f = f − 0.2
    End
Else
    f = f − 0.2
End
```

3.2 Learning and Updating Process

Once the person is identified as the pth individual in the group, face recognition is completed and L new face images from the video sequence are captured and labeled as p while tracking. The set of vectors obtained from the newly collected images is defined as $V_p = \{v_p^1, \cdots, v_p^L\}$. Most of these vectors would be located near the existing sample images (S_p) forming clusters around them. To identify the cluster involving the minimum number of vectors, the number of elements in each cluster is counted by using the following formula:

$$N_M(k) = Count\{n|n \in [1,L], r(v_p^n, S_p) > \tau_3, r(v_p^n, s_p^k) = r(v_p^n, S_p)\}. \quad (3)$$

τ_3 is a value between τ_1 and τ_2, and determines if a sample is inside or outside the existing clusters. If the function "*Count*" returns the number of elements that satisfy the given conditions, there exists α such that $N_M(\alpha) = \inf\{N_M(k)|k \in [1,M]\}$. Using this, the cluster with the minimum number of newly collected face vectors can be found and it is centered by s_p^α. For those l, such that $r(v_p^l, S_p) < \tau_3$, clusters with the most neighbors satisfying $r(v_p^n, v_p^l) > \tau_3$ can be also identified by

$$N_L(l) = Count\{n|n \in [1,L], n \neq l, r(v_p^n, S_p) \leq \tau_3, r(v_p^n, v_p^l) > \tau_3\} \quad (4)$$

Similarly, there exists β that satisfies $N_L(\beta) = \sup\{N_L(l)|l \in [1,L]\}$. If $N_M(\alpha) < N_L(\beta)$, v_p^β is considered more representative than s_p^α, and s_p^α is replaced by v_p^β.

To reduce the effects of atypical external conditions, we perform the replacing strategy while also considering prior information (i.e. the history of $N_M(k)$). For example, if the subject is initially wearing a hat, the first cluster is replaced by a hat-wearing one. However, in the next iteration, if that person removes the hat, there are no collected samples within the newly created hat-wearing cluster. If we only consider the current image, an important cluster may be replaced. To solve this problem, let $N_M^j(k)$ denote the result from the jth iteration calculated by $N_M^j(k) = 0.4N_M^j(k) + 0.6N_M^{j-1}(k)$. This formula provides sufficient consideration of previous results, so that the number of samples located in the atypical cluster (i.e., created by the person wearing a hat) in the following iteration becomes nonzero.

4 Performance Evaluation

For the initial sample images shown in Fig. 4 (left), 20 learning iterations were conducted in different locations in the laboratory to test performance of the algorithm with different lighting conditions and backgrounds. Fig. 4 (right) shows the updated face samples after replacing the cluster with the minimum elements to the newly identified

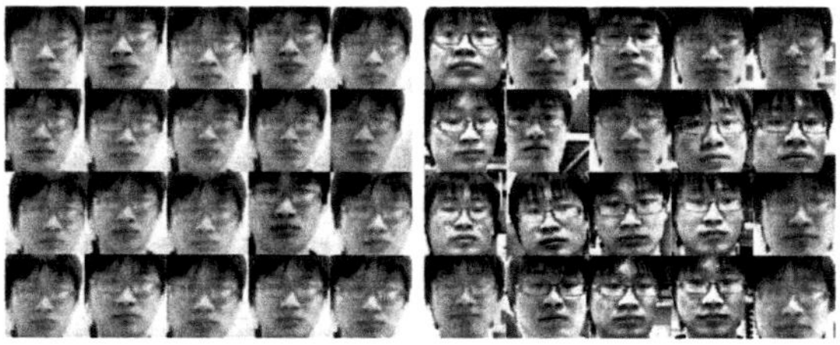

Fig. 4. 20 initial sample face images (left) and 20 updated sample images after 20 learning iterations

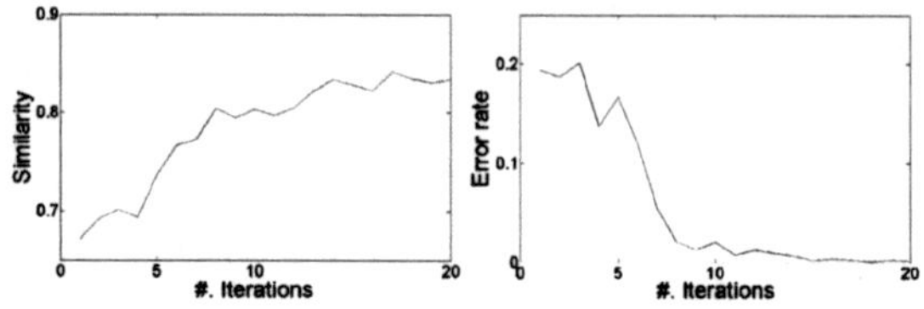

Fig. 5. Changes in similarity and error rate as iterations progress

cluster at each iteration. To evaluate the improvements made by the learning and updating process, 1000 test face images were loaded into the face recognition database. An error is detected when the center of the nearest cluster is not associated with the correct person. For each test face, the average similarity between each test vector and its nearest cluster center associated with the correct person was calculated as shown in Fig. 5 (left). Fig. 5 (right) shows that the error rate quickly drops after 7 iterations and remains low afterward. Similarity did not increase much during the first several iterations because the system had not yet obtained enough representative samples. For example, at the beginning, people may sit in different places for each iteration, so the newly updated cluster may not contribute to improving recognition in the next iteration, resulting in a low average similarity. The similarity increases sharply when the subject returns to the same location as when one of the previous sample sets was obtained.

The presented algorithms have been implemented in a socially interactirve robot, called *Philos* [8]. The design of this robot employs the use of two separate microcontrollers. The first is used for sensor data collecting and processing, wireless communication, and motion control of servo motors and linear actuators. Face detection and recognition were performed on the second controller containing a 720MHz processor and 512MB of RAM (BeagleBoard). Previously developed algorithms, such as [19,12,13], were not applicable for this embedded system due to the hardware limitation; however, the algorithms presented in this paper ran reliably over time.

5 Conclusion and Future Work

This paper presented face tracking and recognition algorithms by utilizing and improving several existing techniques for low-cost, embedded HRI applications. A face detection algorithm based on the AdaBoost classifier was combined with a dynamically resizing potential face window using a skin color filter enabling fast detection with less computing resources. A histogram-based approach was also proposed to improve performance in non-frontal face detection. We also described an incremental face recognition algorithm with an increasing recognition rate during HRI. Our future work includes performance evaluation of the proposed algorithms by comparing with other advanced face detection and recognition algorihtms in terms of accuracy and processing time. For practical HRI applications, efficient and reliable detection and recognition of multiple faces within a single frame is also important area of exploration.

References

1. Aryananda, L.: Recognizing and remembering individuals: online and unsupervised face recognition for humanoid robot. In: IEEE/RSJ International Conference on Intelligent Robots and Systems, vol. 2, pp. 1202–1207 (2002)
2. Bradski, G.R.: Real time face and object tracking as a component of a perceptual user interface. In: Proceedings of the Fourth IEEE Workshop on Applications of Computer Vision, WACV 1998, October 19-21, pp. 214–219 (1998)
3. Christoph, B., Michio, O.: Robotic user interfaces. In: Human and Computer Conf., pp. 130–140 (2001)
4. Feil-Seifer, D., Mataric, M.J.: Defining socially assistive robotics. In: 9th International Conference on Rehabilitation Robotics, ICORR 2005, June 28-July 1, pp. 465–468 (2005)
5. Fieguth, P., Terzopoulos, D.: Color-based tracking of heads and other mobile objects at video frame rates. In: Proceedings of the 1997 IEEE Computer Society Conference on Computer Vision and Pattern Recognition, June 17-19, pp. 21–27 (1997)
6. Ghassabeh, Y.A., Ghavami, A., Moghaddam, H.A.: A New Incremental Face Recognition System. In: 4th IEEE Workshop on Intelligent Data Acquisition and Advanced Computing Systems: Technology and Applications, IDAACS 2007, September 6-8, pp. 335–340 (2007)
7. Hewitt, R., Belongie, S.: Active Learning in Face Recognition: Using Tracking to Build a Face Model. In: Conference on Computer Vision and Pattern Recognition Workshop, CVPRW 2006, June 17-22, p. 157 (2006)
8. Hornfeck, K., Zhang, Y., Lee, K.: Philos: A Sociable Robot for Human Robot Interactions and Wireless health Monitoring. In: Symposium on Applied Computing (SAC 2012), Riva del Garda, Italy, March 26-30 (2012)
9. Huang, W., Lee, B., Rajapakse, M., Li, L.: Face recognition by incremental learning for robotic interaction. In: 2002 IEEE Workshop on Multimedia Signal Processing, December 9-11, pp. 280–283 (2002)
10. Huang, Y., Ao, X., Li, Y.: Real time face detection based on skin tone detector. IJCSNS Intl. J. of Computer Science and Network Security 2(7) (2009)
11. Hunke, M., Waibel, A.: Face locating and tracking for human-computer interaction. In: 1994 Conference Record of the Twenty-Eighth Asilomar Conference on Signals, Systems and Computers, October 31-November 2, vol. 2, pp. 1277–1281 (1994)
12. Liu, Q., Peng, G.: A robust skin color based face detection algorithm. In: 2010 2nd International Asia Conference on Informatics in Control, Automation and Robotics (CAR), March 6-7, vol. 2, pp. 525–528 (2010)
13. Mohamed, A., Weng Y., Jiang, J., Ipson, S.: Face detection based neural networks using robust skin color segmentation. In: 5th International Multi-Conference on Systems, Signals and Devices, IEEE SSD 2008, July 20-22, pp. 1–5 (2008)
14. Murray, M.K.: The nursing shortage: past, present, and future. Journal of Nursing Administration 32(2), 79–84 (2002)
15. Soriano, M., Huovinen, S., Martinkauppi, B., Laaksonen, M.: Using the skin Locus to cope with Changing Illumination Conditions in Color-Based Face Tracking. In: IEEE Nordic Signal Processing Symposium, pp. 383–386 (2000)
16. Triesch, J., Von der Malsburg, C.: Self-organized integration of adaptive visual cues for face tracking. In: Proceedings of the Fourth IEEE International Conference on Automatic Face and Gesture Recognition, pp. 102–107 (2000)
17. Turk, M.A., Pentland, A.P.: Face recognition using eigenfaces. In: Proceedings of the IEEE Computer Society Conference on Computer Vision and Pattern Recognition, CVPR 1991, June 3-6, pp. 586–591 (1991)

18. Vadakkepat, P., Lim, P., De Silva, L.C., Jing, L., Ling, L.L.: Multimodal Approach to Human-Face Detection and Tracking. IEEE Transactions on Industrial Electronics 55(3), 1385–1393 (2008)
19. Viola, P., Jones, M.: Robust real-time face detection. In: Proceedings of the Eighth IEEE International Conference on Computer Vision, ICCV 2001, p. 747 (2001)
20. Vytautas, P.: Distance measures for PCA-based face recognition. Pattern Recognition Letters 25, 711–724 (2004)
21. Wang, K., Li, R., Zhao, L.: Real-time facial expressions recognition system for service robot based-on ASM and SVMs. In: 2010 8th World Congress on Intelligent Control and Automation (WCICA), July 7-9, pp. 6637–6641 (2010)
22. Weng, J., Evans, C.H., Hwang, W.S.: An incremental learning method for face recognition under continuous video stream. In: Proceedings of the Fourth IEEE International Conference on Automatic Face and Gesture Recognition, pp. 251–256 (2000)
23. Wu, X., Gong, H., Chen, P., Zhi, Z., Xu, Y.: Intelligent household surveillance robot. In: IEEE International Conference on Robotics and Biomimetics, ROBIO 2008, February 22-25, pp. 1734–1739 (2009)
24. Yang, F., Paindavoine, M.: Implementation of an RBF neural network on embedded systems: real-time face tracking and identity verification. IEEE Transactions on Neural Networks 14(5), 1162–1175 (2003)

Author Index